Analytical Chemistry

Analytical Chemistry

J. G. DICK   Professor of Chemistry / Sir George Williams University

McGRAW-HILL BOOK COMPANY

New York   Montreal
St. Louis   New Delhi
San Francisco   Panama
Düsseldorf   Rio de Janeiro
Johannesburg   Singapore
Kuala Lumpur   Sydney
London   Toronto
Mexico

Library of Congress Cataloging in Publication Data

Dick, James Gardiner.
    Analytical chemistry.

    1. Chemistry, Analytic--Quantitative.   I.   Title.
QD101.2.D48   545   73-1294
ISBN 0-07-016786-9

Analytical Chemistry

1 2 3 4 5 6 7 8 9 0 K P K P 7 9 8 7 6 5 4 3

This book was set in Times Roman by
Textbook Services, Inc. The editors were
William P. Orr and Joan Stern; the
designer was Wladislaw Finne; and the
production supervisor was John A.
Sabella. The drawings were done by
ANCO Technical Services. Cover
illustration by Stephen Martin.
The printer and binder was Kingsport
Press, Inc.

Contents

Preface

The most obvious aim of a course in quantitative analysis is to provide the groundwork for establishing oneself as a thoroughly qualified analytical chemist. Few students pursuing undergraduate courses in quantitative analysis will have this goal in view. For these students, there is often the feeling that courses of this nature tend to make technicians of them—which is by no means the case. It should be pointed out that the primary aims of courses in quantitative analysis are:

1. To provide a fundamental knowledge of the principles underlying the techniques of quantitative analysis.
2. To provide a laboratory manipulative technique governed by the need to acquire data matching as closely as possible information previously recorded but not available to the student. The requirements of accuracy and precision here tend to stress a disciplined, orderly, and careful technique of laboratory manipulation.
3. To instill a respect for the exact data secured, and to further one's ability to evaluate such data and determine its reliability in the face of the limitations of the techniques of measurement and analysis employed.

It is worthy of note that apart from the general and far-reaching significance of the principles covered by point 1 above, points 2 and 3 are of fundamental importance in subsequent research work and day-to-day pursuits in *any* field of chemistry. With respect to undergraduate courses in chemistry, those involving quantitative analysis are often the first, and subsequently the continuing, introduction of the student to the inexorable demands of honesty and exactitude in data accumulation and evaluation. R. S. Mulliken, a Nobel laureate in chemistry in 1966, has written the following in this connection:

> I think it was in a course in quantitative chemical analysis that an appreciation of the scientific method and its rigours began really to take hold of me. ... There were no short cuts to beat clear thinking, careful technique and endless patience. Later on, I found that the same unnatural methods are always required in those activities commonly called "research."

The tendency today in many universities to eliminate, or to absorb into other generally unrelated courses of study, the specialized and unique discipline of analytical chemistry per se can only be deplored, since it can lead eventually to nothing other than a general lowering of the ability of the research student to appreciate and to perform properly those operations associated with the accurate and precise accumulation of data.

J. G. Dick

1
Introduction

1.1 ANALYTICAL CHEMISTRY

Analytical chemistry is basically the determination of the chemical composition of substances. Until recent years this was, practically speaking, the sole aim of the analytical chemist. Today, however, the field is being extended to embrace to a considerable degree the elucidation of the structural configuration of substances. It remains nevertheless a fact that in its truest sense, analytical chemistry still deals almost exclusively with the compositional aspect of substances, and it is to this aspect that the text is directed.

Analytical chemistry consists principally of two major divisions—qualitative analysis, which deals with the problem of *what* is in a substance; and quantitative analysis, which handles the problem of *how much* of each constituent is present. In the normal course of events, when preparing to analyze a substance of strictly unknown composition, qualitative analysis should precede any attempt at quantitative investigation, since the method of approach selected for the quantitative program may depend on the results of the qualitative analysis. On the other hand, where commercial work in analytical chemistry is involved, when *what* is in a substance is usually known, qualitative analysis is not generally required as a preliminary to any quantitative approach.

The techniques of qualitative analysis that were used for many years, such as in hydrogen-sulfide separations, provided an excellent opportunity to learn much concerning chemical reactions, their applicability under given circumstances, and their limitations. Such acquired information very often proved invaluable when the standard methods of quantitative analysis required modification on the basis of variations in the composition of the substances to be analyzed. In this light, one can perhaps regret the fact that such systematic procedures of qualitative analysis are now but rarely taught or used; but the modern techniques involving use of the emission spectrograph, the x-ray fluorescence spectrometer, reagents specific for many ions and molecules, etc., have provided more than adequate substitutes for the older and more conventional qualitative schemes of analysis.

Both inorganic and organic substances lend themselves to compositional investigation by quantitative analysis. Although many of the general techniques of analysis are applicable to both groups of substances, it is quite often the case that the general method of approach will differ markedly with respect to inorganic and organic substances. A full approach to the subject of quantitative analysis should undoubtedly cover both the inorganic and organic aspects. This text shows a definite bias towards the inorganic aspect, primarily because the author's competence and experience originate from this direction.

1.2   THE METHODS OF QUANTITATIVE ANALYSIS

Quantitative analysis, at its inception, dealt almost exclusively with techniques involving gravimetric procedures. These methods, which require the measurement of the weights of known composition products of the analytical process, or the measurement of weight losses or gains associated with such products, include some of the most accurate and precise techniques that the analytical chemist can employ. Many such methods are still applied today, particularly where the demand for accuracy and precision overrides the inherent disadvantages of these methods—disadvantages which include factors such as the relatively long time lapse for a completed procedure, the need for the careful laboratory techniques of the skilled and experienced worker, and the procedural complications often introduced where the substances being analyzed show unusual compositional complexities.

A second group of techniques which appeared shortly thereafter embraced the procedures of volumetric analysis, those covering the standard procedures of volumetric titration applied without the involvement of instrumental methods for the location of the critical point or points of the titration. In this context such methods require the use of prepared titrant solutions of known concentration and the measurement of the volume of such a solution required to react with a given compositional constituent to a specific point, usually the equivalence point, in the titration. Evidence that the critical point is achieved is provided by an indicator reaction that involves a change in a solution characteristic such as color, turbidity, etc. It should be apparent even at this point that the methods of volumetric analysis have a gravimetric basis, since the preparation of the titrant solution of known composition will nearly always involve at least one accurate weight measurement. The methods of volumetric analysis offer many advantages, not the least of which are the short time lapse to completion for many of the techniques and a somewhat less demanding requirement relative to laboratory manipulative skills.

Over the past 30 years the methods of volumetric analysis have to some considerable extent been subject to a form of instrumentation. By this is meant the increasing use of instrumental techniques, such as the potentiometric, conductometric, amperometric, and absorptiometric techniques, to locate the critical titration point or points or, more generally, to follow the course of the entire titration. In such instances, the volumetric methods are not basically altered from the standard procedures, the instrument functioning as a substitute for some indicator or chemical reaction system for locating a specific titrant volume, such as the equivalence-point volume.

On the other hand, in the past 30 years there has also been an increasing use of what can be described as quantitative analysis performed by instrumental techniques which measure some solution or substance characteristic specifically related directly or indirectly to the concentration of a particular constituent in the substance under investigation. Such techniques include potentiometric, conductometric, coulometric, voltammetric (polarographic), absorptiometric, and fluorometric methods, as well as the methods of flame emission and absorption, emission spectrography, and x-ray emission and absorption spectroscopy. Although such techniques are being applied to a greater and greater extent in the field of industrial and commercial quantitative analysis, it would be a mistake to assume that the standard chemical methods of analysis, such as gravimetric and volumetric methods, will eventually be eliminated from consideration. There are several reasons why this is unlikely ever to occur. First of all, many instrumental methods can not offer the accuracy and precision typical of most chemical methods of analysis. For example, instruments such as the emission spectrograph, the atomic absorption spectrophotometer, and polarographic equipment provide reliable data in the determination of low concentrations of substance constituents but permit only relatively poor accuracy and precision where the higher orders of concentration are involved. Second, such instruments can often be used only after calibration against standard substances whose compositions are accurately known as the result of quantitative analysis by the best of chemical methods of analysis. Third, many instrumental techniques depend, for example, on the measurement of a solution characteristic related to the concentration of a particular constituent, but under conditions where the presence of concentrations of certain other constituents may seriously influence the measurement data secured. In such cases, either chemical methods of separation associated with the general techniques of chemical analysis must be applied in order to eliminate the interfering effects of these constituents or, as is common in many cases of compositional complexity, chemical methods of analysis alone must be applied.

1.3   THE ANALYTICAL CHEMIST AND THE TECHNICIAN

The analytical technician can be considered as someone experienced in laboratory manipulative techniques, capable of following exactly the details of a rigidly defined series of methodic steps in a scheme of chemical or instrumental analysis and able to operate the various instruments used in the modern techniques of quantitative analysis. Such a technician is, in brief, a "cookbook analyst," a term the use of which is not intended to detract in any way from the abilities and value of the analytical technician. The analytical chemist, on the other hand, has an extensive knowledge and understanding of the fundamental principles underlying the methods of chemical and instrumental analysis. He can apply these principles to the modification of existing methods of analysis where required, devise new methods, and adapt developed instrumentation to analytical purposes. He can, where circumstances dictate, devise and modify instrumentation suited to the ends of quantitative analysis. He is fully aware of the limitations imposed by these principles and is prepared to take such limitations into consideration when evaluating the reliability of the analytical data provided by the methods under application.

2
Fundamental Concepts in Quantitative Analysis

2.1 GENERAL

Certain concepts can be considered as fundamental with regard to the principles of quantitative analysis. These include such ideas as solution concentration, chemical equations, chemical equilibrium, stoichiometry, and the basic principles underlying gravimetric, volumetric, and instrumental methods of analysis. In many cases these concepts will receive further detailed treatment in subsequent chapters of the text. The main purpose here is to discuss and review some of the more basic ideas surrounding these concepts.

2.2 SOLUTION CONCENTRATION

The use of solutions in quantitative analysis, whether in association with gravimetric, volumetric, or instrumental methods of analysis, requires that some basis for the expression of solution concentration be used in order to indicate the concentration of whatever solute or solutes have been employed. Although all such systems of concentration expression have a fundamentally similar basis with respect to weight relationships of solute and solvent, circumstances often dictate that the actual method of expression take on some convenient and specific form.

2.2.1 CONCENTRATION AS PERCENT BY WEIGHT

A method of expression of concentration often applied to solutions used in gravimetric work, or in the preliminary procedures of volumetric and instrumental methods, is that of percent by weight. Here the concentration of the solute is expressed as a percentage value representing the grams of solute per 100 g of solution. Thus, the percent by weight for a given solute would be given by

$$P = \frac{W_1}{W_1 + W_2}\, 100$$

where P = percent by weight
W_1 = weight of solute, g
W_2 = weight of solvent, g

Where aqueous solutions are involved, since the weight of 1 ml of water at room temperature is approximately 1 g, it is usually quite adequate to substitute for the weight of water in grams the volume in milliliters.

2.2.2 MOLES (MOL) AND EQUIVALENTS

The reactions of chemistry involve atoms, ions, or molecules. The weight of an individual particle is extremely small, and it is impossible for the chemist to consider directly the weight of a single atom, ion, or molecule in reaction. The basis used for considering substances in reaction is the *gram atom*, the *gram ion*, or the *mole*. These units each involve Avogadro's number, 6.023×10^{23}, of particles, so that we have:

One gram atom = the weight in grams of 6.023×10^{23} atoms
One gram ion = the weight in grams of 6.023×10^{23} ions
One mole = the weight in grams of 6.023×10^{23} molecules

The tendency exists to refer to the weight in grams of Avogadro's number of any particle as a mole, regardless of the nature of the particle. Thus it is common to find the weight in grams of 6.023×10^{23} Na^+ ions referred to as a mole of Na^+; that of 6.023×10^{23} Cu atoms as a mole of Cu.

The weight in grams of Avogadro's number of the molecules of any compound, it will be remembered, is called the *gram-molecular weight* (GMW), is numerically equal to the *molecular weight*, and can be determined as the *formula weight* of the molecule. Thus the mole is identical in concept to the gram-molecular weight.

Similarly, the weight in grams of Avogadro's number of the atoms of any element is called the *gram-atomic weight* (GAW) and is numerically equal to the *atomic weight*.

Chemical reactions occur on the basis of integral numbers of atoms, ions, or molecules, but these reactions often involve more than just a one-particle-for-one-particle basis. In order to determine the general relationships between substances in chemical reactions, the terms *equivalent* and *equivalent weight* are used. In its simplest form of expression, the equivalent of an atom, ion, or molecule in reaction is Avogadro's number of electrons, 6.023×10^{23} e.

Example

$$H^+ + 1e \rightleftharpoons H$$

One mole, or 1.008 g, of hydrogen ions requires 6.023×10^{23} electrons for reduction to hydrogen atoms and is therefore identical to one equivalent of hydrogen ions.

Example

$$Cu^{2+} + 1e \rightleftharpoons Cu^+$$

One mole, or 63.54 g, of copper (II) ions requires 6.023×10^{23} electrons for reduction to copper (I) ions and is therefore identical to one equivalent of copper(II) ions.

Example

$$Cu^{2+} + 2e \rightleftharpoons Cu$$

One mole, or 63.54 g, of copper (II) ions requires $2 \times 6.023 \times 10^{23}$ electrons for reduction to copper atoms, and is therefore identical to two equivalents of copper(II) ions.

The *gram-equivalent weight* (**GEW**) of any substance thus becomes the weight in grams of the atoms, ions, or molecules equivalent in the reaction involved to Avogadro's number of electrons, and it is numerically equal to the equivalent weight. Thus, in the examples just given, the equivalent weight of hydrogen is 1.008, that of copper in the first copper reaction is 63.54, and that of copper in the second copper reaction is 31.77. It is important to note that the equivalent weight of an atom, ion, or molecule is directly related to the reaction in which it is involved, and that it is entirely possible for a substance to show more than one equivalent or equivalent-weight value.

Example

General reaction $H^+ + OH^- \rightleftharpoons H_2O$
Typified by $HCl + NaOH \rightleftharpoons H_2O + NaCl$

GEW HCl $- 6.023 \times 10^{23}$ e $-$ 1 mol OH^- $-$ 1 mol H^+ $-$ 1 mol HCl

GEW HCl $=$ GMW HCl
1 equiv HCl $=$ 1 mol HCl

Example

General reaction $Ag^+ + Cl^- \rightleftharpoons AgCl(s)$
Typified by $AgNO_3 + HCl \rightleftharpoons AgCl(s) + HNO_3$

GEW $AgNO_3 - 6.023 \times 10^{23}$ e $-$ 1 mol Cl^- $-$ 1 mol Ag^+ $-$ 1 mol $AgNO_3$

GEW $AgNO_3 =$ GMW $AgNO_3$
1 equiv $AgNO_3 =$ 1 mol $AgNO_3$

Example

General reaction $2NH_3 + Ag^+ \rightleftharpoons Ag(NH_3)_2{}^+$
Typified by $2NH_3 + AgNO_3 \rightleftharpoons Ag(NH_3)_2{}^+ + NO_3{}^-$

GEW $NH_3 - 6.023 \times 10^{23}$ e $-$ 1 mol Ag^+ $-$ 2 mol NH_3

GEW $NH_3 =$ 2 GMW NH_3
1 equiv $NH_3 =$ 2 mol NH_3

Example

$$K_2Cr_2O_7 + Pb(NO_3)_2 \rightleftharpoons PbCr_2O_7(s) + 2KNO_3$$
$$\text{GEW } K_2Cr_2O_7 = GMW/2 \; K_2Cr_2O_7$$
$$1 \text{ equiv } K_2Cr_2O_7 = \tfrac{1}{2} \text{ mol } K_2Cr_2O_7$$

Example

$$K_2Cr_2O_7 + 6FeCl_2 + 14HCl \rightleftharpoons 2CrCl_3 + 6FeCl_3 + 2KCl + 7H_2O$$
$$\text{GEW } K_2Cr_2O_7 = GMW/6 \; K_2Cr_2O_7$$
$$1 \text{ equiv } K_2Cr_2O_7 = \tfrac{1}{6} \text{ mol } K_2Cr_2O_7$$

2.2.3 CONCENTRATION AS MOLALITY

In disciplines such as physical chemistry, solution concentration is often expressed as *molality*. Although rarely used in the analytical field, it is perhaps worthwhile to consider it briefly. The molality of a solution is given by the number of moles of solute per 1000 g of solvent, and we have

$$m = \frac{W_1}{(MW)(W_2)} \, 1000$$

where m = molality
$\quad\;\; W_1$ = weight of solute, g
$\quad\;\; MW$ = molecular weight of solute, g/mol
$\quad\;\; W_2$ = weight of solvent, g

Example 28.00 g of NaOH is dissolved in 3500 g of water at 25°C. Determine the molality at 25°C in terms of NaOH.

$$m = \frac{28.00}{40.00 \times 3500} \, 1000 = 0.2000$$

Note that the molality of a solution does not change with solution volume changes resulting from temperature changes.

2.2.4 CONCENTRATION AS MOLARITY

One of the important methods of solution-concentration expression used in the field of quantitative analysis is that of *molarity*. The molarity of a solution is given by the number of moles of solute per liter of solution. Since the millimole is one-thousandth of a mole and the milliliter is one-thousandth of a liter, we can use the expression

$$M = \frac{W_1}{(MW)(V)}$$

where M = molarity
$\quad\;\; W_1$ = weight of solute, g or mg
$\quad\;\; MW$ = molecular weight of solute, g/mol, or mg/mmol
$\quad\;\; V$ = volume of solution, l, or ml

Since molarity involves a basis of solution volume, it is apparent that the molarity of a solution will change with volume changes associated with changes in temperature. Thus, a solution prepared as 0.1000 M at 25°C will show a molarity somewhat less than this value at, for example, 50°C.

2.2.5 CONCENTRATION AS FORMALITY

Many substances do not exist in the molecular form, whether in the solid or the solution state. Salts such as NaCl and $FeCl_3$ maintain the ionic form (Na^+ and Cl^-, Fe^{3+} and Cl^-) in the solid state as well as in aqueous solution. The preparation, for example, of a 1-M solution of $FeCl_3$ has thus no logical basis, since the molecular form of $FeCl_3$ does not exist from start to finish of the preparation procedure. Many chemists have therefore insisted that the *formula weight*, rather than the molecular weight, be used in the preparation of solutions, and that the solution concentration then be expressed in *formality*. On this basis one defines formality as being the number of formula weights of a solute per liter of solution. Thus we have

$$F = \frac{W_1}{(FW)(V)}$$

where F = formality
$\quad W_1$ = weight of solute, g
$\quad FW$ = formula weight of solute, grams per formula
$\quad V$ = volume of solution, l

Example   16.22 g of $FeCl_3$ is dissolved in water and diluted to exactly 1000 ml at 25°C. What is the formality of this solution? What is its molarity in terms of Fe^{3+} and in terms of Cl^-?

$$F = \frac{16.22}{162.21 \times 1.000}$$
$$= 0.1000$$
$$M \text{ as } Fe^{3+} = 0.1000$$
$$M \text{ as } Cl^- = 0.3000$$

Again, because of the volume basis of preparation, formal solutions will show changes in formality with volume changes associated with temperature changes.

There is a certain logic where this method of concentration expression is concerned, but in this text, we shall continue to use the molar method of expression as being entirely satisfactory in its application to analytical chemistry.

2.2.6 CONCENTRATION AS NORMALITY

The *normality* of a solution is given by the number of equivalents of solute per liter of solution. It is emphasized most strongly that since the equivalent or equivalent weight of a substance may vary according to the reaction in which it is involved, the use of the normality as a means of expressing a solution concentration should *always* be accompanied by a reference to the associated reaction or the associated molarity.

Since the milliequivalent is one-thousandth of an equivalent, we can use

$$N = \frac{W_1}{(EW)(V)}$$

where N = normality
W_1 = weight of solute, g, or mg
EW = equivalent weight of solute, g/equiv, or mg/mequiv
V = volume of solution, l, or ml

The numerical value of EW in each equation is, of course, identical.

Example NaOH in the reaction

$$NaOH + HCl \rightleftharpoons H_2O + NaCl$$

has an equivalent weight of 40.00. 5.00 g of NaOH is dissolved in water and the solution diluted to exactly 500 ml at 25°C. Determine the normality in terms of NaOH.

$$N = \frac{5.00}{40.00 \times 0.500} = 0.2500$$

Example $K_2Cr_2O_7$ in the reaction

$$K_2Cr_2O_7 + PbCl_2 \rightleftharpoons PbCr_2O_7(s) + 2KCl$$

has an equivalent weight of 147.10, and in the reaction

$$K_2Cr_2O_7 + 6Fe^{2+} + 14H^+ \rightleftharpoons 2Cr^{3+} + 6Fe^{3+} + 2K^+ + 7H_2O$$

it has an equivalent weight of 49.03.

24.52 g of $K_2Cr_2O_7$ is dissolved in water and diluted to exactly 1000 ml at 25°C. Determine the normality of this solution in terms of $K_2Cr_2O_7$ for each reaction.

Reaction with $PbCl_2$: $$N = \frac{24.52}{147.10 \times 1.000} = 0.1667$$

Reaction with Fe^{3+}: $$N = \frac{24.52}{49.03 \times 1.000} = 0.5001$$

Note that the *same* solution has *two* normality values, and that confusion could easily arise when the expression of normality is *not* accompanied by the associated reaction identification or some other positive method of relating normality to solution usage.

Here again, because of the volume basis of preparation, normal solutions will show changes in normality with volume changes associated with changes in temperature.

It is important to note the rather obvious fact that the equivalent weight and the molecular weight are related by the equation

$$EW = \frac{MW}{n}$$

where n is the multiple of Avogadro's number of electrons involved directly or by equivalence in the reaction of 1 mol of substance. Thus, we have

$$N = \frac{nW_1}{(MW)(V)} = nM$$

where N = normality
n = multiple of Avogadro's number of electrons involved in the reaction of one mole of solute
W_1 = weight of solute, g
MW = molecular weight of solute, g/mol
V = volume of solution, l
M = molarity

Example   Using the example from Sec. 2.2.6, we note that the solution of $K_2Cr_2O_7$ prepared involved 24.52 g of the salt in 1000 ml of solution. In the reaction with $PbCl_2$ the multiple of Avogadro's number of electrons is 2, while in the reaction with Fe^{3+} the multiple is 6. The molarity of the solution is given by

$$M = \frac{24.52}{294.19 \times 1.000} = 0.08355$$

and we have

Reaction with $PbCl_2$: $N = nM = 2 \times 0.08335$
$\qquad\qquad = 0.1667$
Reaction with Fe^{3+}: $N = nM = 6 \times 0.08335$
$\qquad\qquad = 0.5001$
identical to the results of the previous normality calculations.

2.2.7   CONCENTRATION AS TITER

In quantitative analytical work involving volumetric methods of analysis in particular, solution concentration is often expressed in the form of the *titer*. This
method of expression is very frequently encountered in laboratories engaged in industrial and commercial analysis. The titer has reference not to the concentration
of the solute in the solution but to the weight of some particular substance with
which the solute in 1 ml of the solution will react. As a case in point, a solution of
potassium dichromate used in the determination of iron by a standard volumetric
oxidation-reduction method shows a titration condition wherein 17.91 ml of the
solution reacts with 0.1000 g of iron exactly, oxidizing the iron quantitatively
from iron(II) to iron(III). The titer of the solution of potassium dichromate in
terms of iron is then given by

$$T = \frac{0.1000 \text{ g}}{17.91 \text{ ml}}$$
$$= 0.005583 \text{ g Fe/ml}$$
$$= 5.583 \text{ mg Fe/ml}$$

Where the normality of the solution is known relative to a specific reaction situation, the titer of the solution can be easily calculated. Such a calculation will be

valid only when the method for the determination of the exact solution normality agrees in detail with the method of analysis in which the solution will be employed. For a specific reaction in which the solution is involved we have

$$N = \frac{mg}{EW \times ml} \qquad T = \frac{mg}{ml}$$

so that

$$T = (N)(EW)$$

where the equivalent weight (EW) is expressed in milligrams/milliequivalent of *the substance involved in the reaction*, and does *not* refer to that of the solute.

Example Calculate the iron titer of a 0.1100-N solution of $K_2Cr_2O_7$ used in the volumetric titration reaction

$$K_2Cr_2O_7 + 6Fe^{2+} + 14H^+ \rightleftharpoons 2Cr^{3+} + 6Fe^{3+} + 2K^+ + 7H_2O$$

In this reaction the equivalent weight of iron is 55.85 and can be expressed as 55.85 g/equiv, or 55.85 mg/meq. We have then

$$T = (N)(EW)$$
$$= 0.1100 \text{ meq/ml} \times 55.85 \text{ mg/meq}$$
$$= 6.144 \text{ mg Fe/ml}$$

As is the case with all methods of concentration expression using a volume basis of preparation, the titer of a solution will change according to volume changes associated with changes in temperature.

2.2.8 IDENTIFICATION OF SOLUTIONS

Solutions used in gravimetric methods, and in the preliminary stages of volumetric or instrumental methods, are commonly concentration-identified by percent by weight and/or molarity relative to the solute. Such solutions should be so marked.

With respect to titrant solutions used in volumetric methods and certain forms of instrumental methods, concentration identification must always show

1. The molarity
2. The normality (or normalities where reactions dictating different solute equivalent weights are involved)
3. The titer (or titers where the solution is capable of being used for the determination of more than one substance)

Such solutions should be marked in accordance with the foregoing.

2.3 CHEMICAL EQUATIONS

Chemical equations represent the reactions between substances. In quantitative analysis most reactions occur in solution and, in inorganic quantitative analysis in particular, in aqueous solution. The form of expression for the chemical reaction as a chemical equation can be selected on the basis of whether the substances involved exist in the ionized or the molecular form. Where highly ionized substances participate, the reaction of the ions is usually shown by the chemical equation rather than the reaction of the stoichiometric formulas for the reacting substances. In the case, for example, of the reaction which takes place when an aqueous solution of silver nitrate is added to an aqueous solution of sodium chloride, we usually write the ionic form of the chemical equation for these two highly ionized substances and show that

$$Ag^+ + Cl^- \rightleftharpoons AgCl(s)$$

although the reaction in terms of the stoichiometric formulas would be

$$AgNO_3 + NaCl \rightleftharpoons AgCl(s) + NaNO_3$$

Where weak electrolytes in reaction are concerned, with corresponding low degree of dissociation, the molecular form is used. For example, in the reaction between acetic acid, a weakly ionized acid in aqueous solution, and sodium hydroxide, a highly ionized base in aqueous solution, we would show

$$OH^- + CH_3{\cdot}COOH \rightleftharpoons H_2O + CH_3{\cdot}COO^-$$

with the stoichiometric formulation for the reaction being

$$NaOH + CH_3{\cdot}COOH \rightleftharpoons H_2O + CH_3{\cdot}COONa$$

The approach to the balancing of chemical equations must take into consideration the fact that two basic forms of chemical reaction may occur. One of these is the reaction of *metathesis*, where the substances involved do not gain or lose electrons. A typical metathetical reaction is

$$Ag^+ + Cl^- \rightleftharpoons AgCl(s)$$

where silver and chlorine maintain the same electronic arrangements before and after the reaction. The second form is the reaction of *oxidation-reduction*, commonly called the *redox* reaction, where the substances involved gain or lose electrons during the reaction and show different electronic arrangements before and after the reaction. A typical redox reaction is

$$Sn^{2+} + 2Fe^{3+} \rightleftharpoons Sn^{4+} + 2Fe^{2+}$$

where tin changes from tin(II) to tin(IV) and iron from iron(III) to iron(II). It is important to avoid confusing the two forms of reaction. Thus the reaction

$$2H_2O \rightleftharpoons H_3O^+ + OH^-$$

is a metathetical reaction, whereas the reaction

$$2H_2O \rightleftharpoons 2H_2 + O_2$$

is a redox reaction. In this latter instance, hydrogen and oxygen have been electronically altered from H^+ to H^0 and from O^{2-} to O^0.

2.3.1 BALANCING OXIDATION-REDUCTION EQUATIONS

The oxidation-numbers method Certain fundamental ideas concerning oxidation-reduction processes and the basis of the oxidation-numbers method of balancing redox equations should be discussed briefly.

Oxidation is the overall process whereby a species in a chemical reaction *loses* one or more electrons and *increases* its *state of oxidation*. *Reduction* is the overall process whereby a species in a chemical reaction *gains* one or more electrons and *decreases* its *state of oxidation*. An *oxidant* is a substance capable of oxidizing a chemical species; it acquires the electron or electrons lost by the species and is itself reduced in the overall process. A *reductant* is a substance capable of reducing a chemical species; it loses the electron or electrons gained by the species and is itself oxidized in the overall process.

The *oxidation state* represents the real or apparent charge on a single atom. Where two or more atoms of the same element exist in different oxidation states in the same formula unit, an average charge is sometimes assigned to a single atom. Oxidation state and valence should not be confused. The oxidation state is the real (or apparent), sometimes arbitrary, charge contributed by an atom in a formula unit to the net charge of the formula unit. The contributory charge may or may not be an integral whole number, but it always carries an associated $+$ or $-$ sign. Valence, on the other hand, is always an integral whole number, has no associated charge sign, and represents the capacity of the atom to form bonds in the combining process. Note that the valences of carbon, hydrogen, and oxygen are 4, 1, and 2, respectively. For acetylene, C_2H_2, the oxidation states are carbon -1 and hydrogen $+1$. For hydrogen peroxide, H_2O_2, the oxidation states are hydrogen $+1$, and oxygen -1, while for maleic acid, $C_4H_4O_4$, the values are carbon $+1$, hydrogen $+1$, and oxygen -2.

The *oxidation number* represents the accumulated total of the oxidation states of all the atoms of the *same* element in the formula unit.

The following indicates, in a general way, the method of assigning oxidation states:

1. The atoms of free elements, whether monatomic or polyatomic in their molecular representation, have oxidation state zero.
2. The atoms in monatomic ions, such as Fe^{2+}, Ag^+, Cl^-, and S^{2-}, show oxidation states given by the ionic charge.
3. The atoms in polyatomic ions involving only one element, such as Hg_2^{2+}, O_2^{2-}, O_2^-, and N_3^-, show oxidation states representing the shared charge on the ion. For the examples given we have oxidation states of $+1$, -1, $-\frac{1}{2}$, and $-\frac{1}{3}$.
4. The atoms of polyatomic ionic or molecular species, such as SO_4^{2-}, $Cr_2O_7^{2-}$, $KMnO_4$, etc., show oxidation states derivable on the basis of certain simple general rules. The oxygen atom is assigned as oxidation state of -2—except in its bondings with fluorine, where the oxidation state is positive (for ex-

ample, $+2$ in OF_2); and except in compounds such as H_2O_2, where the oxidation state is -1. The hydrogen atom is assigned an oxidation state of $+1$, except in hydrides where the oxidation state is generally -1. Periodic group I atoms (Li, Na, K, etc.) are assigned oxidation state $+1$. Periodic group II atoms (Be, Mg, Ca, etc.) are assigned oxidation state $+2$.

The assignment of the oxidation number for an element in a formula unit then becomes, as already noted, the accumulated total of the oxidation states of all the atoms of that element in the formula unit.

Example $NaHSO_4$:

Oxidation states	$Na = +1$
	$H = +1$
	$O = -2$
	$S = +6$
Oxidation numbers	$Na = +1$
	$H = +1$
	$O = -8$
	$S = +6$

It is worthwhile to note that the assignment of oxidation states can be made somewhat easier by the use of net-charge values for many tightly bonded groups or radicals, such as CN^-, SO_4^{2-}, NO_3^-, etc.

We can now consider the general steps required for the balancing of equations of the redox type by the oxidation-numbers method. We shall deal only with equations representing reactions where all of the reactants and products are known. As an example the skeletal (unbalanced) redox equation,

$$K_2Cr_2O_7 + FeCl_2 + HCl \rightleftharpoons CrCl_3 + FeCl_3 + KCl + H_2O$$

is used.

Step 1   Assign oxidation-state values to the elements in those species actually involved in the oxidation-state changes.

$$\overset{+6}{K_2}\overset{+2}{Cr_2}O_7 + \overset{+2}{FeCl_2} + HCl \rightleftharpoons \overset{+3}{CrCl_3} + \overset{+3}{FeCl_3} + KCl + H_2O$$

Step 2   Determine the magnitude of the change in oxidation state for the atom of each reacting species.

| Cr | $+6$ to $+3$ | three per atom |
| Fe | $+2$ to $+3$ | one per atom |

Step 3   Determine the oxidation-number change for each reacting species on the left side.

Cr	three per atom × two atoms	six
Fe	one per atom × one atom	one

Step 4   Balance the gain and loss of electrons represented by the change in the oxidation number for each reacting species by multiplying the associated formula unit on the left side by appropriate whole-number values.

$K_2Cr_2O_7$	one multiple
$FeCl_2$	six multiples

Step 5   Balance the atoms of the redox species on the right side with the associated atoms on the left side by using the proper coefficients to multiply the formula units involved on the right side.

$$K_2Cr_2O_7 + 6FeCl_2 + HCl \rightleftharpoons 2CrCl_3 + 6FeCl_3 + KCl + H_2O$$

Step 6   Balance the species other than the redox species by inspection.

$$K_2Cr_2O_7 + 6FeCl_2 + 14HCl \rightleftharpoons 2CrCl_3 + 6FeCl_3 + 2KCl + 7H_2O$$

The ion-electron, or half-reaction, method   A second technique for the balancing of redox equations is the ion-electron, or half-reaction, method, a method particularly suited to the balancing of equations representing redox reactions taking place in aqueous solution media. Here the redox reaction is divided into two ion-electron half reactions which represent separately the oxidation and reduction processes of the overall redox reaction. Due notice must be taken of the medium in which the reaction is to take place—acidic, ammoniacal, or basic other than ammoniacal. The following indicates the general steps for the balancing of redox equations by this technique. Again we shall consider only those reactions where the reactants and products are known. The skeletal (unbalanced) ionic redox reaction in acidic medium

$$Cr_2O_7{}^{2-} + Fe^{2+} + H^+ \rightleftharpoons Cr^{3+} + Fe^{3+} + H_2O$$

will be used as an example.

Example

Step 1   Write down one of the separate half reactions, either that for the oxidant or that for the reductant.

$$Cr_2O_7{}^{2-} \rightleftharpoons Cr^{3+}$$

Step 2   Balance the half reaction atomically (1) relative to atoms other than oxygen and hydrogen, (2) relative to oxygen atoms, using H_2O molecules for acid media and ammoniacal media and OH^- ions for basic media other than ammoniacal, and (3) relative to hydrogen atoms, using H^+ ions for acid media, $NH_4{}^+$ ions for ammoniacal media (with NH_3 molecules added in equal quantity to the opposite side), and H_2O molecules for basic media other than ammoniacal (with OH^- ions added to the opposite side as a counterbalance).

Atoms other than O and H	$Cr_2O_7^{2-} \rightleftharpoons 2Cr^{3+}$
O atoms	$Cr_2O_7^{2-} \rightleftharpoons 2Cr^{3+} + 7H_2O$
H atoms	$Cr_2O_7^{2-} + 14H^+ \rightleftharpoons 2Cr^{3+} + 7H_2O$

Step 3 Balance the half reaction electronically by adding the required number of electrons to the side in need of negative charge.

$$Cr_2O_7^{2-} + 14H^+ + 6e \rightleftharpoons 2Cr^{3+} + 7H_2O$$

Step 4 Repeat steps 1 to 3 for the second half reaction.

$$Fe^{2+} \rightleftharpoons Fe^{3+}$$

$$Fe^{2+} \rightleftharpoons Fe^{3+} + 1e$$

Step 5 Equalize the electron values for the two half reactions.

$$Cr_2O_7^{2-} + 14H^+ + 6e \rightleftharpoons 2Cr^{3+} + 7H_2O$$

$$6Fe^{2+} \rightleftharpoons 6Fe^{3+} + 6e$$

Step 6 Add the two half reactions to secure the balanced ionic redox equation.

$$Cr_2O_7^{2-} + 6Fe^{2+} + 14H^+ \rightleftharpoons 2Cr^{3+} + 6Fe^{3+} + 7H_2O$$

Example For this example, we have

$$Cu(NH_3)_4^{2+} + CN^- + H_2O \rightleftharpoons Cu(CN)_3^{2-} + CNO^- + NH_4^+ + NH_3$$

in ammoniacal medium.

Step 1 $Cu(NH_3)_4^{2+} \rightleftharpoons Cu(CN)_3^{2-}$

Step 2

Atoms other than O and H	$Cu(NH_3)_4^{2+} + 3CN^- \rightleftharpoons Cu(CN)_3^{2-} + 4NH_3$
O atoms	None required
H atoms	None additional required

Step 3 $Cu(NH_3)_4^{2+} + 3CN^- + 1e \rightleftharpoons Cu(CN)_3^{2-} + 4NH_3$

Step 4 $CN^- \rightleftharpoons CNO^-$

Atoms other than O and H	$CN^- \rightleftharpoons CNO^-$
O atoms	$CN^- + H_2O \rightleftharpoons CNO^-$
H atoms	$CN^- + H_2O + 2NH_3 \rightleftharpoons CNO^- + 2NH_4^+$

$$CN^- + H_2O + 2NH_3 \rightleftharpoons CNO^- + 2NH_4^+ + 2e$$

Step 5 $2Cu(NH_3)_4^{2+} + 6CN^- + 2e \rightleftharpoons 2Cu(CN)_3^{2-} + 8NH_3$

$$CN^- + H_2O + 2NH_3 \rightleftharpoons CNO^- + 2NH_4^+ + 2e$$

Step 6 $2Cu(NH_3)_4^{2+} + 7CN^- + H_2O \rightleftharpoons 2Cu(CN)_3^{2-} + CNO^- + 2NH_4^+ + 6NH_3$

Much of the foregoing has been abstracted from an excellent work by Margolis.* Students wishing to pursue the subject of the balancing of equations in depth are referred to that text.

---

*E. J. Margolis, "Formulation and Stoichiometry: A Review of Fundamental Chemistry." Copyright © 1968. By permission of Appleton,Century,Crofts, Educational Division, Meredith Corporation.

2.4 FUNDAMENTAL PRINCIPLES AND STOICHIOMETRY OF GRAVIMETRIC AND VOLUMETRIC METHODS OF ANALYSIS

2.4.1 GRAVIMETRIC METHODS

The gravimetric method of analysis involves, as the name implies, the measurement of weight as the basic process of the procedure. The following skeletal descriptions indicate the general method.

1. Chemical precipitation; isolation of the precipitate; drying or ignition of the precipitate; cooling and weighing the dried or ignited residue; calculation of the amount of sought-for substance

   Example Determination of silver by precipitation as silver chloride; isolation of the silver chloride by filtration and washing; drying, cooling, and weighing the silver chloride; calculation of silver from the weight of silver chloride

2. Precipitation by electrolytic deposition upon application of a required potential difference between two unattackable electrodes immersed in a solution; isolation of the deposit; drying, cooling, and weighing the deposit; calculation of the amount of sought-for substance

   Example Determination of copper by electrolysis of a copper solution between platinum electrodes; isolation of the copper as a deposit on the previously weighed platinum cathode; drying, cooling, and weighing the electrode and deposit; calculation of copper from the deposit weight

3. Chemical precipitation; isolation of the precipitate; drying or ignition of the precipitate; cooling and weighing the dried or ignited residue; treatment of the residue to expel a compound containing all of the sought-for substance; redrying or reignition; cooling and weighing; determination of the weight loss; calculation of the amount of sought-for substance

   Example Determination of silicon in steel by precipitation as hydrated silicon dioxide; isolation of the hydrated silicon dioxide by filtration and washing; ignition of the hydrated silicon dioxide to silicon dioxide in a platinum crucible; cooling and weighing the crucible and residue; conversion of the silicon dioxide to volatile water and silicon tetrafluoride by treatment with sulfuric and hydrofluoric acids; heating to expel water, silicon tetrafluoride, hydrofluoric, and sulfuric acids; reignition of the crucible and residue; cooling, weighing, and determination of the weight loss as silicon dioxide; calculation of the amount of silicon from the weight loss

   No attempt will be made at this point to detail the theory and procedures of the gravimetric method. These will be covered adequately in a separate chapter. For the moment we shall concern ourselves only with the general stoichiometry of the gravimetric method.

   Having already treated in detail the methods of balancing chemical equations, we can proceed to the use of such balanced equations in gravimetric stoichiometry. It should be noted, first of all, that in general the calculations of gravimetric analysis involve

$$P = \frac{WF}{S} \, 100$$

where P = percent of sought-for substance
$\quad\quad\; W$ = weight of residue or precipitate, g
$\quad\quad\; F$ = gravimetric factor
$\quad\quad\; S$ = sample weight, g

The gravimetric factor represents the factor by which the weight of the pre-cipitate must be multiplied in order to obtain the weight of the sought-for sub-stance. It may also represent, in a more general sense, the factor by which any substance must be multiplied in order to convert its weight to the weight of an as-sociated compound. A few examples will serve to indicate the association of the chemical equation and the gravimetric factor.

Example Assume that 0.4352 g of an unknown substance yields an AgCl precipitate weighing 0.1518 g. Calculate the percentage of silver in the unknown.

The precipitation equation, together with the related mole weights, is

$$Ag^+ + Cl^- \rightleftharpoons AgCl(s)$$
107.87   35.45   143.32

The gravimetric factor for Ag out of AgCl becomes

$$\frac{Ag}{AgCl} = \frac{107.87}{143.32} = 0.7526$$

The percentage of silver in the unknown becomes

$$Ag = \frac{0.1518 \times 0.7526}{0.4352} \, 100 = 26.25\%$$

Example A substance contains, in part, Na_2O and K_2O. 1.8150 g is reacted to convert the Na_2O and K_2O to NaCl and KCl, the weight of the combined chlorides being determined as 0.9226 g. These chlorides are subsequently converted to Na_2SO_4 and K_2SO_4, the mixed sulfates weighing 1.0987 g. Calculate the percentages of Na_2O and K_2O in the substance.

The chemical equations involved, with the associated mole weights, are

$$Na_2O + 2HCl \rightleftharpoons 2NaCl + H_2O$$
61.98   2 × 36.46   2 × 58.44   18.02

$$K_2O + 2HCl \rightleftharpoons 2KCl + H_2O$$
94.20   2 × 36.46   2 × 74.56   18.02

$$2NaCl + H_2SO_4 \rightleftharpoons Na_2SO_4 + 2HCl$$
2 × 58.44   98.08   142.04   2 × 36.46

$$2KCl + H_2SO_4 \rightleftharpoons K_2SO_4 + 2HCl$$
2 × 74.56   98.08   174.27   2 × 36.46

The gravimetric conversion factors are

$$Na_2O \rightarrow NaCl = \frac{2NaCl}{Na_2O} = \frac{2 \times 58.44}{61.98}$$

$$K_2O \to KCl = \frac{2KCl}{K_2O} = \frac{2 \times 74.56}{94.20}$$

$$NaCl \to Na_2SO_4 = \frac{Na_2SO_4}{2NaCl} = \frac{142.04}{2 \times 58.44}$$

$$KCl \to K_2SO_4 = \frac{K_2SO_4}{2KCl} = \frac{174.27}{2 \times 74.56}$$

Let the percentages of Na_2O and K_2O be X and Y, respectively. For the chloride conversion we have

$$1.8150 \times 0.01X \times \frac{2 \times 58.44}{61.98} + 1.8150 \times 0.01Y \times \frac{2 \times 74.56}{94.20} = 0.9226$$

For the sulfate conversion we have

$$1.8150 \times 0.01X \times \frac{2NaCl}{Na_2O} \times \frac{Na_2SO_4}{2NaCl} + 1.8150 \times 0.01Y \times \frac{2KCl}{K_2O} \times \frac{K_2SO_4}{2KCl} = 1.0987$$

$$1.8150 \times 0.01X \times \frac{142.04}{61.98} + 1.8150 \times 0.01Y \times \frac{174.27}{94.20} = 1.0987$$

Simplification and solution simultaneously of the equations for chloride and sufate conversion yield

$Na_2O = 12.84\%$
$K_2O = 16.81\%$

2.4.2 VOLUMETRIC METHODS

General   Volumetric techniques of analysis are in general those which involve as the basic step the measurement of the volume of some *standard solution* of known concentration. This standard solution is called the *titrant*, and the process of determining the volume required is called *titration*. During titration, the active component of the titrant reacts with a quantity of some specific substance in the solution being titrated, and the reaction in question proceeds to some point of completion which has significance in the quantitative analytical sense.

The point of completion is called the *end point* of the titration, and it is usually assumed that this end point will be located at, or as close as possible to, the *equivalence point* of the titration. The equivalence point can be defined as that point in titration where an amount of the active component of the titrant exactly equivalent, stoichiometrically speaking, to the amount of the reacting substance in the titrated solution has been added.

The end point will be signaled by some *indicator* technique, and the volume of the standard solution, or titrant, used to reach this point will be recorded. From this volume, the known concentration of the titrant, and the stoichiometric relationships of the titration reaction, the exact amount of the reacting substance in the titrated solution can be determined. The amount of the substance relative to the sample weight originally taken for the material under analysis provides the percentage value of the substance in the material.

The need for some indicator technique to signal the end point of the titration has been mentioned. The indicator technique applied may be any one of the following:

1. An internal indicator, where one of the components of the titration system functions as the indicator. An example here would be the use of a standard solution of $KMnO_4$ in the redox titration of iron(II). The first excess of $KMnO_4$ solution beyond the equivalence-point volume imparts a pink color to the titrated solution. This is the end-point indication.
2. An internal indicator, where an added substance to the titration system acts as the indicator. An example here would be the use of a weakly acid organic indicator, such as Methyl red, in neutralization titrations. A second example would be the use of potassium chromate in the titration of chloride ion with a standard solution of silver nitrate.
3. An external indicator, where a substance external to the titration system is used to explore reaction conditions between the reacting substances during the course of titration. An example here would be the use of a tannic acid solution as an external indicator in the titration of lead(II) with a standard solution of ammonium molybdate. Such systems are rarely used, particularly today, and have value only when the reaction with the indicator results in an irreversible change. Under such conditions momentary high surface concentrations of the titrant substance can yield premature end-point signals when the indicator is added *to* the titration system.
4. A physicochemical indication of the end point based on some significant change in a physical property at the end point of the titration. Examples of such indicator techniques are the potentiometric, conductometric, amperometric, and absorptiometric methods.

Classification of volumetric methods The *acid-base*, or *neutralization*, methods include the titration of acids or salts of weak bases by standard alkali solutions. The technique is sometimes called *alkalimetry*. Also included is the titration of bases or salts of weak acids by standard acid solutions. This technique is sometimes called *acidimetry*.

The *precipitation* methods include all the methods wherein the reacting substance and the standard solution react to yield a precipitate or a slightly soluble salt as the primary reaction product. This classification, in particular, includes those methods in which silver salts are the slightly soluble products of the titration reaction and in which the silver addition originates from a standard solution of silver(I). Methods of this type are classed under the general subheading of *argentimetric* methods.

The *complexation* methods include all methods wherein the reacting substance and the standard solution react to form a soluble but very slightly dissociated complex substance. These techniques are sometimes referred to as *complexometric* methods.

The *oxidation-reduction*, or *redox*, methods include all methods whereby the reacting substance is oxidized or reduced by the active component of the standard solution.

Reaction requirements for volumetric methods   The reaction should be rapid. This is required in order that, practically speaking, zero time will be needed after each volume addition of titrant for the reaction to reach equilibrium. Slower reaction rates render the associated volumetric processes intolerable with respect to the time consumed for the titration.

The reaction should have a well-defined chemical equation, with no side reactions, so that the stoichiometry of the equation will allow exact calculation of the amount of reacting substance.

The reaction must proceed to completion, when the stoichiometric amount of the standard solution has been added. This criterion permits satisfactory end-point location. The criterion for reaction completion may vary somewhat, depending on the quantity determined. Where significant amounts are under investigation, a concentration maximum of 10^{-6} M for the unreacted portion of the substance involved at the titration equivalence point is usually considered satisfactory. Where lesser amounts are under investigation, a 99.9 percent reaction completion should be realized in order to provide the analytical accuracy customarily desired at the equivalence point of 1 part in 1000.

The titration reaction should be amenable to the use of some simple method of detecting the equivalence point or end point of the titration. As has been previously stated, the equivalence point is that point in the titration where an exact stoichiometric addition of the standard solution has been made. Usually an indicator technique will signal the end point of the titration. Ideally, the end point and the equivalence point should coincide. Where they do not, a titration-error situation may exist.

The requirements outlined in the foregoing are not always realized for every volumetric method. Where a departure exists, it is important to *know* the direction and magnitude of the departure in order to attempt to offset the introduction of a possible error source. Many volumetric titration systems of definite value are applied where one or more of the requirements listed are not ideally met. In these systems, efficient titrations are obtained by knowing the nature and magnitude of the departures from optimum conditions.

Comparison of gravimetric and volumetric methods   Volumetric methods are usually simpler with respect to manipulative details, in that they do not in general involve such operations as isolation, drying or igniting, weighing of residues, etc. This simplicity results in speed in securing results, and in a lessening of the chances for accidental error by a decrease in the number of analytical steps.

Corrections aimed at offsetting problematic situations and nonoptimum conditions are more easily applied with volumetric methods than with the generally more complex gravimetric techniques.

In the majority of instances, volumetric methods are less liable to be affected by interfering actions of certain types. For example, in the gravimetric determination of calcium by isolation as the oxalate and subsequent ignition to calcium oxide, the presence of silicon dioxide causes an error by inclusion with the calcium oxide. This is not the case with the volumetric adaptation of this general technique. Here the isolated oxalate is dissolved in acid and titrated with standard potassium permanganate solution.

By varying the concentration of the titrant within permissible limits, a volumetric method can be made to accommodate a large range of the sought-for substance. The limitation here is associated with reaction characteristics, such as degree of completion and reaction velocity, that may dictate when titration may not be feasible beyond a certain degree of dilution.

The volumetric method is of particular value in the determination of reasonably small quantities of substance. For example, 6 mg of NaCl would require about 10 ml of 0.01 M $AgNO_3$ solution in titration. The error in such a titration would be based on the measurement of the titrant volume; and it would be about 1 part in 500. In the gravimetric precipitation method, such a quantity of NaCl would yield a precipitate of AgCl of about 15 mg. The error in this determination would be based on the measurement of the precipitate weight; and it would be about 1 part in 75.

The gravimetric method does have some advantage, however, in the determination of quantities of substance between 0.1 and 1 g. Here the low relative error in the weight measurement usually provides for a better potentiality for accuracy than would be the case with the corresponding volumetric method.

Because of the ease of correcting or compensating for error in volumetric methods, empirical or nonstoichiometric techniques are sometimes applied (e.g., in the titration of MnO_4^- by $NaAsO_2$ solution, where the degree of completion varies according to the amount of MnO_4^- under investigation, and where the technique requires the simultaneous analysis of a control sample in the same range of MnO_4^- as the unknown). Such techniques are extremely rare in gravimetric analysis. Their use in volumetric analysis is justified only when the method is repeatedly checked to ensure that the system is under proper control.

When a determination is to be based on a reaction that is incomplete quantitatively speaking at the equivalence point (say, for example, a precipitation reaction), the gravimetric method permits the addition of excess amounts of the precipitating substance in order to force an improved degree of completion. Such a procedure, obviously, cannot be applied in any direct volumetric precipitation technique, since the aim is to determine the *exact* equivalence-point volume of titrant. In certain cases relative to volumetric analysis, an excess of the titrant may be deliberately added in order to force the reaction to completion more readily, with back-titration of the excess of titrant then being carried out. However, such a procedure is subject to an increased error as the result of doubling the burette-measurement-uncertainty contribution.

General procedures of volumetric methods   *Standard solution, or titrant*   Very pure reagents of high stability are said to allow preparation of standard solutions without the need for subsequent solution standardization processes. In such cases the amount of the reagent required is accurately weighed out, dissolved, and diluted to the exact volume, and the concentration is assumed on a theoretical basis.

The practice is a poor one. It places too much reliance on the accuracy of a single measurement of weight, and all subsequent results obtained using the standard solution involved depend on the reliability of this single measurement. In addition to this, it is not uncommon to find that the same titrant may require the use of slightly different titer values, depending on the method of analysis in which the titrant is used. For example, a potassium dichromate titrant may yield slightly but significantly different titer values for iron when

1. Standardized against pure iron volumetrically, using an internal indicator such as sodium diphenylbenzidine sulfonate
2. Standardized against pure iron volumetrically, using a technique of analysis identical to that used in 1, but employing a potentiometric first-derivative method for the location of the titration end point

In every technique of standard-solution preparation, consideration must be given, with respect to reagent concentration, solvent nature, etc., to the requirements of the processes of titration and determination in which the solution will be applied.

Standardization process   In the ideal case, the standard solution should be standardized. Again ideally, the process of standardization should be carried out by a technique identical to that to be used in the subsequent analytical process involving the titrant. The greatest care should be taken during all steps of the standardization procedure. Nothing less than triplicate results should be secured, and these must involve three *separate* samples, not three aliquots taken from the *same* sample. The accuracy of the standardization process should be such as to yield a standard value for the titrant suffering from an uncertainty equal to or less than that required for the determination in which the titrant involved is to be used.

Wherever possible, the titrant or standard solution is standardized against a pure form or a pure compound of the substance to be determined. For example, a potassium dichromate titrant should be standardized against pure iron, where iron is the substance to be determined by the titrant. In many cases, however, a standard substance other than that which is to be determined by the standard solution is used in the standardization process. For example, a standard solution of NaOH may be standardized against potassium acid phthalate, $C_6H_4 \cdot COOK \cdot COOH$, and the standard solution may then be used to titrate solutions of H_3O^+ originating from source substances such as HCl, HNO_3, $CH_3 \cdot COOH$, etc.

In general, the standarization process should display the following characteristics:

1. The standardizing substance should be available in the pure state. When it is, the substance is called a *primary standard.*
2. The standardizing substance should be preservable in the pure state without difficulty. This means that the substance should not decompose or be hygroscopic or deliquescent.
3. The standardizing substance should, where practicable, show a high equivalent weight, in order to reduce the effect of weighing uncertainties. Note the following:

$(COOH)_2$ equivalent weight (approximate)	45
Weight for 0.01 equivalents of H_3O^+	0.4500 g
Error in weighing this amount of pure $(COOH)_2$	1:2250

$C_6H_4 \cdot COOK \cdot COOH$ equivalent weight (approximate)	
Weight for 0.01 equivalents of H_3O^+	204
Error in weighing this amount of pure	2.0400 g
$C_6H_4 \cdot COOK \cdot COOH$	1:10,200

Note the decreasing influence of weighing uncertainties on the final result with substances of higher equivalent weight.

4. The standardizing substance should involve a standardization reaction that goes to completion, ideally to a degree not less than that of the determination reaction to follow.
5. Strictly stoichiometric relationships should be observed for the standardization reaction.
6. The standardization reaction must be amenable to the use of some relatively simple indicator technique to determine the end point of the titration.
7. There should be no difference between the end point and the equivalence point. In other words, the titration blank should be zero, or at least small enough to be negligible.
8. A *secondary standard* substance is an impure primary standard. If the degree of purity is known accurately, and if interfering substances are not present, such a secondary standard may be used in a standardization process.

It is not always possible to secure *all* the characteristics listed for a standard substance, but the requirements indicated should be met at all times as closely as possible.

Use of reasonable weight of standardizing substance   The assumption that we shall use here is that a weighing operation is open to a weighing uncertainty of ± 0.0001 g per single weighing, and that two weighings are necessary to establish any weight in a by-difference weighing operation. The following situations should then be noted.

Sample weight, g	Uncertainty, g	Precision
1.0000	±0.0002	1:5000
0.5000	±0.0002	1:2500
0.2000	±0.0002	1:1000
0.1000	±0.0002	1:500

The relative error or precision introduced by the sample weighing operation should ideally be less than the precision and accuracy required for the standardization process. Thus, a weighing precision of better than 1:1000 should generally apply. Under the assumed uncertainty conditions for a single weighing, this requires a sample weight greater than 0.2000 g.

Use of a reasonable titrant volume during standardization   The assumption we shall use is that the volume measuring operation is open to a volume uncertainty of ±0.02 ml per single volume measurement, and that two volume measurements are necessary to establish any titrant volume. The following situations should then be noted.

Titrant volume, ml	Uncertainty, ml	Precision
40.00	±0.04	1:1000
25.00	±0.04	1:625
10.00	±0.04	1:250

The relative error or precision introduced by the volume measurement operation should again ideally be less than the precision and accuracy required for the standardization process. Thus, a volume measuring precision of better than 1:1000 should generally apply. Under the assumed conditions for single-volume measurement uncertainty, this requires a titrant volume of 40.00 ml or greater. It should be understood that with volumes of 40.00 ml or greater, the use of a 50-ml burette for titration purposes is implied. Volumes greater than 50 ml should be avoided, since these require refilling the burette, with subsequent doubling of the volume measurement uncertainties.

Avoid back-titration where possible   As mentioned elsewhere, forward-titration can be carried to some point of excess in order to establish an improved reaction-completion situation. Back-titration with an appropriate titrant then determines the excess volume of the primary titrant beyond the equivalence-point or end-point volume. The following should be noted.

Forward-titration	39.85 ± 0.04 ml
Back-titration	1.30 ± 0.04 ml
Net-titration volume	38.55 ± 0.08 ml
Precisions involved	
Forward-titration	1:1000
Back-titration	1:32
Net titration	1:480

Note that the precision of the net-titration volume is reduced by the additive nature of the volume measuring uncertainties, and that the technique requires the use of two standard solutions, each capable of contributing a source of error through their standard or titer values.

Avoid standardization of one standard solution against another Regardless of the source of advice, this form of standardization should be avoided wherever possible. First, the technique of such a process of standardization may bear little resemblance, relative to the details of the method, to that used in the analytical determination process in which the standard solution is eventually to be used. Sodium thiosulfate solution can be standardized against a standard solution of potassium dichromate in a simple method using dilute hydrochloric acid, potassium iodide, starch solution as an indicator, etc. The use of sodium thiosulfate solution as a titrant in the determination of copper in an ore follows a method which, although somewhat similar, involves several important and significant differences. Thus, the theoretical copper titer calculated from the simple standardization process against standard potassium dichromate solution may not be satisfactory for application in the determination of copper in an ore.

Standardization in triplicate We emphasize again the necessity of standardizing in at least triplicate, the process involving three *separate* samples and not three aliquots taken from the *same* sample. As will be indicated later, the reliability of the average of a series of measurements increases with the number of measurements in the series. The situation applies when three separate samples are used. The use of three aliquots merely serves as a means of reducing the average or standard deviation for the actual process of titration alone; any error in the single sample weight carries through into each of the results secured from the three aliquots.

It is most important to note that the observations made relative to reasonable weight and volume, back-titration, and avoidance of replication of determination on the basis of aliquots applies equally well to **any** *titration process, and not just to those involving a standardization procedure.*

Volumetric standardization and titration processes   In general, the normality of a prepared standard solution, or titrant, can be determined in the standardization process from either

$$N = \frac{W}{(EW)(V \pm V_b)} 1000 \text{ ml}$$

where N = normality, equiv/l
W = weight of standardizing substance, g
EW = equivalent weight of standardizing substance, g/equiv
V = volume of titrant required to reach end point or equivalence point of titration, ml
V_b = indicator blank where applicable, ml

or

$$N = \frac{W}{EW(V \pm V_b)}$$

where N = normality, meq/ml
W = weight of standardizing substance, mg
EW = equivalent weight of standardizing substance, mg/meq
V = volume of titrant required to reach end point or equivalence point of titration, ml
V_b = indicator blank where applicable, ml

The normality value obtained from either expression will, of course, be identical in the arithmetic sense.

Where a standardized solution or titrant is used in the analysis of an unknown, the percentage of the sought-for substance may be found from either of the following equations:

$$P = \frac{(V \pm V_b)(N)(EW)}{1000S} 100$$

$$= \frac{(V \pm V_b)(N)(EW)}{10S}$$

where P = percent of sought-for substance
N = normality of titrant, equiv/l (equiv/1000 ml)
EW = equivalent weight of sought-for substance or substance directly related to sought-for substance in the reaction involved, g/equiv
S = sample weight, g
V = volume of titrant required to reach end point or equivalence point of titration, ml
V_b = indicator blank where applicable, ml

or

$$P = \frac{(V \pm V_b)(N)(EW)}{S} 100$$

where P = percent of the sought-for substance
N = normality of titrant, meq/ml
EW = equivalent weight of sought-for substance or substance directly related to the sought-for substance in the reaction involved, mg/meq
S = sample weight, mg
V = volume of titrant required to reach end point or equivalence point of titration, ml
V_b = indicator blank where applicable, ml

In the above expressions, the molarity of the titrant may be substituted for its normality, where

$$M = \frac{N}{n}$$

as described elsewhere, and where the equivalent weight (EW) is multiplied by the value of n for the titrant substance.

Where the titer of the standard solution is used, the calculations are simplified, since the value of the titrant in terms of the quantity of sought-for substance per milliliter of titrant is known. Here we have

$$P = \frac{(V \pm V_b)(T)}{S} 100$$

where P = percent of sought-for substance
T = titer in terms of the grams of sought-for substance per milliliter of titrant
S = sample weight, g
V = volume of titrant required to reach end point or equivalence point of titration, ml
V_b = indicator blank where applicable, ml

The value of T may also be expressed in milligrams of sought-for substance per milliliter of titrant, providing that the sample weight is expressed in milligrams.

Example A solution of potassium dichromate, $K_2Cr_2O_7$, is prepared and standardized against 0.2690 g of pure iron wire. The indicator used in the titration is sodium diphenylbenzidine sulfonate, and 48.04 ml of the $K_2Cr_2O_7$ titrant is needed to reach the end-point indication. The end point is taken as coinciding with the equivalence point. Determine the normality, the molarity, and the iron titer of the solution.
The balanced equation for the reaction is

$$K_2Cr_2O_7 + 6Fe^{2+} + 14H^+ \rightleftharpoons 2Cr^{3+} + 6Fe^{3+} + 2K^+ + 7H_2O$$

and the equivalent weight of iron in the above reaction is 55.85. Therefore,

$$N = \frac{0.2690}{55.85 \times 48.04} \; 1000 = 0.1002^{\,6}$$

$$= 0.1003$$

$$M = \frac{N}{n} = \frac{0.1002^6}{6} = 0.01671$$

$$T = \frac{0.2690}{48.04} = 0.005600 \text{ g Fe/ml}$$

PROBLEMS

1. 9.84 g of $BaCl_2$ is dissolved in exactly 100.00 ml of water at 25°C. What is the percent by weight of $BaCl_2$ in this solution?

2. Exactly 1.000 l of water at 21°C is used to prepare a 5.67% by weight solution of $AgNO_3$. What weight in grams of $AgNO_3$ is required?

3. Determine the gram-molecular weight (GMW) of potassium zinc ferrocyanide, $K_2Zn_3(Fe(CN)_6)_2$.
Ans. **698.22**

4. Use the *balanced* equations below to determine the following:

(*a*) The titrant is the first substance. Determine its molarity where its titer in the second substance is 1 ml = 0.00500 g.

(*b*) The gram-equivalent weight (GEW) of the first substance in terms of its arithmetic relationship to the gram-molecular weight (GMW).

(1) $2H_3PO_4 + 3Ca(OH)_2 \rightleftharpoons Ca_3(PO_4)_2(s) + 6H_2O$

(2) $Na_2B_4O_7 + 2HCl + 5H_2O \rightleftharpoons 2NaCl + 4H_3BO_3$

(3) $K_2CrO_4 + 2AgNO_3 \rightleftharpoons Ag_2CrO_4(s) + 2KNO_3$

(4) $2Ce(SO_4)_2 + 2FeSO_4 \rightleftharpoons Ce_2(SO_4)_3 + Fe_2(SO_4)_3$

(5) $SnCl_2 + 2HgCl_2 \rightleftharpoons SnCl_4 + Hg_2Cl_2(s)$

(6) $H_2S + I_2 \rightleftharpoons S(s) + 2H^+ + 2I^-$

(7) $CaCl_2 + (NH_4)_2C_2O_4 \rightleftharpoons CaC_2O_4(s) + 2NH_4Cl$

5. Calculate the molarity of the following solutions, prepared as indicated.

(*a*) 0.4320 g of NaOH in 0.075 l of solution

(*b*) 1.967 g of $AgNO_3$ in 127 ml of solution

(*c*) 432.0 mg of $K_2Cr_2O_7$ in 98.0 ml of solution

(*d*) 2.867 g of NaBr in 0.134 l of solution

(*e*) 1526 mg of $KMnO_4$ in 95.2 ml of solution

(*f*) 5.482 g of KCN in 1.875 l of solution

Ans. (*c*) 0.01498 *M*; (*e*) 0.1014 *M*

6. Calculate the normality of each solution in Prob. 5, where the substances involved result in the following reaction products:

(*a*) Na^+ and H_2O

(*b*) $Ag_2CrO_4(s)$

(*c*) Cr^{3+}

(*d*) $AgBr(s)$

(*e*) Mn^{2+}

(*f*) $Ag(CN)_2^-$

7. Calculate the normality of each prepared solution shown in the following, when the substance is involved in the reaction indicated by the balanced chemical equation given.

(a) 29.80 g of NH_3 in 465 ml of solution

$$2NH_3 + Ag^+ \rightleftharpoons Ag(NH_3)_2{}^+$$

(b) 0.6498 g of $Na_2S_2O_3 \cdot 5H_2O$ in 25.45 ml of solution

$$2S_2O_3{}^{2-} + I_2 \rightleftharpoons S_4O_6{}^{2-} + 2I^-$$

(c) 3.691 g of $KMnO_4$ in 985 ml of solution

$$2MnO_4{}^- + 10I^- + 16H^+ \rightleftharpoons 2Mn^{2+} + 5I_2 + 8H_2O$$

(d) 2.134 g of $KBrO_3$ in 56.5 ml of solution

$$BrO_3{}^- + 6I^- + 6H^+ \rightleftharpoons Br^- + 3I_2 + 3H_2O$$

Ans. (c) 0.1186 N

8. Calculate the titers of the following solutions in milligrams per milliliter of the substances mentioned.

(a) 0.1094 N NaOH   In terms of HCl
(b) 0.2039 N $BaCl_2$ In terms of SO_3
(c) 0.0854 N K_2CrO_4 In terms of both Pb and Ag
(d) 0.1102 N $K_2Cr_2O_7$ In terms of both Pb and Fe
(e) 0.0685 N NH_3 In terms of Ag [forming $Ag(NH_3)_2{}^+$]

9. Calculate the molarity of the following solutions:

(a) 24.1 ml of HCl, having a density of 1.18 and containing 37.0% by weight of HCl, is diluted to exactly 1000 ml

(b) H_2SO_4, having a density of 1.28 and containing 30.2% by weight of SO_3

Ans. (b) 4.83 M

10. (a) 75 ml of 0.1104 N HCl is diluted to 450 ml. What is the normality of the new solution?

(b) What volume of water must be added to 24.8 ml of 0.246 N NaOH to yield a 0.0850 N NaOH solution?

(c) 56.4 ml of 0.116 N HCl is added to 47.4 ml of 0.246 N HCl. What is the normality of the resulting solution?

(d) 125.6 ml of 0.153 N NaOH is added with thorough stirring to 234.1 ml of 0.0961 N HCl. What is the normality of the resulting solution relative to what substance?

11. Balance the following skeletal equations (all reactants and products are given) by the oxidation-numbers method.

(a) $K_2Cr_2O_7 + NaCl + H_2SO_4 \rightleftharpoons CrCl_2O_2 + KHSO_4 + NaHSO_4 + H_2O$

(b) $HI + H_2SO_4 \rightleftharpoons I_2 + H_2S + H_2O$

(c) $As_2O_3 + I_2 + H_2O \rightleftharpoons H_3AsO_4 + I^- + H^+$

(d) $KMnO_4 + Na_2C_2O_4 + H_2SO_4 \rightleftharpoons K_2SO_4 + Na_2SO_4 + MnSO_4 + CO_2 + H_2O$

Ans. (b) $8HI + H_2SO_4 \rightleftharpoons 4I_2 + H_2S + 4H_2O$

12. Use the ion-electron method to determine the complete and balanced ionic equation for each of the following skeletal equations. Convert to the molecular formulation, using for each equation the specific data given.

Acidic
(H_2SO_4)
(a) $BrO_3{}^- + Br^- + H^+ \rightleftharpoons Br_2 + H_2O$

Use K to balance the molecular formulation.

Acidic
(HCl)

(b) $BrO_3^- + As^{3+} + H^+ \rightleftharpoons Br^- + As^{5+} + H_2O$

Use K to balance the molecular formulation.

Acidic
(H_2SO_4)

(c) $NO_2^- + MnO_4^- + H^+ \rightleftharpoons NO_3^- + Mn^{2+} + H_2O$

Use K to balance the molecular formulation.

Basic
(NaOH)

(d) $OH^- + S_2O_3^{2-} + I_2 \rightleftharpoons SO_4^{2-} + I^- + H_2O$

Use Na to balance the molecular formulation.

Basic
(NH_3)

(e) $Mn^{2+} + H_2O + NH_3 + O_2 \rightleftharpoons Mn(OH)_3 + NH_4^+$

Use SO_4 to balance the molecular formulation.

13. Samples of impure acidic substances were analyzed by a neutralization method involving over-titration with 0.1049 N NaOH solution, with subsequent back-titration using 0.0998 N HCl. The following values were secured for the samples in question. Calculate the percentage purity relative to the specific substances required.

	Sample weight, g	Volume NaOH, ml	Volume HCl, ml	Sought-for substance
(a)	1.865	31.63	1.84	$C_6H_4 \cdot COOK \cdot COOH$ Potassium, acid phthalate
(b)	2.1398	25.26	2.12	$COONa \cdot COOH$ Sodium hydrogen oxalate
(c)	1.936	34.84	0.68	$COOH \cdot COOH$ Oxalic acid
(d)	1.346	19.49	1.05	$KHSO_4$ Potassium bisulphate

14. Suppose that you are to prepare a silver nitrate titrant to be used in the volumetric determination of KI, so that 1 ml of the titrant would be equal to 1.00 percent of KI when a 0.7500-g sample weight of unknown is used in all analyses. Indicate what the molarity of the $AgNO_3$ solution must be, and how much $AgNO_3$ in grams must be dissolved in water and made up to exactly 1.0000 l.

Ans. 0.04518 M; 7.675 g

15. 15.00 ml of a solution of HNO_3, with density 1.060 g/ml and 11.0% HNO_3 by weight, requires 55.46 ml of an NaOH titrant for complete neutralization. What is the normality of the NaOH solution?

16. Indicate the minimum sample weight that should be taken in order to have a titration volume of not less than 40 ml of 0.1000 N titrant in each of the following analytical titrations. State whether or not you believe the sample-weight minimum requirement to be practical in each case.

(a) Analysis of hematite containing 50.0% Fe_2O_3, the titrant being $KMnO_4$ in the reaction

$$KMnO_4 + 5FeCl_2 + 8HCl \rightleftharpoons MnCl_2 + 5FeCl_3 + KCl + 4H_2O$$

(*b*) Analysis of impure soda ash containing 75.0% Na_2CO_3, the titrant being HCl in the reaction

$$Na_2CO_3 + 2HCl \rightleftharpoons 2NaCl + CO_2 + H_2O$$

(*c*) Analysis of copper ore containing 15.5% copper, the titrant being $Na_2S_2O_3$ in the reaction series:

$$2Cu^{2+} + 4I^- \rightleftharpoons 2CuI + I_2$$
$$2S_2O_3{}^{2-} + I_2 \rightleftharpoons S_4O_6{}^{2-} + 2I^-$$

17. Indicate what weight of sample as a minimum would be required for an ore containing 50.0% iron in order to secure a titrant volume of 40 ml minimum, using 0.1000 N $KMnO_4$ titrant solution in the reaction shown in Prob. 16*a*. What weight would be required for the same situation, but with 1.000 N $KMnO_4$ as the titrant? What weight would be required for the same situation, but with 0.0100 N $KMnO_4$ as the titrant? From the foregoing, draw conclusions as to why titrants in volumetric analysis are usually prepared and used at from 0.1 to 0.2 N.

18. Iron is being determined and reported as FeO in samples of ilmenite ore. The titrant is $K_2Cr_2O_7$ solution in the reaction

$$K_2Cr_2O_7 + 6FeCl_2 + 14HCl \rightleftharpoons 2CrCl_3 + 6FeCl_3 + 2KCl + 7H_2O$$

If 0.5000-g samples are being weighed out for each analysis, indicate what the titrant normality should be in order that the burette volume used for each titration will read the FeO percentage directly.

19. EDTA (disodium ethylenediaminetetraacetic acid) reacts in a complexation titration with magnesium as follows:

$$Mg^{2+} + \underset{\text{(EDTA)}}{Na_2H_2Y} \rightleftharpoons MgY^{2-} + 2Na^+ + 2H^+$$

A standard 0.0100 M solution of Ca^{2+} is prepared and the EDTA solution is standardized against it. The standardization volume is 26.49 ml of the EDTA solution for 25.00 ml of the calcium(II) solution. The reaction between EDTA and calcium(II) is identical to that between EDTA and magnesium(II).

100.0 ml of a sample of natural water is to be analyzed for hardness. This sample requires 34.94 ml of the EDTA solution to react completely with the contained Mg^{2+}. Calculate the magnesium content of the water in parts per million (ppm). Calculate the hardness of the water as ppm of magnesium carbonate, $MgCO_3$. Remember that 1 ppm is 1 mg/l.

Ans. 80.2 ppm Mg; 278 ppm $MgCO_3$

20. A sample of an ore contains calcium as $CaCO_3$. A 0.7564-g sample of this material is treated so as to secure eventually a precipitate of the calcium as calcium oxalate, CaC_2O_4. This precipitate is filtered, washed, and ignited to CaO. The cooled residue is found to weigh 0.1879 g. Calculate the percentage of $CaCO_3$ in the ore.

21. In the sample discussed in Prob. 20, calculate how many milliliters of 0.1091 N $KMnO_4$ solution would be required to react quantitatively with the redissolved calcium oxalate precipitated from 0.9981 g of the ore sample. The titration reaction is

$$2KMnO_4 + 5CaC_2O_4 + 8H_2SO_4 \rightleftharpoons 2MnSO_4 + 10CO_2 + K_2SO_4 + 5CaSO_4 + 8H_2O$$

22. What would be the titer values in grams per milliliter of 0.1069 N $KMnO_4$ solution in terms of iron, titanium, oxalate ion, antimony, and molybdenum for the following titration reactions:

(*a*) $MnO_4^- + 5Fe^{2+} + 8H^+ \rightleftharpoons Mn^{2+} + 5Fe^{3+} + 4H_2O$

(*b*) $MnO_4^- + 5Ti^{3+} + 8H^+ \rightleftharpoons Mn^{2+} + 5Ti^{4+} + 4H_2O$

(*c*) $2MnO_4^- + 5C_2O_4{}^{2-} + 16H^+ \rightleftharpoons 2Mn^{2+} + 10CO_2 + 8H_2O$

(*d*) $4KMnO_4 + 5Sb_2(SO_4)_3 + 24H_2O \rightleftharpoons 10H_3SbO_4 + 2K_2SO_4 + 4MnSO_4 + 9H_2SO_4$

(*e*) $6KMnO_4 + 5Mo_2O_3 + 9H_2SO_4 \rightleftharpoons 10MoO_3 + 3K_2SO_4 + 6MnSO_4 + 9H_2O$

Ans. (*b*) 0.00512 g Ti/ml; (*d*) 0.00651 g Sb/ml

23. Suppose that you are asked to standardize a solution of $KMnO_4$ against 0.3000 g of $Na_2C_2O_4$ and that you find that the titration volume required is 42.54 ml. Work out the titer in terms of $Na_2C_2O_4$ and determine the conversion factors which would then provide from this titer the theoretical titers for iron, titanium, antimony, and molybdenum, as these would react with $KMnO_4$ in the reactions given in Prob. 22.

REFERENCES

1. Margolis, E. J.: "Bonding and Structure," Appleton Century Crofts, New York, 1968.
2. Butler, J. N., B. A. Dunell, and L. G. Harrison: "Problems for Introductory University Chemistry," Addison-Wesley, Reading, Mass., 1967.
3. Day, R. A., Jr., and A. L. Underwood: "Quantitative Analysis," 2d ed., Prentice-Hall, Englewood Cliffs, N.J., 1967.
4. Margolis, E. J.: "Formulation and Stoichiometry," Appleton Century Crofts, New York, 1968.
5. Kolthoff, I. M., E. B. Sandell, E. J. Meehan, and Stanley Bruckenstein: "Quantitative Chemical Analysis," 4th ed., Macmillan, New York, 1969.
6. Fritz, J. S., and G. H. Schenk, Jr.: "Quantitative Analytical Chemistry, 2d ed., Allyn and Bacon, Boston, 1969.
7. Blaedel, W. J., and V. W. Meloche: "Elementary Quantitative Analysis," Harper & Row, New York, 1963.

3
The Treatment of Analytical Data

3.1 GENERAL

No measurement is free from *absolute error*—that is, the difference between the measured value and the true value. Every measurement made, no matter how carefully, will be subject to some margin of absolute error.

In addition to this situation, every measurement made will be subject to some uncertainty in the final value secured, regardless of any relationship to the true value. Thus, we can express any measurement or measured value to a limited number of figures, and the final limiting, or significant, figure will reflect the reliability of the method of measurement.

Although these two aspects of measurement, accuracy and precision, are separate entities, they will be shown later to exhibit a strong degree of interrelationship. An important factor will obviously be to determine the magnitude of the absolute error. An even more important factor will be to determine the reliability of a single measurement and the ability of repetitive measurements of the same dimension or property to agree among each other.

Remember, of course, that we can never *know* the true value. As will be seen later, we can, by a variety of methods, determine a value to which an assigned margin of error will apply. The smaller this margin of error, the more closely will the determined value approach the true value.

3.2 DETERMINATE ERRORS

Errors, the sources of which can be located or defined, are known generally as *determinate*, or *systematic*, errors. A good operator can usually make use of his knowledge and experience to locate sources of determinate error. Errors of this type can be classified, at least partially, as follows:

1. Instrumental errors
    - (*a*) Uncertainty in reading an instrument scale or measuring system
    - (*b*) Faulty balance weights
    - (*c*) Poorly calibrated glassware
    - (*d*) Impurities in reagents
    - (*e*) Poor selection of equipment (e.g., NaOH attack of glass vessels in an analysis to determine SiO_2)
2. Methodic errors
    - (*a*) Solubility of a precipitate
    - (*b*) Incomplete reaction
    - (*c*) Simultaneous precipitation, coprecipitation, or postprecipitation effects
    - (*d*) Decomposition of a precipitate or residue during heating
    - (*e*) Side or auxilliary reactions
3. Operative errors
    - (*a*) Manipulative errors by the analyst
    - (*b*) General lack of laboratory experience
    - (*c*) Uncovered beakers and the inadvertent introduction of foreign materials (e.g., burner SO_2 in SO_3 determination)
    - (*d*) Spilling and loss by bumping during an evaporation process
    - (*e*) Loss of precipitate during filtration (e.g., colloidal state)
    - (*f*) Over- or underwashing during the filtration process
    - (*g*) Poor selection of ignition temperature
    - (*h*) Insufficient drying or ignition
    - (*i*) Weighing hot crucibles and contents
    - (*j*) Poor selection of dessicant or poorly maintained dessicant in the dessicator
4. Personal errors
    - (*a*) Poor color perception
    - (*b*) Color blindness
    - (*c*) Prejudice (e.g., the tendency to constrain a result to agree with a preconceived situation)

It will be worthwhile to discuss at some length the determinate-error source involving the uncertainty in reading an instrument scale or measuring system.

Suppose that a group of individuals is asked to measure a length of 2 ft with a yardstick capable of accurate measurement to the nearest one-sixteenth of an inch. If they are asked to report the measurement to the nearest one-half inch, the possibility of disagreement between the measurements reported by different individuals will be quite remote. On the other hand, suppose that the same group of individuals is asked to weigh a 1-g object to the fourth decimal place on a balance with a limit of measurement in the fourth decimal place. A situation would then arise where the weights reported by different individuals would show some tendency to vary as to the fourth-decimal-place figure or digit.

Every instrument scale or measuring system can be read to a limited degree of certainty. When reading a value from a measuring system, it is considered practical to assign an ability to estimate the value to within one-fifth of the smallest division of the scale involved. Thus, a volume read from a burette with the smallest volume division as 0.1 ml can be estimated to the nearest ± 0.02 ml, and we would, for example, make a single reading of volume as 24.64 ± 0.02 ml. This would imply that a subsequent reading of the same volume could be expected to provide some value between 24.62 and 24.66 ml. Again, a spectrophotometer dial graduated in divisions of percent transmission of one unit, from 0 to 100, can be read to the nearest ± 0.2 percent transmission, and we would report for a single reading, for example, a value of 40.3 ± 0.2 percent transmission. This, as before, implies that if the value were read again, some value between 40.1 and 40.5 percent transmission could be expected.

In point of fact, the question of the uncertainty of a value read from a measuring system can be divided into three categories. The *reading uncertainty*, which we have just discussed, reflects the basic ability to read the scale of the instrument. For most instruments, as previously mentioned, the reading uncertainty will usually be about one-fifth of the smallest scale-division unit, although the physical size of the smallest division may be such as to require higher or lower reading estimates. The reading uncertainty as defined here does *not* take into consideration the possibility of operator error in reading the scale of the instrument, this being an indeterminate-error source.

The *precision uncertainty* refers to the ability of the instrument to replicate readings of the same quantity. Depending on the reliability of the instrument and its condition, the ability to replicate may correspond to or be greater than the reading uncertainty. Obviously, when considering the ability of an instrument to provide values, one should be governed by the precision uncertainty when this exceeds the reading uncertainty.

The *accuracy* of a value read from an instrument has nothing to do with either of the uncertainty factors just described. It represents the difference between the average of replicate instrument readings and the known correct value of the property read. Calibration techniques are required to indicate the level of accuracy of any instrument.

Sources or causes of determinate error may introduce either a *constant-error* situation or a *proportional-error* situation relative to the final results of an analysis.

For example, in an analysis to determine chloride by precipitation as silver chloride, the use of chloride-contaminated nitric acid as a constant volume addition before the separation of silver chloride will introduce a *constant additive* error. Insofar as the final result is concerned, the effect of this error will vary according to the amount of chloride in the sample being analyzed, at least with respect to its relative effect. Suppose, for instance, that the impure nitric acid introduces 1 mg of chloride for each 10 ml of acid used, and that 10 ml of acid is always added. If the sample under analysis contains 0.1000 g of chloride, the relative error will be

0.001 g/0.1000 g   or   + 10 ppt (parts per thousand)

On the other hand, if the sample under analysis contains 0.7000 g of chloride, the relative error will be given as

0.001 g/0.7000 g   or   + 1.4 ppt

Note that this is a *constant*-error situation, regardless of the size of the sample or the amount of chloride under determination, but that the magnitude of the *relative* error is proportional to the amount of chloride being determined.

A constant error is also generally introduced by the reading uncertainty of the instrument or measuring system. For example, where a 25-ml burette is used, the reading uncertainty for a single reading can usually be taken as ± 0.02 ml. Thus, the determinate error introduced by the reading uncertainty is constant regardless of the volume read. The *relative* error will, of course, be proportional to the magnitude of the volume read.

A case in point with respect to a *proportional*-error situation would be an instance where an improper titer value has been assigned to a standard solution used in a volumetric titration. Suppose, for example, that the solution involved is a potassium dichromate solution used in the determination of iron volumetrically, and that the titer used is 1 ml = 0.0050 g of iron, whereas the true titer is 1 ml = 0.0046 g of iron. Here the magnitude of the error introduced will be proportional to the amount of iron determined. Thus, on a 10.00-ml titration, the error will be

+0.004 g of iron   Relative error of 0.004 g/0.046 g, or +87 ppt

On a 20.00-ml titration, the error will be

+0.008 g of iron   Relative error of 0.008 g/0.092 g, or +87 ppt

Observe that this is a *proportional*-error situation, based on the amount of iron determined, but that the *relative* error remains constant.

3.3 INDETERMINATE ERROR

3.3.1 GENERAL

Indeterminate errors are frequently called *accidental*, or *random*, errors. Such errors can be attributed to no known cause, nor can they be predicted as to magnitude or direction for any single measurement in a series or for a single measurement standing alone. The sort of situation that can arise here—and this is given as a single general, but not unique, example—is that wherein an operator misreads an instrument scale value. Once set down and incorporated into the analytical system, there can be no possibility of later attributing this as the source of error. The incident is isolated, random, and accidental.

It is apparent that the possibilities of random error occurring will increase with the number of operations involved in securing a final measurement. Thus, in a complex gravimetric procedure of analysis, where multiple separation, filtration, washing, and re-solution operations may be involved, the possibility of random error occurring is relatively high. In a simpler volumetric titration procedure, with the same analytical aim in view, the reduction in the number of manipulative operations decreases the possibility of accidental error.

Whereas indeterminate errors can not be predicted with exactitude, their patterns of occurrence can be analyzed by the techniques of statistics in order to secure a worthwhile insight into their magnitudes, frequencies of occurrence, and effects on the final expression of results. We can, from such an analysis, arrange for a picture of the probabilities of occurrence of indeterminate error.

3.3.2 FREQUENCY DISTRIBUTION

Table 3.1 shows a series of 74 measurements of percent transmission obtained on solutions of permanganate ion resulting from a spectrophotometric investigation of a steel to determine its manganese content. These values were obtained on 74 separate samples of the same steel dissolved, prepared, and examined in exactly the same manner. The values are arranged in a random manner and vary from a low of 36.9 percent to a high of 38.5 percent. All values are subject to a reading uncertainty of ± 0.2 percent in the recorded percent transmission.

Table 3.2 shows the same data grouped in ascending order of magnitude, with the repetitive values accumulated together to show the frequency of occurrence. It will be observed immediately that there is evidence of a tendency to centralization.

A more efficient method of demonstrating the centralization tendency is to divide the values into cells or groups, with limits or boundaries to each cell or group. The number of cells must be properly chosen, and this will vary with the number of measurements. Usually a number of cells about equal to the root of the number of measurements will suffice. For the 74 values of percent transmission involved here, a division into 9 cells is adequate.

The limits of the cell boundaries are usually set to correspond to a value in the decimal place beyond the last decimal place reported for the values involved,

Table 3.1   Random distribution of percent transmission values*

1. 37.6	20. 37.7	39. 37.9	58. 37.6
2. 38.5	21. 37.8	40. 37.7	59. 37.4
3. 37.8	22. 37.5	41. 37.7	60. 37.7
4. 37.4	23. 37.7	42. 38.1	61. 37.9
5. 37.7	24. 38.0	43. 37.6	62. 38.0
6. 37.5	25. 37.6	44. 37.9	63. 37.7
7. 38.0	26. 37.4	45. 38.1	64. 37.3
8. 37.6	27. 38.1	46. 37.7	65. 37.5
9. 37.3	28. 37.7	47. 38.2	66. 37.7
10. 37.7	29. 37.9	48. 37.6	67. 38.0
11. 37.7	30. 38.3	49. 38.3	68. 37.2
12. 37.2	31. 37.8	50. 37.7	69. 37.8
13. 37.5	32. 38.1	51. 37.3	70. 38.2
14. 37.9	33. 37.7	52. 37.8	71. 37.9
15. 37.0	34. 37.4	53. 37.8	72. 37.1
16. 37.7	35. 37.7	54. 38.4	73. 37.7
17. 38.2	36. 36.9	55. 37.5	74. 37.3
18. 37.1	37. 37.8	56. 37.7	
19. 37.5	38. 37.6	57. 37.4	

*The data were obtained during the spectrophotometric determination of manganese in 74 samples of the same steel.

Table 3.2   Percent transmission values in ascending magnitude and grouped by value

Frequency	Value, percent
1	36.9
1	37.0
2	37.1
2	37.2
4	37.3
5	37.4
6	37.5
7	37.6
18	37.7
7	37.8
6	37.9
4	38.0
4	38.1
3	38.2
2	38.3
1	38.4
1	38.5

or to a figure just beyond the last significant figure for these values. A cell-boundary value about halfway between measured values is appropriate. This technique is customary and does not imply any change in the significance of the figures expressing the results. It merely serves to avoid having any value in the series coincide with the cell-boundary limits, thereby creating confusion as to the proper cell in which to include the value. Table 3.3 shows this new method of distribution.

Two important factors can be noted at this point:

1. Despite the extended range of 36.9 to 38.5 percent, there are few values below 37.2 percent or above 38.2 percent.
2. The highest frequency of occurrence for the series lies near the center of the range.

As a final step in the treatment of the data for this portion of the discussion, note from Fig. 3.1 that it is possible to express the results secured as a series of columns. Each column rests on a base equal to the cell-boundary limits involved, and the columns progress by cell boundaries from the lowest to the highest values. The height of each column is proportional to the frequency of occurrence of the values within the limits of the cell boundaries involved. Note also that the point locating the midcell value at the top of each column is shown. These midcell points are connected to yield a crude frequency-distribution line of a polygonal na-

Table 3.3 Percent transmission values grouped by cells

Cell mid-point, percent	Cell boundary, percent	No. of values
	36.85	
36.95		2
	37.05	
37.15		4
	37.25	
37.35		9
	37.45	
37.55		13
	37.65	
37.75		25
	37.85	
37.95		10
	38.05	
38.15		7
	38.25	
38.35		3
	38.45	
38.55		1
	38.65	

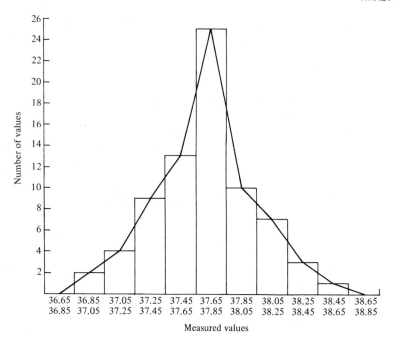

Fig. 3.1 Histogram and frequency distribution for transmission values on 74 replicate steel samples.

ture. It is important to note that either approach stresses the tendency for random values to centralize. The diagram in Fig. 3.1 is called a *histogram*.

3.3.3 DEVIATION

Whereas absolute error represents the departure of a measured value from the true value, *deviation*, or *apparent error*, represents the departure of a measured value from the arithmetic mean or average for the series containing the particular measured value involved. Thus, deviation bears a relationship to the average value of a series similar to that which exists between the absolute error and the true value. Consider a replicate series of measurements involving n values, these being

$$x_1, x_2, x_3, x_4, \ldots, x_n$$

Although the property being measured does not change in value over this series of measurements, random differences will be noted in the individual results constituting the series. As we have seen in the example just given, there will be some repetition of values.

We may determine the *average* value, or *arithmetic mean*, of the series from

$$\bar{x} = \frac{(x_1 + x_2 + x_3 + x_4 + \cdots + x_n)}{n} = \frac{1}{n} \sum_{i=1}^{i=n} x_i$$

As the number of measurements increases, the reliability of the average value $\bar{x}$ improves, and for a series of n measurements, the average value will be $\sqrt{n}$ times as reliable as any single measurement in the series. Thus, the average of a series containing four measurements is two times as reliable as any single measurement. Where the number of measurements is infinite, and where the technique of measurement suffers from indeterminate error only (no sources of determinate error), the average value of the infinite series will be infinitely reliable and will correspond to the true value.

A second method of expressing the results of a series of measurements is to determine the *median M*. The median of an odd number of measurements is the middle value of the orderly array—that is, the results arrayed in ascending order of value with the repetitive values included. Where an even number of measurements is involved, the median is the average of the two middle values of the orderly array. Generally speaking, the median is not as reliable as the mean in expressing the centralizing tendency of the results in a series of measurements. The median can be of some particular value, however, when dealing with a very small number of measurements, since it tends not to be influenced by an odd high or low result in the series.

Note the use of the term $\bar{x}$ for the average value, since it represents the average of a *finite* number of measurements. Where the number of measurements is *infinite*, the average value is sometimes called the *objective mean*, and is represented by the term μ.

The apparent error of each value, or the deviation of each value from the average value, is given by

$$d_i = x_i - \bar{x}$$

where d_i is the deviation for x_i.

In determining the sum of the individual deviations for the series, it is obvious that if sign is used, we have

$$\sum_{i=1}^{i=n} d_i \equiv O$$

If, however, we sum the individual deviations as absolute values, regardless of sign, and determine the *average deviation $\bar{d}$*, we have

$$\bar{d} = \frac{1}{n} \sum_{i=1}^{i=n} |d_i|$$

where $|d_i|$ represents the absolute value of d_i.

Where the number of measurements is infinite, the average deviation is assigned the term δ.

The value of $\bar{d}$, when preceded by a $\pm$ sign, can give us a term indicative, at least in an approximate way, of the range or margin of deviation around the average value of the series.

The average deviation, as we shall see shortly, does not have the statistical significance of other forms of deviation expression, such as the standard deviation, particularly where a small number of measurements is involved. In many instances, however, the average deviation is used in the light of its simplicity of calculation.

The *standard deviation* carries more statistical significance than the average deviation and is determined from

$$s = \sqrt{\frac{\sum\limits_{i=1}^{i=n} (x_i - \bar{x})^2}{n-1}} = \sqrt{\frac{\sum\limits_{i=1}^{i=n} d_i^2}{n-1}}$$

Where an infinite number of measurements is involved, the standard deviation is assigned the term σ.

Another method of expressing the extension of the results of a series of measurements is to calculate the *range R*, which is the difference between the largest and smallest values of the series. The range may be of poor reliability because of the undue influence of any one unusually high or low result where such exists in the series.

Returning now to the 74 percent transmission measurements discussed before, note that we have the following.

Average or mean	$\bar{x}$	37.7%
Median	M	37.7%
Range	R	1.6%
Average deviation	$\bar{d}$	$\pm 0.2^4\%$
Standard deviation	s	$\pm 0.3^2\%$

The average and standard deviations, as means of expressing data relative to the results of the series, are actually ± 0.2 and ± 0.3 percent, respectively. The elevated figures represent digits applicable where the deviations are used in calculation operations. An explanation of this situation will arise out of discussion on significant figures later in this chapter.

Table 3.4 again shows data from the 74 measurement series. In this case, the deviations are grouped in ascending order and accumulated with regard to magnitude. Sign for the deviation has been maintained here, in order that the data may be used in a frequency-of-occurrence plot. This plot, which shows magnitude of deviation with sign or direction vs. frequency of occurrence, is illustrated in Fig. 3.2. If the series under investigation had been extended to an infinite number of measurements, or even to a very large number of measurements, we would have secured a smooth modification of the dashed line shown.

In general, for any infinite series of measurements, the plot of magnitude of plus and minus deviations vs. frequency of occurrence yields a bell-shaped curve.

Table 3.4 Deviation study for 74 transmission values

Deviation from the mean	Number of deviations
0.0	18
+0.1	7
−0.1	7
+0.2	6
−0.2	6
+0.3	4
−0.3	5
+0.4	4
−0.4	4
+0.5	3
−0.5	2
+0.6	2
−0.6	2
+0.7	1
−0.7	1
+0.8	1
−0.8	1

The relationship between the height and practical range of the curve will generally be an indication of the nature of the measuring operation conducted, relative to the general precision and ability to yield repetitive values. Figure 3.3 gives some idea of this situation, and shows the distribution curves for three different series of

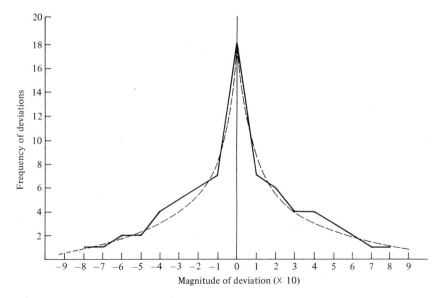

Fig. 3.2 Frequency of deviations—74 transmission-values series.

measurements. In a crude but expressive manner these might be described as follows.

Curve 1   suggests a normal distribution and would tend to indicate an experienced operator using good equipment

Curve 2   suggests a careless operator, an inexact measuring technique, poorly stabilized equipment, or some combination of these or similar factors

Curve 3   suggests a careful operator conducting an extremely precise measuring method on well-stabilized equipment

Regardless of the nature of the basic situation from which replicate measurements are derived, their statistical and graphic analysis will provide a similar curve with respect to general shape. Such a curve is called a *normal-error*, or *deviation-distribution, curve*, and is related to the *Gauss-Laplace law*. It has a mathematical relationship given by the equation

$$y = \frac{1}{\sigma\sqrt{2\pi}} \exp\left[-\frac{1}{2}\left(\frac{x_i - \mu}{\sigma}\right)^2\right]$$

where y = relative frequency

σ = standard deviation for the infinite measurement series and the distance from the zero-deviation axis to either of the points of inflection of the curve

exp = natural logarithmic base

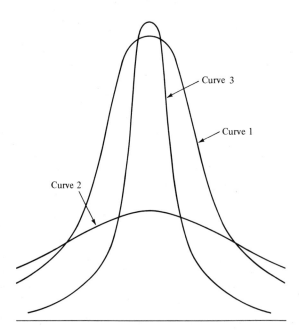

Fig. 3.3   Frequency-vs.-deviation-distribution curves.

x_i = any value in the series

μ = average value for the infinite series of measurements

The equation indicates the following important points:

1. Small deviations occur more frequently than do large ones and are therefore more probable.
2. Plus and minus deviations of equal magnitude will occur with the same frequency.

It is apparent that for any distribution curve, the entire area under the curve to infinity on either side of the zero-deviation axis represents the probability of any deviation occurring from zero to one infinitely large. Since this represents the probability of *some* deviation occurring, from zero to infinitely large, this area can be equated to a 100 percent or 1.00 probability.

Note now that if we cut off any specific deviation range, on both sides of the central axis, the area under the curve and enclosed by the vertical cutoff lines perpendicular to the deviation-magnitude axis, relative to the total area under the curve, represents the probability that any single value in the series will show a deviation of a magnitude within the limiting values assigned.

The standard deviation, for example, which can be shown mathematically to be the deviation on either side of the zero axis that intercepts the curve at the points of inflection, is illustrated relative to area in Fig. 3.4. The area under the curve for the limits of $+\sigma$ to $-\sigma$ is shown as the shaded area, and can be denoted by

Area $(-\sigma$ to $+\sigma)$

The chance, or probability, of a single measurement showing a deviation lying between $\pm\sigma$ is thus given by

$$\frac{\text{Area } (-\sigma \text{ to } + \sigma)}{\text{Total area}}$$

Calculation of the area under the curve for any set of limiting values of the relationship $(x_i-\mu)/\sigma$ can be obtained by integrating the appropriate Gauss-Laplace equation between the limits involved. Such a calculation, for limits given in effect by the standard deviation of $\pm\sigma$, shows a value of 0.6826. This implies that any single measurement in the series has a 0.6826 probability, or a 68.26 percent chance, of showing a deviation lying between $\pm\sigma$. It also implies that any further measurement conducted under exactly the same conditions has the same probability of showing a deviation between $\pm\sigma$. It further implies that in any infinite series of such measurements, 68.26 percent of the values will show deviations from the mean lying between $\pm\sigma$.

Similarly, for a range of deviation of the magnitude of $\pm\delta$, the average deviation, we can show a probability of 0.5762, implying that any single

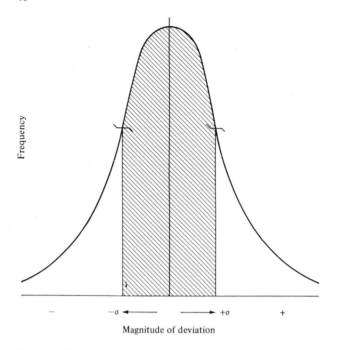

Fig. 3.4 Standard deviation cutoff limits at points of inflection.

measurement in the series as secured, or obtained later by exactly the same technique, has a 0.5762 probability, or 57.62 percent chance, of showing a deviation lying between $\pm\delta$. Again, it implies that for an infinite series, 57.62 percent of the values will show deviations lying between $\pm\delta$.

It will be noted that the average deviation is more restrictive than the standard deviation.

Figure 3.5 indicates graphically, and very generally, the locations on the distribution curve of three common-named deviations, the standard deviation, the average deviation, and the probable deviation, ρ. The latter has a 0.5000 or 50.00 percent probability factor.

It is customary to express the probability that a single value in a series will have a deviation lying between specific plus or minus limits of magnitude by expressing these limits as multiples of the standard deviation. Such a probability value, multiplied by 100, also represents the percentage of the measurements in an infinite series that will show deviations lying within the specific limits selected. The tabulation shown as Table 3.5 indicates this situation.

The conclusions arrived at with respect to the probability values assigned to certain limits of deviation, although valid in the truest sense when applied to an infinite series of measurements, can be used quite adequately to treat data from a finite series of measurements. It will be immediately apparent, even at this point, that the validity of such a treatment will decrease with decreasing numbers of measurements in the series.

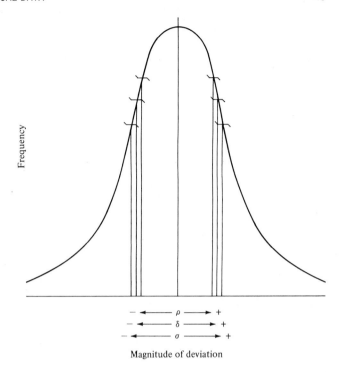

Frequency

Magnitude of deviation

Fig. 3.5 Relative positions at cutoff for standard, average, and probable deviations.

From what has been said so far, it will be realized that if only indeterminate errors are involved, and if a reasonably large series is under examination, the most probable value of the series will be the mean, or average, value. As has already been pointed out, where an infinite series of measurements is involved, and where no sources of determinate error exist, the mean, or average, value of the series is not only the most probable value but also the true value.

Observe that the average value of a large series could be expressed backed by various multiples of the standard deviation. Use of any one of these multiples would then express the probability that any present value of the series, or any subsequent value secured by the same method of measurement, would lie within the range selected. For example, the average value bracketed by $\pm\sigma$ would imply a 68.26 percent chance that any single measurement would lie within this zone around the average value; when bracketed by $\pm\delta$, a 57.62 percent inclusion chance is implied. If the average value is bracketed by a figure representing $\pm3\sigma$, there will be about a 99.5 percent chance that any single value will lie within this zone around the average value. Note also that these figures represent as well the percentages of all values in a large series that can be expected to lie within the limits selected.

Where the method of measurement admits not only to indeterminate error but also to a source of determinate error, the average of a large series of

Table 3.5 Probabilities for certain deviation ranges

Deviation relative to standard deviation $\|x_i - \mu\|/\sigma$	Probability of a single value showing a deviation between $\pm\|x_i - \mu\|/\sigma$
0.20	0.1586
0.50	0.3830
0.67	0.4972
0.80	0.5762 (δ)
1.00	0.6826 (σ)
1.50	0.8664
1.64	0.8990
1.96	0.9500
2.58	0.9902
3.29	0.9990

measurements will show bias relative to the true value. Figure 3.6 shows the distribution curves for two series, each involving a large number of measurements. Both techniques show excellent precision or reproducibility. One technique suffers only from indeterminate error, and the average value $\bar{x}$ and the true value T coincide. The other technique has an additive source of determinate error. Here the value of $\bar{x}$ is higher than T.

3.3.4 DEVIATION AND PRECISION

It was noted that we could express certain deviations by name, or that we could express a deviation magnitude as a multiple of the standard deviation. In order to reach any conclusions about the average, standard, etc., deviation, it was necessary to determine the individual deviation values for the individual measured values in the series. These individual deviation values, expressed as absolute values, were then used in calculations to determine the average deviation or the standard deviation. Once determined in this way, deviation, whether average or standard, or some multiple of the standard, is properly preceded by a $\pm$ sign.

The relative deviation represents the magnitude of the deviation relative to the average value for the series. This expression of the relative deviation results in a statement about the precision for a single value in the series. Precision for single values is seldom used, so that what is called precision usually represents the relative average, standard, etc., deviation. Again, any statement about precision is properly preceded by a $\pm$ sign. The following should be noted.

Precision for a series based on average deviation	$\delta/\mu \times 1000$ (in ppt)
Precision for a series based on standard deviation	$\sigma/\mu \times 1000$ (in ppt)
Precision for a series based on three times the standard deviation	$3\,\sigma/\mu \times 1000$ (in ppt)

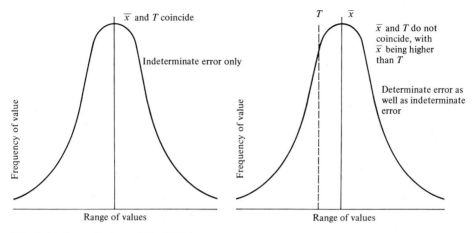

Note that in the graphs shown the horizontal
axis measures the range of values and not
the magnitude of the deviation

Fig. 3.6   Distribution curves for two series.

Note that the precision for the series could also be expressed in parts per hundred (pph) or parts per million (ppm) and that in all cases, what is indicated is an average value bracketed by a range involving a specific probability of inclusion of measured values. Where a finite series of measurements is involved, the notations s and $\bar{x}$ replace σ and μ.

3.4   ABSOLUTE ERROR AND ACCURACY

As opposed to deviation or apparent error, we have *absolute error* or *accuracy*. Absolute error bears a relationship to the true value similar to that of deviation to the average value. The absolute error in a measured value is given by

$$E_i = x_i - T$$

where x_i = a single measured value
$\quad T$ = true value
$\quad E_i$ = absolute error in the value x_i.

E_i is, of course, a measure of the accuracy of the value x_i.

The accuracy can be expressed in the form of the relative error rather than the absolute error by showing

$$\frac{E_i}{T}$$

Accuracy in its relative form of expression can be shown in pph or ppm, although it is usually shown in ppt. These would be expressed in the following forms:

$$\frac{E_i}{T}\,100 = \text{pph}$$

$$\frac{E_i}{T}\,1000 = \text{ppt}$$

$$\frac{E_i}{T}\,10^6 = \text{ppm}$$

We should, in fact, define the accuracy of a measurement in terms of the relative absolute error; and it is important to note that the expression of absolute error or accuracy as relative error in any of the forms indicated will require the use of a plus or a minus sign, except when the values of x_i and T coincide.

It is possible to express the accuracy of a series of measurements by showing the accuracy of the average value of the series. Thus, the absolute error in the average value would be given by

$$E_{(\bar{x})} = \bar{x} - T$$

It is *most* important to note that $E_{(\bar{x})}$ is the absolute error in the average value and not the *average* error. The accuracy of the series of measurements, on this basis, is then given by

$$\frac{E_{(\bar{x})}}{T}\,1000 \quad \text{in ppt}$$

and such a value is always preceded by either a plus or a minus sign, except when the values of $\bar{x}$ and T coincide.

3.5 PRECISION AND ACCURACY COMPARED

We have now arrived at two distinct aspects of the situation relative to the treatment of data. Deviation and precision are the measures of the reproducibility of techniques of measurement; absolute error and accuracy are the measures of the correctness of the values secured as the result of applying such techniques.

It is at once apparent that we can never have true and consistent accuracy without at the same time an order of precision of a similar level of magnitude. Where, for example, a method of measurement applied in a finite series shows a precision of ± 9 ppt, we can hardly expect, on a consistent basis, to achieve for the average values of such series an accuracy of $+0.1$ ppt.

On the other hand, there are occasions where a method of relatively poor precision yields from a series of measurements an average value showing an accuracy of an order much better than the implied precision. Such instances do not occur consistently, and we are justified in assigning to such events a high probability of fluke or accident.

There is no question about the fact, however, that it is possible to obtain in a series of measurements excellent precision but poor accuracy. When such a situation occurs, it is always reasonable to assume the possibility of a contributing source of determinate error of one kind or another.

Therefore, we can, to sum up this situation, note that accuracy without precision is impossible on a consistent basis; precision without accuracy is always a possibility.

3.6 DATA TREATMENT FOR SERIES INVOLVING RELATIVELY FEW MEASUREMENTS

3.6.1 GENERAL

The details of Secs. 3.3 to 3.5 principally involved the treatment of data associated with infinite series, or at least very large numbers, of measurements. The validity of the conclusions arrived at concerning deviation, deviation magnitude, probability of occurrence, inclusion of results, etc., depends largely on the area of application; they will be completely valid only in dealing with an infinite series of measurements. Where the number of measurements is finite, we deal with

s instead of σ
$\bar{d}$ instead of δ
$\bar{x}$ instead of μ

Obviously, the lower the number of measurements in a series, the less reliable will be the average value, the average deviation, the standard deviation, etc., as derived from the series, at least compared with the same data as secured from an application of the same technique of measurement carried to an infinite, or even very large, number of measurements. Since we normally deal with only a few determinations in quantitative analytical work (series involving duplicate to quadruplicate measurements), we should now investigate how data can be interpreted logically when only a limited number of measurements has been made.

3.6.2 STUDENT'S t

W. S. Gosset, writing under the pseudonym of "Student," investigated the possibility of making predictions based on the limitations associated with the statistical analysis of series of measurements involving relatively few values.* He defined a quantity t, sometimes called Student's t, as

$$\pm t = \frac{(\bar{x} - \mu)\sqrt{n}}{s}$$

where $\bar{x}$ = average value for the *finite* series
μ = average value (objective mean) to be expected when the series has been carried to an *infinite* number of measurements
s = standard deviation of the *finite* series
n = number of measurements in the *finite* series

Table 3.6 shows values of t with reference to probability levels of 95 and 99 percent. In the table, "degrees of freedom" refers to one less than n, the number

*Biometrika, **6**: 1 (1908).

of measurements involved in the finite series. The values of t are intended to compensate for the fact that $\bar{x}$ and μ will differ, as well as for the fact that s can only be an approximation of the value of σ. The factor t can be used in the statistical treatment of series that involve only relatively few measurements.

Note that for the infinite series, the value of

$$z = \left| \frac{x_i - \mu}{\sigma} \right|$$

as a multiplier of $\pm\sigma$ is 1.96 for an inclusion probability of 95 percent and 2.58 for an inclusion probability of 99 percent (Table 3.5). Note also, from Table 3.6, that as the number of measurements or determinations increases, the value of t in the 95 and 99 percent columns, respectively, approaches 1.96 and 2.58, attaining these values when the number of measurements is infinite.

Student's t as applied in the expression of results The result of a series of measurements is usually the average value for the acceptable members of the series bracketed by some range value that gives an indication of the probability of inclusion of series values within the range or the probability of deviations up to a particular magnitude. For an infinite series of measurements, for example, the

Table 3.6 Values of Student's t

Degrees of freedom $(n-1)$	95 percent probability	99 percent probability
1	12.71	63.66
2	4.30	9.93
3	3.18	5.84
4	2.78	4.60
5	2.57	4.03
6	2.45	3.71
7	2.37	3.50
8	2.31	3.36
9	2.26	3.25
10	2.23	3.17
11	2.20	3.11
12	2.18	3.06
13	2.16	3.01
14	2.15	2.98
15	2.13	2.95
16	2.12	2.92
17	2.11	2.90
18	2.10	2.88
19	2.09	2.86
20	2.09	2.85
30	2.04	2.75
40	2.02	2.70
∞	1.96	2.58

average value μ could be bracketed by $\pm\delta$, $\pm\sigma$, or $\pm z\sigma$, where z is a multiple for the standard deviation. Again, in a series involving a finite but large number of measurements, the average value $\bar{x}$ could be bracketed by $\pm\bar{d}$, $\pm s$, or $\pm zs$. In this latter instance, the validity of the indicated probability of the inclusion of values within the specific range around $\bar{x}$ would depend upon how many measurements were involved. Where $\bar{x}$ is derived from a very limited number of measurements, the inclusion probability for the method of bracketing $\bar{x}$ will not be maintained. For example, relative to an infinite series, the bracketing of μ by a range of $\pm 1.96\sigma$ indicates that 95 percent of the values will be found to lie around μ within the range of $\pm 1.96\sigma$. Where the number of measurements is finite but large, the bracketing of $\bar{x}$ by $\pm 1.96s$ will still be quite valid with respect to the 95 percent inclusion probability. Where only four measurements are made, however, the lack of reliability in both $\bar{x}$ and s will permit only poor reliability for the 95 percent inclusion probability indicated by the bracketing of $\bar{x}$ by $\pm 1.96s$. In such instances where only a few measurements are made, and where a 95 percent inclusion probability is to be expressed, the value of $\bar{x}$ should be bracketed by the factor $\pm ts$, where the value of t is taken from the 95 percent probability column of Table 3.6 for the number of degrees of freedom involved. Where only four measurements have been made, the number of degrees of freedom will be three, the appropriate value of t will be 3.18, and the use of 3.18 instead of the z value of 1.96 will then provide the larger range around $\bar{x}$ required to yield the inclusion probability of 95 percent.

The same situation applies in the case of the use of the standard deviation for the bracketing of the average value in the expression of the result of a series of measurements. Where the number of measurements is infinite, the result is then expressed as

$$\mu \pm \sigma$$

and implies a 68.26 percent probability of the inclusion of values within the range of $\pm\sigma$ around μ. Where a finite but large number of measurements is involved, the same method of expression of the result will, with reasonable validity, involve

$$\bar{x} \pm s$$

Where, for example, only four measurements have been made, the expression of the result on the same basis will require the use of

$$\bar{x} \pm ts$$

where t has a value of about 1.25.

Student's t and confidence limits It should be noted that the equation for Student's t can be rearranged as

$$\mu = \bar{x} \pm \frac{ts}{\sqrt{n}}$$

The implication here is that the objective mean has a probability of lying within the range of $\pm ts/\sqrt{n}$ around the value of $\bar{x}$ as calculated for a series involving only a few measurements or a low value of n. The probability of finding μ within this range around $\bar{x}$ is given by the selection of t, and if the t value appropriate to the number of degrees of freedom involved is taken from a 95 percent probability column in a table such as Table 3.6, the probability of finding μ in the range assigned is 95 percent. Such a limit around $\bar{x}$ is called a *confidence limit*, or *confidence interval*, and represents the confidence that the objective mean will be found within the indicated interval around $\bar{x}$. In analytical work at least, it is customary to select an acceptable probability area, such as 95 percent, and then determine the associated confidence interval around the value of $\bar{x}$.

It should be observed, of course, that as the number of measurements increases, the magnitude of the confidence interval decreases. This is merely another way of saying that $\bar{x}$ approaches μ as the number of measurements approaches infinity.

The range R can also be used to determine a confidence interval or limit for the inclusion of the objective mean. The expression used in this connection is

$$\mu = \bar{x} \pm C_n R$$

where C_n represents the multiplier of the range R. Table 3.7 shows a series of typical values for C_n for 95 and 99 percent inclusion probabilities for the objective mean. Note that in this table, the number of measurements, and not the degrees of freedom, is used directly. Where the series values do not include an odd high or low value, the confidence interval calculated from the range will agree well with that derived from the t formulation when the same probability or confidence level is used. The possibility of a significant difference between the two intervals arises when the interval from the range has been influenced by an odd high or low value in the series.

Table 3.7   Values of C_n

Number of measurements	95 percent probability	99 percent probability
2	6.35	31.8
3	1.30	3.01
4	0.72	1.32
5	0.51	0.84
6	0.40	0.63
7	0.33	0.51
8	0.29	0.43
9	0.26	0.37
10	0.23	0.33
11	0.21	0.30
12	0.19	0.27

3.6.3 TESTING FOR SIGNIFICANCE

From time to time, particularly in careful analytical work, two different methods are used in the replicate analysis of an unknown, each method yielding a relatively small number of measurements. The average values obtained from the two methods may show a difference. It is not possible by statistical analysis to say which of the average values is the more correct, but it may be possible to determine whether or not the difference between the two values is significant. Such a difference may, for example, be due to random fluctuations (indeterminate error) or directional fluctuations (determinate error).

A technique involving Student's t can be used to determine whether or not it is worthwhile, within a given degree of probability, to look for a determinate-error source underlying the difference between the means of the two methods of measurement. It should be at once apparent that the greater the scatter in the two sets of data or, in other words, the poorer their respective abilities to yield repetitive values, the less likely it will be that the difference between the means is real and significant.

The method of investigating the situation involves what is called a null hypothesis. Here it is assumed that both means are identical. A test which involves Student's t gives a no or yes answer to the correctness of the null hypothesis, with a specific confidence (or probability of being correct) such as 95 or 99 percent. The following indicates the general technique.

The two means under consideration are taken to be $\bar{x}_1$ and $\bar{x}_2$ and the associated standard deviations for the two series are s_1 and s_2. The respective number of measurements made in each series is given by n_1 and n_2. For the two series a value of t is found from

$$t = \frac{|\bar{x}_1 - \bar{x}_2|}{s} \sqrt{\frac{n_1 n_2}{n_1 + n_2}}$$

where the value of s represents here the standard deviation of a single measurement in the *two* methods, and is given by

$$s = \sqrt{\frac{\Sigma(x_{i,1} - \bar{x}_1)^2 + \Sigma(x_{i,2} - \bar{x}_2)^2}{n_1 + n_2 - 2}}$$

Note that the method indicated assumes that the values of s_1 and s_2 will *not* be shown to be significantly different.

Use is then made of a t-value table, such as that shown as Table 3.6. A value of t is selected from the table to correspond to the desired level of probability, and for a degree of freedom corresponding to $(n_1 + n_2 - 2)$. If the value of t selected from the table is greater than the t value calculated according to the equation above, then the null hypothesis is substantiated and $\bar{x}_1$ and $\bar{x}_2$ are the same within the probability level selected. For example, if we select a probability level of 95 percent, and if t from the table is found to be greater than t as calculated, then there is a 95 percent chance that the two means are the same, and that any difference results from random- or indeterminate-error sources.

If, however, the value of t from the table is found to be less than the calculated t value, the null hypothesis is incorrect, and it would be worthwhile to explore both methods for a source or sources of determinate error in order to explain the difference between $\bar{x}_1$ and $\bar{x}_2$. Note that the method of testing for significance, where there is a significant difference between the means, can not indicate which of the two methods may suffer from determinate error. Both methods have to be investigated, unless the experiential background indicates a likelihood that one method is more suspect than the other.

In making the calculation to determine t it was assumed that the two standard deviations s_1 and s_2 were not significantly different. Where these values are significantly different, a much more complicated procedure of testing the difference between two means is required. This procedure will not be discussed here. We shall discuss, instead, a method for testing the two standard deviations to determine whether or not the difference between them is significant. Such a test will indicate whether or not the values of s_1 and s_2 are sufficiently identical to permit the use of the value of s in the t test for significance. The test to determine significance for a difference between two standard deviation values s_1 and s_2 relative to the two series under discussion is usually called the variance ratio, or F test. The *variance* is the square of the standard deviation. The value of the ratio

$$F = \frac{s_1{}^2}{s_2{}^2}$$

is first determined, placing the *larger* of the two s values in the numerator, so that the value of F is always greater than unity. A table of F values, such as that shown as Table 3.8, is now consulted. This table of values covers a 95 percent probability level. If the F value from the table is less than the calculated value, the difference between s_1 and s_2 is significant; if the value from the table is greater than the calculated value, any difference is not significant. At the 95 percent probability level for the F table given, this means that if we find the difference to be significant, there is a 95 percent chance that this decision is valid.

On the basis of the t test for significance it is possible to compare two sets of results, where one set yields a result or average value known to be extremely reliable. Such a result can be secured from a very carefully analyzed sample, such as an analyzed standard from the National Bureau of Standards Standard Institute (NBS). A sample from a standard of this nature can be analyzed for any given component by some specific technique of analysis, and a series of results can be secured from which the values of $\bar{x}$, s, and n would be obtainable. The NBS value for the component involved can then be assigned as the term μ in the equation

$$\mu = \bar{x} \pm \frac{ts}{\sqrt{n}}$$

When t is calculated from this equation, and compared with the value of t shown in Table 3.6 for the number of degrees of freedom given by $(n-1)$, and for the probability level desired, the following observations can then be made:

1. If the t value from the table is greater than the calculated value, then the difference between the value of $\bar{x}$ and the NBS value is due to random or indeterminate error.
2. If the t value from the table is less than the calculated value, the difference between $\bar{x}$ and the NBS value is significant within the probability level used, and the method applied can be presumed to suffer from some determinate source of error.

Note that this latter application of the t test provides a means of exploring methods of analysis to uncover the possibilities of significant sources of determinate error.

3.7   REJECTION OF DOUBTFUL VALUES

3.7.1   GENERAL

An important tool for securing better precision and accuracy in measurement is that which serves as a criterion to determine whether or not a series of measurements contains a value or values which deviate enough from the values remaining to warrant their rejection. Naturally we do not include in any discussion of such criteria the rejection of values on the basis of some *known* error situation.

There are a number of different techniques for the rejection of values that can be applied. What one must be particularly careful to avoid, wherever possible, is the use of techniques which allow the acceptance of values better rejected or the rejection of values better retained.

This requires that consideration be given to the *number* of results in the

Table 3.8   Values of F for a 95 percent probability level

$(n-1)$ for the smaller s^2	$(n-1)$ *for the larger* s^2										
	1	2	3	4	5	6	7	8	9	10	20
1	161	200	216	225	230	234	237	239	241	242	248
2	18.51	19.00	19.16	19.25	19.30	19.33	19.36	19.37	19.38	19.39	19.44
3	10.13	9.55	9.28	9.12	9.01	8.94	8.88	8.84	8.81	8.78	8.66
4	7.71	6.94	6.59	6.39	6.26	6.16	6.09	6.04	6.00	5.96	5.80
5	6.61	5.79	5.41	5.19	5.05	4.95	4.88	4.82	4.78	4.74	4.56
6	5.99	5.14	4.76	4.53	4.39	4.28	4.21	4.15	4.10	4.06	3.87
7	5.59	4.74	4.35	4.12	3.97	3.87	3.79	3.73	3.68	3.63	3.44
8	5.32	4.46	4.07	3.84	3.69	3.58	3.50	3.44	3.39	3.34	3.15
9	5.12	4.26	3.86	3.63	3.48	3.37	3.29	3.23	3.18	3.13	2.93
10	4.96	4.10	3.71	3.48	3.33	3.22	3.14	3.07	3.02	2.97	2.77
20	4.35	3.49	3.10	2.87	2.71	2.60	2.52	2.45	2.40	2.35	2.12

series, using criteria for rejection scaled to the number of results in the series. The higher the reliability of the mean and the standard deviation, which goes hand in hand with increasingly larger numbers of measurements, the more restrictive must be the criteria for result rejection. The lower the reliability of these factors, the wider must be the latitude relative to result-rejection considerations. In point of fact, the criteria for the rejection of results should be directly linked to the number of results or measurements, becoming gradually more restrictive as the number of such measurements increases. The main problem *always* surrounds situations in which the number of results is small, since it is here that the question of insufficient data prevents reliable statistical analysis. In the ordinary course of events, this is exactly the situation that applies in quantitative analytical work, since here we deal usually with analyses performed in anything from duplicate up to quadruplicate, except in those instances of more exact work where replicate analyses involving 10 or more determinations may be conducted.

Indicated in the sections to follow are methods for the rejection of results for series of measurements involving 3 to 10, 11 to 30, and 31 to 100 values or more. These divisions are purely arbitrary and of *no* statistical significance. They merely represent divisions of series numbers that the author has found useful over years of evaluating analytical data.

3.7.2   REJECTION CRITERION FOR A SERIES OF 31 TO 100 RESULTS

We have noted from the deviation-distribution curve that the probability that deviations of a given magnitude range will occur can be worked out. If we have determined the standard deviation for the series involved, we can work out a table which will show the probabilities that deviations higher or lower than the standard deviation will occur. We can, for example, determine the ratio of a result deviation to the standard deviation, and state the probability of deviations of this magnitude occurring.

Suppose, for example, that we determine the average value and the standard deviation for an entire series of from 31 to 100 results. The individual deviations are then compared with the standard deviation. If a doubtful result (or results) is located from this comparison, the average value and the standard deviation are then recalculated with the omission of the doubtful value (or values). The deviation for the doubtful result (or results), relative to the new average value, is now calculated. If the ratio of the deviation for any doubtful result to the standard deviation exceeds some value selected on a specific probability basis, then the result is to be rejected on a confidence basis dictated by the specific probability level selected. If the ratio for the doubtful result is equal to or less than the limiting value involved, the result is accepted and returned to the series. After examination, all acceptable results are used to determine the average value and the standard deviation for the corrected series. A table showing such probability values and deviation ratios is given as Table 3.9.

Table 3.9 Probability of occurrence of deviations relative to the standard deviation $z = |x_i - \bar{x}|/s$

z	Probability of a ratio higher than z	Probability of some deviation equal to or less than $|x_i - \bar{x}|$
0.00	1.00 (100%)	
0.30	0.76 (76%)	0.24 (24%)
0.50	0.62 (62%)	0.38 (38%)
0.67	0.50 (50%)	0.50 (50%)
0.80	0.42 (42%)	0.58 (58%)
1.00	0.32 (32%)	0.68 (68%)
1.50	0.13 (13%)	0.87 (87%)
1.64	0.10 (10%)	0.90 (90%)
1.96	0.05 (5%)	0.95 (95%)
2.58	0.01 (1%)	0.99 (99%)
3.29	0.001 (0.1%)	0.999 (99.9%)

The probability basis of 95 percent is often used by the author to examine series with 31 to 100 results. This implies that for any rejected result, there is only 1 chance in 20 that the result should have been retained. In other words, you would be 95 percent right in rejecting the result involved.

The limiting ratio for this criterion is

$$\frac{\text{Doubtful deviation}}{\text{Standard deviation}} = 2.0$$

and any result yielding a ratio higher than 2.0 must be rejected from the series. If the average deviation, rather than the standard deviation, is used as the examining tool, then since the average deviation is 0.80 times the standard deviation, the limiting ratio becomes

$$\frac{\text{Doubtful deviation}}{\text{Average deviation}} = 2.5$$

Where these criteria are used, the *initial* examination of the values of the *entire* series will consider as doubtful only those values showing deviation–standard deviation ratios in the neighborhood of or higher than 2.0, or deviation–average deviation ratios in the neighborhood of or higher than 2.5.

This criterion, often called the 2.5 d rule, while adequate where the series examined contains 31 to 100 results, tends to reject valid data rather frequently if used in connection with series involving less measurements.

3.7.3 REJECTION CRITERION FOR A SERIES OF 11 TO 30 RESULTS

The general procedure here is identical with that used and outlined in the previous section, except that a limiting ratio of

$$\frac{\text{Doubtful deviation}}{\text{Standard deviation}} = 3.3$$

is set, and any result yielding a ratio higher than 3.3 must be rejected from the series with a 99.9 percent probability that this rejection is valid.

Where the average deviation is used as the examining tool, the limiting ratio then becomes

$$\frac{\text{Doubtful deviation}}{\text{Average deviation}} = 4.0$$

Where these criteria are used the *initial* examination of the values of the *entire* series will consider as doubtful only those values showing deviation/standard deviation, or deviation–average deviation ratios in the neighborhood of or higher than 3.3 and 4.0, respectively. This criterion is often called the $4d$ rule.

3.7.4   REJECTION CRITERION FOR A SERIES OF 3 TO 10 RESULTS

It is an accepted practice to use confidence limits or intervals when data from a series involving 3 to 10 measurements are investigated to determine whether or not values in the series require rejection. The confidence limits used may have a basis in either the standard deviation or the range for the series. This discussion will be limited to rejection based on criteria which involve the use of the range.

We shall assume, first of all, that the series contains more than three values. The values are arranged in ascending order and the range is determined. The difference between the lowest value and the next value to it is then determined. This difference is divided by the range. If the quotient exceeds a stipulated value whose magnitude depends on the number of values under test, then the lowest value is rejected. If it is rejected, the range for the remaining values is calculated, the largest value is then tested, and so on until both lowest and highest values have been accepted. If the original lowest value is accepted when tested, then the largest value is tested, the process being repeated if necessary until both lowest and highest values have been accepted. An example of the method of testing will be given shortly. Where only three values constitute the series of measurements, it is apparent that only one value will be doubtful, and that only one test need be carried out.

The limiting value for the ratio of the difference to the range is called the Q value.*   A table of Q values is given as Table 3.10. These values are based on a 90 percent probability level, implying that a result rejected as the outcome of this method of testing has a 90 percent validity of rejection.

*R. B. Dean and W. J. Dixon, *Anal. Chem.*, **23**: 636 (1951).

Table 3.10   Values of rejection quotient Q at probability level 0.90*

No. of measurements	$Q_{0.90}$
3	0.94
4	0.76
5	0.64
6	0.56
7	0.51
8	0.47
9	0.44
10	0.41
11†	0.39
12†	0.37

*R. B. Dean and W. J. Dixon, *Anal. Chem.*, **23**:636 (1951).
†Added values.

Example   A replicate analysis of an ore for copper involved 10 determinations and yielded the following percent values:

15.42   15.51   15.52   15.52   15.53   15.53   15.54   15.56
15.56   15.68

The results are already set out in ascending order. Determine which results require rejection.
   The lowest value is 15.42 percent, and the range is given by $15.68 - 15.42$. We have therefore for the first test

$$\frac{15.51 - 15.42}{15.68 - 15.42} = \frac{0.09}{0.26} = 0.35 < 0.41 \qquad Q_{0.90} \text{ for } 10$$

The smallest value is therefore *accepted*.
   The highest value is 15.68 percent, with the range as given. We have therefore for the second test

$$\frac{15.68 - 15.56}{15.68 - 15.42} = \frac{0.12}{0.26} = 0.46 > 0.41 \qquad Q_{0.90} \text{ for } 10$$

The highest value is therefore *rejected*.
   For the remaining values the smallest is again 15.42 percent, and the new range is given by $15.56 - 15.42$. We have therefore for the third test

$$\frac{15.51 - 15.42}{15.56 - 15.42} = \frac{0.09}{0.14} = 0.64 > 0.44 \qquad Q_{0.90} \text{ for } 9$$

The smallest value is therefore *rejected*.
   The highest value is now 15.56 percent, and since the next value to it is also 15.56 percent, this value will be *accepted*.
   The smallest value is now 15.51 percent, and the new range is given by $15.56 - 15.51$. We have therefore for the fifth test

$$\frac{15.52 - 15.51}{15.56 - 15.51} = \frac{0.01}{0.05} = 0.20 < 0.47 \qquad Q_{0.90} \text{ for } 8$$

The smallest value is therefore *accepted.*
     The series as corrected now contains eight values ranging from 15.51 percent to 15.56 percent.

When applied to series involving 3 to 10 results, the Q test will allow the rejection of values only when they deviate quite widely from the remaining values, which is particularly the case where the series contains only three or four results. This can lead to the possibility of retaining results that might better be rejected, but the lack of reliability of the data obtainable from a series involving only a small number of measurements precludes anything better in the way of performance.

     Table 3.11 provides a rather interesting tabulation of data relative to the general reliability of the Q test and shows its relationship to the number of values in the series under investigation.

     For a series of three measurements, where the first two values obtained vary as shown in the first two columns of the tabulation, it will be noted that the lowest acceptable third value may range from 8.43 to 9.84 percent, while the highest acceptable third value may range from 10.17 to 11.67 percent. The average values and average deviations for the various sets of three values show significant variations. Indeed, after the first set of three values, the magnitude of the average deviation in each case serves to reduce the number of significant figures for the average value from three to two and from four to three. The difference between the appropriate median and average values increases as the third value tends to differ more and more from the first two values. The median tends to favor the two values of generally similar magnitude.

     On the other hand, for the series involving five measurements, where the maximum difference between the first four values is identical to that given in the first series, it should be noted that the lowest acceptable fifth value varies from 9.82 to 9.94 percent, while the highest acceptable fifth value varies from 10.08 to 10.28 percent. These permissible ranges are much more restrictive than are those for the three measurement series. Here the average values and average deviations show much reduced variations, and the four-significant-figures expression of the average values is not diminished.

     As mentioned earlier, even with the Q test, the validity of the acceptance or rejection of results is a function largely of the number of measurements in the series examined. Where the number is small, such as three or four, the Q test provides for a high acceptance tolerance, permitting the acceptance of values which differ significantly. With an increase to five or more values for the series, the Q test shows improved reliability, and the results accepted show less significant differences from the associated values in the series.

     Again note that the use of the median, where only a few measurements constitute the series and where a result significantly differing from the other results is

Table 3.11 Minimum and maximum acceptable results by Q test

Series $n = 3$ $\qquad\qquad\qquad\qquad\qquad\qquad\qquad\qquad$ $Q_{0.90} = 0.94$

First value	Second value	Third value Min	Max	$\bar{x}$ Min	Max	$\bar{d}$ Min	Max	Median Min	Max	$\bar{d}$ Min	Max
10.00	10.01	9.84	10.17	9.95	10.06	0.07	0.07	10.00	10.01	0.06	0.06
10.00	10.02	9.69	10.33	9.90	10.12	0.14	0.14	10.00	10.02	0.11	0.11
10.00	10.03	9.53	10.50	9.85	10.18	0.22	0.22	10.00	10.03	0.17	0.17
10.00	10.04	9.37	10.67	9.80	10.24	0.29	0.29	10.00	10.04	0.22	0.22
10.00	10.05	9.22	10.83	9.76	10.29	0.36	0.36	10.00	10.05	0.28	0.28
10.00	10.06	9.06	11.00	9.71	10.35	0.43	0.43	10.00	10.06	0.33	0.33
10.00	10.07	8.90	11.17	9.66	10.41	0.50	0.50	10.00	10.07	0.39	0.39
10.00	10.08	8.75	11.33	9.61	10.47	0.57	0.57	10.00	10.08	0.44	0.44
10.00	10.09	8.59	11.50	9.56	10.53	0.65	0.65	10.00	10.09	0.50	0.50
10.00	10.10	8.43	11.67	9.51	10.59	0.72	0.72	10.00	10.10	0.56	0.56

Series $n = 5$ $\qquad\qquad\qquad\qquad\qquad\qquad\qquad\qquad$ $Q_{0.90} = 0.64$

First value	Second value	Third value	Fourth value	Fifth value Min	Max	$\bar{x}$ Min	Max	$\bar{d}$ Min	Max	Median Min	Max	$\bar{d}$ Min	Max
10.00	10.01	10.02	10.03	9.94	10.08	10.00	10.03	0.02	0.02	10.01	10.02	0.02	0.02
10.00	10.02	10.04	10.06	9.90	10.17	10.00	10.06	0.04	0.05	10.02	10.04	0.04	0.04
10.00	10.03	10.06	10.10	9.82	10.28	10.00	10.09	0.07	0.08	10.03	10.06	0.07	0.07

acceptable, provides a somewhat better representative situation than does the use of the average value. In these instances the median tends to line itself up with the more similar values. It should be realized, of course, that on a statistical basis of consideration, and for, say, only three results in the series, there is a very distinct possibility that the result which differs significantly from the other two may be closer to the true value.

3.8 TESTING WHETHER OR NOT A DETERMINATE-ERROR SOURCE MAY EXIST FOR A METHOD OF ANALYSIS

When a new method of analysis is to be applied or when a well-established method is to be used under unusual circumstances, the results for replicate analyses of a given material of known true value can be investigated to determine whether or not the method in question incorporates some form of determinate-error source. In order to accomplish this the true value (or an approximation of the true value of an order of accuracy better than that expected from the method involved) can be used in the technique outlined in the latter half of Sec. 3.6.3. This technique involved the use of t values and a standard material, such as an NBS standard.

A second technique is now outlined. Suppose that we have secured data associated with a series of analyses performed on a known-true-value sample by the specific method under examination and that the series of results involves not less than 10 separate values. Statistical analysis of this data indicates good precision but definite disparity between the average value for the series and the true value. The intention will now be to determine whether or not this disparity is significant and indicative of a contributing source of determinate error with respect to the method.

We shall assume, first of all, that the series has been cleared by the proper rejection criterion of any doubtful values and that the average value represents the corrected series. The absolute error in the average value is first determined as

$$E_{(\bar{x})} = \bar{x} - T$$

the value of $E_{(\bar{x})}$ being obtained as an absolute value (e.g., without sign). The ratio of $E_{(\bar{x})}$ to s, the standard deviation for the corrected series, is now determined, and this ratio is compared with some selected limiting value dictated by the level of probability required. For example, where

$$\frac{E_{(\bar{x})}}{s} > 2.0$$

the probability level has been stipulated as 95 percent by the use of the limiting value of 2.0, which implies in this application that if the value of the ratio is greater than 2.0, there is only a 5 percent chance that the error or disparity noted between the average value and the true value is an accidental or indeterminate one. In other words, there is a 95 percent chance that the method suffers from some source of determinate error.

Where the average deviation is used instead of the standard deviation, the limiting value for the ratio, at the 95 percent probability level, is

$$\frac{E_{(\bar{x})}}{\bar{d}} > 2.5$$

In the event that a higher level of probability is stipulated, with a correspondingly higher order of certainty relative to the decision but a greater allowed latitude for the disparity, the limiting value will be increased. Thus, for the probability level of 99.9 percent, we have

$$\frac{E_{(\bar{x})}}{s} > 3.3 \qquad \text{or} \qquad \frac{E_{(\bar{x})}}{\bar{d}} > 4.0$$

where there is only one chance in a thousand that the error or disparity is of indeterminate origin if the value of the ratio exceeds either 3.3 or 4.0. Here there is a 99.9 percent chance that the disparity represents a determinate-error source in the method under consideration.

3.9 THE PROPAGATION OF ERROR

So far we have discussed basically multiple results series, the accuracy of the average value, and the general order of reproducibility or precision of the series. We have not focused our attention on any individual result, except from the point of view of determining how the individual result acts as part of the series in its contribution to the average value and the average or standard deviation. Where precise and accurate analytical work is concerned, this is the usual focus, and particular attention is not generally concentrated on any individual result in a series providing that the result is acceptable.

For routine industrial or commercial analysis, however, the individual result is of prime importance, since single determinations are the customary basis of analysis in these areas. Ordinarily, such single determinations are considered in relationship to the use of known and proven methods of analysis applied to materials whose component ranges are basically known factors. Here the reliability of the method of analysis as applied to the material type in question is a matter of record.

It will be of interest, however, to determine to what extent a single analytical result is affected by various error sources that will dictate some measure of the final reliability of the result as secured and expressed. One aspect of this particular type of situation is exemplified by the following.

In the determination of silicon in a material the following steps, with their associated data, may be involved:

Weight of impure SiO_2 plus platinum crucible   25.6392 g
Weight of platinum crucible and impurities after
    the conversion and expulsion of SiO_2 25.4439 g

$$\% \text{ silicon} = \frac{(25.6392 - 25.4439) \times Si/SiO_2 \times 100}{\text{sample weight}}$$

Note that in one way or another, each value used on the right-hand side of the equation, except the counting number 100, contributes from its own uncertainty of measurement to the uncertainty in the value for the percent silicon as determined. Thus, if the value for the silicon content stood alone, and without association with replicate results from which statistical data could be secured, we could at least bracket this result with some range situation indicative of the limitations stemming from the contributions of the measuring and other uncertainties involved.

3.9.1 THE PROPAGATION OF DETERMINATE ERROR

In discussing the question of the contribution of measuring and other uncertainties to the uncertainty in a final single result, we involve ourselves in the idea of the propagation of determinate error. It will be recalled that in discussing the uncertainties in the reading of values from instruments, we pointed out that such reading uncertainties are sources of determinate error in that they represent the known limitations in ability to read the scales of the instruments involved. In addition to this, values associated with gravimetric factors, volumetric titers, equivalent weights, etc., have also known areas of uncertainty with respect to the values expressed. Note the following techniques relative to the propagation of determinate error.

Let P be a final single result as secured from the relationship

$$P = A + B + C$$

and let the determinate error values be given by

For A $\pm \Delta A$
For B $\pm \Delta B$
For C $\pm \Delta C$
For P $\pm \Delta P$

The proper equation of the result and the associated error is now

$$P \pm \Delta P = A \pm \Delta A + B \pm \Delta B + C \pm \Delta C$$

so that $\pm \Delta P$ is given by

$$\pm \Delta P = (A + B + C) \pm \Delta A \pm \Delta B \pm \Delta C - P$$
$$= \pm \Delta A \pm \Delta B \pm \Delta C$$

The value of $\pm \Delta P$ now becomes what is commonly referred to as the *maximum possible error* in P, insofar as determinate sources of error are concerned with uncertainties in the measurement of the values of A, B, and C. Where the value of P is given by the equation

$$P = A + B - C$$

the value of $\pm \Delta P$ is still represented by

$$\pm \Delta P = \pm \Delta A \pm \Delta B \pm \Delta C$$

Suppose now that we have the value of P for a final single result given as

$$P = \frac{A \times B}{C}$$

with the determinate-error values for A, B, C, and P being as already noted. The proper expression of the result and the associated error is now given by the following development

$$\log P = \log A + \log B - \log C$$

Differentiating this equation yields

$$\frac{dP}{P} = \frac{dA}{A} + \frac{dB}{B} - \frac{dC}{C}$$

and for finite errors we have

$$\pm \frac{\Delta P}{P} = \pm \frac{\Delta A}{A} \pm \frac{\Delta B}{B} \pm \frac{\Delta C}{C}$$

from which

$$\pm \Delta P = P \left(\pm \frac{\Delta A}{A} \pm \frac{\Delta B}{B} \pm \frac{\Delta C}{C} \right)$$

Again, the value of $\pm \Delta P$ represents the *maximum possible error* in P, insofar as determinate-error sources are concerned with uncertainties in the measurement of the values A, B, and C.

In general the following rules apply:

1. For addition and subtraction operations, the determinate errors are transmitted into the result.
2. For multiplication and division operations, the relative determinate errors are transmitted into the result.

The maximum possible error will not, of course, be achieved for every result obtained on the basis of arithmetic operations involving variables and their associated determinate errors. The probability that the maximum possible error calculated will be achieved for any single result is a function of the number of contributing variables in the expression leading to the result. Where r is the number of contributing variables, the probability that the maximum possible error will be achieved in any single result is given by

$$\text{Probability} = \left(\tfrac{1}{2}\right)^{r-1}$$

for determinate errors of equal occurrence of plus and minus values. Thus, for the equation

$$P = A + B - C$$

the probability of the maximum possible error being achieved for any single value of P is 0.25, or 25 percent.

Example   A volumetric titration involving a forward-titration and a back-titration yielded

Forward-titration
  Burette reading at the start 1.65 ± 0.02 ml
  Burette reading at the finish 24.38 ± 0.02 ml

Back-titration
  Burette reading at the start 0.84 ± 0.02 ml
  Burette reading at the finish 2.95 ± 0.02 ml

  Actual titration volume
  $(24.38 - 1.65) - (2.95 - 0.84) = 20.62$ ml

The maximum possible error in this value is

$$\pm \Delta V = \pm 0.02 \pm 0.02 \pm 0.02 \pm 0.02 = \pm 0.08 \text{ ml}$$

The actual volume used for the titration is now reported as

$$20.62 \pm 0.08 \text{ ml}$$

Example In a gravimetric determination the following data were accumulated:

Crucible plus weight of ignited residue	25.3692 ± 0.0001 g
Crucible alone	25.2346 ± 0.0001 g
Ignited residue weight	0.1346 ± 0.0002 g
Watch glass plus weight of sample	39.8765 ± 0.0001 g
Watch glass alone	38.7867 ± 0.0001 g
Sample weight	1.0898 ± 0.0002 g
Gravimetric factor	0.2394 ± 0.0001

The percentage of the substance involved is given by

$$P = \frac{0.1346 \times 0.2394 \times 100}{1.0898} = 2.9568\ldots\%$$

where the value of 100 is a counting number without error contribution.

$$\pm\frac{\Delta P}{P} = \pm\frac{0.0002}{0.1346} \pm \frac{0.0001}{0.2394} \pm \frac{0.0002}{1.0898}$$

$$= \pm 0.0015 \pm 0.0004 \pm 0.0002$$

$$= \pm 0.0021$$

The value for the maximum possible error now becomes

$$\pm \Delta P = P \times \pm 0.0021$$
$$= \pm 0.006\%$$

and the expression of the result is given as

$$2.957 \pm 0.006\%$$

3.9.2 THE PROPAGATION OF INDETERMINATE ERROR

The previous section provides a general picture relative to the propagation of determinate error in the case of the individual result and indicates the means of using the maximum possible error to bracket the single result secured. Naturally, in any series of measurements, each individual result in the series is subject to a similar situation with respect to the propagation of determinate error, although with series of measurements, determinate-error propagation merely serves to establish the significant-figure limits for the expression of each result in the series. Statistical analysis of the entire series of results is then used to provide a deviation value that can then be properly used to bracket the average value for the series.

It frequently happens, however, that a single result P is secured under conditions where many measurements of the contributing variables have been made

— for example, where P is determined from

$$P = \bar{A} + \bar{B} - \bar{C}$$

and where $\bar{A}$, $\bar{B}$, and $\bar{C}$ each represent the average of multiple measurements of the variables involved. This situation will now be considered.

Determinate-error sources involving reading uncertainties can have definite values assigned (e.g., burette-reading uncertainties, balance-weighing uncertainties, etc.), and as we have seen we can work with these assigned uncertainty values to arrive at a maximum possible error in the final single result. On the other hand, where we have made multiple measurements of each variable, we can average these measurements and determine the standard deviation. This standard deviation may be equal to, smaller than, or larger than the individual measurement uncertainty, and the standard deviation now reflects the indeterminate error in the average value for each variable. Where propagation of such indeterminate errors is concerned, this spread for each variable must be related to the random variation or indeterminate error in the final result.

Again, let P be a single final result secured from the relationship

$$P = \bar{A} + \bar{B} + \bar{C}$$

or from the relationship

$$P = \frac{\bar{A} \times \bar{B}}{\bar{C}}$$

where, in each case, many measurements of A, B, and C have been obtained, so that statistical analysis will yield reasonably accurate information on the standard deviation for each of the average values $\bar{A}$, $\bar{B}$, and $\bar{C}$. Note the following general rules:

1. Where P is obtained by addition and/or subtraction operations, the variances (squares of the standard deviations) for the values of $\bar{A}$, $\bar{B}$, and $\bar{C}$ will be additive in arriving at the variance for the result P.

$$s_P^2 = s_A^2 + s_B^2 + s_C^2$$

2. Where P is obtained by multiplication or division operations, the squares of the *relative* standard deviations additively provide the square of the relative standard deviation for the result P.

$$\left(\frac{s_P}{P}\right)^2 = \left(\frac{s_A}{\bar{A}}\right)^2 + \left(\frac{s_B}{\bar{B}}\right)^2 + \left(\frac{s_C}{\bar{C}}\right)^2$$

Example In the determination of iron in an effluent substance by volumetric titration, the percentage of iron is given by the equation

$$\% \text{ iron} = \frac{\text{ml of titrant} \times \text{titer} \times 100}{\text{sample weight}}$$

Only one sample is analyzed. The sample weight is read in grams five times from the balance as

1.1139   1.1140   1.1140   1.1137   1.1138

The titer value is given as

1 ml = 0.005683 ± 0.000006 g of iron

The titration volume is read in milliliters five times from the burette as

22.62   22.67   22.64   22.65   22.65

Calculate the percent iron and the associated standard deviation (measurements only).

The average sample weight is 1.1138^8 g
The standard deviation is $\pm 0.0001^3$ g
The average titrant volume is 22.64^6 ml
The standard deviation is $\pm 0.01^8$ ml

The iron percentage is given by

$$\% \text{ iron} = \frac{22.64^6 \times 0.005683 \times 100}{1.1138^8} = 11.5539\ldots$$

The standard deviation for the iron value is given by the relationship

$$\left(\frac{s_p}{P}\right)^2 = \left(\frac{0.0001^3}{1.1138^8}\right)^2 + \left(\frac{0.01^8}{22.64^6}\right)^2 + \left(\frac{0.000006}{0.005683}\right)^2$$

$$\frac{s_p}{P} = 0.0013^3$$

$$s_p = \pm 0.02$$

The iron content is then reported as

11.55 ± 0.02%

Suppose that only one sample weight reading was taken at 1.1140 g, subject to a reading uncertainty of ±0.0002 g, and that the titrant volume was read only once at 22.62 ml, subject to a reading uncertainty of ±0.04 ml.

Calculate the percent iron and the associated maximum possible error.

The iron percentage is given by

$$\% \text{ iron} = \frac{22.62 \times 0.005683 \times 100}{1.1140} = 11.5394\ldots$$

The maximum possible error is given by

$$\Delta\% \text{ Fe} = \% \text{ Fe} \left(\pm \frac{0.04}{22.62} \pm \frac{0.000006}{0.005683} \pm \frac{0.0002}{1.1140}\right)$$

$$= \pm 0.03$$

The iron content is then reported as

11.54 ± 0.03%

In general, the question of the propagation of error leads to an interesting and valuable conclusion. If it is possible to locate the weakest link, or the largest error or relative error, in a measurement operation from a group of measurements used in the calculation of a single result, then such datum can be used to avoid the excessive time consumed in determining the other measurements to unnecessary levels of exactitude.

Example Suppose that in the spectrophotometric determination of manganese in steel, the absorbance measurement from the spectrophotometer scale is

0.65 ± 0.02

the value of ± 0.02 representing the reading uncertainty. The final result for percent manganese will be given by

$$\% \ Mn = \frac{absorbance \times Mn \ in \ g/absorbance \ unit \times 100}{sample \ weight}$$

Although the balance used may be capable of weighing to an uncertainty of ± 0.0002 g for a difference weighing, the relative error of 1 part in 33 for the absorbance reading implies that to weigh a 1-g sample to ± 0.0002 g (e.g., a relative error of 1 part in 5000) would be an unnecessary refinement. Weighing the sample to the nearest ± 0.01 g would be adequate for the method involved and would allow significant savings in time when large numbers of such samples must be analyzed.

3.10 SIGNIFICANT FIGURES AND COMPUTATION RULES

Any final computation of a result must indicate the extent of the reliability to be assigned to the result, because the result may be affected by associated determinate and indeterminate sources of error. The exact calculation of a range around the result that is indicative of its reliability is time-consuming. A technique often applied provides a crude but quite adequate means of indicating its reliability by carrying the result computation to a digit at which a level of uncertainty or lack of reliability exists. This is called the method of *significant figures* and represents the result as expressed to a meaningful number of digits. Usually this will involve all the digits of certainty, plus one which represents the area of uncertainty.

Essentially, the use of the proper significant figures avoids the expression of results to numbers of digits that are misleading and maintains for the final result an indication of the precisions or uncertainties of the variables contributing in the calculation of the result. The use of too few or too many digits gives a false impression of the reliability of the result.

In any number expressing the value of a measurement or sequence of measurements, the zeros immediately before and after the decimal point are not significant; they serve merely to locate the decimal point. Enclosed and terminal zeros are, however, significant.

Examples
    0.0164 The two zeros merely serve to locate the decimal point. They are not significant. There are only three significant figures in this number.

0.01064   Again the two zeros on either side of the decimal point serve only to locate it. They are not of significance. The zero enclosed by the digits 1 and 6 is significant. There are four significant figures in this number.

20.0608   All the zeros in this number are significant. There are six significant figures in this number.

5.0300   All the zeros in this number are significant. There are five significant figures in this number.

0.0350   The zeros on either side of the decimal point are not significant; they serve only to locate it. There are three significant figures in this number, since the terminal zero is significant.

24,000   All the zeros in this number are significant. There are five significant figures in this number.

13×10^4 Although the number is multiplied by 10^4 to indicate the order of magnitude, there are only two significant figures in the number.

If we express a balance weight as 1.0738 g, we imply that the last digit is the uncertain one, and we have

1.0738 g subject to an uncertainty of ±0.0001 g

The uncertainty in the last digit might easily be higher than ±0.0001 g, but if this were so, it would have to be clearly indicated. For example, as

1.0738 g   Uncertainty of ±0.0002 g
1.0738 g   Uncertainty of ±0.0003 g

and so on. The expression of volume carries with it the same implication. If we say that we have measured a volume of 28.64 ml, we imply that the last digit is the uncertain one and that we have

28.64 ml subject to an uncertainty of ±0.01 ml

Again remember that the uncertainty in the last digit might have been higher than this, but if this were so, it would have to be clearly indicated. For example, as

28.64 ml   Uncertainty of ±0.02 ml
28.64 ml   Uncertainty of ±0.03 ml

and so on. In the final analysis, the degree of uncertainty in a single weight or volume measurement will be dictated by the equipment used, and this situation will apply to all readings from all instruments. The degree of uncertainty will eventually be transmitted as a magnitude of uncertainty contribution to the final result, and it will be translated by the final significant figure in the eventual form of expression of the result.

The precision of a single measurement is the relative uncertainty in that measurement. If the measurement of 20.05 ml is subject to an uncertainty of

± 0.04 ml, the precision of this single measurement will be given by

$$\frac{\pm 0.04}{20.05}$$

representing a precision of about 1 part in 500. In many cases it is convenient to express the precision in parts per thousand, or ppt. In the instance above, the approximate precision would then be ± 2 ppt.

If we secure the value of 20.05 ml, as we invariably do, by taking the difference between an initial burette reading and a final one, then the uncertainty involved will be ± 0.08 ml. In this case, the precision of the volume value of 20.05 ml would be approximately 1 part in 250, or ± 4 ppt. Since each of the two readings is subject to an uncertainly of ± 0.04 ml, the final volume carries the possibility of an uncertainty double the value of the uncertainty for a single reading. While this is a case of maximum possible error, open to the usual statistical considerations relative to the achieving of the maximum possible error, such reasoning serves adequately at this level.

The same principle applies, of course, to weighing operations; and the uncertainty for a single weighing operation becomes doubled for a weight obtained by adding or subtracting two individual weight measurements. The same principle applies wherever a measured value is obtained as the result of a multiple measuring operation.

All in all, when consideration is given to the question of the precision of a single measurement, concern should not be primarily with the number of decimal places or the size of the number involved. Precision is defined on a relative basis only; it is here that the true magnitude of the uncertainty will show itself.

Examples All these examples cover a single-measurement situation.

1. 40.08 ml	Uncertainty	± 0.04 ml
Approximate precision	1 part in 1000	± 1 ppt
2. 5.05 ml	Uncertainty	± 0.04 ml
Approximate precision	1 part in 125	± 8 ppt
3. 1.0016 g	Uncertainty	± 0.0001 g
Approximate precision	1 part in 10,000	± 0.1 ppt
4. 0.0055 g	Uncertainty	± 0.0001 g
Approximate precision	1 part in 55	± 18 ppt

Note that even though the volume and weight values shown in the foregoing have the same orders of uncertainty, the precisions or relative uncertainties are not in the least comparable. The student should be able to see quite clearly that any final result involving several weight and/or volume measurements will have a number of significant figures governed by the poorest precision with respect to the various measurements involved.

The rules for the computation of results, and for the expression of the significant figures in the final result, can be reduced to certain very simple procedures. While several methods, differing only slightly in their approach, can be applied, the following are suggested as being entirely adequate to the purpose.

1. In expressing any final result, retain no more than one doubtful digit.
2. When rejecting superfluous digits from a calculated result which has been deliberately carried to a number of digits beyond the expected significant figure level,
   (a) Increase by one the last digit retained if the digit rejected is greater than 5.
   (b) Leave the last digit retained unchanged if the digit rejected is less than 5.
   (c) If the rejected digit is exactly 5, increase by one the last digit retained if it is odd, and leave it unchanged if it is even.
3. When adding or subtracting two or more quantities, the sum or difference should contain or include only as many decimal places as there are in the component having the least number of decimal places.
4. In multiplication or division operations, the answer should have a precision of the same general order as that of the least precise component in the operation. While there are several methods of arranging this situation, it is suggested that you adhere to the system that gives the precision of the result of the operation as lying between one-half to five times that of the least precise component.

Example The values 60.3 and 1.05 are to be added and the value 0.162 subtracted from the total.

$$
\begin{array}{r}
60.3 \\
1.05 \\
\hline
61.35 \\
-\ 0.162 \\
\hline
61.188
\end{array}
$$

Express the result as 61.2.

Example A computation operation is to involve the multiplication of the values 21.1, 0.029, and 83.2. The implied uncertainty for each value is a plus or minus unit digit in the final digit expressed.

$$21.1 \times 0.029 \times 83.2 = 50.91008$$

The value of 0.029 is the least precise, with an approximate precision of 1 part in 30.

Five times this is	1 part in 150
One-half this is	1 part in 15
The result is then rounded off to	51

Example A percentage value is computed from the equation

$$\% P = \frac{[(B - A) - (D - C)] \times E \times 100}{(G - F)}$$

where $A = 1.38 \pm 0.02$ ml
$B = 29.82 \pm 0.02$ ml
$C = 0.89 \pm 0.02$ ml
$D = 10.35 \pm 0.02$ ml
$E = 0.00589 \pm 0.00006$ g/ml
$F = 25.8943 \pm 0.0001$ g
$G = 26.7854 \pm 0.0001$ g

The value of % P is given by

$$\% \ P = \frac{[(29.82 - 1.38) - (10.35 - 0.89)] \times 0.00589 \times 100}{26.7854 - 25.8943}$$

$$= \frac{(28.44 - 9.46) \times 0.00589 \times 100}{0.8911}$$

$$= \frac{18.98 \times 0.00589 \times 100}{0.8911}$$

$$= 12.5454. . .$$

Because of the additive and subtractive operations relative to volume, the value of 18.98 ml shows an uncertainty of ± 0.08 ml. This is a precision of about 1 part in 250. The titer value of 0.00589 g/ml has a precision of about 1 part in 100. The sample weight of 0.8911 g has a precision of about 1 part in 4500. The least precise value is the titer at 1 part in 100.

Five times this is	1 part in 500
One-half this is	1 part in 50
The result is expressed as	12.5%

It should be kept firmly in mind, however, that regardless of the significant figures to which any single result in a series of results can be expressed, the average value for the series will be expressed to a number of significant figures which may be

1. The same number of significant figures as the individual results, if the average or the standard deviation as calculated for the series so permits
2. A lesser number of significant figures than the individual results, if the average or standard deviation as calculated for the series permits only this lesser number

Example A series of determinations of chromium in a steel shows percentage values of 16.42, 16.50, 16.46, and 16.40. These values are each expressed to four significant figures, as permitted by the measurement uncertainties for the variables involved in the calculation. The average value for the series is 16.44 percent, and the standard deviation is ± 0.04 percent. Since only four results are involved, the value of s must be multiplied by 1.25 to permit an inclusion probability of about 68 percent. This now allows the result to be expressed as

$16.44 \pm 0.05\%$

and the final value is expressed to the same number of significant figures as the individual results in the series.

A similar series of determinations made in connection with the same steel shows percentage values of 16.35, 16.48, 16.59, and 16.42. Again these values are each expressed to four significant figures, as permitted by the measurement uncertainties for the variables involved in the calculation. The average value for the series is 16.46 percent, the standard deviation is ± 0.1 percent, and the value of s multiplied by 1.25 is still, as rounded off, ± 0.1 percent. This now allows the result to be expressed as

$$16.5 \pm 0.1\%$$

and the final value is expressed to three significant figures, or one less than the individual results in the series.

PROBLEMS

1. A balance has a 1-g weight which actually weighs 1.0009 g. You use this weight each time in weighing the CaO residue from several determinations of calcium in rock samples. Does this situation introduce a constant or proportional source of error? Explain. What is the situation with respect to the relative error introduced?

2. How many more times reliable is the average of 100 determinations than that of 9 determinations?

3. You are to weigh a sample of approximately 50 mg to within ± 3 ppt. What must be the uncertainty of the balance per single reading of weight?   *Ans.* ± 0.08 mg

4. You are to weigh a sample of approximately 0.75 g. The balance to be used has an uncertainty per weighing of ± 5 mg. What will be the uncertainty in weighing the sample?

5. The following measurements of quantity of electricity consumed in an electrochemical process were reported in coulombs:

> 136.4   136.9   135.8   137.1   136.3   136.5   136.3   135.9   135.7   136.5

Consider all the values to be acceptable and determine
   (*a*) The average value
   (*b*) The median
   (*c*) The range
   (*d*) The average deviation
   (*e*) The standard deviation
   (*f*) The value of $\pm ts$ for a 95 percent inclusion of results within the limits of $\bar{x} \pm ts$
   (*g*) The confidence interval $\pm ts/\sqrt{n}$ for a 95 percent probability
   (*h*) The confidence interval $\pm C_n R$ for a 95 percent probability
   (*i*) The average value bracketed by the standard deviation as calculated
   (*j*) The average value bracketed by $\pm ts$ (95 percent inclusion factor)
                          Ans. (*d*) ($\pm 0.3^4$) C, ± 0.3 C; (*g*) ($\pm 0.3^2$) C, ± 0.3 C

6. The same property as that measured in Prob. 5 was determined using a different technique of measurement, with the following results in coulombs:

> 137.3   134.9   136.8   135.5   135.2   136.6   137.1   136.9   136.8   136.4

   (*a*) Consider all the values to be acceptable and determine the same data as outlined for *a* to *h* in Prob. 5.
   (*b*) Compare and comment on the results of the statistical analyses as performed on the data given in Probs. 5 and 6.

7. Suppose, for the technique of measurement used relative to Prob. 6, that only the following three values were secured in coulombs:

> 137.3   134.9   135.2

   (*a*) Consider all the values to be acceptable and determine the same data as outlined for *a* to *h* in Prob. 5.

(b) Compare and comment on the results of the statistical analyses as performed on the data given in Probs. 5, 6, and 7.

8. The following percentage values were obtained in the replicate analysis of a sample of steel for chromium:

16.23   16.28   16.22   16.30   16.25   16.23   16.26   16.25   16.26   16.16

(a) Use the $Q_{0.90}$ test to investigate the series for a value, or values, which must be rejected.
(b) Using the acceptable values for the series determine

The average value
The median
The range
The average deviation
The standard deviation
The precision based on the standard deviation
The confidence interval $\pm ts/\sqrt{n}$ for a 95 percent probability
The confidence interval $\pm C_n R$ for a 95 percent probability
The value of $\pm ts$ for a 95 percent inclusion of results within the limits of $\bar{x} \pm ts$

(c) Given the true value T of the chromium content as 16.28 percent determine
(1) Whether or not the series of acceptable values shows evidence of a source of determinate error. Use both the $t_{0.99}$ test from Sec. 3.6.3 and the $E_{(\bar{x})}/s > 3.3$ test from Sec. 3.8. Comment on the results of applying these tests.
(2) The accuracy of the series of acceptable results.
(d) Comment on your findings relative to the precision and accuracy of the method of analysis.

9. A technique of analysis is applied to determine the percentage of SO_2 in stack gas. The following results are obtained on the basis of 12 replicate analyses:

4.88   4.92   4.90   4.88   4.86   4.85   4.71   4.86   4.87   4.94   4.87   4.99

(a) Use the $2.0s$ ($2.5d$) rule to investigate the series of results for doubtful values. For all acceptable determinations, report

$$\bar{x} \quad M \quad R \quad \bar{d} \quad s \quad \pm ts \ (0.95) \quad \pm ts/\sqrt{n}(0.95) \quad \pm C_n R \ (0.95)$$

(b) Use the $3.3s$ ($4.0d$) rule for the same purpose. For all acceptable determinations report the same data as listed in a.
(c) Use the $Q_{0.90}$ test for the same purpose. For all acceptable determinations report the same data as listed in a.
(d) Give an explanation of the results obtained in a, b, and c.

Ans. (a) Reject 4.71, 4.94, 4.99; (c) Reject 4.71

10. A measurement technique to determine the energy released in a physical process gave the following results in kiloelectronvolts:

39.8   40.1   39.6   41.6

(a) Examine the series of results for the rejection of values by the $Q_{0.90}$ test and the $4.0d$ rule.
(b) Comment on any differences in the rejection of values.

11. A method for the analysis of an ore for aluminum is to be investigated to ascertain its precision and accuracy by analyzing an ore sample of known aluminum content. The known aluminum content is 10.75 percent. Only nine analyses are performed. The percentage values obtained are

10.73   10.73   10.74   10.77   10.77   10.81   10.81   10.82   10.86

(a) Show the results of applying the $2.5d$, the $4.0d$, and the $Q_{0.90}$ criteria for the rejection of values. Comment on any differences and be governed by the result of applying the Q test.

(b) Determine the precision of the method of analysis as based on the standard deviation.

(c) Determine the accuracy of the series of acceptable results.

(d) Using the $E_{(\bar{x})}/s > 2.0$ test, determine whether or not the method suffers from a source of determinate error.

(e) Comment on the precision and accuracy of the method of analysis.

12. A gravimetric method for the determination of aluminum in a zinc-base die-cast alloy involves, finally, the isolation of the aluminum as $Al(OH)_3$ and its ignition to Al_2O_3. The analysis of die-cast alloys of known composition gave the following results.

Aluminum content, g		Error	
Present	Found	Grams	ppt
0.0	0.0	_____	_____
0.0462	0.0470	_____	_____
0.0518	0.0527	_____	_____
0.0602	0.0612	_____	_____
0.0679	0.0692	_____	_____
0.0743	0.0755	_____	_____
0.0860	0.0876	_____	_____

(a) Fill in the two error columns and indicate whether the error source yields a constant or proportional error.

(b) Devise a possible correction to apply to the method of analysis to correct for the error as found. Apply your method of correction, calculate the absolute residual error for each value determined, and show the average absolute residual error.

13. The following data represent the results of two series of analyses for TiO_2 in an extraterrestrial rock sample. Each series represents a different technique of analysis. Consider all results to be acceptable.

Series A TiO₂, percent	Series B TiO₂, percent
8.76	8.89
8.78	8.92
8.74	8.91
8.80	8.92
8.78	8.89
8.77	8.90
8.77	8.88
8.76	8.88
8.79	8.93
8.75	8.94

(*a*) Use the *F* test to determine whether or not any difference between the standard deviations is significant.

(*b*) Determine whether or not the difference between the two means is significant. What is the meaning of your conclusion here?

14. The following results were obtained from a triplicate determination of copper in an aluminum alloy by x-ray fluorescence:

6.78%   6.85%   6.82%

(*a*) What is the highest fourth result that can be obtained without rejection by the $Q_{0.90}$ test?

(*b*) What is the lowest fourth result that can be obtained without rejection by the $Q_{0.90}$ test?

Ans. (*a*) 7.07%; (*b*) 6.56%

15. A series of replicate analyses to determine the tin content of a copper-base alloy involved four determinations and gave the following percentage values:

9.81   9.85   9.80   9.82

If a fifth determination is planned, indicate what the maximum and minimum values would be in order for the determination to be acceptable by the $Q_{0.90}$-test criterion.

16. A single determination of silver, by controlled cathodic potential electrolysis, was made on a copper-silver alloy sample. The following data were obtained:

Weight of watch glass	24.1328 g
Weight of watch glass plus sample	25.1461 g
Weight of platinum cathode	15.5432 g
Weight of platinum cathode plus silver deposit	15.7699 g

All single weighing operations are subject to an uncertainty given by ±0.0002 g. Determine the percentage of silver in the alloy and bracket this value with the maximum possible error for this singlicate determination.   *Ans.* 22.37 ± 0.05%

17. A series of analyses for cobalt in a high-alloy steel involving eight determinations was made by a new method. The average obtained was 10.78 percent. The standard deviation was ±0.04^6. The result was expressed as 10.78 ± 0.05 percent. The true value for the cobalt content of the sample involved was 10.75 percent.

(*a*) Determine by (1) the *t* test (0.95 probability) and (2) the $E_{(\bar{x})}/s > 2.0$ test whether or not the method of analysis used requires investigation to locate a source, or sources, of determinate error.

(*b*) Which test do you think provides the more reliable information? Give reasons for your choice.

18. (*a*) A singlicate determination of nickel in a copper-base alloy was made by an absorptiometric technique. The following data were obtained:

Sample weight	1.0314 ± 0.0002 g
Absorbance	0.462 ± 0.002

The concentration-vs.-absorbance value is 1.000 mg Ni per 0.100 units of absorbance and, for the purposes of this problem, can be taken as an absolute value. Show the percent nickel in the alloy and indicate a range around this value. What does this range represent?

(*b*) Suppose, in this singlicate determination, that the sample weight had been read six times from the balance as

1.0315   1.0316   1.0315   1.0314   1.0313   1.0313   all in grams

and that the absorbance value had been read six times from the spectrophotometer scale as

0.462   0.468   0.460   0.458   0.470   0.465

Now show the percent nickel in the alloy and indicate a range around this value. What does this range represent?

(c) Now suppose that six separate and parallel determinations of nickel were made by the same method of analysis and under exactly the same conditions, with the following results in percent nickel being obtained:

    0.439   0.471   0.446   0.464   0.461   0.450

Now show the percent nickel in the alloy and indicate a range around this value. What does this range represent?

(d) What various conclusions do you reach from a, b, and c?

19. Round off the following numbers to four significant figures:
- (a) 0.034567
- (b) 0.12445
- (c) 1.0005
- (d) 45.395
- (e) 967,856
- (f) 0.00068739
- (g) 240,000
- (h) 21.445

20. Indicate the number of significant figures in the following:
- (a) 0.020430
- (b) 1.06201
- (c) 2.435×10^{-8}
- (d) 1.67×10^{12}
- (e) 405,000
- (f) 0.000439

21. Given the fact that in each of the numbers shown in Prob. 20 the uncertainty in the number is given by ± 1 in the last digit, determine the relative uncertainty in ppt for each number.

22. Given the following calculation systems, determine the result expressed to the proper number of significant figures:
- (a) $127.13 + 4.3104 + 61.2$
- (b) $43.839 - 4.28 + 25.636$
- (c) $41.48 \times 0.0891 \times 0.13 \times 8.46$
- (d) $\dfrac{(0.0865 \times 1.132 \times 4.036)}{0.048}$
- (e) $\dfrac{[24.16 \times (9.85 - 4.3) \times 5.67]}{(15.839 - 14.238)}$

Ans. (b) 65.20; (e) 4.8×10^2

23. Given the equation

$$P = \frac{A \times (B - C) \times Z}{(W - Y)}$$

where

$$A = 4.65 \pm 0.02$$
$$B = 10.98 \pm 0.05$$
$$C = 2.88 \pm 0.05$$
$$Z = 0.668 \pm 0.001$$
$$W = 21.48 \pm 0.02$$
$$Y = 20.21 \pm 0.02$$

calculate P and bracket the value properly.

24. Four separate triplicate series of analyses for iron in the same rock sample gave

Series 1   14.18%, 14.19%, 14.16%
Series 2   14.18%, 14.19%, 14.19%
Series 3   14.10%, 14.17%, 14.24%
Series 4   14.02%, 14.16%, 14.36%

with all results expressed to the number of significant figures allowed by the measurement uncertainties involved. Using the average deviation for each series, express the average value to the proper number of significant figures and bracketed by the appropriate average deviation.

Ans. Series 3, $14.17 \pm 0.05\%$

25. An NBS material is reported to contain 0.564% silicon. Your analysis of this sample yields the following values in percent silicon:

    0.549   0.550   0.547   0.547   0.548

Indicate whether or not your method of analysis suffers from a determinate source of error. Use the $t_{(0.95)}$-test criterion.

26. Two different methods were used to determine the value in ppm of the content of magnesium in a natural water. The following figures were reported.

Method A	Method B
5.65	5.88
5.75	5.78
5.64	5.79
5.73	5.92
5.69	5.95
5.71	5.82

(a) Is the difference between s_A and s_B significant?
(b) Is the difference between $\bar{x}_A$ and $\bar{x}_B$ significant?
(c) What would be your next step?

REFERENCES

1. Kolthoff, I. M., E. B. Sandell, E. J. Meehan, and Stanley Bruckenstein: "Quantitative Chemical Analysis," 4th ed., Macmillan, New York, 1969.
2. Blaedel, W. J., and V. W. Meloche: "Elementary Quantitative Analysis," Harper & Row, New York, 1963.
3. Brown, G. H., and E. M. Sallee: "Quantitative Chemistry," Prentice-Hall, Englewood Cliffs, N.J., 1963.
4. Skoog, D. A., and D. M. West: "Fundamentals of Analytical Chemistry," 2d ed., Holt, New York, 1969.
5. Fritz, J. G., and G. H. Schenk, Jr.: "Quantitative Analytical Chemistry," 2d ed., Allyn and Bacon, Boston, 1969.
6. Barford, N. G.: "Experimental Measurements: Precision, Error and Truth," Addison-Wesley, Reading, Mass., 1967.

7. Pugh, E. M., and G. H. Winslow: "The Analysis of Physical Measurements," Addison-Wesley, Reading, Mass., 1966.
8. Schenck, H., Jr.: "Theories of Engineering Experimentation," McGraw-Hill, New York, 1968.
9. Nalimov, V. V.: "The Application of Mathematical Statistics to Chemical Analysis," Addison-Wesley, Reading, Mass., 1963.

4
Chemical Equilibria

4.1 GENERAL

The reactions of chemical substances involve two characteristics of primary significance—the equilibrium reaction position and the reaction kinetics. Reaction kinetics has to do with the rate at which reactants are converted into products, as well as with the order of the processes, both physical and chemical, by which the conversion takes place. Although of fundamental importance, reaction kinetics does not usually require particular consideration in a course on introductory quantitative analysis because at such a point the main concern is with reactions which are fast, or very fast, and which yield specific products; special attention is usually not paid to the path pursued by the reaction involved beyond ensuring that the required reaction conditions are provided. On the other hand, the question of chemical equilibrium is of primary importance, since a knowledge of the *quantitative* basis of a reaction is fundamental where the reaction is applied in quantitative analytical chemistry.

4.2 THE EQUILIBRIUM CONSTANT

Chemical reactions of the type

$$A + B \rightleftharpoons C + D$$

85

represent reversible reactions, where A and B react at some finite velocity to form C and D. As the amounts of A and B decrease by reaction, the velocity of the reaction A + B falls off. However, as the amounts of C and D increase by reaction, the velocity of the reaction C + D increases. Eventually the A + B and C + D reaction velocities exactly match each other for the temperature condition involved, at which point a dynamic equilibrium is established where the quantities of A, B, C, and D do not change.

The velocity of the reaction A + B or C + D is a function of the activities of the reacting substances and, indeed, is proportional to the product of the activities of the reacting substances, each activity being raised to a power equal to the number of moles of that substance in the reaction. Thus, for the general reaction

$$a\text{A} + b\text{B} + \cdots \rightleftharpoons m\text{M} + n\text{N} + \cdots$$

we have

$$V_1 \ \alpha \ (\alpha_\text{M}{}^m \alpha_\text{N}{}^n \cdots)$$
$$V_1 = K_1(\alpha_\text{M}{}^m \alpha_\text{N}{}^n \cdots)$$
$$V_2 \ \alpha \ (\alpha_\text{A}{}^a \alpha_\text{B}{}^b \cdots)$$
$$V_2 = K_2(\alpha_\text{A}{}^a \alpha_\text{B}{}^b \cdots)$$

where the V values represent reaction velocities and the K values represent the proportionality constants.

At equilibrium we have

$$V_1 = V_2$$

and

$$K_1(\alpha_\text{M}{}^m \alpha_\text{N}{}^n \cdots) = K_2(\alpha_\text{A}{}^a \alpha_\text{B}{}^b \cdots)$$

so that

$$\frac{(\alpha_\text{M}{}^m \alpha_\text{N}{}^n \cdots)}{(\alpha_\text{A}{}^a \alpha_\text{B}{}^b \cdots)} = \frac{K_2}{K_1} = K$$

Here we have $K = K_{\text{eq}}$, the *equilibrium constant*, a constant where T, the temperature of the reaction, is held constant. The expression involving the equilibrium constant represents a statement of the *law of mass action*, or the *law of chemical equilibrium*.

It should be thoroughly understood that the development of the foregoing has been vastly simplified in relationship to factors such as reaction reversibility. Not all reactions are easily reversible, nor can we ever assume, under any circumstances, that the establishment of a state of dynamic equilibrium involves the reduction, singly or collectively, of the activities of the reactant substances to zero.

For the purposes of the temperature and pressure conditions usually encountered in quantitative analytical work, the activity of a pure liquid (i.e., a solvent) can be taken as unity, as can be that of a pure solid.

Under ideal solution conditions the activities of both the solute and the solvent will be given by the mole fractions. Thus the addition of 0.1 mol of NaOH to 1000 g of water yields a 0.1 molal solution each of Na^+ and OH^-. Here we have

$$0.1 \text{ mol} + 0.1 \text{ mol} + \frac{1000}{18} \text{ mol} = 0.2 \text{ mol} + 55.6 \text{ mol}$$
$$= 55.8 \text{ mol}$$

The mole fractions, assuming ideal solution conditions, will yield the activity values, and we have

$$\alpha_{Na^+} = \frac{0.1}{55.8}$$

$$\alpha_{OH^-} = \frac{0.1}{55.8}$$

$$\alpha_{H_2O} = \frac{55.6}{55.8}$$

Note that the value for $\alpha_{H_2O} \approx 1$ and that for dilute solutions of the type usually dealt with in quantitative analysis, the effect of the solute on the activity of the solvent is so small as to be capable of being ignored. Where concentrated solutions are involved, however, the effect of the solute on the solvent activity must be considered.

Under nonideal solution conditions, activity and molality can not be directly related except through the equation

$$\alpha = \gamma m$$

where α = the activity
γ = the activity coefficient
m = the molality

Since the limiting value of the expression α/m is unity, we can equate the ratios of activities and molalities (or molarities as a good approximation to molalities in most cases of dilute solutions), so that we have

$$\frac{\alpha_A}{\alpha_M} = \frac{[A]}{[M]} \qquad \text{etc.}$$

Although α_A and α_M may not be identical to $[A]$ and $[M]$, respectively, in most instances involving analytical work, little in the way of error is normally introduced by assuming that the *ratios* as given above are identical. Under certain circumstances, however, we shall use activity values in order to increase the accuracy of the calculations.

Thus we can rewrite the chemical equilibrium expression just developed to show that

$$aA + bB + \cdots \rightleftharpoons mM + nN + \cdots$$

$$\frac{[M\ ]^m\ [N\ ]^n\ \cdots}{[A\ ]^a\ [B\ ]^b\ \cdots} = K_{eq} \qquad T \text{ constant}$$

Remember that the equilibrium-constant expression, written as shown, is not accurate or exact, but it can yield an excellent and sufficiently accurate first approximation.

We shall be applying the basic expression written as shown in several forms. For example,

$$CH_3 \cdot COOH + H_2O \rightleftharpoons H_3O^+ + CH_3 \cdot COO^-$$

$$\frac{[H_3O^+][CH_3 \cdot COO^-]}{[CH_3 \cdot COOH][H_2O]} = K_{eq}$$

$$\frac{[H_3O^+][CH_3 \cdot COO^-]}{[CH_3 \cdot COOH]} = K_{ion} = K_a$$

Also,

$$AgCl(s) \rightleftharpoons Ag^+ + Cl^-$$

$$\frac{[Ag^+][Cl^-]}{[AgCl]} = K_{eq}$$

$$[Ag^+][Cl^-] = K_{sp}$$

Part 2   Acid-Base Equilibria

4.3   GENERAL

4.3.1   ARRHENIUS' THEORY OF ACIDS AND BASES

The Arrhenius ionic dissociation theory was developed between 1880 and 1890, and was intended to explain the ionic properties exhibited by certain substances in aqueous solution, in particular those of acid and base substances. It was taken, for example, that the following reactions of dissociation took place in aqueous solution:

$$HCl \rightleftharpoons H^+ + Cl^-$$

$$CH_3 \cdot COOH \rightleftharpoons H^+ + CH_3 \cdot COO^-$$

$$KOH \rightleftharpoons OH^- + K^+$$

Thus, the proton H^+ was held responsible for acid properties and the hydroxyl ion OH^-, for base properties. The same general approach was used to explain other dissociation phenomena in aqueous solution. Arrhenius' theory led, however, to certain situations difficult to explain. Two examples will suffice to indicate their general nature.

The true nature of the proton in aqueous solution was still in doubt. Later developments, including contributions by Debye and Hückel, clarified this point with evidence indicating that the proton exists in aqueous solution in close association with a water molecule or water molecules. Thus, the reaction for the dissociation of HCl cannot be shown simply as

$$HCl \rightleftharpoons H^+ + Cl^-$$

but should be shown as

$$HCl + H_2O \rightleftharpoons H_3O^+(aq) + Cl^-(aq)$$

This suggests that an acid is not necessarily a substance which dissociates to yield a proton but rather a molecule capable of transferring a proton to another molecule.

A second problematic situation had to do with the fact that the theory did not explain why certain substances that did not contain the hydroxyl group or did not yield in an obvious way the hydroxyl ion in dissociation were yet capable of acting as bases. For example, in the following reaction,

$$HCl(g) + NH_3(l) \rightleftharpoons NH_4^+ + Cl^-$$

NH_3 acts as a base, since it reacts with HCl, a known acid. NH_3 does not, however, contain the hydroxyl group. Again, $Na_2CO_3(s)$ added to water yields a solution capable of reacting with HCl according to

$$Na_2CO_3 + 2HCl \rightleftharpoons 2NaCl + H_2O + CO_2$$

Here again, Na_2CO_3 does not contain the hydroxyl group nor, on the basis of the Arrhenius theory, does it yield hydroxyl ion in any obvious way when in aqueous solution.

In order to clarify these and other situations, some revision of Arrhenius' theory was obviously required.

4.3.2 BRØNSTED THEORY

The definition of the Brønsted theory relative to acid substances is that they are a species having a tendency to lose or donate a proton. The definition relative to base substances is that they are a species able to accept or receive a proton. Thus, we have

$$\underset{\text{Acid}}{HCl(aq)} + \underset{\text{Base}}{H_2O} \rightleftharpoons H_3O^+(aq) + Cl^-(aq)$$

and, at the same time, we have

$$\underset{\text{Acid}}{H_3O^+(aq)} + \underset{\text{Base}}{Cl^-(aq)} \rightleftharpoons HCl(aq) + H_2O$$

We call HCl and Cl⁻, which differ only by a proton, a conjugate acid-base pair. Again, H_3O^+ and H_2O, which differ only by a proton, represent a second conjugate acid-base pair. The reaction may now be rewritten as

$$HCl + H_2O \rightleftharpoons H_3O^+ + Cl^-$$

Acid 1   Base 2   Acid 2   Base 1

where 1 and 2 represent the conjugate pairs.

We may now explain the reaction of the carbonate ion from Na_2CO_3 as

$$CO_3^{2-} + H_2O \rightleftharpoons HCO_3^- + OH^-$$

Base 1   Acid 2   Acid 1   Base 2

and, for the HCl–NH_3 reaction, we have

$$HCl + NH_3 \rightleftharpoons NH_4^+ + Cl^-$$

Acid 1   Base 2   Acid 2   Base 1

Although, as the best approach, we should use the Brønsted theory in our subsequent discussions of acid-base dissociation reactions, we shall for the sake of simplicity express most of the relations through the use of common Arrhenius' theory equations, using H_3O^+ as the hydrated proton.

4.3.3 THE IONIZATION OF WATER

Water is an amphiprotic substance, which means that it can act as an acid or a base. This is exemplified by the reaction

$$H_2O + H_2O \rightleftharpoons H_3O^+ + OH^-$$

Acid 1   Base 2   Acid 2   Base 1

From this reaction we have

$$\frac{[H_3O^+][OH^-]}{[H_2O]^2} = K_{eq}$$

and

$$[H_3O^+][OH^-] = K_{eq}[H_2O]^2 = K_w$$

The constant K_w is called the ion-product constant, or ionization constant, for water and is a constant at constant temperature.

At room temperature (i.e., 25°C) K_w has a value of 1.01×10^{-14}, so that for pure water at this temperature we have

$$[H_3O^+][OH^-] = 1.01 \times 10^{-14}$$
$$\approx 1.00 \times 10^{-14}$$

therefore, at 25°C,

$$[H_3O^+] = [OH^-] \approx 1.00 \times 10^{-7} \ M$$

Thus, in acid solutions, we have $[H_3O^+]$ greater than $1.00 \times 10^{-7} \ M$, and in basic solutions $[H_3O^+]$ less than $1.00 \times 10^{-7} \ M$.

A shorthand notation for the expression of any ion concentration in solution is to indicate it in the form of the logarithm of the reciprocal of the ion concentration as its molarity. Thus, we have

$$pH = \log \frac{1}{[H_3O^+]} = -\log [H_3O^+]$$

$$pOH = \log \frac{1}{[OH^-]} = -\log [OH^-]$$

$$pCl = \log \frac{1}{[Cl^-]} = -\log [Cl^-]$$

since

$$[H_3O^+][OH^-] = K_w$$
$$pH + pOH = pK_w$$

Example Determine the pH of an aqueous solution $2.00 \times 10^{-4} M$ to $[H_3O^+]$ at 25°C.

$$[H_3O^+] = 2.00 \times 10^{-4} M$$
$$-\log [H_3O^+] = -\log (2.00 \times 10^{-4})$$
$$pH = 4.00 - 0.30$$
$$= 3.70$$

It is most important to note that when an acid solution of high degree of dilution is involved, it may be necessary to consider the contribution of $[H_3O^+]$ from both the acid and the self-dissociation of water.

Example Suppose that an aqueous solution is $6.00 \times 10^{-7} M$ to HCl at 25°C. Determine the pH of this solution neglecting the water-dissociation contribution of H_3O^+. Determine the pH when the water-dissociation contribution is taken into account.

Neglecting the contribution from water dissociation, we have

$$[H_3O^+] = 6.00 \times 10^{-7} M$$

$$pH = 6.22$$

Taking into account the water-dissociation contribution,

$$[H_3O^+]_{total} = [H_3O^+]_{HCl} + [H_3O^+]_{H_2O}$$
$$= 6.00 \times 10^{-7} M + [H_3O^+]_{H_2O}$$

At 25°C, we have

$$(6.00 \times 10^{-7} + [H_3O^+]_{H_2O})([OH^-]_{H_2O}) = 1.01 \times 10^{-14}$$

but

$$[OH^-]_{H_2O} = [H_3O^+]_{H_2O}$$

therefore,

$$[H_3O^+]_{H_2O}^2 + 6.00 \times 10^{-7}[H_3O^+]_{H_2O} - 1.01 \times 10^{-14} = 0$$

Solving the quadratic yields

$$[H_3O^+]_{H_2O} = 0.17 \times 10^{-7} M$$

from which the total value for $[H_3O^+]$ is given as

$$
\begin{aligned}
[H_3O^+]_{total} &= 6.00 \times 10^{-7} + 0.17 \times 10^{-7} \\
&= 6.17 \times 10^{-7} M \\
pH &= 6.21
\end{aligned}
$$

In general, if $[H_3O^+]$ from the dissolved-acid substance is $10^{-6} M$ or greater, the contribution of $[H_3O^+]$ from water dissociation can be ignored. The same situation applies relative to the values of $[OH^-]$ from a dissolved-base substance and from the water dissociation.

4.3.4 THE IONIZATION OF WEAK ACIDS AND WEAK BASES

Strong electrolytes are commonly taken as being 100 percent dissociated in aqueous solution. It is obvious, of course, that no substance can be totally dissociated, since the equilibrium condition demands at least some measure however slight of back-reaction. In general, strong electrolytes can be grouped as

1. Some organic acids, such as HCl, HNO_3, $HClO_4$, etc.
2. Some inorganic bases, such as NaOH, KOH, etc.
3. Most inorganic salts, such as NaCl, KNO_3, etc.
4. Many organic salts, such as $CH_3 \cdot COONa$, $COONa \cdot COONa$, etc.

Weak electrolytes are those dissociated to an extent less than 100 percent. In the majority of cases the extent to which dissociation takes place will be very significantly less than 100 percent. In general, weak electrolytes can be grouped as

1. Some inorganic acids and, in particular, inorganic polybasic acids, such as HCN, H_2SO_4, H_3PO_4, etc.
2. Some inorganic bases, such as $NH_3(aq)$, (NH_4OH), etc.
3. Most organic acids and bases, such as $CH_3 \cdot COOH$, $CH_3 \cdot CH_2OH$, etc.

With strong electrolytes, the determination of an associated-ion concentration from the molar concentration of the electrolyte is quite direct. For example, a 0.1-M solution of HCl yields 0.1 M H_3O^+, a 0.2-M solution of NaOH yields 0.2 M OH^-, a 0.1-M solution of NaCl yields 0.1 M Cl^-, and a 0.2-M solution of Na_2SO_4 yields 0.4 M Na^+.

With weak electrolytes in aqueous solution, the association of the relevant ion concentrations with the molar concentrations of the electrolytes is indirect and frequently quite complex. Examples of such situations relative to aqueous solutions of weak acid and weak base substances will now be given.

4.4 IONIZATION EQUILIBRIA FOR WEAK MONOBASIC ACIDS, BASES, AND RELATED SALTS

4.4.1 IONIZATION EQUILIBRIA FOR A WEAK MONOBASIC ACID HA

It should be understood, first of all, that the weak monobasic acid HA refers only to a general monobasic acid, not to a particular substance. The dissociation reactions of interest for an aqueous solution of HA are

$$HA + H_2O \rightleftharpoons H_3O^+ + A^-$$
$$2H_2O \rightleftharpoons H_3O^+ + OH^-$$

There are five concentrations of importance to be considered in this solution:

$$[HA]_{un} \qquad [A^-] \qquad [H_3O^+] \qquad [OH^-] \qquad C_a$$

where $[HA]_{un}$ represents the concentration of undissociated HA in the solution, and C_a represents the starting or initial molarity of the weak acid HA. The equilibrium equations are

$$\frac{[H_3O^+][A^-]}{[HA]_{un}} = K_a \qquad [H_3O^+][OH^-] = K_w$$

The initial concentration of the acid C_a must be present either as HA_{un} or A^-. We then have the material balance

$$C_a = [HA]_{un} + [A^-] \qquad \text{and} \qquad [HA]_{un} = C_a - [A^-]$$

In addition, the charge balance (negative and positive charges resulting from the ionization reactions) must equate, and we have

$$[H_3O^+] = [A^-] + [OH^-]$$

and

$$[A^-] = [H_3O^+] - [OH^-]$$
$$= [H_3O^+] - \frac{K_w}{[H_3O^+]}$$

Combining the material- and charge-balance equations, we have

$$[HA]_{un} = C_a - \left([H_3O^+] - \frac{K_w}{[H_3O^+]}\right)$$

and substitution of the values for $[HA]_{un}$ and $[A^-]$ into the equilibrium for K_a yields

$$\frac{[H_3O^+]([H_3O^+] - K_w/[H_3O^+])}{C_a - ([H_3O^+] - K_w/[H_3O^+])} = K_a \qquad (1)$$

This yields a cubic equation of the form

$$[H_3O^+]^3 + K_a[H_3O^+]^2 - (K_w + K_aC_a)[H_3O^+] - K_wK_a = 0 \qquad (2)$$

Solutions of cubic equations are rarely simple and only infrequently required in analytical work, so that we shall now examine the general conditions surrounding these equilibria in order to determine what valid assumptions might lead to simpler equations and solutions.

Even where the acid is very weak, the value of $[H_3O^+]$ will generally be greater than $10^{-6}\ M$ and, correspondingly, the value of $K_w/[H_3O^+]$ will be less than $10^{-8}\ M$. Under these conditions the numerator and denominator expression of $([H_3O^+] - K_w/[H_3O^+])$ will reduce to $[H_3O^+]$ approximately. Equation (1) will then modify to

$$\frac{[H_3O^+][H_3O^+]}{C_a - [H_3O^+]} \approx K_a \tag{3}$$

which yields the quadratic

$$[H_3O^+]^2 + K_a[H_3O^+] - K_aC_a = 0 \tag{4}$$

from which we have the solution

$$[H_3O^+] \approx \frac{-K_a + \sqrt{K_a^2 + 4K_aC_a}}{2} \tag{5}$$

Generally, if the acid is weak, the amount of HA dissociating will be small, and the value of $[HA]_{un}$ will very closely approximate the value of C_a. We have then

$$[HA]_{un} \approx C_a$$

and Eq. (3) modifies to

$$\frac{[H_3O^+][H_3O^+]}{C_a} \approx K_a \tag{6}$$

and we have

$$[H_3O^+] \approx \sqrt{K_aC_a} \tag{7}$$

Since we do not intend to employ expressions higher than the quadratic, it remains now to indicate the area of applicability of Eqs. (7) and (5).

In order to obtain Eq. (7), we assumed that the amount of HA dissociating was small compared with C_a. This implies that $[H_3O^+]$, one of the products of this dissociation, is small compared with C_a. From Eq. (7) we note that

$$\frac{\Delta[H_3O^+]}{[H_3O^+]} = \frac{1}{2}\frac{\Delta C_a}{C_a}$$

so that the relative error in the determination of $[H_3O^+]$ from Eq. (7) will be one-half the relative error in C_a resulting from ignoring $[H_3O^+]$ relative to C_a. Thus, if the value of $[H_3O^+]$ as determined is 5 percent of C_a, the relative error in the value of $[H_3O^+]$ determined from Eq. (7) will be about 2.5 percent. The following *general* procedure may be applied.

1. Use Eq. (7) to determine $[H_3O^+]$.
2. Compare this value with C_a. If $[H_3O^+]$ is 5 percent or less of the value of C_a, then the relative error in $[H_3O^+]$ from Eq. (7) will be about 2.5 percent or less. This level of error is satisfactory for most quantitative analytical work.

3. Where $[H_3O^+]$ from Eq. (7) shows a value greater than 5 percent of C_a, or where the relative error level for $[H_3O^+]$ must be less than 2.5 percent, the quadratic Eq. (5) should be solved.

It is important to realize that exactly the same reasoning can be applied to the ionization equilibria associated with aqueous solutions of weak bases. It is suggested that an appropriate student project would be the development of such exact and approximate equations.

Example   For the ionization of acetic acid we have

$$CH_3 \cdot COOH + H_2O \rightleftharpoons H_3O^+ + CH_3 \cdot COO^-$$

with a K_a value of 1.76×10^{-5}. The application of Eq. (7) yields the following $[H_3O^+]$ values and predicted relative errors in $[H_3O^+]$ as determined for various initial acetic acid concentrations.

C_a, M	$[H_3O^+]$, M	$[H_3O^+]/C_a$, %	$\Delta[H_3O^+]/[H_3O^+]$, %
1	4.20×10^{-3}	0.4	0.2
0.01	4.20×10^{-4}	4.2	2.1
0.001	1.33×10^{-4}	13.3	6.7

It is apparent that the initial acid concentration of 0.001 M will require the use of Eq. (5), which gives

$$[H_3O^+] \approx 1.24 \times 10^{-4} M$$

and the relative error in $[H_3O^+]$ from Eq. (7) when compared with the value of $[H_3O^+]$ from Eq. (5), is given by

$$\frac{(1.33 - 1.24) \times 10^{-4}}{1.24 \times 10^{-4}} \cdot 100 = 7.2\%$$

a good approximation to the 6.7 percent predicted value.

4.4.2 IONIZATION EQUILIBRIA FOR THE WEAK MONOBASIC ACID SALT NaA

The action of hydrolysis   By the action of *hydrolysis* is meant the reactions of the anions of weak acids or the cations of weak bases with water. Thus, in aqueous solutions of the salts of weak acids and weak bases, the hydrolysis reactions of the ions involved may dictate to a large extent the final hydrogen or hydroxyl ion concentrations of such solutions.

General examples of solutions of these types, and the hydrolysis actions involved, are given by the following.

Sodium acetate in aqueous solution provides, by dissociation, sodium and acetate ions. Using the abbreviations HOAc for acetic acid, NaOAc for sodium acetate, and OAc^- for the acetate ion, we have

$$NaOAc \rightleftharpoons Na^+ + OAc^-$$

$$OAc^- + H_2O \rightleftharpoons HOAc + OH^-$$

and the overall reaction

$$NaOAc + H_2O \rightleftharpoons HOAc + Na^+ + OH^-$$

Note that a salt of this type in aqueous solution undergoes a reaction where the anion combines with water to form the undissociated weak acid associated with the anion. The resulting solution will be basic. In general, salts of weak acids and strong bases will yield basic aqueous solutions, the degree of basicity depending on the value of K_a for the weak acid involved. The lower the value of K_a, all other factors being equal, the higher the degree of basicity of the salt solution.

An aqueous solution of ammonium chloride dissociates and hydrolyzes

$$NH_4Cl \rightleftharpoons NH_4^+ + Cl^-$$

$$NH_4^+ + 2H_2O \rightleftharpoons NH_4OH + H_3O^+$$

with the overall reaction

$$NH_4Cl + 2H_2O \rightleftharpoons NH_4OH + H_3O^+ + Cl^-$$

and the understanding that we can, for NH_4OH, read $NH_3 + H_2O$. A salt of this type will yield an acidic solution, the cation of the weak base reacting with water to form the undissociated weak base. Salts of weak bases and strong acids will yield acidic aqueous solutions, the degree of acidity increasing, in general, with decreasing value of K_b for the weak base.

Ammonium acetate in aqueous solution dissociates and hydrolyzes

$$NH_4OAc \rightleftharpoons NH_4^+ + OAc^-$$

$$NH_4^+ + 2H_2O \rightleftharpoons NH_4OH + H_3O^+$$

$$OAc^- + H_2O \rightleftharpoons HOAc + OH^-$$

A salt of this type, a product of the reaction between a weak acid and a weak base, may yield either an acidic or a basic solution depending on the relative values of K_a and K_b. If K_a is higher than K_b, the solution will be acidic; if K_b is higher than K_a, the solution will be basic.

The action of hydrolysis is significant in quantitative analytical chemistry. For example, in the titration of a weak acid by a strong base solution, the equivalence point of the titration will involve a solution of the reaction salt, this solution having a pH value dictated largely by the degree of hydrolysis of the weak acid anion of the salt. Again, specific pH conditions in a solution may often be obtained by the use of salts that hydrolyze to predictable extents.

Ionization equilibria for the salt NaA   The salt NaA refers to a generalized sodium salt of a weak monobasic acid, not to a particular substance. The dissociation and hydrolysis reactions of interest for an aqueous solution of NaA are

$$NaA \rightleftharpoons Na^+ + A^-$$
$$A^- + H_2O \rightleftharpoons HA + OH^-$$
$$2H_2O \rightleftharpoons H_3O^+ + OH^-$$

There are five concentrations of importance:

$$[HA]_{un} \quad [A^-] \quad [H_3O^+] \quad [OH^-] \quad C_s$$

where C_s represents the starting or initial molarity of the weak acid salt NaA. The equilibrium equations are

$$\frac{[H_3O^+][A^-]}{[HA]_{un}} = K_a \qquad \frac{[HA]_{un}[OH^-]}{[A^-]} = K_{hyd}$$

$$[H_3O^+][OH^-] = K_w$$

and we observe that the hydrolysis equilibrium constant K_{hyd} can be shown as

$$K_{hyd} = \frac{[HA]_{un}[OH^-]}{[A^-]} = \frac{[HA]_{un}}{[A^-][H_3O^+]}[H_3O^+][OH^-] = \frac{K_w}{K_a}$$

Note from the expression shown that the smaller the value of K_a, the larger the value of K_{hyd} and the greater the action of hydrolysis.

The initial concentration C_s of the salt must be present as either HA_{un} or A^-. We then have the material balance

$$C_s = [A^-] + [HA]_{un} \qquad and \qquad [A^-] = C_s - [HA]_{un}$$

In addition, the charge balance must equate and we have

$$[Na^+] + [H_3O^+] = [A^-] + [OH^-]$$

Note that we have

$$[OH^-] = [OH^-]_{hyd} + [OH^-]_{H_2O}$$
$$= [OH^-]_{hyd} + [H_3O^+]_{H_2O}$$
$$= [HA]_{un} + \frac{K_w}{[OH^-]}$$

so that

$$[HA]_{un} = [OH^-] - \frac{K_w}{[OH^-]}$$

and we have

$$[A^-] = C_s - \left([OH^-] - \frac{K_w}{[OH^-]}\right)$$

Using the hydrolysis equilibrium equation, and substituting these values for $[HA]_{un}$ and $[A^-]$, we have

$$\frac{[OH^-]\left([OH^-] - \dfrac{K_w}{[OH^-]}\right)}{C_s - \left([OH^-] - \dfrac{K_w}{[OH^-]}\right)} = \frac{K_w}{K_a} \tag{8}$$

which provides a cubic equation of the form

$$[OH^-]^3 + \frac{K_w}{K_a}[OH^-]^2 - \frac{K_w K_a + K_w C_s}{K_a}[OH^-] - \frac{K_w{}^2}{K_a} = 0$$

in powers of $[OH^-]$ and

$$[H_3O^+]^3 + (K_a + C_s)[H_3O^+]^2 - K_w[H_3O^+] - K_w K_a = 0 \tag{9}$$

in terms of $[H_3O^+]$.

In order to avoid solutions involving the cubic equation, we shall examine the general conditions surrounding these equilibria to determine what valid assumptions might lead to simpler equations and solutions.

Where reasonable hydrolysis takes place, the value of $[OH^-]$ will be greater than $10^{-6} M$, with a corresponding value of $[H_3O^+] = K_w/[OH^-]$ less than $10^{-8} M$. Under these conditions the value of $([OH^-] - K_w/[OH^-])$ in Eq. (8) will reduce to $[OH^-]$, approximately. Equation (8) will then modify to

$$\frac{[OH^-][OH^-]}{C_s - [OH^-]} \approx \frac{K_w}{K_a} \tag{10}$$

which leads to the quadratic in $[OH^-]$:

$$[OH^-]^2 + \frac{K_w}{K_a}[OH^-] - \frac{K_w C_s}{K_a} = 0$$

and in $[H_3O^+]$:

$$[H_3O^+]^2 - \frac{K_w}{C_s}[H_3O^+] - \frac{K_w K_a}{C_s} = 0 \tag{11}$$

which latter equation has a solution in

$$[H_3O^+] \approx \frac{(K_w/C_s) + \sqrt{(K_w/C_s)^2 + 4K_w K_a/C_s}}{2} \tag{12}$$

Generally speaking, where C_s is of reasonable magnitude, the value of $[OH^-]$ will be much less than C_s. Where valid, this assumption reduces Eq. (10) to

$$\frac{[OH^-][OH^-]}{C_s} \approx \frac{K_w}{K_a} \tag{13}$$

from which we have, in $[OH^-]$,

$$[OH^-] \approx \sqrt{\frac{K_w C_s}{K_a}} \tag{14}$$

and, in $[H_3O^+]$,

$$[H_3O^+] \approx \sqrt{\frac{K_w K_a}{C_s}} \tag{15}$$

Since we shall not employ equations higher than the quadratic, it remains now to indicate the area of applicability of Eqs. (15) and (12).

In order to obtain Eq. (15), we assumed that the value of $[OH^-]$ was much less than C_s. From Eq. (15) we note that

$$\frac{\Delta[H_3O^+]}{[H_3O^+]} = \frac{1}{2}\frac{\Delta C_s}{C_s}$$

so that the relative error in the determination of $[H_3O^+]$ from Eq. (15) will be one-half of the relative error in C_s resulting from ignoring $[OH^-]$ relative to C_s. Thus, if the value of $[OH^-]$, determined as $K_w/[H_3O^+]$ using $[H_3O^+]$ from Eq. (15), is 5 percent of C_s, the relative error in $[H_3O^+]$ from Eq. (15) will be about 2.5 percent. The following *general* procedures may be applied.

1. Use Eq. (15) to determine $[H_3O^+]$.
2. Calculate $[OH^-]$ from $K_w/[H_3O^+]$ and compare with C_s. If $[OH^-]$ is 5 percent or less of C_s, then the relative error in $[H_3O^+]$ from Eq. (15) will be about 2.5 percent or less. This level of error is adequate for most analytical work.
3. Where $[OH^-]$ as calculated in 2 is higher than 5 percent of C_s, or where the relative error level for $[H_3O^+]$ must be less then 2.5 percent, the quadratic Eq. (12) should be solved.

It is important to realize that exactly the same reasoning can be applied to the ionization and hydrolysis equilibria associated with aqueous solutions of the salts of weak bases and strong acids, such as the general salt BCl from BOH and HCl. An appropriate student project would be the development of such exact and approximate equations.

Example   In the ionization and hydrolysis of sodium acetate we have

$$CH_3 \cdot COONa + H_2O \rightleftharpoons CH_3 \cdot COOH + Na^+ + OH^-$$

with a K_a value of 1.76×10^{-5}. The application of Eq. (15) yields the following $[H_3O^+]$ values and predicted relative errors in $[H_3O^+]$ as determined for various initial concentrations of sodium acetate.

C_s, M	$[H_3O^+]$, M	$[OH^-]$, M	$[OH^-]/C_s$, %	$\Delta[H_3O^+]/[H_3O^+]$, %
1	4.20×10^{-10}	2.38×10^{-5}	0.002	0.001
0.01	4.20×10^{-9}	2.38×10^{-6}	0.02	0.01
0.001	1.33×10^{-8}	7.52×10^{-7}	0.08	0.04

It is apparent that for sodium acetate solutions, Eq. (15) should be adequate even with salt concentrations as low as 0.001 M.

4.4.3 IONIZATION EQUILIBRIA FOR MIXTURES OF THE WEAK MONOBASIC ACID HA AND ITS SALT NaA

Buffer action   By *buffer action* is meant the action of a substance, or a mixture of substances, added to solution, usually aqueous in nature, which assists in establishing and maintaining a specific pH value in the solution. Thus a buffered solution tends to resist changes in pH that might ordinarily be brought about by the addition of small amounts of either strong acid or strong base. Although changes in pH do take place in buffered solutions as the result of acid or base additions, these changes are insignificant in comparison with what results under similar circumstances with unbuffered solutions.

Buffered solutions usually involve mixtures of weak acids and the salt of the weak acid, weak bases and the salt of the weak base, or a polybasic acid or base salt such as potassium acid phthalate. The following typifies in general the various buffered solution conditions and associated reactions involved.

An aqueous solution of acetic acid and sodium acetate dissociates as

$$NaOAc \rightleftharpoons Na^+ + OAc^- \qquad \text{strong}$$
$$HOAc + H_2O \rightleftharpoons H_3O^+ + OAc^- \qquad \text{weak}$$

The addition of a small amount of H_3O^+ in the form, say, of HCl results in the reaction

$$OAc^- + H_3O^+ \rightleftharpoons HOAc + H_2O$$

and any resulting change in the pH of the solution will be insignificant providing that the added $[H_3O^+]$ is well under the concentration of NaOAc as C_s.

On the other hand, the addition of a small amount of OH^- in the form, say, of NaOH results in the reaction

$$OH^- + H_3O^+ \rightleftharpoons 2H_2O$$

and any resulting change in the pH of the solution will be insignificant providing that the added $[OH^-]$ is well under the concentration of HOAc as C_a.

Ammonium hydroxide and ammonium chloride in aqueous solution dissociate as

$$NH_4Cl \rightleftharpoons NH_4^+ + Cl^- \qquad \text{strong}$$
$$NH_3 + H_2O \rightleftharpoons NH_4^+ + OH^- \qquad \text{weak}$$

The addition of a small amount of H_3O^+ results in the reaction

$$H_3O^+ + OH^- \rightleftharpoons 2H_2O$$

Any change in the pH of the solution will be insignificant providing that the added $[H_3O^+]$ is well under the concentration of NH_3 as C_b.

The addition of a small amount of OH^- produces the reaction

$$OH^- + NH_4^+ \rightleftharpoons NH_3 + H_2O$$

No significant change in the solution pH will result providing that the added $[OH^-]$ is well under the concentration of NH_4Cl as C_s.

An aqueous solution of the single substitution dibasic acid salt potassium acid phthalate, $C_6H_4 \cdot COOK \cdot COOH$, shown as KHP, dissociates as

$$KHP \rightleftharpoons K^+ + HP^- \qquad \text{strong}$$
$$HP^- + H_2O \rightleftharpoons H_3O^+ + P^{2-} \qquad \text{weak}$$

The addition of a small amount of H_3O^+ results in the reaction

$$H_3O^+ + HP^- \rightleftharpoons H_2P + H_2O$$

and the change in solution pH will be insignificant providing that the added $[H_3O^+]$ is well under the concentration of KHP as C_s. The addition of a small amount of OH^- provides for the reaction

$$OH^- + H_3O^+ \rightleftharpoons 2H_2O$$

and again the pH change will be insignificant providing that the added $[OH^-]$ is well under the concentration of KHP as C_s.

Buffered solutions are very frequently used in quantitative analytical chemistry, wherein they, for example,

1. Ensure the proper progress of a reaction by removing H_3O^+ or OH^- reaction products as these are formed
2. Control the level of pH during a precipitation process
3. Maintain a required level of pH in order to guarantee the proper reaction of an indicator substance
4. Are encountered at certain stages of titration processes, for example, in the titration of a weak acid by a strong base, where buffered solution conditions of some nature exist after the start of the titration and until just before the equivalence point

Buffered solutions are of particular importance in most life processes. Many of the chemical mechanisms of the body are pH-dependent; and as a case in point, the pH of human blood ranges about ± 0.05 pH units around a value of 7.40. Any marked change of a few tenths above or below this range can mean eventual death. Acid and base substances introduced into the system are usually carried by the blood as buffered forms until the excesses (e.g., carbon dioxide and uric acid) can be removed by the lungs or kidneys. Changes in the normal level of the blood pH may provide for conditions which prevent the buffering and transporting of such substances.

Ionization equilibria for mixtures of the weak acid HA and its salt NaA Again the acid HA and salt NaA represent generalized acid and salt substances. The dissociation and hydrolysis reactions of interest for an aqueous solution involving HA and NaA are

$$NaA \rightleftharpoons Na^+ + A^-$$

$$HA + H_2O \rightleftharpoons H_3O^+ + A^-$$

$$A^- + H_2O \rightleftharpoons HA + OH^-$$

$$2H_2O \rightleftharpoons H_3O^+ + OH^-$$

There are six concentrations to be considered:

$$[HA]_{un} \qquad [A^-] \qquad [H_3O^+] \qquad [OH^-] \qquad C_a \qquad C_s$$

where C_a and C_s represent, respectively, the starting or initial molarities of the acid HA and the salt NaA. The equilibrium expressions are

$$\frac{[H_3O^+][A^-]}{[HA]_{un}} = K_a \qquad \frac{[HA]_{un}[OH^-]}{[A^-]} = K_{hyd}$$

$$[H_3O^+][OH^-] = K_w$$

The initial concentrations of C_a and C_s must be present as either A^- or HA_{un}, and we then have the material balance

$$C_s + C_a = [HA]_{un} + [A^-]$$

In addition, the charge balance must equate, and we have

$$[Na^+] + [H_3O^+] = [A^-] + [OH^-]$$

since we have

$$[OH^-] = \frac{K_w}{[H_3O^+]}$$

this yields

$$[A^-] = [Na^+] + [H_3O^+] - \frac{K_w}{[H_3O^+]}$$

From the material-balance equation, using the above value for $[A^-]$, we have

$$C_s + C_a = [HA]_{un} + [Na^+] + [H_3O^+] - \frac{K_w}{[H_3O^+]}$$

so that

$$[HA]_{un} = C_s + C_a - [Na] - \left([H_3O^+] - \frac{K_w}{[H_3O^+]}\right)$$

Since the salt NaA is totally dissociated, for our purposes here

$$[Na^+] = C_s$$

so that we have

$$[HA]_{un} = C_a - \left([H_3O^+] - \frac{K_w}{[H_3O^+]}\right)$$

For the equation

$$[A^-] = [Na^+] + [H_3O^+] - \frac{K_w}{[H_3O^+]}$$

we now have

$$[A^-] = C_s + \left([H_3O^+] - \frac{K_w}{[H_3O^+]}\right)$$

Substitution of these relationships for $[HA]_{un}$ and $[A^-]$ into the acid-dissociation-equilibrium equation yields

$$\frac{[H_3O^+]\{C_s + ([H_3O^+] - K_w/[H_3O^+])\}}{C_a - ([H_3O^+] - K_w/[H_3O^+])} = K_a \tag{16}$$

This provides an exact cubic equation of the form

$$[H_3O^+]^3 + (C_s + K_a)[H_3O^+]^2 - (K_w + K_aC_a)[H_3O^+] - K_wK_a = 0 \tag{17}$$

The following examines the general conditions surrounding these equilibria in order to determine whether or not valid assumptions leading to simpler equations and solutions can be made.

Where the weak acid is reasonably strong and, correspondingly, the salt hydrolysis action is of low order, the $[H_3O^+]$ can be generally expected to be greater than 10^{-6} M, with $[OH^-] = K_w/[H_3O^+]$ less than 10^{-8} M. If this situation exists, Eq. (16) modifies to

$$\frac{[H_3O^+](C_s + [H_3O^+])}{C_a - [H_3O^+]} \approx K_a \tag{18}$$

which leads to the quadratic equation

$$[H_3O^+]^2 + (C_s + K_a)[H_3O^+] - K_aC_a = 0 \tag{19}$$

and a solution in the form

$$[H_3O^+] \approx \frac{-(C_s + K_a) + \sqrt{(C_s + K_a)^2 + 4K_aC_a}}{2} \tag{20}$$

Where the acid is very weak and, correspondingly, the salt hydrolysis is of a high order, it can be generally expected that the value of $[OH^-]$ will be greater than $10^{-6}\ M$, so that $K_w/[H_3O^+]$ will be greater than $10^{-6}\ M$ and $[H_3O^+]$ will be less than $10^{-8}\ M$. If this situation exists, then Eq. (16) modifies to

$$\frac{[H_3O^+](C_s - K_w/[H_3O^+])}{C_a + K_w/[H_3O^+]} \approx K_a \tag{21}$$

which yields the quadratic equation

$$[H_3O^+]^2 - \frac{(K_w + K_aC_a)}{C_s}[H_3O^+] - \frac{K_wK_a}{C_s} = 0 \tag{22}$$

with a solution in the form

$$[H_3O^+] \approx \frac{\{(K_w + K_aC_a)/C_s\} + \sqrt{\{(K_w + K_aC_a)/C_s\}^2 + 4K_wK_a/C_s}}{2} \tag{23}$$

We may now consider, as a second order of assumption, the amount of either $[H_3O^+]$ or $K_w/[H_3O^+]$ relative to C_a and C_s. In general, for Eq. (18), the value of $[H_3O^+]$ will usually be much less than C_a or C_s, and the equation modifies to

$$\frac{[H_3O^+]C_s}{C_a} \approx K_a$$

which yields

$$[H_3O^+] \approx \frac{K_aC_a}{C_s} \tag{24}$$

Again in general, for Eq. (21), the value of $K_w/[H_3O^+]$ or $[OH^-]$ will usually be much less than C_a or C_s, and the equation modifies to

$$\frac{[H_3O^+]C_s}{C_a} \approx K_a$$

which again yields

$$[H_3O^+] \approx \frac{K_aC_a}{C_s} \tag{24}$$

It will be noted that in the derivation of Eq. (24) from Eq. (18), we have, for Eq. (24),

$$\frac{\Delta[H_3O^+]}{[H_3O^+]} = \frac{\Delta C_a}{C_a} + \frac{\Delta C_s}{C_s}$$

so that the relative error in the determination of $[H_3O^+]$ from Eq. (24) will be given by the sum of the relative errors in C_a and C_s resulting from ignoring $[H_3O^+]$ with respect to C_a and C_s. Thus, if the value of $[H_3O^+]$, determined from Eq. (24), is 2.5 percent of C_a and 2.5 percent of C_s, the relative error in $[H_3O^+]$ from Eq. (24) will be about 5 percent.

Again, in the derivation of Eq. (24) from Eq. (21), we have

$$\frac{\Delta[H_3O^+]}{[H_3O^+]} = \frac{\Delta C_a}{C_a} + \frac{\Delta C_s}{C_s}$$

so that the relative error in determination of $[H_3O^+]$ from Eq. (24) will be given by the sum of the relative errors in C_a and C_s resulting from ignoring $K_w/[H_3O^+]$ with respect to C_a and C_s.

The following *general* procedures may be applied.

1. Use Eq. (24) to determine $[H_3O^+]$.
2. (*a*) If the $[H_3O^+]$ is greater than $10^{-7} M$, then practically all the $[H_3O^+]$ in the solution originates from the dissociation of the weak acid HA.
   (*b*) If situation 2*a* is found to be valid, compare the determined value of $[H_3O^+]$ with C_a and C_s. If the total of the percentage values for $[H_3O^+]$ relative to C_a and C_s is 2.5 percent or less, the relative error in $[H_3O^+]$ from Eq. (24) will be about 2.5 percent or less. This level of error will usually be adequate for quantitative analytical work.
   (*c*) If situation 2*a* is found to be valid, and if the value of $[H_3O^+]$ from Eq. (24) shows a predicted relative error greater than 2.5 percent, or if a level of error for $[H_3O^+]$ better than 2.5 percent is required, the quadratic Eq. (20) should be solved.
3. (*a*) If the $[H_3O^+]$ from 1 is less than $10^{-7} M$, then practically all the $[OH^-]$ or $K_w/[H_3O^+]$ originates from the hydrolysis of the salt anion A^-.
   (*b*) If situation 3*a* is found to be valid, determine the value of $K_w/[H_3O^+]$ $= [OH^-]$. Compare this value of $[OH^-]$ with C_a and C_s. If the total of the percentage values for $[OH^-]$ relative to C_a and C_s is 2.5 percent or less, the relative error in $[H_3O^+]$ from Eq. (24) will be about 2.5 percent or less. Again, this will be good enough for most quantitative analytical work.
   (*c*) If situation 3*a* is found to be valid, and if the value of $[H_3O^+]$ from Eq. (24) shows a predicted relative error greater than 2.5 percent, or if a level of error for $[H_3O^+]$ better than 2.5 percent is required, the quadratic Eq. (23) should be solved.

It should be carefully noted that the relative error predicted for $[H_3O^+]$ from Eq. (24) will only agree well with the actual error, determined by comparing $[H_3O^+]$ from Eq. (24) with that from either Eq. (20) or (23), as the case may be, where the percentage of $[H_3O^+]$, on the one hand, and $K_w/[H_3O^+]$ on the other, relative to either C_a or C_s, is not much higher than about 50 percent. The validity of the tests shown is not, however, in question.

In the preparation of buffered solutions, C_a and C_s are usually maintained at reasonably high values in order to provide adequate buffer capacity. These will *generally* lie between 1 and 0.05 M. For the titration of, say, a weak acid by a strong base, it is apparent that C_a and C_s will vary continuously during the course of the titration—C_a being high at the start and decreasing during the titration; and C_s being nonexistent at the start and increasing up to the titration equivalence point.

It is important again to realize that exactly the same reasoning can be applied to the buffered-solution equilibria associated with aqueous solutions of the general weak base BOH and its general salt with a strong acid, BCl. An appropriate student project would be the development of such exact and approximate equations.

Example   For buffered solutions involving acetic acid and sodium acetate we have a K_a value for the acid of 1.76×10^{-5}. Two cases are considered, one involving $C_a = C_s = 1$ M and the other $C_a = C_s = 0.05$ M. In each case the use of Eq. (24) shows the $[H_3O^+]$ to be greater than 10^{-7} M, so that situation 2a is always valid. The application of Eq. (24), together with the predicted relative error in the associated value of $[H_3O^+]$, is shown in the following.

$C_a = C_s$, M	$[H_3O^+]$, M	$[H_3O^+]/C_a$, %	$[H_3O^+]/C_s$, %	$\Delta[H_3O^+]/[H_3O^+]$, %
1	1.76×10^{-5}	0.002	0.002	0.004
0.05	1.76×10^{-5}	0.04	0.04	0.08

Obviously Eq. (24) is adequate for the determination of $[H_3O^+]$ for such buffered solutions.

Example   In the titration of acetic acid by sodium hydroxide, variable C_a and C_s values will be encountered during the course of the titration. The following shows a series of developments in this connection.

Near the start of the titration we shall assume C_a at 0.1 M and C_s at 0.0001 M. Equation (24) yields

$$[H_3O^+] \approx 1.76 \times 10^{-2} M$$

Since $[H_3O^+]$ is greater than $10^{-7} M$, situation 2a is valid. The value of this determined $[H_3O^+]$ relative to C_a and C_s is about 17.6 and 17,600 percent, respectively. Equation (24) cannot obviously be used, and the quadratic Eq. (20) is solved to yield

$$[H_3O^+] \approx 1.27 \times 10^{-3} M$$

Note that this is a case where the predicted and actual errors do not coincide, but where the validity of the test is not in question.

At a point further removed from the titration start we shall assume C_a at 0.1 M and C_s at 0.01 M. Equation (24) gives

$[H_3O^+] \approx 1.76 \times 10^{-4} M$

with this value of $[H_3O^+]$ relative to C_a and C_s being about 0.2 and 1.8 percent, respectively. The total of these is 2.0 percent, so that Eq. (24) is adequate here.

At a point close to the titration equivalence point we shall assume C_a at 0.00005 M and C_s at 0.05 M. Equation (24) gives

$[H_3O^+] \approx 1.76 \times 10^{-8} M$

with situation 3a being valid. The value of $[OH^-]$ calculated as $K_w/[H_3O^+]$ from this $[H_3O^+]$ is $5.68 \times 10^{-7} M$, and this value, relative to C_a and C_s, yields percentages of 1.0 and 0.001, approximately and respectively. Equation (24) is thus adequate here.

It would appear, therefore, that in the titration of 0.1 M acetic acid by 0.1 M sodium hydroxide, the extent of hydrolysis near the equivalence point does not warrant, at least for the approximate titration point investigated, the use of any expression for $[H_3O^+]$ more exact than Eq. (24). For points in the titration near the start, large relative errors are introduced by the use of the simple Eq. (24). Here the use of the quadratic Eq. (20) is required.

Example   For buffered solutions involving hydrogen cyanide and sodium cyanide we have a K_a value for the acid of 5.00×10^{-10}. Two cases are again considered, these being $C_a = C_s = 1 M$ and $C_a = C_s = 0.05 M$. In each case the use of Eq. (24) shows the $[H_3O^+]$ to be less than $10^{-7} M$, so that situation 3a is always valid. The application of Eq. (24), together with the predicted relative error in the associated value of $[H_3O^+]$, is shown in the following.

$C_a = C_s$, M	$[H_3O^+]$, M	$[OH^-]/C_a$, %	$[OH^-]/C_s$, %	$\Delta[H_3O^+]/[H_3O^+]$, %
1	5.00×10^{-10}	0.002	0.002	0.004
0.05	5.00×10^{-10}	0.04	0.04	0.08

Obviously Eq. (24) is adequate for the determination $[H_3O^+]$ for such buffered solutions.

Example   In the titration of hydrogen cyanide by sodium hydroxide we shall assume, for a point near the start of the titration, values of $C_a = 1 M$ and $C_s = 0.001 M$. Equation (24) gives

$[H_3O^+] \approx 5.00 \times 10^{-7} M$

and situation 2a is valid. The value of $[H_3O^+]$ is 0.00005 and 0.05 percent, respectively, of C_a and C_s, so that Eq. (24) is adequate here.

At a point close to the equivalence point of the titration we shall assume $C_a = 0.005 M$ and $C_s = 0.5 M$. Equation (24) yields

$[H_3O^+] \approx 5.00 \times 10^{-12} M$

and situation 3a is valid. This value of $[H_3O^+]$ yields an $[OH^-] = K_w/[H_3O^+]$ value of $2.00 \times 10^{-3} M$, and $[OH^-]$ is 40 and 0.4 percent, respectively, of C_a and C_s. Equation (24) is obviously not adequate, and the quadratic Eq. (23) is solved to yield

$[H_3O^+] \approx 6.55 \times 10^{-12} M$

It appears, therefore, that for points in the titration of 1 M hydrogen cyanide by 1 M sodium hydroxide near the start, Eq. (24) is quite adequate. For points close to the equivalence point, the quadratic Eq. (23) will likely be required.

4.4.4 THE USE OF THE CUBIC EQUATIONS RELATIVE TO WEAK MONOBASIC ACID AND SALT EQUILIBRIA

While the appropriate cubic equation is not really required for any of the weak acid and weak acid salt equilibria so far discussed, it will be noted that during the development of the quadratic equations, assumptions were made that involved expectancies of $[H_3O^+]$ greater than $10^{-6}\,M$ with $[OH^-]$ less than $10^{-8}\,M$, or vice versa. In theory, the relevant cubic equation should be solved for those instances where $[H_3O^+]$ and $[OH^-]$ lie between 10^{-6} and $10^{-8}\,M$. However, for quantitative analytical purposes, the error introduced by extending the ranges involved to $[H_3O^+]$ greater than $10^{-7}\,M$ with $[OH^-]$ less than $10^{-7}\,M$, or vice versa, is relatively insignificant. In fact, it will be recalled that in the general procedures for determining the $[H_3O^+]$ of buffered solutions, the range *was* so extended, situation 2*a* being valid for $[H_3O^+]$ greater than $10^{-7}\,M$ and situation 3*a* being valid for $[H_3O^+]$ less than $10^{-7}\,M$.

4.4.5 BUFFER CAPACITY

The capacity of a buffer is the ability of the buffered solution to resist changes in pH that should result from the addition to the solution of acid or base substances. The *buffer capacity* is usually defined as the number of moles of strong base required to change the pH of one liter of solution by one unit. High buffer capacity will, of course, be obtained where relatively large concentrations of C_a and C_s, C_b and C_s, or C_s alone exist in the buffered solution.

It should be noted that calculations of, for example, the concentrations of weak acid and its salt required to yield a given buffered solution pH can be made on the basis of the equations just developed for buffered-solution conditions. When made in the simplest manner, these calculations can be approximate only. When exact buffered solution pH values are required, such solutions should be approximately prepared and subsequently adjusted under test by a pH meter.

The *general* effectiveness (ability to resist changes in pH that should result from the addition of a strong acid or strong base) of a buffered solution will be greatest where C_a and C_s are equal. Suppose that an aqueous solution involving a moles of the weak acid HA is prepared, and suppose further that NaOH is added slowly to this solution. At any point in this process, let b moles represent the amount of NaOH added. Where $b < a$ only part of the acid has been neutralized, so that $(a - b)$ moles of acid are left and b moles of salt NaA have been formed. If V is the volume of the solution in liters at this point, then by using simple equations as approximations, we have

$$C_a \approx \frac{a - b}{V} \quad \text{and} \quad C_s \approx \frac{b}{V}$$

If we assume that the ionization of the acid HA is slight, we have

$$[HA]_{un} \approx \frac{a - b}{V} \approx C_a$$

And if we assume that the hydrolysis of the salt NaA is slight, we have

$$[A^-] \approx \frac{b}{V} \approx C_s$$

Using the simplest buffer action equation, Eq. (24), we then have

$$[H_3O^+] \approx \frac{K_a C_a}{C_s} \approx \frac{K_a(a-b)/V}{b/V} \approx \frac{K_a(a-b)}{b}$$

From this we have

$$pH = -\log K_a - \log(a-b) + \log b$$

Differentiation of this equation yields

$$\frac{dpH}{db} = \frac{1}{a-b} + \frac{1}{b} = \frac{a}{b(a-b)}$$

For dpH/db to have a minimum value (most effective buffering action), the second derivative d^2pH/db^2 must equal zero; therefore,

$$\frac{d^2pH}{db^2} = -\frac{a(a-2b)}{b^2(a-b)^2} = 0$$

from which

$$b = \frac{a}{2}$$

This indicates that the maximum buffering effectiveness is obtained when $C_a = C_s$.

From the foregoing it is apparent that the selection of the proper weak acid or weak base for a buffered solution of maximum buffering effectiveness will be based on the use of that substance where pK_a or pK_b matches the pH or pOH required for the buffered solution.

Example   A buffered solution is prepared which involves acetic acid $C_a = 0.1\ M$ and sodium acetate $C_s = 0.2\ M$. The total volume is 100 ml and the K_a value is approximately 2.00×10^{-5}.

Determine:
1. The effect of a 10-ml addition of $0.1\ M$ NaOH to this solution
2. The effect of a 10-ml addition of $0.1\ M$ HCl to this solution
3. The buffer capacity of the original solution

Equation (24) gives $[H_3O^+] \approx 1.00 \times 10^{-5}\ M$, and the pH of the buffered solution as prepared will be 5.00.

1. With a 10-ml addition of $0.1\ M$ NaOH we have

$$C_a \approx \frac{100 \times 0.1 - 10 \times 0.1}{110} \approx 8.18 \times 10^{-2}\ M$$

$$C_s \approx \frac{100 \times 0.2 + 10 \times 0.1}{110} \approx 1.91 \times 10^{-1} M$$

so that from Eq. (24) we now have

$$[H_3O^+] \approx 8.56 \times 10^{-6} M$$

and a solution pH of 5.07.

2. With a 10-ml addition of 0.1 M HCl we have

$$C_a \approx \frac{100 \times 0.1 + 10 \times 0.1}{110} = 1.00 \times 10^{-1} M$$

$$C_s \approx \frac{100 \times 0.2 - 10 \times 0.1}{110} = 1.73 \times 10^{-1} M$$

so that from Eq. (24) we have

$$[H_3O^+] \approx 1.16 \times 10^{-5} M$$

and a solution pH of 4.94.

3. The buffer capacity of the original solution is found by determining the number of moles of NaOH required to produce a pH change of one unit per liter of the solution.

Original pH $= 5.00 = 10^{-5} M [H_3O^+]$

New pH $= 6.00 = 10^{-6} M [H_3O^+]$

Therefore,

$$10^{-6} \approx \frac{K_a(C_a - \Delta C_a)}{(C_s + \Delta C_s)}$$

where $\Delta C_a = \Delta C_s =$ the molarity of NaOH. We now have

$$10^{-6} \approx \frac{2.00 \times 10^{-5}(0.1 - \Delta C_a)}{(0.2 + \Delta C_a)}$$

$$\Delta C_a \approx 8.57 \times 10^{-2} M$$

The buffer capacity of the original solution is therefore 0.0857

4.5   IONIZATION EQUILIBRIA FOR POLYBASIC
ACIDS, BASES, AND RELATED SALTS

4.5.1   GENERAL

Polybasic acids are those which dissociate or ionize in solution in stages, each stage yielding H_3O^+ and another acid substance that, in its turn, dissociates to yield H_3O^+, etc. A typical example of this situation would be a dibasic acid such as carbonic acid, H_2CO_3, which dissociates in two stages in the reactions

$$H_2CO_3 + H_2O \rightleftharpoons H_3O^+ + HCO_3^-$$
$$HCO_3^- + H_2O \rightleftharpoons H_3O^+ + CO_3^{2-}$$

A second example would be the tribasic acid phosphoric acid, H_3PO_4, which ionizes in three stages according to the reactions

$$H_3PO_4 + H_2O \rightleftharpoons H_3O^+ + H_2PO_4^-$$
$$H_2PO_4^- + H_2O \rightleftharpoons H_3O^+ + HPO_4^{2-}$$
$$HPO_4^{2-} + H_2O \rightleftharpoons H_3O^+ + PO_4^{3-}$$

It can be expected that with polybasic acids, the extent to which any dissociation stage, after the first, takes place will be relatively low, repression of the secondary, tertiary, etc., stages being brought about by the concentration of H_3O^+ from the primary, secondary, etc., stages.

The dissociation of polybasic bases proceeds along the same general lines, each stage of dissociation yielding OH^- and a base substance that further dissociates in turn to yield OH^-, etc.

Thus, in dealing with polybasic acid or base solutions, we treat with weakly ionized substances, and our treatment of the associated equilibria will reflect this situation.

In treating the various equilibria for polybasic acids and related salts we shall, in the interests of limiting the scope of the discussion, avoid validity proofs for the assumptions made and the explanations of the errors introduced by making the assumptions in question. We shall give general procedures, however, which will permit decision-making processes relative to equations adequate for the determination of the $[H_3O^+]$ in each case; and we shall indicate the approximate relative error to be expected in the $[H_3O^+]$ calculated from any equation used.

4.5.2 IONIZATION EQUILIBRIA FOR THE DIBASIC ACID H_2A

The term H_2A refers to a generalized dibasic acid, not to a particular substance. The dissociation reactions for an aqueous solution of H_2A are

$$H_2A + H_2O \rightleftharpoons H_3O^+ + HA^-$$
$$HA^- + H_2O \rightleftharpoons H_3O^+ + A^{2-}$$
$$2H_2O \rightleftharpoons H_3O^+ + OH^-$$

There are six concentrations of importance to be considered:

$$[H_2A]_{un} \quad [HA^-] \quad [A^{2-}] \quad [H_3O^+] \quad [OH^-] \quad C_a$$

where C_a is the initial or starting concentration of H_2A in molarity. The value of $[H_2A]_{un}$ represents the concentration of undissociated H_2A in the solution.

The equilibrium equations are

$$\frac{[H_3O^+][HA^-]}{[H_2A]_{un}} = K_1 \qquad \frac{[H_3O^+][A^{2-}]}{[HA^-]} = K_2$$
$$[H_3O^+][OH^-] = K_w$$

Note that the K_1 and K_2 equations may be combined to yield

$$\frac{[H_3O^+]^2[A^{2-}]}{[H_2A]_{un}} = K_1K_2$$

The initial concentration of acid C_a must be present as H_2A_{un}, HA^-, or A^{2-}, and we have the material balance

$$C_a = [H_2A]_{un} + [HA^-] + [A^{2-}]$$

$$= [H_2A]_{un} + \frac{[H_2A]_{un}K_1}{[H_3O^+]} + \frac{[H_2A]_{un}K_1K_2}{[H_3O^+]^2}$$

$$= [H_2A]_{un}\left(1 + \frac{K_1}{[H_3O^+]} + \frac{K_1K_2}{[H_3O^+]^2}\right)$$

The charge balance must equate, and we have

$$[H_3O^+] = [HA^-] + 2[A^{2-}] + [OH^-]$$

Note that the coefficient of $[A^{2-}]$ in the charge balance is 2, reflecting the fact that the negative charge contributed by A^{2-} is twice the concentration of the ion. Note also that $[OH^-]$ originates with the water dissociation and is given by $K_w/[H_3O^+]$. The charge-balance equation now yields

$$[H_3O^+] - [OH^-] = \frac{[H_2A]_{un}K_1}{[H_3O^+]} + \frac{2[H_2A]_{un}K_1K_2}{[H_3O^+]^2}$$

$$[H_3O^+] - \frac{K_w}{[H_3O^+]} = [H_2A]_{un}\left(\frac{K_1}{[H_3O^+]} + \frac{2K_1K_2}{[H_3O^+]^2}\right)$$

Dividing the charge balance by the material balance, we have

$$[H_3O^+] - \frac{K_w}{[H_3O^+]} = \frac{C_aK_1[H_3O^+] + 2K_1K_2C_a}{[H_3O^+]^2 + K_1[H_3O^+] + K_1K_2} \tag{25}$$

which yields the quartic equation

$$[H_3O^+]^4 + K_1[H_3O^+]^3 + (K_1K_2 - K_w - K_1C_a)[H_3O^+]^2$$
$$- (K_wK_1 + 2K_1K_2C_a)[H_3O^+] - K_wK_1K_2 = 0 \tag{26}$$

By making various assumptions, we shall now treat these equilibria so as to obtain simpler equations for the determination of $[H_3O^+]$.

The contribution of $[H_3O^+]$ from the water dissociation is represented by $[OH^-] = K_w/[H_3O^+]$ in the charge-balance equation. Generally speaking, the solution will be fairly acid, so that $[H_3O^+]$ will be greater than $10^{-7} M$, with $[OH^-] = K_w/[H_3O^+]$ being less than $10^{-7} M$. Where the value of $K_w/[H_3O^+]$ can thus be ignored relative to $[H_3O^+]$, Eq. (25) modifies to

$$[H_3O^+] \approx \frac{C_aK_1[H_3O^+] + 2K_1K_2C_a}{[H_3O^+]^2 + K_1[H_3O^+] + K_1K_2} \tag{27}$$

which yields the cubic equation

$$[H_3O^+]^3 + K_1[H_3O^+]^2 + (K_1K_2 - K_1C_a)[H_3O^+] - 2K_1K_2C_a = 0 \tag{28}$$

Again we can assume that the secondary dissociation will occur to an extent considerably less than the primary, so that the $[H_3O^+]$ from the secondary dissociation may be ignored. This contribution of $[H_3O^+]$ is represented by the term $[A^{2-}]$ in the material balance and by the term $2[A^{2-}]$ in the charge balance. Where $[H_3O^+]$ from the secondary dissociation can be ignored, the material-balance equation reduces to

$$C_a \approx [H_2A]_{un} \left(1 + \frac{K_1}{[H_3O^+]}\right)$$

and the charge-balance equation reduces to

$$[H_3O^+] \approx \frac{[H_2A]_{un}K_1}{[H_3O^+]}$$

Dividing these modified equations as before, we have

$$[H_3O^+] \approx \frac{C_a K_1}{[H_3O^+] + K_1}$$

which can be expressed as

$$\frac{[H_3O^+]^2}{C_a - [H_3O^+]} \approx K_1 \tag{29}$$

and which yields the quadratic equation

$$[H_3O^+]^2 + K_1[H_3O^+] - K_1 C_a = 0 \tag{30}$$

which has a solution in

$$[H_3O^+] \approx \frac{-K_1 + \sqrt{K_1^2 + 4K_1 C_a}}{2} \tag{31}$$

Note that in this modification of the material- and charge-balance equations, the contribution of $[H_3O^+]$ from the secondary dissociation is approximately given by $[A^{2-}]$. Note also that in the equilibrium equation

$$\frac{[H_3O^+][A^{2-}]}{[HA^-]} = K_2$$

we have $[H_3O^+] \approx [HA^-]$, so that $[A^{2-}] \approx K_2 \approx$ the $[H_3O^+]$ contribution from the secondary dissociation. Observe that the $[H_3O^+]$ from the secondary disso-ciation will then be approximately given by K_2 for the dibasic acid.

In considering now the situation given by Eq. (29), it is apparent that in gen-eral $[H_3O^+]$ is much less than C_a, so that where this condition is valid, we have Eq. (29) reduced to

$$[H_3O^+]^2 \approx K_1 C_a$$

and

$$[H_3O^+] \approx \sqrt{K_1 C_a} \tag{32}$$

The following *general* procedures may be applied. It will be assumed that the relative error in $[H_3O^+]$ as determined should not exceed 5 percent for quantitative analytical purposes.

1. Use Eq. (32) to determine the value of $[H_3O^+]$.
2. Use the value of $[H_3O^+]$ obtained in 1 to determine
   (*a*) The percentage of K_2 relative to $\lceil H_3O^+ \rceil$
   (*b*) The percentage of $[H_3O^+]$ relative to C_a
3. If 2*a* is higher than 5 percent, the relative error in $[H_3O^+]$ from either Eq. (31) or (32) will exceed 5 percent. Equation (28) should be solved.
4. If 2*a* is 5 percent or less, and the sum of 2*a* and one-half of 2*b* is higher than 5 percent, the relative error in $[H_3O^+]$ from Eq. (32) will exceed 5 percent. Equation (31) should be solved to yield an $[H_3O^+]$ with a relative error about equal to the value of 2*a*.
5. If 2*a* is 5 percent or less, and the sum of 2*a* and one-half of 2*b* is less than 5 percent, the relative error in $[H_3O^+]$ from Eq. (32) will be less than 5 percent. Equation (32) should be used with a relative error in $[H_3O^+]$ obtained about equal to the sum of 2*a* and one-half of 2*b*.

Where the relative-error situation for $[H_3O^+]$ must be better than 5 percent, this may require, in order of the relative-error level demanded, the use of Eq. (32), (31), (28), or (26).

Example   An aqueous solution of oxalic acid dissociates

$$COOH \cdot COOH + H_2O \rightleftharpoons H_3O^+ + COOH \cdot COO^-$$

$$COOH \cdot COO^- + H_2O \rightleftharpoons H_3O^+ + (COO^-)_2$$

with $K_1 = 5.90 \times 10^{-2}$ and $K_2 = 6.40 \times 10^{-5}$. The application of Eq. (32) yields the following $[H_3O^+]$ values and predicted relative errors in $[H_3O^+]$ as determined for various initial concentrations of oxalic acid.

C_a, M	$[H_3O^+]$, M	$K_2/[H_3O^+]$, %	$[H_3O^+]/C_a$, %	$\Delta[H_3O^+]/[H_3O^+]$, %
1	2.43×10^{-1}	0.03	24	12
0.1	7.69×10^{-2}	0.08	77	39
0.01	2.43×10^{-2}	0.3	243	122

For a dibasic acid such as oxalic acid, with a relatively high K_1 value, Eq. (32) will be inadequate under the conditions given. Equation (28) is not required, since $K_2/[H_3O^+]$ is less than 5 percent. Equation (31) is good enough and its use yields.

$C_a = 1\ M$ $[H_3O^+] \approx 2.15 \times 10^{-1}\ M$

$C_a = 0.1\ M$ $[H_3O^+] \approx 5.25 \times 10^{-2}\ M$

Example An aqueous solution of carbonic acid, with $K_1 = 4.30 \times 10^{-7}$ and $K_2 = 5.61 \times 10^{-11}$, under the same investigation but with different and appropriate acid concentrations shows the following.

C_a, M	$[H_3O^+], M$	$K_2/[H_3O^+], \%$	$[H_3O^+]/C_a, \%$	$\Delta[H_3O^+]/[H_3O^+], \%$
0.02	9.27×10^{-5}	0.00006	0.5	0.2^5
0.001	2.07×10^{-5}	0.0003	2.1	1.0

For a dibasic acid such as carbonic acid, with a relatively low K_1 value, Eqs. (28) and (31) are not required; Eq. (32) will be adequate under most circumstances.

4.5.3 IONIZATION EQUILIBRIA FOR SOLUTIONS OF THE DIBASIC ACID H_2A AND ITS SALT NaHA

General Aqueous solutions involving mixtures of H_2A and NaHA may represent buffered-solution conditions aimed at establishing and maintaining some specific pH for the solution. They may also represent points in the titration of a dibasic acid by the strong base NaOH, such points lying after the start and before the first-stage equivalence point of the titration.

Solutions of this type may result in acidic or basic conditions, which will depend on the values of K_1 and K_2 for the dibasic acid, as well as on the values of C_a and C_s for the acid and its salt NaHA, respectively.

Ionization equilibria for H_2A–NaHA solutions H_2A and NaHA represent a general dibasic acid and its sodium monosubstitution salt. The dissociation and hydrolysis reactions in aqueous solution are

$$NaHA \rightleftharpoons Na^+ + HA^-$$
$$H_2A + H_2O \rightleftharpoons H_3O^+ + HA^-$$
$$HA^- + H_2O \rightleftharpoons H_3O^+ + A^{2-}$$
$$\text{or} \quad HA^- + H_2O \rightleftharpoons H_2A + OH^-$$
$$2H_2O \rightleftharpoons H_3O^+ + OH^-$$

Here the ion HA^- may react to produce H_3O^+ and A^{2-}, or it may react to yield H_2A_{un} and OH^-. The predominant reaction will be largely dependent on the values of K_1 and K_2, as well as on the values of C_a and $C_{s,NaHA}$. Where the first reaction is predominant, this, together with the reaction for the first-stage dissociation $H_2A + H_2O \rightleftharpoons H_3O^+ + HA^-$, may yield an acidic solution. Where the second reaction is predominant, this may yield a basic solution.

There are seven concentrations to be considered here:

$$[H_2A]_{un} \quad [HA^-] \quad [A^{2-}] \quad [H_3O^+] \quad [OH^-] \quad C_{s,NaHA} \quad C_a$$

The concentration of $C_{s,\text{NaHA}}$ represents the initial molarity of the acid salt and for the purposes of the development of various equilibria equations will be shown as $C_{s,1}$. C_a is the initial molarity of the dibasic acid H_2A. The equilibrium equations applicable here will be identical to those given in Subsec. 4.5.2 for K_1, K_2, K_1K_2, and K_w. The initial concentrations C_a and $C_{s,1}$ must be present as H_2A_{un}, HA^-, or A^{2-}. This yields the material balance

$$C_a + C_{s,1} = [H_2A]_{\text{un}} + [HA^-] + [A^{2-}]$$
$$= [H_2A]_{\text{un}} \left(1 + \frac{K_1}{[H_3O^+]} + \frac{K_1K_2}{[H_3O^+]^2}\right)$$

The charge balance must equate, and we have

$$[H_3O^+] + [Na^+] = [HA^-] + 2[A^{2-}] + [OH^-]$$

Here we have

$$[OH^-] = \frac{K_w}{[H_3O^+]}$$
$$[Na^+] = C_{s,1}$$

and for the charge balance we now show

$$[H_3O^+] - \frac{K_w}{[H_3O^+]} + C_{s,1} = [HA^-] + 2[A^{2-}]$$
$$= [H_2A]_{\text{un}} \left(\frac{K_1}{[H_3O^+]} + \frac{2K_1K_2}{[H_3O^+]^2}\right)$$

Dividing the charge-balance equation by the material-balance equation gives

$$[H_3O^+] - \frac{K_w}{[H_3O^+]}$$
$$= \frac{K_1C_a[H_3O^+] + 2K_1K_2C_a - C_{s,1}[H_3O^+]^2 + K_1K_2C_{s,1}}{[H_3O^+]^2 + K_1[H_3O^+] + K_1K_2} \quad (33)$$

which leads to the quartic equation

$$[H_3O^+]^4 + (C_{s,1} + K_1)[H_3O^+]^3 + (K_1K_2 - K_w - K_1C_a)[H_3O^+]^2$$
$$- (K_wK_1 + 2K_1K_2C_a + K_1K_2C_{s,1})[H_3O^+] - K_wK_1K_2 = 0 \quad (34)$$

We shall now examine the equilibria involved in order to derive simpler equations for the evaluation of $[H_3O^+]$.

It is apparent, first of all, that the value of $([H_3O^+] - K_w/[H_3O^+])$ in Eq. (33) will depend to a considerable extent on the values of K_1 and K_2. Where K_1 and K_2 are relatively high, $[H_3O^+]$ will be high, hydrolysis of HA^- will be low, and any $[OH^-]$ will originate mainly from the water dissociation. In such cases, $[OH^-] = K_w/[H_3O^+]$ can be ignored relative to $[H_3O^+]$. Again, where K_1 and K_2 are relatively low, $[H_3O^+]$ will be low, hydrolysis of HA^- will be high, and $[OH^-]$ will be high and will originate mainly from the hydrolysis action. In such cases, $[H_3O^+]$ can be ignored with respect to $[OH^-] = K_w/[H_3O^+]$.

The solution of H_2A–$NaHA$ *is acidic*   Here we have $[H_3O^+]$ greater than $10^{-7}\,M$ and $[OH^-]$ less than $10^{-7}\,M$. Where $K_w/[H_3O^+]$ is small enough to be ignored relative to $[H_3O^+]$, Eq. (33) yields

$$[H_3O^+] \approx \frac{K_1 C_a [H_3O^+] + 2K_1 K_2 C_a - C_{s,1} [H_3O^+]^2 + K_1 K_2 C_{s,1}}{[H_3O^+]^2 + K_1 [H_3O^+] + K_1 K_2} \quad (35)$$

and this leads to the cubic expression

$$[H_3O^+]^3 + (C_{s,1} + K_1)[H_3O^+]^2 + (K_1 K_2 - K_1 C_a)[H_3O^+]$$
$$- (K_1 K_2 C_{s,1} + 2K_1 K_2 C_a) = 0 \quad (36)$$

Where $[H_3O^+]$ from the secondary dissociation is small enough, this value, represented as before by $[A^{2-}]$ and approximately by K_2, can be ignored. This modifies the material- and charge-balance equations by removing, respectively, $[A^{2-}]$ and $2[A^{2-}]$. The value of $K_w/[H_3O^+]$ in the charge-balance equation has, of course, already been removed. The result of dividing the new charge balance by the new material balance is the equation

$$[H_3O^+] + C_{s,1} \approx \frac{K_1 C_a + K_1 C_{s,1}}{[H_3O^+] + K_1} \quad (37)$$

which can be written as

$$[H_3O^+] \approx K_1 \frac{C_a - [H_3O^+]}{C_{s,1} + [H_3O^+]} \quad (38)$$

yielding the quadratic equation

$$[H_3O^+]^2 + (C_{s,1} + K_1)[H_3O^+] - K_1 C_a = 0 \quad (39)$$

with a solution in

$$[H_3O^+] \approx \frac{-(C_{s,1} + K_1) + \sqrt{(C_{s,1} + K_1)^2 + 4K_1 C_a}}{2} \quad (40)$$

Note the similarity between this equation and Eq. (20).

Finally, in Eq. (38) we note that, very generally, $[H_3O^+]$ will be much less than C_a or $C_{s,1}$. Where valid, this reduces Eq. (38) to

$$[H_3O^+] \approx \frac{K_1 C_a}{C_{s,1}} \quad (41)$$

an equation similar in form to Eq. (24).

The solution of H_2A–$NaHA$ *is basic*   Here we have $[OH^-] = K_w/[H_3O^+]$ greater than $10^{-7}\,M$, with $[H_3O^+]$ less than $10^{-7}\,M$. Where $[H_3O^+]$ is small enough to be ignored relative to $K_w/[H_3O^+]$, Eq. (33) reduces to

$$-\frac{K_w}{[H_3O^+]} \approx \frac{K_1 C_a [H_3O^+] + 2K_1 K_2 C_a - C_{s,1} [H_3O^+]^2 + K_1 K_2 C_{s,1}}{[H_3O^+]^2 + K_1 [H_3O^+] + K_1 K_2} \quad (42)$$

which yields the cubic equation

$$[H_3O^+]^3 - \frac{(K_w + K_1 C_a)}{C_{s,1}} [H_3O^+]^2$$
$$- \left(\frac{K_1 K_2 C_{s,1} + 2K_1 K_2 C_a + K_w K_1}{C_{s,1}} \right) [H_3O^+] - \frac{K_w K_1 K_2}{C_{s,1}} = 0 \quad (43)$$

With respect to this solution, its basicity indicates that K_1 is relatively low and that HA^- hydrolyzes quite strongly. With K_2 less than K_1, the value of $[A^{2-}]$ will be much less than $[HA^-]$. This permits the values $[A^{2-}]$ and $2[A^{2-}]$ to be ignored again in the material- and charge-balance equations, respectively. With the value of $[H_3O^+]$ already removed from the charge-balance equation, division of the modified charge-balance equation by the material-balance equation now yields

$$C_{s,1} - \frac{K_w}{[H_3O^+]} \approx \frac{K_1 C_a + K_1 C_{s,1}}{[H_3O^+] + K_1} \qquad (44)$$

which can be written as

$$[H_3O^+] \approx K_1 \frac{C_a + K_w/[H_3O^+]}{C_{s,1} - K_w/[H_3O^+]} \qquad (45)$$

yielding the quadratic equation

$$[H_3O^+]^2 - \frac{(K_w + K_1 C_a)}{C_{s,1}} [H_3O^+] - \frac{K_w K_1}{C_{s,1}} = 0 \qquad (46)$$

with a solution in

$$[H_3O^+]$$
$$\approx \frac{\{(K_w + K_1 C_a)/C_{s,1}\} + \sqrt{\{(K_w + K_1 C_a)/C_{s,1}\}^2 + 4K_w K_1/C_{s,1}}}{2} \qquad (47)$$

Note the similarity between this equation and Eq. (23). Finally, in Eq. (45), we note that, very generally, $K_w/[H_3O^+]$ will be much less than C_a or $C_{s,1}$. Where valid, this reduces Eq. (45) to

$$[H_3O^+] \approx \frac{K_1 C_a}{C_{s,1}} \qquad (41)$$

General procedures   The following *general* procedures may be applied. It is assumed that the relative error in $[H_3O^+]$ as determined should not exceed 5 percent.

1. Use Eq. (41) to determine $[H_3O^+]$.
2. (*a*) If the $[H_3O^+]$ is greater than $10^{-7} M$, the solution will be acidic.
    (*b*) If situation 2a is found to be valid, use the value of $[H_3O^+]$ from 1 to determine
       (1) The percentage of K_2 relative to $[H_3O^+]$

(2) The percentage of $[H_3O^+]$ relative to C_a

(3) The percentage of $[H_3O^+]$ relative to $C_{s,1}$

(c) If $2b(1)$ is higher than 2.5 percent, the relative error in $[H_3O^+]$ from either Eq. (40) or (41) will exceed 5 percent. Equation (36) should be solved.

(d) If $2b(1)$ is 2.5 percent or less, and twice $2b(1)$ plus the sum of $2b(2)$ and $2b(3)$ is higher than 5 percent, the relative error in $[H_3O^+]$ from Eq. (41) will exceed 5 percent. Equation (40) should be used to yield an $[H_3O^+]$ value with a relative error about equal to twice $2b(1)$.

(e) If $2b(1)$ is 2.5 percent or less, and twice $2b(1)$ plus the sum of $2b(2)$ and $2b(3)$ is less than 5 percent, the relative error in $[H_3O^+]$ from Eq. (41) will be less than 5 percent. Equation (41) should be used with an expected relative error in $[H_3O^+]$ about equal to twice $2b(1)$ plus the sum of $2b(2)$ and $2b(3)$.

3. (a) If the $[H_3O^+]$ from 1 is less than $10^{-7} M$, the solution will be basic.

(b) If situation $3a$ is found to be valid, use the value of $[H_3O^+]$ from 1 to determine

(1) The percentage of K_2 relative to $[H_3O^+]$

(2) The percentage of $K_w/[H_3O^+]$ relative to C_a

(3) The percentage of $K_w/[H_3O^+]$ relative to $C_{s,1}$

(c) If $3b(1)$ is higher than 2.5 percent, the relative error in $[H_3O^+]$ from either Eq. (41) or (47) will exceed 5 percent. Equation (43) should be solved.

(d) If $3b(1)$ is 2.5 percent or less, and twice $3b(1)$ plus the sum of $3b(2)$ and $3b(3)$ is higher than 5 percent, the relative error in $[H_3O^+]$ from Eq. (41) will exceed 5 percent. Equation (47) should be used to yield an $[H_3O^+]$ value with a relative error about twice the value of $3b(1)$.

(e) If $3b(1)$ is 2.5 percent or less, and twice $3b(1)$ plus the sum of $3b(2)$ and $3b(3)$ is less than 5 percent, the relative error in $[H_3O^+]$ from Eq. (41) will be less than 5 percent. Equation (41) should be used with an expected relative error in $[H_3O^+]$ about equal to twice the value of $3b(1)$ plus the sum of $3b(2)$ and $3b(3)$.

Where the relative error situation for $[H_3O^+]$ must be better than 5 percent, this may require, in order of the relative-error level demanded, the use of Eq. (41), (40), (36), or (34) for acidic solutions, or Eq. (41), (47), (43), or (34) for basic solutions.

Example Aqueous solutions of oxalic acid, COOH·COOH, and sodium hydrogen oxalate, COOH·COONa, are to be prepared. The K_1 and K_2 values are 5.90×10^{-2} and 6.40×10^{-5}. The use of Eq. (41) gives for each solution $[H_3O^+]$ greater than $10^{-7} M$, so that situation $2a$ is valid in each case. The values given by Eq. (41), together with the predicted relative error in the associated $[H_3O^+]$, are shown in the following.

$C_{s,1}$ or C_a, M	$[H_3O^+]$, M	$K_2/[H_3O^+]$, %	$[H_3O^+]/C_a$, %	$[H_3O^+]/C_{s,1}$, %	$\Delta[H_3O^+]/[H_3O^+]$, %
$C_{s,1} = 0.1$ $C_a = 0.1$	5.90×10^{-2}	0.1	59	59	118
$C_{s,1} = 0.05$ $C_a = 0.01$	1.18×10^{-2}	0.5	118	24	143

Equation (41) is not adequate for either solution. Since $K_2/[H_3O^+]$ is less than 2.50 percent in each case, Eq. (36) is not required. Equation (40) is used to yield

$$C_{s,1} = 0.1\ M \qquad [H_3O^+] \approx 3.11 \times 10^{-2}\ M$$
$$C_a = 0.1\ M$$

$$C_{s,1} = 0.05\ M \qquad [H_3O^+] \approx 5.15 \times 10^{-3}\ M$$
$$C_a = 0.01\ M$$

Example An aqueous solution of 0.01 M each of carbonic acid, H_2CO_3, and sodium bicarbonate, $NaHCO_3$, is prepared. The K_1 and K_2 values are 4.30×10^{-7} and 5.61×10^{-11}. The use of Eq. (41) gives

$$[H_3O^+] \approx 4.30 \times 10^{-7}\ M$$

Here situation 2a is valid. The value of $K_2/[H_3O^+]$ is 0.01 percent, and this is less than 2.5 percent. $[H_3O^+]/C_a$ and $[H_3O^+]/C_{s,1}$ are each 0.004 percent. Equation (41) is therefore adequate.

Example A second solution involving 0.01 M carbonic acid and 0.5 M sodium bicarbonate was prepared. Equation (41) gives

$$[H_3O^+] \approx 8.60 \times 10^{-9}\ M$$

Situation 3a is valid here. The value of $K_2/[H_3O^+]$ is 0.7 percent, and this is less than 2.5 percent. The value of $K_w/[H_3O^+]$ is $1.16 \times 10^{-6}\ M$, and this value relative to C_a and $C_{s,1}$ is 0.01 and 0.0002 percent, respectively. Equation (41) is again adequate.

4.5.4   IONIZATION EQUILIBRIA FOR A SOLUTION OF THE DIBASIC ACID SALT NaHA

General Such a solution may represent a buffered solution, since aqueous solutions of the monosubstitution salts of dibasic acids and strong bases are capable of accepting, without significant change in pH, small additions of either strong acid or strong base. It may also represent the first-stage equivalence-point-solution condition originating with the titration of a dibasic acid by the strong base NaOH.

Solutions of this type may be acidic or basic; this will depend on the values of K_1 and K_2, as well as on the concentration of the salt NaHA. The anion HA^- of the salt may undergo hydrolysis, thus yielding a basic solution. It may, on the other hand, combine with H_2O, yielding H_3O^+ and an acidic solution.

Ionization equilibria for NaHA solutions NaHA represents the sodium monosubstitution salt of the general dibasic acid H_2A. The dissociation and hydrolysis reactions in aqueous solution are

$$NaHA \rightleftharpoons Na^+ + HA^-$$

$$HA^- + H_2O \rightleftharpoons H_2A + OH^-$$

or

$$HA^- + H_2O \rightleftharpoons H_3O^+ + A^{2-}$$

$$2H_2O \rightleftharpoons H_3O^+ + OH^-$$

Here again the ion HA^- may react to yield H_2A_{un} and OH^-, or H_3O^+ and A^{2-}, the predominant reaction depending on the values of K_1, K_2, and $C_{s,NaHA}$. Predominance relative to the first reaction yields a basic solution, while predominance with respect to the second yields an acidic solution.

There are six concentrations of importance:

$$[H_2A]_{un} \qquad [HA^-] \qquad [A^{2-}] \qquad [H_3O^+] \qquad [OH^-] \qquad C_{s,1}$$

with $C_{s,1}$ representing $C_{s,NaHA}$, the initial molarity of the salt NaHA. The equilibrium equations are identical to those outlined in Subsec. 4.5.2. The initial concentration of $C_{s,1}$ must be present as H_2A_{un}, HA^-, or A^{2-}; and the material balance is

$$C_{s,1} = [H_2A]_{un} + [HA^-] + [A^{2-}]$$

$$= [H_2A]_{un} \left(1 + \frac{K_1}{[H_3O^+]} + \frac{K_1 K_2}{[H_3O^+]^2}\right)$$

The charge balance equates, and we have

$$[H_3O^+] + [Na^+] = [HA^-] + 2[A^{2-}] + [OH^-]$$

which gives

$$[H_3O^+] - \frac{K_w}{[H_3O^+]} + C_{s,1} = [H_2A]_{un}\left(\frac{K_1}{[H_3O^+]} + \frac{2K_1K_2}{[H_3O^+]^2}\right)$$

Dividing the charge balance by the material balance we have

$$[H_3O^+] - \frac{K_w}{[H_3O^+]} = \frac{K_1 K_2 C_{s,1} - C_{s,1}[H_3O^+]^2}{[H_3O^+]^2 + K_1[H_3O^+] + K_1 K_2} \tag{48}$$

which leads to the quartic equation

$$[H_3O^+]^4 + (C_{s,1} + K_1)[H_3O^+]^3 + (K_1 K_2 - K_w)[H_3O^+]^2$$
$$- (K_w K_1 + K_1 K_2 C_{s,1})[H_3O^+] - K_w K_1 K_2 = 0 \tag{49}$$

The equilibria involved will now be examined in order to derive simpler equations for the evaluation of $[H_3O^+]$.

Again it is apparent that the value of the term $([H_3O^+] - K_w/[H_3O^+])$ in Eq. (48) will depend on the values of K_1 and K_2. Where K_1 and K_2 are relatively high, $[H_3O^+]$ will likely be high, and $K_w/[H_3O^+]$ can be ignored relative to $[H_3O^+]$. Where K_1 and K_2 are relatively low, $[OH^-]$ will likely be high as the result of the hydrolysis of HA^-, and $[H_3O^+]$ can be ignored relative to $[OH^-]$ $= K_w/[H_3O^+]$.

The solution of NaHA *is acidic*   Here we have $[H_3O^+]$ greater than $10^{-7}\,M$ and $[OH^-]$ less than $10^{-7}\,M$. Where $K_w/[H_3O^+]$ can be ignored with respect to $[H_3O^+]$, Eq. (48) reduces to

$$[H_3O^+] \approx \frac{K_1K_2C_{s,1} - C_{s,1}[H_3O^+]^2}{[H_3O^+]^2 + K_1[H_3O^+] + K_1K_2} \tag{50}$$

from which we have the cubic equation

$$[H_3O^+]^3 + (C_{s,1} + K_1)[H_3O^+]^2 + K_1K_2[H_3O^+] - K_1K_2C_{s,1} = 0 \tag{51}$$

Equation (50) can be written in the form

$$[H_3O^+]^2 \approx \frac{K_1K_2(C_{s,1} - [H_3O^+])}{(C_{s,1} + K_1) + [H_3O^+]} \tag{52}$$

If we assume that $[H_3O^+]$ will generally be less than $C_{s,1}$, we have Eq. (52) reduced to

$$[H_3O^+]^2 \approx \frac{K_1K_2C_{s,1}}{C_{s,1} + K_1}$$

and

$$[H_3O^+] \approx \sqrt{\frac{K_1K_2C_{s,1}}{C_{s,1} + K_1}} \tag{53}$$

As a final assumption it can be put forward that in most cases, K_1 will be much less than $C_{s,1}$ and, where valid, this reduces Eq. (53) to

$$[H_3O^+] \approx \sqrt{K_1K_2} \tag{54}$$

and we note that where it is possible to apply Eq. (54), the $[H_3O^+]$ of the solution is independent of the value of $C_{s,1}$.

The solution of NaHA *is basic*   Here we have $[H_3O^+]$ less than $10^{-7}\,M$ and $[OH^-]$ greater than $10^{-7}\,M$. Where $[H_3O^+]$ can be ignored relative to $K_w/[H_3O^+] = [OH^-]$, Eq. (48) reduces to

$$-\frac{K_w}{[H_3O^+]} \approx \frac{K_1K_2C_{s,1} - C_{s,1}[H_3O^+]^2}{[H_3O^+]^2 + K_1[H_3O^+] + K_1K_2} \tag{55}$$

which leads to the cubic equation

$$[H_3O^+]^3 - \frac{K_w}{C_{s,1}}[H_3O^+]^2 - \left(\frac{K_1K_2C_{s,1} + K_wK_1}{C_{s,1}}\right)[H_3O^+]$$
$$-\frac{K_wK_1K_2}{C_{s,1}} = 0 \tag{56}$$

Equation (55) can be written in the form

$$[H_3O^+]^2 \approx \frac{K_1K_2(C_{s,1} + K_w/[H_3O^+]) + K_wK_1}{C_{s,1} - K_w/[H_3O^+]} \tag{57}$$

In most cases $K_w/[H_3O^+]$ will be much less than $C_{s,1}$, and we have

$$[H_3O^+] \approx \sqrt{\frac{K_1(K_2 C_{s,1} + K_w)}{C_{s,1}}} \tag{58}$$

Finally, the value of K_w will generally be much less than $K_2 C_{s,1}$ and, where valid, this situation further reduces Eq. (58) to

$$[H_3O^+] \approx \sqrt{K_1 K_2} \tag{54}$$

General procedures   The following *general* procedures may be applied. It is assumed that the relative error in $[H_3O^+]$ as determined should not exceed 5 percent.

1. Use Eq. (54) to determine $[H_3O^+]$.
2. (a) If the $[H_3O^+]$ is greater than $10^{-7} M$, the solution will be acidic.
   (b) If situation 2a is valid, use the value of $[H_3O^+]$ from 1 to determine
      (1) The percentage of $[H_3O^+]$ relative to $C_{s,1}$
      (2) The percentage of $[H_3O^+]$ relative to $C_{s,1} + K_1$
      (3) The percentage of K_1 relative to $C_{s,1}$
   (c) If the sum of 2b(1) and 2b(2) is higher than 10 percent, the relative error in $[H_3O^+]$ from either Eq. (53) or (54) will exceed 5 percent. Equation (51) should be solved.
   (d) If the sum of 2b(1) and 2b(2) is less than 10 percent, and one-half the sum of 2b(1) and 2b(2) plus one-half of 2b(3) is higher than 5 percent, the relative error in $[H_3O^+]$ from Eq. (54) will exceed 5 percent. Equation (53) should be used to yield an $[H_3O^+]$ value with a relative error about equal to one-half the sum of 2b(1) and 2b(2).
   (e) If the sum of 2b(1) and 2b(2) is less than 10 percent, and one-half the sum of 2b(1) and 2b(2) plus one-half of 2b(3) is less than 5 percent, the relative error in $[H_3O^+]$ from Eq. (54) will be less than 5 percent. Equation (54) should be used with an expected relative error in $[H_3O^+]$ about equal to one-half the sum of 2b(1) and 2b(2) plus one-half of 2b(3).
3. (a) If the $[H_3O^+]$ from 1 is less than $10^{-7} M$, the solution will be basic.
   (b) If situation 3a is valid, use the value of $[H_3O^+]$ to determine
      (1) The percentage of $K_w/[H_3O^+]$ relative to $C_{s,1}$
      (2) The percentage of K_w relative to $K_2 C_{s,1}$
   (c) If the value of 3b(1) is higher than 5 percent, the relative error in $[H_3O^+]$ from either Eq. (54) or (58) will exceed 5 percent. Equation (56) should be solved.
   (d) If the value of 3b(1) is less than 5 percent, and the sum of 3b(1) and one-half of 3b(2) is higher than 5 percent, the relative error in $[H_3O^+]$ from Eq. (54) will exceed 5 percent. Equation (58) should be solved with a relative error in $[H_3O^+]$ about equal to 3b(1).

(e) If the value of $3b(1)$ is less than 5 percent, and the sum of $3b(1)$ plus one-half of $3b(2)$ is less than 5 percent, the relative error in $[H_3O^+]$ from Eq. (54) will be less than 5 percent. Equation (54) should be used with an expected relative error in $[H_3O^+]$ about equal to $3b(1)$ plus one-half of $3b(2)$.

Relative-error-level situations for $[H_3O^+]$ better than 5 percent may require, in order of the relative-error level demanded, the use of Eq. (54), (53), (51), or (49) for acidic solutions, or Eq. (54), (58), (56), or (49) for basic solutions.

Example   Aqueous solutions of sodium hydrogen oxalate containing $C_{s,1}$ values of 0.1 and 0.01 M are investigated. In each case the use of Eq. (54) shows a value of $[H_3O^+]$ greater than 10^{-7} M, so that situation $2a$ is valid. The following shows $[H_3O^+]$ from Eq. (54), and the expected relative error involved.

$C_{s,1}$, M	$[H_3O^+]$, M	$[H_3O^+]/C_{s,1}$, %	$[H_3O^+]/(C_{s,1} + K_1)$, %	$K_1/C_{s,1}$, %	$\Delta[H_3O^+]/[H_3O^+]$, %
0.1	1.94×10^{-3}	1.9	1.2	59	31
0.01	1.94×10^{-3}	19	2.8	590	306

In the case of the 0.1-M solution we have $2b(1)$ plus $2b(2)$ less than 10 percent, but one-half of the sum of $2b(1)$ and $2b(2)$ plus one-half of $2b(3)$ exceeds 5 percent. Equation (53) is required and yields

$$[H_3O^+] \approx 1.54 \times 10^{-3} \, M$$

In the case of the 0.01-M solution, we have $2b(1)$ plus $2b(2)$ greater than 10 percent, so that Eq. (51) is solved to yield

$$[H_3O^+] \approx 7.09 \times 10^{-4} \, M$$

Example   Aqueous solutions of sodium bicarbonate containing $C_{s,1}$ values of 0.5 and 0.005 M are investigated. In each case the use of Eq. (54) shows $[H_3O^+]$ less than $10^{-7}M$, so that situation $3a$ is valid. The following shows $[H_3O^+]$ from Eq. (54), and the expected relative error involved.

$C_{s,1}$, M	$[H_3O^+]$, M	$(K_w/[H_3O^+])/C_{s,1}$, %	$K_w/K_2C_{s,1}$, %	$\Delta[H_3O^+]/[H_3O^+]$, %
0.5	4.91×10^{-9}	0.0004	0.04	0.02
0.005	4.91×10^{-9}	0.04	3.6	1.8

Note that Eq. (54) is adequate for the calculation of $[H_3O^+]$ in 0.5- and 0.005-M solutions of $NaHCO_3$.

4.5.5 IONIZATION EQUILIBRIA FOR SOLUTIONS OF THE DIBASIC ACID SALTS NaHA AND Na$_2$A

General A solution of this type can represent a buffered-solution condition aimed at establishing and maintaining a specific solution pH. It can also represent some solution condition between the first- and second-stage equivalence points of a titration between a dibasic acid and the strong base NaOH.

Such a solution condition may also represent a condition between the start and the first-stage equivalence point of a titration between the dibasic acid salt Na$_2$A and a strong acid such as HCl.

The $[H_3O^+]$ of an aqueous solution of NaHA and Na$_2$A will depend on the K_1 and K_2 values for the dibasic acid, as well as on the concentrations of the two salts.

Ionization equilibria for NaHA–Na$_2$A **solutions** The salts NaHA and Na$_2$A represent salts of the general dibasic acid H$_2$A. The dissociation and hydrolysis reactions are

$$NaHA \rightleftharpoons Na^+ + HA^-$$

$$Na_2A \rightleftharpoons 2Na^+ + A^{2-}$$

$$HA^- + H_2O \rightleftharpoons H_3O^+ + A^{2-}$$

or $\quad HA^- + H_2O \rightleftharpoons H_2A + OH^-$

$$A^{2-} + H_2O \rightleftharpoons HA^- + OH^-$$

$$2H_2O \rightleftharpoons H_3O^+ + OH^-$$

Where K_1 and K_2 are relatively low, hydrolysis of both A^{2-} and HA$^-$ may be extensive, yielding a basic solution. Where K_1 and K_2 are relatively high, the reaction of HA$^-$ with H$_2$O to yield H$_3$O$^+$ may result in an acidic solution, particularly where $C_{s,NaHA}$ is high and C_{s,Na_2A} is low. There are seven concentrations of importance:

$$[H_2A]_{un} \qquad [HA^-] \qquad [A^{2-}] \qquad [H_3O^+] \qquad [OH^-] \qquad C_{s,1} \qquad C_{s,2}$$

where $C_{s,1}$ and $C_{s,2}$ represent, respectively, the initial molarities for the salts NaHA and Na$_2$A. These initial molarities must be present as H$_2$A$_{un}$, HA$^-$, or A^{2-}, and we have the material balance

$$C_{s,1} + C_{s,2} = [H_2A]_{un} + [HA^-] + [A^{2-}]$$

$$= [H_2A]_{un} \left(1 + \frac{K_1}{[H_3O^+]} + \frac{K_1 K_2}{[H_3O^+]^2}\right)$$

The charge balance must equate, and we have

$$[H_3O^+] + [Na^+]_{NaHA} + [Na^+]_{Na_2A} = [HA^-] + 2[A^{2-}] + [OH^-]$$

Here we have $[Na^+]_{NaHA} = C_{s,1}$ and $[Na^+]_{Na2A} = 2C_{s,2}$, altering the charge balance to

$$[H_3O^+] - \frac{K_w}{[H_3O^+]} + C_{s,1} + 2C_{s,2} = [HA^-] + 2[A^{2-}]$$

$$= [H_2A]_{un}\left(\frac{K_1}{[H_3O^+]} + \frac{2K_1K_2}{[H_3O^+]^2}\right)$$

Dividing the charge balance by the material balance, we have

$$[H_3O^+] - \frac{K_w}{[H_3O^+]}$$
$$= \frac{K_1K_2C_{s,1} - K_1C_{s,2}[H_3O^+] - C_{s,1}[H_3O^+]^2 - 2C_{s,2}[H_3O^+]^2}{[H_3O^+]^2 + K_1[H_3O^+] + K_1K_2}$$

$$(59)$$

which leads to the quartic equation

$$[H_3O^+]^4 + (C_{s,1} + 2C_{s,2} + K_1)[H_3O^+]^3$$
$$+ (K_1K_2 - K_w + K_1C_{s,2})[H_3O^+]^2 - (K_wK_1 + K_1K_2C_{s,1})[H_3O^+]$$
$$- K_wK_1K_2 = 0 \quad (60)$$

The equilibria involved may now be examined to derive simpler equations for the evaluation of $[H_3O^+]$.

Once again the value of the term $([H_3O^+] - K_w/[H_3O^+])$ in Eq. (59) will depend largely on the values of K_1 and K_2, as well as on the values of $C_{s,1}$ and $C_{s,2}$. Where the value of $[H_3O^+]$ is greater than $K_w/[H_3O^+]$, the solution will be acidic, and $K_w/[H_3O^+]$ may usually be ignored with respect to $[H_3O^+]$. Conversely, where $[H_3O^+]$ is less than $K_w/[H_3O^+]$, the solution will be basic, and $[H_3O^+]$ may usually be ignored relative to $K_w/[H_3O^+]$.

The solution of NaHA–Na$_2$A *is acidic*   Here we have $[H_3O^+]$ greater than $K_w/[H_3O^+]$, and where $K_w/[H_3O^+]$ can be ignored relative to $[H_3O^+]$, Eq. (59) reduces to

$$[H_3O^+] \approx \frac{K_1K_2C_{s,1} - K_1C_{s,2}[H_3O^+] - C_{s,1}[H_3O^+]^2 - 2C_{s,2}[H_3O^+]^2}{[H_3O^+]^2 + K_1[H_3O^+] + K_1K_2}$$

$$(61)$$

which leads to the cubic equation

$$[H_3O^+]^3 + (C_{s,1} + 2C_{s,2} + K_1)[H_3O^+]^2 + (K_1K_2 + K_1C_{s,2})[H_3O^+]$$
$$- K_1K_2C_{s,1} = 0 \quad (62)$$

Equation (61) can be written in the form

$$[H_3O^+] \approx \frac{K_1(K_2C_{s,1} - C_{s,2}[H_3O^+]) - [H_3O^+]^2(C_{s,1} + 2C_{s,2})}{[H_3O^+]^2 + K_1([H_3O^+] + K_2)}$$

$$(63)$$

It is apparent that as a general case, terms involving $[H_3O^+]^2$ will be much less than terms involving only $[H_3O^+]$ and therefore $[H_3O^+]^2(C_{s,1} + 2C_{s,2})$ will be much less than $K_1(K_2C_{s,1} - C_{s,2}[H_3O^+])$ and $[H_3O^+]^2$ will be much less than $K_1([H_3O^+] + K_2)$. Where this situation is valid, Eq. (63) reduces to

$$[H_3O^+] \approx \frac{K_2C_{s,1} - C_{s,2}[H_3O^+]}{[H_3O^+] + K_2} \tag{64}$$

which yields

$$[H_3O^+] \approx K_2\frac{C_{s,1} - [H_3O^+]}{C_{s,2} + [H_3O^+]} \tag{65}$$

and the quadratic equation

$$[H_3O^+]^2 + (C_{s,2} + K_2)[H_3O^+] - K_2C_{s,1} = 0 \tag{66}$$

with a solution in

$$[H_3O^+] \approx \frac{-(C_{s,2} + K_2) + \sqrt{(C_{s,2} + K_2)^2 + 4K_2C_{s,1}}}{2} \tag{67}$$

an equation generally similar to Eqs. (40) and (20).

It can be taken that in most cases, $[H_3O^+]$ will be much less than $C_{s,1}$ or $C_{s,2}$ and, where valid, this reduces Eq. (65) to

$$[H_3O^+] \approx \frac{K_2C_{s,1}}{C_{s,2}} \tag{68}$$

an equation generally similar to Eqs. (41) and (24).

The solution of NaHA−Na$_2$A *is basic* Here we have $[H_3O^+]$ less than $K_w/[H_3O^+]$, and where $[H_3O^+]$ can be ignored with respect to $K_w/[H_3O^+]$, Eq. (59) becomes

$$-\frac{K_w}{[H_3O^+]} \approx \frac{K_1K_2C_{s,1} - K_1C_{s,2}[H_3O^+] - C_{s,1}[H_3O^+]^2 - 2C_{s,2}[H_3O^+]^2}{[H_3O^+]^2 + K_1[H_3O^+] + K_1K_2} \tag{69}$$

which provides the cubic equation

$$[H_3O^+]^3 + \left(\frac{K_1C_{s,2} - K_w}{C_{s,1} + 2C_{s,2}}\right)[H_3O^+]^2 - \left(\frac{K_1K_2C_{s,1} + K_wK_1}{C_{s,1} + 2C_{s,2}}\right)[H_3O^+]$$
$$- \frac{K_wK_1K_2}{C_{s,1} + 2C_{s,2}} = 0 \tag{70}$$

Equation (69) can be written in the form

$$-\frac{K_w}{[H_3O^+]} \approx \frac{K_1(K_2C_{s,1} - C_{s,2}[H_3O^+]) - [H_3O^+]^2(C_{s,1} + 2C_{s,2})}{[H_3O^+]^2 + K_1([H_3O^+] + K_2)} \tag{71}$$

Again, as before, where $[H_3O^+]^2(C_{s,1} + 2C_{s,2})$ is much less than $K_1(K_2C_{s,1} - C_{s,2}[H_3O^+])$ and $[H_3O^+]^2$ is much less than $K_1([H_3O^+] + K_2)$, Eq. (71)

reduces to

$$-\frac{K_w}{[H_3O^+]} \approx \frac{K_2 C_{s,1} - C_{s,2}[H_3O^+]}{[H_3O^+] + K_2} \tag{72}$$

which yields

$$[H_3O^+] \approx K_2 \frac{C_{s,1} + K_w/[H_3O^+]}{C_{s,2} - K_w/[H_3O^+]} \tag{73}$$

and the quadratic equation

$$[H_3O^+]^2 - \left(\frac{K_2 C_{s,1} + K_w}{C_{s,2}}\right)[H_3O^+] - \frac{K_w K_2}{C_{s,2}} = 0 \tag{74}$$

with a solution in

$$[H_3O^+] \approx \frac{\{(K_w + K_2 C_{s,1})/C_{s,2}\} + \sqrt{\{(K_w + K_2 C_{s,1})/C_{s,2}\}^2 + 4 K_w K_2/C_{s,2}}}{2} \tag{75}$$

an equation generally similar to Eqs. (47) and (23).

It can now be taken that $K_w/[H_3O^+]$ will generally be much less than $C_{s,1}$ or $C_{s,2}$, so that where valid, this reduces Eq. (73) to

$$[H_3O^+] \approx \frac{K_2 C_{s,1}}{C_{s,2}} \tag{68}$$

General procedures The following *general* procedures may be applied. A relative error in $[H_3O^+]$ as determined of not more than 5 percent is assumed to be satisfactory.

1. Use Eq. (68) to determine $[H_3O^+]$.
2. (*a*) If $[H_3O^+]$ from 1 is greater than $10^{-7} M$, the solution is acidic.
   (*b*) If situation 2*a* is valid, use $[H_3O^+]$ from 1 to determine
      (1) The percentage of $[H_3O^+]$ relative to $C_{s,1}$
      (2) The percentage of $[H_3O^+]$ relative to $C_{s,2}$
   (*c*) If the sum of 2*b*(1) and 2*b*(2) is 5 percent or less, the relative error in $[H_3O^+]$ from Eq. (68) will be less than 5 percent. Equation (68) should be used to determine $[H_3O^+]$ with a relative error about equal to the sum of 2*b*(1) and 2*b*(2).
   (*d*) If the sum of 2*b*(1) and 2*b*(2) is higher than 5 percent then the relative error in $[H_3O^+]$ from Eq. (68) will exceed 5 percent. Equation (67) is used to calculate $[H_3O^+]$, and this value is then used to determine
      (1) The percentage of $[H_3O^+]^2(C_{s,1} + 2C_{s,2})$ relative to $K_1(K_2 C_{s,1} - C_{s,2}[H_3O^+])$
      (2) The percentage of $[H_3O^+]^2$ relative to $K_1([H_3O^+] + K_2)$
   (*e*) If the sum of 2*d*(1) and 2*d*(2) is 5 percent or less, Eq. (67) will yield a

value of $[H_3O^+]$ with a relative error about equal to the sum of $2d(1)$ and $2d(2)$.

(f) If the sum of $2d(1)$ and $2d(2)$ is higher than 5 percent, Eq. (62) should be used to evaluate $[H_3O^+]$.

3. (a) If $[H_3O^+]$ from 1 is less than $10^{-7} M$, the solution is basic.

(b) If situation $3a$ is valid, use $[H_3O^+]$ from 1 to determine

(1) The percentage of $K_w/[H_3O^+]$ relative to $C_{s,1}$

(2) The percentage of $K_w/[H_3O^+]$ relative to $C_{s,2}$

(c) If the sum of $3b(1)$ and $3b(2)$ is 5 percent or less, Eq. (68) will be adequate for the determination of $[H_3O^+]$ with a relative error about equal to the sum of $3b(1)$ and $3b(2)$.

(d) If the sum of $3b(1)$ and $3b(2)$ is higher than 5 percent, Eq. (75) should be used to calculate $[H_3O^+]$, and this value now used to determine

(1) The percentage of $[H_3O^+]^2(C_{s,1} + 2C_{s,2})$ relative to $K_1(K_2 C_{s,1} - C_{s,2}[H_3O^+])$

(2) The percentage of $[H_3O^+]^2$ relative to $K_1([H_3O^+] + K_2)$

(e) If the sum of $3d(1)$ and $3d(2)$ is 5 percent or less, Eq. (75) will provide a value of $[H_3O^+]$ with a relative error about equal to the sum of $3d(1)$ and $3d(2)$.

(f) If the sum of $3d(1)$ and $3d(2)$ is higher than 5 percent, Eq. (70) should be used to evaluate $[H_3O^+]$.

Example   An aqueous solution contains sodium hydrogen oxalate ($C_{s,1} = 0.1 M$) and sodium oxalate ($C_{s,2} = 0.001 M$). Equation (68) gives

$$[H_3O^+] \approx 6.40 \times 10^{-3} M$$

so that situation $2a$ is valid. This $[H_3O^+]$ is 6.4 and 640 percent, respectively, of $C_{s,1}$ and $C_{s,2}$. Equation (68) is not adequate, and Eq. (67) is used to yield

$$[H_3O^+] \approx 2.05 \times 10^{-3} M$$

The value of $[H_3O^+]^2$ is now found to be 3.4 percent of $K_1([H_3O^+] + K_2)$ and that of $[H_3O^+]^2(C_{s,1} + 2C_{s,2})$ is 167 percent of $K_1(K_2 C_{s,1} - C_{s,2}[H_3O^+])$. Equation (67) is not adequate, and Eq. (62) is required to calculate $[H_3O^+]$ as

$$[H_3O^+] \approx 1.35 \times 10^{-3} M$$

Example   An aqueous solution contains sodium bicarbonate ($C_{s,1} = 0.5 M$) and sodium carbonate ($C_{s,2} = 0.005 M$). Equation (68) gives

$$[H_3O^+] \approx 5.61 \times 10^{-9} M$$

and situation $3a$ is valid. This $[H_3O^+]$ gives a value of $K_w/[H_3O^+] = 1.78 \times 10^{-6} M$, and this is 0.0004 percent of $C_{s,1}$ and 0.04 percent of $C_{s,2}$. Equation (68) is therefore adequate for the calculation of $[H_3O^+]$.

Example   An aqueous solution contains sodium bicarbonate ($C_{s,1} = 0.005 M$) and sodium carbonate ($C_{s,2} = 0.5 M$). Equation (68) gives

$$[H_3O^+] \approx 5.61 \times 10^{-13} M$$

and situation $3a$ is valid. The value of $K_w/[H_3O^+] = 1.78 \times 10^{-2} M$, and this is 356 percent of $C_{s,1}$ and 3.6 percent of $C_{s,2}$. Equation (68) is therefore not adequate, and Eq. (75) is used to yield

$$[H_3O^+] \approx 1.38 \times 10^{-12} M$$

The value of $[H_3O^+]^2$ is less than 0.001 percent of $K_1([H_3O^+] + K_2)$ and the value of $[H_3O^+]^2(C_{s,1} + 2C_{s,2})$ is less than 0.001 percent of $K_1(K_2C_{s,1} - C_{s,2}[H_3O^+])$, so that Eq. (75) is adequate for the calculation of $[H_3O^+]$.

4.5.6 IONIZATION EQUILIBRIA FOR A SOLUTION OF THE DIBASIC ACID SALT Na₂A

General   This is the solution of the double substitution salt Na_2A of the dibasic acid H_2A. Such a solution should be obtained at the second-stage equivalence point of a titration between a dibasic acid and the strong base NaOH, providing, or course, that the relative values of K_1 and K_2 permit such a titration.

The anion A^{2-} of this salt will hydrolyze, yielding a basic solution. With such salts we need only concern ourselves with the basicity of the solution.

Ionization equilibria for Na₂A solutions Na_2A represents the double substitution salt of the general dibasic acid H_2A. The dissociation and hydrolysis reactions in aqueous solution are

$$Na_2A \rightleftharpoons 2Na^+ + A^{2-}$$
$$A^{2-} + H_2O \rightleftharpoons HA^- + OH^-$$
$$HA^- + H_2O \rightleftharpoons H_2A + OH^-$$
$$2H_2O \rightleftharpoons H_3O^+ + OH^-$$

The six concentrations of importance are

$$[H_2A]_{un} \quad [HA^-] \quad [A^{2-}] \quad [H_3O^+] \quad [OH^-] \quad C_{s,2}$$

with $C_{s,2}$ being the initial molarity of the salt Na_2A. The material balance will be given by

$$C_{s,2} = [H_2A]_{un} + [HA^-] + [A^{2-}]$$
$$= [H_2A]_{un} \left(1 + \frac{K_1}{[H_3O^+]} + \frac{K_1K_2}{[H_3O^+]^2}\right)$$

and the charge balance is given by

$$[H_3O^+] + [Na^+]_{Na_2A} = [HA^-] + 2[A^{2-}] + [OH^-]$$

and we have

$$\left([H_3O^+] - \frac{K_w}{[H_3O^+]}\right) + 2C_{s,2} = [H_2A]_{un} \left(\frac{K_1}{[H_3O^+]} + \frac{2K_1K_2}{[H_3O^+]^2}\right)$$

Dividing the charge balance by the material balance, we have

$$[H_3O^+] - \frac{K_w}{[H_3O^+]} = \frac{-K_1 C_{s,2} [H_3O^+] - 2C_{s,2} [H_3O^+]^2}{[H_3O^+]^2 + K_1 [H_3O^+] + K_1 K_2} \tag{76}$$

which leads to the quartic equation

$$[H_3O^+]^4 + (2C_{s,2} + K_1)[H_3O^+]^3$$
$$+ (K_1 K_2 - K_w + K_1 C_{s,2})[H_3O^+]^2$$
$$- K_w K_1 [H_3O^+] - K_w K_1 K_2 = 0 \tag{77}$$

Since the solution will be basic, we have $[H_3O^+]$ generally much less than $K_w/[H_3O^+]$, and Eq. (76) becomes

$$\frac{K_w}{[H_3O^+]} \approx \frac{K_1 C_{s,2} [H_3O^+] + 2C_{s,2} [H_3O^+]^2}{[H_3O^+]^2 + K_1 [H_3O^+] + K_1 K_2} \tag{78}$$

which provides the cubic equation

$$[H_3O^+]^3 + \left(\frac{K_1 C_{s,2} - K_w}{2C_{s,2}}\right)[H_3O^+]^2 - \frac{K_w K_1}{2C_{s,2}}[H_3O^+] - \frac{K_w K_1 K_2}{2C_{s,2}} = 0 \tag{79}$$

Equation (78) can be written as

$$\frac{K_w}{[H_3O^+]} \approx \frac{C_{s,2}(K_1 [H_3O^+] + 2[H_3O^+]^2)}{[H_3O^+]^2 + K_1([H_3O^+] + K_2)} \tag{80}$$

Again, as before, terms involving $[H_3O^+]^2$ will generally be much less than terms involving only $[H_3O^+]$. Where valid, we have $2[H_3O^+]^2$ much less than $K_1[H_3O^+]$ and $[H_3O^+]^2$ much less than $K_1([H_3O^+] + K_2)$. This reduces Eq. (80) to

$$\frac{K_w}{[H_3O^+]} \approx \frac{C_{s,2} [H_3O^+]}{[H_3O^+] + K_2} \tag{81}$$

which can be written as

$$[H_3O^+] \approx \sqrt{\frac{K_w K_2}{C_{s,2}} \left(1 + \frac{[H_3O^+]}{K_2}\right)} \tag{82}$$

which yields the quadratic equation

$$[H_3O^+]^2 - \frac{K_w}{C_{s,2}}[H_3O^+] - \frac{K_w K_2}{C_{s,2}} = 0 \tag{83}$$

with a solution in

$$[H_3O^+] \approx \frac{(K_w/C_{s,2}) + \sqrt{(K_w/C_{s,2})^2 + 4K_w K_2/C_{s,2}}}{2} \tag{84}$$

Finally, where $[H_3O^+]/K_2$ is much less than 1, or $[H_3O^+]$ is much less than K_2, Eq. (82) gives

$$[H_3O^+] \approx \sqrt{\frac{K_w K_2}{C_{s,2}}} \tag{85}$$

The following *general* procedures may be applied. A relative error in $[H_3O^+]$ as determined of not more than 5 percent is assumed to be satisfactory.

1. Use Eq. (85) to determine $[H_3O^+]$.
2. From this $[H_3O^+]$ determine
   (*a*) The percentage of $[H_3O^+]^2$ relative to $K_1([H_3O^+] + K_2)$
   (*b*) The percentage of $2[H_3O^+]^2$ relative to $K_1[H_3O^+]$
   (*c*) The percentage of $[H_3O^+]$ relative to K_2
3. If the sum of 2*a* and 2*b* is higher than 5 percent, Eq. (79) should be used to determine $[H_3O^+]$.
4. If the sum of 2*a* and 2*b* is 5 percent or less, and the sum of 2*a*, 2*b*, and one-half of 2*c* is higher than 5 percent, Eq. (84) is used to yield an $[H_3O^+]$ value with a relative error about equal to the sum of 2*a* and 2*b*.
5. If the sum of 2*a* and 2*b* is 5 percent or less, and the sum of 2*a*, 2*b*, and one-half of 2*c* is 5 percent or less, Eq. (85) will yield an $[H_3O^+]$ value with a relative error about equal to the sum of 2*a*, 2*b*, and one-half of 2*c*.

Example An aqueous solution contains sodium oxalate at 0.2 *M*. Equation (85) yields

$$[H_3O^+] \approx 1.79 \times 10^{-9} M$$

This $[H_3O^+]$ results in $[H_3O^+]^2$ less than 10^{-9} percent of $K_1([H_3O^+] + K_2)$ and $2[H_3O^+]^2$ less than 10^{-5} percent of $K_1[H_3O^+]$. The value of $[H_3O^+]$ is 0.003 percent of K_2. Here Eq. (85) is adequate for the calculation of $[H_3O^+]$.

Example An aqueous solution contains sodium carbonate at 0.5 *M*. Equation (85) gives

$$[H_3O^+] \approx 1.06 \times 10^{-12} M$$

$[H_3O^+]^2$ less than 10^{-4} percent of $K_1([H_3O^+] + K_2)$ and $2[H_3O^+]^2$ less than 10^{-3} percent of $K_1[H_3O^+]$. The value of $[H_3O^+]$ is 1.9 percent of K_2. Equation (85) is adequate here, and yields an $[H_3O^+]$ value with a relative error of about 1 percent.

4.5.7 THE USE OF THE QUARTIC EQUATIONS RELATIVE TO DIBASIC ACID AND DIBASIC ACID SALT EQUILIBRIA

While the appropriate quartic equations are not required for ordinary studies involving dibasic acid and dibasic acid salt equilibria in analytical chemistry, it should be noted that during the development of the cubic equations, assumptions were made relative to $[H_3O^+]$ greater than $10^{-7} M$ with $[OH^-]$ less than $10^{-7} M$, and vice versa. In theory, the relevant quartic equation should be used wherever $[H_3O^+]$ and $[OH^-]$ lie between $10^{-6} M$ and $10^{-8} M$. For quantitative analytical purposes, however, little error is introduced by using the range of $[H_3O^+]$ greater than $10^{-7} M$ with $[OH^-]$ less than $10^{-7} M$, and vice versa.

4.5.8 IONIZATION EQUILIBRIA FOR SOLUTIONS OF A TRIBASIC ACID AND ITS SALTS

General Polybasic acids and bases, beyond the dibasic types, are only occasionally encountered in titration situations for reasons to be outlined in some detail in a subsequent chapter on acid-base titrations.

Tribasic acids are, however, sometimes encountered in titration and other circumstances, and it will be of interest, therefore, to indicate methods for the determination of $[H_3O^+]$ in aqueous solutions involving a tribasic acid and its salts.

In most cases, the development of exact ionization equilibria for such solutions is exhaustive, and somewhat beyond the requirements of a text on introductory quantitative analysis. For this reason, the following discussion covers only the simplest equations for the determination of $[H_3O^+]$, as well as the immediate more accurate equations. Both are given without development or explanation. The student, at this point, should be able to note from the equations involved the similarity between these equations and the corresponding equations developed for the dibasic acid H_2A and its salts. He should find it possible, with the aid of the examples given, to develop for himself those situations which lead to the calculation of the relative error in $[H_3O^+]$ from the simplest equation in each case, from which a decision would be forthcoming as to the necessity of applying the associated more accurate equation.

The following outlines the various general solution conditions for the general tribasic acid H_3A and its salts, together with the appropriate equations for the calculation of the related $[H_3O^+]$.

Aqueous solutions of H_3A *as* C_a

Simplest equation: $\qquad [H_3O^+] \approx \sqrt{K_1 C_a}$ $\qquad\qquad\qquad$ (86)

More accurate equation:

$$[H_3O^+] \approx \frac{-K_1 + \sqrt{K_1^2 + 4K_1 C_a}}{2} \qquad\qquad (87)$$

Aqueous solutions of H_3A *as* C_a *and* NaH_2A *as* $C_{s,1}$

Simplest equation: $\qquad [H_3O^+] \approx \dfrac{K_1 C_a}{C_{s,1}}$ $\qquad\qquad\qquad$ (88)

More accurate equation:

Acid solutions: $\qquad [H_3O^+]$ from Eq. (88) greater than $10^{-7} M$,

$$[H_3O^+] \approx \frac{-(C_{s,1} + K_1) + \sqrt{(C_{s,1} + K_1)^2 + 4K_1 C_a}}{2} \qquad (89)$$

Basic solutions: $\qquad [H_3O^+]$ from Eq. (88) less than $10^{-7} M$,

$$[H_3O^+] \approx \frac{\{(K_w + K_1 C_a)/C_{s,1}\} + \sqrt{\{(K_w + K_1 C_a)/C_{s,1}\}^2 + 4K_w K_1/C_{s,1}}}{2}$$
$$(90)$$

Aqueous solutions of NaH_2A *as* $C_{s,1}$

Simplest equation: $\qquad [H_3O^+] \approx \sqrt{K_1 K_2}$ $\qquad\qquad\qquad$ (91)

More accurate equation:

Acid solutions: $[H_3O^+]$ from Eq. (91) greater than $10^{-7} M$,

$$[H_3O^+] \approx \sqrt{\frac{K_1 K_2 C_{s,1}}{K_1 + C_{s,1}}} \qquad (92)$$

Basic solutions: $[H_3O^+]$ from Eq. (91) less than $10^{-7} M$,

$$[H_3O^+] \approx \sqrt{\frac{K_1 K_2 C_{s,1} + K_w K_1}{C_{s,1}}} \qquad (93)$$

Aqueous solutions of NaH_2A *as* $C_{s,1}$ *and* Na_2HA *as* $C_{s,2}$

Simplest equation: $[H_3O^+] \approx \dfrac{K_2 C_{s,1}}{C_{s,2}}$ $\qquad (94)$

More accurate equation:

Acid solutions: $[H_3O^+]$ from Eq. (94) greater than $10^{-7} M$,

$$[H_3O^+] \approx \frac{-(C_{s,2} + K_2) + \sqrt{(C_{s,2} + K_2)^2 + 4K_2 C_{s,1}}}{2} \qquad (95)$$

Basic solutions: $[H_3O^+]$ from Eq. (94) less than $10^{-7} M$,

$$[H_3O^+]$$
$$\approx \frac{\{(K_2 C_{s,1} + K_w)/C_{s,2}\} + \sqrt{\{(K_2 C_{s,1} + K_w)/C_{s,2}\}^2 + 4K_w K_2/C_{s,2}}}{2}$$
$$(96)$$

Aqueous solutions of Na_2HA *as* $C_{s,2}$

Simplest equation: $[H_3O^+] \approx \sqrt{K_2 K_3}$ $\qquad (97)$

More accurate equation:

Basic solutions: $[H_3O^+]$ from Eq. (97) less than $10^{-7} M$,

$$[H_3O^+] \approx \sqrt{\frac{K_2 K_3 C_{s,2} + K_w K_2}{C_{s,2}}} \qquad (98)$$

Aqueous solutions of Na_2HA *as* $C_{s,2}$ *and* Na_3A *as* $C_{s,3}$

Simplest equation: $[H_3O^+] \approx \dfrac{K_3 C_{s,2}}{C_{s,3}}$ $\qquad (99)$

More accurate equation:

Basi Basic solutions: $[H_3O^+]$ from Eq. (99) less than $10^{-7} M$,

$$[H_3O^+]$$
$$\approx \frac{\{(K_3 C_{s,2} + K_w)/C_{s,3}\} + \sqrt{\{(K_3 C_{s,2} + K_w)/C_{s,3}\}^2 + 4K_w K_3/C_{s,3}}}{2}$$

$$(100)$$

Aqueous solutions of Na_3A *as* $C_{s,3}$

Simplest equation: $[H_3O^+] \approx \sqrt{\dfrac{K_w K_3}{C_{s,3}}}$ 　　　　　　　　(101)

More accurate equation:

Basic solutions: $[H_3O^+]$ from Eq. (101) less than $10^{-7} M$,

$$[H_3O^+] \approx \frac{(K_w/C_{s,3}) + \sqrt{(K_w/C_{s,3})^2 + 4K_w K_3/C_{s,3}}}{2}$$ 　　(102)

Examples   Consider the following solutions involving the tribasic acid H_3PO_4 and its salts. The K values for the acid are

$K_1 = 7.52 \times 10^{-3}$

$K_2 = 6.23 \times 10^{-8}$

$K_3 = 4.80 \times 10^{-13}$

1. A solution involving H_3PO_4 ($C_a = 0.1 M$). Equation (86) yields

    $[H_3O^+] \approx 2.74 \times 10^{-2} M$

    $[H_3O^+]$ is 27 percent of C_a, so that the expected error is about one-half of this value, or 14 percent. Equation (87) is required and gives

    $[H_3O^+] \approx 2.39 \times 10^{-2} M$

2. A solution involving H_3PO_4 ($C_a = 0.1 M$) and NaH_2PO_4 ($C_{s,1} = 0.05 M$). Equation (88) yields

    $[H_3O^+] \approx 1.50 \times 10^{-2} M$

    $[H_3O^+]$ is 15 percent of C_a and 30 percent of $C_{s,1}$. The expected relative error is about the sum of these values, or 45 percent. Equation (89) is required and gives

    $[H_3O^+] \approx 1.10 \times 10^{-2} M$

3. A solution involving NaH_2PO_4 ($C_{s,1} = 0.05 M$). Equation (91) yields

    $[H_3O^+] \approx 2.16 \times 10^{-5} M$

    K_1 is 15 percent of $C_{s,1}$, so that the relative error expected for $[H_3O^+]$ is one-half this value, or about 8 percent. Equation (92) is required and gives

    $[H_3O^+] \approx 2.02 \times 10^{-5} M$

4. A solution involving NaH_2PO_4 ($C_{s,1} = 0.1 M$) and Na_2HPO_4 ($C_{s,2} = 0.05 M$). Equation (94) yields

    $[H_3O^+] \approx 1.25 \times 10^{-7} M$

    $[H_3O^+]$ is 0.0001 percent of $C_{s,1}$ and 0.0003 percent of $C_{s,2}$. Equation (94) is adequate.

5. A solution involving Na_2HPO_4 ($C_{s,2} = 0.05 M$). Equation (97) yields

    $[H_3O^+] \approx 1.73 \times 10^{-10} M$

    $K_w K_2$ is 0.4 percent of $K_2 K_3 C_{s,2}$, so that the relative error in $[H_3O^+]$ is about one-half this value, or 0.2 percent. Equation (97) is adequate.

6. A solution involving Na_2HPO_4 ($C_{s,2} = 0.05$ M) and Na_3PO_4 ($C_{s,3} = 0.1$ M). Equation (99) yields

$$[H_3O^+] \approx 2.40 \times 10^{-13}\ M$$

$K_w/[H_3O^+] \approx 4.17 \times 10^{-2}\ M$ and this is 83 percent of $C_{s,2}$ and 42 percent of $C_{s,3}$. The expected relative error in $[H_3O^+]$ is about equal to the sum of these values, or about 125 percent. Equation (100) gives

$$[H_3O^+] \approx 4.47 \times 10^{-13}\ M$$

7. A solution involving Na_3PO_4 ($C_{s,3} = 0.1$ M). Equation (101) yields

$$[H_3O^+] \approx 2.19 \times 10^{-13}\ M$$

$[H_3O^+]$ is 46 percent of K_3, so that the relative error in $[H_3O^+]$ is about one-half this value, or about 23 percent. Equation (102) is required and gives

$$[H_3O^+] \approx 2.75 \times 10^{-13}\ M$$

Part 3   Precipitation Equilibria

4.6   SLIGHTLY SOLUBLE SALTS

The process of placing a salt in aqueous solution involves basically the removal of ions from the crystalline lattice of the solid to the aqueous solution environment. The organized lattice structure of the crystalline solid represents an equilibrium state, with a relatively low energy situation. To remove ions from such a lattice requires that energy be consumed. On the other hand, energy is released when the removed ions are solvated. The two energies are called the lattice energy and the hydration energy. The extent to which any solid is soluble in aqueous medium depends, therefore, on the difference between the lattice energy and the hydration energy.

With strong electrolytes, the lattice energy increases with the charge on the ions and decreases with the size of the ions. Hydration energy is highest for small ions having a high charge. Note that the two energies are highest for the smallest and most highly charged ions.

With many electrolytes the lattice energy is slightly higher than the hydration energy, so that the solution of these electrolytes is a process requiring some energy in addition to the hydration energy. Thus, such electrolytes are more soluble at elevated solution temperatures, and the solubility increases with increasing temperature.

In all such instances of electrolyte solubility, the action of solvation requires that reactions of dissociation go on, so that all factors that can affect the reactions of dissociation can affect the solubility.

Consider, first of all, the slightly soluble salt BA. This is a general inorganic salt and, like nearly all such salts, a strong electrolyte whose soluble portion can, for practical purposes, be considered to be 100 percent dissociated in aqueous solution. Thus, we have

$$BA(s) \rightleftharpoons B^+(aq) + A^-(aq)$$

For a constant-temperature situation, we have

$$\frac{\alpha_{B^+}\alpha_{A^-}}{\alpha_{BA(s)}} = K_{eq}$$

The activity of $BA(s)$ is a constant, so that

$$\alpha_{B^+}\alpha_{A^-} = K_{eq}\alpha_{BA(s)} = K^\circ_{sp}(BA)$$

We have the relationship

$$\alpha_{B^+} = \gamma_{B^+}[B^+]$$
$$\alpha_{A^-} = \gamma_{A^-}[A^-]$$

where γ_{B^+} and γ_{A^-} are the activity coefficients for $[B^+]$ and $[A^-]$. Therefore,

$$(\gamma_{B^+}[B^+])(\gamma_{A^-}[A^-]) = K^\circ_{sp}(BA)$$

This yields

$$[B^+][A^-] = \frac{K^\circ_{sp}(BA)}{\gamma_{B^+}\gamma_{A^-}}$$

Where the activities are not affected by the presence of relatively large concentrations of diverse ions (ions other than those of the salt itself), and where relatively dilute solutions of BA are involved (always the case with slightly soluble salts), we have

$$\gamma_{B^+} \approx \gamma_{A^-} \approx 1$$

so that, for simple dilute solutions of BA, we have

$$[B^+][A^-] = \frac{K^\circ_{sp}(BA)}{\gamma_{B^+}\gamma_{A^-}} = K_{sp}(BA)$$

In such cases, we have, for a saturated solution of BA alone,

$$[B^+] = [A^-] = S$$

where S is the molar solubility of BA in aqueous solution. From this relationship, we derive

$$[B^+]^2 = [A^-]^2 = S^2 = K_{sp}(BA) = [B^+][A^-]$$

from which

$$[B^+] = [A^-] = S = \sqrt{K_{sp}(BA)}$$

For a slightly soluble salt of the type B_2A, using the same approach,

$$B_2A(s) \rightleftharpoons 2B^+ + A^{2-}$$

and for dilute solutions, where the activities are not affected by relatively large concentrations of diverse ions,

$$[B^+]^2[A^{2-}] = \frac{K^\circ_{sp}(B_2A)}{\gamma_{B^{+2}} \, \gamma_{A^{2-}}} = K_{sp}(B_2A)$$

It will be noted here that

$$[B^+] = 2[A^{2-}] = 2S$$

where S is again the molar solubility of B_2A.

From the foregoing,

$$4[A^{2-}]^3 = \frac{[B^+]^3}{2} = K_{sp}(B_2A)$$

from which we have the relationships

$$[A^{2-}] = \sqrt[3]{\frac{K_{sp}(B_2A)}{4}} \qquad \text{and} \qquad K_{sp}(B_2A) = 4[A^{2-}]^3$$

$$[B^+] = \sqrt[3]{2K_{sp}(B_2A)} \qquad \text{and} \qquad K_{sp}(B_2A) = \frac{[B^+]^3}{2}$$

$$S = \sqrt[3]{\frac{K_{sp}(B_2A)}{4}} \qquad \text{and} \qquad K_{sp}(B_2A) = 4S^3$$

In general, for a slightly soluble salt B_xA_y under the same conditions,

$$B_xA_y(s) \rightleftharpoons xB^{b+} + yA^{a-}$$
$$[B^{b+}]^x[A^{a-}]^y = K_{sp}(B_xA_y)$$

from which we have

$$[A^{a-}] = \sqrt[(x+y)]{\left(\frac{y}{x}\right)^x K_{sp}(B_xA_y)} \qquad \text{and} \qquad K_{sp}(B_xA_y) = \left(\frac{x}{y}\right)^x [A^{a-}]^{(x+y)}$$

$$[B^{b+}] = \sqrt[(x+y)]{\left(\frac{x}{y}\right)^y K_{sp}(B_xA_y)} \qquad \text{and} \qquad K_{sp}(B_xA_y) = \left(\frac{y}{x}\right)^y [B^{b+}]^{(x+y)}$$

$$S = \sqrt[(x+y)]{\frac{K_{sp}(B_xA_y)}{x^x y^y}} \qquad \text{and} \qquad K_{sp}(B_xA_y) = x^x y^y S^{(x+y)}$$

In all the foregoing equations, we have

K°_{sp} = the true or thermodynamic solubility product constant

K_{sp} = the solubility product constant

Examples

1. The solubility of AgCl at 10°C is 0.000089 g/100 ml. The molecular weight of AgCl is 143.32. What is the molar solubility and the solubility product constant at this temperature?

$$S = \frac{0.000089}{143.32} \times 10 = 6.21 \times 10^{-6}\,M$$

$$K_{sp}(AgCl) = S^2 = (6.21 \times 10^{-6})^2 = 3.86 \times 10^{-11}$$

2. The solubility product constant of Ag_2CrO_4 at 25°C is 9.0×10^{-12}. What is the molar solubility at this temperature?

$$S = \sqrt[3]{\frac{K_{sp}(Ag_2CrO_4)}{4}}$$
$$= 1.31 \times 10^{-4}\,M$$

3. The solubility product constant of Fe_2S_3 at 25°C is 1.0×10^{-88}. What is the molar solubility at this temperature?

$$S = \sqrt[(x+y)]{\frac{K_{sp}(Fe_2S_3)}{x^x y^y}}$$
$$= \sqrt[5]{\frac{1.0 \times 10^{-88}}{108}}$$
$$= 9.8 \times 10^{-19}\,M$$

4.7 FACTORS AFFECTING THE SOLUBILITY OF SLIGHTLY SOLUBLE SALTS

4.7.1 GENERAL

There are a number of factors which are capable of affecting the solubility in aqueous solution of slightly soluble salts. In general, these factors are apt to be less important in volumetric than in gravimetric work. Volumetric work usually involves a more closely controlled environment relative to temperature, common- and diverse-ion concentrations, pH, etc., than does gravimetric work. In gravimetric techniques these factors which affect the solubility may vary appreciably from analysis to analysis.

4.7.2 THE COMMON-ION EFFECT

The common-ion effect is that effect on the solubility of slightly soluble salts brought about by the presence in solution of concentrations of an ion in common with one of the ions of the slightly soluble salt. The following two situations may exist with respect to this common-ion effect:

1. In gravimetric work, the precipitating ion is usually added in excess in order to guarantee a quantitative separation of the ion being reacted to form a precipitate.
2. In volumetric work, the precipitating ion is usually added in only that quantity equivalent to the amount of the ion being precipitated.

Note that in the first case, we have an excess of a common ion. In the second, we have only that amount of the common ion left in solution required to satisfy the solubility product constant for the precipitation temperature involved.

As for the relationship between the concentration of the ion being precipitated, the precipitating or common ion, and the molar solubility of the precipitate substance, this is expressed through the solubility product constant for the precipitate substance or slightly soluble salt. Thus, for salts of the BA type we have

$$[Ag^+][Cl^-] = S^2 = K_{sp}(AgCl) = 1.56 \times 10^{-10} \qquad \text{at } 25°C$$

Suppose that it is intended to precipitate silver as AgCl by the addition of Cl^-. After precipitation of AgCl is complete, we have

$$[Ag^+] \text{ remaining in solution} = S(AgCl) = \frac{K_{sp}(AgCl)}{[Cl^-]}$$

If Cl^- had been added so as to have $[Cl^-] = 10^{-1} M$ in solution after the precipitation of AgCl was complete, then

$$[Ag^+] \text{ remaining in solution} = S(AgCl) = \frac{1.56 \times 10^{-10}}{10^{-1}} = 1.56 \times 10^{-9} M$$

Note that for such a separation, the $[Ag^+]$ or $S(AgCl)$ after precipitation is complete can be obtained from

$$[Ag^+] = S(AgCl) = \frac{K_{sp}(AgCl)}{[Cl^-]}$$

Similarly, in the reverse situation, with Ag^+ as the precipitant, we have $[Cl^-]$ or $S(AgCl)$ after precipitation is complete given by

$$[Cl^-] = S(AgCl) = \frac{K_{sp}(AgCl)}{[Ag^+]}$$

It is important to note that the molar solubility and the concentration of the ion being precipitated are inversely proportional to the concentration of the common or precipitating ion, and that each ion has an identical effect in reducing the molar solubility of the slightly soluble salt. Note also that we have, in general, for slightly soluble salts of the BA type,

$$pB + pA = pK_{sp}(BA)$$

where the terms pB, pA, and $pK_{sp}(BA)$ are the logarithms of the reciprocals of the respective ion concentrations or K_{sp} value.

Table 4.1   The effect of either $[Ag^+]$ or $[Cl^-]$ on the molar solubility of AgCl at 25°C

Excess of $[Cl^-]$, M	Solubility of AgCl, M	Excess of $[Ag^+]$, M
0.1	1.56×10^{-9}	0.1
0.01	1.56×10^{-8}	0.01
0.001	1.56×10^{-7}	0.001

Note: The excess of either ion for the same concentration excess has the same effect relative to the molar solubility of AgCl.

Table 4.1 shows the general effect of excesses of either Ag^+ or Cl^- on the molar solubility of AgCl. Figure 4.1 shows the trend of pCl and pS with changing values of pAg, as well as the trend of pAg and pS with changing values of pCl.

In volumetric work, the *theoretical* aim is to secure an exactly equivalent position at the close of the titration, so that for BA-type salts, we should have at the volumetric equivalence point

$$[B^+] = [A^-] = S = \sqrt{K_{sp}(BA)}$$

and for the system just under discussion,

$$[Ag^+] = [Cl^-] = S(AgCl) = \sqrt{K_{sp}(AgCl)}$$
$$= 1.25 \times 10^{-5}\,M \qquad \text{at } 25°C$$

For slightly soluble salts of the B_2A type, such as Ag_2CrO_4, we have the same *general* effect, in that an increase in the concentration of either precipitating ion as common ion results in a decrease in the concentration of the opposing ion and a decrease in the molar solubility of the salt. Thus, we have

$$[Ag^+]^2[CrO_4^{2-}] = K_{sp}(Ag_2CrO_4) = 9.0 \times 10^{-12} \qquad \text{at } 25°C$$

Suppose that we intend to precipitate silver as Ag_2CrO_4 by the addition of CrO_4^{2-}. After any excess of CrO_4^{2-} has been added, we have

$$[Ag^+] \text{ remaining in solution} = 2S(Ag_2CrO_4) = \sqrt{\frac{K_{sp}(Ag_2CrO_4)}{[CrO_4^{2-}]}}$$

If CrO_4^{2-} has been added so as to have in solution after the precipitation of Ag_2CrO_4 a value of $[CrO_4^{2-}] = 10^{-2}\,M$, then,

$$[Ag^+] \text{ remaining in solution} = 2S = \sqrt{\frac{9.0 \times 10^{-12}}{10^{-2}}}$$
$$= 3.0 \times 10^{-5}\,M$$
$$S = 1.5 \times 10^{-5}\,M$$

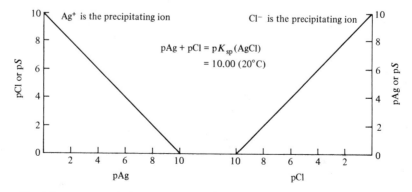

Fig. 4.1 The effect of $[Ag^+]$ or $[Cl^-]$ as the common ion on the molar solubility of AgCl at 20°C.

In the reverse situation, with chromate ion being precipitated by the addition of Ag^+, and with the excess of Ag^+ after the precipitation of Ag_2CrO_4 being $[Ag^+] = 10^{-2} M$, we then have

$$[CrO_4{}^{2-}] \text{ remaining in solution} = S(Ag_2CrO_4) = \frac{K_{sp}(Ag_2CrO_4)}{[Ag^+]^2}$$

$$= \frac{9.0 \times 10^{-12}}{(10^{-2})^2}$$

$$= 9.0 \times 10^{-8} M$$

It is important to note that

1. With Ag^+ as the precipitating or common ion, the molar solubility is inversely proportional to the square of the residual $[Ag^+]$.
2. With $CrO_4{}^{2-}$ as the precipitating or common ion, the molar solubility is inversely proportional to the square root of the residual $[CrO_4{}^{2-}]$.

Thus the effect of $[Ag^+]$ in reducing the molar solubility of Ag_2CrO_4 will be much greater than that of $[CrO_4{}^{2-}]$.

In the above connection, the following may be noted:

$$[Ag^+]_1 = \sqrt{\frac{K_{sp}(Ag_2CrO_4)}{[CrO_4{}^{2-}]_a}} \quad \text{and} \quad [Ag^+]_2 = \sqrt{\frac{K_{sp}(Ag_2CrO_4)}{[CrO_4{}^{2-}]_b}}$$

where

$$[Ag^+]_2 > [Ag^+]_1$$

Again:

$$[CrO_4{}^{2-}]_1 = \frac{K_{sp}(Ag_2CrO_4)}{[Ag^+]_c^2} \quad \text{and} \quad [CrO_4{}^{2-}]_2 = \frac{K_{sp}(Ag_2CrO_4)}{[Ag^+]_d^2}$$

where

$$[CrO_4{}^{2-}]_2 > [CrO_4{}^{2-}]_1$$

When

$$[Ag^+]_1 = [CrO_4{}^{2-}]_1 \quad \text{and} \quad [Ag^+]_2 = [CrO_4{}^{2-}]_2$$

we have

$$\frac{[Ag^+]_1}{[Ag^+]_2} = \frac{[CrO_4{}^{2-}]_1}{[CrO_4{}^{2-}]_2} = \sqrt{\frac{[CrO_4{}^{2-}]_b}{[CrO_4{}^{2-}]_a}} = \frac{[Ag^+]_d^2}{[Ag^+]_c^2}$$

Note, therefore, that in maintaining the same residual concentration of either ion, any change in $[Ag^+]$ equates to this change to the fourth power with respect to the corresponding change in $[CrO_4{}^{2-}]$.

In general, for all slightly soluble salts of the B_2A type,

$$2pB + pA = pK_{sp}(B_2A)$$

Table 4.2 The effect of either $[Ag^+]$ or $[CrO_4]$ on the molar solubility of Ag_2CrO_4 at 25°C

Excess of $[CrO_4{}^{2-}]$, M	Resultant $[Ag^+]$, M	Solubility of Ag_2CrO_4, M	Excess of $[Ag^+]$, $M*$	Resultant $[CrO_4{}^{2-}]$, M
10^{-1}	9.5×10^{-6}	4.7×10^{-6}	1.4×10^{-3}	4.8×10^{-6}
10^{-2}	3.0×10^{-5}	1.5×10^{-5}	7.7×10^{-4}	1.5×10^{-5}
10^{-4}	3.0×10^{-4}	1.5×10^{-4}	2.4×10^{-4}	1.5×10^{-4}
10^{-6}	3.0×10^{-3}	1.5×10^{-3}	7.7×10^{-5}	1.5×10^{-3}
10^{-8}	3.0×10^{-2}	1.5×10^{-2}	2.4×10^{-5}	1.5×10^{-2}

Residual concentration of either $[Ag^+]$ or $[CrO_4{}^{2-}]$	Excess of $[CrO_4{}^{2-}]$		Excess of $[Ag^+]$	
	Required	Multiplier†	Required	Multiplier
$4.7 \times 10^{-6}\,M$	$0.4\,M$	10	$1.4 \times 10^{-3}\,M$	$10^{1/4}$
1.5×10^{-5}	4.0×10^{-2}	10^2	7.7×10^{-4}	$10^{1/2}$
1.5×10^{-4}	4.0×10^{-4}	10^2	2.4×10^{-4}	$10^{1/2}$
1.5×10^{-3}	4.0×10^{-6}	16	7.7×10^{-5}	2
6.0×10^{-3}	2.5×10^{-7}		3.9×10^{-5}	

*Note the effect of $[Ag^+]$ in reducing the molar solubility of Ag_2CrO_4 is significantly greater than that of $[CrO_4{}^{2-}]$.
†Note that the multipliers of the $[CrO_4{}^{2-}]$ concentration differ from those of the $[Ag^+]$ by a fourth-power factor.

Table 4.2 shows the general effect of either $[Ag^+]$ or $[CrO_4{}^{2-}]$ on the molar solubility of Ag_2CrO_4 and on the solubility of the remaining ion. It also shows the excess of either $[Ag^+]$ or $[CrO_4{}^{2-}]$ required to provide a similar residual concentration of the opposing ion. Figure 4.2 shows the trend of pAg and pS with changing $pCrO_4$, as well as the trend of $pCrO_4$ and pS with changing pAg.

In volumetric work, the *theoretical* aim is to secure an exactly equivalent position at the end of the titration, so that for all salts of the B_2A type, the volumetric equivalence point yields

$$[B^+] = 2S = \sqrt[3]{2K_{sp}(B_2A)}$$

$$[A^{2-}] = S = \sqrt[3]{\frac{K_{sp}(B_2A)}{4}}$$

and for the system under discussion this would mean

$$[Ag^+] = 2S = \sqrt[3]{2K_{sp}(Ag_2CrO_4)} \qquad [CrO_4{}^{2-}] = S = \sqrt[3]{\frac{K_{sp}(Ag_2CrO_4)}{4}}$$

4.7.3 THE DIVERSE-ION EFFECT

Many slightly soluble salts show an increased solubility in the presence of increased concentrations of soluble salts having no ion in common with those of

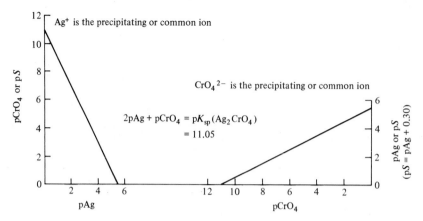

Fig. 4.2 The effect of $[Ag^+]$ or $[CrO_4^{2-}]$ as the common ion on the molar solubility of Ag_2CrO_4 at 25°C.

the slightly soluble salt. Thus, AgCl is more soluble in 0.1 M KNO$_3$ than it is in water, and more soluble in 1 M KNO$_3$ than in 0.1 M KNO$_3$.

This phenomenon is called the diverse-ion, or neutral-salt, effect, and has its basis in the following:

It was pointed out previously that, for the general salt BA, we have the condition

$$[B^+][A^-] = \frac{K^\circ_{sp}(BA)}{\gamma_{B^+}\gamma_{A^-}}$$

and that, where the solutions are highly dilute, the activity coefficients yield

$$\gamma_{B^+} \approx \gamma_{A^-} \approx 1$$

thereby providing

$$[B^+][A^-] = K_{sp}(BA)$$

It should be noted that the proper equation is

$$[B^+][A^-] = \frac{K^\circ_{sp}(BA)}{\gamma_{B^+}\gamma_{A^-}}$$

where $K^\circ_{sp}(BA)$ is a *true* constant at constant temperature.

Where a significant concentration of diverse ions is present, therefore, we have a decrease in the values of γ_{B^+} and γ_{A^-} as a result of the Debye-Hückel-Onsager effect of interionic attraction. Thus γ_{B^+} and γ_{A^-} become less than unity and we have

$$\frac{K^\circ_{sp}(BA)}{\gamma_{B^+}\gamma_{A^-}} \text{ greater than } \frac{K^\circ_{sp}(BA)}{\gamma_{B^+}\gamma_{A^-}}$$
(diverse ion)   (no diverse ion)

Table 4.3   The effect of a diverse-ion salt on the molar solubility of AgCl **and** BaSO₄ *

KNO₃ concentration, M	Ratio (S = molar solubility of AgCl in electrolyte)/ (S_0= molar solubility of AgCl in water)
0.000	1.00
0.001	1.04
0.005	1.08
0.01	1.12

KNO₃ concentration, M	Ratio (S = molar solubility of BaSO₄ in electrolyte)/ (S_0 = molar solubility of BaSO₄ in water)
0.000	1.00
0.001	1.21
0.005	1.48
0.01	1.70

Note: The diverse-ion effect for the substance KNO₃ has, for comparable concentrations, a greater effect on the molar solubility of the divalent ion compound BaSO₄. This is, of course, based on the greater interionic attraction effect of the higher valency ions and the correspondingly greater effect in reducing the activity coefficient for the individual ions involved. In the same way, the effect of a diverse-ion salt having ions of higher valency in increasing the molar solubility of a slightly soluble salt will be greater for comparable concentrations than that of a univalent diverse-ion salt.

*For AgCl: S. Popoff and E. W. Neumann, *J. Phys. Chem.*, **34**:1853 (1930); for BaSO₄: E. W. Neumann, *J. Am. Chem. Soc.*, **55**:879 (1933).

With an increase in $K^{\circ}_{sp}(BA)/(\gamma_{B+}\gamma_{A-})$ we must have an increase in $[B^+][A^-]$. This implies increases in $[B^+]$ and $[A^-]$. Since $[B^+] = [A^-] = S$, then an increase in the molar solubility will result from an increase in the concentration of any diverse ion. Table 4.3 gives an idea of the extent to which the molar solubility of certain slightly soluble salts is affected by the diverse-ion concentration.

The diverse-ion effect does not usually present a serious problem in most of the methods of quantitative analytical chemistry, since the conditions under which the analysis is carried out are normally such as to reduce the effect to the level of insignificance. Where altered conditions of analysis dictate that consideration be given to the diverse-ion effect, the following problem indicates the general method to be pursued.

Example   The Debye-Hückel limiting law for the estimation of activity coefficients is given by

$$-\log \gamma_i = 0.509 Z_i^2 \sqrt{\mu}$$

where μ, the ionic strength, is given by

$$\mu = \tfrac{1}{2}\Sigma C_i Z_i^2$$

with Z_i being the ion charge, C_i the ion concentration, and γ_i the ion-activity coefficient.

Determine the effect of $0.10\ M$ KNO_3 on the molar solubility of Ag_2CrO_4 at 25°C. Note, first of all, that in water at 25°C, we have $K_{sp}(Ag_2CrO_4) = K^{\circ}_{sp}(Ag_2CrO_4)$ $= 9.0 \times 10^{-12}$. Thus, the molar solubility of Ag_2CrO_4 in water at 25°C is

$1.3 \times 10^{-4}\ M$

The ionic strength can be taken on the basis of $0.10\ M$ KNO_3 only, since the values for $[Ag^+]$ and $[CrO_4{}^{2-}]$ will be negligible by comparison; therefore,

$$\mu = \tfrac{1}{2}\{0.1(1)^2 + 0.1(-1)^2\}$$
$$= 0.1$$

The equation for the value of $K_{sp}(Ag_2CrO_4)$ for this solution is

$$\frac{K^{\circ}_{sp}(Ag_2CrO_4)}{(\gamma_{Ag^+})^2 \gamma_{CrO_4{}^{2-}}} = K_{sp}(Ag_2CrO_4)$$

and we have

$$\log K_{sp}(Ag_2CrO_4) = \log K^{\circ}_{sp}(Ag_2CrO_4) - 2\log \gamma_{Ag^+} - \log \gamma_{CrO_4{}^{2-}}$$
$$= -11.05 + (2)(0.509)(1)^2\sqrt{0.1} + (0.509)(-2)^2\sqrt{0.1}$$
$$= -11.05 + 0.32 + 0.64$$
$$= -10.09$$

From this, the $K_{sp}(Ag_2CrO_4)$ at 25°C in $0.1\ M$ KNO_3 is

8.1×10^{-11}

and the molar solubility of Ag_2CrO_4 in $0.1\ M$ KNO_3 is

$2.7 \times 10^{-4}\ M$

Note that this represents an increase of approximately 100 percent in the molar solubility.

It is important to note that whereas the common-ion effect tends to reduce the solubility of a slightly soluble salt, the diverse-ion effect tends to increase it. Thus, as an example, in the precipitation of AgCl from a solution of $AgNO_3$ using KCl as the precipitant, the addition of a planned excess of KCl will serve, through the common-ion effect, to reduce the molar solubility of AgCl. However, through the diverse-ion effect, high excesses of KCl will tend to increase the molar solubility of AgCl. This is one of the reasons why very large excesses of the precipitating ion are avoided in precipitation work.

4.7.4   THE EFFECT OF TEMPERATURE

The action of dissociation for a slightly soluble salt

$$BA(s) \rightleftharpoons B^+(aq) + A^-(aq)$$

will generally be increased as the result of increasing temperature. Correspondingly the molar solubility will increase. For example, the molar solubility of

AgCl is given by

| At 25°C: | 1.78 mg/l | $S = 1.25 \times 10^{-5}\, M$ | $K_{sp}(AgCl) = 1.56 \times 10^{-10}$ |
| At 100°C: | 21.1 mg/l | $S = 1.48 \times 10^{-4}\, M$ | $K_{sp}(AgCl) = 2.17 \times 10^{-8}$ |

In the above, the solubility of AgCl at 100°C yields a $K_{sp}(AgCl)$ value of $2.17 \times 10^{-8}\, M$. Such a level of residual solubility after precipitation would represent a nonquantitative separation of either Ag^+ or Cl^- as AgCl, where no excess of the respective precipitating ion had been added. In order to secure, say, a residual concentration of $[Ag^+] = 10^{-6}\, M$ for a quantitative separation of silver as AgCl at 100°C, we would require in solution after the precipitation of AgCl a chloride ion concentration of

$$[Cl^-]_{min} = \frac{2.17 \times 10^{-8}}{[Ag^+]_{max}} = \frac{2.17 \times 10^{-8}}{10^{-6}}$$
$$= 2.17 \times 10^{-2}\, M$$

At 25°C as the temperature of precipitation, the same residual level of $[Ag^+]$ could be secured with a chloride ion concentration of

$$[Cl^-]_{min} = \frac{1.56 \times 10^{-10}}{10^{-6}}$$
$$= 1.56 \times 10^{-4}\, M$$

It is important to note that a properly adjusted excess of the precipitating ion can be used to offset increases in the solubility of a slightly soluble salt as caused by increases in the temperature at which precipitation takes place.

In gravimetric precipitation work, or in the preliminary precipitation operations preceding completion of an analysis by volumetric titration, the method of precipitation usually involves the use of solutions at temperatures appreciably higher than room temperature. This is usually dictated by conditions involving reaction rate, precipitate particle size, filtration rate, removal of impurities, etc. (see Sec. 11.4 on mechanisms of precipitation). However, in cases involving precipitate substances where these higher temperatures imply molar solubilities that prevent quantitative separation, cooling to room temperature or below prior to filtration, followed by precipitate washing with wash solutions at similar lower temperatures, may be required.

4.7.5 THE EFFECT OF THE NATURE OF THE SOLVENT

The dipolar nature of the water molecule means that water, as a solvent substance, has attractive effects for both anions and cations of a solute and is thus capable of forming hydrated ions. An example of this action is the hydrated hydrogen ion H_3O^+. It is this attractive force which assists in overcoming the forces of attraction that hold anions and cations in the solid crystal-lattice structure, thus allowing aqueous-solution solubility situations.

Organic solvents do not usually provide for the ion-attraction effects which are typical of water, so that most inorganic salts are poorly soluble in organic solvents.

The solubility of a slightly soluble salt in water can often be decreased by the addition of a quantity of some mutually soluble organic solvent.

We might consider as an example of this situation changes in the solubility of lead sulfate brought about by the addition of ethanol to the water medium. The solubility of $PbSO_4$ in water at 25°C is about 45 mg/l. This then represents a residual concentration of Pb^{2+} given by

$$[Pb^{2+}] = 1.48 \times 10^{-4} M$$

and would be indicative of a nonquantitative separation of lead as $PbSO_4$.

An excess of SO_4^{2-} in the solution at the completion of precipitation would help to reduce the residual concentration of Pb^{2+}, but with H_2SO_4 as the precipitating agent, this would be difficult to secure at a reasonable level of concentration of the acid because of its relatively low secondary dissociation. The addition of ethanol to the aqueous solution results in decreases in the solubility of $PbSO_4$ and, of course, in the residual concentration of Pb^{2+}. This effect is shown by the data outlined in Table 4.4. It should be noted that additional decreases in the solubility of $PbSO_4$ would be secured where the solution contained slight excesses of SO_4^{2-} at the completion of precipitation.

In all such instances of the use of organic solvent additions, care must be taken to ensure that a substance other than that desired is not precipitated. A case in point follows from the treatment given relative to the solubility of $PbSO_4$ in water-ethanol mixtures. In a solution containing the sulfates of the metals Pb^{2+}, Sb^{3+}, and Bi^{3+}, the addition of ethanol decreases not only the solubility of $PbSO_4$ but also that of the fairly soluble sulfates of Sb^{3+} and Bi^{3+}. If the level of ethanol is high enough, or if the concentrations of Sb^{3+} and Bi^{3+} are high enough, these latter substances may precipitate as the sulfates, thereby contaminating the precipitate of $PbSO_4$.

4.7.6 THE EFFECT OF pH

General   Consider first of all the slightly soluble salt of a weak acid in aqueous solution containing an appreciable concentration of a strong monobasic acid. Here the anion of the salt will react with H_3O^+ to form undissociated weak acid and to increase the solubility of the salt. Increasing concentrations of the strong monobasic acid will result in increasing salt solubility. The same situation applies, of course, to slightly soluble salts involving the cations of weak bases; the salt solubility will increase with increasing strong monobasic base concentrations in solution.

Similar situations exist with the slightly soluble salts of polybasic acids; again, because of the relatively low degree of dissociation for the secondary and subsequent stages of ionization, increasing values of $[H_3O^+]$ in the solution will

Table 4.4   Solubility of lead sulfate in water-ethanol mixtures*

Concentration of ethanol in volume percent	0	10	20	30	40	50	60	70
Solubility of $PbSO_4$, mg/l	45	17	6.3	2.3	0.77	0.48	0.30	0.09
Solubility of $PbSO_4$, mol/l $\times 10^5$	15	5.6	2.1	0.76	0.25	0.16	0.10	0.03

*I. M. Kolthoff, R. W. Perlich, and D. Weiblen, *J. Phys. Chem.*, **46**:561 (1942).

result in increasing solubility for the salt. The following developments for solution equilibria involving such situations will indicate the general relationship between solution pH and molar solubility of such salts.

Equilibria for an acid aqueous solution of the slightly soluble salt BA of the weak monobasic acid HA   It should be understood that BA and HA refer not to specific substances but to generalized salt and acid.

For the aqueous solution conditions involved, we have

$$BA(s) \rightleftharpoons B^+ + A^-$$

$$A^- + H_3O^+ \rightleftharpoons HA + H_2O$$

$$2H_2O \rightleftharpoons H_3O^+ + OH^-$$

The equilibrium equations associated with the above are

$$[B^+][A^-] = K_{sp}(BA) \qquad \frac{[H_3O^+][A^-]}{[HA]_{un}} = K_a$$

$$[H_3O^+][OH^-] = K_w$$

The material balance is given by

$$S = [A^-] + [HA]_{un} = [B^+] \tag{103}$$

where S is the molar solubility of BA in the acid solution.

The charge balance is given by

$$[H_3O^+] + [B^+] = [OH^-] + [A^-] + [R^-] \tag{104}$$

where $[R^-]$ is the anion concentration stemming from the strong monobasic acid responsible for the value of $[H_3O^+]$ in the solution. We have

$$[B^+] = S$$

so that

$$[A^-] = \frac{K_{sp}}{S}$$

Again, from the material-balance Eq. (103) we have

$$S = [A^-]\left(1 + \frac{[H_3O^+]}{K_a}\right)$$
$$= \frac{K_{sp}}{S}\left(1 + \frac{[H_3O^+]}{K_a}\right)$$

from which

$$S = \sqrt{K_{sp}\left(1 + \frac{[H_3O^+]}{K_a}\right)} \tag{105}$$

Many texts write Eq. (105) as

$$S = \sqrt{\alpha_0 K_{sp}}$$

with $\alpha_0 K_{sp}$ being the effective solubility product constant, or K_{eff}.

Where the solution value of $[H_3O^+]$ at equilibrium is given or is determinable, substitution of this value in Eq. (105) permits its solution for S, the molar solubility of BA in the acid aqueous solution involved.

Note that if C_a, the initial concentration of the strong monobasic acid, is given, then $[H_3O^+] \neq C_a$, since an appreciable amount of the available $[H_3O^+]$ may be used in reaction with the anion of the slightly soluble salt. Where the value of $[H_3O^+]$ at equilibrium can be determined by pH measurement, the problem of determining the molar solubility of the salt identifies itself with the solution of Eq. (105). Where the value of $[H_3O^+]$ can not be so determined, we have

$$[R^-] = C_a$$

Equation (104) then modifies to

$$[H_3O^+] = C_a + [A^-] + [OH^-] - S$$
$$= C_a + [A^-] + \frac{K_w}{[H_3O^+]} - S$$

This leads to a quadratic equation in $[H_3O^+]$. The solution to this equation, substituted into Eq. (105), yields an unwieldy equation for the determination of S.

A simplification can be introduced by assuming that since the aqueous solution has significant acidity, $[H_3O^+]$ is greater than $10^{-7}\ M$ and very likely much greater than $[OH^-]$. Equation (104) then reduces to

$$[H_3O^+] \approx C_a + [A^-] - S$$
$$\approx C_a + \frac{K_{sp}}{S} - S$$

and substitution of this value for $[H_3O^+]$ into Eq. (105) leads to a cubic equation in S given by

$$S^3 + \frac{K_{sp}}{K_a}S^2 - \left(K_{sp} + \frac{C_a K_{sp}}{K_a}\right)S - \frac{K_{sp}^2}{K_a} = 0 \tag{106}$$

Equilibria for an acid aqueous solution of the slightly soluble salt $B_x A_y$ of the polybasic acid $H_a A$ (where $a = 1$, the acid is monobasic)   For the solution conditions involved here, we have

$$B_x A_y(s) \rightleftharpoons xB^{b+} + yA^{a-}$$

$$A^{a-} + H_3O^+ \rightleftharpoons HA^{(a-1)} + H_2O$$

$$HA^{(a-1)-} + H_3O^+ \rightleftharpoons H_2A^{(a-2)-} + H_2O$$

$$\vdots$$

$$2H_2O \rightleftharpoons H_3O^+ + OH^-$$

The equilibria associated with the above are

$$[B^{b+}]^x [A^{a-}]^y = K_{sp}(B_x A_y) \qquad [H_3O^+][OH^-] = K_w$$

$$\frac{[H_3O^+][A^{a-}]}{[HA^{(a-1)-}]} = K_a \qquad \frac{[H_3O^+][HA^{(a-1)-}]}{[H_2A^{(a-2)-}]} = K_{(a-1)}$$

$$\frac{[H_3O^+]^2[A^{a-}]}{[H_2A^{(a-2)-}]} = K_a K_{(a-1)}$$

with the sequence for the polybasic acid continuing to K_1. Note that the general formulation for the polybasic acid also covers the case of the monobasic acid HA.

The material balance is given by

$$S = \frac{[B^{b+}]}{x} = \frac{[A^{a-}] + [HA^{(a-1)-}] + [H_2A^{(a-2)-}] + \cdots + [H_a A]_{un}}{y} \tag{107}$$

where S is the molar solubility of the slightly soluble salt $B_x A_y$ in the acid solution involved.

Since we do not intend to give consideration to any aqueous acid solution other than those for which the equilibrium $[H_3O^+]$ is given or is determinable, we shall omit the charge-balance equation.

We have

$$[B^{b+}] = xS$$

$$[A^{a-}]^y = \frac{K_{sp}}{(xS)^x}$$

From the material-balance Eq. (107), we have

$$S = \frac{[A^{a-}]}{y}\left(1 + \frac{[H_3O^+]}{K_a} + \frac{[H_3O^+]^2}{K_a K_{(a-1)}} + \frac{[H_3O^+]^3}{K_a K_{(a-1)} K_{(a-2)}} + \cdots\right)$$

from which we have

$$S = \sqrt[(x+y)]{\frac{K_{sp}}{x^x y^y}\left(1 + \frac{[H_3O^+]}{K_a} + \frac{[H_3O^+]^2}{K_a K_{(a-1)}} + \cdots\right)^y} \tag{108}$$

where $[H_3O^+]$ is the given or determined equilibrium concentration in the acid solution involved.

Examples

1. Determine the molar solubility of silver acetate, $CH_3 \cdot COOAg$, at $25°C$ in an aqueous solution of 3.00 pH at equilibrium.

   Ignoring any reaction between Ag^+ and H_2O, we have from Eq. (108)

   $$S = \sqrt{K_{sp}\left(1 + \frac{[H_3O^+]}{K_a}\right)} = \sqrt{2.30 \times 10^{-3}\left(1 + \frac{10^{-3}}{1.76 \times 10^{-5}}\right)}$$
   $$= 3.65 \times 10^{-1} M$$

   The molar solubility of $CH_3 \cdot COOAg$ in water at $25°C$ is

   $$S = \sqrt{K_{sp}} = \sqrt{2.30 \times 10^{-3}} = 4.80 \times 10^{-2} M$$

   so that there is an increase in the molar solubility in pH 3.00 solution of about 660 percent, relative to the neutral aqueous solution.

2. Determine the molar solubility at $25°C$ of $CH_3 \cdot COOAg$ in an aqueous solution which initially contained $0.001 M$ HNO_3.

   Here we have $C_a = 0.001 M$ and $[H_3O^+] \neq 0.001 M$ but is much greater than $[OH^-]$. Equation (106) is applicable here; it yields

   $$S \approx 1.21 \times 10^{-1} M$$

   Note that there is an appreciable difference between the molar solubility of $CH_3 \cdot COOAg$ in a solution of 3.00 pH at equilibrium and a solution which is initially $0.001 M$ to HNO_3 before equilibrium is established.

3. Determine the molar solubility of calcium fluoride, CaF_2, at $25°C$ in an aqueous solution of pH 2.00 at equilibrium. Equation (108) yields

   $$S = \sqrt[3]{\frac{K_{sp}}{4}\left(1 + \frac{[H_3O^+]}{K_a}\right)^2}$$
   $$= 2.04 \times 10^{-3} M$$

   The molar solubility of CaF_2 in neutral aqueous solution, ignoring any significant hydrolysis of the fluoride ion, is

   $$S = \sqrt[3]{\frac{K_{sp}}{4}}$$
   $$= 2.15 \times 10^{-4} M$$

   and the relative increase in the molar solubility of CaF_2 for a solution of pH 2.00 at equilibrium is about 850 percent.

4. Determine the molar solubility of calcium oxalate, $COOCa \cdot COOCa$, at $25°C$ in an aqueous solution that is 3.00 pH at equilibrium. Equation (108) gives

   $$S = \sqrt[3]{K_{sp}\left(1 + \frac{[H_3O^+]}{K_2} + \frac{[H_3O^+]^2}{K_1 K_2}\right)}$$
   $$= 2.1 \times 10^{-4} M$$

   The molar solubility of $COOCa \cdot COOCa$ in neutral aqueous solution is given by

   $$S = \sqrt{K_{sp}} = 5.1 \times 10^{-5} M$$

   so that the relative increase in the molar solubility of calcium oxalate in a solution of pH 3.00 at equilibrium is about 310 percent.

5. Determine the molar solubility at $25°C$ of silver sulfide, Ag_2S, in an aqueous solution that is 1.00 pH at equilibrium. Equation (108) gives

$$S = \sqrt[3]{\frac{K_{sp}}{4}\left(1 + \frac{[H_3O^+]}{K_2} + \frac{[H_3O^+]^2}{K_1 K_2}\right)}$$
$$= 1.6 \times 10^{-10} M$$

The molar solubility of Ag_2S in a neutral aqueous solution is about $3.4 \times 10^{-17} M$ so that whereas the molar solubility of Ag_2S is very considerably increased in a solution of pH 1.00 at equilibrium, the separation of silver as Ag_2S from such a solution is still highly quantitative.

Summary In these approaches to the effect of pH on the molar solubility of slightly soluble salts, the equations developed can be solved with comparative ease where the $[H_3O^+]$ of the solution at equilibrium is either given or determined by pH measurement. Where only the initial concentration of the acid is given (without information as to what quantity may subsequently react before equilibrium is established), regardless of whether the acid does or does not involve an anion common with that of the salt, the problem becomes one which involves two variables, the molar solubility of the salt and the $[H_3O^+]$ of the solution at equilibrium. The equations which are then involved are much more complex and, correspondingly, more difficult to solve.

4.7.7 THE EFFECT OF HYDROLYSIS

The previous section indicated the effect of the solution pH or $[H_3O^+]$ on the molar solubility of slightly soluble salts having as anions the anions of weak acid substances. It was noted in general that where the acid responsible for the solution pH did not have a common anion with the salt, increasing the $[H_3O^+]$ by increasing the acid concentration resulted in an increasing molar solubility for the salt.

Where a slightly soluble salt of the general type indicated in the foregoing is added to pure water, there will be some tendency for the salt anion to hydrolyze. The solution *may* become basic, the degree of basicity depending on both the magnitude of the molar solubility of the salt and that of the K value for the dissociation of the acid from which the salt anion originated.

The same equations can be used for hydrolysis phenomena as were developed in the section on the effect of pH on the molar solubility. Basically the problem here centers around the fact that if the $[H_3O^+]$ at equilibrium is not determinable, the equations again involve two variables. In many cases of hydrolysis action involving slightly soluble salts, satisfactory approximate results for molar solubility can be obtained by making assumptions relative to the values of $[H_3O^+]$, $[OH^-]$, K_a, and K_{sp}.

Certain contributory factors can be considered in making assumptions intended to simplify the problem of determining molar solubilities relative to slightly soluble salts capable of hydrolysis actions in pure water medium. These are as follows:

1. The lower the K value of the acid associated with the anion of the slightly soluble salt, the more extensive will be the hydrolysis action involving this anion.

2. The lower the molar solubility of the salt (the K_{sp} value can generally be used as a guide here), the more complete will be the action of anion hydrolysis, but the less significant will be the $[OH^-]$ from hydrolysis relative to the $[OH^-]$ from the water dissociation. Conversely, the higher the molar solubility of the salt, the more significant will be the hydrolysis value of $[OH^-]$ compared with that from water.

The following will serve to illustrate the techniques that can be applied in the determination of the molar solubilities of slightly soluble salts capable of hydrolysis actions.

To determine the molar solubility of silver acetate at 25°C in pure water, giving due consideration to the possibility of anion hydrolysis, we note that we can determine the value of S from Eq. (108) as

$$S = \sqrt{K_{sp}\left(1 + \frac{[H_3O^+]}{K_a}\right)}$$

The charge balance yields

$$[H_3O^+] + [Ag^+] = [OH^-] + [CH_3 \cdot COO^-]$$

Presuming that hydrolysis of the anion does take place, we have $[OH^-]$ greater than $[H_3O^+]$. Again we have $[Ag^+] = S$ and $[CH_3 \cdot COO^-] = K_{sp}/[Ag^+] = K_{sp}/S$. This now gives a charge-balance equation of

$$S \approx \frac{K_w}{[H_3O^+]} + \frac{K_{sp}}{S}$$

which provides a value for $[H_3O^+]$ as

$$[H_3O^+] \approx \frac{K_w S}{S^2 - K_{sp}} \tag{109}$$

Substitution of this value for $[H_3O^+]$ into Eq. (108) as written gives

$$S \approx \frac{K_{sp}}{S}\left(1 + \frac{K_w S}{K_a(S^2 - K_{sp})}\right) \tag{110}$$

from which

$$S \approx 4.80 \times 10^{-2}\ M$$

a value identical to the value of S for silver acetate at 25°C in pure water, neglecting any hydrolysis of the anion. The molar solubility of silver acetate is thus not affected by anion hydrolysis, and this stems mainly from the relatively high value of K_a for acetic acid.

The determination of the molar solubility of copper sulfide, CuS, at 25°C in pure water will be based, first of all, on Eq. (108), which yields

$$S = \sqrt{K_{sp}\left(1 + \frac{[H_3O^+]}{K_2} + \frac{[H_3O^+]^2}{K_1 K_2}\right)}$$

The charge balance is given by

$$[H_3O^+] + 2[Cu^{2+}] = [OH^-] + 2[S^{2-}] + [HS^-]$$

Since the value of $K_{sp}(CuS)$ is very low at 8.0×10^{-36}, we can assume that $[OH^-]$ from S^{2-} hydrolysis will not be significant, and that $[OH^-] \approx 10^{-7}\,M \approx [H_3O^+]$. We have then, from Eq. (108),

$$S \approx \sqrt{8.0 \times 10^{-36}\left(1 + \frac{10^{-7}}{1.1 \times 10^{-15}} + \frac{10^{-14}}{1.0 \times 10^{-22}}\right)}$$

$$\approx 3.9 \times 10^{-14}\,M$$

The molar solubility of CuS in pure water at 25°C, neglecting any hydrolysis of S^{2-}, is given by

$$S = \sqrt{K_{sp}} = 2.8 \times 10^{-18}\,M$$

The determination of the molar solubility of manganese sulfide, MnS, at 25°C in pure water presents a slightly different problem. Again Eq. (108) yields

$$S = \sqrt{K_{sp}\left(1 + \frac{[H_3O^+]}{K_2} + \frac{[H_3O^+]^2}{K_1K_2}\right)}$$

The charge balance gives

$$[H_3O^+] + 2[Mn^{2+}] = [OH^-] + 2[S^{2-}] + [HS^-]$$

Because of the relatively high value of $K_{sp}(MnS)$ at 1.4×10^{-15}, we can assume that $[OH^-]$ is greater than $10^{-7}\,M$ greater than $[H_3O^+]$. In addition to this, we have $[Mn^{2+}] = S$, which modifies the charge balance to

$$2S \approx [OH^-] + 2[S^{2-}] + [HS^-]$$

We can assume that hydrolysis will be complete for the amount of MnS that is soluble. We can also assume that hydrolysis of any HS^- formed by the hydrolysis of S^{2-} will be negligible. We have, therefore,

$$[OH^-] \approx [HS^-]$$
$$[HS^-] \gg [H_2S]_{un}$$

These assumptions modify the derived equation from Eq. (108), as well as the charge balance, to

$$S \approx \sqrt{K_{sp}\left(1 + \frac{[H_3O^+]}{K_2}\right)}$$
$$S \approx [OH^-] + [S^{2-}]$$

and, since we have

$$[S^{2-}] = K_{sp}/[Mn^{2+}] = K_{sp}/S \qquad \text{and} \qquad [OH^-] = K_w/[H_3O^+]$$

the last modification of the charge balance becomes

$$S \approx \frac{K_w}{[H_3O^+]} + \frac{K_{sp}}{S}$$

which yields

$$[H_3O^+] \approx \frac{K_w S}{S^2 - K_{sp}}$$

Substitution of this value for $[H_3O^+]$ into the last modification of the Eq. (108) derivation gives

$$S \approx 2.4 \times 10^{-5} \, M$$

The molar solubility of MnS at 25°C in pure water, neglecting hydrolysis of S^{2-}, is given by

$$S = 3.7 \times 10^{-8} \, M$$

4.7.8 THE EFFECT OF COMPLEXATION

Where complexation reactions are possible with either the anion or the cation of a slightly soluble salt, it is apparent that the concentration of the ligand in solution will have an effect on the solubility of the salt. The ligand may be an ion or molecule either common or foreign to the slightly soluble salt.

A case in point is given by the molar solubility of a salt such as silver chloride in the presence of ammonia, where ammonia reacts with silver(I) ion to form

$$Ag^+ + NH_3 \rightleftharpoons Ag(NH_3)^+$$
$$Ag(NH_3)^+ + NH_3 \rightleftharpoons Ag(NH_3)_2^+$$

There are many situations where slightly soluble salts tend to form complexes with excesses of one or both common ions. For example, silver(I) ion reacts with cyanide ion to form a soluble but very slightly dissociated silver cyanide complex. The complex reacts, in turn, with excess silver(I) ion to form the slightly soluble salt, silver cyanide. Thus we have

$$2CN^- + Ag^+ \rightleftharpoons Ag(CN)_2^-$$
$$Ag(CN)_2^- + Ag^+ \rightleftharpoons 2AgCN$$

The treatment of the effect of complexation on the molar solubility of a slightly soluble salt does not differ markedly from the treatment relative to the effect of pH.

Consider the equilibria for a slightly soluble salt BA in a solution which contains a unidentate ligand L which forms a complex or complexes with the salt cation. The solution conditions yield

$$BA(s) \rightleftharpoons B^+ + A^-$$
$$B^+ + L \rightleftharpoons B(L)^+$$
$$B(L)^+ + L \rightleftharpoons B(L)_2^+$$

. . . .

The equilibria associated with these reactions are

$$[B^+][A^-] = K_{sp}(BA)$$

$$\frac{[B(L)^+]}{[B^+][L]} = K_1 \qquad \frac{[B(L)_2^+]}{[B(L)^+][L]} = K_2 \qquad \frac{[B(L)_2^+]}{[B^+][L]^2} = K_1 K_2$$

and so on to the number of complexes formed between B^+ and L. The material balance is given by

$$S = [A^-] = [B^+] + [B(L)^+] + [B(L)_2^+] + \cdots$$

We have

$$S = [A^-]$$

$$[B^+] = \frac{K_{sp}}{[A^-]} = \frac{K_{sp}}{S}$$

and from the material balance we now have

$$S = \frac{K_{sp}}{S}(1 + K_1[L] + K_1 K_2[L]^2 + \cdots) \tag{111}$$

and

$$S = \sqrt{K_{sp}(1 + K_1[L] + K_1 K_2[L]^2 + \cdots)} \tag{112}$$

In all expressions, the K values represent, of course, the formation or stability constants for the complex formation reaction involved.

Where the value of [L] at equilibrium is given, substitution of this value into Eq. (112) permits the determination of the molar solubility.

Where only the initial concentration of the ligand is given, that is, the value before any reactions leading to equilibrium, the problem of determining the molar solubility becomes more complex; it parallels the situation involved where the molar solubility of a slightly soluble salt of a weak acid is to be determined in a solution for which only the initial concentration of a strong monobasic acid is given. It is suggested that the student endeavor to develop on his own the required equation for the solution of this problem.

Examples

1. Determine the molar solubility at 25°C of silver chloride in a solution which is 1.0 M to NH_3 at equilibrium. Equation (112) gives

$$S = \sqrt{1.56 \times 10^{-10}(1 + 2.0 \times 10^3 \times 1 + 2.0 \times 8.0 \times 10^6 \times 1)}$$

$$= 5.0 \times 10^{-2} M$$

2. Determine the molar solubility of silver chloride at 25°C in a solution which is 1.0 M to NH_3 and 0.10 M to Cl^- at equilibrium.

Here we have $[Ag^+] = K_{sp}/[Cl^-]$ rather that $[Ag^+] = K_{sp}/S$. For Eq. (111) we then have

$$S = \frac{K_{sp}}{[Cl^-]}(1 + K_1[L] + K_1 K_2[L]^2 + \cdots)$$

and this yields

$$S = 2.5 \times 10^{-2} \, M$$

3. Determine the molar solubility of silver chloride at 25°C in a solution that is initially 0.10 M to Cl$^-$, and is then 1.0 M to NH$_3$ at equilibrium. Here we have

$$[Ag^+] = \frac{K_{sp}}{[Cl^-] + [Cl^-]_{AgCl}}$$

where $[Cl^-] = 0.10 \, M$ and $[Cl^-]_{AgCl} = S$, the molar solubility. Equation (111) then becomes

$$S = \frac{K_{sp}}{[Cl^-] + S}(1 + K_1[L] + K_1 K_2[L]^2 + \cdots)$$

and we have

$$S = 2.0 \times 10^{-2} \, M$$

4.8   DIFFERENTIAL PRECIPITATION METHODS

Certain of the methods of quantitative analysis, applicable particularly in gravimetric work but also applicable in a few instances of volumetric work, involve the separation of one ion from another on the basis of a significant difference in the solubilities of the salts formed. Examples of such techniques are

1. The separation of iodide ion from chloride ion, using silver ion as the precipitant
2. The separation of cadmium from manganese, in an acid solution properly adjusted as to pH, using hydrogen sulfide as the precipitant

In many cases, the significant differences in the solubilities of the two or more substances cannot be maintained during the precipitation process unless certain variables such as pH are carefully controlled. In other instances, during precipitation based on differential precipitation phenomena, coprecipitation and other interfering actions sometimes render it impossible to obtain a complete and clean separation of one substance from another, even where the solubility differences indicate that this should be possible. The method can nevertheless be of importance as a separation technique, as the following examples illustrate.

Consider a situation in which one must determine whether or not it is possible to separate iodide ion from chloride ion at 25°C, using silver(I) ion as the precipitant. The initial solution will contain 0.10 M each of the two ions.

The concentration of iodide ion is $10^{-1} \, M$, and the value of $[Ag^+]$ at which AgI is just ready to precipitate is given by

$$[Ag^+] = \frac{K_{sp}}{[I^-]} = \frac{1.50 \times 10^{-16}}{10^{-1}}$$
$$= 1.50 \times 10^{-15} \, M$$

The concentration of chloride ion is also $10^{-1}\,M$, and the $[Ag^+]$ at which AgCl is just ready to precipitate is given by

$$[Ag^+] = \frac{K_{sp}}{[Cl^-]} = \frac{1.56 \times 10^{-10}}{10^{-1}}$$
$$= 1.56 \times 10^{-9}\,M$$

Thus, the AgI separates first, with a controlled addition of Ag^+, and in theory will continue to separate free from AgCl up to a value of $[I^-]$ given by

$$[I^-] = \frac{1.50 \times 10^{-16}}{1.56 \times 10^{-9}} = 9.6 \times 10^{-8}\,M$$

It will obviously not be necessary to carry the precipitation of AgI so close to the point at which AgCl is ready to separate, and once the $[I^-]$ has been reduced to about $10^{-6}\,M$, the separation can be considered as quantitative. This level of $[I^-]$ remaining in solution will leave a residual $[Ag^+]$ of

$$[Ag^+] = \frac{1.50 \times 10^{-16}}{10^{-6}} = 1.50 \times 10^{-10}\,M$$

and this is well under the value of $[Ag^+]$ at which AgCl is ready to precipitate. It is therefore theoretically possible to separate iodide ion quantitatively from chloride ion under the initial solution conditions indicated.

A separation typical of the sulfide-separation technique would be represented by the separation of cadmium from manganese by the use of H_2S as the precipitant. Suppose that the initial concentrations of cadmium(II) ion and manganese(II) ion are 0.10 and 1.0 M, respectively, and that we are interested in determining the maximum value of $[H_3O^+]$ in the solution at which it is possible to separate cadmium quantitatively from manganese. To initiate the precipitation of CdS from a solution containing 0.10 M Cd^{2+}, we require a minimum $[S^{2-}]$ of

$$[S^{2-}]_{min} = \frac{5.0 \times 10^{-27}}{10^{-1}} = 5.0 \times 10^{-26}\,M$$

To separate cadmium quantitatively from this medium, we require a residual $[Cd^{2+}]$ after separation is complete of $10^{-6}\,M$ maximum. This requires a minimum $[S^{2-}]$ of

$$[S^{2-}]_{min} = \frac{5.0 \times 10^{-27}}{10^{-6}_{max}} = 5.0 \times 10^{-21}\,M$$

Using the equilibrium expressions for the dissociation of H_2S, we have

$$\frac{[H_3O^+]^2[S^{2-}]}{[H_2S]_{un}} = 1.0 \times 10^{-22}$$

Assuming that a saturated aqueous solution of H_2S is 0.1 M, we have the maximum desirable $[H_3O^+]$ given by

$$[H_3O^+]_{max}^2 = \frac{1.0 \times 10^{-22} \times 10^{-1}}{5.0 \times 10^{-21}} \quad \text{and} \quad [H_3O^+]_{max} = 4.5 \times 10^{-2}\,M$$

This value of $[H_3O^+]$ represents the maximum solution value at which cadmium can be separated quantitatively. Note that this value of $[H_3O^+]$ would allow a very high concentration of $[Mn^{2+}]$ in solution without separation of MnS, as given by

$$[Mn^{2+}]_{max} = \frac{1.4 \times 10^{-15}}{5.0 \times 10^{-21}}_{min}$$

$$\ggg \quad 1\,M$$

The selection of a value for $[H_3O^+]$, such as $10^{-3}\,M$ less than $4.5 \times 10^{-2}\,M$, allows the residual $[Cd^{2+}]$ after separation to be lower than $10^{-6}\,M$, thus yielding a surer quantitative separation. This value of $[H_3O^+]$ provides an $[S^{2-}]$ value of

$$[S^{2-}] = \frac{1.0 \times 10^{-22} \times 10^{-1}}{(10^{-3})^2}$$

$$= 1.0 \times 10^{-17}\,M$$

At this level of $[S^{2-}]$, we have

$$[Cd^{2+}] \text{ residual at equilibrium} = \frac{5.0 \times 10^{-27}}{1.0 \times 10^{-17}} = 5.0 \times 10^{-10}\,M$$

and an allowable $[Mn^{2+}]$ without precipitation of MnS of

$$[Mn^{2+}]_{max} = \frac{1.4 \times 10^{-15}}{1.0 \times 10^{-17}} = 1.4 \times 10^2\,M \quad \ggg \quad 1\,M$$

It is thus possible to separate cadmium quantitatively from manganese under the solution conditions indicated. The maximum allowable $[H_3O^+]$ would be $4.5 \times 10^{-2}\,M$.

It should again be mentioned that the examples just given represent theoretical situations not always attained under practical conditions, unless precautionary measures are taken. For example, in the separation of iodide ion from chloride ion, high local concentrations of Ag^+ during the addition of this precipitant may allow some precipitate of AgCl to form with the AgI. This separation of AgCl may not be redissolved completely at the time when the precipitate of AgI is filtered off.

Again, in the separation of cadmium from manganese by H_2S, inadequate buffering of the solution before the passage into it of H_2S gas can result in localized separations of MnS with CdS, and it is always possible that such separated MnS may not redissolve prior to filtration of the CdS.

In addition to the forms of interference mentioned, the phenomenon of coprecipitation may intervene to permit a less-than-perfect separation of the two components of the binary complex. The factor of coprecipitation will be discussed in detail in the chapter on the gravimetric methods of analysis.

Part 4   Complexation Equilibria

4.9   GENERAL

Definitions of the terms "complex," "complex ion," or "complex compound" are exceedingly difficult because of the many borderline cases that arise to generate exceptions to the definitions made. With some reservations, therefore, we shall accept the definition of a complex ion as an ion that has at its center an atom or ion surrounded by a number of ions or neutral molecules. In the majority of cases encountered in quantitative analytical chemistry having an inorganic basis, the central atom or ion will be a metal atom or ion. The following examples of complex ions illustrate the situation:

1. Tetraminecopper(II) $Cu(NH_3)_4^{2+}$
2. Dicyanoargentate(I) $Ag(CN)_2^-$
3. Tetrahydroxozincate(II) $Zn(OH)_4^{2-}$
4. Hexanitrocobaltate(III) $Co(NO_2)_6^{3-}$

We shall also define several other terms commonly encountered in complexation work, leaving detailed explanations to arise where applicable out of subsequent treatments.

Complex or coordination compound   A compound which includes a complex ion. An example would be $Ag(NH_3)_2NO_3$.

A ligand   Any atom, ion, or molecule capable of acting as the donor partner in one or more coordinate bonds. An example would be the ligands NH_3 in $Cu(NH_3)_4^{2+}$.

Coordination number   The total number of ions or molecules directly associated with the central atom or ion. For example, in the complex ion $Cu(NH_3)_4^{2+}$, copper has a coordination number of 4.

Unidentate ligands   Ligands that use only one atom at a time as a donor atom, and are thus capable of filling only one coordination position of a given metal ion. An example would be NH_3.

Polydentate ligands   Ligands having two or more atoms which are capable of coordinating to the central atom or ion. An example would be the complex ion triethylenetetraminecopper(II), whose structure can be represented as which is the result of the quadridentate ligand triethylenetetramine satisfying the coordination number of 4 for the copper(II) ion.

Chelates Heterocyclic ring compounds formed by the reaction of a metal ion, as the central ion, with two or more functional groups of the same ligand. The triethylenetetraminecopper(II) compound shown above is an example of a chelate.

Chelons The term proposed for particular chelating ligands capable of forming water-soluble, stable, 1:1 complex ions with metal ions, so that the chelon may be used as a titrant substance in the quantitative determination of the metal ion involved.

The bonding action that exists for complex ions has been the subject of much study and investigation during recent years, and it has been found that it is covalent (i.e., an overlap of ligand and metal ion orbitals, with some degree of ionic interaction).

We shall discuss in this section only complexation situations that involve:

1. Complex ions with unidentate ligands where more than one complex-ion species is commonly encountered
2. Complex ions with polydentate ligands where only one complex-ion species is commonly encountered

4.10 COMPLEX IONS INVOLVING UNIDENTATE LIGANDS— MORE THAN ONE COMPLEX-ION SPECIES POSSIBLE

4.10.1 GENERAL

A general complexation reaction of this type may have its end result typified by the reaction

$$Cu^{2+} + 4NH_3 \rightleftharpoons Cu(NH_3)_4^{2+}$$

The complex ion $Cu(NH_3)_4^{2+}$ has a square planar structure, the Cu^{2+} ion being at the center of the square, each corner of which is occupied by an NH_3 molecule. The bonding action is established between the copper ion and the nitrogen atoms, and the structure may be represented as

For the end reaction above, it should be realized that as is common with most metal ions in aqueous solution, the ion Cu^{2+} does not exist per se but as the hydrated form $Cu(H_2O)_4^{2+}$, and the reaction is more properly written

$$Cu(H_2O)_4^{2+} + 4NH_3 \rightleftharpoons Cu(NH_3)_4^{2+} + 4H_2O$$

In point of fact, the reaction written above, in either form, is only the end result of a series of reactions which can be shown, with their respective equilibrium constants, as

$$Cu^{2+} + NH_3 \rightleftharpoons Cu(NH_3)^{2+} \qquad \frac{[Cu(NH_3)^{2+}]}{[Cu^{2+}][NH_3]} = K_1 = 1.3 \times 10^4$$

$$Cu(NH_3)^{2+} + NH_3 \rightleftharpoons Cu(NH_3)_2^{2+} \qquad \frac{[Cu(NH_3)_2^{2+}]}{[Cu(NH_3)^{2+}][NH_3]} = K_2 = 3.2 \times 10^3$$

$$Cu(NH_3)_2^{2+} + NH_3 \rightleftharpoons Cu(NH_3)_3^{2+} \qquad \frac{[Cu(NH_3)_3^{2+}]}{[Cu(NH_3)_2^{2+}][NH_3]} = K_3 = 8.0 \times 10^2$$

$$Cu(NH_3)_3^{2+} + NH_3 \rightleftharpoons Cu(NH_3)_4^{2+} \qquad \frac{[Cu(NH_3)_4^{2+}]}{[Cu(NH_3)_3^{2+}][NH_3]} = K_4 = 1.3 \times 10^2$$

The overall reaction then becomes, as indicated before,

$$Cu^{2+} + 4NH_3 \rightleftharpoons Cu(NH_3)_4^{2+} \qquad \frac{[Cu(NH_3)_4^{2+}]}{[Cu^{2+}][NH_3]^4} = K_1K_2K_3K_4 = 4.3 \times 10^{12}$$

This latter constant, $K_1K_2K_3K_4$, is called β_4, the overall formation constant for this complex.

In the process of adding NH_3 slowly to a solution of Cu^{2+}, as would occur in the titration of copper(II) with NH_3, all of the complex ions indicated above would be formed in the solution. Only where NH_3 is present in significant excess will the species $Cu(NH_3)_4^{2+}$ be the only complex ion present to an appreciable extent. It is apparent from this that the titration of Cu^{2+} with NH_3 will not be feasible, since the formation of multiple complex ions reduces the rate of change of pCu with volume additions of NH_3 to a level that does not allow exact location of the equivalence point of the titration. If it were possible, which it is not, to add four NH_3 molecules to each Cu^{2+} ion at each addition stage of the titration, then such a titration would be feasible.

Nevertheless, solutions of Cu^{2+} and NH_3, with the concentration of free NH_3 known at equilibrium, are of significance in analytical chemistry, as are others of a similar nature which involve unidentate ligands and ions capable of forming more than one complex-ion species with the ligand involved.

Under the conditions indicated, it will be important to know the distribution of the various complex-ion species for a given equilibrium concentration of the ligand. The following indicates one general approach to this problem.

4.10.2   DISTRIBUTION OF COMPLEX-ION SPECIES FOR THE ION M^{n+} WHERE THE NEUTRAL LIGAND L IS CAPABLE OF FORMING THE COMPLEX-ION SPECIES $M(L)^{n+}$, $M(L)_2^{n+}$, $M(L)_3^{n+}$, ETC.

The solution conditions involved here are

$$M^{n+} + L \rightleftharpoons M(L)^{n+}$$

$$M(L)^{n+} + L \rightleftharpoons M(L)_2^{n+}$$

$$\vdots$$

$$M(L)_{x-1}^{n+} + L \rightleftharpoons M(L)_x^{n+}$$

The respective equilibrium constants, or stepwise stability constants, for the above reactions are

$$\frac{[M(L)^{n+}]}{[M^{n+}][L]} = K_1$$

$$\frac{[M(L)_2^{n+}]}{[M(L)^{n+}][L]} = K_2$$

$$\vdots$$

$$\frac{[M(L)_x^{n+}]}{[M(L)_{x-1}^{n+}][L]} = K_x$$

If we assign a value of C_M as the original or starting concentration of the metal ion, and the values of β_0, β_1, β_2, . . . , β_x as the ratios of the concentrations of the species M^{n+}, $M(L)^{n+}$, . . . , $M(L)_x^{n+}$ to the value of C_M, we have

$$\beta_0 = \frac{[M^{n+}]}{C_M}$$

$$\beta_1 = \frac{[M(L)^{n+}]}{C_M}$$

$$\beta_2 = \frac{[M(L)_2^{n+}]}{C_M}$$

$$\vdots$$

$$\beta_x = \frac{[M(L)_x^{n+}]}{C_M}$$

Using the equilibrium-constant expressions, we have

$$K_1 = \frac{[M(L)^{n+}]}{[M^{n+}][L]} = \frac{\beta_1 C_M}{\beta_0 C_M [L]} = \frac{\beta_1}{\beta_0 [L]}$$

$$K_2 = \frac{[M(L)_2^{n+}]}{[M(L)^{n+}][L]} = \frac{\beta_2 C_M}{\beta_1 C_M [L]} = \frac{\beta_2}{\beta_1 [L]}$$

$$K_3 = \frac{[M(L)_3^{n+}]}{[M(L)_2^{n+}][L]} = \frac{\beta_3 C_M}{\beta_2 C_M [L]} = \frac{\beta_3}{\beta_2 [L]}$$

$$\vdots$$

$$K_x = \frac{[M(L)_x{}^{n+}]}{[M(L)_{x-1}{}^{n+}][L]} = \frac{\beta_x C_M}{\beta_{x-1} C_M [L]} = \frac{\beta_x}{\beta_{x-1}[L]}$$

From the above we have

$$K_1 = \frac{\beta_1}{\beta_0 [L]} \qquad \text{and} \qquad \beta_1 = \beta_0 [L] K_1$$

$$K_2 = \frac{\beta_2}{\beta_1 [L]} \qquad \text{and} \qquad \beta_2 = \beta_0 [L]^2 K_1 K_2$$

$$K_3 = \frac{\beta_3}{\beta_2 [L]} \qquad \text{and} \qquad \beta_3 = \beta_0 [L]^3 K_1 K_2 K_3$$

$$\vdots$$

$$K_x = \frac{\beta_x}{\beta_{x-1} [L]} \qquad \text{and} \qquad \beta_x = \beta_0 [L]^x K_1 K_2 K_3, \ldots, K_x$$

The material-balance expression is given by

$$C_M = [M^{n+}] + [M(L)^{n+}] + [M(L)_2{}^{n+}] + \cdots + [M(L)_x{}^{n+}] \tag{113}$$

and we have

$$1 = \beta_0 + \beta_1 + \beta_2 + \beta_3 + \cdots + \beta_x \tag{114}$$

from which

$$1 = \beta_0 + \beta_0 [L] \, K_1 + \beta_0 [L]^2 K_1 K_2 + \beta_0 [L]^3 K_1 K_2 K_3 + \cdots$$
$$+ \beta_0 [L]^x K_1 K_2 K_3, \ldots, K_x \tag{115}$$

and we have

$$\beta_0 = \frac{1}{1 + [L] K_1 + [L]^2 K_1 K_2 + [L]^3 K_1 K_2 K_3 + \cdots + [L]^x K_1 K_2 K_3, \ldots, K_x}$$

$$\beta_1 = \frac{[L] K_1}{1 + [L] K_1 + [L]^2 K_1 K_2 + \cdots + [L]^x K_1 K_2 K_3, \ldots, K_x}$$

$$\beta_2 = \frac{[L]^2 K_1 K_2}{1 + [L] K_1 + [L]^2 K_1 K_2 + \cdots + [L]^x K_1 K_2 K_3, \ldots, K_x}$$

$$\vdots$$

$$\beta_x = \frac{[L]^x K_1 K_2 K_3, \ldots, K_x}{1 + [L] K_1 + [L]^2 K_1 K_2 + \cdots + [L]^x K_1 K_2 K_3, \ldots, K_x} \tag{116}$$

Where the concentration of the ligand at equilibrium is known, the ratio value for each species can be determined and, using the original concentration of the M^{n+} ion, the actual concentrations of each complex-ion species secured.

Where the concentration of the ligand at equilibrium is not known, assumptions can be made to allow some reduction in the complexity of the problem of determining the ratio value for each species. Under most circumstances which develop in analytical chemistry, the ligand will be present in excess at equilibrium.

It can be assumed, therefore, that the metal ion is in its most highly complexed form, and that the concentrations of all the other complex-ion species are negligible. It can be further assumed that the free-ligand concentration at equilibrium will be identical to the excess concentration.

Note that where charged ligands rather than neutral ligands are involved, such as CN^-, the general situation does not change, and the solution to the concentration-ratio problem will be the same as the foregoing. The reaction situations, for a charged ligand of the type L^{p-}, would yield the complex-ion species

$$M(L)_x^{(n-xp)}$$

where x has the values $1, 2, 3, \ldots, x$, and where x is the coordination number for M in the particular system under study. The value of n represents the charge for the ion M^{n+}.

Example Determine for 25°C the concentrations of all species in a mixture of Cu^{2+} and NH_3 where the equilibrium concentration of NH_3 is 0.10 M. Consider only the five possible species Cu^{2+}, $Cu(NH_3)^{2+}, \ldots, Cu(NH_3)_4^{2+}$. Suppose the starting or initial concentration of Cu^{2+} to have been 0.01 M.

The formation constants from the tables are

$$K_1 Cu(NH_3)^{2+} = 1.3 \times 10^4 \qquad K_2 Cu(NH_3)_2^{2+} = 3.2 \times 10^3$$

$$K_3 Cu(NH_3)_3^{2+} = 8.0 \times 10^2 \qquad K_4 Cu(NH_3)_4^{2+} = 1.3 \times 10^2$$

The denominator of Eq. (116) gives a value of

$$4.7 \times 10^8$$

and the calculated values of β are then given by Eq. (116) as

$$\beta_0 = 2.1 \times 10^{-9} \qquad \beta_1 = 2.8 \times 10^{-6}$$

$$\beta_2 = 9.0 \times 10^{-4} \qquad \beta_3 = 7.1 \times 10^{-2}$$

$$\beta_4 = 9.2 \times 10^{-1}$$

The initial value of $[Cu^{2+}]$ is 0.01 M, and we then have

$$[Cu^{2+}] = 2.1 \times 10^{-11} M \qquad [Cu(NH_3)^{2+}] = 2.8 \times 10^{-8} M$$

$$[Cu(NH_3)_2^{2+}] = 9.0 \times 10^{-6} M \qquad [Cu(NH_3)_3^{2+}] = 7.1 \times 10^{-4} M$$

$$[Cu(NH_3)_4^{2+}] = 9.2 \times 10^{-3} M$$

4.11 COMPLEX IONS INVOLVING POLYDENTATE LIGANDS— ONLY ONE COMPLEX-ION SPECIES POSSIBLE

4.11.1 METAL ION COMPLEXATION EQUILIBRIA NOT DIRECTLY AFFECTED BY SOLUTION pH

A typical polydentate ligand capable of forming only a single complex-ion species under normal conditions with copper(II) ion is triethylenetetramine, commonly called "trien." The ligand is quadridentate with Cu^{2+}, yielding the structural formula shown on page 161. Since only the single 1:1 complex-ion species is formed, and since the stability constant for triethylenetetraminecopper(II) is high, as given by

$$\frac{[\text{Cu(trien)}^{2+}]}{[\text{Cu}^{2+}][\text{trien}]} = 2.5 \times 10^{20}$$

the use of trien as a titrating agent for copper(II) can be expected to yield a quantitative reaction, a satisfactory rate of change of pCu versus trien around the equivalence point of the titration, and an easily determinable volume of the titrant at the equivalence point. Since the ligand and the complex with copper(II) are both water-soluble, and since the reaction occurs rapidly, the trien compound is suitable for aqueous solution titrations of Cu^{2+}.

4.11.2 METAL ION COMPLEXATION EQUILIBRIA DIRECTLY AFFECTED BY SOLUTION pH

Many complexing agents or ligands are affected by the pH of the solution, particularly where these ligands are the anions of weak acid substances or the cations of weak base substances. Using the former case as an example, it is apparent that the presence of appreciable hydrogen ion in solution results in a reaction where the ligand combines with hydrogen ion in one or more steps to yield undissociated forms of the weak acid from which the ligand was derived. A typical situation is represented by the substance ethylenediaminetetraacetic acid, commonly called EDTA. The fully dissociated anion of this acid is a chelating agent, reacting with many metal ions to form the appropriate complex ions. With four-carboxyl groups and two neutral nitrogen atoms, the EDTA substance is capable of acting as a quadridentate, quinquedentate, or sexidentate ligand. For the sake of convenience, the free acid form of EDTA is often represented as H_4Y.

While it is possible that in solutions of fairly high $[H_3O^+]$, species such as MHY^- could exist, and while it is possible again that in solutions of very low $[H_3O^+]$, hydroxyl ions may provide complex ions such as $M(OH)Y^{3-}$, we shall ignore these special cases and consider only the condition where one complex-ion species is secured for any metal ion–EDTA ligand complexation reaction.

Using EDTA or H_4Y as the example of a pH-affected polydentate ligand, we note that

$$H_4Y + H_2O \rightleftharpoons H_3O^+ + H_3Y^-$$
$$H_3Y^- + H_2O \rightleftharpoons H_3O^+ + H_2Y^{2-}$$
$$H_2Y^{2-} + H_2O \rightleftharpoons H_3O^+ + HY^{3-}$$
$$HY^{3-} + H_2O \rightleftharpoons H_3O^+ + Y^{4-}$$

The rearranged equilibrium-constant expressions are given by

$$[HY^{3-}] = \frac{[H_3O^+][Y^{4-}]}{K_4}$$

$$[H_2Y^{2-}] = \frac{[H_3O^+][HY^{3-}]}{K_3} = \frac{[H_3O^+]^2[Y^{4-}]}{K_3K_4}$$

$$[H_3Y^-] = \frac{[H_3O^+][H_2Y^{2-}]}{K_2} = \frac{[H_3O^+]^3[Y^{4-}]}{K_2K_3K_4}$$

$$[H_4Y] = \frac{[H_3O^+][H_3Y^-]}{K_1} = \frac{[H_3O^+]^4[Y^{4-}]}{K_1K_2K_3K_4}$$

If we now define as C_Y the total concentration of the free or uncomplexed EDTA anion, we have as the material balance

$$C_Y = [Y^{4-}] + [HY^{3-}] + [H_2Y^{2-}] + [H_3Y^-] + [H_4Y] \tag{117}$$

Substitution of the four equilibrium-constant expressions above into Eq. (117) yields

$$C_Y = [Y^{4-}]\left(1 + \frac{[H_3O^+]}{K_4} + \frac{[H_3O^+]^2}{K_3K_4} + \frac{[H_3O^+]^3}{K_2K_3K_4} + \frac{[H_3O^+]^4}{K_1K_2K_3K_4}\right)$$

$$\tag{118}$$

If we designate as α the term in the brackets, we then have

$$[Y^{4-}] = \frac{C_Y}{\alpha} \tag{119}$$

which represents the availability of the ligand Y^{4-} for complexation purposes. It is apparent that this availability will be a function of the magnitude of α, itself a function of $[H_3O^+]$. The higher the value of $[H_3O^+]$, the lower the availability of Y^{4-}.

Suppose that we consider the complexation reaction of a 1:1 ligand–metal ion basis where we have

$$M^{n+} + Y^{4-} \rightleftharpoons M(Y)^{(n-4)}$$

with an equilibrium constant of

$$\frac{[M(Y)^{(n-4)}]}{[M^{n+}][Y^{4-}]} = K_{\text{stability}} \tag{120}$$

If we substitute for the value of $[Y^{4-}]$ the equality of Eq. (119), we have

$$\frac{[M(Y)^{(n-4)}]}{[M^{n+}]C_Y} = \frac{K}{\alpha}$$

Note that the value of K/α, often referred to as the conditional formation constant, varies with pH.

As the pH decreases, the value of α becomes greater, the conditional formation constant smaller, and the reaction less quantitative. With increasing pH, the value of α approaches unity and the reaction approaches a degree of completion dictated by the magnitude of the stability constant K. Where the pH of the solution is low enough to permit a predominance of forms of the EDTA anion other than Y^{4-}, such as, for example, H_2Y^{2-}, we are faced with complexation reactions of the type

$$M^{n+} + H_2Y^{2-} + 2H_2O \rightleftharpoons M(Y)^{(n-4)} + 2H_3O^+$$

where the pH of the solution tends to decrease as the complex ion $M(Y)^{(n-4)}$ is formed. Under these circumstances, it is apparent that the complexation reaction must take place under properly buffered solution conditions in order to prevent pH decreases in the solution as the reaction proceeds.

Example Determine at 25°C the possible degree of quantitative completion of a reaction between Mg^{2+} and EDTA with a total uncomplexed EDTA ligand concentration at equilibrium given by $C_Y = 10^{-2} M$. Consider solution conditions at (1) pH = 6 and (2) pH = 10.

The K values for EDTA are

$$K_1 = 1.0 \times 10^{-2} \qquad K_2 = 2.1 \times 10^{-3}$$
$$K_3 = 6.9 \times 10^{-7} \qquad K_4 = 5.5 \times 10^{-11}$$

and the formation constant for Mg(EDTA), $Mg(Y)^{2-}$, is

$$K = 4.9 \times 10^8$$

The value of α from Eq. (118) is given by

$$\alpha \approx 4.5 \times 10^4$$

for a solution pH of 6. This yields a value of K/α of about 1.1×10^4. For the expression

$$\frac{[Mg(Y)^{2-}]}{[Mg^{2+}]} = \frac{KC_Y}{\alpha}$$

we have a value then of 1.1×10^2, approximately. Thus, the value of the ratio of $[Mg(Y)^{2-}]$ to $[Mg^{2+}]$ at equilibrium is much less than 10^6, and we can assume that the complexation reaction will by no means be quantitatively complete at a solution condition of pH = 6.

The value of α from Eq. (118) is given by

$$\alpha \approx 2.8$$

for a solution pH of 10. This yields a K/α of about 1.8×10^8. The ratio of $[Mg(Y)^{2-}]$ to $[Mg^{2+}]$ at equilibrium is then about 1.8×10^6. This value is greater than 10^6, and we can assume that the reaction is quantitatively complete at a solution condition of pH = 10.

Note that if we were dealing with a titration of $0.10 M$ Mg^{2+} by $0.10 M$ EDTA, the value of $[Mg(Y)^{2-}]$ at the equivalence for a quantitative reaction would be about $0.050 M$, and we would have $[Mg^{2+}]$ at equivalence approximately equal to C_Y. If the titration is conducted in a solution buffered so as to maintain a pH value of 10.0 during the titration, we then have

$$\frac{[Mg(Y)^{2-}]}{[Mg^{2+}]^2} \approx \frac{K}{\alpha} \approx 1.8 \times 10^8$$

from which we have

$$[Mg^{2+}] \approx 1.7 \times 10^{-5} M$$

at the equivalence point. This value is slightly higher than $10^{-6} M$, so that the titration is not exactly complete in the quantitative sense. For such a titration, however, involving, say, 100 ml of $0.10 M$ Mg^{2+} at the start, the residual amount of Mg^{2+} at the equivalence point would be given by 3.4×10^{-6} mol/200 ml, this volume being the final-solution volume at the equivalence point. This is about 0.0001 g of Mg^{2+} and would represent for this titration a quantitative reaction at the equivalence point.

Part 5   Oxidation-Reduction Equilibria

4.12   GENERAL

Reactions involving oxidation-reduction processes, or redox reactions, are encountered very frequently in quantitative analytical chemistry, and are of particular importance in volumetric methods. The terms oxidation and reduction have been defined elsewhere in this volume; basically, oxidation refers to the loss of one or more electrons by a species during reaction, and reduction refers to the gain of one or more electrons by a species during reaction. Typical of reactions associated with these basic definitions are

$$Fe^{2+} - 1e \rightleftharpoons Fe^{3+} \qquad \text{oxidation}$$

$$Sn^{4+} + 2e \rightleftharpoons Sn^{2+} \qquad \text{reduction}$$

$$Cu^{2+} + 2e \rightleftharpoons Cu \qquad \text{reduction}$$

and we may indicate as a typical oxidation-reduction reaction

$$Ce^{4+} + Fe^{2+} \rightleftharpoons Ce^{3+} + Fe^{3+}$$

where cerium(IV) is reduced to cerium(III), while iron(II) is oxidized to iron(III).

While the definitions above constitute proper identifications of the terms oxidation and reduction, extension of these definitions in a rather unique way is possible in discussing solution conditions. Certain reactions not identifiable as oxidation-reduction reactions can lead to conditions in solution during the reaction that are capable of being interpreted in a somewhat related manner. For example, the general acid-base neutralization reaction

$$H_3O^+ + OH^- \rightleftharpoons 2H_2O$$

can be regarded as the removal, by the formation of the stable species H_2O, of H_3O^+ and OH^- from the solution just as effectively as would the imaginary reduction reaction $H_3O^+ + 1e \rightleftharpoons H_3O^\circ$ and the imaginary oxidation reaction $OH^- - 1e \rightleftharpoons OH^\circ$.

Similarly, precipitation reactions typified by

$$Ag^+ + Cl^- \rightleftharpoons AgCl(s)$$

can be looked upon as the removal from the solution of Ag^+ and Cl^- just as effectively as the imagined oxidation-reduction reactions $Ag^+ + 1e \rightleftharpoons Ag^\circ$ and $Cl^- - 1e \rightleftharpoons Cl^\circ$.

In the same way, a complexation reaction, yielding a stable complex, such as

$$Ag^+ + 2CN^- \rightleftharpoons Ag(CN)_2^-$$

can be thought of as removing Ag^+ and CN^- from solution just as effectively as the imagined oxidation-reduction reactions $Ag^+ + 1e \rightleftharpoons Ag^\circ$ and $CN^- - 1e \rightleftharpoons CN^\circ$.

There is no intention here of implying that, for example, Ag exists as Ag° in AgCl. What we can picture is that in the solution medium, the precipitation of Ag^+ as AgCl is similar in effect to the formation of Ag° from Ag^+. In actuality, Ag in AgCl exists as Ag^+, just as it does in the aqueous solution; but in the form of AgCl, Ag is not available as Ag^+, except to a very limited extent, in the solution environment.

The ability of an atom, ion, or molecule to be reduced or oxidized can be described in terms of its reduction or oxidation potential; and when two species interact in an oxidation-reduction reaction, the difference between their respective oxidation or reduction potentials, as standard values, can be related to the equilibrium constant for the reaction involved. The following discussion provides some ideas concerning the general development of oxidation-reduction potentials for simple half-cells or single-electrode systems.

4.13 HALF-CELL POTENTIALS

Consider a situation where a strip or electrode of pure copper metal is immersed in a solution of copper(II) ions. The general system involved is shown in Fig. 4.3. We shall discuss this system from some points of view that might be challenged on the basis of not being strictly in accordance with theory but still serve to explain quite clearly how potential may be developed between an electrode and an aqueous solution of its ions.

When the copper electrode is immersed in the solution of copper(II), or Cu^{2+}, two reaction tendencies manifest themselves. The first of these is a reaction in which copper atoms leave the electrode and appear in solution at the electrode-

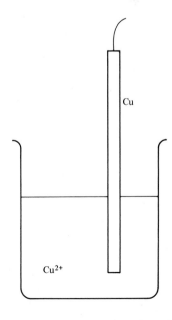

Fig. 4.3 Copper electrode immersed in a solution of Cu^{2+}.

solution interface as copper(II) ions. This reaction is shown as

$$Cu - 2e \rightleftharpoons Cu^{2+}$$

Such a reaction tends to leave near the electrode surface, but in the electrode, an excess of two electrons per copper atom entering solution. This, in effect, provides for an electrode having a higher electron density than electrical neutrality implies.

The second reaction tendency is that in which copper(II) ions leave the solution and appear on the electrode as copper atoms. The reaction is

$$Cu^{2+} + 2e \rightleftharpoons Cu$$

which tends to provide an electrode deficiency of two electrons per copper(II) ion leaving the solution. This, in effect, would provide for an electrode having a lower electron density than electrical neutrality implies.

Obviously, in the first reaction, the electrode becomes negative relative to the solution, and in the second reaction, the electrode becomes positive relative to the solution.

If we now consider a very highly dilute solution of Cu^{2+}, the likelihood is that the reaction

$$Cu - 2e \rightleftharpoons Cu^{2+}$$

will be predominant. We might then represent the two reaction velocities by

$$Cu - 2e \longrightarrow Cu^{2+} \tag{1}$$
$$Cu^{2+} + 2e \rightarrow Cu \tag{2}$$

denoting the reactions as (1) and (2). If we visualize little hindrance to the progress of these reactions (and this is highly imaginative), we can picture a situation where the velocity of reaction (1) decreases as the result of increasing $[Cu^{2+}]$ in the solution, while the velocity of reaction (2) increases. In time and under a condition of constant temperature, regardless of whether or not an equilibrium is established at equal reaction velocities, because reaction (1) was predominant, we will have established an electrode negatively charged relative to the solution. The magnitude of the difference in potential between the electrode and the solution will be a function of the degree of predominance of reaction (1) over reaction (2) at the start, and will therefore be a function of the starting concentration of Cu^{2+}. The more dilute the solution, the higher the potential difference established between the electrode and its solution.

The same argument can be used in relation to highly concentrated solutions of Cu^{2+}, where reaction (2) will, at the start, have a higher velocity than reaction (1). Here again, while the signs of the electrode and solution potentials, and that of their difference in potential, will be reversed, the difference in their potentials will increase in magnitude with increasing solution concentration of Cu^{2+}.

One can imagine, of course, a particular and unique concentration of Cu^{2+}

wherein equal reaction velocities will be established immediately and no potential difference will be established between the electrode and this solution.

If we indicate positive electrode or solution potentials by + signs and negative potentials by − signs, and if we indicate the electrode potential by ψ_e and the solution potential by ψ_s, we can show the potential difference as

$$\psi_e - \psi_s$$

and we can state that

1. The more dilute the solution, below the concentration of Cu^{2+} at which equal reaction velocities are established immediately, the more negative the value of $\psi_e - \psi_s$
2. The more concentrated the solution, above the concentration of Cu^{2+} at which equal reaction velocities are established immediately, the more positive the value of $\psi_e - \psi_s$

Within the limitations of this situation, if the sign and the magnitude of the potential difference $\psi_e - \psi_s$ is known, it is possible to determine the value of the $[Cu^{2+}]$ at the start. Conversely, where the $[Cu^{2+}]$ at the start is known, it is possible to predict the sign and magnitude of the potential difference $\psi_e - \psi_s$ that would be established. Apparently we can anticipate a range of values for $\psi_e - \psi_s$, depending on the range of solution $[Cu^{2+}]$ values.

Even though the foregoing discussion is imaginative, particularly in view of the fact that half-cells cannot continue in reaction to equilibrium unless some means of removing charge accumulations at the electrode-solution interface is provided, it serves to illustrate the development of potential differences between electrodes and solutions of their ions and the dependency of this potential difference, as to sign and magnitude, on ion concentration in the solution.

4.14 THE NERNST EQUATION

We shall now consider briefly the mechanism which permits the establishment of half-cell or single-electrode potentials as $\psi_e - \psi_s$ for any situation involving an oxidation-reduction reaction.

It is apparent that in order for a potential difference to be established between an electrode and its solution, some change in the free energy of the system must occur. Since the potential difference implies the ability to do electrical work external to the system, the energy to do this work must originate with a change in the energy balance of the electrode-solution system. It should be at once apparent from the preceding discussion that this energy change is directly related to the change in the chemical composition of the solution. In the example just used, the compositional change was a change in the $[Cu^{2+}]$ of the solution.

The concentration change yields a change in the free energy of the electrode-solution system. The free energy of this system is given by

$$G = H - TS \tag{121}$$

where G = free energy
H = enthalpy
T = absolute temperature
S = entropy

All systems, regardless of their nature, tend wherever possible to achieve spontaneously lower values of G, and we have, for such systems,

$$\Delta G = \Delta H - T\,\Delta S \tag{122}$$

For a spontaneous reaction, ΔH is negative and ΔS positive, so that ΔG is negative. Here both the ΔH and ΔS factors favor spontaneity.

For a general reaction of the type

$$aA + bB + \cdots \rightleftharpoons mM + nN + \cdots$$

we have

$$\Delta G = \Delta G^\circ + 2.303\,RT \log \frac{\alpha_M{}^m \alpha_N{}^n \cdots}{\alpha_A{}^a \alpha_B{}^b \cdots} \tag{123}$$

where ΔG° is the free-energy change for the reaction indicated when the activity of each product and reactant is unity, with R being the gas constant and T the absolute temperature.

This can be shown as

$$\Delta G = \Delta G^\circ + 2.303\,RT \log \frac{[M]^m [N]^n \cdots}{[A]^a [B]^b \cdots} \times \frac{(\gamma_M)^m (\gamma_N)^n \cdots}{(\gamma_A)^a (\gamma_B)^b \cdots} \tag{124}$$

As outlined in the general discussion on chemical equilibria, we can assume that for solutions of normal dilution,

$$\gamma_M \approx \gamma_N \approx \gamma_A \approx \gamma_B \approx 1$$

so that we then have

$$\Delta G = \Delta G^\circ + 2.303\,RT \log \frac{[M]^m [N]^n \cdots}{[A]^a [B]^b \cdots} \tag{125}$$

The change in free energy resulting from the transfer of Avogadro's number of electrons (Ne) through a given potential difference E is given by $(Ne)E$ and

$$\Delta G = -nFE$$

where n represents the moles of electrons involved in the reaction, or the electron transfer per mole, and F is the Faraday, or 96,490 C/equiv, and is equal to (Ne).

This leads to the equation

$$-nFE = -nFE° + 2.303 \; RT \log \frac{[M]^m [N]^n \cdots}{[A]^a [B]^b \cdots} \tag{126}$$

and

$$E = E° - 2.303 \frac{RT}{nF} \log \frac{[M]^m [N]^n \cdots}{[A]^a [B]^b \cdots} \tag{127}$$

At 298°K or 25°C, we have

$$E = E° - \frac{0.059}{n} \log \frac{[M]^m [N]^n \cdots}{[A]^a [B]^b \cdots} \tag{128}$$

Here $E°$ is the potential developed for the reaction involved where the activity of each product and reactant is unity. For the Nernst equation, as shown in the form of Eq. (128), molarities are substituted for activities, and unity values are represented by any substance in the solid state and soluble substances at the 1-M concentration in solution.

Thus, for any simple oxidation-reduction reaction, we have

$$Ox + ne \rightleftharpoons Red$$

and from Eq. (128),

$$E_{Ox/Red} = E°_{Ox/Red} - \frac{0.059}{n} \log \frac{[Red]}{[Ox]}$$

$$= E°_{Ox/Red} + \frac{0.059}{n} \log \frac{[Ox]}{[Red]} \tag{129}$$

$$E_{Red/Ox} = E°_{Red/Ox} - \frac{0.059}{n} \log \frac{[Ox]}{[Red]} \tag{130}$$

Note the general practice of maintaining the oxidized state in the numerator of the logarithmic expression.

It should be apparent, even at this point, that for a fixed situation of $[Ox]$ and $[Red]$ we have

$$E_{Red/Ox} = -E_{Ox/Red}$$

and for unit activities for the Ox and Red substances, or appropriately assigned unit values for $[Ox]$ and $[Red]$, we have

$$E°_{Red/Ox} = -E°_{Ox/Red}$$

4.15   THE CONVENTION OF SIGNS

Confusion arises from time to time—although less today than in the past—concerning the proper sign to be placed before a difference in potential for a half-cell or single-electrode system. Earlier texts may show one of the following:

$$Cu^{2+} + 2e \rightleftharpoons Cu \qquad E^\circ_{Cu^{2+}/Cu} = 0.340 \text{ V}$$
$$Cu^{2+} + 2e \rightleftharpoons Cu \qquad E^\circ_{Cu^{2+}/Cu} = -0.340 \text{ V}$$

The explanation is quite simple. It was indicated earlier that the potential difference between an electrode and a solution of its ions could be shown as

$$\psi_e - \psi_s$$

It would have been just as valid to show it as

$$\psi_s - \psi_e$$

whereupon it is apparent that we would have

$$\psi_e - \psi_s = -(\psi_s - \psi_e)$$

thus explaining the equal magnitude but opposite sign of the E° values in some of the earlier texts.

Modern texts generally tend to adhere to the International Union of Pure and Applied Chemistry (IUPAC) recommendations which, in effect, stipulate the potential difference as

$$\psi_e - \psi_s$$

so that increasing negative charge on the electrode implies increasingly negative values for the potential difference between the electrode and its solution. Under this convention the copper half-cell, given by the reaction

$$Cu^{2+} + 2e \rightleftharpoons Cu$$

gives a potential value

$$E^\circ_{Cu^{2+}/Cu} = 0.340 \text{ V}$$

4.16   THE MEASUREMENT OF HALF-CELL OR SINGLE-ELECTRODE POTENTIALS

As mentioned elsewhere, it is not possible to measure in any direct manner the potential difference between an electrode and its solution, mainly because of the accumulation of charges at the electrode-solution interface. Where such charge accumulations can be removed, potential differences can be measured. The establishment of external electrical connection and internal ion-exchange connection between one half-cell and another provides a means whereby such charge accumulations can be removed and the potential difference between the two half-cell systems measured. If the potential difference for one of the interconnected half-cells is known, that of the other can be determined. The reference half-cell to which the potentials of other half-cell systems are referred is the *normal*, or *standard, hydrogen electrode*, or *SHE*. This electrode is shown in simplified form in

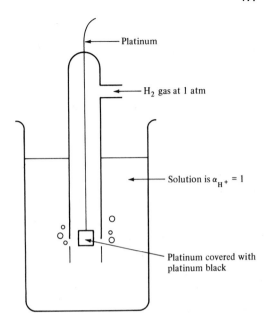

Platinum

H_2 gas at 1 atm

Solution is $\alpha_{H^+} = 1$

Platinum covered with platinum black

Fig. 4.4 The standard hydrogen electrode (SHE).

Fig. 4.4; it is a standard electrode by virtue of having a unit activity situation for both reactant and product, for either reaction direction, in the half-cell reaction

$$2H^+ + 2e \rightleftharpoons H_2$$

The electrode consists of a platinum wire sealed into a vessel somewhat similar to an inverted test tube, with a side-arm inlet and perforations in the walls near the open end. The wire terminates in a platinum strip coated with platinum black, a reduced form of platinum. The tube is immersed in a solution which has $\alpha_{H^+} = 1$. Hydrogen gas, under 1 atm, is introduced through the side arm and provides for $\alpha_{H_2} = 1$. The platinum black coating is exposed to both hydrogen gas and hydrogen ion as the gas escapes through the perforations. Such an electrode is assigned a potential difference value of 0.000 V at all temperatures, implying that the reactions

$$2H^+ + 2e \rightleftharpoons H_2$$

are in equilibrium. Thus, this reference electrode presents no resistance to the spontaneous reaction direction for any linked half-cell under test.

Suppose, for example, that we are to determine the standard potential of the copper-copper(II) half-cell previously discussed. By the standard potential is meant the half-cell potential relative to the reduction reaction

$$Cu^{2+} + 2e \rightleftharpoons Cu$$

when the situation is $\alpha_{Cu^{2+}} = 1$ and $\alpha_{Cu} = 1$. For our purposes we can assume that $\alpha_{Cu^{2+}} = 1$ is approximately secured in a solution where $[Cu^{2+}] = 1\,M$. The solid

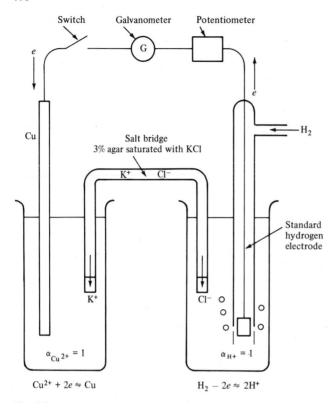

Fig. 4.5 Measurement of the standard reduction potential of the copper half-cell.

Cu electrode, in common with solid substances, provides $\alpha_{Cu} = 1$. A simple form of the equipment required is shown in Fig. 4.5. The copper half-cell and the standard hydrogen electrode solutions are linked through a conducting bridge, and the electrodes are connected to a potentiometer capable of measuring potential difference accurately.* Closing the switch momentarily results in a galvanometer deflection, the direction of which then establishes the current or electron-flow direction in the external or measuring circuit. For the cell combination of half-cells under discussion, the flow will be from the hydrogen electrode to the copper electrode. This implies immediately that the spontaneous reaction direction for the standard copper half-cell is

$$Cu^{2+} + 2e \rightleftharpoons Cu$$

The reaction direction at this time for the standard hydrogen electrode is

$$H_2 - 2e \rightleftharpoons 2H^+$$

*That is, without significant current flow, in order to prevent compositional changes in the solutions over the period of potential measurement.

The potentiometer is now adjusted until no deflection occurs when the galvanometer switch is closed momentarily. At this point the potential of the potentiometer exactly balances the potential difference between the copper half-cell and the hydrogen half-cell. The potentiometer potential required to secure this balance is read and is equal to

$$E^\circ_{Cu^{2+}/Cu} + E^\circ_{H_2/2H^+} = E_{measured}$$

from which we have

$$E^\circ_{Cu^{2+}/Cu} = E_{measured} - E^\circ_{H_2/2H^+}$$

Since $E^\circ_{H_2/2H^+} = 0.000$ V, we have determined directly as the value of $E_{measured}$ the standard reduction potential of the copper half-cell. What we have actually determined is the reduction potential of the copper half-cell when $[Cu^{2+}] = 1\ M$, but this will very closely approximate the standard reduction potential which involves $\alpha_{Cu^{2+}} = 1$. This value will be found to be 0.340 V at 25°C. It is apparent that the value of $E^\circ_{Cu/Cu^{2+}}$ will be -0.340 V, indicating that the oxidation reaction

$$Cu - 2e \rightleftharpoons Cu^{2+}$$

is not spontaneous for $[Cu^{2+}] = 1\ M$.

In an identical manner, wherever side reactions of the electrode with the solution do not occur, the standard reduction potentials for other oxidation-reduction half-cell systems may be developed. A table of standard reduction potentials may be found in the Appendixes.

4.17 HALF-CELL COMBINATIONS AS GALVANIC CELLS

Consider an example of a combination of half-cells to produce a cell that is sometimes called a galvanic cell. Such a combination is shown in Fig. 4.6. The half-cell combination of zinc-zinc(II) and copper-copper(II) is involved, with the solutions linked by a salt or conductive bridge and the electrodes connected externally through a potentiometer. The concentration values for Cu^{2+} and Zn^{2+} are, respectively,

$$[Cu^{2+}] = 10^{-1}\ M \qquad \text{and} \qquad [Zn^{2+}] = 10^{-2}\ M$$

The reduction potential for the copper half-cell system is given by

$$E_{Cu^{2+}/Cu} = E^\circ_{Cu^{2+}/Cu} + \frac{0.059}{2} \log \frac{[Cu^{2+}]}{[Cu]}$$

$$= 0.340 + \frac{0.059}{2} \log 10^{-1}$$

$$= 0.310\ V \qquad \text{at } 25°C$$

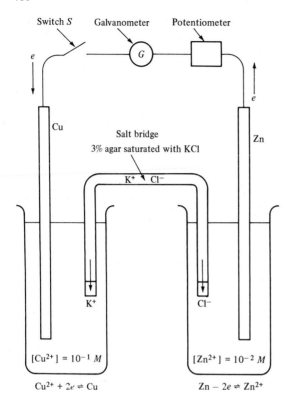

Fig. 4.6 Measurement of the cell emf for the galvanic cell consisting of combined copper and zinc half-cells.

Similarily, for the zinc half-cell system in reduction we have

$$E_{Zn^{2+}/Zn} = -0.763 + \frac{0.059}{2} \log 10^{-2}$$

$$= -0.82^2 V \qquad \text{at } 25°C$$

It is apparent that the copper half-cell has the higher reduction potential, so that the interlinked half-cells will result in the reactions

$$Cu^{2+} + 2e \rightleftharpoons Cu$$

$$Zn - 2e \rightleftharpoons Zn^{2+}$$

and an overall cell reaction of

$$Zn + Cu^{2+} \rightleftharpoons Zn^{2+} + Cu$$

and that a voltage between the electrodes is established at

$$E_{\text{measured}} = E_{Cu^{2+}/Cu} + E_{Zn/Zn^{2+}}$$

which is usually written

$$E_{\text{measured}} = E_{\text{Cu}^{2+}/\text{Cu}} - E_{\text{Zn}^{2+}/\text{Zn}}$$

and yields

$$E_{\text{measured}} = 0.310 - (-0.82^2)$$
$$= 1.13^2 \text{V} \qquad \text{at } 25°\text{C}$$

Although, in the potentiometric measurement device shown in Fig. 4.6, additional sources of potential, such as those involving the salt bridge and solution junctions, might have to be considered, the measurement system would show

1. Electron flow externally from the zinc to the copper electrode
2. The zinc electrode as the negative pole and the copper electrode as the positive pole
3. The half-cell reactions as

$$\text{Cu}^{2+} + 2e \rightleftharpoons \text{Cu} \qquad \text{Zn} - 2e \rightleftharpoons \text{Zn}^{2+}$$

4. The voltage or emf between the electrodes as 1.13^2V at 25°C

Such a galvanic cell will continue for a time interval to do electrical work external to the system, with a decrease in $[\text{Cu}^{2+}]$ and an increase in $[\text{Zn}^{2+}]$. Obviously the value of $E_{\text{Zn}^{2+}/\text{Zn}}$ must increase, while that of $E_{\text{Cu}^{2+}/\text{Cu}}$ must decrease. At the point where we have

$$E_{\text{Cu}^{2+}/\text{Cu}} = E_{\text{Zn}^{2+}/\text{Zn}} \tag{131}$$

the cell ceases to be capable, at 25°C, of generating a potential difference and of doing further electrical work external to the system. We have, at this point, an equilibrium established where, from Eq. (131),

$$E°_{\text{Cu}^{2+}/\text{Cu}} + \frac{0.059}{2} \log [\text{Cu}^{2+}]_{\text{eq}} = E°_{\text{Zn}^{2+}/\text{Zn}} + \frac{0.059}{2} \log [\text{Zn}^{2+}]_{\text{eq}} \tag{132}$$

and

$$\frac{0.059}{2} \log \frac{[\text{Zn}^{2+}]_{\text{eq}}}{[\text{Cu}^{2+}]_{\text{eq}}} = E°_{\text{Cu}^{2+}/\text{Cu}} - E°_{\text{Zn}^{2+}/\text{Zn}}$$
$$= 0.340 - (-0.763)$$
$$= 1.103 \text{ V}$$

from which

$$\frac{[\text{Zn}^{2+}]_{\text{eq}}}{[\text{Cu}^{2+}]_{\text{eq}}} = 10^{37.4} = K_{\text{eq}} \qquad \text{at } 25°\text{C} \tag{133}$$

where the value of $10^{37.4}$ represents the equilibrium constant at 25°C for the reaction

$$\text{Zn} + \text{Cu}^{2+} = \text{Zn}^{2+} + \text{Cu}$$

and is properly written

$$\frac{[Zn^{2+}]_{eq}[Cu]_{eq}}{[Zn]_{eq}[Cu^{2+}]_{eq}} = 10^{37.4} = K_{eq} \qquad \text{at } 25°C$$

Note the general reaction of the simple oxidation-reduction type:

$$aOx_2 + n_2e \rightleftharpoons bRed_2 \tag{134}$$

$$cRed_1 - n_1e \rightleftharpoons dOx_1 \tag{135}$$

with an overall reaction of

$$n_2cRed_1 + n_1aOx_2 \rightleftharpoons n_2dOx_1 + n_1bRed_2 \tag{136}$$

This reaction can be treated so as to yield equations applicable in all specific instances of the reaction type.

Here we have

$$E_{Ox_1/Red_1} = E°_{Ox_1/Red_1} + \frac{0.059}{n_1} \log \frac{[Ox_1]^d}{[Red_1]^c}$$

and

$$E_{Ox_2/Red_2} = E°_{Ox_2/Red_2} + \frac{0.059}{n_2} \log \frac{[Ox_2]^a}{[Red_2]^b}$$

Where E_{Ox_2/Red_2} is obviously greater than E_{Ox_1/Red_1}, the cell voltage or emf is given by

$$E_{cell} = E_{Ox_2/Red_2} - E_{Ox_1/Red_1}$$

$$= E°_{Ox_2/Red_2} - E°_{Ox_1/Red_1} + \frac{0.059}{n_1 n_2} \log \frac{[Ox_2]^{n_1 a}[Red_1]^{n_2 c}}{[Red_2]^{n_1 b}[Ox_1]^{n_2 d}} \tag{137}$$

In the special case where $n_1 = n_2$ we have, for the logarithmic term and its coefficient,

$$+\frac{0.059}{n_1} \log \frac{[Ox_2]^a[Red_1]^c}{[Red_2]^b[Ox_1]^d} \tag{138}$$

At equilibrium we have, in general,

$$E_{Ox_2/Red_2} = E_{Ox_1/Red_1}$$

and

$$E°_{Ox_2/Red_2} + \frac{0.059}{n_2} \log \frac{[Ox_2]^a}{[Red_2]^b} = E°_{Ox_1/Red_1} + \frac{0.059}{n_1} \log \frac{[Ox_1]^d}{[Red_1]^c}$$

from which

$$\frac{0.059}{n_1 n_2} \log \frac{[Ox_1]^{n_2 d}[Red_2]^{n_1 b}}{[Red_1]^{n_2 c}[Ox_2]^{n_1 a}} = E°_{Ox_2/Red_2} - E°_{Ox_1/Red_1}$$

and the equilibrium constant given by

$$\frac{[Ox_1]^{n_2 d}[Red_2]^{n_1 b}}{[Red_1]^{n_2 c}[Ox_2]^{n_1 a}} = {}_{10}\left(\frac{E^\circ_{Ox_2/Red_2} - E^\circ_{Ox_1/Red_1}}{0.059/n_1 n_2}\right) \tag{139}$$

Again, where $n_1 = n_2$, we have

$$\frac{[Ox_1]^{d}[Red_2]^{b}}{[Red_1]^{c}[Ox_2]^{a}} = {}_{10}\left(\frac{E^\circ_{Ox_2/Red_2} - E^\circ_{Ox_1/Red_1}}{0.059/n_1}\right) \tag{140}$$

With the use of the value of 0.059, and with E° values for 25°C, the above equations will yield the cell voltage or emf and the equilibrium constant at 25°C.

If we apply these general equations to galvanic cells, we can determine the cell emf and the equilibrium constant for the overall reaction associated with each case.

It is important to note that in each case given in the foregoing, including the general case, the use of a single solution—for example, copper and zinc electrodes immersed in a solution with $[Cu^{2+}] = 10^{-1}\,M$ and $[Zn^{2+}] = 10^{-2}\,M$—would result in the identical values being secured for the cell emf, electrode polarity, and equilibrium constant. In addition to this, it should be remembered that in the case where solid zinc metal is immersed in a solution of copper(II) at 25°C, for example, the overall reaction will be identical to the overall cell reaction; and where sufficient zinc exists to allow the reactions to go to equilibrium, the same equilibrium-constant value will govern the ratio of $[Zn^{2+}]/[Cu^{2+}]$ at equilibrium.

Example The following represents an example of the application to half-cell combinations of the general equations just developed.

For combination of silver and copper half-cells at 25°C, where $[Ag^+] = 10^{-2}\,M$ and $[Cu^{2+}] = 10^{-1}\,M$,

$$E_{Ag^+/Ag} = 0.800 + 0.059 \log 10^{-2}\,M = 0.68^2\,V$$

$$E_{Cu^{2+}/Cu} = 0.340 + \frac{0.059}{2} \log 10^{-1}\,M = 0.310\,V$$

so that the half-cell and overall cell reactions are

$$Ag^+ + 1e \rightleftharpoons Ag$$
$$Cu - 2e \rightleftharpoons Cu^{2+}$$
$$Cu + 2Ag^+ \rightleftharpoons Cu^{2+} + 2Ag$$

This sets up values for the general equations of

$Ox_2 = Ag^+$	$Red_2 = Ag$	$n_2 = 1$	$a = 1$	$b = 1$
$Ox_1 = Cu^{2+}$	$Red_1 = Cu$	$n_1 = 2$	$c = 1$	$d = 1$

The cell emf, E_{cell}, is given by Eq. (137) as

$$E_{cell} = 0.800 - 0.340 + \frac{0.059}{2} \log \frac{[Ag^+]^2[Cu]}{[Ag]^2[Cu^{2+}]}$$
$$= 0.37^2\,V \quad \text{at } 25°C$$

The equilibrium constant is given by Eq. (139) as

$$\frac{[Cu^{2+}][Ag]^2}{[Cu][Ag^+]^2} = {}_{10}\left(\frac{0.800 - 0.340}{0.059/2}\right) = 10^{15.6} = K_{eq} \quad \text{at } 25°C$$

4.18 EXTENSION OF HALF-CELL AND CELL CONDITIONS TO MORE GENERAL OXIDATION-REDUCTION REACTIONS

A general oxidation-reduction reaction was indicated in Eq. (136) as

$$n_2 c \, \text{Red}_1 + n_1 a \text{Ox}_2 \rightleftharpoons n_2 d \text{Ox}_1 + n_1 b \, \text{Red}_2$$

and the equilibrium constant for this reaction was given by either Eq. (139) or (140). Suppose that the general reaction above represented a titration-reaction situation, with Red_1 being titrated by Ox_2. It is apparent, from Eq. (136), that at any point in such a titration, after the start, we have

$$[\text{Red}_2] = \frac{n_1 b}{n_2 d}[\text{Ox}_1] \tag{141}$$

At the equivalence point *only* we have, in addition to this equality,

$$[\text{Ox}_2] = \frac{n_1 a}{n_2 c}[\text{Red}_1] \tag{142}$$

Under these conditions, and for the titration equivalence point *only*, Eq. (139) becomes

$$\frac{[\text{Ox}_1]^{n_2 d}\{(n_1 b/n_2 d)[\text{Ox}_1]\}^{n_1 b}}{[\text{Red}_1]^{n_2 c}\{(n_1 a/n_2 c)[\text{Red}_1]\}^{n_1 a}} = {}_{10}\left(\frac{E°_{\text{Ox}_2/\text{Red}_2} - E°_{\text{Ox}_1/\text{Red}_1}}{0.059/n_1 n_2}\right)$$

and

$$\frac{(n_1 b/n_2 d)^{n_1 b}[\text{Ox}_1]^{(n_2 d + n_1 b)}}{(n_1 a/n_2 c)^{n_1 a}[\text{Red}_1]^{(n_2 c + n_1 a)}} = {}_{10}\left(\frac{E°_{\text{Ox}_2/\text{Red}_2} - E°_{\text{Ox}_1/\text{Red}_1}}{0.059/n_1 n_2}\right)$$

with

$$\frac{[\text{Ox}_1]^{(n_2 d + n_1 b)}}{[\text{Red}_1]^{(n_2 c + n_1 a)}} = \frac{(n_1 a/n_2 c)^{n_1 a}}{(n_1 b/n_2 d)^{n_1 b}} {}_{10}\left(\frac{E°_{\text{Ox}_2}/\text{Red}_2 - E°_{\text{Ox}_1}/\text{Red}_1}{0.059/n_1 n_2}\right) \tag{143}$$

When $n_1 = n_2$, we have

$$\frac{[\text{Ox}_1]^{(d+b)}}{[\text{Red}_1]^{(c+a)}} = \frac{(a/c)^a}{(b/d)^b} {}_{10}\left(\frac{E°_{\text{Ox}_2/\text{Red}_2} - E°_{\text{Ox}_1/\text{Red}_1}}{0.059/n_1}\right) \tag{144}$$

From the appropriate equation, the degree of quantitative completion of the overall reaction at the equivalence point of the titration can be determined.

The cell or solution potential at any point in the titration, after the start, will also be of interest. For such a titration situation, where high reaction velocities are a normal requirement, an immediate equilibrium is established at any stage in the titration, unlike the situation that applies in the case of galvanic cell reactions. At any stage in the titration after the start, therefore, the cell or solution potential will be given by

$$E_{cell} = E_{Ox_2/Red_2} = E_{Ox_1/Red_1}$$

with

$$E_{Ox_2/Red_2} = E^{\circ}_{Ox_2/Red_2} + \frac{0.059}{n_2} \log \frac{[Ox_2]^a}{[Red_2]^b}$$

$$E_{Ox_1/Red_1} = E^{\circ}_{Ox_1/Red_1} + \frac{0.059}{n_1} \log \frac{[Ox_1]^d}{[Red_1]^c}$$

from which

$$n_2 E_{Ox_2/Red_2} = n_2 E^{\circ}_{Ox_2/Red_2} + 0.059 \log \frac{[Ox_2]^a}{[Red_2]^b} \tag{145}$$

$$n_1 E_{Ox_1/Red_1} = n_1 E^{\circ}_{Ox_1/Red_1} + 0.059 \log \frac{[Ox_1]^d}{[Red_1]^c} \tag{146}$$

and we have

$$(n_1 + n_2)E_{cell} = n_2 E_{Ox2/Red2} + n_1 E_{Ox1/Red1}$$

$$= n_2 E^{\circ}_{Ox2/Red2} + n_1 E^{\circ}_{Ox1/Red1} + 0.059 \log \frac{[Ox_2]^a [Ox_1]^d}{[Red_2]^b [Red_1]^c} \tag{147}$$

Solution of this equation for E_{cell} will yield the solution potential for any point in the titration after the start.

At the equivalence point *only*, there is a special relationship, given by Eqs. (141) and (142) as

$$\frac{[Ox_2]^a}{[Red_2]^b} = \frac{\{(n_1 a/n_2 c)[Red_1]\}^a}{\{(n_1 b/n_2 d)[Ox_1]\}^b}$$

Substitution of this ratio for $[Ox_2]^a/[Red_2]^b$, or for $[Ox_1]^d/[Red_1]^c$ determined in a similar manner, into Eq. (147) will lead to the value of E_{cell} at the equivalence point of the titration.

The general equations developed in this section can now be applied to ox-idation-reduction titration systems having reactions of the type illustrated by Eq. (136).

Example   For the titration of Fe^{2+} in aqueous solution at 25°C by the addition of Ce^{4+} solution, where the half-cell and overall cell reactions are

$$Fe^{2+} - 1e \rightleftharpoons Fe^{3+}$$

$$Ce^{4+} + 1e \rightleftharpoons Ce^{3+}$$

$$Fe^{2+} + Ce^{4+} \rightleftharpoons Fe^{3+} + Ce^{3+}$$

and where the $[Fe^{3+}]$ at the equivalence point is approximately 0.05 M, we have

$Ox_2 = Ce^{4+}$ $Red_2 = Ce^{3+}$ $n_2 = 1$ $a = 1$ $b = 1$

$Ox_1 = Fe^{3+}$ $Red_1 = Fe^{2+}$ $n_1 = 1$ $c = 1$ $d = 1$

The equilibrium constant is given by Eq. (140) as

$$\frac{[Fe^{3+}][Ce^{3+}]}{[Fe^{2+}][Ce^{4+}]} = {}_{10}\left(\frac{1.459 - 0.770}{0.059}\right) = 10^{11.7} = K_{eq} \qquad \text{at } 25°C$$

The degree of quantitative completion of the reaction at the equivalence point is given by Eq. (144) as

$$\frac{[Fe^{3+}]^2}{[Fe^{2+}]^2} = 10^{11.7} \qquad \text{and} \qquad \frac{[Fe^{3+}]}{[Fe^{2+}]} \approx 10^{5.8}$$

therefore, as an approximation,

$$[Fe^{2+}] \approx \frac{5 \times 10^{-2}}{10^{5.8}} \approx 7.9 \times 10^{-8} M$$

Since the value of $[Fe^{2+}]$ at the equivalence point is less than $10^{-6} M$, this may be taken as indicative of a quantitatively complete reaction at the equivalence point.

The cell or solution potential at the equivalence point is given by Eq. (147) as

$$2E_{cell,eq\ pt} = 1.459 + 0.770 + 0.059 \log \frac{[Ce^{4+}][Fe^{3+}]}{[Ce^{3+}][Fe^{2+}]}$$

but, at the equivalence point *only*, from Eqs. (141) and (142), we have

$$\frac{[Ce^{4+}]}{[Ce^{3+}]} = \frac{[Fe^{2+}]}{[Fe^{3+}]}$$

We have, therefore,

$$E_{cell,eq\ pt} = \frac{1.459 + 0.770}{2} = 1.115 \text{ V} \qquad \text{at } 25°C$$

4.19 REFERENCE ELECTRODE SYSTEMS

4.19.1 GENERAL

It has already been indicated that the potential values for oxidation-reduction systems are generally referred to the standard hydrogen electrode, or SHE. Because of difficulties associated with the maintenance of standard conditions for this electrode, as well as the dangers inherent in the use of free-flowing hydrogen gas, most measurements of potential for half-cells and cells are made against reference electrodes other than the SHE. The essential characteristics for the proper functioning of such electrodes are as follows.

1. The potential of the electrode relative to the SHE must be accurately known.
2. The potential of the electrode must not change, except with changes in temperature, whether the electrode is in use or not.

Although many such electrode systems exist, only two will be discussed here.

4.19.2 THE CALOMEL ELECTRODE

This reference electrode system consists of mercury, the slightly soluble salt mercurous chloride, Hg_2Cl_2, commonly called calomel, and a solution of potassium chloride, KCl. Figure 4.7 shows a typical calomel electrode system. The potential developed by this half-cell system will vary according to the concentration of KCl. Our discussion here will center around a calomel electrode with a saturated KCl solution, this being referred to as the *saturated KCl calomel electrode*, or *SCE*.

The half-cell reaction involves the system

$$Hg_2^{2+} + 2e \rightleftharpoons 2Hg$$

which can also be written

$$Hg^+ + 1e \rightleftharpoons Hg$$

The Nernst equation applied to this reaction yields

$$E_{Hg^+/Hg} = E^\circ_{Hg^+/Hg} + 0.059 \log [Hg^+]$$
$$= 0.796 + 0.059 \log [Hg^+]$$

The saturation value for KCl in water at 25°C is about 3.5 M, so that $[Cl^-]$ in solution will be 3.5 M. The solubility product constant for Hg_2Cl_2 at 25°C is 1.3×10^{-18}, so that

$$[Hg^+]^2 [Cl^-]^2 = K_{sp}(Hg_2Cl_2)$$
$$= 1.3 \times 10^{-18}$$

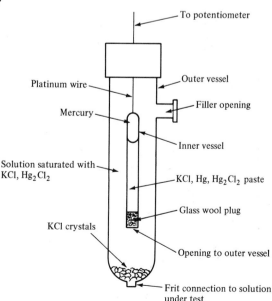

Fig. 4.7   Saturated KCl calomel electrode.

In a saturated solution of KCl at 25°C, the $[Hg^+]$ value will be

$$[Hg^+] = \sqrt{\frac{1.3 \times 10^{-18}}{(3.5)^2}}$$

$$= 3.3 \times 10^{-10} \, M$$

Substitution of this value for $[Hg^+]$ into the Nernst equation above yields

$$E_{Hg^+/Hg} = 0.796 + 0.059 \log 3.3 \times 10^{-10}$$

$$= 0.796 - 0.560$$

$$= 0.236 \, V \qquad \text{at } 25°C$$

The best measured value for the SCE, without the addition of the junction potential correction, is 0.2412 V at 25°C. The difference between the calculated and the measured value reflects the various approximations used in the former, such as molarities for activities, less than exact values for the molar solubility of KCl, the $K_{sp}(Hg_2Cl_2)$, etc.

Observe that the concentration of Cl^- from KCl controls the $[Hg^+]$ value, and that the electrode maintains a constant potential, at constant temperature, regardless of the reaction direction for

$$Hg_2^{2+} + 2e \rightleftharpoons 2Hg$$

during electrode use. Table 4.5 shows the potential values for different KCl concentration calomel electrodes, all at 25°C. The advantage of the saturated KCl electrode lies in the fact that with solid KCl present, the solution concentration of KCl is easily maintained. Its disadvantage is associated with its relatively high rate of change of potential with change in temperature, that is, its $\Delta E/\Delta T$ compared with that of either the 1 or the 0.1 M KCl electrode. It does not, in other words, have the temperature stability of potential of these latter electrodes.

Table 4.5   KCl calomel reference electrode potentials at 25°C

KCl	E, volts	E_j, volts*	$E + E_j$, volts
Saturated	0.2412	0.0033	0.2445
1 M	0.2801	0.0029	0.2830
0.1 M	0.3337	0.0019	0.3356

*The value of E_j refers to the potential developed at the junction or interface of the electrode and the solution in which it is immersed. This will vary with the nature of the solution (and depending on whether or not a salt bridge is used). The values shown are representative of those for the dilute solutions normally encountered in analytical work.

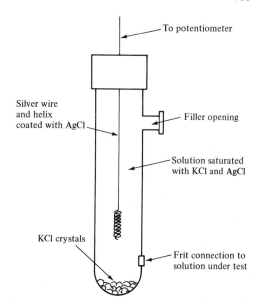

To potentiometer

Silver wire
and helix
coated with AgCl

Filler opening

Solution saturated
with KCl and AgCl

KCl crystals

Frit connection to
solution under test

Fig. 4.8   Saturated KCl Ag/AgCl electrode.

4.19.3   THE SILVER–SILVER CHLORIDE ELECTRODE

This system consists of silver, the slightly soluble salt silver chloride, AgCl, and a solution of potassium chloride, KCl. Figure 4.8 illustrates a typical silver–silver chloride electrode. Since nearly all such electrodes involve a saturated solution of KCl, and are referred to as *saturated KCl Ag/AgCl electrodes*, we shall confine our discussion to this particular electrode system. By a treatment similar to that given for the saturated KCl calomel electrode, we can show that this Ag/AgCl electrode yields

$$E_{Ag^+/Ag} = 0.189 \text{ V} \qquad \text{at } 25°C$$

The measured value for the saturated KCl silver–silver chloride electrode is 0.1956 V at 25°C, with a value of 0.1989 V as the value of $E + E_j$. Again, the discrepancy between the calculated and the measured potential stems from the approximate nature of the data used in the calculations.

When using the SCE as a reference electrode in the measurement of the potential of a copper-copper(II) half-cell at 25°C, with $[Cu^{2+}] = 10^{-1} M$, we would note that $E_{Cu^{2+}/Cu} = 0.310$ V and $E_{SCE, reduction} = 0.245$ V, the latter value being the value of $E + E_j$ for the reference electrode. This indicates a stronger reduction tendency for the copper system, therefore we would have

$$E_{measured} = 0.310 - E_{SCE, reduction}$$
$$= 0.310 - 0.245$$
$$= 0.065 \text{ V} \qquad \text{at } 25°C$$

In the measurement of the potential of a zinc-zinc(II) half-cell at 25°C, with

$[Zn^{2+}] = 10^{-2} M$, we would find the $E_{Zn^{2+}/Zn} = -0.82^2$ V and, of course, $E_{SCE,reduction} = 0.245$ V. Here we have a stronger reduction tendency for the SCE electrode, and we thus have

$$E_{measured} = E_{SCE,reduction} - (-0.82^2)$$
$$= 0.245 + 0.82^2$$
$$= 1.06^7 \text{ V} \qquad \text{at } 25°C$$

Since the copper system was in reduction, linked with the SCE in oxidation, we have determined the reduction potential of the copper system relative to SCE as 0.065 V at 25°C. Since the zinc system was in oxidation, linked with the SCE in reduction, we have determined the oxidation half-cell emf of the zinc system (the IUPAC convention reserves the term "potential" for use in describing the half-cell emf in reduction), relative to SCE as 1.06^7 V at 25°C. The half-cell emf of the copper system in oxidation will, of course, be -0.065 V relative to SCE; the reduction potential of the zinc system relative to SCE will be -1.06^7 V. Note that oxidation for the copper system as given, and reduction for the zinc system as given, are nonspontaneous reactions.

The reduction potential for any half-cell or cell, measured relative to SCE, can be converted to the reduction potential relative to SHE by using the following equation:

$$E \text{ vs. SHE} = E \text{ vs. SCE} + E_{SCE,reduction}$$
$$= E \text{ vs. SCE} + 0.245 \text{ V} \qquad \text{at } 25°C$$

For the copper and zinc systems just discussed, this yields the following.

For the copper system: E vs. SHE $= 0.065 + 0.245 = 0.310$ V   at 25°C
For the zinc system: E vs. SHE $= -1.06^7 + 0.245 = -0.82^2$ V   at 25°C

Similarly, for the conversion of reduction potentials for half-cells or cells relative to the saturated KCl Ag/AgCl electrode, to reduction potentials relative to SHE, we have

$$E \text{ vs. SHE} = E \text{ vs. Ag/AgCl}_{sat\ KCl} + E_{Ag/AgCl,reduction}$$
$$= E \text{ vs. Ag/AgCl}_{sat\ KCl} + 0.199 \text{ V} \qquad \text{at } 25°C$$

The values used for E_{SCE} and $E_{Ag/AgCl}$ are the $E + E_j$ values at 25°C. For the conversion of reduction potentials for half-cells or cells relative to Ag/AgCl$_{sat\ KCl}$ electrode, to reduction potentials relative to SCE, we have

$$E \text{ vs. SCE} = E \text{ vs. Ag/AgCl}_{sat\ KCl} - (E_{SCE} - E_{Ag/AgCl})$$

the E_{SCE} and $E_{Ag/AgCl}$ values being the reduction potential values. This gives

$$E \text{ vs. SCE} = E \text{ vs. Ag/AgCl}_{sat\ KCl} - (0.245 - 0.199)$$
$$= E \text{ vs. Ag/AgCl}_{sat\ KCl} - 0.046 \text{ V} \qquad \text{at } 25°C$$

4.20 INDICATING, OR WORKING, ELECTRODE SYSTEMS

It is apparent that to measure any half-cell or solution potential, a two-electrode arrangement is required. Of the pair, one will be the reference electrode and the other the indicator or working electrode. This latter electrode must be capable of responding to the concentrations for the particular oxidation-reduction situation involved, by changing its potential to correspond to a changing ratio of [Ox]/[Red] in the solution system. We can, for example, use a silver electrode in a silver ion solution, where the silver electrode will assume a potential related through the Nernst equation to the [Ag^+] in the solution. Similarly, we can use a hydrogen electrode or, as we shall see later, a glass electrode, immersed in a solution of H_3O^+ to report a potential dictated by the value of [H_3O^+] in the solution. In all these instances, the relationship of indicator electrode potential to ion concentration or ion-concentration ratios, as given by the Nernst equation, will hold rigidly only where the measurement of potential is made under conditions that closely approximate a zero current flow in the measuring circuit.

Many oxidation-reduction solution situations involve different oxidation states for the same substance—for example, Fe^{2+} and Fe^{3+}, where the ratio [Fe^{3+}]/[Fe^{2+}] changes, as it would during a titration of iron(II) by an appropriate oxidant. In such cases, an inert electrode can be used to report the changing potential implied by the changing ratio of [Ox]/[Red]. A case in point here is the use of a platinum metal electrode under these circumstances. Such an electrode reports the changing potential associated with the changing ratio [Ox]/[Red], without itself being involved in any reaction. Platinum is not easily attacked chemically by most solutions and, for this reason, such platinum electrodes are called "inert" electrodes.

4.21 FACTORS INFLUENCING ELECTRODE POTENTIALS

Several factors are capable of influencing the potential values measured for half-cells and cells. Some of the more important of these factors are outlined in the following.

4.21.1 THE EFFECT OF THE CONCENTRATION OF THE ACTIVE ION OR IONS

The effect of the concentration of the active ion or ions has already been generally treated, and we have noted that for

$$E_{Ox/Red} = E^\circ_{Ox/Red} + \frac{0.059}{n} \log \frac{[Ox]}{[Red]}$$

an increase in the value of [Ox]/[Red] results in an increase in the value of $E_{Ox/Red}$. In general, where an oxidation-reduction titration system involves the reaction

$$Red - ne \rightleftharpoons Ox$$

this means an increase in $E_{Ox/Red}$ during the titration; and for the titration reaction

$$Ox + ne \rightleftharpoons Red$$

this implies a decrease in $E_{Ox/Red}$ as the titration proceeds.

4.21.2 THE EFFECT OF DIVERSE-ION CONCENTRATION

By diverse ions we mean any ion in solution other than those directly involved in the oxidation-reduction reaction. From the Nernst equation in the form

$$E_{Ox/Red} = E_{Ox/Red}^{\circ} + \frac{0.059}{n} \log \frac{\alpha_{Ox}}{\alpha_{Red}}$$

where activities, instead of molarities, are properly used it is apparent that the presence of any diverse ion in significant concentration will result in a decrease in the activities of the Ox and Red states, where each state represents an ion species. Note that

$$\alpha_{Ox} = [Ox]\gamma_{Ox}$$
$$\alpha_{Red} = [Red]\gamma_{Red}$$

and that a decrease in each activity implies a decrease in the respective activity coefficient. The Nernst equation above can be written

$$
\begin{aligned}
E_{Ox/Red} &= E_{Ox/Red}^{\circ} + \frac{0.059}{n} \log \frac{[Ox]\gamma_{Ox}}{[Red]\gamma_{Red}} \\
&= E_{Ox/Red}^{\circ} + \frac{0.059}{n} \log \frac{\gamma_{Ox}}{\gamma_{Red}} + \frac{0.059}{n} \log \frac{[Ox]}{[Red]} \\
&= E_{f,Ox/Red}^{\circ} + \frac{0.059}{n} \log \frac{[Ox]}{[Red]}
\end{aligned}
$$

where $E_{f,Ox/Red}^{\circ}$ is called the formal reduction potential of the half-cell, varying according to the diverse-ion concentration or, totally, as the ionic strength of the solution. Thus the presence of diverse ions can cause a shift in the value of $E_{f,Ox/Red}^{\circ}$ and consequently a shift in the value of $E_{Ox/Red}$. Since most applications of oxidation-reduction reaction systems in volumetric analysis involve titration conditions, where the variation of $E_{Ox/Red}$ during the titration with additions of oxidant or reductant is used to locate the titration equivalence point; and since the variations of the ionic strength of the solution during the titration are usually small enough to be ignored, such variations in $E_{f,Ox/Red}^{\circ}$ are of comparatively little importance in volumetric techniques. Where potentiometric measurements are made, from which direct concentration calculations result through the Nernst equation, such changes in the value of $E_{f,\ Ox/Red}^{\circ}$ with changes in the ionic strength of the solutions tested are of importance and must be given due consideration. This situation will be discussed in greater detail in the section on potentiometry and potentiometric titrations (Sec. 12.2).

It is to be noted that variations in solution $[H_3O^+]$ represent ionic strength variations and can result in altered values for $E^\circ_{f,\text{Ox/Red}}$. Thus, the value of $E^\circ_{f,\text{Fe}^{3+}/\text{Fe}^{2+}}$ is 0.700 V in 1 M HCl, 0.680 V in 1 M H$_2$SO$_4$, and 0.730 V in 1 M HClO$_4$.

4.21.3 THE EFFECT OF pH

The general effect of $[\text{H}_3\text{O}^+]$ on $E^\circ_{f,\text{Ox/Red}}$ has been discussed in the foregoing section, but it must be emphasized at this point that such effects refer strictly to the case where $[\text{H}_3\text{O}^+]$ acts only within the scope of the diverse-ion effect. For many oxidation-reduction systems, H$_3$O$^+$ actually enters as part of the reaction system. Where this occurs, the effect of $[\text{H}_3\text{O}^+]$ on the reaction characteristics is far more profound.

The following general oxidation-reduction reaction of the pH-dependent type can be used to provide general equations for critical values relative to such systems:

$$a\text{Ox}_2 + y\text{H}_3\text{O}^+ + n_2 e \rightleftharpoons b\text{Red}_2 + w\text{H}_2\text{O}$$

$$c\,\text{Red}_1 - n_1 e \rightleftharpoons d\text{Ox}_1$$

with an overall reaction

$$n_2 c\,\text{Red}_1 + n_1 a\text{Ox}_2 + n_1 y\text{H}_3\text{O}^+ \rightleftharpoons n_2 d\text{Ox}_1 + n_1 b\,\text{Red}_2 + n_1 w\text{H}_2\text{O} \quad (148)$$

The equilibrium constant is found by noting that at equilibrium we have

$$E_{\text{Ox}_1/\text{Red}_1} = E_{\text{Ox}_2/\text{Red}_2}$$

therefore,

$$E^\circ_{\text{Ox}_1/\text{Red}_1} + \frac{0.059}{n_1} \log \frac{[\text{Ox}_1]^d}{[\text{Red}_1]^c} = E^\circ_{\text{Ox}_2/\text{Red}_2} + \frac{0.059}{n_2} \log \frac{[\text{Ox}_2]^a [\text{H}_3\text{O}^+]^y}{[\text{Red}_2]^b [\text{H}_2\text{O}]^w}$$

and

$$\frac{0.059}{n_1 n_2} \log \frac{[\text{Red}_2]^{n_1 b} [\text{Ox}_1]^{n_2 d} [\text{H}_2\text{O}]^{n_1 w}}{[\text{Ox}_2]^{n_1 a} [\text{Red}_1]^{n_2 c} [\text{H}_3\text{O}^+]^{n_1 y}} = E^\circ_{\text{Ox}_2/\text{Red}_2} - E^\circ_{\text{Ox}_1/\text{Red}_1}$$

The equilibrium constant is given by

$$\frac{[\text{Red}_2]^{n_1 b} [\text{Ox}_1]^{n_2 d} [\text{H}_2\text{O}]^{n_1 w}}{[\text{Ox}_2]^{n_1 a} [\text{Red}_1]^{n_2 c} [\text{H}_3\text{O}^+]^{n_1 y}} = 10 \left(\frac{E^\circ_{\text{Ox}_2/\text{Red}_2} - E^\circ_{\text{Ox}_1/\text{Red}_1}}{0.059/n_1 n_2} \right) \quad (149)$$

Where $n_1 = n_2$, we have

$$\frac{[\text{Red}_2]^b [\text{Ox}_1]^d [\text{H}_2\text{O}]^w}{[\text{Ox}_2]^a [\text{Red}_1]^c [\text{H}_3\text{O}^+]^y} = 10 \left(\frac{E^\circ_{\text{Ox}_2/\text{Red}_2} - E^\circ_{\text{Ox}_1/\text{Red}_1}}{0.059/n_1} \right) \quad (150)$$

From Eq. (149), for normally dilute solutions, we have

$$\frac{[\text{Red}_2]^{n_1 b} [\text{Ox}_1]^{n_2 d}}{[\text{Ox}_2]^{n_1 a} [\text{Red}_1]^{n_2 c}} = 10 \left(\frac{E^\circ_{\text{Ox}_2/\text{Red}_2} - E^\circ_{\text{Ox}_1/\text{Red}_1}}{0.059/n_1 n_2} \right) \times [\text{H}_3\text{O}^+]^{n_1 y} \quad (151)$$

At any point in a titration involving this system, after the start, we have

$$[\text{Red}_2] = \frac{n_1 b}{n_2 d}[\text{Ox}_1] \tag{152}$$

At the equivalence point *only* we have, in addition to this equality,

$$[\text{Ox}_2] = \frac{n_1 a}{n_2 c}[\text{Red}_1] \tag{153}$$

for which Eq. (151) yields, for the equivalence point *only*,

$$\frac{[\text{Ox}_1]^{n_2 d}\{(n_1 b/n_2 d)[\text{Ox}_1]\}^{n_1 b}}{[\text{Red}_1]^{n_2 c}\{(n_1 a/n_2 c)[\text{Red}_1]\}^{n_1 a}} = 10^{\left(\dfrac{E^{\circ}_{\text{Ox}_2/\text{Red}_2} - E^{\circ}_{\text{Ox}_1/\text{Red}_1}}{0.059/n_1 n_2}\right)} \times [\text{H}_3\text{O}^+]^{n_1 y}$$

and

$$\frac{(n_1 b/n_2 d)^{n_1 b}[\text{Ox}_1]^{(n_2 d + n_1 b)}}{(n_1 a/n_2 c)^{n_1 a}[\text{Red}_1]^{(n_2 c + n_1 a)}} = 10^{\left(\dfrac{E^{\circ}_{\text{Ox}_2/\text{Red}_2} - E^{\circ}_{\text{Ox}_1/\text{Red}_1}}{0.059/n_1 n_2}\right)} \times [\text{H}_3\text{O}^+]^{n_1 y} \tag{154}$$

From the above, we obtain

$$\frac{[\text{Ox}_1]^{(n_2 d + n_1 b)}}{[\text{Red}_1]^{(n_2 c + n_1 a)}} = \frac{(n_1 a/n_2 c)^{n_1 a}}{(n_1 b/n_2 d)^{n_1 b}} 10^{\left(\dfrac{E^{\circ}_{\text{Ox}_2/\text{Red}_2} - E^{\circ}_{\text{Ox}_1/\text{Red}_1}}{0.059/n_1 n_2}\right)} \times [\text{H}_3\text{O}^+]^{n_1 y} \tag{155}$$

Where $n_1 = n_2$, we have

$$\frac{[\text{Ox}_1]^{(d+b)}}{[\text{Red}_1]^{(c+a)}} = \frac{(a/c)^a}{(b/d)^b} 10^{\left(\dfrac{E^{\circ}_{\text{Ox}_2/\text{Red}_2} - E^{\circ}_{\text{Ox}_1/\text{Red}_1}}{0.059/n_1}\right)} \times [\text{H}_3\text{O}^+]^y \tag{156}$$

From either of these equations, as may be appropriate, the approximate degree of quantitative completion of the reaction at the equivalence point may be determined for a specific solution pH.

The solution or cell potential at any point in the titration after the start is given by

$$E_{\text{cell}} = E_{\text{Ox}_2/\text{Red}_2} = E_{\text{Ox}_1/\text{Red}_1}$$

with

$$E_{\text{Ox}_2/\text{Red}_2} = E^{\circ}_{\text{Ox}_2/\text{Red}_2} + \frac{0.059}{n_2} \log \frac{[\text{Ox}_2]^a [\text{H}_3\text{O}^+]^y}{[\text{Red}_2]^b [\text{H}_2\text{O}]^w}$$

$$n_2 E_{\text{Ox}_2/\text{Red}_2} = n_2 E^{\circ}_{\text{Ox}_2/\text{Red}_2} + 0.059 \log \frac{[\text{Ox}_2]^a [\text{H}_3\text{O}^+]^y}{[\text{Red}_2]^b [\text{H}_2\text{O}]^w}$$

$$E_{\text{Ox}_1/\text{Red}_1} = E^{\circ}_{\text{Ox}_1/\text{Red}_1} + \frac{0.059}{n_1} \log \frac{[\text{Ox}_1]^d}{[\text{Red}_1]^c}$$

$$n_1 E_{\text{Ox}_1/\text{Red}_1} = n_1 E^{\circ}_{\text{Ox}_1/\text{Red}_1} + 0.059 \log \frac{[\text{Ox}_1]^d}{[\text{Red}_1]^c}$$

Thus, after the start, we have the solution or cell potential during titration as

$$(n_1 + n_2)E_{cell} = n_2 E^\circ_{Ox_2/Red_2} + n_1 E^\circ_{Ox_1/Red_1}$$
$$+ 0.059 \log \frac{[Ox_2]^a [Ox_1]^d [H_3O^+]^y}{[Red_2]^b [Red_1]^c [H_2O]^w} \quad (157)$$

from which, for normally dilute solutions, we have

$$(n_1 + n_2)E_{cell} = n_2 E^\circ_{Ox_2/Red_2} + n_1 E^\circ_{Ox_1/Red_1} + 0.059 \log [H_3O^+]^y$$
$$+ 0.059 \log \frac{[Ox_2]^a [Ox_1]^d}{[Red_2]^b [Red_1]^c} \quad (158)$$

Substitution of the ratio relationship for $[Ox_2]^a/[Red_2]^b$ or $[Ox_1]^d/[Red_1]^c$, as derived from the equalities given by Eq. (152) or (153), into Eq. (158) will provide the solution or cell potential which is specific for the equivalence point of the titration.

Example   Consider the titration of a solution of Fe^{2+} by a solution of $Cr_2O_7^{2-}$ at 25°C. Determine

1. The equilibrium constant for the reaction.
2. The degree of completion of the reaction at the equivalence point of the titration where we have (1) $[H_3O^+] = 1\,M$ and (2) $[H_3O^+] = 10^{-4}\,M$. Suppose $[Fe^{3+}]$ at the equivalence point in each case to be given by 0.05 M.
3. The minimum $[H_3O^+]$ for a quantitative reaction at the equivalence point when $[Fe^{3+}] = 0.05\,M$.
4. The solution or cell potential at the equivalence point when (1) $[H_3O^+] = 1\,M$ and (2) $[H_3O^+] = 10^{-4}\,M$.
      The half-cell and overall cell reactions are

$$Cr_2O_7^{2-} + 14H_3O^+ + 6e \rightleftharpoons 2Cr^{3+} + 21H_2O$$
$$Fe^{2+} - 1e \rightleftharpoons Fe^{3+}$$
$$6Fe^{2+} + Cr_2O_7^{2-} + 14H_3O^+ \rightleftharpoons 6Fe^{3+} + 2Cr^{3+} + 21H_2O$$

and we have

$Ox_1 = Fe^{3+}$	$Red_1 = Fe^{2+}$	$n_1 = 1$	$a = 1$	$b = 2$	$y = 14$
$Ox_2 = Cr_2O_7^{2-}$	$Red_2 = Cr^{3+}$	$n_2 = 6$	$c = 1$	$d = 1$	$w = 21$

1. The equilibrium constant is given by Eq. (149) as

$$\frac{[Cr^{3+}]^2 [Fe^{3+}]^6 [H_2O]^{21}}{[Cr_2O_7^{2-}][Fe^{2+}]^6 [H_3O^+]^{14}} = 10\left(\frac{1.330 - 0.770}{0.059/6}\right) = 10^{56.9} \quad \text{at } 25°C$$

2. The degree of quantitative completion is given by Eq. (155) as

$$\frac{[Fe^{3+}]^8}{[Fe^{2+}]^7} = \frac{(1/6)}{(2/6)^2} 10^{56.9} [H_3O^+]^{14}$$
$$= (1.5 \times 10^{56.9}) [H_3O^+]^{14}$$
$$[Fe^{2+}] = \sqrt[7]{\frac{[Fe^{3+}]^8}{(1.5 \times 10^{56.9}) [H_3O^+]^{14}}}$$

For $[H_3O^+] = 1\ M$ and $[Fe^{3+}] = 0.05\ M$, we have

$$[Fe^{2+}] = 2.29 \times 10^{-10}\ M$$

This is very much less than $10^{-6}\ M$, and the titration at $1\ M\ [H_3O^+]$ will be quantitatively complete at the equivalence point.

For $[H_3O^+] = 10^{-4}\ M$ and $[Fe^{3+}] \approx 0.05\ M$, we have

$$[Fe^{2+}] \approx 2.29 \times 10^{-2}\ M$$

This is very much higher than $10^{-6}\ M$, and the titration at $10^{-4}\ M\ [H_3O^+]$ will not be quantitatively complete at the equivalence point.

3. The minimum $[H_3O^+]$ for a quantitative reaction at the equivalence point will be given by Eq. (155) as

$$\frac{(5 \times 10^{-2})^8}{(10^{-6}\ M_{max})^7} = (1.5 \times 10^{56.9})\,[H_3O^+]^{14}_{(min)}$$

$$[H_3O^+]_{(min)} = 1.51 \times 10^{-2}\ M$$

$$pH_{(max)} = 1.8$$

4. The solution potential at the equivalence point is given by Eq. (158) as

$$7E_{cell} = 6 \times 1.330 + 0.770 + 0.059\ \log\ [H_3O^+]^{14} + 0.059\ \log\ \frac{[Cr_2O_7^{2-}]\,[Fe^{3+}]}{[Cr^{3+}]^2\,[Fe^{2+}]}$$

but, at the equivalence point, we have the relation

$$\frac{[Fe^{3+}]}{[Fe^{2+}]} = \frac{[Cr^{3+}]}{2[Cr_2O_7^{2-}]}$$

as given by Eqs. (152) and (153). Thus, we have

$$7E_{cell} = 7.980 + 0.770 + 0.059\ \log\ [H_3O^+]^{14} + 0.059\ \log\ \frac{1}{2[Cr^{3+}]}$$

Before going on with this problem, it is important to note that in this case, the equivalence-point potential will depend on both the starting concentration of Fe(II), since this will dictate the $[Cr^{3+}]$ at the equivalence point, and on the $[H_3O^+]$ at the equivalence point. The former situation will be encountered wherever we have an oxidation-reduction reaction in which the values of a, b, c, and d respond to the test

$$a \neq b \qquad and/or \qquad c \neq d$$

When the value of $[Fe^{3+}]$ at the equivalence point is 0.05 M, then from Eq. (152), the value of $[Cr^{3+}]$ at the equivalence point will be $\tfrac{1}{3} \times 0.05\ M$, or 0.0167 M. For $[H_3O^+] = 1\ M$, this yields a value of E_{cell}:

$$E_{cell} = 1.26^2\ V \qquad at\ 25°C$$

For $[H_3O^+] = 10^{-4}\ M$, this yields an approximate value of

$$E_{cell} \approx 0.79^0\ V \qquad at\ 25°C$$

Example   The particular reduction reaction for As(V) to As(III) in aqueous solution depends on the value of the solution pH, and may be one of the following.

pH 0 to 2.3: $H_3AsO_4 + 2H_3O^+ + 2e \rightleftharpoons H_3AsO_3 + 3H_2O$

pH 2.3 to 6.8: $H_2AsO_4^- + 3H_3O^+ + 2e \rightleftharpoons H_3AsO_3 + 4H_2O$

pH 6.8 to 9.2: $HAsO_4^{2-} + 4H_3O^+ + 2e \rightleftharpoons H_3AsO_3 + 5H_2O$

pH 9.2 to 11.0: $AsO_4^{3-} + 5H_3O^+ + 2e \rightleftharpoons H_3AsO_3 + 6H_2O$

Using the proper combination of Ka values for the acid H_3AsO_4, a standard potential of $E^\circ_{H_3AsO_4/H_3AsO_3} = 0.58$ V at 25°C, and considering that at all times the ion-concentration ratio involving any form of $[As(V)]/[As(III)]$ will be approximated by unity, prepare a graph of reduction potential for As(V)/As(III) versus pH for the pH values, 0, 1.0, 2.0, 2.3, 3.0, 4.0, 5.0, 6.0, 6.8, 7.0, 8.0, 9.0, 9.2, 10.0, and 11.0.

For the pH zone 0 to 2.3 we have, for the indicated reaction,

$$E_{As(V)/As(III)} = 0.58 + \frac{0.059}{2} \log \frac{[H_3AsO_4][H_3O^+]^2}{[H_3AsO_3]}$$

$$= 0.58 + \frac{0.059}{2} \log [H_3O^+]^2$$

At pH = 0: $E_{As(V)/As(III)} = 0.58$ V

At pH = 1.0: $E_{As(V)/As(III)} = 0.58 + \frac{0.059}{2} \log (10^{-1})^2$

$$= 0.58 - 0.06 = 0.52 \text{ V}$$

At pH = 2.0: $E_{As(V)/As(III)} = 0.58 + \frac{0.059}{2} \log (10^{-2})^2$

$$= 0.58 - 0.12 = 0.46 \text{ V}$$

At pH = 2.3: $E_{As(V)/As(III)} = 0.58 + \frac{0.059}{2} \log (10^{-2.3})^2$

$$= 0.58 - 0.14 = 0.44 \text{ V}$$

For the pH zone 2.3 to 6.8 we have, for the indicated reaction equation,

$$E_{As(V)/As(III)} = 0.58 + \frac{0.059}{2} \log \frac{[H_3AsO_4][H_3O^+]^2}{[H_3AsO_3]}$$

but, from the acid dissociation equation, we have

$$H_3AsO_4 + H_2O \rightleftharpoons H_3O^+ + H_2AsO_4^-$$

and

$$\frac{[H_3O^+][H_2AsO_4^-]}{[H_3AsO_4]} = K_1 \quad \text{with } [H_3AsO_4] = \frac{[H_3O^+][H_2AsO_4^-]}{K_1}$$

Substitution of this latter equality into the potential equation yields

$$E_{As(V)/As(III)} = 0.58 + \frac{0.059}{2} \log \frac{[H_2AsO_4^-][H_3O^+]^3}{[H_3AsO_3]K_1}$$

$$= 0.58 + \frac{0.059}{2} \log \frac{[H_3O^+]^3}{K_1}$$

At pH = 3.0: $E_{As(V)/As(III)} = 0.58 + \frac{0.059}{2} \log \frac{(10^{-3})^3}{5.62 \times 10^{-3}}$

$$= 0.58 - 0.20 = 0.38 \text{ V}$$

At pH = 4.0: $E_{As(V)/As(III)} = 0.58 + \frac{0.059}{2} \log \frac{(10^{-4})^3}{5.62 \times 10^{-3}}$

$$= 0.58 - 0.29 = 0.29 \text{ V}$$

At pH $= 5.0$: $E_{As(V)/As(III)} = 0.58 + \dfrac{0.059}{2} \log \dfrac{(10^{-5})^3}{5.62 \times 10^{-3}}$

$= 0.58 - 0.38 = 0.20 \text{ V}$

At pH $= 6.0$: $E_{As(V)/As(III)} = 0.58 + \dfrac{0.059}{2} \log \dfrac{(10^{-6})^3}{5.62 \times 10^{-3}}$

$= 0.58 - 0.46 = 0.12 \text{ V}$

At pH $= 6.8$: $E_{As(V)/As(III)} = 0.58 + \dfrac{0.059}{2} \log \dfrac{(10^{-6.8})^3}{5.62 \times 10^{-3}}$

$= 0.58 - 0.54 = 0.04 \text{ V}$

For the pH zone 6.8 to 9.2, we have again

$$E_{As(V)/As(III)} = 0.58 + \frac{0.059}{2} \log \frac{[H_3AsO_4][H_3O^+]^2}{[H_3AsO_3]}$$

From the acid dissociation for this pH zone, we have

$$H_2AsO_4^- + H_2O \rightleftharpoons H_3O^+ + HAsO_4^{2-}$$

from which we have

$$\frac{[H_3O^+][HAsO_4^{2-}]}{[H_2AsO_4^-]} = K_2$$

which gives, with the K_1 expression,

$$\frac{[H_3O^+]^2[HAsO_4^{2-}]}{[H_3AsO_4]} = K_1 K_2$$

and we have

$$[H_3AsO_4] = \frac{[H_3O^+]^2[HAsO_4^{2-}]}{K_1 K_2}$$

Substitution of this latter equality into the potential equation yields

$$E_{As(V)/As(III)} = 0.58 + \frac{0.059}{2} \log \frac{[HAsO_4^{2-}][H_3O^+]^4}{[H_3AsO_3]K_1 K_2}$$

from which

$$E_{As(V)/As(III)} = 0.58 + \frac{0.059}{2} \log \frac{[H_3O^+]^4}{K_1 K_2}$$

At pH $= 7.0$: $E_{As(V)/As(III)} = 0.58 + \dfrac{0.059}{2} \log \dfrac{(10^{-7})^4}{9.55 \times 10^{-10}}$

$= 0.58 - 0.56 = 0.02 \text{ V}$

At pH $= 8.0$: $E_{As(V)/As(III)} = 0.58 + \dfrac{0.059}{2} \log \dfrac{(10^{-8})^4}{9.55 \times 10^{-10}}$

$= 0.58 - 0.68 = -0.10 \text{ V}$

At pH $= 9.0$: $E_{As(V)/As(III)} = 0.58 - 0.80 = -0.22 \text{ V}$

At pH $= 9.2$: $E_{As(V)/As(III)} = 0.58 - 0.82 = -0.24 \text{ V}$

For the pH zone 9.2 to 11.0, using the next dissociation equation for the acid, and the same general reasoning, we have

$$E_{As(V)/As(III)} = 0.58 + \frac{0.059}{2} \log \frac{[H_3O^+]^5}{K_1 K_2 K_3}$$

At pH = 10.0: $E_{As(V)/As(III)} = 0.58 + \dfrac{0.059}{2} \log \dfrac{(10^{-10})^5}{3.77 \times 10^{-21}}$

$$= 0.58 - 0.87 = -0.29 \text{ V}$$

At pH = 11.0: $E_{As(V)/As(III)} = 0.58 - 1.02 = -0.44 \text{ V}$

The resulting plot of reduction potential for the couple As(V)/As(III) versus pH is shown in Fig. 4.9. Note that it shows a separate slope for each pH zone investigated. It will be obvious, of course, that all values determined for the reduction potential are versus SHE.

Example   What is the minimum difference for $(E^\circ_{Ox_2/Red_2} - E^\circ_{Ox_1/Red_1})$ at 25°C, when $n_1 = n_2 = 2$, and $a = b = c = d = 1$, in order to have a quantitatively complete reaction (99.9 percent complete) at the equivalence point of the system.

$$Ox_2 + n_2 e \rightleftharpoons Red_2$$

$$Red_1 - n_1 e \rightleftharpoons Ox_1$$

$$Red_1 + Ox_2 \rightleftharpoons Ox_1 + Red_2$$

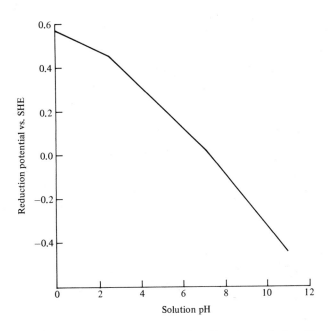

Fig. 4.9   Reduction potential As(V)/As(III) versus solution pH.

The equilibrium constant is

$$\frac{[Ox_1][Red_2]}{[Red_1][Ox_2]} = 10^{\left(\frac{E^\circ_{Ox_2/Red_2} - E^\circ_{Ox_1/Red_1}}{0.059/2}\right)} = K_{eq}$$

For a reaction 99.9 percent complete at the equivalence point,

$$\frac{[Ox_1]}{[Red_1]} = 1000/1 \text{ minimum} \qquad \text{and} \qquad \frac{[Red_2]}{[Ox_2]} = 1000/1 \text{ minimum}$$

therefore,

$$10^6_{min} = 10^{\left(\frac{E^\circ_{Ox_2/Red_2} - E^\circ_{Ox_1/Red_1}}{0.059/2}\right)}$$

and we have

$$(E^\circ_{Ox_2/Red_2} - E^\circ_{Ox_1/Red_1})_{min} = 6 \times \frac{0.059}{2} = 0.177 \text{ V}$$

Example   We have the half-cell reactions

$$Cu^{2+} + 2e \rightleftharpoons Cu$$
$$Cu^+ + 1e \rightleftharpoons Cu$$

where

$$E^\circ_{Cu^{2+}/Cu} = 0.340 \text{ V} \qquad E^\circ_{Cu^+/Cu} = 0.522 \text{ V} \qquad \text{at } 25°C$$

Determine the value of $E^\circ_{Cu^{2+}/Cu^+}$ at 25°C. Is the reaction shown below then spontaneous at 25°C and for unit activities of Cu^{2+} and Cu?

$$Cu^{2+} + 1e \rightleftharpoons Cu^+$$

$$E_{Cu^{2+}/Cu} = 0.340 + \frac{0.059}{2} \log \frac{[Cu^{2+}]}{[Cu]}$$

$$2E_{Cu^{2+}/Cu} = 0.680 + 0.059 \log \frac{[Cu^{2+}]}{[Cu]} \tag{A}$$

$$E_{Cu^+/Cu} = 0.522 + 0.059 \log \frac{[Cu^+]}{[Cu]} \tag{B}$$

Subtracting (B) from (A) we have

$$(2E_{Cu^{2+}/Cu} - E_{Cu^+/Cu}) = 0.680 - 0.522 + 0.059 \log \frac{[Cu^{2+}]}{[Cu^+]} = E_{Cu^{2+}/Cu^+}$$

Where unit activities are implied for Cu^{2+} and Cu^+, we have

$$E^\circ_{Cu^{2+}/Cu^+} = 0.680 - 0.522$$
$$= 0.158 \text{ V} \qquad \text{at } 25°C$$

The reaction

$$Cu^{2+} + 1e \rightleftharpoons Cu^+$$

is definitely spontaneous at 25°C and for unit activities of the two ions.

PROBLEMS

Part 2

1. Determine the $[H_3O^+]$ of the following solutions. Where the predicted relative error is less than 5 percent, use the simplest equation for the solution of each problem. Show the corresponding pH values.

    (a) 0.268 g of HNO_3 in 1.00 l of solution

    (b) 0.167 g of $HClO_4$ in 546 ml of solution

    (c) 0.221 g of $CH_3 \cdot COOH$ in 295 ml of solution

    (d) 0.198 g of $H \cdot COOH$ in 300 ml of solution

    (e) 15.6 g of NaOH in 1.24 l of solution

    (f) 1.89 g NH_3 in 146 ml of solution

    (g) 0.444 g of $H \cdot COONa$ in 125 ml of solution

    (h) 0.398 g of NH_4Cl in 95 ml of solution

    (i) 8.0×10^{-7} mol of HCl in 1 l of solution

    (j) 4.10 g of $COOH \cdot COOH$ in 1.00 l of solution

    (k) 10.00 g $COOH \cdot COONa$ in 950 ml of solution

    (l) 1.85 g of $COONa \cdot COONa$ in 220 ml of solution

Ans. (a) $4.25 \times 10^{-3} M$, 2.37 pH; (e) $3.18 \times 10^{-14} M$, 13.50 pH; (h) $6.62 \times 10^{-6} M$, 5.18 pH; (k) $1.52 \times 10^{-3} M$, 2.82 pH

2. Determine the $[H_3O^+]$ of the following solutions with a relative error not exceeding 2.5 percent. Show the corresponding pH values.

    (a) 25 ml of 0.200 M NH_3 and 20 ml of 0.150 M HCl

    (b) 35 ml of 0.100 M NaOH and 45 ml of 0.120 M $CH_3 \cdot COOH$

    (c) 45 ml of 0.100 M NaOH and 30 ml of 0.120 M $CH_3 \cdot COOH$

    (d) 100.0 ml of HCl at a pH of 2.00 and 50.0 ml of water

    (e) 75.0 ml of HCl at a pH of 1.40 and 20.0 ml of NaOH at a pH of 10.60

3. Calculate the concentrations of all species present in the following solutions. Base your calculations on a relative error of 2.5 percent maximum.

    (a) 50.0 ml of 0.100 M NH_3 and 50.0 ml of 0.100 M HCl

    (b) 100 ml of 0.100 M NH_3 and 50.0 ml of 0.050 M HCl

    (c) 50 ml of 0.0100 M C_6H_5COOH and 50 ml of 0.0050 M KOH

4. Calculate the dissociation constant for a weak monobasic acid HA, where the molecular weight is 70 and a solution resulting from 2.10 g of the acid in 300 ml of solution has a pH of 3.15.

Ans. 5.0×10^{-6}

5. A weak monobasic acid shows 1.5 percent ionization in a 0.10-M solution.

    (a) Calculate the percentage ionization in a 0.050-M solution.

    (b) Calculate the percentage ionization in a 0.50-M solution.

    (c) Determine at what point of solution concentration relative to weak acid there is an ionization percentage of 1.

Use the simplest equations in all calculations.

6. Using a 0.100-M solution of NH_3, indicate how many grams of NH_4Cl must be added (no volume change when added) to obtain 1 l of solution having a pH of 8.20.

7. What volumes of 0.100 M $CH_3 \cdot COOH$ and 0.200 M NaOH should be added together to obtain 1 l of solution with a pH of 5.20. Assume that the volumes used are strictly additive.

Ans. 731 ml $CH_3 \cdot COOH$, 269 ml NaOH

8. (a) Determine the $[H_3O^+]$ and the pH of a 1.00-M solution of $CHCl_2 \cdot COOH$, dichloracetic acid, using the simplest equation.

    (b) Calculate the predicted relative error in a above.

    (c) Calculate the $[H_3O^+]$ and the pH using the appropriate quadratic equation, and compare the actual relative error in the $[H_3O^+]$ from a with the predicted relative error.

9. (a) Determine the $[H_3O^+]$ and the pH of a 0.0100-M solution of NH_3, using both the simplest and quadratic equations.

(b) Compare the predicted relative error and the actual relative error in the $[H_3O^+]$ from the simplest equation.

(c) Comment on which equation should, in your opinion, be used to determine $[H_3O^+]$ in this solution.

10. (a) Determine the $[H_3O^+]$ and the pH for solutions of $NaNO_2$, sodium nitrite, at the following concentrations:

(1) 0.100 M

(2) 0.0010 M

Use only the simplest equation in each case.

(b) Calculate the predicted relative error for $[H_3O^+]$ for both solutions

(c) Comment on the validity of using the simplest equation in each case.

Ans. (a) (1) 6.8×10^{-9} M, 8.17 pH; (a) (2) 6.8×10^{-8} M, 7.17 pH

11. (a) Determine the $[H_3O^+]$ and the pH for solutions of NH_4Cl at the following concentrations:

(1) 0.100 M

(2) 0.0010 M

Use both the simplest and quadratic equations in each case.

(b) What are the predicted relative errors for both solutions relative to $[H_3O^+]$ as determined from the simplest equation?

(c) Comment on the validity of using the simplest equation for the $[H_3O^+]$ determination for each solution.

12. (a) Determine the $[H_3O^+]$ and the pH for solutions resulting from the titration of 0.0100 M HIO_4, periodic acid, and 0.0100 M NaOH at points where

(1) $C_a = 0.0100$ M

   $C_s = 0.0010$ M

(2) $C_a = 0.00010$ M

   $C_s = 0.0050$ M

Use both the simplest and quadratic equations in each case.

(b) What are the predicted and actual relative errors in $[H_3O^+]$ as determined from the simplest equation in each case? Comment on any significant discrepancies.

13. (a) Determine the $[H_3O^+]$ and the pH for solutions resulting from the titration of 0.100 M NH_3 and 0.100 M HCl at points where

(1) $C_b = 0.100$ M

   $C_s = 0.00050$ M

(2) $C_b = 0.0010$ M

   $C_s = 0.0500$ M

Use the simplest and quadratic equations in each case.

(b) Calculate the predicted and actual relative errors in $[H_3O^+]$ as determined from the simplest equation. Comment on any discrepancies.

(c) Comment on the proper equation to be used for the determination of $[H_3O^+]$ in each case.

14. Determine the $[H_3O^+]$ and the pH of the following solutions. The value of $[H_3O^+]$ should have a relative error not exceeding 5 percent.

(a) H_2SO_3	0.100 M
(b) H_2SO_3	0.050 M
NaHSO_3$	0.010 M
(c) $NaHSO_3$	0.020 M
(d) $NaHSO_3$	0.030 M
Na_2SO_3$	0.020 M
(e) Na_2SO_3	0.010 M

Ans. (c) 2.4×10^{-5} M, 4.62 pH quadratic; (e) 2.5×10^{-10} M, 9.60 pH simplest

15. It is desired to change the pH of 100 ml of 0.100 M HCl from 1.00 to 4.40 by the addition of sodium acetate, $CH_3 \cdot COONa \cdot 3H_2O$. How much solid sodium acetate salt must be added in grams? Assume no volume change for the solution as the result of the addition.

16. What is the buffer capacity of a solution which is 0.100 M to NH_3 and 0.200 M to NH_4Cl?

Part 3

17. Calculate the molar solubility in water of each of the following. Do not consider any hydrolysis effects relative to the anion or cation.

    (*a*) AgBr, solubility 1.33 mg/l
    (*b*) $BaSO_4$, solubility 0.000243 g/100 ml
    (*c*) CaF_2, solubility 0.001337 g/80 ml
    (*d*) $Mg(OH)_2$, solubility 1.59 mg/150 ml   *Ans.* (*b*) $1.04 \times 10^{-5} M$; (*d*) $1.82 \times 10^{-4} M$

18. Determine the K_{sp} values of each of the slightly soluble salts in Prob. 17. Calculate the concentration of each ion in a saturated aqueous solution of the salt.

19. Calculate the molar solubility in a saturated aqueous solution of each of the following. Do not consider any hydrolysis effects relative to the anion or cation.

    (*a*) $K_{sp}(Ca_3(PO_4)_2)$ 1.0×10^{-26}
    (*b*) $K_{sp}(Fe(OH)_3)$ 4.5×10^{-37}
    (*c*) $K_{sp}(MnS)$ 1.4×10^{-15} *Ans.* (*b*) $3.6 \times 10^{-10} M$

20. Calculate the residual concentrations, and the molar solubilities, of the indicated substances in the following saturated and equilibrated aqueous solutions. Ignore any hydrolysis effects relative to the anion or cation.

    (*a*) Residual $[Ag^+]$ and $S(AgI)$ in a solution containing $[I^-]$ at a value of $10^{-2} M$
    (*b*) Residual $[SO_4^{2-}]$ and $S(BaSO_4)$ in a solution containing $[Ba^{2+}]$ at a value of $1.6 \times 10^{-3} M$
    (*c*) Residual $[OH^-]$ and $S(Al(OH)_3)$ in a solution containing $[Al^{3+}]$ at a value of $1.5 \times 10^{-2} M$
    (*d*) Residual $[Zn^{2+}]$ and $S(Zn_3(PO_4)_2)$ in a solution containing $[PO_4^{3-}]$ at a value of $10^{-2} M$
    (*e*) Residual $[Ca^{2+}]$ and $S(CaC_2O_4)$ in a solution containing $[C_2O_4]$ at a value of 0.05 M

21. Determine the molar solubility of CaF_2 in an aqueous solution which is 0.10 M to KNO_3. Ignore any hydrolysis effects of the anion or cation of the salt CaF_2, and consider the ionic strength of the solution to have its origin in the concentration of KNO_3 alone.   *Ans.* $4.50 \times 10^{-4} M$

22. Compare the molar solubilities of FeS in aqueous solutions which are

    (*a*) 0.10 M to KNO_3
    (*b*) 0.10 M to K_2SO_4

Ignore any hydrolysis action of the anion or cation of FeS, and consider that the ionic strength of the solution originates with either the KNO_3 or K_2SO_4 concentration alone.

23. Determine the molar solubility of AgCN in an aqueous solution which is pH 5.00 to a strong monobasic acid at equilibrium. Compare this value with the molar solubility in pure water, neglecting any hydrolysis effects relative to the anion or cation of AgCN.

24. Calculate the molar solubility of CaC_2O_4 in an aqueous solution which is pH 2.00 to a strong monobasic acid at equilibrium. Compare with the molar solubility in pure water, neglecting any anion or cation hydrolysis.   *Ans.* pH 2.00: $6.87 \times 10^{-4} M$; water: $5.07 \times 10^{-5} M$

25. What is the molar solubility of CaF_2 in an aqueous solution which is pH 3.00 to a strong monobasic acid at equilibrium? Compare with the molar solubility in pure water, ignoring any anion or cation hydrolysis effects.

26. What is the molar solubility of AgCN in a solution which is $10^{-5} M$ to $[H_3O^+]$ from a strong monobasic acid before equilibrium is established?

27. What is the molar solubility of $BaSO_4$ in the following solutions?

    (*a*) Pure water (ignore any anion or cation hydrolysis effects)
    (*b*) A solution in which a strong monobasic acid yields an equilibrium pH of 2.00
                                Ans. (*a*) $1.0 \times 10^{-5} M$; (*b*) $1.4 \times 10^{-5} M$

28. Show the molar solubility of $Ca_3(PO_4)_2$ in the following solutions:
      (a) Pure water (ignore any anion or cation hydrolysis effects)
      (b) A solution which is pH 2.00 at equilibrium because of the presence of a strong monobasic acid

29. Determine the concentration of Ca^{2+} in a saturated solution of CaF_2 with a pH of 1.00 at equilibrium. What concentration of F^- must be added to this solution to yield the same $[Ca^{2+}]$ value which would be obtained in a saturated pure water solution of CaF_2, ignoring any anion or cation hydrolysis effects?

30. The value of $K_{sp}(AgCl)$ at 100°C is 2.17×10^{-8}. A saturated aqueous solution of AgCl at 100°C will be obtained as the result of an analytical procedure. What $[Cl^-]$ should be present in this solution at equilibrium to give a value of $[Ag^+]$ equal to that representative of a saturated solution of AgCl in pure water at 25°C?   *Ans.* $1.74 \times 10^{-3} M$

31. Determine the molar solubilities for CdS and ZnS in pure water, giving consideration to anion hydrolysis effects.

32. (a) Determine the molar solubility of AgBr in a solution 1.0 M to NH_3 at equilibrium, and compare this value to the molar solubility in pure water, ignoring any anion or cation hydrolysis effects.
      (b) Calculate the molar solubility of AgBr in a solution 0.10 M to Br^- and 1.0 M to NH_3 at equilibrium.

33. A metal hydroxide, $M(OH)_2$, which is a slightly soluble salt in aqueous medium, has a K_{sp} value of 4.0×10^{-15}. To a 0.10-M solution of the metal ion, NaOH is added. Assuming no volume change to occur, calculate the pH of the solution when
      (a) One percent of the metal has been precipitated.
      (b) Fifty percent of the metal has been precipitated.
      (c) Ninety-nine percent of the metal has been precipitated.
                                     Ans. (a) 7.30 pH; (b) 7.45 pH; (c) 8.3 pH

34. Water is saturated with both $AgIO_3$ and AgOH at 25°C. Neglecting any hydrolysis effects, calculate the molar solubility of each slightly soluble salt in the solution.

Part 4

35. Determine the concentration distribution of each species involving cadmium in a solution which starts off as 0.1 M to Cd^{2+} and, after the addition of NH_3 without volume change, ends up with an equilibrium value of $[NH_3] = 0.2 M$.

36. 100 ml of 0.1 M Cu^{2+} is titrated with a trien solution which is 0.2 M. Calculate the $[Cu^{2+}]$ in solution at the equivalence point of this titration. Is this titration suitable for use in the quantitative determination of copper?   *Ans.* $1.6 \times 10^{-11} M$

37. You are considering the titration of 100 ml each of 0.2-M solutions of Ba^{2+}, Ca^{2+}, Co^{2+}, Hg^{2+}, and Mg^{2+}. These titrations will involve 0.2 M EDTA and will be carried out in solutions buffered to a pH of 5.00. Indicate what will be the expected approximate concentration of each of these ions at the equivalence points of the respective titrations. Indicate also whether or not each titration will be feasible from the point of view of a quantitative determination.

38. Using the same general conditions as indicated in Prob. 37 but with a buffered solution condition of pH 11.00, indicate what will be the expected approximate concentrations of Ba^{2+}, Ca^{2+}, and Mg^{2+} at the equivalence points of their respective titrations. Will these titrations be feasible from the point of view of a quantitative determination? Can the titration of Ba^{2+} be made quantitative at all?
      Ans. $[Ba^{2+}] \approx 4.4 \times 10^{-5} M$; $[Ca^{2+}] \approx 1.1 \times 10^{-6} M$; $[Mg^{2+}] \approx 1.6 \times 10^{-5} M$

39. Determine the concentration of all species involving cadmium where 100 ml of 0.1 M Cd^{2+} is added to 900 ml of a 1-M solution of KCN. Why is the value of $[Cd(CN)_4^{2-}]$ so high?

Part 5

40. Determine for 25°C the half-cell potentials, or half-cell emf values, for the following half-cells:
      (a) Cu^{2+} to Cu $[Cu^{2+}] = 10^{-4} M$
      (b) Au^{3+} to Au $[Au^{3+}] = 2 M$

(c) Ag^+ to Ag $[Ag^+] = 2.5 \times 10^{-3} M$
(d) Sn^{4+} to Sn^{2+} (in 1 M HCl) $[Sn^{4+}] = 4.3 \times 10^{-2} M$
                                    $[Sn^{2+}] = 2.6 \times 10^{-1} M$
(e) Fe^{3+} to Fe^{2+} (in 1 M HCl) $[Fe^{3+}] = 4.6 \times 10^{-4} M$
                                    $[Fe^{2+}] = 3.7 \times 10^{-2} M$
(f) Pb to Pb^{2+} $[Pb^{2+}] = 1.7 \times 10^{-3} M$
(g) Bi to Bi^{3+} $[Bi^{3+}] = 4.8 \times 10^{-4} M$

Ans. (c) 0.64^6; (e) 0.58^8; (g) -0.135

41. Determine for 25°C
    (a) The half-cell reactions
    (b) The overall cell reaction
    (c) The electrode polarity
    (d) The cell emf
    (e) The equilibrium constant
for the following galvanic cells:
    (1) Zinc and silver electrodes in a solution of $[Zn^{2+}] = 1.9 \times 10^{-4} M$ and $[Ag^+] = 5.3 \times 10^{-2} M$
    (2) Cadmium and copper electrodes in a solution of $[Cd^{2+}] = 7.8 \times 10^{-1} M$ and $[Cu^{2+}] = 1.5 M$
    (3) Iron and copper electrodes in a solution of $[Fe^{2+}] = 8.4 \times 10^{-1} M$ and $[Cu^{2+}] = 1.2 \times 10^{-2} M$

42. Granulated metals, or powdered metals of mesh size 30 or greater, when added in excess to solutions of certain cations, will either precipitate the metal represented by the cation or will reduce the cation to a soluble but lower oxidation state form. Indicate what cation reduction reactions would take place with the following combinations. Indicate the concentration of the cation in solution at equilibrium, given the starting concentration shown in each case, and assuming no solubility problems relative to the ions resulting from solution of the added metal. Are the reactions capable of achieving a quantitative degree of completion at equilibrium, assuming an equilibrium concentration of the cation of $10^{-6} M$ maximum to represent, in each case, a degree of completion that is quantitative?

Cation	*Metal addition to excess*
(a) $[Cu^{2+}] = 5.0 \times 10^{-2} M$	Fe ($\rightarrow Fe^{2+}$)
(b) $[Fe^{3+}] = 1.00 M$ (1 M H_2SO_4)	Zn
(c) $[Sn^{4+}] = 0.50 M$ (1 M HCl)	Ni
(d) $[Ag^+] = 0.20 M$	Cd

43. A Jones reductor is often used for the quantitative reduction of certain ions to soluble ionic species of lower state of oxidation. This latter species is subsequently titrated in a standard oxidation-reduction volumetric procedure. The Jones reductor is essentially a long glass tube, mounted vertically, containing 20-30 mesh zinc powder. The solution containing the substance or substances to be reduced is passed through the column.
    A 2.5-M H_2SO_4 solution of the following ions is passed through such a reductor. Of the possible reactions indicated for each substance, prove which will actually occur, assuming the $E°$ value in each case to be applicable to the appropriate reduction reaction, and assuming a value for $E_{Zn^{2+}/Zn}$ based on an average value of $[Zn^{2+}] = 10^{-3} M$ accumulated in solution during passage though the column. What substance or substances present in the initial solution will lead to difficulties in the operation of the reductor and why?
    (a) Fe^{3+} ($Fe^{3+} + 1e \rightleftharpoons Fe$)($Fe^{2+} + 2e \rightleftharpoons Fe$)

(b) $Cr_2O_7^{2-}$ $(Cr_2O_7^{2-} + 14H_3O^+ + 6e \rightleftharpoons 2Cr^{3+} + 21H_2O)(Cr^{3+} + 1e \rightleftharpoons Cr^{2+})$

$(Cr^{2+} + 2e \rightleftharpoons Cr)$

(c) Ce^{4+} $(Ce^{4+} + 1e \rightleftharpoons Ce^{3+})(Ce^{3+} + 3e \rightleftharpoons Ce)$

(d) MnO_4^- $(MnO_4^- + 8H_3O^+ + 5e \rightleftharpoons Mn^{2+} + 12H_2O)(Mn^{2+} + 2e \rightleftharpoons Mn)$

(e) TiO^{2+} $(TiO^{2+} + 2H_3O^+ + 1e \rightleftharpoons Ti^{3+} + 3H_2O)(Ti^{3+} + 1e \rightleftharpoons Ti^{2+})$

(f) VO_2^+ $(VO_2^+ + 2H_3O^+ + 1e \rightleftharpoons VO^{2+} + 3H_2O)(VO^{2+} + 2H_3O^+ + 1e \rightleftharpoons$

$V^{3+} + 3H_2O)(V^{3+} + 1e \rightleftharpoons V^{2+})$

(g) Ni^{2+} $(Ni^{2+} + 2e \rightleftharpoons Ni)$

(h) Cu^{2+} $(Cu^{2+} + 2e \rightleftharpoons Cu)$

44. If silver powder instead of zinc powder were used in the reductor column in Prob. 43, what reduction reactions would then take place in the case of each substance in the solution? What would be the advantage of this type of column under certain circumstances? *Note*: 1 *M* HCl in place of 2.5 *M* H_2SO_4.

45. The indicator electrodes mentioned below are immersed at 25°C in the solutions involved, and are linked through a potentiometer with the SHE. The potentiometer gives the emf and electrode polarity values shown.

(a) A silver-indicating electrode in a solution of Ag^+ shows polarity + and a measured emf with the SHE of 0.548 V. Determine the solution $[Ag^+]$.

(b) A silver-indicating electrode in a saturated solution of AgCl shows a polarity of + and a measured emf with the SHE of 0.511 V. Determine the $K_{sp}(AgCl)$ at 25°C.

(c) A silver-indicating electrode in a solution of Ag^+ and $Ag(NH_3)_2^+$ with $[NH_3] = 2$ *M* shows polarity + and an emf with the SHE of 0.245 V. Assuming that only the $[Ag(NH_3)_2^+]$ is significant, that of $[Ag(NH_3)^+]$ being negligible, determine the value of $[Ag(NH_3)_2^+]$ in the solution.

(d) An inert platinum electrode in a solution of Fe^{3+}, Fe^{2+}, and 1 *M* H_2SO_4, where the $[Fe^{2+}]$ is known to be 0.050 *M*, shows a polarity of + and a measured emf with the SHE of 0.700 V. Determine the value of $[Fe^{3+}]$.

(e) An inert platinum electrode is to be used to follow the course of a titration of Fe^{2+} in 1 *M* H_2SO_4 by the addition without volume change of solid $KMnO_4$. If the starting concentration of Fe^{2+} is 0.100 *M*, what will be the solution potential when

   (1) 25 percent
   (2) 50 percent
   (3) 90 percent
   (4) 99 percent
   (5) 99.9 percent

of the Fe^{2+} has been converted to Fe^{3+}. If the solution potential given by the inert electrode and SHE is 1.372 V, with the inert electrode as + polarity, use the Nernst equation *only* to determine whether or not the reaction $Fe^{2+} - 1e \rightleftharpoons Fe^{3+}$ is likely to be quantitatively complete at this point in the titration, which, incidentally, represents the equivalence point.

(f) A hydrogen electrode is immersed in a solution of acetic acid, $CH_3 \cdot COOH$, which is known to be exactly 0.100 *M*. The electrode shows negative polarity and gives with the SHE a measured emf of 0.170 V. What is the pH of the solution and what is the K_a value for acetic acid at 25°C?   *Ans.* (a) 5.4×10^{-5} *M*; (c) 0.025 *M*; (f) 2.88 pH, $K_a = 1.74 \times 10^{-5}$

46. A silver-indicating electrode is immersed in a solution saturated with silver carbonate, Ag_2CO_3, at 25°C. When linked with an SCE electrode through a potentiometer, a measured emf is noted, with the SCE showing − polarity. Determine the value of this emf, and determine the value relative to both the SHE and $Ag/AgCl_{sat\ KCl}$ electrode.

47. A zinc electrode shows negative polarity and a measured emf value of 1.148 V when linked to a $Ag/AgCl_{sat\ KCl}$ electrode through a potentiometer, both electrodes being immersed at 25°C in a solution saturated with zinc phosphate, $Zn_3(PO_4)_2$. Calculate the $K_{sp}(Zn_3(PO_4)_2)$ at 25°C.

Ans. 1.39×10^{-32}

48. A silver-indicator electrode, used in a titration between Ag^+ and Cl^-, gave a series of values of emf, reference SHE, during the titration. The values are indicated below for the titration at 25°C, and it was noted that the SHE always showed negative polarity.

(1) 0.741 V
(2) 0.623 V
(3) 0.505 V
(4) 0.387 V
(5) 0.269 V

Determine the value of $[Ag^+]$ in the solution at each titration point. Show the emf values that would have been secured at the same titration points with the $Hg/Hg_2SO_{4(sat\,K_2SO_4)}$ electrode as the reference electrode. Show the polarity of the electrodes for each value of emf. The reduction potential for $Hg/Hg_2SO_{4(sat\,K_2SO_4)}$ at 25°C is 0.624 V.

49. Calculate the value of the reduction potential for the mercury–mercurous sulfate electrode, with saturated K_2SO_4 solution. The saturation value of K_2SO_4 in water at 25°C is 0.70 M.

50. A solution of Sn(II) can be used to reduce Fe(III) in solution to Fe(II). Prior to titration of the Fe(II) so obtained, the slight excess of Sn(II) is oxidized by the addition of Hg(II) solution. The following half-cell and overall cell reactions can be assumed to be involved:

$$2Hg^{2+} + 2e \rightleftharpoons Hg_2^{2+}$$
$$Sn^{2+} - 2e \rightleftharpoons Sn^{4+}$$
$$Sn^{2+} + 2Hg^{2+} \rightleftharpoons Sn^{4+} + Hg_2^{2+}$$

If the solution is 0.10 M to HCl, and the $[Sn^{4+}]$ at equilibrium after the addition of an amount of Hg(II) equivalent to the amount of Sn(II) is 0.050 M, calculate for 25°C,

(a) The equilibrium constant
(b) The $[Sn^{2+}]$ at equilibrium after the Hg(II) addition
(c) The solution potential versus SHE
(d) The solution potential versus SCE

51. Compare the values secured for Prob. 50(a) to 50(d) above with those obtainable when the half-cell and overall cell reactions are assumed to be

$$Hg^{2+} + 1e \rightleftharpoons Hg^+$$
$$Sn^{2+} - 2e \rightleftharpoons Sn^{4+}$$
$$Sn^{2+} + 2Hg^{2+} \rightleftharpoons Sn^{4+} + 2Hg^+$$

52. In an Example from Subsec. 4.21.3, it was found that the minimum difference for $(E^\circ_{Ox_2/Red_2} - E^\circ_{Ox_1/Red_1})$ at 25°C, when $n_1 = n_2 = 2$, and $a = b = c = d = 1$, was 0.177 V in order to have a quantitatively complete reaction (99.9 percent complete) at the equivalence point.

Determine the minimum difference for this same value for the conditions outlined below. All calculations are to be based on a 25°C temperature. In all cases it is assumed that $a = b = c = d = 1$.

(a) $n_2 = 1$ $n_1 = 1$
(b) $n_2 = 1$ $n_1 = 2$
(c) $n_2 = 2$ $n_1 = 2$
(d) $n_2 = 2$ $n_1 = 3$
(e) $n_2 = 3$ $n_1 = 3$ *Ans.* (b) 0.266 V; (d) 0.148 V

53. Given the half-cell reactions, and the associated standard potentials at 25°C,

$$Sn^{4+} + 2e \rightleftharpoons Sn^{2+} \qquad E^\circ_{Sn^{4+}/Sn^{2+}} = 0.139 \text{ V}$$
$$Sn^{2+} + 2e \rightleftharpoons Sn \qquad E^\circ_{Sn^{2+}/Sn} = -0.136 \text{ V}$$

calculate the standard potential at 25°C for the half-cell reaction

$$Sn^{4+} + 4e \rightleftharpoons Sn$$

54. Given the half-cell reactions, and the associated standard potentials at 25°C,

$$MnO_2 + 4H_3O^+ + 2e \rightleftharpoons Mn^{2+} + 6H_2O \qquad E°_{MnO_2/Mn^{2+}} = 1.208 \text{ V}$$

$$MnO_4^- + 4H_3O^+ + 3e \rightleftharpoons MnO_2 + 6H_2O \qquad E°_{MnO_4^-/MnO_2} = 1.679 \text{ V}$$

calculate the standard potential at 25°C for the half-cell reaction

$$MnO_4^- + 8H_3O^+ + 5e \rightleftharpoons Mn^{2+} + 12H_2O$$

55. The half-cell reaction

$$MnO_4^- + 4H_3O^+ + 3e \rightleftharpoons MnO_2 + 6H_2O$$

is pH-dependent. Given the standard potential at 25°C as 1.679 V, use the pH dependency to calculate the standard potential at 25°C for the half-cell reaction

$$MnO_4^- + 2H_2O + 3e \rightleftharpoons MnO_2 + 4OH^- \qquad\qquad\qquad \textit{Ans.} \; 0.578 \text{ V}$$

56. Fe^{2+} can be titrated by the addition of VO_2^+. The reaction involved is

$$VO_2^+ + Fe^{2+} + 2H_3O^+ \rightleftharpoons VO^{2+} + Fe^{3+} + 3H_2O$$

Determine, for a titration carried out at 25°C in 1 M H_2SO_4,

    (a) The equilibrium constant.

    (b) The $[Fe^{2+}]$ unconverted at the equivalence point of the titration when $[H_3O^+] = 1 \, M$ and $[Fe^{3+}] = 0.020 \, M$. Is this titration then quantitatively complete at the equivalence point?

    (c) The maximum pH for a quantitatively complete reaction at the equivalence point when $[Fe^{3+}] = 0.020 \, M$.

    (d) The solution potential versus $Ag/AgCl_{(sat\,KCl)}$ at the equivalence point, when $[H_3O^+] = 1 \, M$.

    (e) The solution potential versus $Ag/AgCl_{(sat\,KCl)}$ at the equivalence point, when $[H_3O^+]$ has the value determined from c above.

57. As(III) can be titrated with I_2 in slightly acid to fairly basic solutions, with the appropriate reactions

At pH 2.3 − 6.8, approximately: $H_3AsO_3 + I_2 + 4H_2O \rightleftharpoons H_2AsO_4^- + 2I^- + 3H_3O^+$

At pH 6.8 − 9.2 approximately: $H_3AsO_3 + I_2 + 5H_2O \rightleftharpoons HAsO_4^{2-} + 2I^- + 4H_3O^+$

From the graph prepared in an Example from Subsec. 4.21.3, determine the approximate minimum pH for a 99.9 percent conversion of H_3AsO_3 to $H_2AsO_4^-$ at the equivalence point.

Ans. Approximately, pH minimum 3.2

REFERENCES

Parts 2 to 5

1. Robbins, O., Jr.: "Ionic Reactions and Equilibria," Macmillan, New York, 1967.
2. Margolis, E. M.: "Ionic Equilibria," Macmillan, New York, 1966.
3. Freiser, N., and Q. Fernando: "Ionic Equilibria in Analytical Chemistry," Wiley, New York, 1963.
4. Fleck, G. M.: "Equilibrium in Solution," Holt, New York, 1966.
5. Bard, A. J.: "Chemical Equilibrium," Harper & Row, New York, 1966.
6. Lee, T. S., and L. G. Sillen, "Chemical Equilibrium in Analytical Chemistry," reprint, Interscience, New York, 1959.

Additional references for Part 5 only

7. Lingane, J. J.: "Electroanalytical Chemistry," 2d ed., Interscience, New York, 1958.
8. Ives, D. J. G., and G. J. Janz: "Reference Electrodes," Academic, New York, 1961.

5
Generalized Titration Equations and Theory of Titration Curves

5.1 GENERALIZED TITRATION EQUATIONS

Any volumetric titration, regardless of type, is carried out in accordance with a specific pattern of changes in the solution concentrations of the reacting substances, and the products, occurring during the titration process. Consider a general reaction

$$R + T \rightleftharpoons A + B \tag{159}$$

taking place under titration conditions. Here R is the reacting substance and T is the titrant substance. The equilibrium constant is given by

$$\frac{[A][B]}{[R][T]} = K_{eq} \qquad \text{temperature constant}$$

where the molarities of the products and reactants are used rather than their activities, which would be more exact. At the start of the titration we have, let us say, a volume of V_R milliliters of reactant solution at a molar concentration of M_R. At the start of the titration we thus have

$$\frac{V_R M_R}{1000} \text{ moles of R}$$

Consider now that the titration has been carried out to some point *between* the start and the equivalence point, where we have added a volume of V_T milliliters of titrant solution at a titrant molarity of M_T. The number of moles of R left unti-trated at this point in the titration will be given by

$$\frac{V_R M_R}{1000} - \frac{V_T M_T}{1000}$$

and the molar concentration of R in the solution at this point will be given by

$$[R]_{V_T} \approx \frac{(V_R M_R - V_T M_T)/1000}{V_R + V_T} \, 1000$$

$$\approx \frac{V_R M_R - V_T M_T}{V_R + V_T} \tag{160}$$

The approximation sign is used here because there are actually two sources of $[R]$ in the solution at any such point in the titration. These are $[R]$ left unreacted at the point V_T and $[R]$ from the equilibrium equation. This latter value of $[R]$ represents $[R]$ from the back-reaction. Where the equilibrium constant K_{eq} is large (the equilibrium for the titration reaction lies far to the right), the reaction will be quite complete and quantitative, and the value of $[R]$ from the back-reaction will be very small.

Thus we have, on the most accurate basis,

$$[R]_{V_T} = [R]_{\text{unreacted}} + [R]_{\text{back-reaction}}$$

$$= \frac{V_R M_R - V_T M_T}{V_R + V_T} + [R]_{\text{back-reaction}}$$

The value of $[R]_{\text{back-reaction}}$, or $[R]_{br}$, will obviously be identical to the value of $[T]_{\text{back-reaction}}$, or $[T]_{br}$, for the general reaction given, and we have

$$[T]_{br} = \frac{[A][B]}{K_{eq}[R]_{V_T}} = [R]_{br} \tag{161}$$

Therefore,

$$[R]_{V_T} = \frac{V_R M_R - V_T M_T}{V_R + V_T} + \frac{[A][B]}{K_{eq}[R]_{V_T}} \tag{162}$$

which involves a quadratic equation in $[R]_{V_T}$, under conditions where generally $[A] = [B]$ at each titration point, and can be estimated for titrations starting under known conditions of concentration and volume of reactant solution and concentration of titrant.

For the majority of such titration points, where V_T lies after the start of the titration and appreciably before the equivalence point, and where the titration reaction goes well towards completion, we will have

$$\frac{[A][B]}{K_{eq}[R]_{V_T}} \ll \frac{V_R M_R - V_T M_T}{V_R + V_T}$$

and, as a good approximation of $[R]_{V_T}$ at these points, we have

$$[R]_{V_T} \approx \frac{V_R M_R - V_T M_T}{V_R + V_T} \tag{160}$$

However, where points of V_T near the equivalence point are involved, or where significantly incomplete titration reactions are the case (low values of K_{eq}), the exact Eq. (162) should be solved for $[R]_{V_T}$.

At the equivalence point of the titration we have, of course,

$$V_R M_R = V_T M_T$$

and the value of $[R]$ at the equivalence point is given by

$$[R]_{equiv\ point} = [R]_{ep} = \frac{[A][B]}{K_{eq}[R]_{ep}} \tag{163}$$

and we have

$$[R]_{ep} = \sqrt{\frac{[A][B]}{K_{eq}}} \tag{164}$$

For any point in the titration *after* the equivalence point, where T will be in excess, we will have

$$[T]_{V_T} = [T]_{excess} + [T]_{br} \tag{165}$$

and

$$[T]_{V_T} = \frac{V_T M_T - V_R M_R}{V_T + V_R} + \frac{[A][B]}{K_{eq}[T]_{V_T}} \tag{166}$$

Again, for points in the titration well beyond the equivalence point, we have

$$[T]_{br} \ll [T]_{excess}$$

and

$$[T]_{V_T} \approx \frac{V_T M_T - V_R M_R}{V_T + V_R} \tag{167}$$

For Eqs. (166) and (167), the value of $[R]_{V_T}$ in either case can be found by substituting the value for $[T]_{V_T}$ into the equation

$$[R]_{V_T} = \frac{[A][B]}{K_{eq}[T]_{V_T}} \tag{168}$$

We should now take a closer look at the relationships developed so far. Note that the number of moles of R at the start of the titration was given by

$$\frac{V_R M_R}{1000}$$

and that at a point in the titration V_T between the start and the equivalence point,

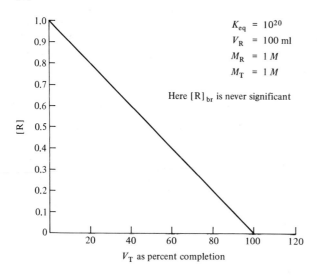

Fig. 5.1 $[R]$ vs. V_T with dilution effect corrected.

we had, neglecting for the moment the back-reaction contribution,

$$[R]_{V_T} \approx \frac{V_R M_R - V_T M_T}{V_R + V_T} \tag{160}$$

On the basis of the *original* volume of R solution at the start of V_R milliliters, the molarity of R at the point V_T as described above would be, neglecting the back-reaction contribution, given by

$$[R]_{V_T} \approx \frac{V_R M_R - V_T M_T}{V_R} \tag{169}$$

Note that apart from the *reaction* effect in decreasing $[R]$, the influence of V_T is to provide a *dilution* effect in decreasing $[R]$ as the titration progresses between the start and the equivalence point.

Equation (169) could, for example, be written as

$$[R]_{V_T} \approx \frac{V_R M_R}{V_R} - \frac{V_T M_T}{V_R}$$

$$\approx M_R - \frac{M_T}{V_R} V_T \tag{170}$$

and this is of the form

$$y = b + mx$$

a linear equation, indicating that $[R]$ varies linearly as V_T. If the back-reaction contribution of $[R]$ is not neglected, Eq. (170) becomes

$$[R]_{V_T} = \left(M_R - \frac{M_T}{V_R} V_T \right) + \frac{[A][B]}{K_{eq}[R]_{V_T}} \tag{171}$$

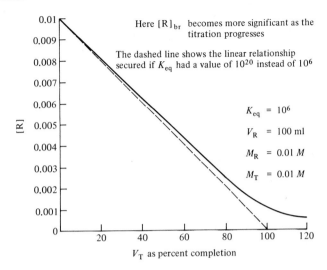

Fig. 5.2 $[R]$ vs. V_T with dilution effect corrected.

and near the equivalence point, where the second term on the right may become significant relative to the term in parentheses, a departure from linearity may take place, the point of departure and the magnitude of departure depending on the values of M_R, M_T, and K_{eq}. Figures 5.1 and 5.2 show graphically the general situation involved. Note that in these graphs, the value of V_T is shown on a percent completion basis for the titration rather than on an actual volume basis.

On the other hand, Eq. (160), which calculates $[R]_{V_T}$ on the basis of the dilution effect introduced by V_T but neglects the back-reaction effect, yields

$$[R]_{V_T} = \frac{V_R M_R}{V_R + V_T} - \frac{V_T M_T}{V_R + V_T} \tag{172}$$

This equation is nonlinear with respect to the relationship between $[R]$ and V_T. The dilution effect of V_T during a titration is shown in Fig. 5.3. Note the increase in the departure from linearity as the effect of dilution by V_T increases, using as a measure of the dilution effect the ratio

$$\frac{V_{T,ep}}{V_R}$$

Where $V_{T,ep} \leq V_R$, we can usually ignore the dilution effect in most standard volumetric titrations. Note that the ratio above of $V_{T,ep}$ and V_R can also be shown as

$$\frac{M_R}{M_T} = \frac{V_{T,ep}}{V_R} \leq 1$$

as applied *specifically* to the reaction type shown in Eq. (159). Certain types of volumetric titrations, such as amperometric, conductometric, and spectrophotometric titrations, require dilution-effect corrections under more restrictive con-

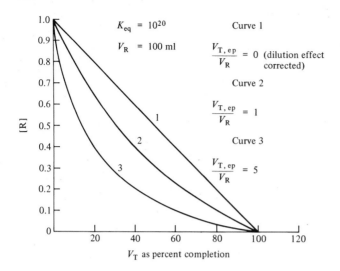

Fig. 5.3   [R] vs. V_T with dilution effect uncorrected.

ditions, for example, whenever we have

$$\frac{M_R}{M_T} = \frac{V_{T,ep}}{V_R} > \frac{1}{20}$$

Such dilution-effect corrections can be made by noting that Eq. (160) can be converted into Eq. (169) by multiplication of the former by a dilution-correction factor

$$\left(\frac{V_R M_R - V_T M_T}{V_R + V_T}\right)\left(\frac{V_R + V_T}{V_R}\right) = [R]_{V_T, \text{ dilution effect}} \left(\frac{V_R + V_T}{V_R}\right)$$

$$= [R]_{V_T, \text{ dilution effect corrected}} \qquad (173)$$

the factor $(V_R + V_T)/V_R$ being called the *dilution-correction factor.*

In developing this situation for correction of the dilution effect during titration, we have used the expression for $[R]_{V_T}$ that neglected the back-reaction contribution of $[R]$ at V_T. The extension of the same reasoning to cover the more exact expression should be obvious.

The following summarizes the titration expressions for the calculation of $[R]$ for any point in a titration involving the general reaction shown by Eq. (159). The general equations are uncorrected for the dilution effect, in anticipation of the fact that most standard volumetric titrations will involve the condition

$$\frac{M_R}{M_T} = \frac{V_{T,ep}}{V_R} \leq 1$$

for Eq. (159) type reactions.

1. At the start of the titration, $[R] = M_{R, \text{ substance}}$. It should be noted that this relationship applies *only* in those cases where 1 mol of R is secured from 1 mol

of the substance from which R is derived, and where $[R]$ results from the 100 percent dissociation (practically speaking) of this substance. The application of the relationship for other than the mole-for-mole situation is obvious. Instances where less than 100 percent dissociation exists will be discussed under specific titration groups (i.e., the titration of a weak acid by a strong base).

2. Between the start and the equivalence point of the titration, for any point in titration V_T, Eq. (160) is used, except when quite close to the equivalence point and/or where K_{eq} is relatively low, under which conditions Eq. (162) may be required.

3. At the equivalence point, Eq. (164) is used.

4. After the equivalence point, for any point in titration V_T, Eq. (167) is used, except when quite close to the equivalence point and/or where K_{eq} is relatively low, under which conditions Eq. (166) may be required. These equations solve for $[T]_{V_T}$ and in each case, the appropriate $[R]_{V_T}$ is determined by the substitution of the value for $[T]_{V_T}$ into Eq. (168).

Where a more generalized reaction occurs during the titration, such as,

$$rR + tT \rightleftharpoons aA + bB \qquad (174)$$

we can develop the following generalized titration equations which, of course, also encompass those already outlined.

1. At the start of the titration, $[R] = M_{R,\text{substance}}$. Again the special conditions referred to apply here, relative to the mole-for-mole relationship and the 100 percent dissociation condition.

2. Between the start and the equivalence point, for any point V_T, the equation

$$[R]_{V_T} \approx \frac{V_R M_R - r/t\, V_T M_T}{V_R + V_T} \qquad (175)$$

is used, except when quite close to the equivalence point and/or where K_{eq} is relatively low, under which conditions the equation

$$[R]_{V_T} = \frac{V_R M_R - r/t\, V_T M_T}{V_R + V_T} + \frac{r}{t}\sqrt[t]{\frac{[A]^a[B]^b}{K_{eq}[R]_{V_T}{}^r}} \qquad (176)$$

may be required.

3. At the equivalence point,

$$[R]_{ep} = \frac{r}{t}\sqrt[t]{\frac{[A]^a[B]^b}{K_{eq}[R]_{ep}{}^r}} \qquad (177)$$

$$= {}^{(r+t)}\sqrt{\left(\frac{r}{t}\right)^t \frac{[A]^a[B]^b}{K_{eq}}} \qquad (178)$$

4. After the equivalence point, for any point in titration V_T, the equation

$$[T]_{V_T} \approx \frac{V_T M_T - t/r\, V_R M_R}{V_T + V_R} \tag{179}$$

is used, except when quite close to the equivalence point and/or where K_{eq} is relatively low, under which conditions the expression

$$[T]_{V_T} = \frac{V_T M_T - t/r V_R M_R}{V_T + V_R} + \frac{t}{r} \sqrt[r]{\frac{[A]^a [B]^b}{K_{eq}[T]_{V_T}{}^t}} \tag{180}$$

may be required. These equations solve for $[T]_{V_T}$ and in each case, the appropriate $[R]_{V_T}$ is determined by substitution of the value for $[T]_{V_T}$ into the equilibrium constant equation

$$[R]_{V_T} = \sqrt[r]{\frac{[A]^a [B]^b}{K_{eq}[T]_{V_T}{}^t}} \tag{181}$$

As a matter of note, it should be pointed out that Eqs. (176) and (180), and therefore by implication Eq. (181), do not permit simple solutions for $[R]_{V_T}$ when powers of $[R]_{V_T}$, $[T]_{V_T}$, $[A]$, and/or $[B]$ capable of yielding equations of a degree higher than that of the quadratic exist. Such equations can be solved by approximation or other techniques, but in the absence of computer programming, such solutions are both tedious and time-consuming. This situation is generally of minor importance because with titrations requiring the use of Eqs. (176) and (180), the expectation is that the titration reaction will not be quantitative at the equivalence point. Thus the determination of the lack of quantitative completion at the equivalence point would be usually the only important factor, and the need to determine the values of $[R]_{V_T}$ at points before and after the equivalence point would exist on the basis of academic interest alone.

Although we have dealt with a strictly general titration reaction

$$rR + tT \rightleftharpoons aA + bB$$

there are many titration reactions which can be generalized in a similar manner, but with modifications leading to a somewhat simpler approach. Typical of such titration reactions, classified by specific groups, are the following:

Neutralization:	$H_3O^+ + OH^- \rightleftharpoons 2H_2O$	(a)
Precipitation:	$Ag^+ + Cl^- \rightleftharpoons AgCl(s)$	(b)
Complexation:	$Cu^{2+} + trien \rightleftharpoons Cu(trien)^{2+}$	(c)

Such reactions generalized would then lead to

$$rR + tT \rightleftharpoons aA \tag{182}$$

where titration conditions are concerned, and from the above we would have

$$[R]^r[T]^t = \frac{[A]^a}{K_{eq}} = K[A]^a \tag{183}$$

where $K[A]^a$ has the values, for the typical titration reactions given in the foregoing, of

$$[H_3O^+][OH^-] = K_w \qquad (a)$$
$$[Ag^+][Cl^-] = K_{sp}(AgCl) \qquad (b)$$
$$[Cu^{2+}][trien] = K_{instability}[Cu(trien)^{2+}] \qquad (c)$$

In Eq. (c), $K_{instability}$ is, of course, the inverse of the value of $K_{stability}$.

Since we have applied the law of chemical equilibrium to the reaction in the reverse direction to that occurring during the titration, it is apparent that the *lower* the value of $K[A]^a$, the more complete will be the titration reaction.

The following generalized titration equations may be derived, again on a basis of not having been corrected for dilution effects.

1. At the start of the titration, $[R] = M_{R,substance}$. Again, the special conditions referred to previously apply here relative to the mole-for-mole relationship and the 100 percent dissociation condition. For titrations similar to the examples referred to, we would have

$$[H_3O^+] = M_{acid} \qquad \text{for example, } M_{HCl}$$
$$[Ag^+] = M_{Ag,substance} \qquad \text{for example, } M_{AgNO_3}$$
$$[Cu^{2+}] = M_{Cu,substance} \qquad \text{for example, } M_{CuSO_4}$$

2. Between the start and the equivalence point, for any point V_T, the equation

$$[R]_{V_T} \approx \frac{V_R M_R - r/t \, V_T M_T}{V_R + V_T} \qquad (184)$$

is used, except when quite close to the equivalence point and/or where $K[A]^a$ is relatively *high*, under which conditions the equation

$$[R]_{V_T} = \frac{V_R M_R - r/t \, V_T M_T}{V_R + V_T} + \frac{r}{t}\sqrt[t]{\frac{K[A]^a}{[R]_{V_T}{}^r}} \qquad (185)$$

may be required. Note that, for the titration examples used, Eq. (185) gives

$$[H_3O^+]_{V_{OH^-}} = \frac{V_{H_3O^+} M_{H_3O^+} - V_{OH^-} M_{OH^-}}{V_{H_3O^+} + V_{OH^-}} + \frac{K_w}{[H_3O^+]_{V_{OH^-}}}$$

$$[Ag^+]_{V_{Cl^-}} = \frac{V_{Ag^+} M_{Ag^+} - V_{Cl^-} M_{Cl^-}}{V_{Ag^+} + V_{Cl^-}} + \frac{K_{sp}(AgCl)}{[Ag^+]_{V_{Cl^-}}}$$

$$[Cu^{2+}]_{V_{trien}} = \frac{V_{Cu^{2+}} M_{Cu^{2+}} - V_{trien} M_{trien}}{V_{Cu^{2+}} + V_{trien}} + \frac{K_{inst}[Cu(trien)^{2+}]}{[Cu^{2+}]_{V_{trien}}}$$

since, in these examples, $r = t = 1$.

3. At the equivalence point,

$$[R]_{ep} = \sqrt[(r+t)]{\left(\frac{r}{t}\right)^t K[A]^a} \qquad (186)$$

4. After the equivalence point, for any point in titration V_T, the equation

$$[T]_{V_T} \approx \frac{V_T M_T - t/r \, V_R M_R}{V_T + V_R} \qquad (187)$$

is used, except when quite close to the equivalence point and/or where $K[A]^a$ is relatively *high*, under which conditions the equation

$$[T]_{V_T} = \frac{V_T M_T - t/r \, V_R M_R}{V_T + V_R} + \frac{t}{r} \sqrt[r]{\frac{K[A]^a}{[T]_{V_T}^t}} \qquad (188)$$

may be required. These equations solve for $[T]_{V_T}$ and, as before, $[R]_{V_T}$ can be found, in each case, by substitution of the value of $[T]_{V_T}$ into the equation

$$[R]_{V_T} = \sqrt[r]{\frac{K[A]^a}{[T]_{V_T}^t}} \qquad (189)$$

Example   Consider the titration system where Sn(II) is titrated by Fe(III) in an aqueous medium $1\,M$ to HCl. The reactions are

$$Sn^{2+} - 2e \rightleftharpoons Sn^{4+}$$

$$Fe^{3+} + 1e \rightleftharpoons Fe^{2+}$$

$$Sn^{2+} + 2Fe^{3+} \rightleftharpoons Sn^{4+} + 2Fe^{2+}$$

Suppose the solution to be 100.00 ml of $0.1000\,M$ $SnCl_2$ at the start, and the titrant to be $0.2000\,M$ $FeCl_3$. Consider that the $[H_3O^+]$ of the solution remains unchanged during the titration. For a titration at 25°C determine

(a) $[Sn^{2+}]$ at the start
(b) $[Sn^{2+}]$ when 98.00 ml of $FeCl_3$ solution is added
(c) $[Sn^{2+}]$ when 99.90 ml of $FeCl_3$ solution is added
(d) $[Sn^{2+}]$ at the equivalence point
(e) $[Sn^{2+}]$ when 100.10 ml of $FeCl_3$ solution is added
(f) $[Sn^{2+}]$ when 110.00 ml of $FeCl_3$ solution is added

This reaction is of the general type

$$rR + tT \rightleftharpoons aA + bB$$

where $r = 1$

$\qquad t = 2$

$\qquad a = 1$

$\qquad b = 2$

or of the general type

$$n_2 c \, Red_1 + n_1 a Ox_2 \rightleftharpoons n_2 d Ox_1 + n_1 b Red_2$$

where $n_1 = 2$

$\qquad n_2 = 1$

$\qquad a = b = c = d = 1$

The equilibrium constant is given by Eq. (139) as

$$\frac{[Sn^{4+}][Fe^{2+}]^2}{[Sn^{2+}][Fe^{3+}]^2} = 10\left(\frac{0.700 - 0.139}{0.059/2}\right) = 10^{19.0} = K_{eq}$$

(a) $[Sn^{2+}]$ at the start $= M_{SnCl_2} = 1.00 \times 10^{-1}\, M$.

(b) For $[Sn^{2+}]$ when 98.00 ml of $FeCl_3$ solution is added, Eq. (176) gives

$$[Sn^{2+}] = \frac{100.00 \times 0.1 - \frac{1}{2} \times 98.00 \times 0.2}{198.00} + \frac{1}{2}\sqrt{\frac{[Sn^{4+}][Fe^{2+}]^2}{10^{19.0}[Sn^{2+}]}}$$

$$= 0.00101\, M + \frac{1}{2}\sqrt{\frac{[Sn^{4+}][Fe^{2+}]^2}{10^{19.0}[Sn^{2+}]}}$$

Using the following approximate equations,

$$[Sn^{2+}] \approx 0.00101\, M \qquad \text{Eq. (175)}$$

$$[Sn^{4+}] \approx \frac{\frac{1}{2} \times 98.00 \times 0.2}{198.00} \approx 0.0495\, M$$

$$[Fe^{2+}] = 2[Sn^{4+}] \approx 0.0990\, M$$

we have

$$\frac{1}{2}\sqrt{\frac{[Sn^{4+}][Fe^{2+}]^2}{10^{19.0}[Sn^{2+}]}} \approx 1.10 \times 10^{-10}M \ll 1.01 \times 10^{-3}M$$

so that Eq. (175) is adequate for the determination of $[Sn^{2+}]$ and gives

$$[Sn^{2+}] \approx 1.01 \times 10^{-3}\, M$$

(c) For $[Sn^{2+}]$ when 99.90 ml of $FeCl_3$ solution is added, Eq. (176) gives

$$[Sn^{2+}] = 0.0000500\, M + \frac{1}{2}\sqrt{\frac{[Sn^{4+}][Fe^{2+}]^2}{10^{19.0}[Sn^{2+}]}}$$

Using approximate equations as in b, we have

$$\frac{1}{2}\sqrt{\frac{[Sn^{4+}][Fe^{2+}]^2}{10^{19.0}[Sn^{2+}]}} \approx 5.00 \times 10^{-10}M \ll 5.00 \times 10^{-5}M$$

so that again Eq. (175) is adequate at

$$[Sn^{2+}] \approx 5.00 \times 10^{-5}\, M$$

(d) For $[Sn^{2+}]$ at the equivalence point, from Eq. (143), we have

$$\frac{[Sn^{4+}]}{[Sn^{2+}]} = 10^{6.33}$$

and we can assume that we have, at the equivalence point,

$$[Sn^{2+}] + [Sn^{4+}] = [Sn]_{total}$$

$$= \frac{100.00 \times 0.1}{200.00} = 0.0500\, M$$

so that

$$[Sn^{2+}] + 10^{6.33}[Sn^{2+}] = 0.0500\, M$$

and

$$[Sn^{2+}] \approx \frac{5\,00 \times 10^{-2}\,M}{10^{6.33}} \approx 2.34 \times 10^{-8}\,M$$

The use of Eq. (178) gives the identical value.

(e) For $[Sn^{2+}]$ when 100.10 ml of $FeCl_3$ solution is added, Eq. (180) gives

$$[Fe^{3+}] = \frac{100.10 \times 0.2 - 2 \times 100.00 \times 0.1}{200.10} + 2\frac{[Sn^{4+}][Fe^{2+}]^2}{10^{19.0}[Fe^{3+}]^2}$$

$$= 0.000100\,M + 2\frac{[Sn^{4+}][Fe^{2+}]^2}{10^{19.0}[Fe^{3+}]^2}$$

Again using approximate equations, we have

$$[Fe^{3+}] \approx 0.000100\,M \qquad \text{Eq. (179)}$$

$$[Sn^{4+}] \approx \frac{100.00 \times 0.1}{200.10} \approx 0.0500\,M$$

$$[Fe^{2+}] = 2[Sn^{4+}] \approx 0.1000\,M$$

and we have

$$2\frac{[Sn^{4+}][Fe^{2+}]^2}{10^{19.0}[Fe^{3+}]^2} \approx 1.00 \times 10^{-14}\,M \ll 1.00 \times 10^{-4}\,M$$

so that Eq. (179) is adequate and gives

$$[Fe^{3+}] \approx 1.00 \times 10^{-4}\,M$$

Substitution of this value into Eq. (181) yields

$$[Sn^{2+}] \approx 5.00 \times 10^{-15}\,M$$

(f) For $[Sn^{2+}]$ when 110.00 ml of $FeCl_3$ solution is added, Eq. (180) gives

$$[Fe^{3+}] = 0.00952\,M + 2\frac{[Sn^{4+}][Fe^{2+}]^2}{10^{19.0}[Fe^{3+}]^2}$$

Using approximate equations as in e, we have

$$2\frac{[Sn^{4+}][Fe^{2+}]^2}{10^{19.0}[Fe^{3+}]^2} \approx 9.52 \times 10^{-19}\,M \ll 9.52 \times 10^{-3}\,M$$

so that Eq. (179) is adequate and yields

$$[Fe^{3+}] \approx 9.52 \times 10^{-3}\,M$$

Substitution of this value into Eq. (181) gives

$$[Sn^{2+}] \approx 4.76 \times 10^{-19}\,M$$

Example   You are to titrate 80.00 ml of 0.1000 M HCl with 0.2100 M NaOH. If the titration is carried out at 25°C, determine

(a) $[H_3O^+]$ at the start
(b) $[H_3O^+]$ when 15.00 ml of NaOH solution is added
(c) $[H_3O^+]$ when 25.00 ml of NaOH solution is added

(d) $[H_3O^+]$ when 38.00 ml of NaOH solution is added
(e) $[H_3O^+]$ at the equivalence point
(f) $[H_3O^+]$ when 38.30 ml of NaOH solution is added
(g) $[H_3O^+]$ when 45.00 ml of NaOH solution is added

The reaction involved is

$$HCl + NaOH \rightleftharpoons NaCl + H_2O$$

or more generally,

$$H_3O^+ + OH^- \rightleftharpoons 2H_2O$$

The ionization constant equation is

$$[H_3O^+][OH^-] = K_w = 1.00 \times 10^{-14} \, M$$

The equivalence point will occur at

$$V_{T,ep} = \frac{80.00 \times 0.1}{0.2100} = 38.10 \text{ ml}$$

The reaction is of the general type

$$rR + tT \rightleftharpoons aA$$

where

$$r = t = 1 \qquad K[A]^a = K_w$$

(a) For $[H_3O^+]$ at the start,

$$[H_3O^+] = M_{HCl} = 0.1000 \, M = 1.00 \times 10^{-1} \, M$$

(b) For $[H_3O^+]$ when 15.00 ml of NaOH solution is added, Eq. (185) gives

$$[H_3O^+] = \frac{80.00 \times 0.1 - 15.00 \times 0.21}{95.00} + \frac{K_w}{[H_3O^+]}$$
$$= 0.0510 \, M + \frac{K_w}{[H_3O^+]}$$

It is apparent the $K_w/[H_3O^+]$ will be much less than 0.0510 M, so that Eq. (184) will be adequate to give

$$[H_3O^+] \approx 5.10 \times 10^{-2} \, M$$

(c) For $[H_3O^+]$ when 25.00 ml of NaOH solution is added, Eq. (185) gives

$$[H_3O^+] = 0.0262 \, M + \frac{K_w}{[H_3O^+]}$$

and, again, $K_w/[H_3O^+]$ is much less than 0.0262 M, so that we have Eq. (184) adequate to give

$$[H_3O^+] \approx 2.62 \times 10^{-2} \, M$$

(d) For $[H_3O^+]$ when 38.00 ml of NaOH solution is added, Eq. (185) gives

$$[H_3O^+] = 0.000169 \, M + \frac{K_w}{[H_3O^+]}$$

and, again, $K_w/[H_3O^+]$ is much less than 0.000169 M, so that we have Eq. (184) adequate to give

$$[H_3O^+] \approx 1.69 \times 10^{-4}\, M$$

(e) For $[H_3O^+]$ at the equivalence point, Eq. (186) gives

$$[H_3O^+] = \sqrt{K_w} = 1.00 \times 10^{-7}\, M$$

(f) For $[H_3O^+]$ when 38.30 ml of NaOH solution is added, Eq. (188) gives

$$[OH^-] = \frac{38.30 \times 0.21 - 80.00 \times 0.1}{118.30} + \frac{K_w}{[OH^-]}$$
$$= 0.000363\, M + \frac{K_w}{[OH^-]}$$

Here we can show that $K_w/[OH^-]$ is much less than 0.000363 M, so that Eq. (187) is adequate as

$$[OH^-] \approx 3.63 \times 10^{-4}\, M$$

Substituted in Eq. (189) this gives

$$[H_3O^+] \approx 2.75 \times 10^{-11}\, M$$

(g) For $[H_3O^+]$ when 45.00 ml of NaOH solution is added, Eq. (188) gives

$$[OH^-] = 0.0116\, M + \frac{K_w}{[OH^-]}$$

Again $K_w/[OH^-]$ can be shown to be much less than 0.0116 M, so that Eq. (187) is again adequate as

$$[OH^-] \approx 1.16 \times 10^{-2}\, M$$

with the substitution of this value into Eq. (189) yielding

$$[H_3O^+] \approx 8.62 \times 10^{-13}\, M$$

5.2   THE THEORY OF TITRATION CURVES

5.2.1   GENERAL

In the previous section it was noted that for several types of titration reaction, general equations could be developed for the calculation of values of [R] during the progress of the titration. One of the reaction types discussed was

$$r\mathrm{R} + t\mathrm{T} \rightleftharpoons a\mathrm{A} \qquad\qquad (182)$$

For the purposes of easier manipulation in developing titration-curve theory, we shall show this reaction as

$$\mathrm{R} + \mathrm{T} \rightleftharpoons \mathrm{A} \qquad\qquad (190)$$

The method of applying the law of chemical equilibrium, used in the previous section for reactions of this general type, would lead to

$$[R][T] = \frac{[A]}{K_{eq}} = K[A]$$

so that, the lower the value of $K[A]$ or K, the more complete the reaction.

Although for simplicity's sake we shall, as mentioned before, concentrate on this reaction in our approach to titration-curve theory, it should be remembered that subsequent developments in this section will apply *in the same general sense* to any form of titration reaction, due consideration being given to the specific characteristics of the reaction involved.

Let us consider, first of all, a titration involving the reaction shown in Eq. (190). We shall assume, for the preliminary treatment, that the effect of dilution during titration has been compensated for, and that the value of $K[A]$ or K is so low that the back-reaction contribution of $[R]_{br}$ or $[T]_{br}$ is negligible at all practical titration points after the start, with the exception of the equivalence point itself.

Here we would have, for all titration points after the start and before the equivalence point,

$$[R]_{V_T} \approx \frac{V_R M_R - V_T M_T}{V_R} \tag{169}$$

and

$$-\log [R]_{V_T} = -\log (V_R M_R - V_T M_T) + \log V_R$$

so that

$$pR_{V_T} = -\log (V_R M_R - V_T M_T) + \log V_R$$

Differentiating with respect to V_T leads to

$$\frac{dpR_{V_T}}{dV_T} = \frac{M_T}{V_R M_R - V_T M_T} \tag{191}$$

Note that this idealized version of the course of the titration between the start and the equivalence point indicates that

1. The slope of the curve for the plot of pR_{V_T} vs. V_T is represented by dpR_{V_T}/dV_T.
2. Since the value of $(V_R M_R - V_T M_T)$ approaches zero as a limit at the equivalence point, the slope of the curve increases to a maximum value at the equivalence point.
3. In the idealized version shown by Eq. (191), the slope becomes infinitely great at the equivalence point.

If consideration is now given to the back-reaction situation, this leads to

$$[R]_{V_T} = \frac{V_R M_R - V_T M_T}{V_R} + [R]_{br} \tag{192}$$

for the range of titration points just considered. Through the same process as applied before, this equation leads to

$$\frac{d\text{pR}_{V_T}}{dV_T} = \frac{M_T - V_R(d[\text{R}]_{br}/dV_T)}{(V_R M_R - V_T M_T) + [\text{R}]_{br} V_R} \tag{193}$$

For this less idealized version of the course of the titration between the start and the equivalence point we note that

1. As before, the slope of the curve for the plot of pR_{V_T} vs. V_T is represented by $d\text{pR}_{V_T}/dV_T$.
2. Since the value of $(V_R M_R - V_T M_T)$ approaches zero as a limit at the equivalence point, the slope of the curve increases to a maximum at the equivalence point.
3. Because of the limitations imposed by the back-reaction contribution of $[\text{R}]_{br}$, the slope attains a practical or finite maximum value at the equivalence point. Note that the magnitude of this finite maximum value will be a function of the magnitude of the back-reaction contribution and will decrease with increasing value of this contribution. The back-reaction contribution, of course, increases with increasing values of $K[\text{A}]$ or K.

If further consideration is now given to this reaction, this time from the point of view of the dilution effect, where this factor has not been compensated for, we have

$$[\text{R}]_{V_T} = \frac{V_R M_R - V_T M_T}{V_R + V_T} + [\text{R}]_{br} \tag{194}$$

again for the same range of titration points previously discussed. Application of the same process leads to

$$\frac{d\text{pR}_{V_T}}{dV_T} = \frac{M_T - \{(d[\text{R}]_{br}/dV_T)(V_R + V_T) + [\text{R}]_{br}\}}{(V_R M_R - V_T M_T) + [\text{R}]_{br}(V_R + V_T)} + \frac{1}{V_R + V_T} \tag{195}$$

For this nonidealized and, indeed, very standard version of the course of the titration between the start and the equivalence point, we note that

1. As before, the slope of the curve for the plot of pR_{V_T} vs. V_T is represented by $d\text{pR}_{V_T}/dV_T$.
2. Since the value of $(V_R M_R - V_T M_T)$ approaches zero as a limit at the equivalence point, the slope of the curve increases to a maximum at the equivalence point.
3 Because of the limitations imposed by the back-reaction contribution of $[\text{R}]_{br}$, we have the identical situation as that noted in 3 of the discussion just completed.
4. The effect of dilution during titration is to decrease the slope for comparable points in the titration and to render less acute the rate of change of pR_{V_T} rela-

tive to V_T, particularly in the zone close to the equivalence point. These effects are introduced in Eq. (195) by the terms in the right-hand member involving $(V_R + V_T)$.

For equations associated with titration conditions after the equivalence point, we have a choice, on the same progressive basis as that used before, of the equations

$$[T]_{V_T} \approx \frac{V_T M_T - V_R M_R}{V_R} \tag{196}$$

$$[T]_{V_T} = \frac{V_T M_T - V_R M_R}{V_R} + [T]_{br} \tag{197}$$

$$[T]_{V_T} = \frac{V_T M_T - V_R M_R}{V_T + V_R} + [T]_{br} \tag{198}$$

These equations lead, respectively, to

$$pT_{V_T} = -\log (V_T M_T - V_R M_R) + \log V_R$$

and since

$$pR_{V_T} + pT_{V_T} = pK[A]$$

we have

$$pR_{V_T} = pK[A] - pT_{V_T}$$

so that

$$\frac{dpR_{V_T}}{dV_T} = \frac{M_T}{V_T M_T - V_R M_R} \tag{199}$$

Using the same approach, we can show that the remaining two equations are

$$\frac{dpR_{V_T}}{dV_T} = \frac{M_T + V_R(d[T]_{br}/dV_T)}{(V_T M_T - V_R M_R) + [T]_{br} V_R} \tag{200}$$

$$\frac{dpR_{V_T}}{dV_T} = \frac{M_T + \{(d[T]_{br}/dV_T)(V_T + V_R) + [T]_{br}\}}{(V_T M_T - V_R M_R) + [T]_{br}(V_T + V_R)} - \frac{1}{V_T + V_R} \tag{201}$$

Note that, for these equations, we have the following general situations progressively.

1. The slope of the curve for the plot of pR_{V_T} vs. V_T is given by dpR_{V_T}/dV_T for points after the equivalence point, just as for points before the equivalence point.
2. Since the value of $(V_T M_T - V_R M_R)$ increases after the equivalence point, the slope of the curve decreases after the equivalence point.
3. The limitations imposed by the back-reaction contribution of $[T]_{br}$ or $[R]_{br}$, and the dilution effect, tend to render less acute the rate of change of pR_{V_T} relative to V_T.

In summing up the developments so far, it is to be noted that the slope of the curve for the plot of pR_{V_T} vs. V_T increases to a maximum at the equivalence point of the titration and then decreases with increasing excess of titrant after the equivalence point. Thus, the value of dpR_{V_T}/dV_T passes through a point of inflection at the titration equivalence point.

An examination of Eqs. (194) and (198) indicates that as the values of M_R and/or M_T are decreased for comparable starting volumes of V_R, the value of $[R]_{V_T}$ will be decreased for similar points in the titration before the equivalence point, and increased for similar points of titrant excess after the equivalence point. This leads, for such situations, to higher values of pR_{V_T} before the equivalence point and lower values after the equivalence point, all for comparable points of titration completion and titrant excess. Table 5.1 shows this effect, by a comparison of similar titration completion points for, respectively, the titration of 100 ml of 0.1 M strong acid by 0.1 M strong base, 100 ml of 0.01 M strong acid by 0.1 M strong base, and 100 ml of 0.001 M identical acid by 0.001 M identical base.

Do not miss the significance of the situation outlined in Table 5.1. Note that the range of change of pH around the equivalence point decreases with increasing degree of dilution of either or both the reactant (H_3O^+) and the titrant (OH^-) solutions. This implies that the accuracy of locating the equivalence-point volume value for the titration decreases with increasing degree of dilution at the start for the solutions titrated together.

The curves for the plots of pH vs. volume of titrant V_T, for each of the titrations covered by Table 5.1, are shown in Fig. 5.4. Note the decrease in the range of change of pH around the equivalence point with increasing degree of dilution at

Table 5.1 Strong acid–strong base titrations at various starting molarities*

Titration completion, %	pH		
	100 *ml* 0.1 *M acid* vs. 0.1 *M base*	100 *ml* 0.01 *M acid* vs. 0.1 *M base*	100 *ml* 0.001 *M acid* vs. 0.001 *M base*
0	1.0	2.0	3.0
10	1.1	2.1	3.1
25	1.2	2.2	3.2
50	1.5	2.4	3.5
75	1.9	2.6	3.9
90	2.3	3.0	4.3
95	2.6	3.3	4.6
99	3.3	4.0	5.3
101	10.7	10.0	8.7
110	11.7	11.0	9.7
120	12.0	11.3	10.0

*The titration data have been corrected, in each instance, for the back-reaction effect where required but not corrected for the titration-dilution effect.

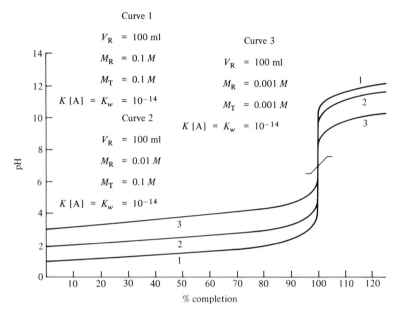

Fig. 5.4   The effect of starting molarities on the range of change of pR (pH) around the equivalence point.

the start for either or both the reactant and titrant solutions. Appreciate the fact that these titration curves exemplify in general the situation relative to plots of pR_{V_T}, or pR, vs. V_T.

The statement that the value of dpR_{V_T}/dV_T passes through a maximum at the equivalence point indicates that a plot of finite values of dpR_{V_T}/dV_T vs. V_T, or $\Delta pR_{V_T}/\Delta V_T$ vs. V_T, must increase to a maximum value and subsequently decrease. Such a set of plots for $\Delta pH/\Delta V_B$ vs. V_B is shown in Figs. 5.5, 5.6, and 5.7, and covers the data for the titrations outlined in Table 5.1.

Again, since dpR_{V_T}/dV_T passes through a point of inflection at the equivalence point, the second derivative $d^2pR_{V_T}/dV_T^2$ must pass through zero at the equivalence point. For the titrations under discussion, since a rising curve of pR_{V_T} vs. V_T is involved, this implies that $d^2pR_{V_T}/dV_T^2$ must increase to a maximum, pass through zero at the equivalence point, attain a minimum immediately after, and subsequently increase again with increasing excess of titrant   A set of plots for finite values of $d^2pR_{V_T}/dV_T^2$ vs. V_T, or $\Delta^2pH/\Delta V_B^2$ vs. V_B, is shown in Figs. 5.8, 5.9, and 5.10. Again, the data involved is taken from the titrations outlined in Table 5.1.

Note that for the general reaction

$$R + T \rightleftharpoons A$$

if we had plotted pT_{V_T} vs. V_T, we would have had opposing concentration change

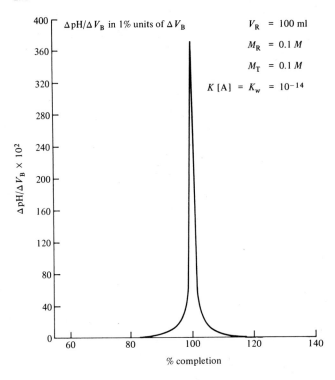

Fig. 5.5 $\Delta pH/\Delta V_B$ vs. V_B.

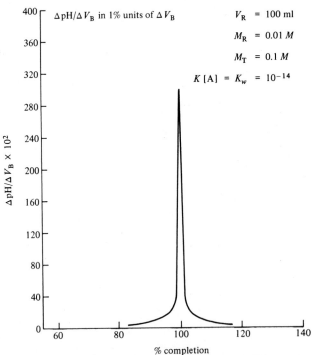

Fig. 5.6 $\Delta pH/\Delta V_B$ vs. V_B.

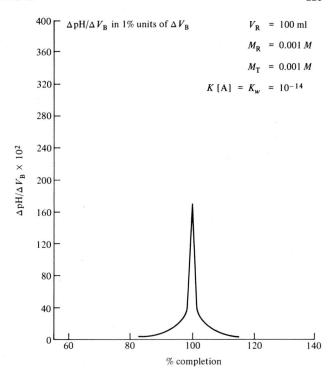

$\Delta pH/\Delta V_B$ in 1% units of ΔV_B

$V_R = 100$ ml

$M_R = 0.001\ M$

$M_T = 0.001\ M$

$K[A] = K_w = 10^{-14}$

% completion

Fig. 5.7 $\Delta pH/\Delta V_B$ vs. V_B.

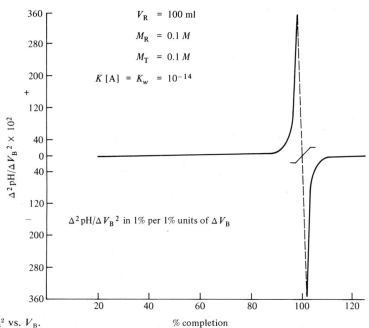

$V_R = 100$ ml

$M_R = 0.1\ M$

$M_T = 0.1\ M$

$K[A] = K_w = 10^{-14}$

$\Delta^2 pH/\Delta V_B^2$ in 1% per 1% units of ΔV_B

% completion

Fig. 5.8 $\Delta^2 pH/\Delta V_B^2$ vs. V_B.

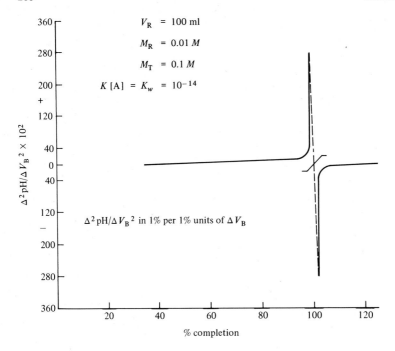

Fig. 5.9 $\Delta^2 pH / \Delta V_B^2$ vs. V_B.

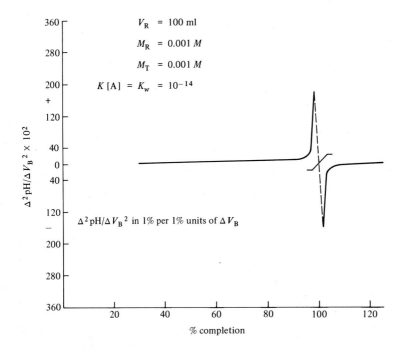

Fig. 5.10 $\Delta^2 pH / \Delta V_B^2$ vs. V_B.

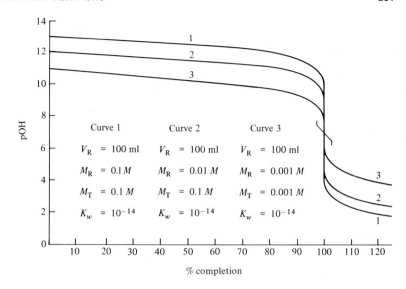

Fig. 5.11   The effect of starting molarities on the range of change of pT (pOH) around the equivalence point.

situations to those outlined for the plots involving pR_{V_T}. The slope of the curve would have, in other words, decreased to a minimum at the equivalence point and subsequently increased. Such plots are shown in Fig. 5.11 and again involve data from Table 5.1. The same general situation would, of course, apply in the case of plots of $\Delta pT_{V_T}/\Delta V_T$ vs. V_T and $\Delta^2 pT_{V_T}/\Delta V_T{}^2$ vs. V_T.

It is of prime importance to give consideration to the effect of the magnitude of $K[A]$, or K, the factor which bears on the degree of reaction completion, on the curve forms for titrations associated with the general reaction given by Eq. (190). Obviously, higher values of $K[A]$, or K, in the equation

$$[R]_{V_T} = \frac{V_R M_R - V_T M_T}{V_R + V_T} + [R]_{br}$$
$$= \frac{V_R M_R - V_T M_T}{V_R + V_T} + \frac{K}{[R]_{V_T}}$$

provide for higher concentrations of $[R]$ from the back-reaction, and therefore increased values of $[R]_{V_T}$ for any comparable point in titration before the equivalence point.

Again, with $[R]$ at the equivalence point given generally by the equation

$$[R]_{ep} = \sqrt{K[A]}$$

increasing values of $K[A]$ or K will permit higher values of $[R]_{ep}$.

For the situation after the equivalence point, it is apparent that we will have, for any point V_T,

$$[R]_{V_T} = \frac{K[A]}{[T]_{V_T}}$$

Table 5.2 The effect of the value of K on the range of change of pR_{V_T} around the equivalence point*

Titration completion, %	All titrations are 100 ml of 0.1 M R vs. 0.1 MT		
	pR_{V_T} $K = 10^{-10}$	pR_{V_T} $K = 10^{-8}$	pR_{V_T} $K = 10^{-6}$
0	1.0	1.0	1.0
10	1.1	1.1	1.1
25	1.2	1.2	1.2
50	1.5	1.5	1.5
75	1.9	1.9	1.8
90	2.3	2.3	2.2
95	2.6	2.6	2.5
99	3.3	3.3	2.9
101	6.7	4.7	3.1
110	7.7	5.7	3.7
120	8.0	6.0	4.0

*The titration data have been corrected for the back-reaction effect where required but not corrected for the dilution effect during titration.

so that any increase in the value of $K[A]$, or K, will result, for comparable points after the equivalence point, in higher values of $[R]_{V_T}$.

All the foregoing discussion indicates that with increasing values of $K[A]$, or K, the range of change of $[R]$ around the equivalence point will be decreased, with a corresponding decrease in the range of change of pR_{V_T} or pR. Thus, the ability to locate accurately the equivalence-point volume of the titration will be adversely affected.

This effect is shown in Table 5.2, which includes data associated with titrations involving identical values of V_R, M_R, and M_T, but with varying values of K. The effect, for the same data, is demonstrated graphically by the combined curves of Fig. 5.12.

If one now considers the combined effect of an increasing degree of dilution at the start of the reactant and/or titrant solutions and an increasing value of $K[A]$, or K, it is at once apparent that the effect of increasing values of $K[A]$, or K, will be to increase the sensitivity of a titration to increasing degree of dilution at the start of reactant and/or titrant solutions. In other words, increasing values of $K[A]$, or K, will severely limit the extent to which the starting concentrations of reactant and/or titrant solutions can be diluted, always bearing in mind the necessity of locating the equivalence-point volume with a sensible degree of accuracy.

Table 5.3 and the associated Figs. 5.13 and 5.14 show the effects of both degree of dilution of reactant and titrant solutions at the start and the magnitude of K, on the range of change of pR_{V_T} around the equivalence point and on the curve forms for the plots of pR_{V_T} vs. V_T. Note the loss of accuracy in locating the equivalence-point volume of titrant that is implied by these curves as they progress from low to high degrees of dilution and from low to high values of K.

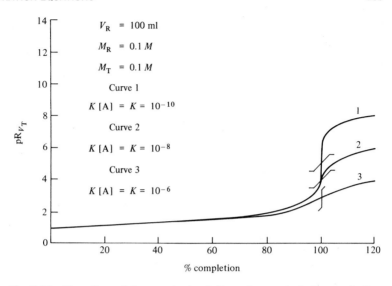

Fig. 5.12 The effect of the magnitude of K on the range of change of pR_{V_T} around the equivalence point.

5.2.2 LOCATION OF THE EQUIVALENCE POINT: pR_{V_T}-VS.-V_T CURVES

The determination of the equivalence-point volume of titrant from the curve resulting from the plot of pR_{V_T} vs. V_T involves first of all the location of the point of inflection of this curve. For mole-for-mole titration reactions the point of inflection and the equivalence point will coincide. For other than mole-for-mole reactions there will be a lack of coincidence between these points. This phenomenon will be discussed in Sec. 5.2.6.

Several methods of curve analysis can be applied in the location of the point of inflection. The three outlined in the following are perhaps the most commonly encountered.

The method of bisection This technique, shown under application in Fig. 5.15, is usually applied to reasonably symmetrical curves for pR_{V_T} vs. V_T, where curve sections showing good approximations to straight lines are obtained before and after the equivalence-point zone of high rate of change of the slope, and where a mole-for-mole titration reaction is involved.

Tangential method using parallel tangents One of the methods of applying this technique involves the use of a wide plastic sheet, marked with paired-by-color parallel lines on either side of a central parallel slot wide enough to accommodate a pencil point. The device is shown in Fig. 5.16, and its method of application is shown in Fig. 5.17. The technique can be applied to symmetrical curve plots

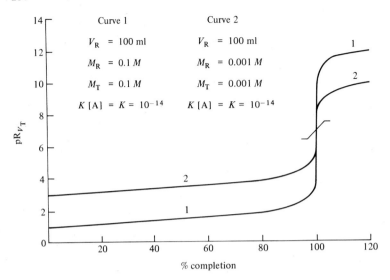

Fig. 5.13 The effect of both magnitude of K and the degree of dilution of reactant and titrant solutions on the range of change of pR_{V_T} around the equivalence point.

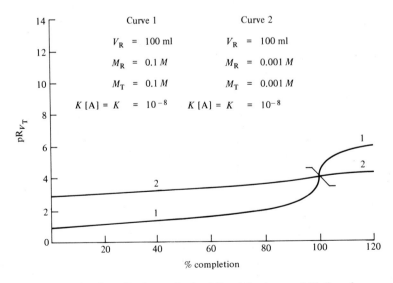

Fig. 5.14 The effect of both magnitude of K and the degree of dilution of reactant and titrant solutions on the range of change of pR_{V_T} around the equivalence point.

Table 5.3   The effect of both magnitude of K and the degree of dilution of reactant and titrant solutions on the range of change of pR_{V_T} around the equivalence point*

	$K = 10^{-14}$	
Titration completion, %	100 *ml* of 0.1 *M* R vs. 0.1 *M* T	100 *ml* of 0.001 *M* R vs. 0.001 *M* T
	pR_{V_T}	pR_{V_T}
0	1.0	3.0
10	1.1	3.1
25	1.2	3.2
50	1.5	3.5
75	1.9	3.9
90	2.3	4.3
95	2.6	4.6
99	3.3	5.3
101	10.7	8.7
110	11.7	9.7
120	12.0	10.0
	$K \times 10^{-8}$	
0	1.0	3.0
10	1.1	3.1
25	1.2	3.2
50	1.5	3.5
75	1.9	3.7
90	2.3	3.9
95	2.6	3.9^5
99	3.3	3.9^8
101	4.7	4.0^2
110	5.7	4.1
120	6.0	4.2

*The titration data have been corrected for the back-reaction effect where required but not corrected for the dilution effect during titration.

which do not have approximately straight-line sections before and after the equivalence-point zone, and where a mole-for-mole titration reaction is involved. It can also, of course, be applied to curve plots which do show straight-line sections before and after the zone of the equivalence point.

Circle-fit method   This technique involves the use of a thin but rigid plastic sheet marked with circles of varying diameter, the center of each circle being indicated

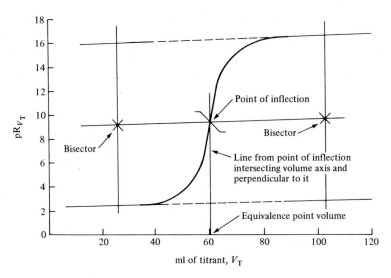

Fig. 5.15 Location of the equivalence-point volume—pR_{V_T}-vs.-V_T curve: method of bisection.

by a hole through the sheet capable of accommodating the point of a pencil. The device is shown in Fig. 5.18. Those circles having the proper circumferential curvatures to match the curve sections just before and just after the equivalence-point zone are identified, and their centers are marked on the plot sheet while the circles are lined up with the curves. These centers are then joined by a straight line which intersects the plot of pR_{V_T} vs. V_T at the *equivalence point* of the titration, regardless of whether this point does or does not coincide with the point of inflection. The application is shown in Fig. 5.19. The method can be applied to symmetrical or asymmetrical curves representing mole-for-mole or other than mole-for-mole titration reactions.

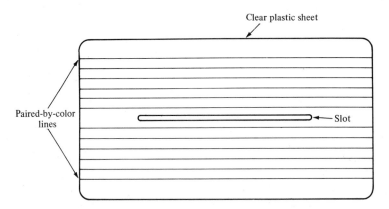

Fig. 5.16 Paired-by-color parallel lines for tangential method.

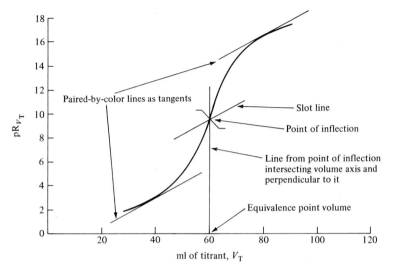

Fig. 5.17 Location of the equivalence-point volume—pR_{V_T}-vs.-V_T curve: tangential method.

5.2.3 LOCATION OF THE EQUIVALENCE POINT: $\Delta pR_{V_T}/\Delta V_T$-VS.-$V_T$ CURVES

Insofar as the analysis of $\Delta pR_{V_T}/\Delta V_T$-vs.-$V_T$ curves is concerned, a typical example of this form of analysis is shown in Fig. 5.20. The assumption is made that the extrapolated lines on either side of the approximate location of the equivalence-point zone will meet at a point on that line perpendicular to the volume axis and dropped from the point where the curve attains its theoretical maximum value. A similar approach will apply in the case of a $\Delta pR_{V_T}/\Delta_{V_T}$-vs.-$V_T$ plot

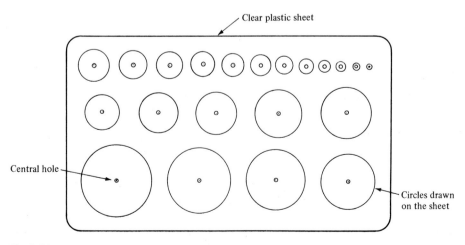

Fig. 5.18 Varying circle sheet for circle-fit method.

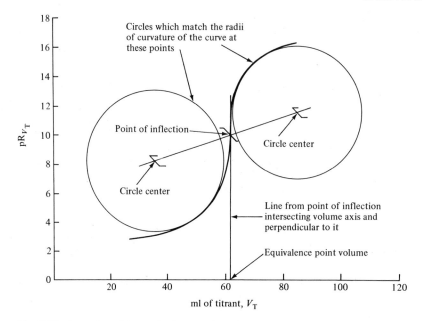

Fig. 5.19 Location of the equivalence-point volume—pR_{V_T}-vs.-V_T curve: circle-fit method.

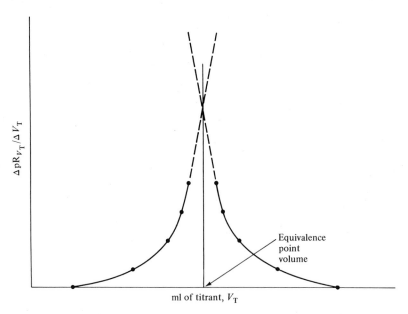

Fig. 5.20 Location of the equivalence-point volume—$\Delta pR_{V_T}/\Delta V_T$-vs.-$V_T$ curve.

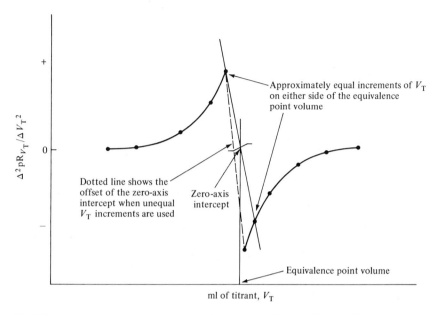

Fig. 5.21   Location of the equivalence-point volume—$\Delta^2 pR_{V_T}/\Delta V_T^2$-vs.-$V_T$ curve.

which attains a minimum value at the equivalence point. Properly applied, the method will locate the equivalence-point volume for titration reactions involving mole-for-mole or other than mole-for-mole situations.

5.2.4 LOCATION OF THE EQUIVALENCE POINT: $\Delta^2 pR_{V_T}/\Delta V_T^2$-VS.-$V_T$ CURVES

A typical example of the analysis of a curve for the plot of $\Delta^2 pR_{V_T}/\Delta V_T^2$ vs. V_T is shown in Fig. 5.21. Note that where values of $\Delta^2 pR_{V_T}/\Delta V_T^2$ do not represent values of V_T symmetrically located on either side of the equivalence point, some sort of adjustment should be made in order that the equivalence point be accurately located. In practice, such adjustments are difficult to achieve, but the error introduced where the volume increments are small is insignificant. Plots of this type are rarely made; but where they are, and where the volume increments before and after the equivalence-point zone are small, the equivalence-point volume can be located with reasonable accuracy for titration reactions of the mole-for-mole or other than mole-for-mole type.

5.2.5 LOCATION OF THE EQUIVALENCE POINT: TABULATION METHOD

It can be appreciated that graphic methods, and the analysis of the related curves, may or may not be capable of accurate equivalence-point-volume location. Much depends on the general shape of any given curve and since this will depend, for

Table 5.4   Data from a titration of 50.00 ml of 0.1 M strong acid vs. 0.09 M strong base*

Volume of base, ml	pH	$(\Delta pH/\Delta V_B) \times 10^3$, $\Delta V_B = 0.1$ ml	$(\Delta^2 pH/\Delta V_B^2) \times 10^3$, $\Delta V_B^2 = 0.1/0.1$ ml
0.00	1.00		
5.00	1.08		
10.00	1.16		
20.00	1.34		
30.00	1.54		
40.00	1.81		
		3.8 (42.50)	
45.00	2.00		0.044 (45.00)
		6 (47.50)	
50.00	2.30		0.1 (49.00)
		9 (50.50)	
51.00	2.39		0.2 (51.00)
		11 (51.50)	
52.00	2.50		0.4 (52.00)
		15 (52.50)	

*The titration data have been corrected for the back-reaction effect where required but not corrected for the dilution effect during titration. The volume values in parentheses after each $\Delta pH/\Delta V_B$ and $\Delta^2 pH/\Delta V_B^2$ value associate the average slope, or rate of change of the slope, with the average volume value between the titration points involved.

hand-drawn curves, on the number of V_T points investigated around the equivalence-point zone and on the ability of the operator to fit a curve properly to these points, there is much latitude for inaccuracy in the location of the equivalence-point volume. In addition to this, the processes of recording data, plotting points, and fitting the curve are tedious and time-consuming.

Our main reason for discussing curve analysis as a means of locating the equivalence-point volume has been because of the fact that many laboratories obtain, from automatic potentiographic titration equipment, machine-drawn curves representing the continuous plot of pR_{V_T} vs. V_T, or $\Delta pR_{V_T}/\Delta V_T$ vs. V_T, during the titration. Such curves are smooth, and by the methods just discussed, reasonably accurate location of the equivalence-point volume is nearly always possible within the limitations dictated by the characteristics of the titration and the titration reaction, such as magnitude of $K[A]$, or K, degree of dilution at the start of the reactant and/or titrant solutions, and the mole relationships between the reactant and titrant substances.

It should now be apparent that the tabulation of titration data, and in particular the data for points just before and just after the equivalence point, can lead to simple methods of identifying the equivalence-point volume. Table 5.4 shows the

Table 5.4 *(Continued)*

Volume of base, ml	pH	$(\Delta \text{pH}/\Delta V_B) \times 10^3$, $\Delta V_B = 0.1$ ml	$(\Delta^2 \text{pH}/\Delta V_B^2) \times 10^3$, $\Delta V_B^2 = 0.1/0.1$ ml
53.00	2.65		0.7 (53.00)
		22 (53.50)	
54.00	2.87		2.3 (54.00)
		45 (54.50)	
55.00	3.32		8.2 (54.78)
		90 (55.05)	
55.10	3.41		20 (55.10)
		110 (55.15)	
55.20	3.52		30 (55.20)
		140 (55.25)	
55.30	3.66		80 (55.30)
		220 (55.35)	
55.40	3.88		220 (55.40)
		440 (55.45)	
55.50	4.32		4820 (55.50)
		5260 (55.55)	
55.60	9.58		−4750 (55.60)
		510 (55.65)	
55.70	10.09		−280 (55.70)
		230 (55.75)	
55.80	10.32		−80 (55.80)
		150 (55.85)	
55.90	10.47		−40 (55.90)
		110 (55.95)	
56.00	10.58		−11 (56.23)
		51 (56.50)	
57.00	11.09		−2.9 (57.00)
		22 (57.50)	
58.00	11.31		−0.7 (58.00)
		15 (58.50)	
59.00	11.46		−0.5 (59.00)
		10 (59.50)	
60.00	11.56		−0.1 (61.00)
		6 (62.50)	
65.00	11.87		
75.00	12.15		
90.00	12.35		

data tabulated for a titration of a strong acid by a strong base where

$$R = H_3O^+ \quad T = OH^- \quad M_R = 0.1\,M \quad M_T = 0.09\,M$$
$$V_R = 50.00\text{ ml} \quad K[A] = K_w = 10^{-14}$$

Note carefully how the values of $\Delta \text{pR}_{V_T}/\Delta V_T$ ($\Delta \text{pH}/\Delta V_B$) and $\Delta^2 \text{pR}_{V_T}/\Delta V_T^2$

$(\Delta^2 pH/\Delta V_B^2)$ are determined. The volume figures in parentheses approximate the average V_T appropriate to the associated average slope or rate of change of slope determined. For this titration, the finite interval of ΔV_T (ΔV_B) selected was 0.1 ml. This could have been any appropriate value, such as 0.2 ml, 0.05 ml, etc. The selection of the value for ΔV_T is a matter of judgment; it is based on the range of change of pR_{V_T} around the equivalence point. Where this is expected to be small, a value of ΔV_T at 0.2 or 0.1 ml will often be suitable; where the rate of change is expected to be large, a ΔV_T value of 0.05 or 0.02 ml may be required. The volume intervals or increments of V_T around the equivalence point are uniform in size and arranged with an eye to as much symmetry as possible. Those volume points far removed from the equivalence-point zone are not uniform, neither are they symmetrically arranged, and the method of treating these values is obvious.

The methods of determining the equivalence-point volume for two aspects of the data tabulation in Table 5.4 are described in the following.

1. We have indicated earlier that where pR_{V_T} increases with V_T, the $\Delta pR_{V_T}/\Delta V_T$-vs.-$V_T$ plot reaches a maximum value at the equivalence point for mole-for-mole titration reactions. Thus, for such reactions, the titration volume at which $\Delta pR_{V_T}/\Delta V_T$ attains its maximum value is identified as the equivalence-point volume. For the titration now under consideration, $\Delta pH/\Delta V_B$ attains its maximum value at an equivalence-point volume of 55.55 ml of OH^- solution.

2. When discussing earlier the plot of $\Delta^2 pR_{V_T}/\Delta V_T^2$ vs. V_T it was noted that the value of $\Delta^2 pR_{V_T}/\Delta V_T^2$ passed through zero at the equivalence point for titration reactions on a mole-for-mole basis. A value of zero is infrequently obtained in tabulations of data secured from a titration. What is more common is to encounter a situation similar to that exemplified by the Table 5.4 data, where $\Delta^2 pR_{V_T}/\Delta V_T^2$, in this case $\Delta^2 pH/\Delta V_B^2$, attains a high positive value and, for the next recorded titration point, takes on a high negative value. The equivalence-point volume can be determined by using these $\pm$ values for $\Delta^2 pH/\Delta V_B^2$ as absolute values in the equation

Magnitude of last positive value = 4820   corresponds to 55.50 ml

Magnitude of first negative value = 4750   corresponds to 55.60 ml

$$\text{Equivalence-point volume} = 55.50 + \frac{482}{482 + 475}(55.60 - 55.50)$$
$$= 55.50 + 0.05 = 55.55 \text{ ml}$$

It should be pointed out again that as was the case with the $\Delta pH/\Delta V_B$ tabulation, the approach of $\Delta^2 pH/\Delta V_B^2$ to zero from the positive side is a reflection of the fact that for the titration under discussion, pH increases with V_B.

5.2.6 COINCIDENCE OF THE POINT OF INFLECTION AND THE EQUIVALENCE POINT

It is of prime importance to note that with respect to most of the graphical and tabulation methods mentioned in this section, the assumption is made that there will be a coincidence between the point of location of the equivalence point on the titration curve and the point of inflection of the curve. This assumption is, in point of fact, only valid for those symmetrical titration curves derived for titration reactions where the reactant and the titrant combine in a one-for-one ratio. Such a reaction would be typified by

$$Ag^+ + Cl^- \rightleftharpoons AgCl(s)$$

In the case of asymmetrical curves derived with respect to titration reactions where the reactant and titrant combine in other than a one-for-one ratio, the point of inflection of the curve and the equivalence point will not coincide. Such a reaction would be exemplified by

$$5Fe^{2+} + MnO_4^- + 8H_3O^+ \rightleftharpoons 5Fe^{3+} + Mn^{2+} + 12H_2O$$

where the reactant and titrant combine in the ratio of five for one. A curve typical of this situation may be found in Chap. 9, Fig. 9.6. In the majority of cases involving asymmetrical titration curves, however, the range of change of pR around the equivalence point will be large enough to permit this discrepancy without the introduction of a significant titration error.

An exception to this general situation is the curve-fit method for the analysis of pR-vs.-V_T or E_{cell}-vs.-V_T curves. Where properly applied, this technique will permit a good approximation to the equivalence-point volume, even where the equivalence point and the point of inflection of the curve do not coincide. Again, in the analysis of first- and second-derivative-vs.-titrant-volume curves, the methods of curve analysis, properly applied, will usually serve to permit the location of the equivalence-point volume, even where coincidence between the equivalence point and the point of inflection or its attributes does not exist.

PROBLEMS

1. A titration system involves a solution of Sn(II) titrated by Ce(IV) in $1\,M\,H_2SO_4$. The solution at the start is 50.00 ml of 0.1000 M SnSO$_4$, and the titrant is 0.2000 M Ce(IV). If the titration is carried out at 25°C, determine

    (a) The [Sn^{2+}] at the start
    b) The [Sn^{2+}] after 15.00 ml of Ce(IV) is added
    (c) The [Sn^{2+}] 0.05 ml before the equivalence-point volume of Ce(IV)
    (d) The [Sn^{2+}] at the equivalence point
    (e) The [Sn^{2+}] 0.05 ml after the equivalence-point volume of Ce(IV)
    (f) The [Sn^{2+}] after 60.00 ml of Ce(IV) is added

Ans. (b) $5.38 \times 10^{-2}\,M$; (e) $1.00 \times 10^{-40}\,M$

2. Consider the titration system where Fe(II) is titrated by VO_2^+ in $2\ M\ H_2SO_4$. The solution is 100.00 ml of $0.0500\ M\ FeSO_4$ at the start, and the titrant is $0.1500\ M\ VO_2^+$. For a titration conducted at 25°C, and assuming the concentration of H_2SO_4 to remain constant during the titration, determine

(a) The $[Fe^{2+}]$ at the start

(b) The $[Fe^{2+}]$ 0.50 ml before the equivalence-point volume of VO_2^+

(c) The $[Fe^{2+}]$ 0.05 ml before the equivalence-point volume of VO_2^+

(d) The $[Fe^{2+}]$ at the equivalence point

(e) The $[Fe^{2+}]$ 0.05 ml after the equivalence-point volume of VO_2^+

(f) The $[Fe^{2+}]$ 0.50 ml after the equivalence-point volume of VO_2^+

Note: For the purposes of determining the $[H_3O^+]$ for the solution, you may consider the H_2SO_4 to be totally dissociated for the first stage.

3. You are required to titrate 25.00 ml of $0.0100\ M\ KOH$ at 25°C by the addition of a titrant involving HNO_3. Suppose

(a) The titrant to be $0.0100\ M\ HNO_3$

(b) The titrant to be $0.0050\ M\ HNO_3$

Determine for each titration the range of change of pH for ±0.05 ml of titrant around the equivalence-point volume. Comment on any difference between these values.

4. Consider the possibility of titrating Ag^+ at 25°C by the use of one of the titrants Cl^-, Br^-, or I^-. Suppose the starting solution in each case to be 50.00 ml of $0.1000\ M\ AgNO_3$, and the titrant, regardless of the type chosen, to be $0.2000\ M$. Tabulate the values of pAg for the following titration points in milliliters of titrant solution added:

    24.00   24.50   24.70   24.90   24.95   25.00   25.05   25.10   25.30

    25.50   26.00

Indicate in each case the range of change of pAg for ±0.05 ml of titrant solution around the equivalence point.

5. In the titration discussed in Prob. 4, assume that you have an indicator which gives an excellent end-point indication between pAg values of 6.50 and 7.50. Assuming the critical indicator change point to occur at $pAg = 7.00$, determine for each type of titrant the titrant volume at which the indicator reacts and gives a positive end-point indication. Do not attempt to determine this volume closer than the nearest 0.05 ml. Comment on your findings.

Ans. Cl^- 25.60 ml, Br^- 25.00 ml, I^- 25.00 ml

6. A titration is to be conducted at 20°C involving 50.00 ml of $0.2000\ M\ MgCl_2$ and $0.2500\ M$ EDTA. The EDTA solution will involve the disodium dihydrogen ethylenediaminetetraacetic acid substance, commonly written as Na_2H_2Y. During the entire titration, the solution will be buffered to yield a pH of 10.00. Determine the values of

(a) The $[Mg^{2+}]$ at the start

(b) The $[Mg^{2+}]$ after 25.00 ml of EDTA is added

(c) The $[Mg^{2+}]$ after 39.00 ml of EDTA is added

(d) The $[Mg^{2+}]$ after 39.90 ml of EDTA is added

(e) The $[Mg^{2+}]$ at the equivalence point

(f) The $[Mg^{2+}]$ after 40.10 ml of EDTA is added

(g) The $[Mg^{2+}]$ after 41.00 ml of EDTA is added

Note. See Chap. 4, Part 4, for the equations required to determine necessary complexation relationships.

7. 25.00 ml of $FeSO_4$ solution is titrated with $0.1000\ M\ Ce(IV)$ solution at $pH = 1.00$ and 25°C. The data given below shows the course of the titration in terms of volume of titrant and the corresponding $[Fe^{2+}]$ as calculated from potentiometric monitoring. Tabulate the values of $\Delta pFe/\Delta V_{Ce}$ and $\Delta^2 pFe/\Delta V_{Ce}^2$, and determine the equivalence-point volume of Ce(IV) solution relative to each tabulation. From these equivalence-point volumes determine the starting molarity of the $FeSO_4$ solution. Use 0.05 ml increments of ΔV_{Ce}.

Volume Ce (IV) added, ml	$[Fe^{2+}]$
39.20	7.79×10^{-4}
39.40	5.82×10^{-4}
39.50	4.84×10^{-4}
39.60	3.87×10^{-4}
39.70	2.90×10^{-4}
39.80	1.93×10^{-4}
39.85	1.44×10^{-4}
39.90	9.63×10^{-5}
39.95	4.79×10^{-5}
40.00	5.43×10^{-8}
40.05	6.14×10^{-11}
40.10	3.08×10^{-11}
40.20	1.54×10^{-11}
40.30	1.02×10^{-11}
40.40	7.64×10^{-12}
40.50	6.10×10^{-12}

8. 60.00 ml of a solution of $PbCl_2$ is titrated at 25°C by $0.1000 M$ K_2CrO_4. Determine the equivalence-point volume of titrant and the starting molarity of the $PbCl_2$ solution from the titration data provided below. Use only the $\Delta^2 pPb/\Delta V^2_{CrO_4^{2-}}$ tabulation.

Volume K_2CrO_4 added, ml	$[Pb^{2+}]$
25.00	5.88×10^{-3}
26.00	4.65×10^{-3}
27.00	3.45×10^{-3}
27.50	2.88×10^{-3}
28.50	1.69×10^{-3}
29.50	5.59×10^{-4}
30.50	3.3×10^{-11}
31.50	1.1×10^{-11}
32.00	8.2×10^{-12}
33.00	5.6×10^{-12}
34.00	4.3×10^{-12}
35.00	3.4×10^{-12}

9. A titration of 50.00 ml of NaOH against $0.1000 M$ HCl was carried out at 25°C under the following two conditions:

(a) Where the volume increments of titrant added around the equivalence point were likely to be fairly evenly spaced around this point

(b) Where the volume increments of titrant added around the equivalence point were unlikely to be evenly spaced around this point

The following indicates the titration data secured in each case. Using a tabulation of $\Delta^2 pH/\Delta V^2_{HCl}$ at a ΔV_{HCl} value of 0.10 ml and multiplying the ratio of $\Delta^2 pH/\Delta V^2_{HCl}$ by 10^3, determine the equivalence-point-volume values for each titration. Note the difference between these values and comment on them. To support your argument, make an approximate graph, for each titration, of $\Delta^2 pH/\Delta V^2_{HCl}$ vs. V_{HCl}.

Titration (a) form		Titration (b) form	
Volume HCl added, ml	pH	Volume HCl added, ml	pH
43.00	11.45	42.00	11.60
43.50	11.36	43.00	11.45
44.00	11.24	44.00	11.24
44.50	11.09	45.00	10.84
45.00	10.84	46.00	3.44
45.50	10.19	47.00	2.85
45.60	9.72	48.00	2.62
45.70	4.28	49.00	2.47
45.80	3.80	50.00	2.36
45.90	3.58		
46.00	3.44		
46.50	3.06		
47.00	2.85		

Ans. (a) 45.65 ml; *(b)* 45.50 ml

10. Consider a system where a solution of Fe(II) is titrated by a solution of Ce(IV). The starting solution is 100.00 ml of 0.1000 M FeSO$_4$, and the titrant is 0.1000 M Ce(SO$_4$)$_2$. Consider the values of $E^\circ_{Ce^{4+}/Ce^{3+}}$ and $E^\circ_{Fe^{3+}/Fe^{2+}}$ at 25°C to be 1.459 and 0.770 V, respectively. For a titration carried out at 25°C determine

(a) $[Fe^{2+}]$ at the start
(b) $[Fe^{2+}]$ at 25.00 ml of added Ce(SO$_4$)$_2$ solution
(c) $[Fe^{2+}]$ at 50.00 ml of added Ce(SO$_4$)$_2$ solution
(d) $[Fe^{2+}]$ at 95.00 ml of added Ce(SO$_4$)$_2$ solution
(e) $[Fe^{2+}]$ at 99.90 ml of added Ce(SO$_4$)$_2$ solution
(f) $[Fe^{2+}]$ at the equivalence point
(g) $[Fe^{2+}]$ at 100.10 ml of added Ce(SO$_4$)$_2$ solution
(h) $[Fe^{2+}]$ at 110.00 ml of added Ce(SO$_4$)$_2$ solution

11. Consider the titration of 100.00 ml of 0.1000 M FeSO$_4$ by 0.0200 M KMnO$_4$ where the solution is $[H_3O^+] = 1\ M$ throughout the titration. Consider the values of $E^\circ_{MnO_4^-/Mn^{2+}}$ and $E^\circ_{Fe^{3+}/Fe^{2+}}$ to be, respectively, 1.510 and 0.770 V at 25°C. The titration reaction is

$$5Fe^{2+} + MnO_4^- + 8H_3O^+ \rightleftharpoons 5Fe^{3+} + Mn^{2+} + 12H_2O$$

For a titration at 25°C determine

(a) $[Fe^{2+}]$ at the start
(b) $[Fe^{2+}]$ at 50.00 ml of added KMnO$_4$ solution
(c) $[Fe^{2+}]$ at 99.90 ml of added KMnO$_4$ solution

(d) $[Fe^{2+}]$ at the equivalence point

(e) $[Fe^{2+}]$ at 100.10 ml of added $KMnO_4$ solution

(f) $[Fe^{2+}]$ at 110.00 ml of added $KMnO_4$ solution

Ans. (b) $3.33 \times 10^{-2} M$; (e) $5.75 \times 10^{-14} M$

12. Consider the same situation as outlined in Prob. 11, but assume that the solution is $[H_3O^+] = 10^{-6} M$ throughout the titration. For a titration carried out at 25°C determine

(a) $[Fe^{2+}]$ at the equivalence point

(b) The *form* of the equation leading to the determination of $[Fe^{2+}]$ when 99.90 ml of $KMnO_4$ solution has been added

13. Consider a titration of silver(I) ion by $CrO_4{}^{2-}$ where we have 100.00 ml of 0.0750 M $AgNO_3$ at the start, and a titrant of 0.0750 M K_2CrO_4. Under 25°C titration conditions determine

(a) $[Ag^+]$ at the start

(b) $[Ag^+]$ at 40.00 ml of added K_2CrO_4 solution

(c) $[Ag^+]$ at 49.90 ml of added K_2CrO_4 solution

(d) $[Ag^+]$ at the equivalence point

(e) $[Ag^+]$ at 55.00 ml of added K_2CrO_4 solution

REFERENCES

1. Day, R. A., Jr., and A. L. Underwood, "Quantitative Analysis," 2d ed., Prentice-Hall, Englewood Cliffs, N.J., 1967.
2. Kolthoff, I. M., E. B. Sandell, E. J. Meehan, and Stanley Bruckenstein, "Quantitative Chemical Analysis," 4th ed., Macmillan, New York, 1969.
3. Fritz, J. S., and G. H. Schenk, Jr., "Quantitative Analytical Chemistry," 2d ed., Allyn and Bacon, Boston, 1969.
4. Blaedel, W. J., and V. W. Meloche, "Elementary Quantitative Analysis," Harper & Row, New York, 1963.
5. Brown, G. H., and E. M. Sallee, "Quantitative Chemistry," Prentice-Hall, Englewood Cliffs, N.J., 1963.
6. Skoog, D. A., and D. M. West, "Fundamentals of Analytical Chemistry," 2d ed., Holt, New York, 1969.
7. Ayres, Gilbert H., "Quantitative Chemical Analysis," 2d ed., Harper & Row, New York, 1968.

6
Neutralization Titrations

6.1 GENERAL

Neutralization or acid-base titrations are generally classified as those in which H_3O^+ in solution is titrated by OH^-, or vice versa, and this classification holds regardless of whether strong acids, strong bases, weak acids, weak bases, weak acid salts, or weak base salts are involved in the titration reaction. The following provides examples of some typical neutralization titration reactions:

1. $\quad\quad\quad HCl + NaOH \rightleftharpoons NaCl + H_2O$

2. $\quad CH_3 \cdot COOH + NaOH \rightleftharpoons CH_3 \cdot COONa + H_2O$

3. $\quad\quad\quad NH_4OH + HCl \rightleftharpoons NH_4Cl + H_2O$

4. $\quad CH_3 \cdot COONa + HClO_4 \rightleftharpoons CH_3 \cdot COOH + NaClO_4$

5. $\quad\quad\quad H_3PO_4 + NaOH \rightleftharpoons NaH_2PO_4 + H_2O$

6. $\quad\quad NaH_2PO_4 + NaOH \rightleftharpoons Na_2HPO_4 + H_2O$

Note that the titration reactions shown above have all been written in the molecular rather than the ionic form since as ionic reactions, all of them basically involve

$$H_3O^+ + OH^- \rightleftharpoons 2H_2O$$

In the examples above we have typically (1) a strong acid–strong base titration, (2) a weak acid–strong base titration, (3) a weak base–strong acid titration, (4) a weak acid salt where the hydrolysis product OH^- is titrated by a strong acid, and (5) and (6) the two-stage titration of a tribasic acid by a strong base.

Such titrations can be carried out under standard volumetric titration conditions, using an internal indicator substance to locate the equivalence point of the titration. They can also be carried out under potentiometric conditions, where the change in $[H_3O^+]$ during the titration process results in a change in the potential for the half-cell:

$$2H^+ + 2e \rightleftharpoons H_2$$

This technique will be discussed subsequently in detail in Chap. 12, Sec. 12.2 on potentiometry and potentiometric titrations. In the potentiometric titration technique, the potentiometer, or pH meter, provides the means of locating the equivalence point of the titration and, as such, substitutes for the internal indicator substance used in the standard volumetric method. Other instrumental methods of locating the equivalence point, such as the spectrophotometric and conductometric methods, may also be applied.

We shall discuss in the sections to follow the general method of locating the titration equivalence point through the use of internal indicator substances which change color in the zone of the equivalence-point volume of titrant.

6.2 NEUTRALIZATION TITRATION INTERNAL INDICATORS

The indicator substances used internally to locate the equivalence points and/or end points of neutralization titrations are usually complex organic compounds which act as weak acid or weak base substances in aqueous solution. Before we proceed further with this discussion, attention is drawn specifically to the statement just made—"equivalence points and/or end points." Where the indicator substance used reacts by changing its color at the exact equivalence point of the titration, then the end point and the equivalence point coincide; where the indicator reacts appreciably before or after the exact equivalence point, we have determined the end point of the titration which, in this instance, does not coincide with the equivalence point. Where there exists a lack of coincidence between the end-point volume and the equivalence-point volume of titrant, the titration is said to suffer from a *titration error* and requires the use of an *indicator-blank* correction in order to correct the volume of titrant for the end point to the titrant volume corresponding to the equivalence point. Such a correction, depending on the direction of the titration error, will be added to or subtracted from the end-point volume of titrant. As we shall see later, indicator-blank corrections can only be applied where the difference between the end-point and equivalence-point volumes of titrant is relatively small.

The organic substances used as indicators have a color for the dissociated or ionized form which is different from that for the undissociated or un-ionized form. An indicator substance which acts like a weak acid in solution can be depicted as HIn; it shows, in an acid solution, the color of the HIn or undissociated form, since $[H_3O^+]$ from the acid solution will repress the dissociation of the indicator substance. In an alkaline solution this same indicator will show the color of the In⁻ or dissociated form, since the low $[H_3O^+]$ of the alkaline solution will promote the dissociation of the indicator. The indicator substance HIn can be considered as dissociating according to

$$HIn + H_2O \rightleftharpoons H_3O^+ + In^-$$ (202)

and the equilibrium constant is given by

$$\frac{[H_3O^+][In^-]}{[HIn][H_2O]} = K_{eq}$$ (203)

and, for normally dilute solutions, we have

$$\frac{[H_3O^+][In^-]}{[HIn]} = K_a$$ (204)

with K_a being the dissociation constant for the indicator HIn.

For the logarithmic relationship this becomes

$$pH = pK_a - \log\frac{[HIn]}{[In^-]}$$ (205)

Note that the color of the indicator in solution will at all times be a function of the ratio of the concentrations of the undissociated and dissociated forms of the indicator. Thus we have

$$Color = \frac{[HIn]}{[In^-]} = \frac{\text{color of undissociated form}}{\text{color of dissociated form}} = \frac{[H_3O^+]}{K_a}$$

It is thus apparent that any such indicator will change color over some definite range of $[H_3O^+]$ or pH and that, in general, this will be a specific range for a given value of K_a.

The eye can not detect all the subtle shifts in color that take place during the transition from the extreme undissociated color to the extreme dissociated color, and vice versa. We can therefore determine the approximate limiting values for the ratio $[HIn]/[In^-]$ above which and below which the eye will see only the undissociated and dissociated colors.

Practically speaking, we can say that for most indicator substances which act like weak acids in solution, the eye sees only the undissociated color when we have

$$\frac{[H_3O^+]}{K_a} = \frac{[HIn]}{[In^-]} \geq 10 \tag{206}$$

and only the dissociated color when we have

$$\frac{[H_3O^+]}{K_a} = \frac{[HIn]}{[In^-]} \leq \frac{1}{10} \tag{207}$$

It is immediately apparent that the transition zone of color change, sometimes called the "indicator color change range," lies in the area of

$$\frac{[H_3O^+]}{K_a} = \frac{1}{10} \text{ to } 10 \tag{208}$$

which yields

$$[H_3O^+] = \frac{K_a}{10} \text{ to } 10K_a \tag{209}$$

and in the logarithmic form,

$$pH = pK_a \pm 1 \tag{210}$$

Thus ΔpH for the range of change of pH corresponding to the indicator color change range is given as ± 1 unit around the value of pK_a. The same reasoning can, of course, be applied to indicator substances which act like weak bases in solution, in which case the identical conclusion is reached.

Table 6.1 lists some of the common neutralization indicator substances and their indicator color change range in pH units. Note that, in general, all these indicators show a change range of 1 to 2 pH units. It should be pointed out that some of these indicators may have ranges of change that are not symmetrical around pK_a. Both the variation in ΔpH and the lack of symmetry depend largely on the nature of the dissociated and undissociated colors, since the eye is *not* capable of detecting *all* colors with the same degree of sensitivity.

From the values listed in Table 6.1 it is apparent that the selection of the proper indicator substance for a given neutralization titration will be of prime importance, if the indicator color change range in pH is to coincide with the range of change of pH around the equivalence point or close to it, thus allowing exact agreement between the end point and the equivalence point. In making the indicator selection it may also be of importance, under certain circumstances, to give consideration to whether the titration approaches its equivalence point pH value from the higher or lower pH side.

Table 6.1 Neutralization indicator substances

Common name	Color change	Color change range in pH
Picric acid	Colorless to yellow	0.1–0.8
Methyl violet	Yellow to blue	0.0–1.6
Crystal violet	Yellow to blue	0.0–1.7
Ethyl violet	Yellow to blue	0.0–2.3
Methyl green	Yellow to blue	0.3–1.8
Cresol red*	Red to yellow	1.0–2.0
Para-methyl red	Red to yellow	1.0–3.0
Thymol blue†	Red to yellow	1.2–2.8
Meta-cresol purple‡	Red to yellow	1.2–2.8
2,6-Dinitrophenol	Colorless to yellow	2.0–4.0
Methyl yellow	Red to yellow	2.9–4.0
Bromophenol blue	Yellow to blue	3.0–4.6
Congo red	Blue to red	3.0–5.0
Methyl orange	Red to yellow	3.1–4.4
Ethyl orange	Red to yellow	3.5–4.8
Bromocresol green	Yellow to blue	3.8–5.4
Methyl red	Red to yellow	4.2–6.2
Litmus	Red to blue	4.5–8.3
Propyl red	Red to yellow	4.8–6.4
Methyl purple	Purple to green	4.8–5.4
Chlorophenol red	Yellow to red	4.8–6.4
Para-nitrophenol	Colorless to yellow	5.0–7.0
Bromocresol purple	Yellow to purple	5.2–6.8
Alizarin⌀	Yellow to red	5.5–6.8
Bromothymol blue	Yellow to blue	6.0–7.6
Brilliant yellow	Yellow to orange	6.5–7.8
Neutral red	Red to yellow	6.8–8.0
Phenol red	Yellow to red	6.8–8.4
Para-α-naphthalein	Yellow to blue	7.0–9.0
Meta-cresol purple‡	Yellow to purple	7.4–9.0
Phenolphthalein	Colorless to red	8.3–10.0
Thymolphthalein	Colorless to blue	9.3–10.6
Alizarin yellow R	Yellow to violet	10.1–12.0
2,4,6-Trinitrotoluene	Colorless to orange	11.5–13.0
1,3,5-Trinitrobenzene	Colorless to orange	12.0–14.0

*Cresol red also has a color change of yellow to red at pH 7.0 to 8.8.
†Thymol blue also has a color change of yellow to blue at pH 8.0 to 9.6.
‡Note the two color change intervals for meta-cresol purple.
⌀Alizarin also has a color change of red to purple at pH 11.0 to 12.4.

6.3   NEUTRALIZATION INDICATOR CLASSIFICATIONS

The majority of the organic indicator substances are aromatic compounds with two or more benzene rings joined by one or more carbon or nitrogen atoms. The solution reaction involves a change from a benzenoid to a quinoid structure

Benzenoid   Quinoid

or vice versa, the quinoid structure being one usually associated with color. The three most common indicator categories relative to structure and reaction are the phthaleins, the sulfonphthaleins, and the azos. Examples of these, with their characteristic solution reaction, are shown.

6.3.1. THE PHTHALEINS

In the main, these indicators are colorless in the undissociated form and colored in the dissociated form. A typical example is phenolphthalein.

Colorless   Red

The color change takes place over the pH range of 8.3 to 10.0.

6.3.2 THE SULFONPHTHALEINS

Many of these indicators show two distinct and separate ranges of color change. One of these usually occurs in solutions of relatively low pH, with the other occurring in neutral or alkaline solutions. The color changes are varied for this group. A typical example is phenol red, or phenolsulfonphthalein.

Red   Yellow

This color change takes place over the range of pH of approximately 1.0 to 2.0

with this color change taking place over the pH range of 6.8 to 8.4. Most of the instances where these indicators are used involve the second area of color change, that is, the neutral to alkaline solution change.

6.3.3 THE AZOS

Most of these indicators have a color change from red to yellow. A typical example is methyl orange

with the color shift occurring at a pH range of 3.1 to 4.4.

6.4 FACTORS AFFECTING INDICATOR USE

There are several factors which have a bearing on the selection and functionability of indicator substances in neutralization titrations. The more significant of these are outlined in brief.

6.4.1 INDICATOR SELECTION

The indicator substance selected should have a color change range in pH which coincides with the pH range which occurs close to and around the equivalence point pH of the titration. As will be discussed later in somewhat greater detail, a lack of coincidence in this area can lead to a significant difference between the titration end-point volume, as given by the indicator reaction and color change, and the equivalence-point volume.

6.4.2 QUANTITY OF INDICATOR

In order to avoid the involvement of an appreciable amount of titrant in the indicator color change reaction, relative to the equivalence-point volume, the amount of the indicator used should be as small as is conveniently possible. The high color intensity of most indicator substances helps toward achieving this goal, and usually the final indicator concentration in the solution being titrated will be approximately 0.0001 to 0.0004 percent (e.g., two to eight drops of a 0.1 percent solution of the indicator per 100 ml of solution).

6.4.3 COLOR CHANGE DETECTION

The *first* color change that can be clearly detected is taken as the indicator color change. Any operator should rely on his *own* ability to judge color change, since this is very apt to differ to a greater or lesser extent from one individual to another. Many individuals suffer from deficiencies in color perception which range all the way from generally poor perception of color to outright color blindness. Where this latter situation exists, the form of color blindness (i.e., red-green, yellow-blue, etc.) should serve as a guide in avoiding the use of certain indicators. As a case in point, a red-green color-blind person should avoid the use of those indicators which display red and/or green color change patterns, substituting for such indicators on an appropriate basis those which show color changes involving blue, yellow, and/or purple.

When an operator of generally poor color perception and/or color memory is conducting a series of neutralization titrations involving an indicator, a common practice is to carry out one of the titrations using a pH meter and to determine the indicator color corresponding to the titration equivalence-point pH value. Subsequent titrations in the series are then carried out to the identical color.

6.4.4 THE EFFECT OF TEMPERATURE

Solution temperature changes can be expected to affect neutralization titration indicators and their end-point indications from two directions. There will be, first of all, the effect of changing solution temperature on the $[H_3O^+]$ or pH of the solution being titrated. For example, because of the effect of elevated temperature on the degree of dissociation, we would expect that in the titration of a solution of a weak acid by a strong base, the $[H_3O^+]$ values will generally be higher for comparable titration completion stages with higher solution temperatures within defined limitations. Second, we have the effect of change in temperature on the degree of dissociation of the indicator substance itself. Here again we have a somewhat similar situation, although the indicator type will to a certain extent dictate the magnitude of the response to changing temperature situations. Nevertheless, a shift in the indicator color change range can be expected with changing temperature of the solution environment. All in all the situation is such that for a given neutralization titration employing a specific indicator, titrant standardization and normal titration conditions should involve an appropriate *fixed* temperature.

6.4.5 THE EFFECT OF THE PRESENCE OF COLLOIDS IN SOLUTIONS UNDER TITRATION

The ability of colloidal particles to adsorb ions at their surfaces provides for the possibility of premature or delayed indicator color change situations. Thus, the use of indicators in neutralization titrations where colloids are formed or already exist in the solution medium can lead to considerable error. For example, in carrying out a titration of the neutralization type using an alkaline titrant under conditions where a colloid such as the hydrous metal oxide $Al_2O_3 \cdot xH_2O$ is formed, it can be expected that OH^- ions from the solution may be adsorbed at the surface of the colloid. This will provide for localized concentrations which may be high enough to permit the indicator to give its color change at the colloid surface even when the solution concentration of hydroxyl ion is not yet high enough to permit the indicator change in the solution.

We shall now give consideration to various neutralization titration systems with the understanding that since previous chapters and sections have dealt with acid-base equilibria, general titration equations, and titration-curve theory, we shall be using, with the necessary references, equations already developed for the calculation of data pertinent to the titration points considered.

6.5 STRONG ACID–STRONG BASE TITRATIONS

Remember first of all that by the terms "strong acid" and "strong base," we imply acid and base substances such as HCl, HNO_3, NaOH, and KOH, which in the practical sense can be considered as 100 percent dissociated in aqueous solution. For the purposes of this discussion we shall consider the titration direction as being that given by the titration of the strong acid solution by the addition of the strong base titrant. No attempt will be made to elaborate on the details associated with the opposite titration direction; this is left to the student to consider, although problems covering this aspect are included at the end of this section.

Since the reaction under consideration is in general that of a strong monobasic acid with a strong monobasic base, we have, as an example,

$$HCl + NaOH \rightleftharpoons NaCl + H_2O$$

or

$$H_3O^+ + Cl^- + Na^+ + OH^- \rightleftharpoons 2H_2O + Na^+ + Cl^-$$

or simply

$$H_3O^+ + OH^- \rightleftharpoons 2H_2O$$

This latter reaction is typical of Eq. (182), with $r = t = 1$, and we have

$$[H_3O^+][OH^-] = K_w$$

so that we can proceed to use those equations associated with the general reaction of Eq. (182) to determine the value of $[H_3O^+]$ or pH for any point in such a titration.

Let us consider, first of all, the titration of 100.00 ml of $0.1000 M$ HCl by $0.1000 M$ NaOH. Here we have

$M_R = M_A$ = molarity of acid at the start = $0.1000 M$

$M_T = M_B$ = molarity of base = $0.1000 M$

$V_R = V_A$ = volume of acid at the start = 100.00 ml

$V_T = V_B$ = volume of base added

$[R]_{V_T} = [H_3O^+]_{V_E} \qquad [T]_{V_T} = [OH^-]_{V_B} \qquad pR_{V_T} = pH_{V_B} = pH$

$r = t = 1 \qquad K[A]^a = K_w$

The values of $[H_3O^+]$ and pH for the various points in this titration, with an indication of the equations used to calculate them, are shown in Table 6.2. Note that by any of the methods outlined in Chap. 5, the equivalence-point volume is 100.00 ml of $0.1000 M$ NaOH. Determination of this titration point has been purposely simplified by choosing equal concentrations for HCl and NaOH, by having these concentrations as simple numbers, and by using uniform volume intervals or increments of NaOH around the equivalence point.

Consideration should now be given to the selection of an indicator substance for this titration. The equivalence point pH is 7.00, this value being, of course, common to all simple strong acid–strong base titration systems. The range of change of pH for the points ± 0.05 ml around this equivalence-point volume is given by $9.40 - 4.60$, or 4.80 pH units. With the pH change for this zone starting at 4.60 and finishing at 9.40, the following indicators selected from Table 6.1 provide indicator-blank values of less than $\pm$ml as adjustments to the end-point volume of titrant to correct it to the equivalence-point volume.

Indicator	Range pH
Propyl red	4.8–6.4
Methyl purple	4.8–5.4
Chlorophenol red	4.8–6.4
Para-nitrophenol	5.0–7.0
Bromocresol purple	5.2–6.8
Alizarin	5.5–6.8
Bromothymol blue	6.0–7.6
Brilliant yellow	6.5–7.8
Neutral red	6.8–8.0
Phenol red	6.8–8.4
Para-α-naphthalein	7.0–9.0
Meta-cresol purple	7.4–9.0

Since the titration error can be found by taking the indicator blank, with the opposite sign, dividing it by the equivalence-point volume, and multiplying this value by 100, each of the indicators listed above would provide for a titration error pos-

Table 6.2 Strong-acid-vs.-strong-base titration*
100.00 ml of 0.1000 M HCl vs. 0.1000 M NaOH

Base added, ml	Total volume, ml	$[H_3O^+]$	Equation used	pH	$(\Delta pH/\Delta V_B) \times 10^3$ $0.1\ ml = \Delta V_B$		$(\Delta^2 pH/\Delta V_B^2) \times 10^3$ $0.1\ ml/0.1\ ml$	
0.00	100.00	1.00×10^{-1}	$[H_3O^+] = M_A$	1.00				
1.00	101.00	9.80×10^{-2}	(184)	1.01				
5.00	105.00	9.05×10^{-2}	(184)	1.04				
15.00	115.00	7.39×10^{-2}	(184)	1.13				
25.00	125.00	6.00×10^{-2}	(184)	1.22				
35.00	135.00	4.81×10^{-2}	(184)	1.32				
45.00	145.00	3.80×10^{-2}	(184)	1.42				
55.00	155.00	2.90×10^{-2}	(184)	1.54				
65.00	165.00	2.12×10^{-2}	(184)	1.67				
75.00	175.00	1.43×10^{-2}	(184)	1.84				
85.00	185.00	8.11×10^{-3}	(184)	2.09				
					5	(90.00)		
95.00	195.00	2.56×10^{-3}	(184)	2.59			+0.2	(93.50)
					18	(97.00)		
99.00	199.00	5.02×10^{-4}	(184)	3.30			+2	(98.13)
					60	(99.25)		
99.50	199.50	2.51×10^{-4}	(184)	3.60			+13	(99.40)
					100	(99.55)		
99.60	199.60	2.00×10^{-4}	(184)	3.70			+20	(99.60)
					120	(99.65)		
99.70	199.70	1.50×10^{-4}	(184)	3.82			+60	(99.70)
					180	(99.75)		
99.80	199.80	1.00×10^{-4}	(184)	4.00			+120	(99.80)
					300	(99.85)		
99.90	199.90	5.00×10^{-5}	(184)	4.30			+400	(99.89)
					600	(99.92)		
99.95	199.95	2.50×10^{-5}	(184)	4.60			+8400	(99.95)
					4800	(99.98)		
100.00	200.00	1.00×10^{-7}	(186)	7.00			0.0	(100.00)
					4800	(100.02)		
100.05	200.05	4.00×10^{-10}	(187),(189)	9.40			−8400	(100.05)
					600	(100.08)		
100.10	200.10	2.00×10^{-10}	(187),(189)	9.70			−400	(100.12)
					300	(100.15)		
100.20	200.20	1.00×10^{-10}	(187),(189)	10.00			−120	(100.20)
					180	(100.25)		
100.30	200.30	6.67×10^{-11}	(187),(189)	10.18			− 60	(100.30)
					120	(100.35)		

*$[H_3O^+]$ and pH values are not corrected for the dilution effect. The volume values in parentheses after each $\Delta pH/\Delta V_B$ and $\Delta^2 pH/\Delta V_B^2$ value associate the average slope, or rate of change of the slope, with the average volume value between the titration points involved. Before the equivalence-point back-reaction, contributions are ignored if they constitute 2.5 percent or less of the contribution of $[H_3O^+]$ from the acid not yet neutralized. After the equivalence-point back-reaction, contributions are ignored if they constitute 2.5 percent or less of the contribution of $[OH^-]$ from the excess base.

Table 6.2   (Continued)

Base added, ml	Total volume, ml	$[H_3O^+]$	Equation used	pH	$(\Delta pH/\Delta V_B) \times 10^3$ 0.1 ml $= \Delta V_B$	$(\Delta^2 pH/\Delta V_B^2) \times 10^3$ 0.1 ml/0.1 ml
100.40	200.40	5.01×10^{-11}	(187),(189)	10.30		-20 (100.40)
					100 (100.45)	
100.50	200.50	4.01×10^{-11}	(187),(189)	10.40		
101.00	201.00	2.01×10^{-11}	(187),(189)	10.70		
105.00	205.00	4.08×10^{-12}	(187),(189)	11.39		
110.00	210.00	2.10×10^{-12}	(187),(189)	11.68		
115.00	215.00	1.43×10^{-12}	(187),(189)	11.84		
125.00	225.00	9.01×10^{-13}	(187),(189)	12.05		
135.00	235.00	6.71×10^{-13}	(187),(189)	12.17		

sibility of less than ± 0.05 percent, or less than 0.5 ppt. Figure 6.1 shows the location of the color change range, or transition range, in pH for two other indicators, methyl orange and thymolphthalein, relative to the general titration curve. There are two simple methods of determining the indicator blank and the titration error for such indicators, where the transition range does not coincide exactly with the pH change around the equivalence point.

The first of these requires that a titration of the same type be conducted using both the indicator involved and a pH meter. The data from the pH meter is

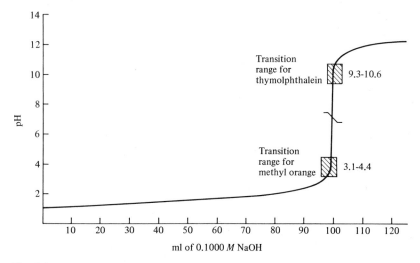

Fig. 6.1   Strong-acid-vs.-strong-base titration: 100.00 ml of 0.1000 *M* HCl vs. 0.1000 *M* NaOH.   Transition ranges for methyl orange and thymolphthalein indicators.

used to determine the exact equivalence point and the associated volume of titrant, and the first color change of the indicator is used to establish the end-point titrant volume. The difference between the two volume values then yields the indicator blank, or volume to be added to or subtracted from the end-point volume to correct it to the equivalence-point volume. Note that indicator-blank values will always carry an associated sign, indicating addition to or subtraction from the end-point volume. The titration error is then found by taking the indicator-blank value, with opposing sign, dividing it by the equivalence-point-volume value, and multiplying the result by 100. This provides the titration error as a percent value, which can be applied in a calculation converting any end-point-volume value in a similar titration situation to the appropriate equivalence-point volume.

The second method assumes, first of all, that the indicator or indicators in question, in this case methyl orange and thymolphthalein, change color at some point in the titration reasonably close to the equivalence-point pH value, and that the color change occurs in a portion of the titration curve where the rate of change of pH with titrant volume is very high. Where these criteria are met, as they are in the present example, a simple method of calculation will provide the *approximate* indicator blank and titration error for each indicator. Once the titration error has been determined it can, as noted before, be applied to determine the indicator blank for other similar titrations involving different end-point- and equivalence-point-volume values. The method is outlined below.

It must be determined, first of all, whether the indicator in question changes color before or after the equivalence-point pH value. In the case of methyl orange, the color change takes place between 3.1 and 4.4 pH and therefore *before* the equivalence point. For the purposes of approximation, it can be assumed that the central value of the color change range, 3.7^5 pH, represents that point at which the color change will be definitely detected. We have already noted, from the data outlined in Table 6.2, that in this zone of the titration, it is not necessary to give consideration to any back-reaction contribution of $[H_3O^+]$. Equation (184) was adequate for calculation purposes, and we can, for methyl orange, equate the $[H_3O^+]$ corresponding to the central value of the transition range to the right-hand side of Eq. (184). This yields

$$10^{-3.75} \approx \frac{100.00 \times 0.1000 - V_B \times 0.1000}{100.00 + V_B}$$

so that

$$0.100178 V_B \approx 9.9822 \text{ ml}$$

and

$$V_B \approx 99.64 \text{ ml}$$

With V_B calculated at 99.64 ml, instead of the equivalence-point volume of 100.00 ml, the end-point volume differs by -0.36 ml from the equivalence-point volume. The indicator blank would therefore be $+0.36$ ml or $+0.35$ ml, rounded off to the nearest 0.05 ml in conformity with the approximate nature of the calculation as dictated by the use of the central value of the color change range for the indicator. The titration error here would then be given by

$$\text{Titration error } \% \approx \frac{-(\text{indicator blank in ml})}{\text{equivalence-point volume in ml}} \, 100$$

$$\approx \frac{-0.35 \text{ ml}}{100.00 \text{ ml}} \, 100 \approx -0.35$$

Note that the titration error has both sign and magnitude. The sign represents the direction of departure of the end-point volume from the equivalence-point volume; thus negative values indicate titrations where the end-point volume is less than the equivalence-point volume, and vice versa.

In the case of thymolphthalein, with a color change range in pH of 9.3 to 10.6, the indicator changes color *after* the equivalence point. Again, for such points in the titration, Table 6.2 data indicate that the back-reaction contribution of $[OH^-]$ need not be considered, and that Eq. (187) is adequate. Since this expression yields the value for $[OH^-]$, we can convert the central value of the color change range to a value of $[OH^-]$. For thymolphthalein, the central value of the range is 9.9^5 pH, and this yields 4.0^5 pOH. Used properly in Eq. (187), we have

$$10^{-4.05} \approx \frac{0.1000V_B - 100.00 \times 0.1000}{V_B + 100.00}$$

from which

$$0.0999109V_B \approx 10.00891 \text{ ml}$$

and

$$V_B \approx 100.18 \text{ ml}$$

This gives a rounded-off indicator blank of -0.20 ml, and a titration error of $+0.20$ percent.

Figure 6.2 should now be carefully noted. It shows the same general situation as Fig. 6.1 but it covers two other indicators, para-methyl red and alizarin yellow R. Both these indicators change color at titration-point zones far removed from the equivalence point, and these changes occur in areas of the curve of the titration where the rate of change of pH with titrant volume is relatively low, so that over the interval of the indicator color change in each case a very considerable change in titrant volume is taking place. In instances involving indicators of this type, no valid or useful calculations can be made with respect to the determination of the indicator blank or titration error.

It has been stated already that once determined, the titration error for a specific indicator used in a strong acid–strong base titration can be used to approximate the indicator blanks for other *similar* titrations using the same general

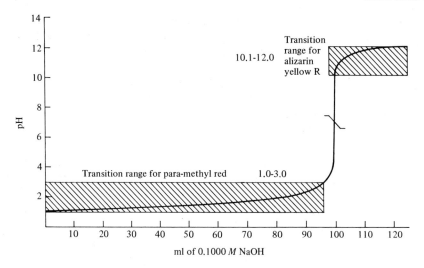

Fig. 6.2 Strong-acid-vs.-strong-base titration: 100.00 ml of 0.1000 M HCl vs. 0.1000 M NaOH. Transition ranges for para-methyl red and alizarin yellow R.

strength of titrant, regardless of the magnitude of the end-point volume. For example, using the titration error of -0.35 percent determined for methyl orange, we note the following.

End-point volume, ml 0.1000 M NaOH	Titration error, %	Indicator blank, ml	Equivalence-point volume, ml 0.1000 M NaOH
56.70	-0.35	$+0.20$	56.90
28.56	-0.35	$+0.10$	28.66

The basis on which the above calculations were made is

56.70 ml $\approx$ 99.65% $=$ end-point volume

56.90 ml $\approx$ 100.00% $=$ equivalence-point volume

$+$ 0.20 ml $\approx$ indicator blank

Since only approximate values are involved, the indicator blanks so calculated should be rounded off to the nearest ± 0.05 ml.

Consider, finally, the titration of 100.00 ml of 0.0010 M HCl with 0.0010 M NaOH. Table 6.3 shows the associated titration data. In spite of the decreased concentrations for both reactant and titrant solutions, the volume increment around the equivalence point has been maintained at 0.10 ml, with the usual one drop, or 0.05 ml, just before and just after the equivalence point. Note, however, that the ΔV_B and $\Delta V_B{}^2$ increments have been increased, respectively, to 0.2 ml and 0.2 ml/0.2 ml. The equivalence-point volume is 100.00 ml of 0.0010 M NaOH.

It will be noted that because of the reduced concentrations involved, the back-reaction contributions near the equivalence point require consideration. Thus, between 99.70 ml of added base and the equivalence point, Eq. (185) has been used; and between the equivalence point and 100.30 ml of added base, Eqs. (188) and (189) have been used.

The range of change of pH between the points ±0.05 ml around the equivalence-point volume is 7.45 to 6.55, or 0.90 pH units, a considerably smaller pH change than that encountered previously. This smaller change results from the higher degree of dilution of reactant and titrant solutions. The pH change for this zone starts at 6.55 and finishes at 7.45, and there are *no* indicators in the Table 6.1 list having a color change range fitting exactly inside this zone. There are, however, four indicators which have color change ranges so close to this zone that calculations based on the methods already outlined, using the proper equations giving consideration to back-reaction contributions and taking the central value for each indicator range as the critical value, show indicator blanks ±0.05 ml or less. The associated titration errors would be, of course, ±0.05 percent or less, or 0.5 ppt or less. These indicators are as follows.

Indicator	Range pH
Bromothymol blue	6.0–7.6
Brilliant yellow	6.5–7.8
Neutral red	6.8–8.0
Phenol red	6.8–8.4

The following shows the calculation of the approximate indicator blank for the neutral red indicator. The central value of the range is 7.4 pH, and this shows that the indicator changes color *after* the equivalence point. The zone in which the color change takes place is one in which the back-reaction contribution of OH^- must be considered, as will be noted from Table 6.3. The pOH value for the center of the indicator range is 6.6. We have then, for Eq. (188),

$$10^{-6.6} = \frac{0.0010 V_B - 100.00 \times 0.0010}{100.00 + V_B} + \frac{K_w}{10^{-6.6}}$$

and

$$V_B = 100.04 \text{ ml}$$

$$\text{Indicator blank} = -0.05 \text{ ml}$$

$$\text{Titration error} = +0.05\%$$

Table 6.3 Strong-acid-vs.-strong-base titration*

100.00 ml of 0.0010 M HCl vs. 0.0010 M NaOH

Base added, ml	Total volume, ml	$[H_3O^+]$	Equation used	pH	$(\Delta pH/\Delta V_B) \times 10^3$ 0.2 ml $= \Delta V_B$	$(\Delta^2 pH/\Delta V_B^2) \times 10^3$ 0.2 ml/0.2 ml
0.00	100.00	1.00×10^{-3}	$[H_3O^+] = M_A$	3.00		
1.00	101.00	9.80×10^{-4}	(184)	3.01		
5.00	105.00	9.05×10^{-4}	(184)	3.04		
15.00	115.00	7.39×10^{-4}	(184)	3.13		
25.00	125.00	6.00×10^{-4}	(184)	3.22		
35.00	135.00	4.81×10^{-4}	(184)	3.32		
45.00	145.00	3.80×10^{-4}	(184)	3.42		
55.00	155.00	2.90×10^{-4}	(184)	3.54		
65.00	165.00	2.12×10^{-4}	(184)	3.67		
75.00	175.00	1.43×10^{-4}	(184)	3.84		
85.00	185.00	8.11×10^{-5}	(184)	4.09		
95.00	195.00	2.56×10^{-5}	(184)	4.59		
					36 (97.00)	
99.00	199.00	5.02×10^{-6}	(184)	5.30		+8 (98.12)
					120 (99.25)	
99.50	199.50	2.51×10^{-6}	(184)	5.60		+53 (99.40)
					200 (99.55)	
99.60	199.60	2.00×10^{-6}	(184)	5.70		+80 (99.60)
					240 (99.65)	
99.70	199.70	1.51×10^{-6}	(185)	5.82		+240 (99.70)
					360 (99.75)	
99.80	199.80	1.01×10^{-6}	(185)	6.00		+400 (99.80)
					560 (99.85)	
99.90	199.90	5.20×10^{-7}	(185)	6.28		+1386 (99.89)
					1080 (99.92)	
99.95	199.95	2.85×10^{-7}	(185)	6.55		+2880 (99.95)
					1800 (99.98)	
100.00	200.00	1.00×10^{-7}	(186)	7.00		0.0 (100.00)
					1800 (100.02)	
100.05	200.05	3.51×10^{-8}	(188),(189)	7.45		−2880 (100.05)
					1080 (100.08)	
100.10	200.10	1.92×10^{-8}	(188),(189)	7.72		−1386 (100.12)
					560 (100.15)	
100.20	200.20	9.91×10^{-9}	(188),(189)	8.00		−400 (100.20)
					360 (100.25)	
100.30	200.30	6.62×10^{-9}	(188),(189)	8.18		−240 (100.30)
					240 (100.35)	

*$[H_3O^+]$ and pH values are not corrected for the dilution effect. The values in parentheses after each $\Delta pH/\Delta V_B$ and $\Delta^2 pH/\Delta V_B^2$ value associate the average slope, or rate of change of the slope, with the average volume value between the titration points involved. Before the equivalence-point back-reaction, contributions are ignored if they constitute 2.5 percent or less of the contribution of $[H_3O^+]$ from the acid not yet neutralized. After the equivalence-point back-reaction, contributions are ignored if they constitute 2.5 percent or less of the contribution of $[OH^-]$ from the excess base.

Table 6.3 *(Continued)*

Base added, ml	Total volume, ml	$[H_3O^+]$	Equation used	pH	$(\Delta pH/\Delta V_B) \times 10^3$ 0.2 ml $= \Delta V_B$	$(\Delta^2 pH/\Delta V_B^2) \times 10^3$ 0.2 ml/0.2 ml
100.40	200.40	5.01×10^{-9}	(187),(189)	8.30		-80 (100.40)
					200 (100.45)	
100.50	200.50	4.01×10^{-9}	(187),(189)	8.40		-53 (100.60)
					120 (100.75)	
101.00	201.00	2.01×10^{-9}	(187),(189)	8.70		
105.00	205.00	4.08×10^{-10}	(187),(189)	9.39		
110.00	210.00	2.10×10^{-10}	(187),(189)	9.68		
115.00	215.00	1.43×10^{-10}	(187),(189)	9.84		
125.00	225.00	9.01×10^{-11}	(187),(189)	10.05		
135.00	235.00	6.71×10^{-11}	(187),(189)	10.17		

Because of this restricted indicator selection situation, Table 6.4 has been arranged to show the indicator blanks, titration errors, and the equations used to calculate the indicator blanks. From this data it is apparent that chlorophenol red and phenolphthalein represent indicators which, because of the magnitude of the corrections required, should not be used in such a titration. Meta-cresol purple is, in fact, just barely feasible. Figure 6.3 shows the general titration curve, to which has been matched the transition ranges for chlorophenol red and phenolphthalein.

Again therefore, for the same general titration, but using more dilute reactant and titrant solutions, a much restricted indicator selection would be available, if the intent is to try to maintain a level of 0.5 ppt in the location of the equivalence-point volume. It is apparent that any titration involving still higher degrees of dilution will, even with the best indicator possible, and because of the restricted change of pH in the neighborhood of the equivalence point, permit the location of the equivalence-point volume with less than 1 ppt reliability.

Table 6.4 **Strong-acid-vs.-strong-base titration: indicator blanks and titration errors**
100.00 ml of 0.0010 M HCl vs. 0.0010 M NaOH

Indicator	Color change range, pH	Central value of range, pH	Equation used	Indicator blank, ml	Titration error, %
Chlorophenol red	4.8–6.4	5.6	(184)	+0.50	−0.50
Alizarin	5.5–6.8	6.1⁵	(185)	+0.15	−0.15
Meta-cresol purple	7.4–9.0	8.2	(188)	−0.30	+0.30
Phenolphthalein	8.3–10.0	9.1⁵	(187)	−2.8	+2.8

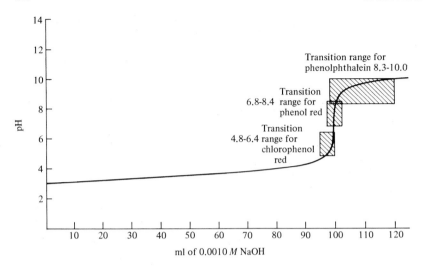

Fig. 6.3   Strong-acid-vs.-strong-base titration: 100.00 ml of 0.0010 M HCl vs. 0.0010 M NaOH.   Transition   ranges   for   chlorophenol   red,   phenol   red,   and phenolphthalein.

6.6   WEAK ACID—STRONG BASE TITRATIONS

"Weak acid" here means a monobasic acid substance dissociated in aqueous solution to an extent significantly less than 100 percent, and typified by substances such as formic acid H·COOH, acetic acid CH_3·COOH, hydrocyanic acid HCN, etc.   We shall deal only with such monobasic weak acids in titration with monobasic strong bases.   For the purposes of this discussion, only the normal direction of such a titration will be considered, that is, with the acid being titrated by the base.   No attempt will be made to discuss in detail the titration of a weak monobasic base by a strong monobasic acid, since the general approach is the same and the problems at the end of this section will include this aspect.   The student should familiarize himself with the process of determining $[H_3O^+]$ values for the various stages in a titration of this type.

The approach to weak acid—strong base titrations must combine equations, or modified equations, from both Chap. 4, Part 2, on acid-base equilibria, and Chap. 5, on generalized titration equations. Perhaps the problem is best presented by discussing the various points or zones of such a titration separately and generally. One point of importance should be noted. Regardless of the titration point or zone involved, we shall not give consideration to equations higher than the quadratic form.

6.6.1   THE WEAK ACID SOLUTION AT THE START OF THE TITRATION

Here the $[H_3O^+]$ of the solution will be determined from Eq. (5) or (7).   The equation chosen will depend on the accuracy required for the determination of

$[H_3O^+]$, and for the titration calculations involved in this section, we shall consider the value of $[H_3O^+]$ of adequate accuracy if the relative error is ± 5 percent or less.

6.6.2 TITRATION POINTS BETWEEN THE START AND THE EQUIVALENCE POINT

During the titration the value of C_a, the weak acid concentration, will decrease and that of C_s, the concentration of the salt of the weak acid and the strong base, will increase, for points in the titration between the start and the equivalence point. This indicates that a buffered solution condition, with varying values of C_a and C_s, will develop in this zone. The determination of $[H_3O^+]$ will thus require the use of Eq. (20) or (23), depending on the relative weakness of the acid or, if adequate, the simpler Eq. (24). Again, the choice of a simple or a quadratic equation will be based on the necessity of expressing the $[H_3O^+]$ with a relative error of ± 5 percent or less. Regardless of the equation used, the values of C_a and C_s will be determined by modifications of equations developed in Chap. 5. These will be

$$C_a = \frac{V_A M_A - V_B M_B}{V_A + V_B} \tag{211}$$

$$C_s = \frac{V_B M_B}{V_A + V_B} \tag{212}$$

6.6.3 AT THE EQUIVALENCE POINT OF THE TITRATION

At the equivalence point the solution will be that of the salt of the weak acid and the strong base. Depending on the relative weakness of the acid, this will lead to more or less hydrolysis action. Calculation of the $[H_3O^+]$ at the equivalence point will then be based on Eq. (12) or (15), the choice being based on obtaining a relative error in the $[H_3O^+]$ determined of ± 5 percent or less. Regardless of the equation required, the value of C_s will be determined from

$$C_s = \frac{V_A M_A}{V_A + V_B} \tag{213}$$

since this represents the maximum amount of C_s that can be formed.

6.6.4 TITRATION POINTS AFTER THE EQUIVALENCE POINT

For points after the equivalence point two situations may develop. Where such points lie after the equivalence point, but close to it, the contribution of $[OH^-]$ from hydrolysis of the salt and that from the excess base may both have to be considered. Again, where such points lie well beyond the equivalence point, the contribution of $[OH^-]$ from the salt hydrolysis may be insignificant compared with that from the excess base, thus permitting the former to be ignored. Here again an accuracy level permitting a relative error of ± 5 percent or less is maintained, and if the hydrolysis contribution is higher than 5 percent of the $[OH^-]$ from the excess base, the two sources must be considered. For example, the initial equation to be used will be

$$[OH^-] = [OH^-]_{\text{salt hydrolysis}} + [OH^-]_{\text{base excess}}$$
$$= \frac{K_w}{\sqrt{\dfrac{K_w K_a}{C_s}}} + \frac{V_B M_B - V_A M_A}{V_B + V_A} \tag{214}$$

where C_s is always given by Eq. (213), using the appropriate values of V_A, M_A, and V_B. If the first member on the right-hand side is higher than 5 percent of the second member, Eq. (214) must be solved properly. If it is 5 percent or less of the second member, the $[OH^-]$ value will be given adequately by the second member alone. In the data tabulations to follow, the term Eq. (214) corrected refers to the full Eq. (214); the term Eq. (214) uncorrected refers to Eq. (214) where only the second member on the right is required.

6.6.5 THE TITRATION OF ACETIC ACID AND SODIUM HYDROXIDE SOLUTIONS

Table 6.5 shows the data for the titration of 50.00 ml of 0.1000 M $CH_3 \cdot COOH$ by 0.1000 M NaOH. The K_a value for the acid was taken as 1.76×10^{-5}. The equations required to determine the $[H_3O^+]$ for each titration point are shown, and the student should verify the equation requirements indicated. By any of the methods outlined in Chap. 5, the equivalence-point volume is found to be 50.00 ml of 0.1000 M NaOH.

The equivalence-point pH value is 8.73. For different concentrations of reactant and/or titrant solutions in this titration system, the equivalence point pH will vary according to the value of C_s at the equivalence point. The value will, however, always be higher than 7.00 pH, reflecting the salt hydrolysis situation at this titration point. Because of the buffering action common before the equivalence point with titrations of this type, the range of change of pH around the equivalence point is generally much reduced compared with that which would be obtained for comparable concentration titrations for strong acid–strong base systems. For this reason, the range of change of pH for points ± 0.10 ml around the equivalence point, instead of ± 0.05 ml, will be considered with respect to indicator selection. This range is 10.02 to 7.45, or 2.57 pH units. Table 6.1 shows that the indicators meta-cresol purple (7.4 to 9.0) and phenolphthalein (8.3 to 10.0) would fit well into this pH change range, and would provide for indicator blanks of ± 0.10 ml or less, with titration errors of ± 0.20 percent or less, or 2 ppt or less.

Other indicators from Table 6.1 show, however, color change ranges so close to the range of change of pH for ± 0.10 ml around the equivalence point as to indicate that they might also be suitable for use, with low indicator blanks and titration errors. The two methods of determining the indicator blank and titration error that were described before apply here also. Using the calculation method, and applying it to the indicators neutral red, phenol red, para-α-naphthalein, thymolphthalein and alizarin yellow R, we can determine the feasibility of using

Table 6.5 Weak-acid-vs.-strong-base titration*
50.00 ml of 0.1000 M CH$_3$·COOH vs. 0.1000 M NaOH

Base added, ml	Total volume, ml	$[H_3O^+]$	Equation used	pH	$(\Delta pH/\Delta V_B) \times 10^3$ 0.1 ml $= \Delta V_B$	$(\Delta^2 pH/\Delta V_B^2) \times 10^3$ 0.1 ml/0.1 ml
0.00	50.00	1.33×10^{-3}	(7)	2.88		
0.10	50.10	1.22×10^{-3}	(20)	2.91		
0.50	50.50	9.05×10^{-4}	(20)	3.04		
2.50	52.50	4.64×10^{-4}	(20)	3.33		
5.00	55.00	1.55×10^{-4}	(24)	3.88		
10.00	60.00	7.04×10^{-5}	(24)	4.15		
15.00	65.00	4.11×10^{-5}	(24)	4.39		
25.00	75.00	1.76×10^{-5}	(24)	4.75		
35.00	85.00	7.54×10^{-6}	(24)	5.12		
45.00	95.00	1.96×10^{-6}	(24)	5.71		
					18.5 (47.00)	
49.00	99.00	3.59×10^{-7}	(24)	6.45		+1.8 (48.12)
					60 (49.25)	
49.50	99.50	1.78×10^{-7}	(24)	6.75		+13 (49.40)
					100 (49.55)	
49.60	99.60	1.42×10^{-7}	(24)	6.85		+30 (49.60)
					130 (49.65)	
49.70	99.70	1.06×10^{-7}	(24)	6.98		+40 (49.70)
					170 (49.75)	
49.80	99.80	7.07×10^{-8}	(24)	7.15		+130 (49.80)
					300 (49.85)	
49.90	99.90	3.53×10^{-8}	(24)	7.45		+990 (49.90)
					1290 (49.95)	
50.00	100.00	1.88×10^{-9}	(15)	8.73		0.0 (50.00)
					1290 (50.05)	
50.10	100.10	9.52×10^{-11}	(214)†	10.02		−1010 (50.10)
					280 (50.15)	
50.20	100.20	5.00×10^{-11}	(214)‡	10.30		−100 (50.20)
					180 (50.25)	
50.30	100.30	3.34×10^{-11}	(214)‡	10.48		−60 (50.30)
					120 (50.35)	
50.40	100.40	2.51×10^{-11}	(214)‡	10.60		−20 (50.40)
					100 (50.45)	
50.50	100.50	2.01×10^{-11}	(214)‡	10.70		
51.00	101.00	1.01×10^{-11}	(214)‡	11.00		
55.00	105.00	2.10×10^{-12}	(214)‡	11.68		
60.00	110.00	1.10×10^{-12}	(214)‡	11.96		
70.00	120.00	5.99×10^{-13}	(214)‡	12.22		

*$[H_3O^+]$ and pH values are not corrected for the dilution effect. The values in parentheses after each $\Delta pH/\Delta V_B$ and $\Delta^2 pH/\Delta V_B^2$ value associate the average slope, or rate of change of the slope, with the average volume value between the titration points involved.
†Corrected.
‡Uncorrected.

these indicators in this titration. Table 6.6 shows the indicators mentioned, together with the associated indicator blanks, titration errors, and equations used to calculate the indicator blanks. Shown immediately following are two sample calculations for indicator blanks and titration errors as associated with neutral red and alizarin yellow R.

Neutral red has its central value for the color change range at pH 7.4. The indicator therefore changes color *before* the equivalence point and, as shown by Table 6.5, Eq. (24) can be used satisfactorily here to give

$$10^{-7.4} \approx K_a \frac{V_A M_A - V_B M_B}{V_B M_B} \approx 1.76 \times 10^{-5} \frac{5.000 - 0.1000 V_B}{0.1000 V_B}$$

and

$$0.100226 V_B \approx 5.000$$
$$V_B \approx 49.89 \text{ ml}$$
$$\text{Indicator blank} = +0.10 \text{ ml}$$
$$\text{Titration error} = -0.20\%$$

Alizarin yellow R has its central value at 11.0^5 pH. The indicator changes color *after* the equivalence point and, as given by Table 6.5, the uncorrected Eq. (214) can be used. Thus we have a central value as 2.9^5 pOH, which provides

$$10^{-2.95} \approx \frac{V_B M_B - V_A M_A}{V_B + V_A} \approx \frac{0.1000 V_B - 5.000}{V_B + 50.00}$$

and

$$0.09888 V_B \approx 5.056$$
$$V_B \approx 51.13 \text{ ml}$$
$$\text{Indicator blank} = -1.1 \text{ ml}$$
$$\text{Titration error} = +2.2\%$$

Table 6.6 Weak-acid-vs.-strong-base titration: indicator blanks and titration errors
50.00 ml of 0.1000 M CH$_3$·COOH vs. 0.1000 M NaOH

Indicator	Color change range, pH	Central value of range, pH	Equation used	Indicator blank, ml	Titration error, %
Neutral red	6.8–8.0	7.4	(24)	+0.10	−0.20
Phenol red	6.8–8.4	7.6	(24)	+0.05	−0.10
para-α-Naphthalein	7.0–9.0	8.0	(24)	+0.05	−0.10
Thymolphthalein	9.3–10.6	9.9^5	(214)*	−0.10	+0.20
Alizarin yellow R	10.1–12.0	11.0^5	(214)†	−1.1	+2.2

*Corrected.
†Uncorrected.

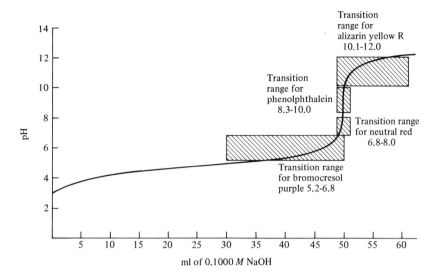

Fig. 6.4 Weak-acid-vs.-strong-base titration: 50.00 ml of 0.1000 M CH$_3$·COOH vs. 0.1000 M NaOH. Transition ranges for bromocresol purple, neutral red, phenolphthalein, and alizarin yellow R.

Obviously alizarin yellow R should not be used in the titration under discussion. Figure 6.4 shows the general titration curve and, matched to it, the indicators bromocresol purple, neutral red, phenolphthalein, and alizarin yellow R. Note that the color change ranges for bromocresol purple and alizarin yellow R occur in titration zones where the rate of change of pH with volume of base solution is relatively low, thus rendering these indicators, practically speaking, useless for the accurate location of the equivalence-point volume.

As was the case with strong acid–strong base titrations, dilution of the acid and/or base solution will result in titrations having, for comparable plus or minus volume values around the equivalence point, a lower range of change of pH. Table 6.7 provides an example of this situation in the titration of 50.00 ml of 0.0100 M acetic acid by 0.0100 M sodium hydroxide. Note that the range of change of pH for ±0.10 ml around the equivalence-point volume is now only 1.62 pH units. Table 6.8 shows the same selection of indicators as was found to perform adequately in the previous titration. Four of these indicators are still capable of location of the equivalence-point volume with an error of 2 ppt or less. Note the phenolphthalein would give an error of about 3 ppt, while thymolphthalein, which was an adequate indicator for the 0.1000-M titration, would not be satisfactory at all for the 0.0100-M titration. Figure 6.5 shows the general curve of the titration, with the indicators neutral red, phenolphthalein, and thymolphthalein matched thereto. The extended block represents the general position of the color change range for thymolphthalein in a titration zone of relatively low rate of

Table 6.7   Weak-acid-vs. strong-base titration*
50.00 ml of 0.0100 M CH$_3$·COOH vs. 0.0100 M NaOH

Base added, ml	Total volume, ml	[H$_3$O$^+$]	pH	Equation used
49.75	99.75	8.85 × 10^{-8}	7.05	(24)
49.80	99.80	7.07 × 10^{-8}	7.15	(24)
49.85	99.85	5.30 × 10^{-8}	7.28	(24)
49.90	99.90	3.53 × 10^{-8}	7.45	(24)
49.95	99.95	1.76 × 10^{-8}	7.76	(24)
50.00	100.00	5.93 × 10^{-9}	8.23	(15)
50.05	100.05	1.50 × 10^{-9}	8.82	(214)†
50.10	100.10	8.55 × 10^{-10}	9.07	(214)†
50.15	100.15	6.00 × 10^{-10}	9.22	(214)†
50.20	100.20	4.61 × 10^{-10}	9.34	(214)†
50.25	100.25	4.00 × 10^{-10}	9.40	(214)‡

*[H$_3$O$^+$] and pH values are not corrected for the dilution effect.
†Corrected.
‡Uncorrected.

change of pH with volume of base solution. Further reductions in the concentrations of reactant and/or titrant solutions lead to lower values of ΔpH around the equivalence point, with correspondingly greater difficulty in locating this point with relatively low error.

6.6.6   THE EFFECT OF THE MAGNITUDE OF K_a

With weak acid–strong base titrations the effect of decreasing value of K_a for the acid titrated, considering identical reactant and titrant solution concentrations, will be to decrease the magnitude of ΔpH around the equivalence point for compa-

Table 6.8   Weak-acid-vs.-strong-base titration: indicator blanks and titration errors
50.00 ml of 0.0100 M CH$_3$· COOH vs. 0.0100 M NaOH

Indicator	Color change range, pH	Central value of range, pH	Equation used	Indicator blank, ml	Titration error, %
Neutral red	6.8–8.0	7.4	(24)	+0.10	−0.20
Phenol red	6.8–8.4	7.6	(24)	+0.05	−0.10
Para-α-naphthalein	7.0–9.0	8.0	(24)	+0.05	−0.10
Meta-cresol purple	7.4–9.0	8.2	(24)	<+0.05	<−0.10
Phenolphthalein	8.3–10.0	9.1^5	(214)*	−0.15	+0.30
Thymolphthalein	9.3–10.6	9.9^5	(214)†	−0.90	+1.8

*Corrected.
†Uncorrected.

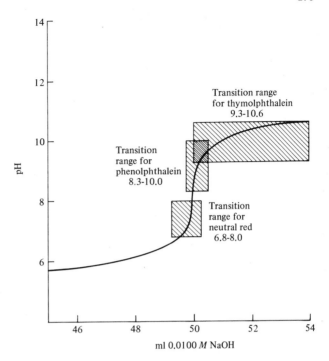

Fig. 6.5 Weak-acid-vs.-strong-base titration: 50.00 ml of 0.0100 M $CH_3 \cdot COOH$ vs. 0.0100 M NaOH. Transition ranges for neutral red, phenolphthalein, and thymolphthalein.

rable plus or minus base solution volume values around this point. This will render it more difficult to locate the equivalence point with low error. Table 6.9 shows titration data for the titration of 50.00 ml of 0.1000 M hydrocyanic acid, HCN, vs. 0.1000 M NaOH. The K_a value for HCN is taken as 5.00×10^{-10}.

The equivalence point pH is 11.00, and the value of ΔpH for ± 2.50 ml around the equivalence-point volume of 50.00 ml of 0.1000 M NaOH is only 11.54 to 10.52, or 1.02 pH units. The best indicator from Table 6.1 would be alizarin yellow R, with a color change range of 10.1 to 12.0 pH, which implies that alizarin yellow R will start to change color at 10.1 pH (about 43 ml of base added) and will finish its color change at 12.0 pH (about 61 ml of base added). Between 43 and 61 ml of added base the indicator changes color very gradually. Even with the use of a color standard for the indicator change, the titration using alizarin yellow R is unlikely to be able to locate the equivalence-point volume with an error better than about four parts per hundred, or 40 ppt. With a pH meter, the error might be reduced to about two parts per hundred, or 20 ppt.

Table 6.10 shows a tabulation of ΔpH for weak acids of varying K_a value from 10^{-3} to 10^{-10}, using acid and base solution strengths of 0.1000 M, a starting volume of the acid solution of 50.00 ml, and considering titration points at 0.10 ml

Table 6.9 Weak-acid-vs.-strong-base titration*
50.00 ml of 0.1000 M HCN vs. 0.1000 M NaOH

Base added, ml	Total volume, ml	$[H_3O^+]$	Equation used	pH	$(\Delta pH/\Delta V_B) \times 10^3$ 0.1 ml = ΔV_B	$(\Delta^2 pH/\Delta V_B^2) \times 10^3$ 0.1 ml/0.1 ml
0.00	50.00	7.07×10^{-6}	(7)	5.15		
0.10	50.10	2.50×10^{-7}	(24)	6.60		
0.50	50.50	4.95×10^{-8}	(24)	7.31		
2.50	52.50	9.50×10^{-9}	(24)	8.02		
5.00	55.00	4.50×10^{-9}	(24)	8.35		
25.00	75.00	5.00×10^{-10}	(24)	9.30		
					4.7 (35.00)	
45.00	95.00	5.77×10^{-11}	(23)	10.24		+0.06 (40.62)
					11.2 (46.25)	
47.50	97.50	3.00×10^{-11}	(23)	10.52		+0.21 (47.37)
					16 (48.50)	
49.50	99.50	1.44×10^{-11}	(23)	10.84		+1.2 (49.12)
					32 (49.75)	
50.00	100.00	1.00×10^{-11}	(15)	11.00		0.0 (50.00)
					32 (50.25)	
50.50	100.50	6.92×10^{-12}	(214)†	11.16		−1.2 (50.87)
					17 (51.50)	
52.50	102.50	2.91×10^{-12}	(214)†	11.54		

*$[H_3O^+]$ and pH values are not corrected for the dilution effect. The volume values in parentheses after each $\Delta pH/\Delta V_B$ and $\Delta^2 pH/\Delta V_B^2$ value associate the average slope, or the rate of change of the slope, with the average volume value between the titration points involved.
†Corrected.

Table 6.10 Weak-acid-vs.-strong-base titration*
50.00 ml of 0.1000 M acid vs. 0.1000 M NaOH
ΔpH for ± 0.10 ml around the equivalence-point volume for varying K_a

K_a	0.10 ml before pH	Equation used	Equivalence point, pH	0.10 ml after pH	Equation used†	ΔpH, ±0.10 ml
10^{-3}	5.70	(24)	7.85	10.00	(214)	4.30
10^{-4}	6.70	(24)	8.35	10.01	(214)	3.31
10^{-5}	7.70	(24)	8.85	10.03	(214)	2.33
10^{-6}	8.68	(23)	9.35	10.09	(214)	1.41
10^{-7}	9.56	(23)	9.85	10.23	(214)	0.67
10^{-8}	10.25	(23)	10.35	10.51	(214)	0.26
10^{-9}	10.82	(23)	10.85	10.91	(214)	0.09
10^{-10}	11.33	(23)	11.35	11.37	(214)	0.04

*$[H_3O^+]$ and pH values are not corrected for the dilution effect.
†Corrected.

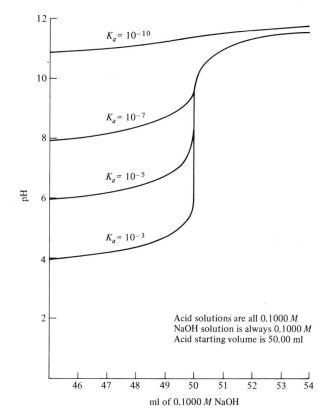

Fig. 6.6 Weak-acid-vs.-strong-base titrations: varying value of acid K_a.

before the equivalence point, the equivalence point and 0.10 ml after the equivalence point. It should be understood that the pH values shown were secured by calculation and do not represent data capable of being read off a pH meter dial. Figure 6.6 gives a graphic representation of the titrations for K_a values of 10^{-3}, 10^{-5}, 10^{-7}, and 10^{-10}.

Table 6.11 shows the effect of dilution of the acid and/or base solutions for the three K_a values of 10^{-3}, 10^{-5}, and 10^{-7}. The general effect is given by the value of ΔpH for ±0.10 ml around the equivalence-point volume. For each titration, the starting volume of the acid was 50.00 ml. Note that ΔpH decreases with both decreasing K_a value and decreasing concentration of the acid and/or base solution. It can be expected that the titration of 0.1-M solutions of acid and base, with an acid having a K_a value of 10^{-7}, will show an error in the location of the equivalence-point volume of about 5 to 10 ppt. Figure 6.7 shows graphically the situation for the titrations covered by Table 6.11.

Table 6-11   Weak-acid-vs.-strong-base titration*
50.00 ml of acid vs. NaOH
ΔpH for ±0.10 ml around the equivalence-point volume, varying K_a and solution strengths

Acid-base strength, M	K_a	0.10 ml before pH	Equation used	Equivalence point, pH	0.10 ml after pH	Equation used†	ΔpH, ±0.10 ml
1	10^{-3}	5.70	(24)	8.35	11.00	(214)	5.30
0.1	10^{-3}	5.70	(24)	7.85	10.00	(214)	4.30
1	10^{-5}	7.70	(24)	9.35	11.01	(214)	3.31
0.1	10^{-5}	7.70	(24)	8.85	10.03	(214)	2.33
1	10^{-7}	9.70	(23)	10.35	11.09	(214)	1.39
0.1	10^{-7}	9.56	(23)	9.85	10.23	(214)	0.67

*$[H_3O^+]$ and pH values are not corrected for the dilution effect.
†Corrected.

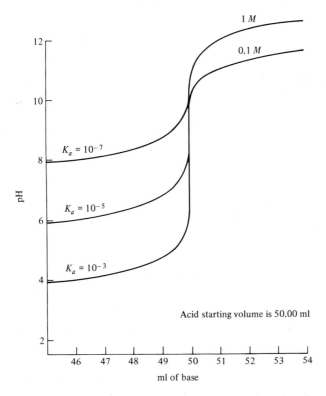

Fig. 6.7   Weak-acid-vs.-strong-base titrations: varying value of acid K_a and solution molarity.

6.7 THE FEASIBILITY OF STRONG ACID–STRONG BASE AND WEAK ACID–STRONG BASE TITRATIONS

We shall deal here only with the situation of an acid substance titrated by a base substance. It will be understood that the same general argument can be applied in the case of titrations of base substances by acids.

For equivalence-point volumes of about 50 ml, it is apparent that as long as we have a ΔpH value of 1 to 2 units for ± 0.10 ml around the equivalence-point volume, the location of this volume, using the best indicator fit available, will be in error by only a few parts per thousand. Lower values of ΔpH, resulting from either or both reactant and/or titrant dilution or decreased values of K_a, will increase the magnitude of the error, as will equivalence-point-volume values appreciably less than 50 ml. For many titrations, such a situation may be quite acceptable, and ΔpH values yielding errors as large as a few parts per hundred may be tolerated under these circumstances. We do not intend to generalize here, since considerable variation exists for the allowed values of K_a, equivalence-point volume, reactant-solution concentration, and titrant-solution concentration. In any given circumstance, the limiting values of these variables, relative to securing a ΔpH of 1 to 2 units for ± 0.10 ml (or whatever volume latitude is tolerable) around the equivalence-point volume can be easily calculated.

Example Give some idea of the feasibility of a titration between 50.00 ml of $0.0001\,M$ HCl and $0.0001\,M$ NaOH.

The equivalence point pH is 7.00. Considering a minimum ΔpH of 1.00 unit around the equivalence-point volume, this implies a range of 6.50 to 7.50 pH. Brilliant yellow (6.5 to 7.8) fits reasonably well into this range. Using the exact Eqs. (185), (188), and (189), we have

49.71 ml of base added for 6.50 pH
50.28 ml of base added for 7.50 pH

With brilliant yellow, the equivalence-point volume might be located within ± 0.10 to ± 0.20 ml, or with an error of about 2 to 4 ppt. The color change (yellow to orange) is not advantageous, so that the larger error possibility would likely apply, with an error of about 4 ppt being anticipated.

Example Comment on the feasibility of a titration between 50.00 ml of $0.0100\,M$ benzoic acid and $0.0100\,M$ potassium hydroxide.

The K_a value for benzoic acid is 6.46×10^{-5}. Equation (15) gives an equivalence point pH at 7.95. Assuming a minimum ΔpH around this value of 1.00 unit, we have a range of 7.45 to 8.45 pH, into which the indicator meta-cresol purple fits reasonably well with its color change interval of 7.4 to 9.0. Using Eqs. (24) and (214) corrected, we have

49.95 ml of added base for 7.45 pH
50.05 ml of added base for 8.45 pH

This indicator can be expected to locate the equivalence-point volume within about ± 0.05 ml, or with an error of about 1 ppt.

6.8 WEAK ACID–WEAK BASE TITRATIONS

Titrations of weak acids by weak bases, and vice versa, using indicator substances, are not feasible processes because of the very gradual change of solution
pH with volume of titrant. This gradual change carries through the entire titration, before and after the equivalence point, and results from the double
hydrolysis (both anion and cation) of the salt of the weak acid and the weak base,
and the very low degree of dissociation of both reactant and titrant substances.
Under these circumstances, the change of pH around the equivalence point is
very gradual, and location of the equivalence-point volume by indicator color
change methods can be nothing other than highly inaccurate, even where a color
standard has been previously prepared for the calculated equivalence point pH.
Even in cases where a pH meter, or a potentiometer, is used this does little to
improve the accuracy. Automatic potentiometric titrators capable of providing a
first-derivative plot give a somewhat lower error but still do not render such titrations generally feasible. Conductometric titration techniques can, under the
proper circumstances, eliminate or at least minimize this problem. All in all,
however, except in certain most specific instances, such titrations are not usually
necessary, since any weak acid substance can be titrated by a strong base, or vice
versa.

   Obviously, for such media, if $K_a > K_b$, the solution of the salt of the corresponding weak acid and weak base will show a pH value less than 7.00. Where
$K_b > K_a$, the solution pH will be greater than 7.00. It is required, from time to
time, to determine the pH for solutions of salts of weak acids and weak bases.
Such a salt can be represented by BA, where B^+ is the cation of the weak base and
A^- the anion of the weak acid. We then have

$$BA \rightleftharpoons B^+ + A^-$$
$$B^+ + A^- + H_2O \rightleftharpoons BOH + HA$$

and we have

$$\frac{[BOH][HA]}{[B^+][A^-]} = K_{hyd} = \frac{[BOH]}{[B^+][OH^-]} \cdot \frac{[HA]}{[H_3O^+][A^-]} \cdot [H_3O^+][OH^-] \quad (215)$$

so that

$$K_{hyd} = \frac{K_w}{K_b K_a} \quad (216)$$

On the left-hand side of Eq. (215) where, as a general rule, we have K_a and K_b
reasonably similar, we have

$$[BOH] \approx [HA] \quad \text{and} \quad [B^+] \approx [A^-]$$

so that

$$\frac{[HA]^2}{[A^-]^2} \approx K_{hyd} \approx \frac{K_w}{K_b K_a} \approx \frac{[H_3O^+]^2}{K_a^2}$$

and

$$[H_3O^+] \approx \sqrt{\frac{K_w K_a}{K_b}} \qquad (217)$$

with

$$pH \approx \tfrac{1}{2}(pK_w + pK_a - pK_b) \qquad (218)$$

Note that, as a reasonable approximation, the pH of a solution of the salt of a weak acid and a weak base is independent of the concentration of the salt, and dependent only on the values of K_b and K_a.

6.9 TITRATIONS OF SOLUTIONS OF WEAK ACID– STRONG BASE SALTS BY STRONG ACIDS

As before, only the titration of weak monobasic acid salts with strong acids will be discussed; it is assumed that the student will appreciate the necessity of extending the same reasoning to the titration of weak base salts. Titrations of weak acid salts by strong acids can be carried out with ease providing that the weak acid is sufficiently weak to permit extensive hydrolysis of the salt anion, since it is the OH^- product of anion hydrolysis that is titrated by the strong acid. For such a titration process, the various critical titration zones are

1. At the start the solution consists only of the salt involved. The $[H_3O^+]$ for such a solution can be determined from Eq. (15) or (12), the choice of equation being dictated by its ability to provide an $[H_3O^+]$ with ± 5 percent or less relative error.
2. After the start and before the equivalence point, the solution involves buffering conditions of varying C_s and C_a, and the $[H_3O^+]$ can be determined from either Eq. (20) or (23), depending on the relative weakness of the weak acid, or from Eq. (24) if this simpler equation is capable of yielding an $[H_3O^+]$ value with ± 5 percent or less relative error.
3. At the equivalence point the solution is entirely one of the weak acid and the salt of the strong acid, which latter substance does not have any effect on the situation. Thus Eq. (5) or (7) may be used to calculate the $[H_3O^+]$ value, the choice again being based on maintaining ± 5 percent or less relative error in $[H_3O^+]$.
4. After the equivalence point the solution consists of a mixture of the weak acid, the strong acid salt, and the excess of strong acid. Here again the strong acid salt exerts no influence on the solution pH. Under these conditions a close approximation to the value of $[H_3O^+]$ can be obtained from

$$[H_3O^+] = [H_3O^+]_{\text{weak acid}} + [H_3O^+]_{\text{strong acid}}$$

$$= \sqrt{K_a C_a} + \frac{V_A M_A - V_S M_S}{V_A + V_S}$$

$$= \sqrt{K_a \frac{V_S M_S}{V_A + V_S}} + \frac{V_A M_A - V_S M_S}{V_A + V_S} \qquad (219)$$

where V_S = starting volume of salt solution, ml

M_S = starting molarity of salt solution

Soon after the equivalence point, the contribution of $[H_3O^+]$ from the weak acid becomes less than 5 percent of the contribution from the strong acid excess. Wherever this situation exists Eq. (219) will involve only the second term in the right-hand member.

It is at once apparent that the weaker the parent acid of the salt, the more extensive the hydrolysis of the salt anion and the more pronounced the change of $[H_3O^+]$ or pH around the equivalence point. Thus, in the titration of 0.1-M solutions, it is found that, generally speaking, salts derived from weak acids having K_a values increasingly larger than 10^{-7} will result in equivalence-point-volume location with correspondingly larger errors. A 0.1-M solution of the salt sodium cyanide, NaCN, from the weak acid HCN, $K_a = 5.00 \times 10^{-10}$, can be titrated with 0.1 M HCl with an equivalence-point-volume location error of only one or two parts per thousand for a 50-ml titrant volume. It will be recalled that on the other hand, the titration of 0.1 M HCN by 0.1 M NaOH did *not* permit equivalence-point-volume location with any worthwhile accuracy.

An important point should be covered here. The generalization just made is in association with the usual form of volumetric neutralization titration technique, with the equivalence point being located through the use of a color change indicator. It is possible to titrate salts of weak acids, with K_a values greater than 10^{-7}, using slightly modified methods for the location of the equivalence point. For example, it is possible to titrate a 0.05-M solution of sodium acetate, $CH_3 \cdot COONa$, $K_a = 1.76 \times 10^{-5}$, with 1 M perchloric acid, $HClO_4$, using methyl orange as the indicator and a recording spectrophotometer set to examine the solution during titration at the appropriate wavelength, to provide a plot of absorbance vs. volume of 1 M $HClO_4$ from which the equivalence-point volume can be located. It should be remembered, however, that although this method is more accurate than visual equivalence-point location, the small volume of titrant used (on the basis of the titrant being 20 times the concentration of the solution titrated) provides for a fairly large relative error, and the equivalence-point volume in such a titration is rarely located with an error better than about 10 ppt or 1 pph.

Table 6.12 shows titration data for the titration of 50.00 ml of 0.1000 M NaCN by 0.1000 M HCl, using a K_a value of 5.00×10^{10}. Note that the ΔpH for ± 0.10 ml around the equivalence point is 2.62 units, and that the indicator methyl red (6.2 to 4.2) would fit exactly into this pH change, yielding an equivalence-point-volume location with an error of 2 ppt or less. Figure 6.8 shows the general titration curve, with the color change ranges for methyl yellow, methyl red, and neutral red matched to the curve. Note that methyl yellow and neutral red have their transition ranges in those portions of the curve where the rate of change of pH with titrant volume is low enough to allow a significant titration error. Methods of calculating the indicator blank or titration error, for any indicator with

Table 6.12   Weak-acid salt-vs.-strong-acid titration*
50.00 ml of 0.1000 M NaCN vs. 0.1000 M HCl

Acid added, ml	Total volume, ml	$[H_3O^+]$	Equation used	pH	$(\Delta pH/\Delta V_A) \times 10^3$ 0.1 $ml = \Delta V_A$	$(\Delta^2 pH/\Delta V_A^2) \times 10^3$ 0.1 ml/0.1 ml
0.00	50.00	7.07×10^{-12}	(15)	11.15		
					-62 (0.25)	
0.50	50.50	1.44×10^{-11}	(23)	10.84		-4 (0.87)
					-16 (1.50)	
2.50	52.50	3.00×10^{-11}	(23)	10.52		-0.2 (2.62)
					-11.2 (3.75)	
5.00	55.00	5.77×10^{-11}	(23)	10.24		-0.06 (9.37)
					-4.7 (15.00)	
25.00	75.00	5.00×10^{-10}	(24)	9.30		0.0 (25.00)
					-4.7 (35.00)	
45.00	95.00	4.50×10^{-9}	(24)	8.35		-0.08 (40.62)
					-13.2 (46.25)	
47.50	97.50	9.50×10^{-9}	(24)	8.02		-0.6 (47.37)
					-36 (48.50)	
49.50	99.50	4.95×10^{-8}	(24)	7.31		-12 (49.10)
					-178 (49.70)	
49.90	99.90	2.50×10^{-7}	(24)	6.60		-449 (49.82)
					-1300 (49.95)	
50.00	100.00	5.00×10^{-6}	(7)	5.30		-20 (50.00)
					-1320 (50.05)	
50.10	100.10	1.05×10^{-4}	(219)†	3.98		$+460$ (50.17)
					-170 (50.30)	
50.50	100.50	4.98×10^{-4}	(219)‡	3.30		$+24$ (50.52)
					-60 (50.75)	
51.00	101.00	9.90×10^{-4}	(219)‡	3.00		$+1.9$ (51.87)
					-17 (53.00)	
55.00	105.00	4.76×10^{-3}	(219)‡	2.32		

*$[H_3O^+]$ and pH values are not corrected for the dilution effect. The values of volume in parentheses after each $\Delta pH/\Delta V_A$ and $\Delta^2 pH/\Delta V_A^2$ value associate the average slope, or the rate of change of the slope, with the average volume value between the titration points involved. The titration data, more extensive for the first and second derivatives, shows two zero points for the second derivative. The second of these is the equivalence point; the first zero point location is shown on Fig. 6.8. The equivalence-point volume for the titration is given by

$$49.82 \text{ ml} + \left\{ \frac{449}{449 + 460} (50.17 - 49.82) \text{ ml} \right\} = 49.99 \text{ ml}$$

the very slight discrepancy between the calculated value and the theoretical value being due to the minor lack of symmetry between 49.82 ml and 50.17 ml as points on either side of the equivalence point. This minor lack of symmetry occurs as the result of rounding off calculated values of volume.

†Corrected
‡Uncorrected.

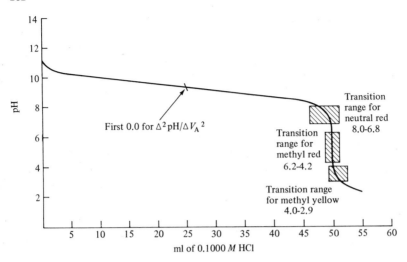

Fig. 6.8   Weak-acid salt-vs.-strong-acid titration: 50.00 ml of 0.1000 M NaCN vs. 0.1000 M HCl. Transition ranges for methyl yellow, methyl red, and neutral red.

a color change range close enough to the ΔpH for ± 0.10 ml around the equivalence point to make it worthy of consideration, will be similar to those already outlined for the same purpose in weak-acid-vs-strong-base titrations.

6.10   POLYBASIC ACID–STRONG BASE TITRATIONS

6.10.1   GENERAL

By polybasic acids is meant those acid substances which react with water in dissociation reaction steps, each step yielding H_3O^+. This situation was already outlined in Chap. 4, Part 2. Such substances are not strong electrolytes, the extent of dissociation being significantly less than 100 percent, since each dissociation stage after the primary stage is repressed by the $[H_3O^+]$ from the primary dissociation. Thus, in effect, and in nearly every instance, each stage of dissociation of a polybasic acid represents the ionization of a weak electrolyte or weak acid.

It is apparent that under the proper circumstances, it might be possible to titrate such acids in steps, the number of steps depending on the polybasic nature of the acid; two steps and two equivalence points for a dibasic acid; three steps and three equivalence points for a tribasic acid; etc. It will be our purpose here to investigate the *progress* and *feasibility* of such multistage titrations. Obviously the same generalizations could be made with respect to polybasic base substances. It should be understood that the student will be responsible for the extension of the arguments in this section to cover situations involving the titrations of polybasic bases by strong acids.

6.10.2 DIBASIC ACID–STRONG BASE TITRATIONS

In order to outline the general progress of a polybasic titration, we shall limit the preliminary discussion to the titration of a dibasic acid by a strong base. The method of approach to such a titration will be to combine equations, or modified equations, from Chap. 4, Part 2, on acid-base equilibria, Chap. 5, on generalized titration equations, and the material developed so far in this chapter. There are certain points of importance that should be stressed. First, regardless of the titration zone involved, no consideration will be given to any equation for $[H_3O^+]$ higher than the quadratic. Second, the value of $[H_3O^+]$ will be considered adequate with respect to accuracy when the relative error is ± 5 percent or less. Third, the simplest equation will always be attempted first; a more complicated equation will be used only where required to maintain the accuracy level outlined.

1. At the start of the titration, the $[H_3O^+]$ will be determined from Eq. (31) or (32).

2. Between the start and the first-stage equivalence point, we have a progressive decrease in the dibasic acid concentration C_a and an increase in the concentration of the monohydrogen salt concentration $C_{s,1}$. Equation (41) is the simplest equation that can be used here and if it yields an unsatisfactory value of $[H_3O^+]$, it will be substituted by either Eq. (40) or (47). The choice from these last two equations will depend, respectively, on whether $[H_3O^+]$ from Eq. (41) is greater than $10^{-7} M$ or less than $10^{-7} M$. Regardless of the equation used, the values of C_a and $C_{s,1}$ will be given by Eqs. (211) and (212), respectively, as

$$C_a = \frac{V_A M_A - V_B M_B}{V_A + V_B} \qquad C_{s,1} = \frac{V_B M_B}{V_A + V_B}$$

3. At the first-stage equivalence point, the solution will be that of the monohydrogen salt of the dibasic acid. The value of $[H_3O^+]$ may be found from the simple Eq. (54). If this value proves insufficiently accurate, Eq. (53) or (58) may be used, the choice depending, respectively, on whether the $[H_3O^+]$ from Eq. (54) is greater than $10^{-7} M$ or less than $10^{-7} M$. Regardless of the equation used, the value of $C_{s,1}$ will be found from Eq. (213) as

$$C_{s,1} = \frac{V_A M_A}{V_A + V_B}$$

4. For points between the first- and second-stage equivalence points, we shall consider that at the first-stage equivalence point, the solution was $V_{S,1}$ in volume, with a $C_{s,1}$ molarity given by $M_{S,1}$. For this titration zone we have variable concentrations of the two dibasic acid salts as $C_{s,1}$ and $C_{s,2}$. The simple Eq. (68) can be used and, if inadequate, Eq. (67) or (75) may be substituted, the choice depending, respectively, on whether the $[H_3O^+]$ from Eq. (68) is greater than $10^{-7} M$ or less than $10^{-7} M$. Regardless of the equation used, the values of $C_{s,1}$ and $C_{s,2}$ are determined, respectively, from modifications

of Eqs. (211) and (212) as

$$C_{s,1} = \frac{V_{S,1}M_{S,1} - V_BM_B}{V_{S,1} + V_B} \qquad C_{s,2} = \frac{V_BM_B}{V_{S,1} + V_B}$$

where V_B is now taken as the volume of base solution added starting from the first-stage equivalence point.

5. At the second-stage equivalence point, the solution involves only the double replacement salt of the dibasic acid in a concentration of $C_{s,2}$, and the $[H_3O^+]$ can be determined from the simple Eq. (85) or, if required, the more accurate Eq. (84). The value of $C_{s,2}$ may be determined from Eq. (213) as

$$C_{s,2} = \frac{V_{S,1}M_{S,1}}{V_{S,1} + V_B}$$

where $V_{S,1}$, $M_{S,1}$, and V_B have the origins indicated in step 4 above.

6. After the second-stage equivalence point, two situations may occur. For titration points after the second-stage equivalence point, but close to it, $[OH^-]$ contributions from both the salt hydrolysis and the excess base may have to be considered. Where these points lie well after the equivalence point, $[OH^-]$ from the salt hydrolysis may be insignificant compared with that from the excess base, allowing the former to be ignored. The general equation to be used will be

$$[OH^-] = \frac{K_w}{\sqrt{K_wK_2/C_{s,2}}} + \frac{V_BM_B - V_{S,1}M_{S,1}}{V_B + V_{S,1}}$$

where, again, $V_{S,1}$, $M_{S,1}$, and V_B have the origins indicated in step 4. Where the first member on the right is 5 percent or less of the second member, only the second member need be considered. The equation is, of course, a modification of Eq. (214).

Table 6.13 shows the data for a titration of 50.00 ml of 0.1000 M maleic acid, $C_2H_2{\cdot}COOH{\cdot}COOH$, by 0.1000 M NaOH. The K values for the acid are taken as $K_1 = 1.2 \times 10^{-2}$ and $K_2 = 6.0 \times 10^{-7}$. It is strongly suggested that the student verify the use of the equations indicated in Table 6.13. The first-stage equivalence point pH is given as 4.12, and the ΔpH for ± 1.00 ml around this point is 0.84 units. The indicators methyl orange (3.1 to 4.4), ethyl orange (3.5 to 4.8), and bromocresol green (3.8 to 5.4) would be capable of providing indicator blanks of about ± 1.00 ml or less, with ethyl orange being the best-suited substance. The titration errors would be about ± 2.0 percent or less, or about 20 ppt or less. It is apparent that the first-stage equivalence point will not be located with a high order of accuracy in the quantitative sense. A discussion of this will follow shortly.

The second-stage equivalence point has a value of 9.38 pH, and the ΔpH for ± 0.10 ml around this point is 1.07 units. The indicators phenolphthalein (8.3 to 10.0) and thymolphthalein (9.3 to 10.6) would locate the equivalence-point volume with indicator blanks of about ± 0.10 ml or less and titration errors of approximately ± 0.10 percent or less, or 1 ppt or less. Phenolphthalein would obviously

Table 6.13 Dibasic-acid-vs.-strong-base titration*
50.00 ml of 0.1000 M maleic acid vs. 0.1000 M NaOH

Base added, ml	Total volume, ml	$[H_3O^+]$	Equation used	pH	$(\Delta pH/\Delta V_B) \times 10^3$ $0.1\ ml = \Delta V_B$	$(\Delta^2 pH/\Delta V_B^2) \times 10^3$ $0.1\ ml/0.1\ ml$
0.00	50.00	2.9×10^{-2}	(31)	1.54		
2.00	52.00	2.6×10^{-2}	(40)	1.58		
30.00	80.00	5.4×10^{-3}	(40)	2.27		
					7.5 (39.50)	
49.00	99.00	2.0×10^{-4}	(40)	3.70		+0.34 (44.50)
					42 (49.50)	
50.00	100.00	7.6×10^{-5}	(53)	4.12		0.0 (50.00)
					42 (50.50)	
51.00	101.00	2.9×10^{-5}	(68)	4.54		−0.24 (58.00)
					6.3 (65.50)	
80.00	130.00	4.3×10^{-7}	(68)	6.37		
					8.0 (89.50)	
99.00	149.00	1.2×10^{-8}	(68)	7.92		+0.5 (94.38)
					60 (99.25)	
99.50	149.50	6.1×10^{-9}	(68)	8.22		+26 (99.48)
					168 (99.70)	
99.90	149.90	1.3×10^{-9}	(68)	8.89		+141 (99.82)
					490 (99.95)	
100.00	150.00	4.2×10^{-10}	(85)	9.38		+110 (100.00)
					580 (100.05)	
100.10	150.10	1.1×10	(214)†	9.96		−195 (100.18)
					148 (100.30)	
100.50	150.50	2.8×10^{-11}	(214)†	10.55		−22 (100.52)
					56 (100.75)	
101.00	151.00	1.5×10^{-11}	(214)†	10.83		−9 (103.12)
					11 (105.50)	
110.00	160.00	1.6×10^{-12}	(214)‡	11.80		

*$[H_3O^+]$ and pH values are not corrected for the dilution effect. The volume values in parentheses after each $\Delta pH/\Delta V_B$ and $\Delta^2 pH/\Delta V_B^2$ value associate the average slope, or the rate of change of the slope, with the average volume value between the titration points involved. The second-stage equivalence point is given by

$$99.82\ ml + \left(\frac{141}{141 + 195}\ 0.36\ ml\right) = 99.97\ ml$$

The first-stage equivalence point does not involve symmetrical values of volume around the equivalence-point volume.
†Corrected.
‡Uncorrected.

be the better choice of the two. It should be noted that the accuracy of location of the second-stage equivalence-point volume, at least for this particular titration system, far exceeds that for the location of the first-stage equivalence-point volume.

Despite the fact that the first-stage equivalence-point volume may frequently be located with significant error, this may be a critical point in the titration, particularly where the value of K_2 is so low that the error in the location of the second-stage equivalence-point volume is even greater than that relative to the first-stage volume location.

It is not our intention to discuss in any detail the determination of indicator blanks and titration errors for dibasic acid titrations as exemplified by that for maleic acid, since the methods involved do not differ in principle from those already discussed and applied earlier in this chapter. We shall only point out that the use of meta-cresol purple for the location of the second-stage equivalence-point volume of the titration in hand provides for an indicator blank of $+0.50$ ml and a titration error of -0.50 percent, or 5 ppt. Equation (68) is used in this derivation. Where the indicator alizarin yellow R is used for the same purpose, the indicator blank is about -1.7 ml and the titration error is about $+1.7$ percent, or 17 ppt. Equation (214) uncorrected is used in this derivation. Neither indicator gives the accuracy associated with phenolphthalein.

Figure 6.9 shows the general curve of the titration, matched to which are the transition ranges for the indicators ethyl orange, phenolphthalein, meta-cresol purple, and alizarin yellow R.

6.10.3 FEASIBILITY OF DIBASIC ACID–STRONG BASE TITRATIONS IN PARTICULAR AND POLYBASIC ACID–STRONG BASE TITRATIONS IN GENERAL

Using a dibasic acid titration as an example, we note that the titration reaction to the first-stage equivalence point involves

$$H_2A + OH^- \rightleftharpoons HA^- + H_2O$$

The degree of completion of this reaction at the equivalence point can be approximated by considering the extent to which the possible reactions

$$HA^- + H_2O \rightleftharpoons H_3O^+ + A^{2-}$$

$$HA^- + H_2O \rightleftharpoons H_2A + OH^-$$

can take place. The reaction completion situation can thus *in general* be indicated by the combination reaction

$$H_2A + A^{2-} \rightleftharpoons 2HA^- \tag{220}$$

for which the equilibrium constant equation is

$$\frac{[HA^-]^2}{[H_2A]_{un}[A^{2-}]} = K_{eq} \tag{221}$$

For the dibasic acid we have

$$\frac{[H_3O^+][HA^-]}{[H_2A]_{un}} = K_1 \quad \text{and} \quad \frac{[H_3O^+][A^{2-}]}{[HA^-]} = K_2$$

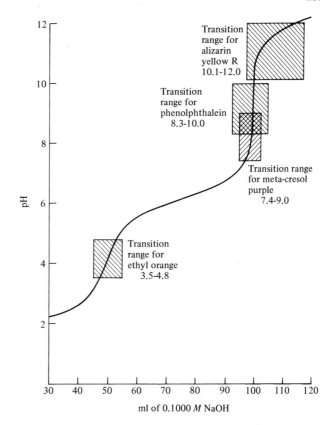

Fig. 6.9 Dibasic-acid-vs.-strong-base titration: 50.00 ml of 0.1000 M maleic acid vs. 0.1000 M NaOH. Transition ranges for ethyl orange, meta-cresol purple, phenolphthalein, and alizarin yellow R.

from which

$$\frac{[\text{HA}^-]^2}{[\text{H}_2\text{A}]_{\text{un}}[\text{A}^{2-}]} = \frac{K_1}{K_2} = K_{\text{eq}} \tag{222}$$

where K_{eq} is the equilibrium constant for the reaction of Eq. (220). If we consider that the loss of HA^- to A^{2-} or $\text{H}_2\text{A}_{\text{un}}$ represents the degree to which the reaction is incomplete at the first-stage equivalence point, and if we assign a value of $[\text{HA}^-] = 1$ and $[\text{A}^{2-}] \approx [\text{H}_2\text{A}]_{\text{un}} \approx x$, we can then show that from Eq. (222)*

$$\frac{1}{x^2} \approx \frac{K_1}{K_2} \quad \text{and} \quad \frac{1}{x} \approx \sqrt{\frac{K_1}{K_2}} \tag{223}$$

---

*Equality of $[\text{A}^{2-}]$ and $[\text{H}_2\text{A}]_{\text{un}}$ is a singular case, but the general case often approximates this ideal.

where x can be taken as indicative of the approximate error in locating the first-stage equivalence-point volume. For example, where K_1/K_2 has a value of 10^4, the value of x will be about 10^{-2}, and the location of the first-stage equivalence-point volume, with the best indicator selection possible, will be subject to an error of about one part per hundred. If K_1/K_2 is 10^6, the value of x will be 10^{-3}, and the error will be about one part per thousand. Obviously, for any worthwhile accuracy, say of a few parts per thousand, the ratio K_1/K_2 should not be less than 10^5. For the titration of maleic acid discussed in Sec. 6.10.2, the value of x is about 0.007 as determined from the K_1/K_2 of 2.0×10^4. This indicates an error of about seven parts per thousand. With ethyl orange as the best-suited indicator, the expected error in the location of the first-stage equivalence-point volume of this titration was estimated to be about 20 ppt or less.

It will be recalled that when discussing the titration of weak acids and strong bases, the limiting value of K_a, for a 50.00-ml equivalence-point volume and 0.1000-M reactant- and titrant-solution concentrations, was estimated to be about 10^{-7} where the error in locating the equivalence-point volume was to be held to a few parts per thousand, and about 10^{-8} for an error of a part or two per hundred. If this situation is now considered in conjunction with the suggested minimum value of 10^5 for the ratio K_1/K_2, we note that for $K_2 \geq 10^{-8}$ we should have $K_1 \geq 10^{-3}$ in order for the first- and second-stage equivalence points to be located with errors in the neighborhood of a part per hundred or a few parts per thousand.

The foregoing presents a rather interesting picture with respect to polybasic acids of higher order than the dibasic. If, for example, we set a limiting value of about 10^{-8} on K_3 for a tribasic acid, this would effectively remove any tribasic acid from the possibility of showing *three* separate and distinct breaks in the titration curve capable of locating the associated equivalence-point volumes with any worthwhile accuracy. Again, if the tribasic acid has a K_1 value of 10^{-2}, this will require K_2 and K_3 values of 10^{-7} and 10^{-12}, respectively, if the first- and second-stage equivalence-point volumes are to be located with significant accuracy. It is immediately apparent that with a K_3 value of 10^{-12}, the third-stage equivalence-point volume will not be capable of determination except in the crudest sense. In general, the values of K_1, K_2, K_3, etc., will decide how many distinct break points can be expected in the titration curve, as well as where such break points will be located. Where the final K value is too low to permit accurate location of the equivalence-point volume for this stage, it will be necessary to consider one of the earlier equivalence points as a possible critical point in the titration. It is always possible that the K values may be such that the acid involved can not be titrated so as to locate *any* equivalence-point volume with worthwhile accuracy.

Some rather interesting and circuitous methods have been used in the titration of polybasic acid substances. For example, in the titration of H_3PO_4 with NaOH, the solution at the second-stage equivalence point involves only Na_2HPO_4, and the value of K_3 for this acid at 4.80×10^{-13} precludes the possibility of anything in the way of accuracy in the location of the third-stage equivalence point. The addition of an adequate amount of $CaCl_2$ to the second-stage equivalence-point solution results in the reaction

$$2Na_2HPO_4 + 3CaCl_2 + 2H_2O \rightleftharpoons Ca_3(PO_4)_2(s) + 4Na^+ + 2H_3O^+ + 6Cl^-$$

and the liberated H_3O^+ can then be titrated with standard NaOH solution in a simulated "third-stage titration," where the equivalence-point volume can be located with good accuracy.

6.10.4 TRIBASIC ACID–STRONG BASE TITRATIONS

Since the progress of the titration of a tribasic acid follows closely along the lines of that of a dibasic acid, subject to the limitations outlined in the previous section on the feasibility of such titrations, we shall avoid completely a discussion of this subject in depth. The method of approach to the theoretical aspect of such a titration will be to combine equations, or modified equations, from Chap. 4, Part 2, on acid-base equilibria, Chap. 5, on generalized titration equations, and material developed so far in this chapter. Again we stress three significant points. First, no equation higher than a quadratic will *ordinarily* be given consideration. Second, the value of $[H_3O^+]$ will be of adequate accuracy when the relative error is ± 5 percent or less. Third, the simplest equation will always be attempted first.

Table 6.14 shows the data for the titration of 50.00 ml of 0.1000 M H_3PO_4 with 0.1000 M NaOH. The K values for the acid are taken as $K_1 = 7.52 \times 10^{-3}$, $K_2 = 6.23 \times 10^{-8}$, and $K_3 = 4.80 \times 10^{-13}$. The equations required to yield a relative error in $[H_3O^+]$ of ± 5 percent or less are shown. Because of the low value for K_3, no attempt has been made to carry the titration calculations much beyond the second-stage equivalence point.

The feasibility of determing the first-stage equivalence-point volume would, in theory, be given by x as determined from the K_1/K_2 ratio at about 0.003. The expected error in locating this equivalence-point volume is therefore about 3 ppt. With respect to the second-stage equivalence-point volume, the feasibility of its location would be given by x as determined from the K_2/K_3 at a value of about 0.003 again. The expected error in locating the second-stage equivalence-point volume would also be about 3 ppt. Because of the lower value of K_2 relative to K_1, the second-stage equivalence-point-volume location will be slightly less accurate than that of the first stage.

From Table 6.14 the first-stage equivalence point pH is noted to be 4.69, and the ΔpH for ± 0.50 ml around this point is given as 1.03 pH units. The indicators bromocresol green (3.8 to 5.4) and methyl red (4.2 to 6.2) would then provide indicator-blank values of about ± 0.50 ml or less, with bromocresol green being the best choice. The titration error for this indicator would be about ± 1.0 percent or less, or 10 ppt or less. This is relatively close to the predicted zone of feasibility.

The second-stage equivalence point pH is noted to be 9.66, and the ΔpH for ± 0.50 ml around this point is given as 0.92 pH units. The indicator thymolphthalein (9.3 to 10.6) would be capable of providing for an indicator blank of about ± 0.50 ml or less, with a titration error of about ± 0.50 percent or less, or 5 ppt,

Table 6.14 Tribasic-acid-vs.-strong-base titration*
50.00 ml of 0.1000 M H_3PO_4 vs. 0.1000 M NaOH

Base added, ml	Total volume, ml	$[H_3O^+]$	Equation used	pH	$(\Delta pH/\Delta V_B) \times 10^3$ $0.1\ ml = \Delta V_B$	$(\Delta^2 pH/\Delta V_B^2) \times 10^3$ $0.1\ ml/0.1\ ml$
0.00	50.00	2.39×10^{-2}	(87)	1.62		
2.00	52.00	2.13×10^{-2}	(89)	1.67		
30.00	80.00	3.85×10^{-3}	(89)	2.41		
					7.9 (39.50)	
49.00	99.00	1.20×10^{-4}	(89)	3.92		+0.5 (44.38)
					52 (49.25)	
49.50	99.50	6.60×10^{-5}	(89)	4.18		+10 (49.50)
					102 (49.75)	
50.00	100.00	2.02×10^{-5}	(92)	4.69		+0.4 (50.00)
					104 (50.25)	
50.50	100.50	6.16×10^{-6}	(94)	5.21		−8.4 (50.50)
					62 (50.75)	
51.00	101.00	3.05×10^{-6}	(94)	5.52		−5.5 (55.62)
					7.9 (60.50)	
70.00	120.00	9.35×10^{-8}	(94)	7.03		−0.02 (70.25)
					3.9 (80.00)	
90.00	140.00	1.56×10^{-8}	(94)	7.81		+0.06 (87.25)
					12 (94.50)	
99.00	149.00	1.28×10^{-9}	(94)	8.89		+1.1 (96.88)
					62 (99.25)	
99.50	149.50	6.25×10^{-10}	(94)	9.20		+6 (99.50)
					92 (99.75)	
100.00	150.00	2.21×10^{-10}	(98)	9.66		0 (100.00)
					92 (100.25)	
100.50	150.50	7.59×10^{-11}	(100)	10.12		−6 (100.50)
					62 (100.75)	
101.00	151.00	3.72×10^{-11}	(100)	10.43		−1.7 (101.88)
					23 (103.00)	
105.00	155.00	4.32×10^{-12}	(99)	11.36		−0.4 (105.25)
					7.2 (107.50)	
110.00	160.00	1.92×10^{-12}	(99)	11.72		

*$[H_3O^+]$ and pH values are not corrected for the dilution effect. The values of volume in parentheses after each $\Delta pH/\Delta V_B$ and $\Delta^2 pH/\Delta V_B^2$ value associate the average slope, or the rate of change of the slope, with the average volume value between the titration points involved. The first-stage equivalence-point volume is given by

$$49.50\ ml + \left(\frac{10}{10 + 8.4}\ 1.00\ ml \right) = 50.04\ ml$$

The second-stage equivalence-point volume is given by

$$99.50\ ml + \left(\frac{6}{6 + 6}\ 1.00\ ml \right) = 100.00\ ml$$

again close to the predicted zone of feasibility. The lower value for the titration error here arises out of the higher equivalence-point volume of base solution (double that of the first-stage equivalence point) associated with the same increment of volume around the equivalence point.

Note that bromocresol green and thymolphthalein could *not* be used together to locate both the first- and second-stage equivalence points, since each indicator has a color change shift to blue. Where both equivalence points are to be located, methyl red and thymolphthalein could be used together to yield a color shift from red to yellow for the first stage and yellow to green for the second stage.

The student interested in the calculation of indicator blanks and titration errors, associated with indicators fitting close to but not exactly into the ΔpH change around the equivalence point in each case, should be able, from the material already presented in this chapter, to devise equations to suit these situations.

Figure 6.10 shows the general curve of the titration covered by Table 6.14.

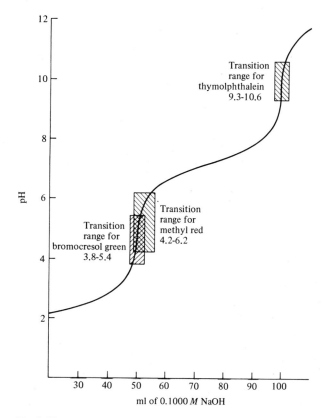

Fig. 6.10 Tribasic-acid-vs.-strong-base titration: 50.00 ml of 0.1000 M H_3PO_4 vs. 0.1000 M NaOH. Transition ranges for bromocresol green, methyl red, and thymolphthalein.

Matched to this curve are the transition ranges for the indicators methyl red, bromocresol green, and thymolphthalein.

6.11 TITRATIONS OF THE SALTS OF POLYBASIC ACIDS AND STRONG BASES BY STRONG ACIDS

As was the case with a titration involving the salt of a weak acid and a strong base, the salts of polybasic acids may be titrated with strong acids, providing that the K value of the polybasic acid permits adequate hydrolysis action for the salt involved. This would normally require that the K_1 value be not greater than about 10^{-7}, if the equivalence-point volume for the final titration stage is to be located with an error of about one part per hundred. As for intermediate equivalence points, these will again, relative to the general feasibility of determining the associated equivalence-point volumes within a few parts per thousand to a part per hundred, depend on the ratios K_1/K_2, K_2/K_3, etc., being not less than 10^5. Thus, we have the following.

1. With boric acid, H_3BO_3, having values of $K_1 = 7.3 \times 10^{-10}$, $K_2 = 1.8 \times 10^{-13}$, and $K_3 = 1.6 \times 10^{-14}$, in the titration of sodium borate, Na_3BO_3, with a strong acid such as HCl, the first-stage equivalence-point break will be imperceptible, since $K_2/K_3 \ll 10^5$. The second-stage equivalence-point break will be determinable in a crude way only, since $K_1/K_2 < 10^5$. The third-stage equivalence-point break will be strongly defined, since the value of $K_1 \ll 10^{-7}$.
2. With phosphoric acid, H_3PO_4, having values of $K_1 = 7.52 \times 10^{-3}$, $K_2 = 6.23 \times 10^{-8}$, and $K_3 = 4.80 \times 10^{-13}$, the titration of Na_3PO_4 with HCl will yield a well-defined first-stage break, since $K_2/K_3 > 10^5$. The second-stage equivalence-point break will also be determinable with a value of $K_1/K_2 > 10^5$, but the associated volume value may be determined with slightly less accuracy than that of the first stage, since the value of K_2 at 6.23×10^{-8} is pushing the limit of feasibility. The third-stage equivalence point will not be determinable, since $K_1 \gg 10^{-7}$.

Again we shall not outline in detail the methods of determining $[H_3O^+]$ for various titration points, since the techniques already discussed provide the general direction to be taken in each case. Table 6.15 shows a tabulation of data typical of the titration of 50.00 ml of 0.1000 M sodium carbonate, Na_2CO_3, by 0.1000 M HCl. The K values for H_2CO_3 are $K_1 = 4.30 \times 10^{-7}$ and $K_2 = 5.61 \times 10^{-11}$. It is apparent at once that for the first-stage equivalence point, the value of K_1/K_2 at less than 10^5 implies that the volume associated with this point will be determined with an error of about 1 part per hundred. With a K_1 value of 4.30×10^{-7}, the second-stage equivalence point will be determined with an error of about 5 ppt or greater.

In Table 6.15 cubic equations have been used in certain instances in order to

Table 6.15 Titration of a polybasic acid salt by a strong acid*
50.00 ml of 0.1000 M Na$_2$CO$_3$ by 0.1000 M HCl

Acid added, ml	Total volume, ml	[H$_3$O$^+$]	Equation used	pH	(ΔpH/ΔV_A) $\times$ 10^3 0.1 ml = ΔV_A	(Δ^2pH/ΔV_A^2) $\times$ 10^3 0.1 ml/0.1 ml
0.00	50.00	2.37 $\times$ 10^{-12}	(85)	11.63		
2.00	52.00	3.98 $\times$ 10^{-12}	(75)	11.40		
20.00	70.00	3.74 $\times$ 10^{-11}	(68)	10.43		
					3.5 (25.00)	
30.00	80.00	8.47 $\times$ 10^{-11}	(68)	10.08		+0.008 (30.00)
					4.3 (35.00)	
40.00	90.00	2.24 $\times$ 10^{-10}	(68)	9.65		+0.036 (38.75)
					7.0 (42.50)	
45.00	95.00	5.05 $\times$ 10^{-10}	(68)	9.30		+0.20 (44.75)
					16 (47.00)	
49.00	99.00	2.19 $\times$ 10^{-9}	(70)	8.66		+0.71 (48.12)
					32 (49.25)	
49.50	99.50	3.18 $\times$ 10^{-9}	(70)	8.50		+2.0 (49.50)
					42 (49.75)	
50.00	100.00	5.10 $\times$ 10^{-9}	(54)	8.29		−2.0 (50.00)
					32 (50.25)	
50.50	100.50	7.48 $\times$ 10^{-9}	(43)	8.13		−2.4 (50.50)
					20 (50.75)	
51.00	101.00	9.33 $\times$ 10^{-9}	(43)	8.03		
56.00	106.00	5.86 $\times$ 10^{-8}	(41)	7.23		
60.00	110.00	1.08 $\times$ 10^{-7}	(41)	6.97		

* [H$_3$O$^+$] and pH values are not corrected for the dilution effect. The values of volume in parentheses after each ΔpH/ΔV_A and Δ^2pH/ΔV_A^2 value associate the average slope, or rate of change of the slope, with the average volume value between the titration points involved. The first-stage equivalence-point volume is given by

$$49.50 \text{ ml} + \left(\frac{2.0}{2.0 + 2.0} \, 0.50 \text{ ml}\right) = 49.75 \text{ ml}$$

although, in view of the Δ^2pH/ΔV_A^2 value at 50.50 ml, a closer approximation can be given by

$$49.50 \text{ ml} + \left(\frac{2.0}{2.0 + 2.4} \, 1.00 \text{ ml}\right) = 49.95 \text{ ml}$$

The second-stage equivalence-point volume is given by

$$99.50 \text{ ml} + \left(\frac{8.8}{8.8 + 14} \, 1.00 \text{ ml}\right) = 99.90 \text{ ml}$$

The discrepancies between these values and the theoretical values of 50.00 and 100.00 ml are due to approximations associated with the calculated pH and ΔpH/ΔV_A, Δ^2pH/ΔV_A^{2} values for titration points in the critical zones. These discrepancies anticipate errors in the determination of the respective equivalence-point volumes of about 1 part per thousand.

Table 6.15 *(Continued)*

Acid added, ml	Total volume, ml	$[H_3O^+]$	Equation used	pH	$(\Delta pH/\Delta V_A) \times 10^3$ $0.1\ ml = \Delta V_A$	$(\Delta^2 pH/\Delta V_A^2) \times 10^3$ $0.1\ ml/0.1\ ml$
70.00	120.00	2.87×10^{-7}	(41)	6.54		
					3.5 (75.00)	
80.00	130.00	6.45×10^{-7}	(41)	6.19		+0.008 (80.00)
					4.3 (85.00)	
90.00	140.00	1.72×10^{-6}	(41)	5.76		+0.036 (88.75)
					7.0 (92.50)	
95.00	145.00	3.87×10^{-6}	(41)	5.41		+0.25 (94.75)
					18 (97.00)	
99.00	149.00	2.11×10^{-5}	(41)	4.68		+1.6 (98.12)
					54 (99.25)	
99.50	149.50	3.85×10^{-5}	(40)	4.41		+8.8 (99.50)
					98 (99.75)	
100.00	150.00	1.20×10^{-4}	(32)	3.92		+3.6 (100.00)
					116 (100.25)	
100.50	150.50	4.53×10^{-4}	(219)†	3.34		−14 (100.50)
					46 (100.75)	
101.00	151.00	7.80×10^{-4}	(219)†	3.11		−1.4 (101.88)
					16 (103.00)	
105.00	155.00	3.22×10^{-3}	(219)‡	2.49		−0.21 (105.25)
					5.8 (107.50)	
110.00	160.00	6.25×10^{-3}	(219)‡	2.20		

†Modified and corrected.
‡Modified only.

obtain sufficient accuracy in $[H_3O^+]$. In addition, Eq. (219) used for the determination of $[H_3O^+]$ for points after the second-stage equivalence point has been modified and corrected for some points and modified alone for others. The ΔpH for ± 1.00 ml around the first-stage equivalence point pH of 8.29 is given as 0.63 pH unit. The indicator meta-cresol purple (9.0 to 7.4) would fit reasonably well into this ΔpH, yielding an equivalence-point-volume location with an error of about 2 parts per hundred or less. Phenolphthalein (10.0 to 8.3) can also be used with about the same order of error. Although the value of pH at which the color change for this indicator starts is quite high, the nature of the change, red to colorless, provides for the disappearance of the red color at a point in volume relatively close to the equivalence-point value. The ΔpH for ± 0.50 ml around the second-stage equivalence point pH of 3.92 is given as 1.07 pH units. Methyl orange (4.4 to 3.1) would fit almost exactly into this ΔpH, yielding an equivalence-point-volume location with an error of about 5 ppt or less. Because of possible variations in the concentration of H_2CO_3 around this equivalence point, due to variable loss of CO_2 from the solution, there exists a fair level of uncertainty relative to the location of the equivalence-point volume, and the expected error may be higher than the estimated value of 5 ppt.

When one attempts to locate both equivalence-point volumes, the indicators phenolphthalein and methyl orange should be used so as to permit proper color change definition at each titration break point, the methyl orange indicator being added to the solution only after the first-stage equivalence point color change of red to colorless has been achieved. Again we shall not discuss indicator-blank and titration-error calculations related to indicators fitting less exactly the ΔpH change ranges around each equivalence point. The student interested in such calculations can use the material already presented in this chapter for the purpose of devising the necessary equations.

We noted that at the second-stage equivalence point, some lack of sharpness in definition would result from the two factors of possible variation in H_2CO_3 concentration and relatively high value of K_1 for H_2CO_3. A method of increasing the definition, and therefore the accuracy of locating the equivalence-point volume, is to add the strong acid titrant to an excess beyond that for the methyl orange color change. The solution is then boiled to expel CO_2 and subsequently back-titrated with a standard solution of a strong base such as NaOH. While providing for a sharper definition of the end point, the method has the disadvantage common to all back-titration techniques, namely, that of increasing the total uncertainty in the measured volume of strong acid titrant representing the second-stage equivalence-point volume.

Figure 6.11 shows the general curve of the titration covered by the data provided in Table 6.15. Matched to this curve are the transition ranges for phenolphthalein, meta-cresol purple, and methyl orange.

From the foregoing it is apparent that it will be possible to titrate a mixture of $NaHCO_3$ and Na_2CO_3 in solution by using a strong acid titrant and to determine from this titration the amount of each substance present in the mixture. Using phenolphthalein as the indicator, the solution mixture is titrated to the first-stage equivalence point, and the volume of titrant required is noted as V_1 milliliters. The addition of methyl orange at this point is followed by titration to the second-stage equivalence point, and the *total* volume of titrant used is noted as V_2 milliliters. The volume of titrant required to titrate the Na_2CO_3 in the mixture is then V_1 milliliters, while that required to titrate the $NaHCO_3$ component of the mixture is given by (V_2 milliliters $- 2V_1$ milliliters). A similar form of titration can, of course, be applied to mixtures of the salts of any polybasic acid, subject to the usual limitations relative to titration feasibility.

A further development arising out of the foregoing permits the qualitative and quantitative identification of solutions of substances such as HCl, H_3PO_4, and NaH_2PO_4 present singly or in the allowable paired combinations. Of the three substances above, only the paired combination of HCl and NaH_2PO_4 can not be so investigated, since this results in a reaction forming H_3PO_4 which continues until the solution consists largely of either HCl and H_3PO_4 or NaH_2PO_4 and H_3PO_4. Such a series of investigations can be carried out on the solutions involved by titrating with standard NaOH solution separate and equal aliquots, using bromocresol green as the indicator for one titration and thymolphthalein for the other. The following situations develop:

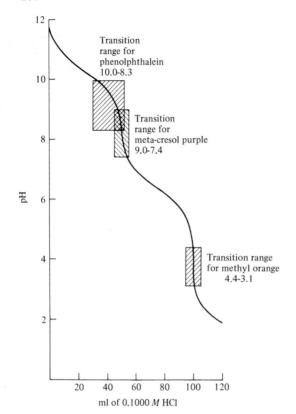

Fig. 6.11 Titration of a polybasic acid salt by strong acid: 50.00 ml of 0.1000 M Na$_2$CO$_3$ vs. 0.1000 M HCl. Transition ranges for phenolphthalein, meta-cresol purple, and methyl orange.

1. Solutions of HCl alone yield the same volume for each titration:

   Volume with thymolphthalein = volume with bromocresol green

2. Solutions of H$_3$PO$_4$ alone titrate to NaH$_2$PO$_4$ with bromocresol green, then to Na$_2$HPO$_4$ with thymolphthalein:

   Volume with thymolphthalein = 2 × volume with bromocresol green

3. Solutions of HCl and H$_3$PO$_4$ will titrate to NaCl and NaH$_2$PO$_4$ with bromocresol green, with the NaH$_2$PO$_4$ being titrated to Na$_2$HPO$_4$ with thymolphthalein:

   Volume with thymolphthalein < 2 × volume with bromocresol green

4. Solutions of H$_3$PO$_4$ and NaH$_2$PO$_4$ will titrate the H$_3$PO$_4$ portion to NaH$_2$PO$_4$ with bromocresol green, with the combined NaH$_2$PO$_4$ being titrated to Na$_2$HPO$_4$ with thymolphthalein:

   Volume with thymolphthalein > 2 × volume with bromocresol green

5. Solutions of NaH_2PO_4 alone will yield an immediate color change with bromocresol green, with the NaH_2PO_4 being titrated to Na_2HPO_4 with thymolphthalein:

Volume with bromocresol green = zero
Volume with thymolphthalein = volume for NaH_2PO_4

6.12 TITRATIONS OF MIXTURES OF ACIDS BY STRONG BASES

The titration of a mixture of acids will follow generally the course of the titration of a polybasic acid, since the latter can be thought of as a solution of single acids of different dissociation constants. Under these circumstances it will be clearly seen that the titration of a mixture of acids, relative to feasibility, will be governed by the general conditions already expressed in Sec. 6.10.3. To recapitulate, we note that mixtures of more than two acids will not likely be capable of titration so as to yield distinct equivalence-point breaks for each acid (the limitation relative to tribasic acid titrations); the K_a value for the weaker acid should not be less than 10^{-8}, and preferably not less than 10^{-7} (the usual limitation relative to reasonably accurate location of the associated equivalence point); and the values of K_{a_1}/K_{a_2} should not be less than 10^5 (the limitation stipulated for a distinct equivalence-point break for the titration of the stronger acid).

Thus, the titration of a strong acid–weak acid mixture will be feasible, relative to two distinct equivalence-point breaks being obtained, providing that the dissociation constant of the weak acid is not greater than 10^{-5} and not less than 10^{-8}, but preferably not less than 10^{-7}. The value of $[H_3O^+]$ at the start of such a titration will be given by the concentration of the strong acid, since the $[H_3O^+]$ contributed by the weak acid will be negligible as the result of repression of the weak acid ionization by the high $[H_3O^+]$ from the strong acid. Between the start and the first-stage equivalence point the situation changes progressively. For points located close to the start, only the $[H_3O^+]$ from the remaining strong acid need be considered; but for points located close to the first equivalence point, the contribution of $[H_3O^+]$ from both the residual strong acid and the weak acid should be taken into account. In most cases, however, a reasonable approximation to the $[H_3O^+]$ value for all such points will be given by considering only the $[H_3O^+]$ from the remaining strong acid. In such instances, Eq. (185) or (184) may be used. Since the solution at the first equivalence point will consist of the weak acid at a concentration dictated by the titration dilution effect to this point, and the neutral strong acid salt, the $[H_3O^+]$ at this point will be given by either Eq. (5) or (7), the equation used being dependent on the relative error allowed for $[H_3O^+]$—usually ± 5 percent maximum. Note also that this first-stage value of $[H_3O^+]$ sets the limiting value for the calculation of $[H_3O^+]$ from Eq. (185) or (184). After the first-stage equivalence point, the titration proceeds in a manner identical to that for the titration of a weak acid, and the equations required for the calculation of $[H_3O^+]$ for the various titration zones involved will be identical to those outlined in Sec. 6.6. Table 6.16 shows data associated with such a titration,

Table 6.16 Titrations of strong acid–weak acid mixtures with strong base*

50.00 ml of 0.1000 M HCl and 0.1500 M CH$_3$·COOH vs. 0.2000 M NaOH

Base added, ml	Total volume, ml	$[H_3O^+]$	Equation used	pH	$(\Delta pH/\Delta V_B) \times 10^3$ 0.1 ml $= \Delta V_B$	$(\Delta^2 pH/\Delta V_B^2) \times 10^3$ 0.1 ml/0.1 ml
0.00	50.00	1.00×10^{-1}	Strong acid	1.00		
2.00	52.00	8.86×10^{-2}	(184)	1.05		
					3.0 (3.50)	
5.00	55.00	7.27×10^{-2}	(184)	1.14		+0.005 (5.50)
					3.2 (7.50)	
10.00	60.00	5.00×10^{-2}	(184)	1.30		+0.02 (10.00)
					4.2 (12.50)	
15.00	65.00	3.08×10^{-2}	(184)	1.51		+0.048 (15.00)
					6.6 (17.50)	
20.00	70.00	1.43×10^{-2}	(184)	1.84		+0.25 (19.50)
					18 (22.00)	
24.00	74.00	2.70×10^{-3}	(184)	2.57		+0.52 (23.25)
					31 (24.50)	
25.00	75.00	1.33×10^{-3}	(7)	2.88		+0.80 (25.00)
					39 (25.50)	
26.00	76.00	5.35×10^{-4}	(20)	3.27		−0.80 (26.50)
					17 (28.00)	
30.00	80.00	1.14×10^{-4}	(24)	3.94		−0.15 (31.50)
					6.4 (35.00)	
40.00	90.00	2.64×10^{-5}	(24)	4.58		−0.016 (40.00)
					4.8 (45.00)	
50.00	100.00	8.80×10^{-6}	(24)	5.06		+0.074 (50.50)
					13 (56.00)	
62.00	112.00	2.30×10^{-7}	(24)	6.64		+2.6 (59.10)
					172 (62.20)	
62.40	112.40	4.70×10^{-8}	(24)	7.33		+515 (62.32)
					1460 (62.45)	

*$[H_3O^+]$ and pH values are not corrected for the dilution effect. The values of volume in parentheses after each $\Delta pH/\Delta V_B$ and $\Delta^2 pH/\Delta V_B^2$ value associate the average slope, or rate of change of the slope, with the average volume value between the titration points involved. The first-stage equivalence-point volume is given by

$$25.00 \text{ ml} + \left(\frac{0.80}{0.80 + 0.80} \, 1.50 \text{ ml}\right) = 25.75 \text{ ml}$$

The discrepancy in the calculated value relative to the theoretical value of 25.00 ml reflects an error in the calculation of $[H_3O^+]$ at the 24.00-ml point of the titration. Here Eq. (184) is used, and no consideration is given to the contribution of $[H_3O^+]$ from the weak acid. The second-stage equivalence-point volume is given by

$$62.32 \text{ ml} + \left(\frac{515}{515 + 529} \, 0.36 \text{ ml}\right) = 62.50 \text{ ml}$$

Table 6.16 *(Continued)*

Base added, ml	Total volume, ml	$[H_3O^+]$	Equation used	pH	$(\Delta pH/\Delta V_B) \times 10^3$ 0.1 ml $= \Delta V_B$		$(\Delta^2 pH/\Delta V_B^2) \times 10^3$ 0.1 ml/0.1 ml	
62.50	112.50	1.62×10^{-9}	(15)	8.79			$+30$	(62.50)
					1490	(62.55)		
62.60	112.60	5.62×10^{-11}	(214)†	10.28			-529	(62.68)
					168	(62.80)		
63.00	113.00	1.13×10^{-11}	(214)†	10.95			-11	(63.40)
					34	(64.00)		
65.00	115.00	2.30×10^{-12}	(214)†	11.64			-0.73	(65.75)
					8.6	(67.50)		
70.00	120.00	8.00×10^{-13}	(214)†	12.07				

†Uncorrected.

in this case 50.00 ml of 0.1000 M HCl and 0.1500 M acetic acid, $CH_3 \cdot COOH$, vs. 0.2000 M NaOH. The K_a value for acetic acid is 1.76×10^{-5}.

The pH at the first-stage equivalence point is 2.88, the ΔpH for ± 1.00 ml around this point being $3.27 - 2.57$, or 0.70 pH unit. The indicator 2,6-dinitrophenol, color change range 2.0 to 4.0, can be used here to determine the first-stage equivalence-point volume with an error of about ± 4 parts per hundred or less. It is apparent that because of the relatively high K_a value of 1.76×10^{-5} for $CH_3 \cdot COOH$, the first equivalence-point volume will be poorly located. K_a values less than 10^{-5} would permit a more accurate location of this volume.

The pH at the second equivalence point is 8.79, and the ΔpH for ± 0.10 ml around this point is given by $10.28 - 7.33$, or 2.95 pH units. As noted in Sec. 6.6.5, the indicator phenolphthalein would provide for location of the associated volume with an error of about 1 part per thousand.

It is important to observe that the two indicators suggested can only be used in *separate* first- and second-stage titration situations, since the color changes do not adequately permit simultaneous use.

Figure 6.12 shows the titration curve, matched to which are the two indicators selected. In the titration of a mixture of two weak acids, the situation is somewhat more complicated. No attempt will be made to discuss the details involved, since the student may determine, from previous data, the methods for calculating the critical points of such a titration.

Figure 6.13 shows the titration curve for 50.00 ml of a solution 0.1000 M to dichloracetic acid, $CHCl_2 \cdot COOH$, and 0.1000 M to acetic acid, $CH_3 \cdot COOH$, the titration being carried out by 0.1000 M NaOH. The values of K_a for these acids are 3.32×10^{-2} and 1.76×10^{-5}, respectively. As predicted by the ratio of $K_{a_1}/K_{a_2} < 10^5$, the first equivalence-point break is poorly defined. For example, the ΔpH for ± 3.00 ml around the first equivalence point pH of 3.12 is found to be

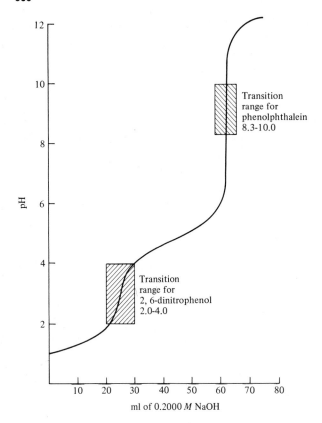

Fig. 6.12 Titration of a strong acid–weak acid mixture with strong base: 50.00 ml of 0.1000 M HCl and 0.1500 M CH$_3$·COOH vs. 0.2000 M NaOH. Transition ranges for 2, 6-dinitrophenol and phenolphthalein.

about 0.90 unit. With an indicator such as 2,6-dinitrophenol, color change range 2.0 to 4.0, this equivalence-point volume might be located with an error of about 6 parts per hundred or less.

The value of $K_{a_2} > 10^{-7}$ provides for a well-defined break at the second equivalence point. The ΔpH for ±0.10 ml around this equivalence point pH of 8.64 is about 2.40 units, and phenolphthalein, color change range 8.3 to 10.0, locates the associated volume with an error of about a part or two per thousand.

6.13   ERROR IN LOCATING THE EQUIVALENCE–POINT VOLUMES IN MULTISTAGE TITRATIONS

In the titration of a polybasic acid solution or the solution of a salt of a polybasic acid, any equivalence point may in theory yield the volume data required for quantitative evaluation. In practice, as pointed out in Sec. 6.10.3, the critical equiva-

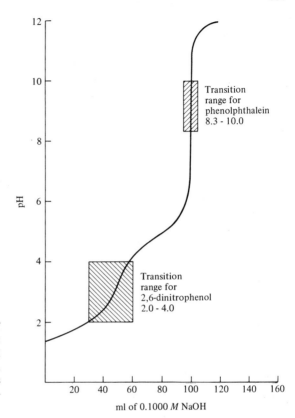

Fig. 6.13 Titration of a mixture of two weak acids using a strong base: 50.00 ml of 0.1000 M CHCl$_2$·COOH and 0.1000 M CH$_3$·COOH vs. 0.1000 M NaOH. Transition ranges for 2, 6-dinitrophenol and phenolphthalein.

lence point is always that which provides the best definition or volume locational accuracy.

In titrations involving at the start solutions of acid mixtures, or solution mixtures of the salts of weak or polybasic acids, each equivalence point provides specific and unique analytical data and must be considered individually. In such instances, any uncertainty in an earlier equivalence-point-volume location must be reflected in the location of a subsequent equivalence-point volume, regardless of the sharpness of definition relative to this latter volume. For example, in a two-stage titration by sodium hydroxide solution of a mixture of a strong and a weak acid, the first-stage equivalence-point volume yields data associated with the quantity of strong acid, while the second-stage volume provides data relative to the weak acid quantity. It is apparent, however, that

Second-stage volume = total volume − first-stage volume

and any uncertainty in locating the first-stage volume will be carried over to the determination of the second-stage volume, regardless of the clarity of definition of the total-volume value.

6.14 DETERMINATION OF NITROGEN BY THE KJELDAHL METHOD

The analysis of organic materials (pure compounds, foods, fertilizers, etc.) for nitrogen is carried out very frequently by the Kjeldahl method. For example, estimations of protein in food products are customarily based on Kjeldahl nitrogen determinations. The technique involves several steps, including a neutralization titration. A generalization of these follows.

The usual first step is an oxidation process using hot concentrated sulfuric acid to oxidize the organic material to carbon dioxide and water, the nitrogen being converted to ammonium hydrogen sulfate, NH_4HSO_4. This is undoubtedly the most critical step in the entire technique, and because of its time-consuming nature, catalysts such as mercury, copper, selenium, or compounds of these elements are often used to hasten the process. A normal modification, again aimed at decreasing the time for the oxidation step, is to add potassium sulfate as a means of increasing the temperature during digestion. The conversion of nitrogen to ammonium hydrogen sulfate proceeds well, providing that the element is bound to amide or amine groups. Where it is bound in azo, nitro, etc., groups conversion is often incomplete because of formation of elemental nitrogen or oxides of nitrogen. Where organic compounds suspected of containing such groups are involved, a prior reduction step must be included. This can usually be accomplished by the addition of a reductant such as sodium thiosulfate, $Na_2S_2O_3$, to the sulfuric acid medium before digestion is started.

Upon completion of the oxidation process the solution is cooled and diluted, and a concentrated solution of sodium hydroxide is floated as a separate layer on top of the sulfuric acid solution. If mercury or copper has been used for catalytic purposes, these ions will be precipitated by the addition of potassium sulfide prior to floating the sodium hydroxide layer. The flask is now connected to a distillation apparatus and the layers are mixed by agitation. Sulfuric acid is neutralized by sodium hydroxide in excess, and ammonia is liberated according to

$$NH_4HSO_4 + 2OH^- \rightleftharpoons NH_3(g) + H_2O + SO_4{}^{2-}$$

Heating the flask now evolves ammonia and water, which are absorbed in a receiver flask containing either a measured amount of standard hydrochloric acid solution or a saturated solution of boric acid, H_3BO_3. In the former case, the ammonia is neutralized as ammonium chloride, NH_4Cl; in the latter, it is converted to ammonium dihydrogen borate, $NH_4H_2BO_3$.

Where the hydrochloric acid absorption method is used, the excess of standard acid is titrated with standard sodium hydroxide solution. Where the boric acid method has been applied, the ammonium salt of the weak boric acid is titrated with standard hydrochloric acid. Since, in each case, the solution at the equivalence point contains ammonium chloride, an indicator having a color change pH interval in the acid zone, such as methyl red or bromocresol green, is required. Obviously the boric acid absorption method has the advantage of requiring the preparation and use of only one standard solution.

Calculation of the amount of ammonia and therefore of nitrogen follows from either the amount of standard hydrochloric acid used in the absorption process

$$HCl + NH_3 \rightleftharpoons NH_4Cl$$

or from the amount of standard hydrochloric acid used in the titration of the borate ion from the ammonium dihydrogen borate

$$NH_4H_2BO_3 \rightleftharpoons NH_4^+ + H_2BO_3^-$$
$$H_2BO_3^- + HCl \rightleftharpoons H_3BO_3 + Cl^-$$

6.15 DETERMINATION OF ALCOHOLS

Neutralization methods can be used in the determination of organic compounds such as alcohols, ketones, etc., by making use of reactions which quantitatively convert such substances to forms which can be titrated by standard acid or base solutions. As a case in point, alcohols can be reacted with acetic anhydride in pyridine or ethyl acetate solution according to*

$$(CH_3 \cdot CO)_2O + ROH \rightleftharpoons CH_3 \cdot COOR + CH_3 \cdot COOH$$

    Anhydride   Alcohol   Ester   Acid

When the reaction is complete, the excess of anhydride is reacted with water:

$$(CH_3 \cdot CO)_2O + H_2O \rightleftharpoons 2CH_3 \cdot COOH$$

The total quantity of acetic acid formed is now titrated with standard sodium hydroxide solution:

$$CH_3 \cdot COOH + NaOH \rightleftharpoons CH_3 \cdot COONa + H_2O$$

Exactly the same quantity of acetic anhydride is now reacted with water, and the resulting acetic acid is titrated with standard sodium hydroxide solution. The difference between the second and first titration volumes represents the quantity of ester produced in the alcohol reaction, and for the example used, this is on a mole-for-mole basis with the alcohol reacting.

PROBLEMS

1. (a) With a relative error in $[H_3O^+]$ of not more than ± 5 percent, determine the values of $[H_3O^+]$ and pH for the following titration points in three separate titrations involving
        (1) 50.00 ml of 1.000 M KOH and 1.000 M HCl
        (2) 50.00 ml of 0.1000 M KOH and 0.1000 M HCl
        (3) 50.00 ml of 0.00100 M KOH and 0.00100 M HCl
Milliliters of HCl are

0.00	5.00	10.00	20.00	30.00	40.00	45.00	49.00	49.50	49.80
49.90	49.95	50.00	50.05	50.10	50.20	50.50	51.00	55.00	
60.00									

*J. S. Fritz and G. H. Schenk, *Anal. Chem.*, **31**: 1808 (1959).

(b) Sketch the titration curves involved on the same graph, and comment on that portion of the general curve most critically affected by changes in KOH and HCl concentrations.

(c) Was it necessary at any point to use the back-reaction correction Eqs. (185) and (188)? If so, indicate the points and the titrations involved.

(d) List the indicators that might be chosen for each titration from Table 6.1 so that each titration will show an indicator blank not exceeding ± 0.05 ml and, therefore, an equivalence-point volume in error by not more than 1 ppt.

(e) Comment on the range of indicator choice for each titration.

(f) Suppose that the following indicators were used

1.000-M titration	2,6-Dinitrophenol and alizarin yellow R
0.1000-M titration	Ethyl orange and thymolphthalein
0.00100-M titration	Para-nitrophenol and phenol red

Determine the approximate indicator blanks (to the nearest 0.05 ml) and the titration errors (to the nearest 0.05 percent). Comment on the general applicability of each indicator.

2. The following data were recorded during the titration of 50.00 ± 0.02 ml of a solution of KOH with 0.1045 ± 0.0001 M HCl, using a pH meter to follow the titration

Volume HCl, ml	pH
40.00	11.86
45.00	11.14
45.50	10.90
45.70	10.75
45.90	10.54
46.10	10.12
46.30	4.06
46.50	3.52
46.70	3.28
47.00	3.08
47.50	2.86
48.00	2.72

Use a $\Delta^2 \text{pH}/\Delta V_A^2$ tabulation to determine the equivalence-point volume of HCl solution. Determine the molarity of the KOH solution. Show the uncertainty in the KOH molarity based on the propagation of determinate error, assuming a burette-reading error or uncertainty of ± 0.02 ml per reading. *Ans.* 46.20 ml, 0.0966 M

3. (a) The titration of the following weak acid solutions is projected:

(1) 50.00 ml of 0.1000 M acid vs. 0.1000 M NaOH		$K_a = 1.00 \times 10^{-7}$
(2) 50.00 ml of 0.1000 M acid vs. 0.1000 M NaOH		$K_a = 1.00 \times 10^{-8}$
(3) 50.00 ml of 1.000 M acid vs. 1.000 M NaOH		$K_a = 1.00 \times 10^{-8}$

Calculate the pH value for each titration at milliliters of added base:

49.50	49.70	49.90	50.00	50.10	50.30	50.50

Use equations capable of yielding a relative error in $[H_3O^+]$ of ± 5 percent or less, and indicate the equations used by number.

(b) Using only the data accumulated above, determine for each titration a ΔpH, for symmetrical points around the equivalence-point volume, of approximately one pH unit. Based on these ΔpH values, select the *best-suited* indicator for each titration from Table 6.1.

(c) Using the indicators selected in b, determine the approximate error in parts per thousand for locating the equivalence-point volume for each titration.

(d) Give your general conclusions, relative to the feasibility of weak acid–strong base titrations, as based on the foregoing data.

4. (a) A titration of 50.00 ± 0.02 ml of a weak monobasic acid solution with $0.0965\ M$ NaOH is followed by a pH meter and yields the following data.

Volume NaOH, ml	pH
56.50	4.36
56.70	4.58
56.80	4.77
56.90	5.08
57.00	7.50
57.10	9.98
57.20	10.27
57.50	10.66

Prepare a $\Delta^2 \mathrm{pH}/\Delta V_B{}^2$ tabulation and locate the equivalence-point volume.

(b) Determine the molarity of the weak acid solution, and assuming a burette-reading uncertainty of ± 0.02 ml per reading, use the propagation of determinate errors to find the uncertainty in the acid-solution molarity.

(c) Using only the simplest equation, determine the value of K_a for the weak monobasic acid.

(d) What indicators from Table 6.1 would be capable of locating the equivalence-point volume with an error of about 2 ppt or less? How does this volume locational error compare with the burette-reading uncertainty for the equivalence-point volume?

(e) What indicator blanks and titration errors (to the nearest 0.05 ml and 0.05 percent, respectively) would be applicable if the following indicators were used:

(1) Bromocresol green

(2) Alizarin yellow R

(f) Would the use of these indicators be practical?

5. (a) A solid substance contains neutral salts and benzoic acid, $C_6H_5 \cdot COOH$. The molecular weight of benzoic acid is 122.12. 753 ± 1 mg of the solid is weighed out, dissolved in water, and diluted to 25.00 ± 0.04 ml. This solution is titrated with $0.0500 \pm 0.0001\ M$ NaOH, using para-α-naphthalein as the indicator. The end point occurs at 32.05 ml, subject to a burette-reading uncertainty of ± 0.02 ml per reading. The error in locating the equivalence-point volume, as based on the use of para-α-naphthalein, is about 2 ppt. What is the percentage of benzoic acid in the solid?

(b) What is the *approximate* K_a value for benzoic acid (a monobasic acid), using the simplest equation and taking the pH at the equivalence point as the central value of the pH color change range for the para-α-naphthalein indicator? How does this value compare with the tabled value for the K_a of benzoic acid?

(c) How does the indicator error in locating the equivalence-point volume compare with

(1) The weighing error

(2) The error in the starting volume of solution

(3) The burette-reading uncertainty for the equivalence-point volume

(4) The error in the molarity of the NaOH solution

Is the error in (2) important in the calculation of the percentage of benzoic acid in the solid?

(d) Show that you have expressed your value for the percentage of benzoic acid to the proper significant figures.

    (e) Based on determinate-error propagation calculations, show the percentage of the benzoic acid component followed by the correct indication of uncertainty.

Ans. (a) 26.0%; (b) 3×10^{-4}; (e) $26.0 \pm 0.1\%$

6. (a) The following titrations are to be conducted with 0.1000 M HCl:
  - (1) 50.00 ml of 0.1000 M diethylamine, $(C_2H_5)_2NH$
  - (2) 50.00 ml of 0.1000 M hydroxylamine, H_2NOH
  - (3) 50.00 ml of 0.1000 M ammonia, NH_3

Using equations capable of yielding values of $[OH^-]$ showing relative errors of ± 10 percent or less, and indicating the resulting pH values, calculate the pH at the following levels of HCl solution addition in milliliters:

0.00	10.00	20.00	30.00	40.00	45.00	49.00	49.50	49.80
50.00	50.20	50.50	51.00	55.00	60.00			

    (b) Sketch the titration curves on the same graph and comment on their general appearance.

    (c) Using only the data accumulated for each titration, determine for each titration a ΔpH for symmetrical points around the equivalence-point volume of about one pH unit minimum. Make a list, from Table 6.1, of the indicators best suited for each titration.

    (d) What would be the approximate error in each case in locating the equivalence-point volume, based on the conditions found for c?

7. (a) In the titration of 26.00 ± 0.05 ml of a solution of the weak monobasic base ethylamine, $C_2H_5NH_2$, with 0.1505 M HCl, a pH meter yielded the following data.

Volume HCl, ml	pH
15.00	10.33
17.00	10.09
19.00	9.71
20.00	9.31
20.50	8.80
20.60	8.55
20.70	5.96
20.80	3.65
20.90	3.27
21.00	3.06
21.50	2.62
22.00	2.40
22.50	2.26
23.00	2.16

Use a $\Delta^2 pH / \Delta V_A^2$ tabulation to locate the equivalence-point volume.

    (b) Determine the molarity of the ethylamine solution.

    (c) From Table 6.1 make a list of those indicators which will allow location of the equivalence-point volume with an error of about 5 ppt or less.

    (d) Considering the burette-reading uncertainty of ± 0.02 ml per reading for the equivalence-point-volume location by the tabulation method, use the determinate-error propagation method to show the uncertainty in the molarity expressed for b.

    (e) Considering the equivalence-point-volume locational error of 5 ppt when using the indicator method, use the same error propagation method to show the uncertainty in the molarity for b.

    (f) Explain how the situation in e might be bettered.

8. In the titration of 0.1000 M ammonia as outlined in Prob. 6a (3), determine the indicator blanks (to the nearest 0.05 ml) associated with the use of the following indicators. The simplest equations will be adequate.

(*a*) Methyl orange
(*b*) Bromothymol blue
(*c*) Neutral red

Indicate if the use of these indicators would be feasible.

9. A material contains only neutral salts and calcium oxide, CaO. 1.0000 g of this substance is dissolved in 50.00 ml of 0.2560 M HCl. The resulting solution contains an excess of HCl, which is then titrated with 0.3311 M NaOH to an equivalence-point volume of 9.54 ml. The indicator error is 1 ppt and the burette-reading uncertainty is ±0.02 ml per reading. What is the percentage of calcium oxide in the material?   *Ans.* 27.0%

10. A solid substance contains only CaO and MgO. 1.0000 g is dissolved in 50.00 ml of 1.000 M HCl. The excess HCl is titrated with 0.4800 M NaOH and requires 24.22 ml to reach the equivalence point. What are the percentages of CaO and MgO?

11. Most indicator substances used in neutralization work are prepared solutions ranging around 0.1 percent by weight in water, alcohol, or another appropriate solvent. Since the average molecular weight of many indicators would be given by a value of about 300, this indicates a solution strength of about $3 \times 10^{-3} M$.

Where four drops (0.20 ml) of the indicator solution are used for a titration which involves about 100 ml total solution volume at the equivalence point, the indicator concentration at this point will be about $6 \times 10^{-6} M$. Where the titrant is 0.1 M, the indicator substance consumes approximately 0.006 ml of titrant for the indicator reaction. Thus, the normal indicator concentration level does not create any problem relative to the consumption of titrant.

Consider that 0.05 ml of titrant consumed in this way would be significant, and calculate from this the indicator volume addition which would permit this consumption where the titrant is 0.1 M and the equivalence point occurs at a total solution volume of about 100 ml.   *Ans.* 1.7 ml

12. A 0.10-M solution of ammonium formate, H·COONH$_4$, is prepared. What is the approximate pH of this solution? What would be the approximate pH values for 0.05- and 1.0-M solutions of the same salt? Explain the fact that the pH appears to be independent of the concentration of the salt.

13. (*a*) A titration of the weak acid salt NaA is projected, with the associated weak monobasic acid having a K_a value of 2.06×10^{-9}. This titration will involve 40.00 ml of a 0.0875-M solution of the salt, with a titrant of 0.1000 M HCl. Determine the $[H_3O^+]$ for the titration (with a relative error of ±10 percent maximum) at the volume addition levels in milliliters of HCl solution:

0.00	10.00	20.00	25.00	30.00	34.00	34.50	34.80	35.00
35.20	35.50	36.00	40.00	45.00	50.00			

and show the pH values involved.

(*b*) Considering a ΔpH for ±0.20 ml around the equivalence-point volume, make a list from Table 6.1 of suitable indicators.

(*c*) Show the approximate error expected in the location of the equivalence volume using the indicators selected in *b*.

(*d*) Show the indicator blank and the titration error (to the nearest 0.05 ml and 0.05 percent, respectively) where bromothymol blue is used as the indicator.

14. (*a*) A titration is to be conducted between 50.00 ml of 0.1000 M sulfurous acid, H_2SO_3, and 0.1000 M KOH in a closed system. The K values for the acid are $K_1 = 1.7 \times 10^{-2}$ and $K_2 = 6.2 \times 10^{-8}$. Using equations that yield a relative error in $[H_3O^+]$ of ±10 percent or less, and showing the equations by number, determine the $[H_3O^+]$ and pH for the following levels of KOH solution addition in milliliters:

0.00	10.00	20.00	30.00	40.00	49.00	49.50	49.80	50.00
50.20	50.50	51.00	60.00	70.00	80.00	90.00	99.00	99.50
99.80	100.00	100.20	100.50	101.00	105.00	110.00		

(b) Sketch the curve for this titration.

(c) Show the calculated feasibility of determining the first-stage equivalence-point volume, based on the values of K_1 and K_2.

(d) Using a ΔpH for ± 0.50 ml around each equivalence-point volume, select from Table 6.1 indicators suited to the location of the first- and second-stage equivalence-point volumes.

(e) What will be the approximate errors in locating these volumes with the indicators selected in d?

(f) Can the indicators selected be used to locate the two equivalence-point volumes in a single titration?

15. (a) 25.00 ml of 0.1000 M Na_3PO_4 is to be titrated with 0.2000 M HCl. Using equations capable of yielding a relative error of ± 10 percent or less for $[H_3O^+]$, calculate the pH for the following volume points in milliliters relative to the HCl solution, and show by number the equations used:

0.00	10.00	12.00	12.25	12.50	12.75	13.00	15.00	20.00
22.00	24.00	24.50	24.75	25.00	25.25	25.50	26.00	28.00
30.00	35.00	36.00	37.00	37.50	38.00	39.00		

(b) Sketch the titration curve.

(c) Show the feasibility of locating the three possible equivalence points for this titration, using the following criteria:

ΔpH for ± 0.25 ml around the first-stage equivalence point
ΔpH for ± 0.25 ml around the second-stage equivalence point
ΔpH for ± 0.50 ml around the third-stage equivalence point

(d) Using the same criteria as in c, show from the Table 6.1 selection what indicators could be used to locate the first- and second-stage equivalence-point volumes.

(e) Considering the indicators selected in d and the related ΔpH criteria, show the approximate anticipated errors in locating the first- and second-stage equivalence-point volumes.

16. (a) 50.00 ml of a solution of HCl is titrated with 0.1049 M NaOH solution. Bromothymol blue is used as the indicator and yields a volume of 59.48 ml of NaOH solution at the end point. The indicator will yield an approximate accuracy of equivalence-point-volume location of better than 1 ppt. What is the molarity of the HCl solution?

(b) If alizarin yellow R had been used as the indicator, what end-point volume (to the nearest 0.05 ml) would have been secured?

(c) Calculate the indicator blank (to the nearest 0.05 ml) and the titration error (to the nearest 0.1 percent) in the case of b above.

(d) Calculate the molarity of the HCl solution based on the titration volume determined for b above. What would have been the relative error in this molarity value?

Ans. (a) 0.1248 M; (b) 60.65 ml; (c) -1.15 ml, $+2.0\%$

17. (a) Calculate the equivalence-point-volume values for 0.1025 M HCl used as the titrant in the following titrations:

    (1) 50.00 ml of 0.0875 M NaCN
    (2) 45.00 ml of 0.0988 M NaOH
    (3) 61.00 ml of 0.0865 M Na_3PO_4 to Na_2HPO_4
    (4) 50.00 ml of 0.1000 M Na_2HPO_4 to NaH_2PO_4
    (5) 60.00 ml of 0.0675 M NH_4OH $\{NH_3(aq)\}$

(b) What are the pH values at each of the above equivalence points?

18. The titration of a weak acid–strong acid solution is projected. Both acids are monobasic, and the titrant is 0.0875 M KOH. In the first titration, methyl orange is used as the indicator, and a 50.00-ml aliquot of the acid solution mixture requires a volume of 22.68 ml of base for the end point. A second 50.00-ml aliquot is titrated, using phenolphthalein as the indicator, and 49.87 ml of base is needed to reach the end point. The accuracy of locating the equivalence-point volume for the methyl orange ti-

tration is about 2 pph; that for the phenolphthalein titration is about 1 ppt. Calculate the molarities of the strong acid and weak acid in the acid solution mixture to the proper significant figures.

19. Solutions are to be investigated which contain HCl, H_3PO_4, or NaH_2PO_4 either singly or in any allowable paired combination. 30.00-ml aliquots of each solution were titrated with 0.2000 M NaOH using bromocresol green for one of the aliquots and thymolphthalein for the other. The following volumes were recorded.

Solution investigated	Bromocresol green, ml	Thymolphthalein, ml
(1)	22.5	45.0
(2)	0.0	27.8
(3)	23.7	32.2
(4)	14.8	14.8
(5)	10.1	28.4

Since the indicator-volume locational accuracy for most of these titrations is about 8 ppt, the burette readings were rounded off to the nearest 0.1 ml. What are the components of each solution and their respective molarities?   *Ans.* (3) 0.101 M HCl, 0.0567 M H_3PO_4

20. A series of solutions contain NaOH, Na_2CO_3, and $NaHCO_3$ either singly or in any allowable paired combination. 25.00-ml aliquots of each solution were titrated with 0.1500 M HCl, using phenolphthalein as the indicator for one of the aliquots and methyl orange for the other. The following volumes were recorded.

Solution investigated	Phenolphthalein, ml	Methyl orange, ml
(1)	21.3	21.3
(2)	18.4	28.8
(3)	14.0	28.0
(4)	0.0	18.3
(5)	11.2	35.6

Since the indicator-volume locational accuracy for most of these titrations varies from 5 ppt to 2 ppt approximately, the burette readings were rounded off to the nearest 0.1 ml. What are the components of each solution and their respective molarities?

21. 50.00-ml aliquots of a solution containing Na_3PO_4 and Na_2HPO_4 were titrated with 0.0780 M HCl, one aliquot to a thymolphthalein end point and the other to a bromocresol green end point. The thymolphthalein end point occurred at 14.6 ml of acid, while that of bromocresol green occurred at 48.7 ml. What are the concentrations of Na_3PO_4 and Na_2HPO_4 in milligrams per milliliter of the original solution?

22. A sample of a food product is to be analyzed for nitrogen by the Kjeldahl method. A 0.9865-g sample of the product properly treated is eventually distilled into a receiver flask containing 50.00 ml of 0.1085 M HCl. The unreacted HCl, after distillation is complete, is titrated with 0.1108 M NaOH and requires 31.65 ml of the base to reach a methyl red end point. What is the nitrogen content of the food product?   *Ans.* 2.72%

23. An H_2SO_4 steel-sheet pickling solution requires a range of pH of from 2.0 to 3.0 for efficient operation. 50.00 ml of this solution, when titrated with 0.1000 M NaOH to a bromothymol blue end point, requires 45.66 ml of the base. What is the pH of the solution? What addition of solid NaOH in pounds per thousand gallons of pickling solution must be made in order to increase the pH to a value of 2.5? (Brit. gal.)

24. A method for the determination of aluminum in magnesium alloys involves the dissolving of a sample of the alloy in HCl. The excess HCl is then partially neutralized with 1:2 NH_4OH and then completely neutralized with NaOH solution., The solution is then titrated at 80°C with standard NaOH solution according to

$$2AlCl_3 + 5NaOH \rightleftharpoons Al_2(OH)_5Cl + 5NaCl$$

A 1.0562-g sample of a magnesium alloy is dissolved in HCl. The excess of acid is neutralized according to the technique above. The titration with 1.062 M NaOH requires 12.75 ml to reach an end point located by pH meter pointer deflection. The uncertainty in the weighing operation is ± 0.0002 g and that of the burette volume is ± 0.04 ml. What is the percentage of aluminum in the magnesium alloy?

25. 1.0625 g of a material containing calcium carbonate and inert substances is treated with 50.00 ml of 0.1000 M HCl. After treatment, the CO_2 is expelled by boiling briefly. The excess HCl is then titrated with 0.1055 M NaOH, 36.80 ml being required to reach a neutral red end point. Assuming no loss of HCl during the expulsion of CO_2 and a volume locational accuracy for neutral red of better than 1 ppt, calculate the percentage of $CaCO_3$ in the material. *Ans.* 5.27%

26. Comment on the feasibility of the following:
   (*a*) The titration of 50.00 ml of 0.0001 M HCl by 0.0001 M NaOH
   (*b*) The titration of 50.00 ml of 0.0100 M benzoic acid by 0.0100 M KOH

27. Comment on the feasibility of the following:
   (*a*) The titration of 0.100 M H_2SO_4 by 0.100 M NaOH in two distinct stages
   (*b*) The titration of 0.100 M boric acid, H_3BO_3, by 0.100 M NaOH in three distinct stages
   (*c*) The titration of 0.100 M H_3PO_4 by 0.100 M NaOH in three distinct stages

REFERENCES

1. Böttger, W., ed., "Newer Methods of Volumetric Analysis," R. E. Oesper, transl., D. Van Nostrand, Princeton, N.J., 1938.
2. Kolthoff, I. M., P. J. Elving, and E. B. Sandell, "Treatise on Analytical Chemistry," Wiley, New York, 1959 (continuing series).
3. Kolthoff, I. M., E. B. Sandell, E. J. Meehan, and Stanley Bruckenstein, "Quantitative Chemical Analysis," 4th ed., Macmillan, New York, 1969.
4. Kolthoff, I. M., V. A. Stenger, and R. Belcher, "Volumetric Analysis," 3 vols., 2d. ed., Interscience, New York, 1957.
5. Laitinen, H. A., "Chemical Analysis," McGraw-Hill, New York, 1960.
6. Tomiček, O., "Chemical Indicators," Butterworth, London, 1951.

7
Precipitation Titrations

7.1 GENERAL

Precipitation titrations as a group can be described as titrations in which the titration reaction results in the formation of a precipitate or slightly soluble salt. It is apparent that any precipitation reaction could be in theory adapted to a volumetric technique, providing that (1) the precipitation reaction reaches equilibrium very rapidly after each addition of titrant, (2) no interfering situations occur (coprecipitation, occlusion of foreign ions, adsorption, etc.), (3) an indicator capable of locating the stoichiometric equivalence point with reasonable accuracy is available.

In general, there are few precipitation methods used in volumetric analysis, in comparison with the diversity of methods based on neutralization, oxidation-reduction, and complexation processes. Some of these methods are among the oldest and most frequently applied of the volumetric techniques, however, and of these, the methods that involve the titration of the halides Cl^-, Br^-, and I^- by silver(I), often called argentometric methods, are perhaps the most prominent.

Examples of these and other precipitation methods are

1. $\qquad Cl^- + AgNO_3 \rightleftharpoons AgCl(s) + NO_3^-$

2. $\qquad Ag^+ + KSCN \rightleftharpoons AgSCN(s) + K^+$

3. $\qquad SO_4^{2-} + BaCl_2 \rightleftharpoons BaSO_4(s) + 2Cl^-$

4. $\qquad 3Zn^{2+} + 2K_4Fe(CN)_6 \rightleftharpoons K_2Zn_3\{Fe(CN)_6\}_2(s) + 6K^+$

5. $C_2O_4^{2-} + (CH_3 \cdot COO)_2Pb \rightleftharpoons PbC_2O_4(s) + 2CH_3 \cdot COO^-$

6. $\qquad Pb^{2+} + (NH_4)_2MoO_4 \rightleftharpoons PbMoO_4(s) + 2NH_4^+$

In each instance, the second substance on the left is the titrant salt.

The indicator substances used to locate the equivalence point in precipitation titrations do not belong to a general class of substances, as is the case with neutralization titration indicators. Precipitation titration indicators include

1. Those which form a second precipitate with the titrant at or near the titration equivalence point, this second precipitate being clearly detectable relative to color, etc; such techniques involve a form of differential precipitation (see Sec. 4.8)
2. Those which are involved, in one way or another, in adsorption processes at the precipitate surface, and which provide a color change at or near the equivalence point as a result of these processes
3. Those which form with the titrant a clearly detectable colored soluble complex at or near the equivalence point
4. Those which undergo a change in color associated with a change in the solution potential at or near the equivalence point (oxidation-reduction indicators)

Rather than generalize on the problems associated with each indicator technique, we shall give particularized discussions centered around some representative methods using these techniques.

7.2   DIFFERENTIAL PRECIPITATION INDICATOR TECHNIQUES

7.2.1   TITRATION OF CHLORIDE AND BROMIDE BY SILVER(I) ION: MOHR METHOD

The Mohr method, one of the oldest of volumetric techniques, involves the titration of either chloride or bromide with a standard solution of silver nitrate, $AgNO_3$, using a soluble chromate salt such as potassium chromate, K_2CrO_4, as the indicator substance. The reactions are

$\qquad Cl^- + Ag^+ \rightleftharpoons AgCl(s)$

or

$\qquad Br^- + Ag^+ \rightleftharpoons AgBr(s) \qquad$ and $\qquad CrO_4^{2-} + 2Ag^+ \rightleftharpoons Ag_2CrO_4(s)$

The secondary precipitate, silver chromate, Ag_2CrO_4, has a distinct red color, the first appearance of which, in combination with the basic color of the solution being titrated, signals the titration end point. Usually a buff color in solution is secured. The determination of iodide or thiocyanate can not be so conducted, since adsorption processes prevent the appearance of a clear indication of the end point.

Table 7.1 shows the appropriate values for $[Cl^-]$, pCl, $[Ag^+]$, and pAg for the titration of 50.00 ml of $0.1000\,M$ NaCl by $0.1000\,M$ $AgNO_3$, the $K_{sp}(AgCl)$ being 1.56×10^{-10}. The equations used for calculation purposes are shown by number. These were selected to provide $[Cl^-]$ and $[Ag^+]$ values in error by not more than ± 5 percent. The indicator is K_2CrO_4, used as a 2-ml addition, before the start of the titration, of a 5 percent K_2CrO_4 solution. This addition yields a value for $[CrO_4^{2-}]$ of about $5 \times 10^{-3}\,M$ at the titration end point. Note that the ΔpCl for ± 0.10 ml around the equivalence point is 1.91 units, a value matched by that for ΔpAg.

For such a differential indicator technique to work perfectly, the end point and the stoichiometric equivalence point should coincide, implying that the chloride ion should be removed as AgCl within the limits of $\sqrt{K_{sp}(AgCl)}$, just when the first color of Ag_2CrO_4 is observed. Several factors combine to render this ideal situation somewhat difficult to achieve. First, some appreciable quantity of Ag_2CrO_4 precipitate must be present before the red color becomes visible. Second, the chromate ion imparts a fairly intense yellow color to solutions having the $[CrO_4^{2-}]$ required for any reasonable possibility that the end point and equivalence point will coincide. This color tends to obscure the point of solution darkening caused by the appearance of Ag_2CrO_4 color, and to further increase the quantity of silver(I) solution needed to achieve the end point. Third, the masking effect of the generally heavy white precipitate of AgCl requires the addition of appreciable silver(I) in order to produce enough Ag_2CrO_4 to overcome this effect. The entire situation may be summarized as follows relative to the titration of chloride ion.

The calculation of the required concentration of CrO_4^{2-} at the equivalence point, for end-point and equivalence-point coincidence, involves

$$[Cl^-] \text{ at the equivalence point} = \sqrt{K_{sp}(AgCl)} = \sqrt{1.56 \times 10^{-10}}$$
$$= 1.24 \times 10^{-5}\,M$$
$$= [Ag^+] \text{ at the equivalence point}$$

The value of $[CrO_4^{2-}]$ at which this concentration of Ag^+ will be just ready to yield a precipitate of Ag_2CrO_4 is given by

$$[CrO_4^{2-}] = \frac{K_{sp}(Ag_2CrO_4)}{[Ag^+]^2} = \frac{9.0 \times 10^{-12}}{(1.24 \times 10^{-5})^2} = 5.8 \times 10^{-2}\,M$$

and it should be noted that additional silver(I) solution, beyond the equivalence-point volume, must now be added to provide sufficient Ag_2CrO_4 precipitate for end-point detection. The value of $5.8 \times 10^{-2}\,M$ for $[CrO_4^{2-}]$ at the equivalence point is high enough to impart a yellow color to the solution of sufficient intensity

Table 7.1   Titration of chloride ion by the Mohr method*
50.00 ml of 0.1000 M NaCl vs. 0.1000 M AgNO$_3$
[CrO$_4{}^{2-}$] at the equivalence point about 5×10^{-3} M

Volume AgNO$_3$, ml	[Cl⁻]	pCl	[Ag⁺]	pAg	Equation used	$(\Delta pCl/\Delta V_{Ag})\times10^3$ 0.1 ml $= \Delta V_{Ag}$	$(\Delta^2 pCl/\Delta V_{Ag}{}^2)\times10^3$ 0.1 ml/0.1 ml
0.00	1.00×10^{-1}	1.00					
						+1.8 (5.00)	
10.00	6.67×10^{-2}	1.18	2.34×10^{-9}	8.63	(184)		+0.001 (10.00)
						+1.9 (15.00)	
20.00	4.28×10^{-2}	1.37	3.63×10^{-9}	8.44	(184)		+0.004 (20.00)
						+2.3 (25.00)	
30.00	2.50×10^{-2}	1.60	6.17×10^{-9}	8.21	(184)		+0.013 (30.00)
						+3.6 (35.00)	
40.00	1.11×10^{-2}	1.96	1.41×10^{-8}	7.85	(184)		+0.037 (38.75)
						+6.4 (42.50)	
45.00	5.26×10^{-3}	2.28	2.95×10^{-8}	7.53	(184)		+0.18 (44.50)
						+13.6 (46.50)	
48.00	2.04×10^{-3}	2.69	7.59×10^{-8}	7.12	(184)		+0.87 (47.50)
						+31 (48.50)	
49.00	1.01×10^{-3}	3.00	1.55×10^{-7}	6.81	(184)		+3.9 (48.88)
						+60 (49.25)	
49.50	5.02×10^{-4}	3.30	3.09×10^{-7}	6.51	(184)		+18 (49.45)
						+133 (49.65)	
49.80	2.00×10^{-4}	3.70	7.76×10^{-7}	6.11	(184)		+84 (49.75)
						+300 (49.85)	
49.90	1.00×10^{-4}	4.00	1.56×10^{-6}	5.81	(184)		+600 (49.90)
						+900 (49.95)	
50.00	1.24×10^{-5}	4.90	1.24×10^{-5}	4.90	(186)		+10 (50.00)
						+910 (50.05)	
50.10	1.56×10^{-6}	5.81	9.99×10^{-5}	4.00	(187)		-600 (50.10)
						+310 (50.15)	
50.20	7.76×10^{-7}	6.11	2.00×10^{-4}	3.70	(187)		-89 (50.25)
						+133 (50.35)	
50.50	3.13×10^{-7}	6.51	4.98×10^{-4}	3.30	(187)		-18 (50.55)
						+60 (50.75)	
51.00	1.56×10^{-7}	6.81	9.90×10^{-4}	3.00	(187)		-4.1 (51.12)
						+29 (51.50)	
52.00	7.94×10^{-8}	7.10	1.96×10^{-3}	2.71			

*[Cl⁻], pCl, [Ag⁺], and pAg values are not corrected for the dilution effect. The values in parentheses after each $\Delta pCl/\Delta V_{Ag}$ and $\Delta^2 pCl/\Delta V_{Ag}{}^2$ value associate the average slope, or rate of change of the slope, with the average volume value between the titration points involved. The equivalence-point volume is 50.00 ml of 0.1000 M AgNO$_3$.

to present difficulty relative to end-point detection, and the normal value for [CrO$_4{}^{2-}$] at the end point is about 5×10^{-3} M. This is the value obtained in the titration covered by Table 7.1.

Note that this latter value for [CrO$_4{}^{2-}$] results in an [Ag⁺] at the point where the precipitate of Ag_2CrO_4 is just ready to form

$$[Ag^+] = \sqrt{\frac{K_{sp}(Ag_2CrO_4)}{[CrO_4^{2-}]}} = \sqrt{\frac{9.0 \times 10^{-12}}{5 \times 10^{-3}}} = 4.2 \times 10^{-5} \, M$$

with a corresponding $[Cl^-]$ at this point of

$$[Cl^-] = \frac{K_{sp}AgCl)}{[Ag^+]} = \frac{1.56 \times 10^{-10}}{4.2 \times 10^{-5}} = 3.7 \times 10^{-6} \, M$$

These values of concentration represent pAg and pCl values of 4.38 and 5.43, respectively, indicating a point in the titration slightly beyond the stoichiometric equivalence point. The actual point of added 0.1000 M AgNO$_3$ solution would be about 50.04 ml—0.04 ml beyond the equivalence point, with additional AgNO$_3$ solution still required to yield a detectable level of Ag_2CrO_4.

By using 100 ml of a 5×10^{-3}-M solution of K$_2$CrO$_4$, and adding 0.01 M AgNO$_3$ dropwise, the color darkening due to Ag_2CrO_4 formation becomes detectable at a level of 0.40 ml of the AgNO$_3$ solution. The sensitivity level of $5 \times 10^{-3} \, M \, CrO_4^{2-}$ to silver(I) is thus

$$\frac{0.40 \text{ ml} \times 0.01 \, M}{1000} = 4.0 \times 10^{-6} \text{ mole/100 ml}$$

In the titration under discussion, this would require about 0.04 ml additional 0.1000 M AgNO$_3$, and the end point would occur at a value of approximately 50.08 ml of the AgNO$_3$ solution. Correction of this volume value for the sensitivity of CrO_4^{2-} would involve an indicator blank of -0.04 ml, and it would return the end-point volume to 50.04 ml, differing from the stoichiometric equivalence-point volume by $+0.04$ ml and representing a titration error of about 1 ppt.

For all 0.1-M titrations corrected for indicator sensitivity by the use of the indicator blank, the titration error of about 1 ppt remains constant. Decreases in the molarity of reactant and/or titrant solutions result in increases in the titration error. The indicator-blank value changes, of course, with changing molarity of the silver(I) titrant, increasing with decreasing molarity; as well as with changing volume of the solution at the end point, increasing with increasing volume at the end point.

Actually, the sensitivity determination just outlined is not usually in itself adequate, since it does not take into consideration the masking effect of the AgCl precipitate. A more practical determination of the indicator blank can be made by taking about 0.2 g of chloride-free calcium carbonate, CaCO$_3$, and suspending it in 100 ml of water containing 2 ml of a 5 percent solution of K$_2$CrO$_4$. Titration of this suspension with standard silver nitrate solution to an observable color change provides an indicator-blank value applicable to chloride titrations involving generally similar conditions of end-point volume and silver(I) solution strength. The CaCO$_3$ suspension provides a masking effect paralleling that of the AgCl precipitate. With 0.1000 M AgNO$_3$ as the titrant, an indicator blank of about -0.08 ml is commonly found for the above test.

As indicated in Sec. 5.2.1, the effect of increasing values of the K_{sp} is to reduce the magnitude of the ΔpR changes for symmetrical titrant volumes around the equivalence-point volume and to increase the sensitivity of the titration to dilution of the starting concentrations of reactant and/or titrant solutions. The $K_{sp}(AgCl)$ at 1.56×10^{-10} is low enough to provide for an adequate ΔpCl (or ΔpAg) range for ± 0.10 ml around the equivalence point in the 0.1-M titration. Titration of 0.01-M solutions of Cl^- and $AgNO_3$ would not be too feasible, however, the reduction of the ΔpCl for ± 0.10 ml around the equivalence-point volume and the increase in the magnitude of the indicator blank being such as to permit only poor equivalence-point-volume locational accuracy.

By the same token, a decrease in the K_{sp} value, as, for example, in the titration of bromide ion by silver(I), where the $K_{sp}(AgBr) = 5.0 \times 10^{-13}$, provides for a larger ΔpBr (or ΔpAg) for ± 0.10 ml around the equivalence-point volume for the 0.1-M titration, with correspondingly less sensitivity to dilution of the starting bromide ion and/or silver(I) solution concentrations.

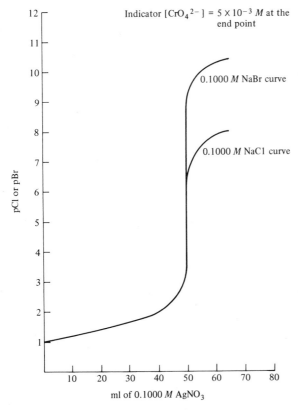

Fig. 7.1 Titration of 50.00 ml of 0.1000 M NaCl or NaBr with 0.1000 M AgNO$_3$: Mohr method.

Figure 7.1 shows the curve for the titration data outlined in Table 7.1. On the same graph is shown the curve for the titration of 50.00 ml of 0.1000 M NaBr under the same general conditions. The extended ΔpBr range around the equivalence point will be noted.

In all titrations of the Mohr type, the solution acidity is of primary importance. Too high an acidity leads to the reactions

$$2CrO_4^{2-} + 2H_3O^+ \rightleftharpoons 2HCrO_4^- + H_2O \rightleftharpoons Cr_2O_7^{2-} + H_2O$$

The solubilities of $AgHCrO_4$ and $Ag_2Cr_2O_7$ are appreciably higher than that of Ag_2CrO_4, so that an acid solution requires in titration higher additions of silver(I) to yield any end-point indication.

Solutions that are too basic result in the reaction

$$2Ag^+ + 2OH^- \rightleftharpoons 2AgOH \rightleftharpoons Ag_2O + H_2O$$

with significant silver(I) being precipitated as the oxide instead of entering into the normal Mohr titration reactions.

The Mohr titration should be carried out in solutions having pH values not lower than 7 and not higher than 10.

The increasing solubility of Ag_2CrO_4 with increasing solution temperature requires that these titrations be carried out at room temperature.

7.3 ADSORPTION INDICATOR TECHNIQUES

7.3.1 TITRATION OF CHLORIDE BY SILVER(I) ION: FAJANS' METHOD

The titration of chloride with a standard solution of silver nitrate, using the weak organic acid substance fluorescein as the indicator, is one of the adsorption indicator techniques developed by Fajans and his associates.* During the first stage of the titration, up to but not including the equivalence point, the precipitate of AgCl tends to adsorb the lattice ion that is in excess in the solution—in this case the chloride ion. At this stage the negative fluoresceinate ion is repelled by the negatively charged precipitate surface and the solution maintains the yellow-green color imparted by fluorescein. As the equivalence point is approached, the adsorbed chloride ions are removed by the precipitating silver(I) ions, and with the first excess of silver(I), the precipitated AgCl surface adsorbs Ag^+ ions and becomes positively charged. The fluoresceinate ions are now attracted strongly to this positively charged surface. At the surface, silver fluoresceinate is formed, which yields a deep red color *at the precipitate surface*. Since some excess of silver(I) must be present to permit the surface adsorption of Ag^+, the end point occurs slightly beyond the stoichiometric equivalence point, but under normal conditions, the indicator blank is negligibly small.

*K. Fajans and O. Hassel, Z. Elektrochem., **29**:495(1923). K. Fajans and H. Wolff, Z. anorg. allgem. Chem., **137**:221(1924).

Several important factors, in some cases particularized for the foregoing titration process but generally applicable to adsorption indicator techniques, should be noted.

1. Since the colored compound formed between the indicator and the primary adsorbed ion at the precipitate surface (e.g., silver fluoresceinate at the AgCl surface) is itself a substance of limited solubility, the concentration of the indicator should be kept low enough to avoid precipitation of the compound.
2. The development of the colored compound, formed between the indicator and the primary adsorbed ion, occurs at the precipitate surface *only* and is the result of an adsorption not a precipitation reaction. Since the color change is a surface effect, the best sensitivity is secured with the highest dispersion of the precipitate. The tendency at the equivalence point (or isoelectric point of equality of positive and negative charges in solution) is for the precipitate particles to coagulate or flocculate, with a corresponding reduction in the surface area. It is therefore customary, in certain titrations such as the chloride-silver(I) titration, to add a substance such as dextrin to maintain high particle dispersion. On the other hand, high dispersion tends to enhance the photo-decomposition effect of precipitates such as AgCl or AgBr, etc.; and to minimize this situation, titrations involving the formation of such precipitates should be carried out fairly rapidly under diffuse lighting. Since certain of the adsorption indicators tend to promote photo-decomposition reactions, such as dichlorfluorescein, titration speed and lighting conditions are even more important when such indicators are used.
3. High concentrations of electrolytes in the solution being titrated should be avoided: first, because of their flocculation effect on the precipitate; and second, because the anions may compete with the indicator anion in the adsorption process, thus reducing the amount of adsorbed indicator and lessening the intensity of the color change.
4. Excessive stirring or agitation may increase the rate of precipitate coagulation, as will titration at elevated temperatures. Limited but adequate stirring and titration at room temperature are important requirements.
5. The titration of low concentration solutions of reactant and/or titrant should be avoided, since the total surface area of the resultant precipitate may be inadequate for proper color development. As an example, in the case of the chloride-silver(I) titration, concentrations of chloride ion less than 0.005 M are rarely titrated. It is important to note, however, that the adsorption indicator method is less sensitive to reactant and/or titrant solution dilution than is the Mohr method.
6. Since most adsorption indicators are weak organic acids, or salts of such acids (for example, sodium dichlorfluoresceinate), solutions which are too acid may result in the reaction

$$In^- + H_3O^+ \rightleftharpoons HIn + H_2O$$

with loss of effective concentration of the indicator anion. Although the tolerance towards $[H_3O^+]$ in solution varies with the indicator used (fluorescein greater than 7 pH, dichlorfluorescein greater than 4 pH, eosin or tetrabromofluorescein greater than 2 pH), most adsorption indicator titrations are conducted in neutral or slightly alkaline solution. Several adsorption indicators where the effective ion is a cation are available, and these can be used in the titration of solutions that are strongly acid.

7. The indicator ion must be attracted as a counter ion to the primary adsorbed ion and must not replace this latter ion if the color development reaction is to occur to an adequate extent. Eosin, for example, can not be used in the chloride-vs.-silver(I) titration for this reason.

8. Since the color development process is a reversible one, titration in the opposite sense is often feasible. It is possible, for example, to titrate silver(I) with a standard solution of a halide, such as a standard sodium chloride solution.

9. The following lists several titration situations involving adsorption indicators.

Determined	Titrant	Indicator
Cl^-, Br^-, SCN^-	$AgNO_3$	Fluorescein
		Dichlorfluorescein
Ag^+	$NaCl$	Fluorescein
		Dichlorfluorescein
Br^-, I^-, SCN^-	$AgNO_3$	Eosin

Adsorption indicators have also been adapted to many other precipitation titrations. As a case in point, the determination of lead(II) by titration with a standard solution of ammonium molybdate, $(NH_4)_2MoO_4$, in the reaction

$$Pb^{2+} + (NH_4)_2MoO_4 \rightleftharpoons PbMoO_4(s) + 2NH_4^+$$

was originally carried out through the use of tannic acid solution as an external white-spot plate indicator, the end point occurring with the first appearance of a yellow-brown color in the spot plate test. In its modern application, the adsorption indicators eosin or erythrosine may be used as internal end-point indicators.

7.4 COMPLEX–FORMATION INDICATOR TECHNIQUES

7.4.1 THE VOLHARD METHOD

The Volhard method involves basically the titration of silver(I) in acid solution with a standard solution of potassium thiocyanate, KSCN, using iron(III) as the indicator substance. The titration reaction results in the formation of the slightly soluble salt silver thiocyanate, AgSCN,

$$Ag^+ + SCN^- \rightleftharpoons AgSCN(s)$$

with the indicator reaction resulting in the formation of the intensely red soluble complex $Fe(SCN)^{2+}$

$$Fe^{3+} + SCN^- \rightleftharpoons Fe(SCN)^{2+}$$

Although the Volhard method can be used in the volumetric determination of silver(I), one of its more general applications is in the determination indirectly of the halides Cl^-, Br^-, and I^-. In such applications a measured excess of a standard solution of silver nitrate is added to the acid solution of the halide (the solution is usually about $3 \times 10^{-1} M$ to HNO_3 at the equivalence point or end point) and the excess is determined by back-titration with standard potassium thiocyanate solution, using iron(III) as the indicator. As carried out, this is a form of differential precipitation titration. The fact that the titration is conducted in a highly acid solution is an advantage. Neutral or alkaline solution titrations for the halides encounter difficulties where the unknown contains substances slightly soluble in such media, such as oxalates or carbonates. The Volhard method provides a means of avoiding these difficulties.

We shall discuss, first of all, the direct determination of silver by titration of silver(I) with standard potassium thiocyanate solution. The sensitivity of iron(III) as the indicator, usually at a concentration of about $[Fe^{3+}] = 10^{-2} M$ at the end point in a solution approximately $3 \times 10^{-1} M$ to HNO_3, is given by the fact that a visible color will be imparted when the $[Fe(SCN)^{2+}]$ in solution is about $6.5 \times 10^{-6} M$. At the stoichiometric equivalence point (not the end point) the value of $[SCN^-]$ is given by

$$[SCN^-] = \sqrt{K_{sp}(AgSCN)} = \sqrt{1.07 \times 10^{-12}} = 1.03 \times 10^{-6} M$$

and at this point the concentration of $[Fe(SCN)^{2+}]$ in solution will be given by

$$\frac{[Fe(SCN)^{2+}]}{[Fe^{3+}][SCN^-]} = 1.38 \times 10^2$$

the stability-constant equation. From this we have

$$\begin{aligned}[Fe(SCN)^{2+}] &= (1.38 \times 10^2)[Fe^{3+}][SCN^-] \\ &= 1.38 \times 10^2 \times 10^{-2} \times 1.03 \times 10^{-6} \\ &= 1.42 \times 10^{-6} M\end{aligned}$$

a concentration of $Fe(SCN)^{2+}$ appreciably lower than that at which the reddish color of the complex is visible. The equivalence point thus occurs before the end point. At the end point, where the required concentration of $6.5 \times 10^{-6} M$ is attained for $[Fe(SCN)^{2+}]$, we have

$$[SCN^-] = \frac{[Fe(SCN)^{2+}]}{[Fe^{3+}] K_{stab}} = \frac{6.5 \times 10^{-6}}{10^{-2} \times 1.38 \times 10^2} = 4.7 \times 10^{-6} M$$

The additional concentration of KSCN required after the equivalence point can be found from

$$[KSCN] = \underset{\substack{\text{Formed after the} \\ \text{equivalence point}}}{[Fe(SCN)^{2+}]} + [SCN^-] + \underset{\substack{\text{Formed after the} \\ \text{equivalence point}}}{[AgSCN]}$$

The first expression on the right is given by

$$6.5 \times 10^{-6}\,M - 1.4^2 \times 10^{-6}\,M = 5.1 \times 10^{-6}\,M$$

while the third expression on the right is given by

$$[Ag^+]\text{ at the equivalence point} - [Ag^+]\text{ at the end point}$$

$$1.03 \times 10^{-6}\,M - \frac{K_{sp}(AgSCN)}{[SCN^-]}$$

$$1.03 \times 10^{-6}\,M - \frac{1.07 \times 10^{-12}}{4.7 \times 10^{-6}}$$

$$1.03 \times 10^{-6}\,M - 2.3 \times 10^{-7}\,M$$

$$8.0 \times 10^{-7}\,M$$

We now have

$$[KSCN] = 5.1 \times 10^{-6}\,M + 4.7 \times 10^{-6}\,M + 8.0 \times 10^{-7}\,M = 1.1 \times 10^{-5}\,M$$

and this represents, for an approximate 100-ml end-point total volume of solution, a volume of 0.1 M KSCN solution of

$$\frac{V_{KSCN} \times 0.1\,M}{100\text{ ml}} = 1.1 \times 10^{-5}\,M$$

$$V_{KSCN} = 0.01\text{ ml}$$

so that the indicator blank in the Volhard titration will be −0.01 ml. Based on approximately 50.00 ml of 0.1 M KSCN solution being used in the titration, this represents a titration error of about 0.2 ppt. The indicator blank is indeed negligible, being significantly less than the burette-reading uncertainty of ±0.02 ml per reading.

Because of a tendency for silver thiocyanate to adsorb silver(I) ions from solution, the first color change is apt to show up prematurely, and it will fade upon vigorous agitation. For this reason the titration should be continued, with vigorous agitation, until the reddish color change indicating the end point is permanent.

The application of the Volhard method to the determination of the halides proceeds without difficulty where bromide and iodide are involved, since the solubilities of AgBr and AgI are appreciably lower than that of AgSCN. The fact that the solubility of AgCl is higher than that of AgSCN results in the reaction

$$AgCl + SCN^- \rightleftharpoons AgSCN + Cl^-$$

Since the titration involves the addition of excess standard silver(I) solution to the chloride solution and back-titration of the excess with standard potassium thiocyanate solution, the occurrence of this reaction to a significant extent near the end point results in the addition of KSCN solution beyond what should be theoretically required. It is possible to estimate the extent to which overconsumption of KSCN solution will take place in the chloride application of the Volhard titration. At the end point of the titration we have

$$[Ag^+] = \frac{K_{sp}(AgCl)}{[Cl^-]} = \frac{K_{sp}(AgSCN)}{[SCN^-]}$$
$$= \frac{1.56 \times 10^{-10}}{[Cl^-]} = \frac{1.07 \times 10^{-12}}{[SCN^-]}$$

so that

$$[Cl^-] = \frac{1.56 \times 10^{-10} \times 4.7 \times 10^{-6}}{1.07 \times 10^{-12}}$$
$$= 6.8 \times 10^{-4} \, M$$

The value of $[Cl^-]$ at the equivalence point of the titration reaction between Cl^- and Ag^+ is given by $\sqrt{K_{sp}(AgCl)} = 1.24 \times 10^{-5} \, M$. In a titration involving 0.1 M KSCN, 0.1 M Cl^-, and 0.1 M $AgNO_3$, with an end-point total solution volume of about 100 ml, the excess volume of KSCN solution required to reach the end point, where $[Cl^-] = 6.8 \times 10^{-4} \, M$, is

$$\frac{V_{KSCN} \times 0.1 \, M}{100 \text{ ml}} = 6.8 \times 10^{-4} \, M$$

$$V_{KSCN} = 0.68 \text{ ml}$$

or 0.70 ml rounded off to the nearest 0.05 ml, and indicative of a significantly large titration error of about 14 ppt. Because of other factors, the titration error is not infrequently considerably higher in practical titrations.

Several techniques have been devised to avoid this situation. The most obvious of these is the removal of the precipitated AgCl by filtration, the filtrate or an aliquot thereof being subsequently titrated with standard KSCN solution. Although coagulation of the precipitated AgCl by digestion prior to filtration decreases filtration and washing time, the process has the disadvantage of prolonging the total elapsed time for the analysis.

A second technique has been used with considerable success. This technique is one devised by Caldwell and Moyer* and involves shaking the solution, after precipitation of the AgCl and before titration with KSCN solution, with a small volume of nitrobenzene. Coagulation of the AgCl and the coating of the coagulate with a layer of nitrobenzene prevents it from reacting with the aqueous solution layer during the back-titration process. Other immiscible organic liquids, such as benzene or ligroin, may also be used.

It is possible by increasing the concentration of the Fe^{3+} indicator in solution

*J. R. Caldwell and H. V. Moyer, *Ind. Eng. Chem., Anal. Ed.*, **7**:38(1935).

at the end point to offset the titration error introduced by the higher solubility of AgCl relative to that of AgSCN. Swift and others* have proposed the use of higher Fe^{3+} concentrations to this end. It is possible to calculate the optimum $[Fe^{3+}]$ required to offset exactly the error in the chloride determination. The method is as follows.

Exact correspondence between the end point and the equivalence point should exist for zero titration error. At the equivalence point (not the end point) of the Volhard adaptation for the chloride determination, the concentration of Ag^+ in the solution is given by:

$$[Ag^+] = [Cl^-] + [SCN^-] + [Fe(SCN)^{2+}]$$

We have, for a visible color of $Fe(SCN)^{2+}$, a value of $[Fe(SCN)^{2+}] = 6.5 \times 10^{-6}$ M and

$$\frac{[Fe(SCN)^{2+}]}{[Fe^{3+}][SCN^-]} = 1.38 \times 10^2$$

with

$$[SCN^-] = \frac{6.5 \times 10^{-6}}{[Fe^{3+}] \, 1.38 \times 10^2} \tag{a}$$

We also have

$$[Ag^+] = \frac{K_{sp}(AgSCN)}{[SCN^-]} = \frac{1.07 \times 10^{-12}}{[SCN^-]}$$

and substitution of (a) yields

$$[Ag^+] = \frac{1.07 \times 10^{-12} [Fe^{3+}] \, 1.38 \times 10^2}{6.5 \times 10^{-6}} \tag{b}$$

Again,

$$[Cl^-] = \frac{K_{sp}(AgCl)}{[Ag^+]} = \frac{1.56 \times 10^{-10}}{[Ag^+]}$$

and substitution of (b) now yields

$$[Cl^-] = \frac{1.56 \times 10^{-10} \times 6.5 \times 10^{-6}}{1.07 \times 10^{-12} [Fe^{3+}] \, 1.38 \times 10^2} \tag{c}$$

Substitution of the values for (a), (b), and (c), along with the value for $[Fe(SCN)^{2+}]$, into the equality expression for $[Ag^+]$ now gives

$$\frac{1.07 \times 10^{-12} [Fe^{3+}] \, 1.38 \times 10^2}{6.5 \times 10^{-6}} = \frac{1.56 \times 10^{-10} \times 6.5 \times 10^{-6}}{1.07 \times 10^{-12} [Fe^{3+}] \, 1.38 \times 10^2}$$
$$+ \frac{6.5 \times 10^{-6}}{[Fe^{3+}] \, 1.38 \times 10^2} + 6.5 \times 10^{-6}$$

*E. H. Swift, G. M. Arcand, R. Lutwack, and D. J. Meier, *Anal. Chem.*, **22**:306(1950).

Solving this equation for $[Fe^{3+}]$ yields a value of

$$[Fe^{3+}] = 0.71 \ M$$

The establishment of a concentration level of 0.7 M for $[Fe^{3+}]$ at the titration end point should offset the error in the chloride determination and provide for adequate correspondence between the end point and equivalence point. Some initial difficulty may be encountered in locating the end-point color shift because of interference originating with the background color of Fe^{3+} at this concentration, but this can be overcome by experience.

The Volhard method, with modifications where required, can also be used in the determination of cyanide, CN^-, oxalate, $C_2O_4^{2-}$, sulfide, S^{2-}, chromate, CrO_4^{2-}, carbonate, CO_3^{2-}, thiocyanate, SCN^-, etc., ions.

PROBLEMS

1. Calculate the molarities of the following solutions:

    (a) 15.839 g of $AgNO_3$ in 978 ml of solution

    (b) A solution of NaBr where 36.45 ml requires 40.18 ml of 0.0995 M $AgNO_3$ to reach the equivalence point

    (c) A solution of $K_4Fe(CN)_6$ where 50.00 ml requires 37.65 ml of 0.1008-M zinc(II) solution to reach the equivalence point

    (d) A solution of $AgNO_3$ where 46.70 ml is required to titrate a solution containing 0.2162 g of NaCl.   *Ans.* (b) 0.1097 M

2. Calculate the titers of the following solutions:

    (a) 0.0964 M $AgNO_3$ in terms of chloride and bromide

    (b) 0.1109 M KSCN in terms of silver(I)

    (c) 0.1412 M $K_4Fe(CN)_6$ in terms of zinc(II)

3. A titration involving 38.00 ml of 0.0985 M Br^- and 0.1025 M $AgNO_3$ is to be carried out by the Mohr method. Ignoring the indicator reaction, determine the pBr and pAg values for the following additions in milliliters of $AgNO_3$ solution:

    0.00   1.00   5.00   10.00   20.00   30.00   35.00   36.00   36.30   36.40   36.50
    36.60   36.70   37.00   38.00   43.00

Use equations that yield not more than ±5 percent error for the ion concentration calculated.

4. In the titration outlined in Prob. 3 the stoichiometric equivalence point is achieved at 36.52 ml of $AgNO_3$ solution. If 2.00 ml of a 5.00 percent solution of K_2CrO_4 is added at the start of the titration show

    (a) The $[CrO_4^{2-}]$ at the equivalence point

    (b) The $[Br^-]$ when Ag_2CrO_4 is just ready to precipitate

    (c) The end-point volume of $AgNO_3$ solution under these conditions

    (d) The indicator blank and the titration error

5. In the titration covered by Probs. 3 and 4, consider that the sensitivity of the $[CrO_4^{2-}]$ in solution at the end point (or equivalence point) is represented by 4.6×10^{-6} mole of Ag^+ per 100 ml of solution. Ignoring the masking effect of the AgBr precipitate, show

    (a) The end-point volume of $AgNO_3$ solution under these conditions

    (b) The indicator blank associated with the correction for the Ag^+ sensitivity of CrO_4^{2-} only.

                                                   Ans. (a) 36.58 ml; (b) −0.04 ml

6. 40.55 ml of a solution of NaCl is titrated with $0.1000\ M$ AgNO$_3$ in a Mohr titration. A total volume of AgNO$_3$ solution of 49.85 ml is required to reach the end point. The $[\text{CrO}_4{}^{2-}]$ at the end point is $5.0 \times 10^{-3}\ M$, and sensitivity at this concentration is given by 4.0×10^{-6} mole of Ag$^+$ per 100 ml of solution. Ignoring the masking effect of the AgCl precipitate, show the molarity of the NaCl solution for

      (*a*) Titration volume uncorrected
      (*b*) Titration volume corrected for the CrO$_4{}^{2-}$ sensitivity to Ag$^+$ alone

7. Water used in an industrial process is adjusted to contain 20.0 g of chloride ion per British gallon. A 50.00-ml sample of this water is titrated in a Mohr titration, using $0.1500\ M$ AgNO$_3$. If 25.80 ml of AgNO$_3$ solution is needed to reach the end point and the indicator blank is -0.08 ml, calculate the required addition of NaCl per gallon of solution in order to meet the optimum level of chloride ion concentration.   *Ans.* 12.5 g

8. A tank contains a solution of NaBr. 25.00 ml of this solution requires 29.65 ml of $0.0975\ M$ AgNO$_3$ to reach the end point in a Mohr method titration. 275 g of NaBr is now added to the tank and thoroughly mixed in. 25.00 ml of this new solution now requires 42.18 ml of the AgNO$_3$ solution. If the indicator blank in each titration is -0.10 ml, calculate the number of British gallons of solution in the tank. Ignore any volume changes based on salt addition or volume removed for titration.

9. A soldering alloy of lead and tin was analyzed to determine its lead content. A sample of 1.0876 g was weighed out and dissolved in dilute HNO$_3$. After subsequent treatment, a final solution containing all the lead as lead(II) was titrated with $0.0969\ M$ (NH$_4$)$_2$MoO$_4$ solution, using eosin as the adsorption indicator. The end-point volume was 26.89 ml. Using uncertainties of ±0.0001 g and ±0.02 ml per weight and volume reading, calculate the percentage of lead in the solder.   *Ans.* 49.6%

10. (*a*) A material contains KCl and an inert substance. A 0.6051-g sample is weighed out and dissolved in 50.00 ml of $1\ M$ HNO$_3$. A 25.00-ml aliquot is treated with 35.00 ml of $0.1000\ M$ AgNO$_3$, and after digestion, the precipitated AgCl is filtered off and washed. The filtrate is then titrated with $0.1000\ M$ KSCN after adding 5.0 ml of a 10.0 percent solution of NH$_4$Fe(SO$_4$)$_2 \cdot$ 12H$_2$O in 6 M HNO$_3$. The end point is detected at 8.81 ml of KSCN solution. What is the percentage of KCl in the material?

    (*b*) A second 25.00-ml aliquot is treated with 35.00 ml of $0.1000\ M$ AgNO$_3$, 5.0 ml of the NH$_4$Fe(SO$_4$)$_2 \cdot$12H$_2$O solution is added, and the solution is titrated with $0.1000\ M$ KSCN. On the basis of the discussion given in the chapter on precipitation titrations; calculate

      (1) The end-point volume of $0.1000\ M$ KSCN, rounded off to the nearest 0.05 ml
      (2) The percentage of KCl in the material based on this volume
      (3) The relative error in parts per thousand for this value of percent KCl

11. A magnesium alloy contains zinc as the only substance capable of reacting with potassium ferrocyanide. The alloy is supposed to contain 5.5 percent of zinc. In the process of producing the alloy, the molten material is sampled and the sample analyzed for zinc by titration with K$_4$Fe(CN)$_6$ solution. The sample weight taken was 2.7851 g and the titration required 36.75 ml of $0.0359\ M$ K$_4$Fe(CN)$_6$ solution. The indicator blank was -0.08 ml. Determine how many pounds of zinc metal must be added per 100 lb of molten alloy in order to obtain the required composition of 5.5% zinc.

REFERENCES

1. Böttger, W., ed., "Newer Methods of Volumetric Analysis," R. E. Oesper, transl., D. Van Nostrand, Princeton, N. J., 1938.
2. Kolthoff, I. M., P. J. Elving, and E. B. Sandell, "Treatise on Analytical Chemistry," Wiley, New York, 1959 (continuing series).

3. Kolthoff, I. M., E. B. Sandell, E. J. Meehan, and Stanley Bruckenstein, "Quantitative Chemical Analysis," 4th ed., Macmillan, New York, 1969.
4. Kolthoff, I. M., V. A. Stenger, and R. Belcher, "Volumetric Analysis," 3 vols., 2d ed., Interscience, New York, 1957.
5. Laitinen, H. A., "Chemical Analysis," McGraw-Hill, New York, 1960.
6. Tomiček, O., "Chemical Indicators," Butterworth, London, 1951.

8

Complexation Titrations

8.1 GENERAL

Complexation titrations involve titration reactions that result in the formation of a soluble complex or coordination compound. Any such complexation reaction could in theory be applied as a volumetric technique providing that (1) the reaction reaches equilibrium very rapidly following each addition of titrant, (2) interfering situations do not intervene (such as stepwise formation of various complexes resulting in more than one complex existing in solution in significant concentrations during the titration process), and (3) an indicator reaction capable of locating the stoichiometric equivalence point with reasonable accuracy is available.

In point of fact, many complexation reactions suitable to standard volumetric titration techniques can not be so adapted because of the lack of indicators capable of giving color-shift situations at or near the equivalence points involved. In a number of such instances, however, instrumental methods of locating the equivalence point can be applied, such as the potentiometric, conductometric, amperometric, etc., methods. We shall for the moment treat only complexation titrations where indicator techniques are employed.

The student is referred to Chap. 4, Part 4, covering complexation equilibria, for much of the background to complexation titrations.

8.2 TITRATIONS INVOLVING UNIDENTATE LIGANDS

In the majority of instances unidentate ligands, such as NH_3, form successively a number of complexes with the metal ion involved, these complexes being present during the titration process in significant concentrations. For example, in the case of copper(II) titrated with NH_3, the solution at the theoretical equivalence point for the reaction

$$Cu^{2+} + 4NH_3 \rightleftharpoons Cu(NH_3)_4^{2+}$$

will show appreciable concentrations of the additional species $Cu(NH_3)^{2+}$, $Cu(NH_3)_2^{2+}$, and $Cu(NH_3)_3^{2+}$. As discussed in Chap. 4, Part 4, location of the equivalence point with any worthwhile accuracy is not possible. Only where a significant excess of NH_3 is present will the species $Cu(NH_3)_4^{2+}$ be the only complex ion present to any considerable extent. Because of this tendency of unidentate ligands, few of such substances are capable of acting as one of the components in a complexation titration system. There are some notable exceptions, two of which follow.

8.2.1 THE TITRATION OF CYANIDE BY SILVER(I) ION: LIEBIG METHOD

The Liebig method involves the titration of cyanide with a standard solution of silver nitrate, the titration reaction being

$$2CN^- + Ag^+ \rightleftharpoons Ag(CN)_2^-$$

resulting in the formation of the highly stable complex dicyanoargentate ion, $Ag(CN)_2^-$. Almost coinciding with the stoichiometric equivalence point of the titration, Ag^+ from the silver(I) solution reacts with $Ag(CN)_2^-$ to form the white slightly soluble salt silver dicyanoargentate, $Ag\{Ag(CN)_2\}$, according to the reaction

$$Ag(CN)_2^- + Ag^+ \rightleftharpoons Ag\{Ag(CN)_2\}(s)$$

The resulting first permanent turbidity imparted to the solution signals the end point of the titration. The coincidence of the end point and the equivalence point is almost exact, as indicated by the following.

Consider the titration of 50.00 ml of 0.2000 M NaCN by 0.1000 M AgNO$_3$. At the equivalence point (or the end point) the value of $[Ag(CN)_2^-]$ will be represented by $5.0 \times 10^{-2} M$ as a satisfactory approximation. At the end point, therefore, where $Ag\{Ag(CN)_2\}$ is just ready to precipitate, we have

$$[Ag^+] = \frac{K_{sp}(Ag\{Ag(CN)_2\})}{[Ag(CN)_2^-]} \quad \text{and} \quad [Ag^+] = \frac{[Ag(CN)_2^-]}{[CN^-]^2 K_{stab}}$$

so that

$$[CN^-] = \sqrt{\frac{[Ag(CN)_2^-]^2}{K_{sp}(Ag\{Ag(CN)_2\}K_{stab}}}$$

$$= \sqrt{\frac{(5.0 \times 10^{-2})^2}{2.0 \times 10^{-12} \times 7.1 \times 10^{19}}}$$

$$= 4.2 \times 10^{-6} \ M$$

At the equivalence point the value of $[CN^-]$ is given by

$$[CN^-] = 2[Ag^+] \quad \text{and} \quad \frac{[Ag(CN)_2^-]}{[Ag^+][CN^-]^2} = 7.1 \times 10^{19}$$

with

$$[CN^-] = \sqrt[3]{\frac{5.0 \times 10^{-2}}{0.5 \times 7.1 \times 10^{19}}}$$

$$= 1.1 \times 10^{-7} \ M$$

This latter concentration of CN^- is not achieved, however, since precipitation of $Ag\{Ag(CN)_2\}$ begins as soon as the value of $[CN^-] = 4.2 \times 10^{-6} \ M$ is attained. The end point thus occurs very slightly before the equivalence point, although the difference is insignificant, particularly in consideration of the fact that some further addition of $AgNO_3$ solution is required to provide sufficient $Ag\{Ag(CN)_2\}$ for visual detection.

Table 8.1 shows the data for the titration in question. After the start of the titration and up to but not including the 50.00-ml point of $AgNO_3$ solution addition, the values for $[CN^-]$ were calculated using Eq. (184), with the corresponding $[Ag^+]$ calculated from the stability-constant equation. At the 50.00-ml point, the $[CN^-]$ was calculated as shown for the end point, with $[Ag^+]$ determined from the stability-constant equation. After the 50.00-ml point, the values for $[Ag^+]$ were determined from the solubility-constant equation, with $[CN^-]$ being determined from the stability-constant equation alone.

Note that to a reasonable level of $AgNO_3$ solution added after the end-point volume, there is no practical change in either $[Ag^+]$ or $[CN^-]$; this on the basis of the continued precipitation of $Ag(CN)_2^-$ as $Ag\{Ag(CN)_2\}$. Figure 8.1 indicates the general form of the titration curve.

There is a tendency for a precipitate of $Ag\{Ag(CN)_2\}$ to form just before the actual end point as the result of high localized concentrations of Ag^+. Because of the slow rate of redissolving of this precipitate, vigorous agitation of the solution, with slow addition of titrant, is required in order to avoid premature end-point indication.

Liebig titrations are usually carried out in a medium alkaline to sodium or potassium hydroxide (about 0.05 M).

Table 8.1   Titration of cyanide ion by the Liebig method*
50.00 ml of 0.2000 M NaCN vs. 0.1000 M AgNO$_3$

Volume AgNO$_3$, *ml*	[CN$^-$]	pCN	[Ag$^+$]	pAg	$(\Delta pCN/\Delta V_{Ag}) \times 10^3$ 0.1 ml $= \Delta V_{Ag}$		$(\Delta^2 pCN/\Delta V^2_{Ag}) \times 10^3$ 0.1 ml/0.1 ml	
0.00	2.0×10^{-1}	0.70						
10.00	1.3×10^{-1}	0.89	1.4×10^{-20}	19.85				
					$+1.8$	(15.00)		
20.00	8.6×10^{-2}	1.07	5.4×10^{-20}	19.27			$+0.005$	(20.00)
					$+2.3$	(25.00)		
30.00	5.0×10^{-2}	1.30	2.1×10^{-19}	18.68			$+0.013$	(30.00)
					$+3.6$	(35.00)		
40.00	2.2×10^{-2}	1.66	1.3×10^{-18}	17.89			$+0.043$	(38.75)
					$+6.8$	(42.50)		
45.00	1.0×10^{-2}	2.00	6.7×10^{-18}	17.17			$+0.25$	(44.75)
					$+18$	(47.00)		
49.00	2.0×10^{-3}	2.70	1.8×10^{-16}	15.74			$+1.9$	(48.12)
					$+60$	(49.25)		
49.50	1.0×10^{-3}	3.00	7.0×10^{-16}	15.16			$+18$	(49.45)
					$+133$	(49.65)		
49.80	4.0×10^{-4}	3.40	4.4×10^{-15}	14.36			$+84$	(49.75)
					$+300$	(49.85)		
49.90	2.0×10^{-4}	3.70	1.8×10^{-14}	13.74			$+400$	(49.89)
					$+600$	(49.92)		
49.95	1.0×10^{-4}	4.00	7.0×10^{-14}	13.16			$+4320$	(49.95)
					$+2760$	(49.98)		
50.00	4.2×10^{-6}	5.38	4.0×10^{-11}	10.40			-5520	(50.00)
					0	(50.02)		
50.05	4.2×10^{-6}	5.38	4.0×10^{-11}	10.40			0	(50.05)
					0	(50.08)		
50.10	4.2×10^{-6}	5.38	4.0×10^{-11}	10.40			0	(50.12)
					0	(50.15)		
50.20	4.2×10^{-6}	5.38	4.0×10^{-11}	10.40			0	(50.25)
					0	(50.35)		
50.50	4.2×10^{-6}	5.38	4.0×10^{-11}	10.40				

*[CN$^-$], pCN, [Ag$^+$], and pAg are not corrected for the dilution effect. The values in parentheses after each $\Delta pCN/\Delta V_{Ag}$ and $\Delta^2 pCN/\Delta V_{Ag}^2$ value associate the average slope, or rate of change of the slope, with the average volume value between the titration points involved. The equivalence-point volume is 50.00 ml of 0.1000 M AgNO$_3$ solution.

A clearer end-point indication can be obtained by a modification of the Liebig method known as the Liebig-Denigés method. In this technique, the titration is carried out in a medium alkaline to NH$_3$, using potassium iodide, KI, as the indicator substance. The presence of NH$_3$ prevents the formation of Ag{Ag(CN)$_2$} by the reaction

$$Ag\{Ag(CN)_2\} + 4NH_3 \rightleftharpoons 2Ag(NH_3)_2^+ + 2CN^-$$

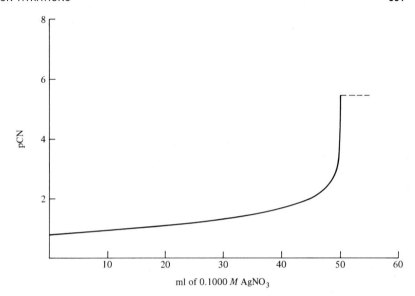

Fig. 8.1 Titration of 50.00 ml of 0.2000 M NaCN with 0.1000 M AgNO$_3$: Liebig method.

forming the complex diaminesilver(I), Ag(NH$_3$)$_2^+$, and keeping the silver ion concentration low. The silver iodide salt, AgI, has a much lower solubility than Ag{Ag(CN)$_2$}, however, and the appearance of a solution turbidity due to precipitated AgI signals the titration end point. The correspondence between the end point and the equivalence point is good providing that the NH$_3$ and KI concentrations are kept within the proper limits.

Some importance must be attached to the concentrations of NH$_3$ and KI, or I$^-$, at the end point, since improper concentrations can adversely affect the coincidence of end point and equivalence point. Values of [NH$_3$] between 0.3 M and 0.6 M, and [I$^-$] between 0.01 and 0.03 M, will be found satisfactory.

A combination of the Liebig and Volhard methods can be used to determine cyanide and chloride in the same solution. The solution is first titrated with silver(I) solution to the first turbidity (the Liebig end point). Silver(I) solution is next added in an amount in excess of that required to convert all Ag(CN)$_2^-$ to Ag{Ag(CN)$_2$} and all Cl$^-$ to AgCl. The combined precipitates are filtered off and washed, and the filtrate titrated with standard KSCN solution to determine the excess of silver(I) (the Volhard end point). A similar method can be applied in the analysis of solutions containing cyanide and bromide, cyanide and iodide, and cyanide and thiocyanate.

A standard solution of potassium cyanide, KCN, can be used as a titrant in the determination of copper(II) and nickel(II). In the case of the former, the titration takes place in a medium alkaline to NH$_3$ and we have

$$2Cu(NH_3)_4^{2+} + 7CN^- + H_2O \rightleftharpoons 2Cu(CN)_3^{2-} + CNO^- + 2NH_4^+ + 6NH_3$$

the end point being signaled by the disappearance of the blue color due to the $Cu(NH_3)_4{}^{2+}$ complex ion.

In the determination of nickel(II), the titration takes place in a medium alkaline to NH_3 to which a small concentration of potassium iodide has been added. A measured volume of standard $AgNO_3$ solution is added, and a solution turbidity due to precipitated AgI is obtained. Titration with standard KCN solution is now carried out to the disappearance of the turbidity. Back-titration with standard $AgNO_3$ solution is carried out to the appearance of the first turbidity. The volume of KCN solution is corrected for the total volume of $AgNO_3$ solution used. The titration reaction with KCN is

$$Ni(NH_3)_4{}^{2+} + 4CN^- \rightleftharpoons Ni(CN)_4{}^{2-} + 4NH_3$$

8.2.2 THE TITRATION OF CHLORIDE BY MERCURY(II)

The reactions of chloride with mercury(II) can be summarized as

$$\begin{aligned}
Cl^- + Hg^{2+} &\rightleftharpoons HgCl^+ & K_1 &= 5.5 \times 10^6 \\
Cl^- + HgCl^- &\rightleftharpoons HgCl_2 & K_2 &= 3.0 \times 10^6 \\
Cl^- + HgCl_2 &\rightleftharpoons HgCl_3{}^- & K_3 &= 7.1 \\
Cl^- + HgCl_3{}^- &\rightleftharpoons HgCl_4{}^{2-} & K_4 &= 1.0 \times 10
\end{aligned}$$

The low stability of the last two complexes implies that a titration of chloride by mercury(II) could be carried out with a very considerable increase in the value of $[Hg^{2+}]$ around the equivalence point for the reaction forming mercuric chloride, $HgCl_2$.

The method involves the titration of a dilute nitric acid solution of the chloride with a standard solution of mercury(II) in the form of mercuric nitrate, $Hg(NO_3)_2$. A solution of sodium nitroprusside, $Na_2Fe(CN)_5NO$, is used as the indicator, the volume added being adjusted to yield a value of $[Fe(CN)_5NO^{2-}]$ of about $1.5 \times 10^{-3} M$ in solution at the titration end point. The end point occurs with the appearance of the first permanent solution turbidity resulting from the reaction of Hg^{2+} with $Fe(CN)_5NO^{2-}$ to form the white slightly soluble salt mercuric nitroprusside, $HgFe(CN)_5NO$:

$$Hg^{2+} + Fe(CN)_5NO^{2-} \rightleftharpoons HgFe(CN)_5NO(s)$$

Several factors of importance exist with respect to this titration. The end point occurs after the equivalence point, and in part excess mercury(II) solution reacts with mercuric chloride according to

$$Hg^{2+} + HgCl_2 \rightleftharpoons 2HgCl^+$$

so that the actual $[Hg^{2+}]$ in solution is appreciably less than the value expected from the excess of mercury(II) solution added. The extent to which this reaction takes place depends on the concentration of $HgCl_2$ present. The end point will occur when sufficient mercury(II) solution has been added to provide a visually detectable precipitate of $HgFe(CN)_5NO$. It is apparent that the excess of mercury(II) solution required to reach this point will vary according to the $HgCl_2$ concentration, so that the indicator blank will not be constant in value. The determination of the indicator blank can be accomplished by first calculating the approximate $HgCl_2$ concentration at the end point from the volume of mercury(II) solution consumed. A solution of $HgCl_2$ of the same strength is then prepared, and a volume of this solution similar to the titration end-point volume, and containing the same concentration of the nitroprusside indicator, is titrated with the mercury(II) solution to the first permanent turbidity. The volume required is then applied as the indicator blank to the actual titration volume. A general picture of such corrections is given by Kolthoff and Stenger.* As an example, the titration of 50 ml of 0.1 M chloride with 0.05 M mercury(II) would require an indicator blank of about -0.2 ml.

The magnitude of the indicator-blank correction is also a function to some extent of the acidity of the solution and the rate of addition of the titrant, so that these variables must be kept within reasonable control. The value of the method lies in the fact that chloride can be determined in acid solution, as well as at concentration levels as low as 10 ppm.

Organic substances capable of forming colored complexes with mercury(II) have been used as indicators in place of sodium nitroprusside. Of these, diphenylcarbazide (colorless to violet) and diphenylcarbazone (orange to violet) have been found satisfactory. Control of the solution pH is required in order to obtain the best coincidence of end point and equivalence point. Roberts found a solution pH of 1.5 to 2.0 necessary for diphenylcarbazide, while Clark reported a range of 3.2 to 3.3 as best for diphenylcarbazone.†

Although titration methods involving mercury(II) for the determination of bromide, cyanide, and thiocyanate can be applied, these do not provide advantages over the standard methods involving silver(I) as titrant. Iodide has also been determined by titration with mercury(II), but with limited accuracy.

8.3 TITRATIONS INVOLVING POLYDENTATE LIGANDS

8.3.1 GENERAL

Several polydentate ligands are capable of forming water-soluble complexes with many metal ions, where these complexes are of the 1:1 type. Such complexation reactions occur as a single step, as opposed to the stepwise formation of complexes common with many unidentate ligands. As a result of this single-step formation and the usually high stability of the complexes formed, such polydentate ligands, used as titrants, are capable of yielding large enough changes in the metal-ion concentration around the equivalence point to permit, where adequate

---

*I. M. Kolthoff and V. A. Stenger, "Volumetric Analysis," 2d. ed., vol 2, Interscience, New York, 1947, pp. 332–333.
†I. Roberts, *Ind. Eng. Chem., Anal. Ed.*, **8**:365(1936); F. E. Clark, *Anal. Chem.*, **22**:553(1950).

indicator situations are available, detection of the end point of the titration and reasonable coincidence between the end point and the equivalence point.

The magnitude of the change in the metal-ion concentration, or the pM, around the equivalence point will be a function of the stability constant for the metal-ligand complex (or as we shall see later, of the value of this constant modified by the effect of solution pH), the larger the value of the constant the greater the ΔpM around the equivalence point.

Such ligands, in many cases, are chelons; and the titrations involving them are sometimes referred to as chelometric titrations.

Although several such polydentate ligands are known and used, we shall consider as a fundamental example the ligand ethylenediaminetetraacetic acid

$$\begin{matrix} \text{HOOC}\cdot\text{CH}_2 & & & & \text{CH}_2\cdot\text{COOH} \\ & \diagdown & & \diagup & \\ & \text{N}-\text{CH}_2\cdot\text{CH}_2-\text{N} & \\ & \diagup & & \diagdown & \\ \text{HOOC}\cdot\text{CH}_2 & & & & \text{CH}_2\cdot\text{COOH} \end{matrix}$$

commonly called EDTA and usually represented in the acid form as H_4Y. A discussion of this ligand and its dissociation reactions, reactions with metal ions, influence of solution pH on these reactions, etc., is given in Chap. 4, Part 4. Ordinarily the acid form of the ligand is not used in the preparation of titrant solutions because of its relatively low solubility in water, and the disodium dihydrate salt $Na_2H_2Y\cdot 2H_2O$ is most commonly used in this connection.

8.3.2 TITRATIONS OF METAL IONS WITH EDTA

These titrations are usually carried out in buffered solutions of the metal ion involved. The use of the proper pH is important; it will involve a consideration of the stability constant of the metal-EDTA complex. As outlined in Chap. 4, Part 4, the ability to form the metal-EDTA complex will be a function of the value of α as calculated from Eq. (118); and for each metal-EDTA complex, there will be a minimum pH value below which the titration will not be feasible.

Buffering of the solution is required, since the EDTA substance may exist predominantly in one of the several protonated ion states such as HY^{3-}, H_2Y^{2-}, or H_3Y^-, rather than as Y^{4-}, depending on the solution pH at which the titration is carried out. Such conditions could lead to reactions such as

$$M^{n+} + H_2Y^{2-} + 2H_2O \rightleftharpoons M(Y)^{(n-4)} + 2H_3O^+$$

where the production of H_3O^+ during the course of the titration could, unless the solution is properly buffered, result in a decrease in the pH and a nonquantitative conversion of M^{n+} to $M(Y)^{(n-4)}$ at the equivalence point.

Again, the pH of the solution to be titrated may have to be adjusted and controlled to a value permitting the proper functioning of the indicator substance to be used; and this is a second area of consideration relative to solution pH conditions during titration. Obviously, the final pH to which the solution will be adjusted and buffered will represent a proper balance between the demands for a quantitatively complete reaction at the equivalence point and those relative to the requirements for the indicator to be used.

Leaving the situation with respect to indicators aside for the moment, we note that very generally alkaline solutions will be required for metal-EDTA complexes of relatively low stability constant, such as calcium or magnesium, while less alkaline to mildly acid solutions might be considered for those of high stability constant, such as copper or zinc.

The titration of magnesium(II) ion by EDTA Consider the titration of 50.00 ml of 0.1000 M magnesium(II) by 0.1000 M EDTA. Because of the relatively low value of the stability constant for the magnesium-EDTA complex, the titration should be carried out at as high a pH as possible, practically speaking. However, in order to avoid the competitive reaction resulting from the formation of magnesium hydroxide, $Mg(OH)_2$, at high pH values, the direct titration for magnesium(II) alone usually involves an NH_3-NH_4Cl buffer with a solution pH of about 10.0

The value of α as calculated from Eq. (118), omitting the third, fourth, and fifth terms within the parentheses, is 2.8. The value of $[Mg(Y)^{2-}]$ at the equivalence point is given by $5.0 \times 10^{-2} M$ as a good approximation, and we have

$$C_Y \approx [Mg^{2+}]$$

so that Eq. (120) leads to

$$\frac{5.0 \times 10^{-2}}{[Mg^{2+}]^2} \approx \frac{K_{stab}}{\alpha} \approx \frac{4.9 \times 10^8}{2.8}$$

and

$$[Mg^{2+}] \approx 1.7 \times 10^{-5} M$$

at the equivalence point.

Table 8.2 shows the calculated data for the titration involved. The values of $[Mg^{2+}]$ before the equivalence point are calculated from Eq. (184), although at the levels of 49.90 and 49.95 ml of EDTA solution, Eq. (185) could have been more properly applied. After the equivalence point C_Y is calculated from the simplest Eq. (187), with $[Mg(Y)^{2-}]$ being determined on a solution dilution basis from

$$[Mg(Y)^{2-}] = \frac{V_{Mg^{2+}} M_{Mg^{2+}}}{V_{Mg^{2+}} + V_{EDTA}}$$

and $[Mg^{2+}]$ being finally calculated from Eq. (120).

Note that ΔpMg for ±0.05 ml around the equivalence-point volume of 50.00 ml of 0.1000 M EDTA solution is 0.95 units.

Figure 8.2 shows the curve for this titration, as well as those for similar titrations conducted in solutions buffered at 8 and 9 pH. The decreasing range of

Table 8.2 Titration of magnesium (II) by EDTA*
50.00 ml of 0.1000 M Mg²⁺ vs. 0.1000 M EDTA at pH = 10.0

Volume EDTA, ml	$[Mg^{2+}]$	pMg	Equation used	$(\Delta pMg/\Delta V_{EDTA})\times10^3$ 0.1 $ml = \Delta V_{EDTA}$	$(\Delta^2 pMg/\Delta V^2_{EDTA})\times10^3$ 0.1 ml/0.1 ml
0.00	1.00×10^{-1}	1.00			
5.00	8.18×10^{-2}	1.09	(184)		
				+1.8 (7.50)	
10.00	6.67×10^{-2}	1.18	(184)		+0.0013 (11.25)
				+1.9 (15.00)	
20.00	4.28×10^{-2}	1.37	(184)		+0.004 (20.00)
				+2.3 (25.00)	
30.00	2.50×10^{-2}	1.60	(184)		+0.013 (30.00)
				+3.6 (35.00)	
40.00	1.11×10^{-2}	1.96	(184)		+0.084 (39.75)
				+11.6 (44.50)	
49.00	1.01×10^{-3}	3.00	(184)		+1.0 (46.88)
				+60 (49.25)	
49.50	5.02×10^{-4}	3.30	(184)		+18 (49.45)
				+133 (49.65)	
49.80	2.00×10^{-4}	3.70	(184)		+84 (49.75)
				+300 (49.85)	
49.90	1.00×10^{-4}	4.00	(184)		+400 (49.89)
				+600 (49.92)	
49.95	5.00×10^{-5}	4.30	(184)		+680 (49.95)
				+940 (49.98)	
50.00	1.7×10^{-5}	4.77	(118),(120)		+40 (50.00)
				+960 (50.02)	
50.05	5.6×10^{-6}	5.25	(187),(120)		−720 (50.05)
				+600 (50.08)	
50.10	2.8×10^{-6}	5.55	(187),(120)		−400 (50.12)
				+300 (50.15)	
50.20	1.4×10^{-6}	5.85	(187),(120)		−82 (50.25)
				+137 (50.35)	
50.50	5.5×10^{-7}	6.26	(187),(120)		−20 (50.55)
				+58 (50.75)	
51.00	2.8×10^{-7}	6.55	(187),(120)		−0.99 (53.12)
				+11 (55.50)	
60.00	2.8×10^{-8}	7.55	(187),(120)		

*$[Mg^{2+}]$ and pMg values are not corrected for the dilution effect. The values in parentheses after each $\Delta pMg/\Delta V_{EDTA}$ and $\Delta^2 pMg/\Delta V^2_{EDTA}$ value associate the average slope, or rate of change of the slope, with the average volume value between the titration points involved. The equivalence-point volume is 50.00 ml of 0.1000 M EDTA solution.

ΔpMg with decreasing pH will be noted. Based on a desirable ΔpMg of about 1.0 unit, the titrations involved show potential accuracies in the location of the equivalence points of 1 ppt, 3 ppt, and 10 ppt for the pH 10, pH 9, and pH 8 conditions, respectively.

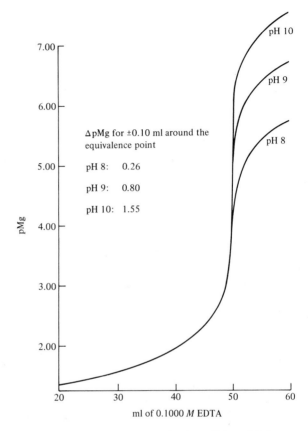

Fig. 8.2 Titration of 50.00 ml of 0.1000 M Mg^{2+} vs. 0.1000 M EDTA.

The general effect of the solution pH on the reaction between metal ions and EDTA has been discussed elsewhere, together with its end result in increasing the value of α and decreasing the conditional formation constant, given as K in

$$K = \frac{K_{\text{stab}}}{\alpha}$$

Table 8.3 shows the approximate minimum desirable solution pH values for various metal ions titrated with EDTA under conditions where no competitive complexing situation exists in solution.

Reactions competitive with the complexation process The necessity of conducting many EDTA titrations in solutions of relatively high pH frequently results in difficulties relative to the hydrolysis of certain metal ions, since the hydrolysis process, where it leads to the formation of insoluble hydroxides or

Table 8.3 Minimum solution pH for satisfactory end-point location metalion–EDTA titrations*

Metal ion	Minimum solution pH
Fe^{3+}	1.5
Hg^{2+}	2.2
Cu^{2+}	3.2
Ni^{2+}	3.2
Pb^{2+}	3.3
Cd^{2+}	4.0
Co^{2+}	4.1
Zn^{2+}	4.1
Fe^{2+}	5.1
Ca^{2+}	7.3
Mg^{2+}	10.0

*Minimum solution pH based on a conditional formation constant value

$$K = \frac{K_{stab}}{\alpha}$$

of not less than 10^8, and no competitive complexation condition in solution.

basic salts, may compete with the complexation process. In some instances, this situation can be put to a useful purpose, for example, where calcium and magnesium are present together in solution. Titration with EDTA at pH 10 results in both ions being titrated. The addition of NaOH to raise the pH to a value higher than 12 causes the precipitation of magnesium(II) as $Mg(OH)_2$, and titration with EDTA serves to determine calcium alone.

In other instances, particularly with respect to the heavier metals, such as copper or zinc, the situation is frequently complicated by the necessity of buffering the solution to a pH compatible with both the magnitude of the stability constant and the indicator color change requirements and, at the same time, avoiding the separation of the hydrolysis products of such metals. The common use of ammonia with ammonium salts as buffering agents to avoid such hydrolysis phenomena can result in complex formation reactions capable of competing with the EDTA titration reaction. In certain cases such competitive complexing substances are added with a special objective in view, for example, in the titration of lead by EDTA in the presence of nickel, where cyanide is added to form the highly stable complex ion $Ni(CN)_4^{2-}$, thus permitting the titration of the lead without interference from nickel. An example of an EDTA titration in the presence of a competitive complexing situation follows.

The titration of copper(II) ion by EDTA Consider the titration of 50.00 ml of 0.1000 M copper(II) by 0.1000 M EDTA where the copper(II) solution is buffered to 9.0 pH by the use of $[NH_3] = 0.1\ M$ and $[NH_4Cl] = 0.2\ M$. Assume that the EDTA solution has the same $[NH_3]$ and $[NH_4Cl]$ values, so that the titration process does not result in a change in the solution pH.

The titration reaction is basically

$$Cu(NH_3)_4^{2+} + HY^{3-} + H_2O \rightleftharpoons Cu(Y)^{2-} + 4NH_3 + H_3O^+$$

It should be noted that the extent to which the reaction proceeds now depends on the NH_3 concentration, as well as the solution pH. It is apparent that too high a concentration of NH_3 should be avoided, since this will decrease the ΔpCu around the equivalence point by decreasing the degree of quantitative completion of the copper-EDTA reaction.

Using developments arising out of Chap. 4, Part 4, we may arrive at methods for calculating the value of $[Cu^{2+}]$ at various points in such a titration. Note that we can assign a value of C_M to the total concentration of copper(II) not associated with EDTA, and that this will include $[Cu^{2+}]$, $[Cu(NH_3)^{2+}]$, $[Cu(NH_3)_2^{2+}]$, $[Cu(NH_3)_3^{2+}]$, and $[Cu(NH_3)_4^{2+}]$. The value of $[Cu^{2+}]$ then becomes the amount of C_M not involved in the various aminecopper(II) complexes. We now have

$$\beta_0 = \frac{[Cu^{2+}]}{C_M}$$

where β_0 is given by Eq. (116) as

$$\beta_0 = \frac{1}{1 + [NH_3]K_1 + [NH_3]^2K_1K_2 + [NH_3]^3K_1K_2K_3 + [NH_3]^4K_1K_2K_3K_4}$$

K_1, K_2, etc., being the stability constants for the various aminecopper(II) complexes. The calculation of β_0 now permits the determination of $[Cu^{2+}]$ for any point in titration where C_M represents the amount of copper(II) in any form other than that of the copper-EDTA complex.

Equation (120) now yields

$$\frac{[Cu(Y)^{2-}]}{[Cu^{2+}]C_Y} = \frac{K_{stab}}{\alpha}$$

and, with $[Cu^{2+}] = \beta_0 C_M$, we have

$$\frac{[Cu(Y)^{2-}]}{C_M C_Y} = \frac{K_{stab}\,\beta_0}{\alpha}$$

from which C_M at the equivalence point can be determined and $[Cu^{2+}]$ at this point calculated.

The value of α based on a solution pH of 9.0 is given by Eq. (118) as

$$\alpha = 19$$

while the value of β_0 calculated from Eq. (116) is

$$\beta_0 = 2.2 \times 10^{-9}$$

For points in the titration before the equivalence point, C_M is determined from Eq. (184) as

$$C_M = \frac{50.00 \times 0.1000 - V_{EDTA} M_{EDTA}}{50.00 + V_{EDTA}}$$

For the 49.90-ml point in the titration we have

$$C_M = 1.00 \times 10^{-4} \, M$$

and, ignoring any dissociation of the several copper complexes in solution,

$$\begin{aligned}
[Cu^{2+}] &= C_M \beta_0 \\
&= 1.00 \times 10^{-4} \, M \times 2.2 \times 10^{-9} \\
&= 2.2 \times 10^{-13} \, M
\end{aligned}$$

Note that without consideration of the competitive complexing situation introduced by the NH_3, the value of $[Cu^{2+}]$ at this point would be given by $1.00 \times 10^{-4} \, M$.

At the equivalence point we have

$$C_M = C_Y$$

and we have

$$\begin{aligned}
[Cu(Y)^{2-}] &= 0.05 \, M - C_M \\
&\approx 0.05 \, M
\end{aligned}$$

since $C_M \ll [Cu(Y)^{2-}]$. We now have

$$\frac{0.05}{C_M^2} \approx \frac{K_{stab}\beta_0}{\alpha} \approx \frac{6.2 \times 10^{18} \times 2.2 \times 10^{-9}}{19}$$

and

$$C_M \approx 8.4 \times 10^{-6} \, M$$

with

$$\begin{aligned}
[Cu^{2+}] &\approx 8.4 \times 10^{-6} \, M \times \beta_0 \\
&\approx 1.8 \times 10^{-14} \, M
\end{aligned}$$

Again, without the competitive situation, we would have determined $[Cu^{2+}]$ from Eq. (120) as

$$[Cu^{2+}] = 3.9 \times 10^{-10} \, M$$

After the equivalence point we have, from Eq. (187),

$$[EDTA] \approx C_Y \approx \frac{V_{EDTA} M_{EDTA} - 50.00 \times 0.1000}{V_{EDTA} + 50.00}$$

For the 50.10-ml point in the titration this yields

$$[\text{EDTA}] \approx C_Y \approx 9.99 \times 10^{-5} \; M$$

and we have

$$[\text{Cu(Y)}^{2-}] = \frac{V_{\text{Cu}^{2+}} M_{\text{Cu}^{2+}}}{V_{\text{Cu}^{2+}} + V_{\text{EDTA}}}$$
$$= 5.00 \times 10^{-2} \; M$$

from which

$$C_M \approx \frac{[\text{Cu(Y)}^{2-}]\,\alpha}{C_Y K_{\text{stab}} \beta_0}$$
$$\approx 7.0 \times 10^{-7} \; M$$

with

$$[\text{Cu}^{2+}] \approx 7.0 \times 10^{-7} \; M \times \beta_0$$
$$\approx 1.5 \times 10^{-15} \; M$$

The competitive situation plays no part after the equivalence point, and the value of $[\text{Cu}^{2+}]$ at the 50.10-ml point would be $1.5 \times 10^{-15} \; M$ regardless of the presence or absence of NH_3.

Note that for ± 0.10 ml around the equivalence-point volume, the ΔpCu is given by $14.82 - 12.66$, or 2.16 units, where the solution is buffered with the NH_3–NH_4Cl mixture, and $14.82 - 4.00$, or 10.82 units, where there is no competitive complexing situation. Understand fully that this latter situation is purely imaginative, and in the absence of NH_3–NH_4Cl the copper-EDTA titration would be carried out at a much lower solution pH.

Table 8.4 shows the data for the titration described at the start of this section. Figure 8.3 shows the titration curves for this titration, and for a similar titration carried out with $[\text{NH}_3] = 0.01 \; M$ and pH $= 9.0$. Note the effect of increasing $[\text{NH}_3]$ on the ΔpCu around the equivalence point.

8.3.3 OTHER POLYDENTATE LIGAND OR CHELON TITRANTS

1. 1,2-Diaminocyclohexane-N,N,N',N'-tetraacetic acid, commonly called DCYTA, forms highly stable complexes with metal ions, and can be used in titrations similar to those where EDTA is applied.
2. Ethyleneglycol bis (β-aminoethylether)-N,N,N',N'-tetraacetic acid, known as EGTA, has a low stability constant for the magnesium(II) complex, and thus permits the determination of calcium in the presence of magnesium without the necessity of separating the magnesium as Mg(OH)_2.
3. Nitrilotriacetic acid, known as NTA, and generally used as the trisodium salt, can be used for many of the titrations to which EDTA is normally applied. The generally lower values of its stability constants for the metal complexes does not permit a satisfactory determination of calcium, magnesium, and barium.
4. Triethylenetetramine, commonly called trien, has been used in the titration of copper(II), nickel(II), cadmium(II), zinc(II), and mercury(II).

Table 8.4 Titration of copper (II) by EDTA in the presence of ammonia*
50.00 ml of 0.1000 M Cu^{2+} vs. 0.1000 M EDTA
Solution pH $= 9.0$ $[NH_3] = 0.1\,M$ $[NH_4Cl] = 0.2\,M$

Volume EDTA, ml	$[Cu^{2+}]$	pCu	$(\Delta pCu/\Delta V_{EDTA}) \times 10^3$ $0.1\ ml = \Delta V_{EDTA}$	$(\Delta^2 pCu/\Delta V^2_{EDTA}) \times 10^3$ $0.1\ ml/0.1\ ml$
0.00	2.2×10^{-10}	9.66		
5.00	1.8×10^{-10}	9.74		
			$+1.6$ (7.50)	
10.00	1.5×10^{-10}	9.82		$+0.0067$ (11.25)
			$+2.1$ (15.00)	
20.00	9.4×10^{-11}	10.03		$+0.002$ (20.00)
			$+2.3$ (25.00)	
30.00	5.5×10^{-11}	10.26		$+0.013$ (30.00)
			$+3.6$ (35.00)	
40.00	2.4×10^{-11}	10.62		$+0.084$ (39.75)
			$+11.6$ (44.50)	
49.00	2.2×10^{-12}	11.66		$+1.0$ (46.88)
			$+60$ (49.25)	
49.50	1.1×10^{-12}	11.96		$+18$ (49.45)
			$+133$ (49.65)	
49.80	4.4×10^{-13}	12.36		$+84$ (49.75)
			$+300$ (49.85)	
49.90	2.2×10^{-13}	12.66		$+400$ (49.89)
			$+600$ (49.92)	
49.95	1.1×10^{-13}	12.96		$+1920$ (49.95)
			$+1560$ (49.98)	
50.00	1.8×10^{-14}	13.74		-40 (50.00)
			$+1540$ (50.02)	
50.05	3.1×10^{-15}	14.51		-1840 (50.05)
			$+620$ (50.08)	
50.10	1.5×10^{-15}	14.82		-426 (50.12)
			$+300$ (50.15)	
50.20	7.6×10^{-16}	15.12		-90 (50.25)
			$+120$ (50.35)	
50.50	3.1×10^{-16}	15.51		-14 (50.55)
			$+62$ (50.75)	
51.00	1.5×10^{-16}	15.82		-1.0 (53.12)
			$+11$ (55.50)	
60.00	1.5×10^{-17}	16.82		

*$[Cu^{2+}]$ and pCu values are not corrected for the dilution effect. The values in parentheses after each $\Delta pCu/\Delta V_{EDTA}$ and $\Delta^2 pCu/\Delta V^2_{EDTA}$ value associate the average slope, or rate of change of the slope, with the average volume value between the titration points involved. The equivalence-point volume is 50.00 ml of EDTA 0.1000-M solution.

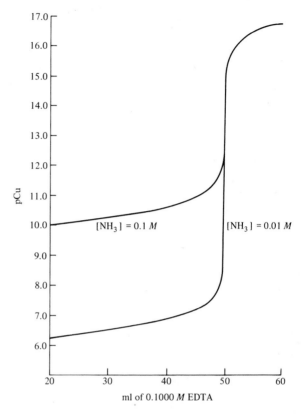

Fig. 8.3 Titrations of 50.00 ml of Cu^{2+} vs. 0.1000 M EDTA at pH = 9.0 and in the presence of ammonia.

8.3.4 INDICATORS FOR TITRATIONS INVOLVING POLYDENTATE LIGANDS

While it is possible in many instances to locate the end point or equivalence point in complexation titrations by the application of instrumental techniques, several indicator substances are available for use in the standard volumetric techniques. These indicators, often called *metallochromic indicators*, are organic substances capable of forming intensely colored complexes with many metal ions. Since these substances may also react with H_3O^+ to form colored compounds, often of a similar color to that of the metal-indicator complex, pH control of the solution is important from this aspect alone in order to prevent competition between the metal ion and H_3O^+ for the indicator substance.

Obviously the indicator must form the metal-indicator complex at a concentration of the metal ion appropriate to that expected throughout the titration up to very close to the equivalence point, and this complex must be destroyed at the equivalence point, with the appropriate color change, otherwise indicator-blank corrections of significant magnitude would be implied.

Indicators of the metallochromic type act on the basis of a color-shift situation similar to that associated with acid-base indicators, although as mentioned already a complication is introduced by the fact that color shifts may occur independently on the basis of pH and pM changes.

As a case in point we shall discuss the indicator Eriochrome black T. This indicator shows color shifts based on pH changes as follows:

H_2In^- (red)

HIn^{2-} (blue)

HIn^{2-} (blue)

In^{3-} (orange)

The K_2 value for the red-to-blue shift is 5.0×10^{-7}, while that involved in the blue-to-orange shift is given by $K_3 = 2.5 \times 10^{-12}$. In the range of pH of 7 to 11, the indicator assumes the blue color due to HIn^{2-}. The addition of metal ions under the proper conditions (Mg^{2+} can not be added to solutions with a pH in excess of about 10.5 due to precipitation of $Mg(OH)_2$), such as calcium(II), magnesium(II), zinc(II), etc., results in the formation of the metal-indicator complex according to reactions typified by those for zinc and magnesium:

$$Mg^{2+} + HIn^{2-} + H_2O \rightleftharpoons MgIn^- + H_3O^+$$
$$Zn^{2+} + 2HIn^{2-} + 2H_2O \rightleftharpoons ZnIn_2^{4-} + 2H_3O^+$$

These metal-indicator complexes are red in color, resulting in the solution prior to titration taking on a red color. The addition of a polydentate ligand titrant such as EDTA results in the formation of the more stable metal-ligand complex and the eventual practical destruction of the metal-indicator complex, at which point the solution color shifts from red to blue.

The selection of the proper indicator for a given polydentate ligand titration requires a knowledge of the stability constant of the metal-ligand complex and that of the metal-indicator complex. Taken in conjunction with the solution pH required to be applied, the best indicator selection, where an indicator is available, can be made.

An excellent treatment of end-point detection and indicator selection for complexation titrations adaptable to the use of metallochromic indicators is given by Reilley and Schmid.* For the purposes of this discussion we shall show only a method for determining the approximate indicator blank associated with the magnesium(II)-vs.-EDTA titration covered by Table 8.2, where Eriochrome black T is used as the indicator.

The titration was carried out at a pH of 10.0, a low enough value to avoid the formation of $Mg(OH)_2$, and where the indicator is largely in the HIn^{2-} form. We have the metal-indicator reaction as

$$Mg^{2+} + HIn^{2-} + H_2O \rightleftharpoons MgIn^- + H_3O^+$$

and we have

$$\frac{[MgIn^-][H_3O^+]}{[Mg^{2+}][HIn^{2-}]} = K$$

The third-stage dissociation constant for the indicator is given by

$$\frac{[H_3O^+][In^{3-}]}{[HIn^{2-}]} = K_3 = 2.5 \times 10^{-12}$$

and we have

$$\frac{[MgIn^-]}{[Mg^{2+}][In^{3-}]} = \frac{K}{K_3} = K_{stab,\ metal-indicator}$$

*C. N. Reilley and R. W. Schmid, *Anal. Chem.*, **31**:887(1959).

Thus $K = K_3 K_{\text{stab, metal-indicator}}$ and, with $K_{\text{stab}} = 1.0 \times 10^7$, the value of K is given by 2.5×10^{-5}. We now have

$$\frac{[\text{MgIn}^-]}{[\text{Mg}^{2+}][\text{HIn}^{2-}]} = \frac{K}{[\text{H}_3\text{O}^+]}$$

and

$$\frac{[\text{MgIn}^-]}{[\text{HIn}^{2-}]} = \frac{K}{[\text{H}_3\text{O}^+]}[\text{Mg}^{2+}]$$

Considering a range of $[\text{MgIn}^-]/[\text{HIn}^{2-}]$ of $\frac{1}{10}$ to 10 for the detectability by the eye of the color shift from red to blue (similar to the situation relative to acid-base indicators), we have

$$\frac{1}{10} \text{ to } 10 = \frac{K}{[\text{H}_3\text{O}^+]}[\text{Mg}^{2+}]$$

For pH $= 10.0$ and $K = 2.5 \times 10^{-5}$ we have

$$\text{pMg} = 5.4 \pm 1$$

corresponding to the color-shift range of the indicator.

It will be noted from Table 8.2 that the pMg at the equivalence point is 4.77, so that the end point as detected by Eriochrome black T can be expected to occur slightly after the equivalence point. Using the central value of the color-shift range in pMg at pH 10.0, we have a value of 5.4. The end point thus occurs at a value of the EDTA titrant given by

$$\frac{[\text{Mg}(\text{Y})^{2-}]}{[\text{Mg}^{2+}]C_{\text{Y}}} = \frac{K_{\text{stab, metal-EDTA}}}{\alpha}$$
$$= \frac{4.9 \times 10^8}{2.8}$$

where

$$[\text{Mg}(\text{Y})^{2-}] = \frac{50.00 \times 0.1000}{V_{\text{EDTA}} + 50.00}$$

$$C_{\text{Y}} = \frac{V_{\text{EDTA}} \times 0.1000 - 50.00 \times 0.1000}{V_{\text{EDTA}} + 50.00}$$

$$[\text{Mg}^{2+}] = 4.0 \times 10^{-6} \ M \qquad \text{from pMg} = 5.4$$

and we have

$$V_{\text{EDTA}} = 50.07 \text{ ml}$$

and an indicator blank of about -0.07 ml, corresponding to a titration error of about $+1.4$ ppt.

Some other indicators available for use in titrations involving polydentate ligands, together with some of the metal ions with which they have been used, are

Metalphthalein	MPT	Ca, Ba, Sr
1-2-(Pyridylazo)-2-naphthol	PAN	Cu, Zn, Cd, In
Pyrocatechol violet	PYV	Bi, Cu, Zn, Al
Xylenol orange	XYO	Bi, Sc, Pb, Zn, Cd, Hg
Calcon	CAL	Ca
Murexide	MUR	Ni, Cu, Zn, Co, Ca
Calgamite	CMG	Ca, Mg

8.3.5 TITRATION METHODS INVOLVING POLYDENTATE LIGANDS

No attempt will be made to outline these methods in detail. A general approach will serve to illustrate the basic techniques. For a more extensive approach to the subject, the student is referred to an excellent treatment by Schwarzenbach.*

Determination by direct titration   The method here is identical to that described for the determination of magnesium by EDTA, using Eriochrome black T as the indicator. The solution temperature and pH are adjusted to appropriate values, the indicator is added, and the titration proceeds to the indicator color change. The volume of titrant used, corrected for any indicator blank, permits direct calculation of the quantity of the metal involved.

Determination by indirect titration   The determination of certain substances which do not react with polydentate ligands is sometimes possible by a technique of indirect titration. Sulfate can, for example, be determined by adding a measured excess of barium(II). The barium sulfate precipitate is filtered off and washed, the excess barium in the filtrate being subsequently titrated under the proper solution conditions by EDTA.

Determination by back-titration   Reactions of metal ions with polydentate ligands which take place slowly do not permit the use of the direct titration technique. In such cases a measured excess of standard polydentate ligand solution is added, and a sufficient time is allowed for the reaction to go to completion. The solution temperature and pH are adjusted to appropriate values, the proper indicator is added, and the excess of ligand titrant is determined by titration with a standard solution of a metal ion which reacts rapidly with the ligand.

Determination by displacement titration   In certain instances of metal ions the solution conditions involved may be such that a titration indicator is not available. If the metal ion involved forms a much more stable complex with the ligand intended to be used as the titrant than, say, magnesium, a solution of the ligand as the magnesium-ligand complex can then be added to the metal-ion solution. The displacement reaction

$$Mg(Y)^{2-} + M^{2+} \rightleftharpoons M(Y)^{2-} + Mg^{2+}$$

*G. Schwarzenbach, "Complexometric Titrations," Methuen, London, 1960.

then takes place, and the Mg^{2+} resulting from the reaction can be titrated with a standard EDTA solution.

Determination by masking effect titration   In the case of a solution containing more than one metal ion it usually happens that titration with a polydentate ligand results in all of the metal-ligand complexes being formed, so that the volume of titrant consumed reflects the sum of the metals present. It is sometimes possible to use a competitive ligand to prevent the reaction of all the metal ions, except one, with the polydentate ligand titrant, thereby allowing the determination of the single metal ion unaffected. Such a procedure is called masking, and it can be applied, for example, in the determination of magnesium in the presence of nickel. In this case, cyanide ion is added prior to titration and the very stable tetracyanonickelate(II) ion, $Ni(CN)_4^{2-}$ is formed. This complex ion does not react with EDTA, since the nickel-EDTA complex is very much less stable, and the nickel is said to be masked by cyanide. The subsequent titration with standard EDTA solution reports a titration volume associated with magnesium alone.

PROBLEMS

1. An electroplating bath specification calls for the solution to contain from 0.20 to 0.30 lb of NaCN per British gallon. A sample of an operative solution is taken and 25.00 ml is titrated with 0.1025 M $AgNO_3$ using the Liebig-Denigés method. The end point was reached at 37.20 ml of the $AgNO_3$ solution. The indicator blank for this titration is $+0.05$ ml. Determine the weight in pounds of NaCN to be added, without volume change, per British gallon of electroplating solution in order to bring the NaCN content to the upper-level requirement of the specification.   *Ans.* 0.15 lb

2. A homogeneous powder contains only NaCN and KCN. A sample weight of 0.5000 g is dissolved and titrated by the Liebig-Denigés method using 0.0985 M $AgNO_3$. The end-point volume is 42.15 ml, with an indicator blank of $+0.05$ ml to be used. What are the percentages of NaCN and KCN in the powder?

3. A solution contains only NaCN and NaCl. A 50.00-ml sample of the solution is titrated under the proper conditions with 0.1000 M $AgNO_3$ in a Liebig titration. The end point occurs with the first turbidity at 16.85 ml of the $AgNO_3$ solution. 100.00 ml additional 0.1000-M $AgNO_3$ solution is now added and, after digestion, the precipitated substances are filtered off and washed. The filtrate is now diluted to exactly 200.00 ml. A 50.00-ml aliquot is titrated under the proper conditions with 0.1000 M KSCN in a Volhard titration, the end point occurring at 11.45 ml of KSCN solution. Ignoring any indicator blanks, calculate the molarities of NaCN and NaCl in the original solution.

4. A steel containing nickel is analyzed by titration with standard KCN solution according to the method given in the text. A 0.5000 $\pm$ 0.0002 g sample of the steel is dissolved and treated, and eventually 0.1000-M $AgNO_3$ solution is added to yield the first turbidity due to AgI. Titration with 0.1000-M KCN solution is now carried out to the complete disappearance of this turbidity. This occurs at 9.65 $\pm$ 0.04 ml of KCN solution. Back-titration with the $AgNO_3$ solution returns the first turbidity. A total volume of 1.75 $\pm$ 0.04 ml of 0.1000-M $AgNO_3$ solution was consumed in both operations. Neglecting any indicator-blank situations, determine the percentage of nickel in the steel to the proper significant figure.   *Ans.* 1.80%

5. 50.00 ml of 0.1000 M calcium(II) is titrated with 0.1000 M EDTA under a buffered pH 10.0 condition. Using the simplest equations, determine the $[Ca^{2+}]$ and pCa at the following levels of EDTA solution addition in milliliters:

0.00   10.00   30.00   40.00   49.00   49.50   49.90   50.00   50.10   50.50   51.00

6. What would be the indicator blank and titration error where the indicator Eriochrome black T is used in the titration for Prob. 5? Since many such determinations of calcium(II) involve relatively dilute solutions, show the indicator blank and titration error for 50.00 ml of 0.0100 M calcium(II) vs. 0.0100 M EDTA using the same indicator. Comment on this entire situation.

7. Suppose that you were to add to 50.00 ml of a 0.0100-M solution of calcium(II) enough EDTA and magnesium(II) to make the solution, without volume change, 0.00010 M to each substance prior to titration at pH 10.0 with 0.0100 M EDTA. Calculate the indicator blank and titration error in the calcium determination where Eriochrome black T is used as the indicator in the subsequent titration. Comment on this situation relative to that of Prob. 6.   *Ans.* +0.01 ml, −0.2 ppt

8. Suppose that the titration given in Prob. 5 was carried out in (*a*) a solution buffered to pH 8.0 and (*b*) a solution buffered to pH 6.0. Determine in each case the [Ca^{2+}] and pCa for the following titration points in milliliters of EDTA solution:

    49.50   50.00   50.50

Use the simplest equations as being adequate for the purpose. What are your general conclusions?

9. A powder contains only a soluble sulfate salt and neutral salts. A sample weight of 1.2400 g is dissolved in water and subsequently treated with 50.00 ml of 0.2000-M $BaCl_2$ solution. The precipitate is eventually filtered off and washed. The filtrate is brought to the proper buffered pH condition and diluted to exactly 100.00 ml. A 50.00-ml aliquot is titrated under the proper conditions with 0.1000 M EDTA. The end point is reached at 22.35 ml of EDTA solution. Ignoring any indicator blank, report the percentage of sufate in the powder as Na_2SO_4.   *Ans.* 63.3%

10. A 50.00-ml sample of a natural water is titrated under the proper conditions with 0.0102-M EDTA solution. 29.85 ml of EDTA solution is required to reach the end point which, for this purpose, coincides with the equivalence point. What is the hardness of the water expressed as parts per million of $CaCO_3$?

REFERENCES

1. Kolthoff, I. M., P. J. Elving, and E. B. Sandell, "Treatise on Analytical Chemistry," Wiley, New York, 1959 (continuing series).
2. Kolthoff, I. M., E. B. Sandell, E. J. Meehan, and Stanley Bruckenstein, "Quantitative Chemical Analysis," 4th ed., Macmillan, New York, 1969.
3. Kolthoff, I. M., V. A. Stenger, and R. Belcher, "Volumetric Analysis," 3 vols., 2d ed., Interscience, New York, 1957.
4. Ringbom, A., "Complexation in Analytical Chemistry," Interscience, New York, 1963.
5. Schwarzenbach, G., "Complexometric Titrations," Methuen, London, 1960.

9
Oxidation-reduction Titrations

9.1 OXIDATION-REDUCTION REACTIONS

9.1.1 GENERAL

Oxidation-reduction, or redox, titrations involve titration reactions where there is a transfer of electrons between the reactant and titrant reaction systems. The balancing of such reactions was discussed in Chap. 2, while the general theory of oxidation-reduction equilibria was outlined in considerable detail in Chap. 4, Part 5.

As is generally the case where adaptability to volumetric titration analysis is concerned, many redox reactions could in theory be so applied, providing that equilibrium is established very rapidly following each addition of titrant, and providing that an indicator technique capable of locating the stoichiometric equivalence point with reasonable accuracy is available. Because of the large number of oxidation-reduction reactions that are capable of being so adapted, there are many more redox titration methods used in volumetric analysis than there are neutralization, precipitation, or complexation titration techniques.

Many standard redox titrations are carried out using indicator substances which, by color change situations, locate with good accuracy the equivalence points of such titrations. The two half reactions for any redox titration system are, however, in equilibrium at all points after the start of the titration, so that the reduction potentials for the half-cells involved are identical for all such points. Thus the cell potential, or E_{cell}, changes during the titration process and provides a characteristic and relatively large ΔE_{cell} in the neighborhood of the equivalence point for all titrations conducted on a sound basis of quantitative conversion. While it is possible to plot the logarithm of the reactant concentration reciprocal, or pR, vs. the volume of titrant to obtain a typical titration curve, it is also possible to obtain just as typical a curve by plotting E_{cell} vs. the volume of titrant.

The variation of E_{cell} with volume of titrant indicates that for most redox titration systems, a potentiometric technique of following the course of the titration would be most suitable, and this technique would then serve, in one way or another, to locate the equivalence point of the titration. The use of the potentiometric method eliminates the necessity of finding an indicator substance where the color change interval corresponds to the range of change of the reactant concentration, or the range of change of E_{cell}, around the equivalence point, thus rendering more general the adaptability of redox reactions to volumetric titration analysis.

The methods of potentiometric titration will be discussed fully in a subsequent chapter. For the moment we shall restrict our treatment of redox titrations to a general outline of the titration process and the application thereto of indicator end-point detection methods.

Redox titration reactions can in general be classed in two main groups, with three important subgroups for each main group. These, together with the associated typical characteristics, are given in the section following.

9.1.2 TITRATIONS WHERE H_3O^+ OR OH^- DO NOT PARTICIPATE DIRECTLY IN THE REDOX REACTION

1. An equimolar relationship exists between the reactant and product for both the oxidant and reductant half reactions. This implies that for a given $[H_3O^+]$, the value of E_{cell} at the titration equivalence point will generally be independent of the starting concentrations of the reactant and/or titrant solutions, and that the curve of E_{cell} vs. volume of titrant will be identical for titrations of $1\ M, 0.1\ M, 0.05\ M$, etc., concentrations of reactant and titrant. Variation in $[H_3O^+]$ in solution will, through the diverse-ion effect on $E^{\circ}_{f,\ Ox/Red}$ for each half-cell, result in variation in the value of E_{cell} at all titration points; but for the usual run of such titrations, these variations will not be significant relative to the magnitude of ΔE_{cell} around the equivalence point in each case. In general, then, E_{cell} for all comparable titration points will be independent of both $[H_3O^+]$ and the starting concentrations of reactant and/or titrant.

An equimolar relationship exists between the oxidant and reductant substances. This implies that the curve of E_{cell} vs. volume of titrant will be symmetrical around the equivalence point.

Example

$$Fe^{2+} + Ce^{4+} \rightleftharpoons Fe^{3+} + Ce^{3+}$$

2. An equimolar relationship exists between the reactant and product for both the oxidant and reductant half reactions. This provides for the characteristic that E_{cell} for all comparable titration points will in general be independent of both $[H_3O^+]$ and the starting concentrations of reactant and/or titrant.

   A nonequimolar relationship exists between the oxidant and the reductant substances. This implies a curve of E_{cell} vs. volume of titrant that will be asymmetrical around the equivalence point.

Example

$$Sn^{2+} + 2Fe^{3+} \rightleftharpoons Sn^{4+} + 2Fe^{2+}$$

3. A nonequimolar relationship exists between the reactant and product for one or both of the oxidant and reductant half reactions. This implies that although E_{cell} for comparable titration points will generally be independent of $[H_3O^+]$, it will be dependent on the starting concentrations of reactant and/or titrant.

   A nonequimolar relationship exists between the oxidant and reductant substances. This again implies that the curve of E_{cell} vs. volume of titrant will be asymmetrical around the equivalence point.

Example

$$2S_2O_3{}^{2-} + I_2 \rightleftharpoons S_4O_6{}^{2-} + 2I^-$$

9.1.3   TITRATIONS WHERE H₃O⁺ OR OH⁻ DO PARTICIPATE DIRECTLY IN THE REDOX REACTION

1. An equimolar relationship exists between the reactant and product for both the oxidant and reductant half reactions. This implies that for a given $[H_3O^+]$ in solution, the value of E_{cell} at the equivalence point will generally be independent of the starting concentrations of reactant and/or titrant solutions, and the curve of E_{cell} vs. volume of titrant will be identical for titrations of $1\,M$, $0.1\,M$, $0.05\,M$, etc., concentrations of reactant and titrant.

   Since the extent to which the main redox reaction goes to quantitative completion will depend on the $[H_3O^+]$, the value of E_{cell} at all comparable titration points will be dependent on the value of $[H_3O^+]$. For all such titrations there will be a critical value of $[H_3O^+]$ beyond which the main redox reaction will not attain a quantitative conversion situation at the titration equivalence point.

   An equimolar relationship exists between the oxidant and reductant substances. This implies that the curve of E_{cell} vs. volume of titrant will be symmetrical around the equivalence point.

Example

$$Fe^{2+} + VO_4{}^{3-} + 6H_3O^+ \rightleftharpoons Fe^{3+} + VO^{2+} + 9H_2O$$

2. An equimolar relationship exists between the reactant and product for both the oxidant and reductant half reactions. For a given value of $[H_3O^+]$ in solution, E_{cell} for all comparable titration points will in general be independent of the starting concentrations of reactant and/or titrant. Again the value of E_{cell} for all comparable titration points will be dependent on the $[H_3O^+]$ in the solution, and there will be a critical $[H_3O^+]$ beyond which the main redox reaction will not attain a quantitative conversion situation at the titration equivalence point.

A nonequimolar relationship exists between the oxidant and reductant substances, and the curve of E_{cell} vs. volume of titrant will be asymmetrical around the equivalence point.

Example

$$5Fe^{2+} + MnO_4^- + 8H_3O^+ \rightleftharpoons 5Fe^{3+} + Mn^{2+} + 12H_2O$$

3. A nonequimolar relationship exists between the reactant and product for one or both of the oxidant and reductant half reactions. For a given $[H_3O^+]$ in solution, the value of E_{cell} at all comparable titration points will be dependent on the starting concentrations of reactant and/or titrant. Again the value of E_{cell} for all comparable titration points will be dependent on the $[H_3O^+]$ in the solution, and there will be a critical $[H_3O^+]$ beyond which the main redox reaction will fail to attain a quantitative conversion situation at the titration equivalence point.

A nonequimolar relationship exists between the oxidant and reductant substances, and the curve of E_{cell} vs. volume of titrant will be asymmetrical around the equivalence point.

Example

$$6Fe^{2+} + Cr_2O_7^{2-} + 14H_3O^+ \rightleftharpoons 6Fe^{3+} + 2Cr^{3+} + 21H_2O$$

9.2 TYPICAL OXIDATION-REDUCTION TITRATION SYSTEMS

9.2.1 GENERAL

We shall not attempt a full discussion of a representative system for each of the six important subgroups outlined in Secs. 9.1.2 and 9.1.3. We shall discuss two subgroup systems from each main group, and it will be understood that, in general, the treatment of any other systems within each main group will follow along similar lines.

For *all* six of the important subgroup systems, however, we shall show tables providing typical titration data and the associated curves for these data. Where tabled data covers a system *not* discussed in detail in the text, it is suggested that the student verify the data for the various titration points by making the appropriate calculations.

9.2.2 H_3O^+ DOES NOT PARTICIPATE DIRECTLY IN THE REDOX REACTION

The titration of iron(II) with cerium(IV)   This is a redox titration typical of subgroup 1 in Sec. 9.1.2. An acid solution of iron(II) is titrated with a standard solution of cerium(IV). The titration reaction

$$Fe^{2+} + Ce^{4+} \rightleftharpoons Fe^{3+} + Ce^{3+}$$

corresponds in form to Eq. (136), where

$$n_1 = n_2 = 1 \qquad a = b = c = d = 1$$

The acid medium is required to prevent hydrolysis of the reactants and products, so that $[H_3O^+]$ affects the value of $E^\circ_{f,Ox/Red}$, but H_3O^+ does not participate directly in the redox reaction.

Consider the titration of 50.00 ml of 0.1000 M Fe^{2+} with 0.1000 M Ce^{4+} in a medium 0.5 M with respect to H_2SO_4. The values of $E^\circ_{f,Fe^{3+}/Fe^{2+}}$ and $E^\circ_{f,Ce^{4+}/Ce^{3+}}$ at 25°C in this medium can be taken as 0.700 and 1.459 V, respectively, relative to the SHE.

The equilibrium constant is given by Eq. (140):

$$\frac{[Fe^{3+}][Ce^{3+}]}{[Fe^{2+}][Ce^{4+}]} = 10^{12.9} = 7.9 \times 10^{12}$$

At the titration equivalence point we have from Eq. (144)

$$\frac{[Fe^{3+}]^2}{[Fe^{2+}]^2} = 7.9 \times 10^{12}$$

$$\frac{[Fe^{3+}]}{[Fe^{2+}]} = 2.8 \times 10^{6}$$

Thus, at the equivalence point for the 0.1000 M reactant and titrant concentrations titration, we have

$$[Fe^{3+}] + [Fe^{2+}] = 0.0500 \ M$$
$$[Fe^{3+}] = 2.8 \times 10^6 [Fe^{2+}]$$

so that

$$[Fe^{2+}] = 1.8 \times 10^{-8} \ M$$

and this represents a quantitative conversion of Fe^{2+} to Fe^{3+} at the equivalence point. For such quantitatively complete reactions, the calculation of $[Fe^{2+}]$ and E_{cell} at the various titration points proceeds generally according to the following. At the start of the titration, the value of $[Fe^{2+}]$ in solution is given by

$$[Fe^{2+}] = 1.00 \times 10^{-1} \ M$$

while the value of E_{cell} is given by

$$E_{\text{cell}} = E^{\circ}_{f,\text{Fe}^{3+}/\text{Fe}^{2+}} + 0.059 \log \frac{[\text{Fe}^{3+}]}{[\text{Fe}^{2+}]}$$

Since the value of $[\text{Fe}^{3+}]$ before the titration starts is not precisely known (it will depend on the method of preparing the solution of Fe^{2+}, the extent to which oxidation of Fe^{2+} by air has taken place, etc.), it is customary to leave as undefined the value of E_{cell} at the start.

At any titration point after the start and up to but not including the equivalence point, the value of $[\text{Fe}^{2+}]$ can be determined exactly from Eq. (176). A method of determining whether or not this exact equation is required was outlined in the Examples in Sec. 5.1. In most cases the simpler Eq. (175) will be adequate. Again, for such titration points, the value of E_{cell} will be found from

$$E_{\text{cell}} = E^{\circ}_{f,\text{Fe}_{3+}/\text{Fe}_{2+}} + 0.059 \log \frac{[\text{Fe}^{3+}]}{[\text{Fe}^{2+}]}$$

where $[\text{Fe}^{3+}]$ can be found exactly from

$$[\text{Fe}^{3+}] = [\text{Ce}^{3+}] = \frac{V_{\text{Ce}^{4+}} M_{\text{Ce}^{4+}}}{V_{\text{Fe}^{2+}} + V_{\text{Ce}^{4+}}} - [\text{Ce}^{4+}]$$

$$= \frac{V_{\text{Ce}^{4+}} M_{\text{Ce}^{4+}}}{V_{\text{Fe}^{2+}} + V_{\text{Ce}^{4+}}} - \frac{[\text{Ce}^{3+}][\text{Fe}^{3+}]}{K_{\text{eq}}[\text{Fe}^{2+}]}$$

Usually $[\text{Ce}^{4+}]$ will be small enough to be ignored, and we have

$$[\text{Fe}^{3+}] = [\text{Ce}^{3+}] \approx \frac{V_{\text{Ce}^{4+}} M_{\text{Ce}^{4+}}}{V_{\text{Fe}^{2+}} + V_{\text{Ce}^{4+}}}$$

Thus, at the 10.00-ml point of added Ce^{4+} solution, we have

$$[\text{Fe}^{2+}] \approx \frac{50.00 \times 0.1000 - 10.00 \times 0.1000}{50.00 + 10.00} \approx 6.67 \times 10^{-2} \, M$$

$$[\text{Fe}^{3+}] = [\text{Ce}^{3+}] \approx \frac{10.00 \times 0.1000}{50.00 + 10.00} \approx 1.67 \times 10^{-2} \, M$$

$$E_{\text{cell}} = 0.700 + 0.059 \log \frac{[\text{Fe}^{3+}]}{[\text{Fe}^{2+}]}$$

$$\approx 0.66^5 \, V$$

The value of $[\text{Fe}^{2+}]$ at the equivalence point is $1.8 \times 10^{-8} \, M$, found by the method already outlined, and the value of E_{cell} at this point is given by Eq. (147) as

$$E_{\text{cell}} = \frac{1.459 + 0.700}{2} = 1.080 \, V$$

For titration points after the equivalence point, the value of E_{cell} is determined from

$$E_{\text{cell}} = E^{\circ}_{f,\text{Ce}^{4+}/\text{Ce}^{3+}} + 0.059 \log \frac{[\text{Ce}^{4+}]}{[\text{Ce}^{3+}]}$$

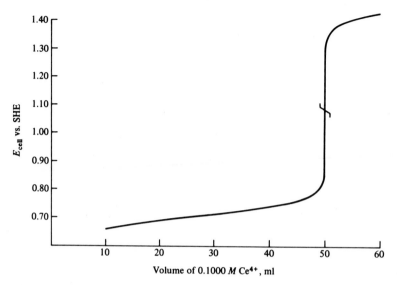

Fig. 9.1 Titration of 50.00 ml of 0.1000 M Fe^{2+} vs. 0.1000 M Ce^{4+} in 0.5 M H_2SO_4.

where $[Ce^{4+}]$ can be determined exactly from Eq. (180) and more usually from the simpler Eq. (179). The $[Ce^{3+}]$ value can generally be determined from the simple expression

$$[Fe^{3+}] = [Ce^{3+}] \approx \frac{V_{Fe^{2+}} M_{Fe^{2+}}}{V_{Ce^{4+}} + V_{Fe^{2+}}}$$

The value of E_{cell} determined in this way is then substituted in the expression

$$E_{cell} = E^{\circ}_{f,Fe^{3+}/Fe^{2+}} + 0.059 \log \frac{[Fe^{3+}]}{[Fe^{2+}]}$$

and the value of $[Fe^{2+}]$ for these titration points is determined. Thus, for the point where 60.00 ml of Ce^{4+} solution has been added, we have

$$[Ce^{4+}] \approx \frac{60.00 \times 0.1000 - 50.00 \times 0.1000}{60.00 + 50.00} \approx 9.09 \times 10^{-3}\,M$$

$$[Fe^{3+}] = [Ce^{3+}] \approx \frac{50.00 \times 0.1000}{60.00 + 50.00} \approx 4.54 \times 10^{-2}\,M$$

$$E_{cell} = 1.459 + 0.059 \log \frac{[Ce^{4+}]}{[Ce^{3+}]}$$

$$\approx 1.41^8\,V$$

and for $[Fe^{2+}]$ at this point

$$1.41^8 \approx 0.700 + 0.059 \log \frac{[Fe^{3+}]}{[Fe^{2+}]} \qquad \text{and} \qquad [Fe^{2+}] \approx 3.1 \times 10^{-14}\,M$$

Table 9.1 shows data typical of the titration involved. It will be noted that the $\Delta pFe(II)$ for ± 0.05 ml around the equivalence-point volume is 6.86, corresponding to a ΔE_{cell} of 0.40 V.

Figure 9.1 shows the titration curve for E_{cell} vs. the volume of 0.1000 M Ce^{4+} solution. The symmetry of the curve around the equivalence point will be noted.

Table 9.1 Titration of iron(II) by cerium(IV)*
50.00 ml of 0.1000 M Fe^{2+} vs. 0.1000 M Ce^{4+} in 0.5 M H_2SO_4

Volume Ce^{4+}, ml	$[Fe^{2+}]$	pFe(II)	E_{cell} vs. SHE	$(\Delta E_{cell}/\Delta V_{Ce^{4+}}) \times 10^3$ 0.1 ml = $\Delta V_{Ce^{4+}}$		$(\Delta^2 E_{cell}/\Delta V^2_{Ce^{4+}}) \times 10^3$ 0.1 ml/0.1 ml	
0.00	1.00×10^{-1}	1.00					
10.00	6.67×10^{-2}	1.18	0.66				
20.00	4.28×10^{-2}	1.37	0.69				
				+0.20	(25.00)		
30.00	2.50×10^{-2}	1.60	0.71			+0.0010	(30.00)
				+0.30	(35.00)		
40.00	1.11×10^{-2}	1.96	0.74			+0.0013	(38.75)
				+0.40	(42.50)		
45.00	5.26×10^{-3}	2.28	0.76			+0.013	(44.75)
				+1.0	(47.00)		
49.00	1.01×10^{-3}	3.00	0.80			+0.13	(48.12)
				+4.0	(49.25)		
49.50	5.02×10^{-4}	3.30	0.82			+1.3	(49.48)
				+10	(49.70)		
49.90	1.00×10^{-4}	4.00	0.86			+14	(49.81)
				+40	(49.92)		
49.95	5.00×10^{-5}	4.30	0.88			+720	(49.95)
				+400	(49.98)		
50.00	1.8×10^{-8}	7.74	1.08			0	(50.00)
				+400	(50.02)		
50.05	6.9×10^{-12}	11.16	1.28			−720	(50.05)
				+40	(50.08)		
50.10	3.4×10^{-12}	11.47	1.30			−14	(50.19)
				+10	(50.30)		
50.50	6.7×10^{-13}	12.17	1.34			−1.3	(50.52)
				+4.0	(50.75)		
51.00	3.2×10^{-13}	12.50	1.36			−0.13	(51.88)
				+1.0	(53.00)		
55.00	6.6×10^{-14}	13.18	1.40			−0.013	(55.25)
				+0.40	(57.50)		
60.00	3.1×10^{-14}	13.51	1.42				

*$[Fe^{2+}]$ and pFe(II) are not corrected for the dilution effect. The values in parentheses after each $\Delta E_{cell}/\Delta V_{Ce^{4+}}$ and $\Delta^2 E_{cell}/\Delta V^2_{Ce^{4+}}$ value associate the average slope, or rate of change of the slope, with the average volume value between the titration points involved. The equivalence-point volume is 50.00 ml of 0.1000 M Ce^{4+} solution.

Figure 9.2 shows the curves for pFe(II) vs. the volume of Ce^{4+} solution for titrations involving $0.1000\ M$, $0.0100\ M$, and $0.0010\ M$ concentrations of Fe^{2+} and Ce^{4+}. Note the shift in the curves resulting from the changes in the starting concentrations of reactant and titrant. E_{cell} vs. volume of Ce^{4+} yields, of course, identical curves for these titrations.

The titration of tin(II) with iron(III)   This redox titration is typical of subgroup 2 in Sec. 9.1.2. An acid solution of tin(II) is titrated with a standard solution of iron(III) under an inert gas. The titration reaction

$$Sn^{2+} + 2Fe^{3+} \rightleftharpoons Sn^{4+} + 2Fe^{2+}$$

corresponds in form to Eq. (136), where

$$n_1 = 2 \qquad n_2 = 1 \qquad a = b = c = d = 1$$

The acid medium is again required to prevent hydrolysis of the reactants and products. $[H_3O^+]$ affects only the $E^\circ_{f,Ox/Red}$ values and H_3O^+ does not participate directly in the redox reaction. The inert gas is required to prevent air oxidation of Sn^{2+} and Fe^{2+}. Table 9.2 shows typical data for the titration of 50.00 ml of $0.0500\ M$ Sn^{2+} with $0.1000\ M$ Fe^{3+} in a medium 2 M to HCl. The values of $E^\circ_{f,Fe^{3+}/Fe^{2+}}$ and $E^\circ_{f,Sn^{4+}/Sn^{2+}}$ at 25°C in this medium can be taken as 0.700 and 0.139 V, respectively, relative to the SHE.

   Note that the value of $\Delta pSn(II)$ for ± 0.05 ml around the equivalence-point volume is 10.00, involving 3.36 units for 0.05 ml before the equivalence point to

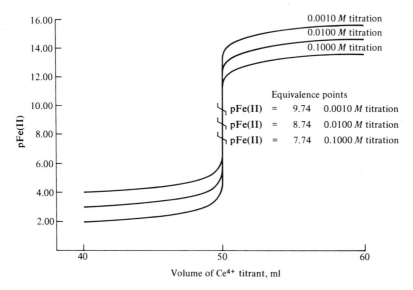

Fig. 9.2   Titrations of 50.00 ml Fe^{2+} vs. Ce^{4+} in 0.5 M H_2SO_4.

Table 9.2 Titration of tin(II) by iron(III)*
50.00 ml of 0.0500 M Sn^{2+} vs. 0.1000 M Fe^{3+} in 2 M HCl

Volume Fe³⁺, ml	[Sn²⁺]	pSn(II)	E_{cell} vs. SHE	$(\Delta E_{cell}/\Delta V_{Fe^{3+}}) \times 10^3$ 0.1 ml $= \Delta V_{Fe^{3+}}$	$(\Delta^2 E_{cell}/\Delta V^2_{Fe^{3+}}) \times 10^3$ 0.1 ml/0.1 ml
0.00	5.00×10^{-2}	1.30			
10.00	3.33×10^{-2}	1.48	0.12		
20.00	2.14×10^{-2}	1.67	0.13		
30.00	1.25×10^{-2}	1.90	0.14		
40.00	5.55×10^{-3}	2.26	0.16		
				+0.20 (42.50)	
45.00	2.63×10^{-3}	2.58	0.17		+0.0067 (44.75)
				+0.50 (47.00)	
49.00	5.05×10^{-4}	3.30	0.19		+0.067 (48.12)
				+2.0 (49.25)	
49.50	2.51×10^{-4}	3.60	0.20		+0.67 (49.48)
				+5.0 (49.70)	
49.90	5.00×10^{-5}	4.30	0.22		+6.7 (49.81)
				+20 (49.92)	
49.95	2.50×10^{-5}	4.60	0.23		+360 (49.95)
				+200 (49.98)	
50.00	1.1×10^{-8}	7.96	0.33		+360 (50.00)
				+380 (50.02)	
50.05	2.5×10^{-15}	14.60	0.52		−680 (50.05)
				+40 (50.08)	
50.10	5.8×10^{-16}	15.24	0.54		−14 (50.19)
				+10 (50.30)	
50.50	2.5×10^{-17}	16.60	0.58		−1.3 (50.52)
				+4.0 (50.75)	
51.00	5.8×10^{-18}	17.24	0.60		−0.13 (51.88)
				+1.0 (53.00)	
55.00	2.4×10^{-19}	18.62	0.64		−0.013 (55.25)
				+0.40 (57.50)	
60.00	5.3×10^{-20}	19.28	0.66		

*[Sn²⁺] and pSn(II) are not corrected for the dilution effect. The values in parentheses after each $\Delta E_{cell}/\Delta V_{Fe^{3+}}$ and $\Delta^2 E_{cell}/\Delta V^2_{Fe^{3+}}$ value associate the average slope, or rate of change of the slope, with the average volume value between the titration points involved. The equivalence-point volume is 50.00 ml of 0.1000 M Fe^{3+} solution.

the equivalence point, and 6.64 units for the equivalence point to 0.05 ml after. The corresponding values for E_{cell} are +0.29, +0.10, and +0.19 V. The asymmetry around the equivalence point is further exemplified by Fig. 9.3, which shows the curve for E_{cell} vs. Fe^{3+} solution volume.

The titration of iodine with thiosulfate This redox titration is typical of subgroup 3 in Sec. 9.1.2. A slightly acid solution of iodine, I_2, containing iodide ion, is titrated with a standard solution of sodium thiosulfate, $Na_2 S_2 O_3$. Although the

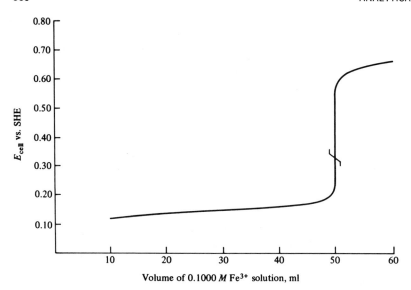

Fig. 9.3 Titration of 50.00 ml of 0.0500 M Sn^{2+} vs. 0.1000 M Fe^{3+} in 2 M HCl.

iodine is present in solution as the triiodide ion, I_3^-, we shall, for the moment, consider the solution as involving only iodine. The titration reaction

$$2S_2O_3^{2-} + I_2 \rightleftharpoons S_4O_6^{2-} + 2I^-$$

again corresponds in form to Eq. (136), where

$$n_1 = n_2 = 2 \qquad a = d = 1 \qquad b = c = 2$$

H_3O^+ does not participate directly in the redox reaction, and $[H_3O^+]$ affects only the values of $E^\circ_{f,Ox/Red}$.

Table 9.3 shows typical titration data for such a system. Note that ΔpI_2 for 0.05 ml before the equivalence point to the equivalence point is 1.91, that for the equivalence point to 0.05 ml after is 3.79, giving a ΔpI_2 for ± 0.05 ml around the equivalence point of 5.70. The corresponding values for ΔE_{cell} are -0.05, -0.11, and -0.16 V. The asymmetry will be noted.

Figure 9.4 shows the curve of E_{cell} vs. volume of $Na_2S_2O_3$ solution for the titration under discussion, as well as that for a titration involving 0.0050 M I_2 and 0.0100 M $Na_2S_2O_3$. The displacement of the curve with changing starting concentration of reactant and titrant is well marked, as is the asymmetry around the equivalence point.

9.2.3 H_3O^+ DOES PARTICIPATE DIRECTLY IN THE REDOX REACTION

The titration of vanadate with iron(II) This is a redox titration typical of subgroup 1 in Sec. 9.1.3. An acid solution of vanadate ion, VO_4^{3-}, is titrated with

Table 9.3 Titration of iodine by thiosulfate*

50.00 ml of 0.0500 M I_2 vs. 0.1000 M $Na_2S_2O_3$ at pH $= 6.0$ and 10^{-2} M I^-

Volume $Na_2S_2O_3$, ml	$[I_2]$	pI_2	E_{cell} vs. SHE	$(\Delta E_{cell}/\Delta V_{S_2O_3{}^{2-}}) \times 10^3$ $0.1\ ml = \Delta V_{S_2O_3{}^{2-}}$	$(\Delta^2 E_{cell}/\Delta_{S_2O_3{}^{2-}}^2) \times 10^3$ $0.1\ ml/0.1\ ml$
0.00	5.00×10^{-2}	1.30			
10.00	3.33×10^{-2}	1.48	0.60		
20.00	2.14×10^{-2}	1.67	0.58		
30.00	1.25×10^{-2}	1.90	0.56		
				-0.10 (35.00)	
40.00	5.55×10^{-3}	2.26	0.55		-0.0013 (38.75)
				-0.20 (42.50)	
45.00	2.63×10^{-3}	2.58	0.54		-0.0067 (44.75)
				-0.50 (47.00)	
49.00	5.05×10^{-4}	3.30	0.52		-0.16 (48.12)
				-4.0 (49.25)	
49.50	2.51×10^{-4}	3.60	0.50		-0.22 (49.48)
				-5.0 (49.70)	
49.90	5.00×10^{-5}	4.30	0.48		-6.6 (49.81)
				-20 (49.92)	
49.95	2.50×10^{-5}	4.60	0.47		-160 (49.95)
				-100 (49.98)	
50.00	3.1×10^{-7}	6.51	0.42		-240 (50.00)
				-220 (50.02)	
50.05	5.0×10^{-11}	10.30	0.31		$+360$ (50.05)
				-40 (50.08)	
50.10	1.4×10^{-12}	11.85	0.29		$+13$ (50.19)
				-10 (50.30)	
50.50	4.6×10^{-13}	12.34	0.25		$+1.3$ (50.52)
				-4.0 (50.75)	
51.00	1.1×10^{-13}	12.96	0.23		$+0.13$ (51.88)
				-1.0 (53.00)	
55.00	4.5×10^{-15}	14.35	0.19		$+0.013$ (55.25)
				-0.40 (57.50)	
60.00	1.0×10^{-15}	15.00	0.17		

*$[I_2]$ and pI_2 are not corrected for the dilution effect. The values in parentheses after each $\Delta E_{cell}/\Delta V_{S_2O_3{}^{2-}}$ and $\Delta^2 E_{cell}/\Delta V_{S_2O_3{}^{2-}}^2$ value associate the average slope, or rate of change of the slope, with the average volume value between the titration points involved. The equivalence-point volume is 50.00 ml of 0.1000 M $Na_2S_2O_3$ solution.

a standard solution of ferrous sulfate, $FeSO_4$, or ferrous ammonium sulfate, $FeSO_4 \cdot (NH_4)_2SO_4$. The titration reaction

$$Fe^{2+} + VO_4{}^{3-} + 6H_3O^+ \rightleftharpoons Fe^{3+} + VO^{2+} + 9H_2O$$

corresponds in form to Eq. (148), where

$$n_1 = n_2 = 1 \qquad a = b = c = d = 1 \qquad y = 6 \qquad w = 9$$

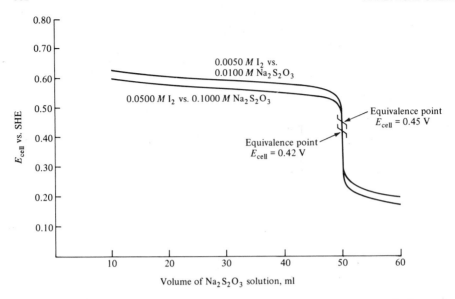

Fig. 9.4 Titrations of 50.00 ml I_2 solutions vs. $Na_2S_2O_3$ solutions at pH 6.0 and $[I^-] = 10^{-2}\ M$.

H_3O^+ is essential if the reaction

$$VO_4^{3-} + 6H_3O^+ + 1e \rightleftharpoons VO^{2+} + 9H_2O$$

is to be driven to the right, and without a sufficiently high $[H_3O^+]$ in solution, this reaction and the redox reaction for the titration can not proceed to quantitative completion at the titration equivalence point. Thus, regardless of the minor effect of $[H_3O^+]$ on $E^\circ_{f,Ox/Red}$, H_3O^+ participates directly in the redox reaction.

Consider the titration of 50.00 ml of 0.1000 M VO_4^{3-} with 0.1000 M Fe^{2+} in a solution 5.0 M to H_2SO_4. The values of $E^\circ_{f,VO_4^{3-}/VO^{2+}}$ and $E^\circ_{f,Fe^{3+}/Fe^{2+}}$ at 25°C can be taken as 1.200 and 0.700 V, respectively, relative to the SHE.

The equilibrium constant is given by Eq. (150) as

$$\frac{[VO^{2+}][Fe^{3+}][H_2O]^9}{[VO_4^{3-}][Fe^{2+}][H_3O^+]^6} = 10^{8.5}$$

At 5.0 M H_2SO_4 the value of $[H_3O^+]$ can be taken as approximately 5 M, so that we have

$$\frac{[VO^{2+}][Fe^{3+}]}{[VO_4^{3-}][Fe^{2+}]} = 10^{8.5}\,[H_3O^+]^6 = 5.0 \times 10^{12}$$

At the equivalence point of the titration we have, from Eq. (156),

$$\frac{[VO^{2+}]^2}{[VO_4^{3-}]^2} = 5.0 \times 10^{12} \quad \text{and} \quad \frac{[VO^{2+}]}{[VO_4^{3-}]} = 2.2 \times 10^6$$

Thus, at the equivalence point, we have

$$[VO^{2+}] + [VO_4^{3-}] = 0.0500 \, M \qquad \text{and} \qquad [VO^{2+}] = 2.2 \times 10^6 \, [VO_4^{3-}]$$

so that

$$[VO_4^{3-}] = 2.3 \times 10^{-8} \, M$$

This represents a quantitative conversion of VO_4^{3-} to VO^{2+} at the equivalence point, and for such quantitatively complete reactions, the following outlines the calculations for various titration points.

At the start of the titration $[VO_4^{3-}] = 1.00 \times 10^{-1} \, M$, with E_{cell} undefined. At the 10.00-ml point of added Fe^{2+} solution,

$$[VO_4^{3-}] \approx \frac{50.00 \times 0.1000 - 10.00 \times 0.1000}{50.00 + 10.00} \approx 6.67 \times 10^{-2} \, M$$

$$[VO^{2+}] = [Fe^{3+}] \approx \frac{10.00 \times 0.1000}{50.00 + 10.00} \approx 1.67 \times 10^{-2} \, M$$

$$E_{cell} = 1.200 + 0.059 \log \frac{[VO_4^{3-}][H_3O^+]^6}{[VO^{2+}]}$$

$$\approx 1.48^2 \, V$$

At the equivalence point, the $[VO_4^{3-}]$ is $2.3 \times 10^{-8} \, M$, and E_{cell} is given by Eq. (158) as

$$2E_{cell} = 1.200 + 0.700 + 0.059 \log [H_3O^+]^6$$

$$E_{cell} = 1.07^4 \, V$$

At the point where 60.00 ml of Fe^{2+} solution has been added, we have

$$[Fe^{2+}] \approx \frac{60.00 \times 0.1000 - 50.00 \times 0.1000}{60.00 + 50.00} \approx 9.09 \times 10^{-3} \, M$$

$$[Fe^{3+}] = [VO^{2+}] \approx \frac{50.00 \times 0.1000}{60.00 + 50.00} \approx 4.54 \times 10^{-2} \, M$$

$$E_{cell} = 0.700 + 0.059 \log \frac{[Fe^{3+}]}{[Fe^{2+}]}$$

$$\approx 0.74^1 \, V$$

and for $[VO_4^{3-}]$ at this point,

$$0.74^1 \approx 1.200 + 0.059 \log \frac{[VO_4^{3-}][H_3O^+]^6}{[VO^{2+}]}$$

and

$$[VO_4^{3-}] \approx 4.9 \times 10^{-14} \, M$$

Table 9.4 shows data typical of this titration. The ΔpVO_4 for ± 0.05 ml around the equivalence point is 6.66, corresponding to a ΔE_{cell} value of $-0.39 \, V$. Figure 9.5

Table 9.4   Titration of vanadate by iron(II)*

50.00 ml of 0.1000 M VO_4^{3-} vs. 0.1000 M Fe^{2+} in 5.0 M H_2SO_4

Volume Fe^{2+}, ml	$[VO_4^{3-}]$	pVO_4	E_{cell} vs. SHE	$(\Delta E_{cell}/\Delta V_{Fe^{2+}}) \times 10^3$ 0.1 $ml = \Delta V_{Fe^{2+}}$		$(\Delta^2 E_{cell}/\Delta V_{Fe^{2+}}^2) \times 10^3$ 0.1 $ml/0.1$ ml	
0.00	1.00×10^{-1}	1.00					
10.00	6.67×10^{-2}	1.18	1.48				
20.00	4.28×10^{-2}	1.37	1.46				
				-0.20	(25.00)		
30.00	2.50×10^{-2}	1.60	1.44			-0.0010	(30.00)
				-0.30	(35.00)		
40.00	1.11×10^{-2}	1.96	1.41			-0.0013	(38.75)
				-0.40	(42.50)		
45.00	5.26×10^{-3}	2.28	1.39			-0.013	(44.75)
				-1.0	(47.00)		
49.00	1.01×10^{-3}	3.00	1.35			-0.13	(48.12)
				-4.0	(49.25)		
49.50	5.02×10^{-4}	3.30	1.33			-1.3	(49.48)
				-10	(49.70)		
49.90	1.00×10^{-4}	4.00	1.29			-14	(49.81)
				-40	(49.92)		
49.95	5.00×10^{-5}	4.30	1.27			-720	(49.95)
				-400	(49.98)		
50.00	2.3×10^{-8}	7.64	1.07			$+40$	(50.00)
				-380	(50.02)		
50.05	1.1×10^{-11}	10.96	0.88			$+680$	(50.05)
				-40	(50.08)		
50.10	5.4×10^{-12}	11.27	0.86			$+14$	(50.19)
				-10	(50.30)		
50.50	1.1×10^{-12}	11.96	0.82			$+1.3$	(50.52)
				-4.0	(50.75)		
51.00	5.3×10^{-13}	12.28	0.80			$+0.13$	(51.88)
				-1.0	(53.00)		
55.00	1.0×10^{-13}	13.00	0.76			$+0.013$	(55.25)
				-0.40	(57.50)		
60.00	4.9×10^{-14}	13.31	0.74				

* $[VO_4^{3-}]$ and pVO_4 are not corrected for the dilution effect. The values in parentheses after each $\Delta E_{cell}/\Delta V_{Fe^{2+}}$ and $\Delta^2 E_{cell}/\Delta V_{Fe^{2+}}^2$ value associate the average slope, or rate of change of the slope, with the average volume value between the titration points involved. The equivalence-point volume is 50.00 ml of 0.1000 M Fe^{2+} solution.

shows the curve for E_{cell} vs. volume of Fe^{2+} solution, and it should be noted that the curve is symmetrical around the equivalence point.

The minimum $[H_3O^+]$ for a quantitative conversion of VO_4^{3-} to VO^{2+} at the equivalence point of this titration can be found by modifying Eq. (156). Thus, for a 0.1000-M titration, and in consideration of 10^{-6} M as the maximum allow-

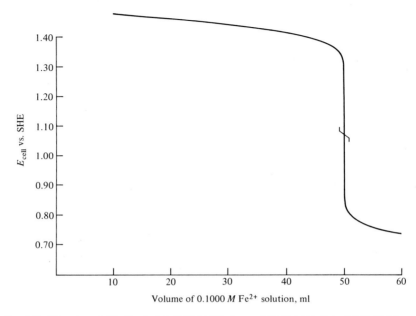

Fig. 9.5 Titration of 50.00 ml of 0.1000 M VO_4^{3-} vs. 0.1000 M Fe^{2+} in 5.0 M H_2SO_4.

able $[VO_4^{3-}]$ at the equivalence point, we have

$$\frac{[VO^{2+}]^2}{[VO_4^{3-}]^2} = 10^{8.5}[H_3O^+]^6 \quad \text{and} \quad \frac{(0.0500 \, M)^2}{(10^{-6} \, M_{max})^2} = 10^{8.5}[H_3O^+]_{min}^6$$

$$[H_3O^+]_{min} = 1.4 \, M$$

Titrations of vanadate ion and iron(II) are normally carried out in media with $[H_3O^+]$ very significantly higher than the calculated minimum value.

The titration of iron(II) with permanganate This redox titration is typical of subgroup 2 in Sec. 9.1.3. An acid solution of iron(II) is titrated with a standard solution of potassium permanganate, $KMnO_4$. The titration reaction

$$5Fe^{2+} + MnO_4^- + 8H_3O^+ \rightleftharpoons 5Fe^{3+} + Mn^{2+} + 12H_2O$$

corresponds in form to Eq. (148), where

$$n_1 = 1 \quad n_2 = 5 \quad a = b = c = d = 1 \quad y = 8 \quad w = 12$$

Again H_3O^+ is essential if the reaction

$$MnO_4^- + 8H_3O^+ + 5e \rightleftharpoons Mn^{2+} + 12H_2O$$

is to go to practical completion, and if a sufficiently high $[H_3O^+]$ does not exist in

Table 9.5 Titration of iron(II) by potassium permanganate*
50.00 ml of 0.1000 M Fe^{2+} vs. 0.0200 M $KMnO_4$ at pH = 1.00

Volume $KMnO_4$, ml	[Fe^{2+}]	pFe(II)	E_{cell} vs. SHE	$(\Delta E_{cell}/\Delta V_{MnO_4^-}) \times 10^3$ 0.1 ml = $\Delta V_{MnO_4^-}$	$(\Delta^2 E_{cell}/\Delta V^2_{MnO_4^-}) \times 10^3$ 0.1 ml/0.1 ml
0.00	1.00×10^{-1}	1.00			
10.00	6.67×10^{-2}	1.18	0.74		
20.00	4.28×10^{-2}	1.37	0.76		
30.00	2.50×10^{-2}	1.60	0.78		
				+0.20 (35.00)	
40.00	1.11×10^{-2}	1.96	0.80		+0.0053 (38.75)
				+0.60 (42.50)	
45.00	5.26×10^{-3}	2.28	0.83		+0.0088 (44.75)
				+1.0 (47.00)	
49.00	1.01×10^{-3}	3.00	0.87		+0.13 (48.12)
				+4.0 (49.25)	
49.50	5.02×10^{-4}	3.30	0.89		+1.3 (49.48)
				+10 (49.70)	
49.90	1.00×10^{-4}	4.00	0.93		+13 (49.81)
				+40 (49.92)	
49.95	5.00×10^{-5}	4.30	0.95		+1360 (49.95)
				+720 (49.98)	
50.00	3.8×10^{-11}	10.42	1.31		−1160 (50.00)
				+140 (50.02)	
50.05	2.3×10^{-12}	11.64	1.38^0		−264 (50.05)
				+8.0 (50.08)	
50.10	2.0×10^{-12}	11.70	1.38^4		−2.6 (50.19)
				+2.0 (50.30)	
50.50	1.4×10^{-12}	11.85	1.39^2		−0.26 (50.52)
				+0.80 (50.75)	
51.00	1.2×10^{-12}	11.92	1.39^6		−0.026 (51.88)·
				+0.20 (53.00)	
55.00	8.7×10^{-13}	12.06	1.40^4		−0.0031 (55.25)
				+0.060 (57.50)	
60.00	7.2×10^{-13}	12.14	1.40^7		

*[Fe^{2+}] and pFe(II) are not corrected for the dilution effect. The values in parentheses after each $\Delta E_{cell}/\Delta V_{MnO_4^-}$ and $\Delta^2 E_{cell}/\Delta V^2_{MnO_4^-}$ value associate the average slope, or rate of change of the slope, with the average volume value between the titration points involved. The equivalence-point volume is 50.00 ml of 0.0200 M $KMnO_4$ solution.

solution, this reaction and the titration redox reaction can not proceed to quantitative completion at the titration equivalence point.

Table 9.5 shows data typical of the titration of 50.00 ml of 0.1000 M Fe^{2+} with 0.0200 M $KMnO_4$ in a solution 0.10 M to H_2SO_4, that is, at pH = 1.00. The values of $E^\circ_{f,Fe^{3+}/Fe^{2+}}$ and $E^\circ_{f,MnO_4^-/Mn^{2+}}$ at 25°C can be taken, respectively, as 0.770 and 1.510 V relative to the SHE. The ΔpFe(II) for − 0.05 ml to the equivalence point is 6.12, that for the equivalence point to +0.05 ml is 1.22, yielding a

ΔpFe(II) for ± 0.05 ml around the equivalence-point volume of 7.34. The corresponding ΔE_{cell} values are $+0.36$, $+0.07$, and $+0.43$ V. Figure 9.6 shows the curve of E_{cell} vs. volume of KMnO$_4$ solution for this titration. The asymmetrical nature of the curve around the equivalence point is apparent.

The minimum $[H_3O^+]$ for a quantitative conversion of Fe^{2+} to Fe^{3+} in this titration at the equivalence point can be shown to be

$$[H_3O^+]_{min} = 4.9 \times 10^{-5} \; M$$

Titrations of iron(II) and permanganate ion are carried out in media much higher in $[H_3O^+]$ than the minimum value shown.

The titration of iron(II) with dichromate   This redox titration typifies subgroup 3 in Sec. 9.1.3. An acid solution of iron(II) is titrated with a standard solution of potassium dichromate, K$_2$Cr$_2$O$_7$. The titration reaction

$$6Fe^{2+} + Cr_2O_7^{2-} + 14H_3O^+ \rightleftharpoons 6Fe^{3+} + 2Cr^{3+} + 21H_2O$$

corresponds in form to Eq. (148), where

$$n_1 = 1 \qquad n_2 = 6 \qquad a = c = d = 1 \qquad b = 2 \qquad y = 14 \qquad w = 21$$

H$_3$O$^+$ is essential if the reaction

$$Cr_2O_7^{2-} + 14H_3O^+ + 6e \rightleftharpoons 2Cr^{3+} + 21H_2O$$

is to go to the right to practical completion, and again, if the $[H_3O^+]$ is not high

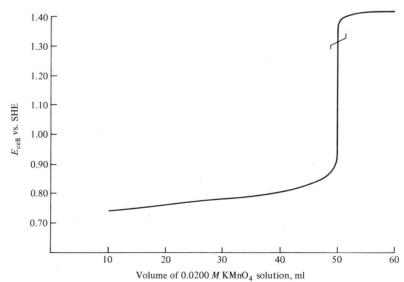

Fig. 9.6   Titration of 50.00 ml of 0.1000 M Fe^{2+} vs. 0.0200 M KMnO$_4$ solution at pH $= 1.00$.

Table 9.6 Titration of iron(II) by potassium dichromate*

50.00 ml of 0.1000 M Fe^{2+} vs. 0.0167 M K$_2$Cr$_2$O$_7$ at pH $= 1.00$

Volume K$_2$Cr$_2$O$_7$, ml	[Fe^{2+}]	pFe(II)	E_{cell} vs. SHE	($\Delta E_{cell}/\Delta V_{Cr_2O_7{}^{2-}}$) × 10^3 0.1 $ml = \Delta V_{Cr_2O_7{}^{2-}}$	($\Delta^2 E_{cell}/\Delta V^2_{Cr_2O_7{}^{2-}}$) × 10^3 0.1 ml/0.1 ml
0.00	1.00 × 10^{-1}	1.00			
10.00	6.67 × 10^{-2}	1.18	0.74		
20.00	4.28 × 10^{-2}	1.37	0.76		
30.00	2.50 × 10^{-2}	1.60	0.78		
				+0.20 (35.00)	
40.00	1.11 × 10^{-2}	1.96	0.80		+0.0053 (38.75)
				+0.60 (42.50)	
45.00	5.26 × 10^{-3}	2.28	0.83		+0.0088 (44.75)
				+1.0 (47.00)	
49.00	1.01 × 10^{-3}	3.00	0.87		+0.13 (48.12)
				+4.0 (49.25)	
49.50	5.02 × 10^{-4}	3.30	0.89		+1.3 (49.48)
				+10 (49.70)	
49.90	1.00 × 10^{-4}	4.00	0.93		+13 (49.81)
				+40 (49.92)	
49.95	5.00 × 10^{-5}	4.30	0.95		+680 (49.95)
				+380 (49.98)	
50.00	2.3 × 10^{-8}	7.64	1.14		−604 (50.00)
				+76 (50.02)	
50.05	6.2 × 10^{-9}	8.21	1.17^8		−140 (50.05)
				+6.0 (50.08)	
50.10	5.4 × 10^{-9}	8.27	1.18^1		−2.0 (50.19)
				+1.5 (50.30)	
50.50	4.2 × 10^{-9}	8.38	1.18^7		−0.20 (50.52)
				+0.60 (50.75)	
51.00	3.8 × 10^{-9}	8.42	1.19^0		−0.019 (51.88)
				+0.17 (53.00)	
55.00	2.8 × 10^{-9}	8.55	1.19^7		−0.0024 (55.25)
				+0.060 (57.50)	
60.00	2.3 × 10^{-9}	8.64	1.20^0		

*[Fe^{2+}] and pFe(II) are not corrected for the dilution effect. The values in parentheses after each $\Delta E_{cell}/\Delta V_{Cr_2O_7{}^{2-}}$ and $\Delta^2 E_{cell}/\Delta V^2_{Cr_2O_7{}^{2-}}$ value associate the average slope, or rate of change of the slope, with the average volume value between the titration points involved. The equivalence-point volume is 50.00 ml of 0.0167 M K$_2$Cr$_2$O$_7$ solution.

enough, this reaction and the titration redox reaction can not attain quantitative completion at the titration equivalence point.

Table 9.6 shows data typical of this titration. The ΔpFe(II) for -0.05 ml to the equivalence point is 3.34, that for the equivalence point to $+0.05$ ml is 0.57, giving a ΔpFe(II) for ± 0.05 ml around the equivalence point of 3.91. The corresponding ΔE_{cell} values are $+0.19$, $+0.04$, and $+0.23$ V.

Figure 9.7 shows the curve of E_{cell} vs. volume of K$_2$Cr$_2$O$_7$ solution. The asymmetric nature of the curve around the equivalence point is obvious.

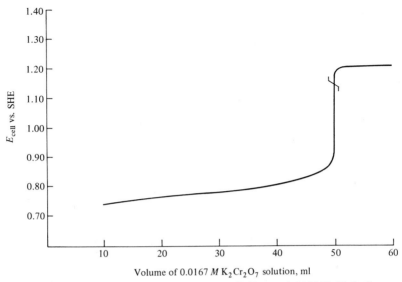

Fig. 9.7 Titration of 50.00 ml of 0.1000 M Fe^{2+} vs. 0.0167 M K$_2$Cr$_2$O$_7$ at pH = 1.00.

The minimum [H$_3$O$^+$] for a quantitative conversion of Fe^{2+} to Fe^{3+} at the equivalence point of this titration is given by

$$\frac{[Fe^{3+}]^8}{[Fe^{2+}]^7} = \tfrac{3}{2} \times 10^{56.9} [H_3O^+]^{14}$$

$$\frac{(5.00 \times 10^{-2}\ M)^8}{(10^{-6}\ M_{max})^7} = \tfrac{3}{2} \times 10^{56.9} [H_3O^+]^{14}_{min}$$

$$[H_3O^+]_{min} = 1.5 \times 10^{-2}\ M$$

The greater sensitivity of the dichromate titration of iron(II) to [H$_3$O$^+$], relative to that for the permanganate titration, is worthy of note. Titrations of iron(II) by potassium dichromate are normally conducted in media much higher in [H$_3$O$^+$] than the minimum calculated above.

9.3. MULTIPLE REDOX SYSTEM TITRATIONS

It is possible to titrate, for example, a solution of two reductants with a standard solution of a suitable oxidant. The curve of E_{cell} vs. volume of titrant will show two distinct breaks, one at each equivalence point, providing that the separation of the $E^{\circ}_{f,Ox/Red}$ values for the two reductants is large enough and that the $E^{\circ}_{f,Ox/Red}$ value for the oxidant is sufficiently higher than the larger $E^{\circ}_{f,Ox/Red}$ value of the reductant systems. Naturally, fundamental criteria for the success of such a titration will also include rapid establishment of equilibria for both redox systems and availability of an indicator system or systems capable of locating each equivalence point with reasonable accuracy.

   Such a multiple-system titration is analogous to that of a neutralization titration where two acids in solution are titrated with a standard base solution, where the K_a values for the acids are sufficiently different, or to that of a precipitation titration where two substances in solution are titrated with a standard solution of a common precipitant, and the K_{sp} values for the two slightly soluble substances formed differ by a sufficiently large margin.

   An example of such a multiple redox titration system is the titration of an acid solution of titanium(III) and iron(II) by a standard solution of cerium(IV). The various reactions involved are

$$Ti^{3+} + Ce^{4+} + 3H_2O \rightleftharpoons TiO^{2+} + Ce^{3+} + 2H_3O^+$$

$$TiO^{2+} + Fe^{2+} + 2H_3O^+ \rightleftharpoons Ti^{3+} + Fe^{3+} + 3H_2O$$

$$Fe^{2+} + Ce^{4+} \rightleftharpoons Fe^{3+} + Ce^{3+}$$

The equilibrium for the second reaction lies far to the left, so that only just before or at the first equivalence point, where $[Ti^{3+}]$ is very low, will the value of $[Fe^{3+}]$ be of any real significance relative to $[Ti^{3+}]$.

   Consider the titration of 50.00 ml of a solution 0.0500 M to Ti^{3+} and 0.1000 M to Fe^{2+} by 0.1000 M Ce^{4+} in a medium 2 M to H_2SO_4. The value of $[H_3O^+]$ here will be approximately 2 M. The values for $E^\circ_{f,TiO^{2+}/Ti^{3+}}$, $E^\circ_{f,Fe^{3+}/Fe^{2+}}$, and $E^\circ_{f,Ce^{4+}/Ce^{3+}}$ at 25°C and in 2 M H_2SO_4 can be taken as 0.10, 0.700, and 1.459 V, respectively, relative to the SHE.

   In such a system, the titanium(III) is titrated first, with the $[Fe^{2+}]$ in solution being relatively unaffected except by the dilution effect, until the first equivalence point is reached. Subsequent to this, the iron(II) is titrated, the titration being complete at the second equivalence point.

   For quantitative conversions of Ti^{3+} to TiO^{2+} and Fe^{2+} to Fe^{3+} at the first and second equivalence points, respectively, the following general methods hold relative to the calculation of E_{cell} values at the various titration points.

   At the start of the titration, $[Ti^{3+}] = 5.00 \times 10^{-2} M$, with E_{cell} undefined.

   Between the start and up to but not including the first equivalence point, the values of $[Ti^{3+}]$ and $[TiO^{2+}]$ are found quite accurately from

$$[Ti^{3+}] = \frac{V_{Ti^{3+}} M_{Ti^{3+}} - V_{Ce^{4+}} M_{Ce^{4+}}}{V_{start} + V_{Ce^{4+}}} + [Ce^{4+}] + [Fe^{3+}]$$

the second and third terms on the right representing generally the loss of TiO^{2+} in the back-reaction for the titration and in the reaction

$$TiO^{2+} + Fe^{2+} + 2H_3O^+ \rightleftharpoons Ti^{3+} + Fe^{3+} + 3H_2O$$

The values of $[Ce^{4+}]$ and $[Fe^{3+}]$ will be low enough at these titration points to be ignored, and we have $[Ti^{3+}]$ given by the first expression on the right, which is, of course, Eq. (175). The value of $[TiO^{2+}]$ for such titration points is given quite accurately by

$$[TiO^{2+}] = \frac{V_{Ce^{4+}} M_{Ce^{4+}}}{V_{start} + V_{Ce^{4+}}} - [Ce^{4+}] - [Fe^{3+}]$$

but, again, we may ignore the second and third terms on the right and determine the value of $[TiO^{2+}]$ from the first equation. E_{cell} for these points is then given by

$$E_{cell} = E^{\circ}_{f,TiO^{2+}/Ti^{3+}} + 0.059 \log \frac{[TiO^{2+}] \, [H_3O^+]^2}{[Ti^{3+}]}$$

At the first equivalence point, the situation relative to E_{cell} will be governed by the reaction

$$TiO^{2+} + Fe^{2+} + 2H_3O^+ \rightleftharpoons Ti^{3+} + Fe^{3+} + 3H_2O$$

and we have

$$E_{cell} = E^{\circ}_{f,TiO^{2+}/Ti^{3+}} + 0.059 \log \frac{[TiO^{2+}] \, [H_3O^+]^2}{[Ti^{3+}]}$$

$$E_{cell} = E^{\circ}_{f,Fe^{3+}/Fe^{2+}} + 0.059 \log \frac{[Fe^{3+}]}{[Fe^{2+}]}$$

so that

$$2E_{cell} = E^{\circ}_{f,TiO^{2+}/Ti^{3+}} + E^{\circ}_{f,Fe^{3+}/Fe^{2+}} + 0.059 \log \frac{[TiO^{2+}] [Fe^{3+}] [H_3O^+]^2}{[Ti^{3+}] [Fe^{2+}]}$$

In general, at the first equivalence point, we have

$$[Fe^{3+}] \approx [Ti^{3+}]$$

$$[Fe^{2+}] \approx \frac{50.00 \times 0.1000}{50.00 + 25.00} \approx 6.67 \times 10^{-2} M$$

$$[TiO^{2+}] \approx \frac{50.00 \times 0.0500}{50.00 + 25.00} \approx 3.33 \times 10^{-2} M$$

and we have for the titration in question

$$2E_{cell} = 0.10 + 0.700 + 0.059 \log \frac{[TiO^{2+}] \, [H_3O^+]^2}{[Fe^{2+}]}$$

$$\approx 0.82 \text{ V}$$

$$E_{cell} \approx 0.41 \text{ V}$$

With an E_{cell} value of 0.41 V and $[TiO^{2+}] = 3.33 \times 10^{-2} M$ at the equivalence point, we have:

$$0.41 \approx 0.10 + 0.059 \log \frac{[TiO^{2+}] \, [H_3O^+]^2}{[Ti^{3+}]}$$

and

$$[Ti^{3+}] \approx 7.6 \times 10^{-7} \, M$$

at the first equivalence point, representing at this point a quantitative conversion of Ti^{3+} to TiO^{2+} for the titration conditions involved.

After the first equivalence point and up to but not including the second equivalence point, the titration proceeds as if the iron(II) and cerium(IV) systems were titrated alone, and for each such point we have

$$[Fe^{2+}] \approx \frac{V_{Fe^{2+}} M_{Fe^{2+}} - V_{Ce^{4+}} M_{Ce^{4+}}}{V_{Fe^{2+}} + V_{Ce^{4+}}}$$

$$[Fe^{3+}] \approx \frac{V_{Ce^{4+}} M_{Ce^{4+}}}{V_{Fe^{2+}} + V_{Ce^{4+}}}$$

$$E_{cell} = E^{\circ}_{f,Fe^{3+}/Fe^{2+}} + 0.059 \log \frac{[Fe^{3+}]}{[Fe^{2+}]}$$

where $V_{Fe^{2+}}$ and $M_{Fe^{2+}}$ are the volume and molarity of the solution at the first equivalence point relative to Fe^{2+}, and $V_{Ce^{4+}}$ is the volume of Ce^{4+} solution added *after* the first equivalence point.

At the second equivalence point we have, from Eq. (155),

$$2E_{cell} = E^{\circ}_{f,Ce^{4+}/Ce^{3+}} + E^{\circ}_{f,Fe^{3+}/Fe^{2+}} + 0.059 \log \frac{[Ce^{4+}][Fe^{3+}]}{[Ce^{3+}][Fe^{2+}]}$$

Although we can use the approximate relationship

$$[Ce^{4+}] \approx [Fe^{2+}]$$

because of the prior titration of Ti^{3+}, the value of $[Ce^{3+}]$ will be given by

$$[Ce^{3+}] \approx \frac{V_{Fe^{2+}} M_{Fe^{2+}}}{V_{start} + V_{Ce^{4+}}} + \frac{V_{Ti^{3+}} M_{Ti^{3+}}}{V_{start} + V_{Ce^{4+}}}$$

and, for $[Fe^{3+}]$, we have

$$[Fe^{3+}] \approx \frac{V_{Fe^{2+}} M_{Fe^{2+}}}{V_{start} + V_{Ce^{4+}}}$$

where $V_{Fe^{2+}}$ and $M_{Fe^{2+}}$ are the starting volume and molarity for Fe^{2+}; $V_{Ti^{3+}}$ and $M_{Ti^{3+}}$ are the starting volume and molarity for Ti^{3+}; and $V_{Ce^{4+}}$ is the *total* volume of Ce^{4+} solution added. Thus we have

$$2E_{cell} = 1.459 + 0.700 + 0.059 \log \frac{[Fe^{3+}]}{[Ce^{3+}]}$$

$$E_{cell} \approx 1.07 \, V$$

The value for $[Fe^{2+}]$ at the second equivalence point is thus

$$1.07 \approx 0.700 + 0.059 \log \frac{[Fe^{3+}]}{[Fe^{2+}]} \qquad \text{and} \qquad [Fe^{2+}] \approx 2.1 \times 10^{-8} \, M$$

representing a quantitative conversion of Fe^{2+} to Fe^{3+} at the equivalence point. After the second equivalence point the value of E_{cell} is determined from

$$E_{cell} = E^{\circ}_{f,\,Ce^{4+}/Ce^{3+}} + 0.059 \log \frac{[Ce^{4+}]}{[Ce^{3+}]}$$

and we have

$$[Ce^{4+}] \approx \frac{V_{Ce^{4+}} M_{Ce^{4+}} - V_{Fe^{2+}} M_{Fe^{2+}}}{V_{Ce^{4+}} + V_{Fe^{2+}}}$$

where $V_{Fe^{2+}}$ and $M_{Fe^{2+}}$ are the volume and molarity of Fe^{2+} at the first equivalence point, and $V_{Ce^{4+}}$ is the volume of Ce^{4+} solution added *after* the first equivalence point. For $[Ce^{3+}]$ we have

$$[Ce^{3+}] \approx \frac{V_{Fe^{2+}} M_{Fe^{2+}}}{V_{start} + V_{Ce^{4+}}} + \frac{V_{Ti^{3+}} M_{Ti^{3+}}}{V_{start} + V_{Ce^{4+}}}$$

where $V_{Fe^{2+}}$, $M_{Fe^{2+}}$, $V_{Ti^{3+}}$, and $M_{Ti^{3+}}$ are the *starting* volumes and *starting* molarities of Fe^{2+} and Ti^{3+}, and $V_{Ce^{4+}}$ is the *total* volume of Ce^{4+} solution added.

Figure 9.8 gives data typical of such a titration in the form of the curve of

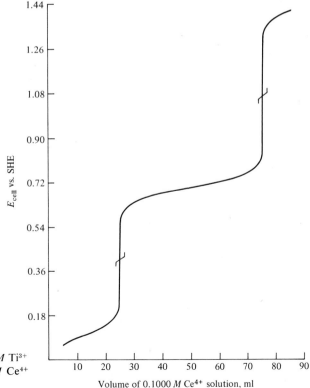

Fig. 9.8 Titration of 0.0500 M Ti^{3+} and 0.1000 M Fe^{2+} vs. 0.1000 M Ce^{4+} in 2 M H_2SO_4.

E_{cell} vs. volume of cerium(IV) solution. The symmetric nature of the curve around each equivalence point, and the excellent definition of these equivalence points, will be noted. As a matter of interest, the ΔE_{cell} value for ± 0.05 ml around the first and second equivalence points is approximately 0.26 and 0.39 V, respectively.

9.4 FEASIBILITY OF OXIDATION-REDUCTION TITRATIONS

The feasibility of oxidation-reduction titrations depends largely on the separation of the respective $E^{\circ}_{f,Ox/Red}$ values and the characteristics of the redox reaction. Consider the general reaction

$$n_2 c\mathrm{Red}_1 + n_1 a\mathrm{Ox}_2 \rightleftharpoons n_2 d\mathrm{Ox}_1 + n_1 b\mathrm{Red}_2$$

At the equivalence point of a titration between the reductant and the oxidant, we have from Eq. (143) by rearrangement

$$(E^{\circ}_{f,Ox_2/Red_2} - E^{\circ}_{f,Ox_1/Red_1})_{min} = \frac{0.059}{n_1 n_2} \log \frac{(n_1 b/n_2 d)^{n_1 b} [\mathrm{Ox}_1]^{(n_2 d + n_1 b)}}{(n_1 a/n_2 c)^{n_1 a} [\mathrm{Red}_1]^{(n_2 c + n_1 a)}_{max}}$$

$$\text{at } 25^{\circ}\text{C} \qquad (224)$$

or, where $n_1 = n_2$,

$$(E^{\circ}_{f,Ox_2/Red_2} - E^{\circ}_{f,Ox_1/Red_1})_{min}$$

$$= \frac{0.059}{n_1} \log \frac{(b/d)^b [\mathrm{Ox}_1]^{(d+b)}}{(a/c)^a [\mathrm{Red}_1]^{(c+a)}_{max}} \qquad \text{at } 25^{\circ}\text{C} \qquad (225)$$

Table 9.7 Minimum difference for $(E^{\circ}_f, Ox_2/Red_2 - E^{\circ}_f, Ox_1/Red_1)$ for a quantitative conversion of Red$_1$ to Ox$_1$ at the equivalence point. [Ox$_1$] at the equivalence point = 0.05 M and [Red$_1$]$_{max}$ at this point = 10^{-6} M

n_1	n_2	a	b	c	d		$(E^{\circ}_f, Ox_2/Red_2 - E^{\circ}_f, Ox_1/Red_1)$, volts minimum, at 25°C
1	1	1	1	1	1		0.555
2	1	1	1	1	1		0.416
1	2	1	1	1	1		0.416
2	2	1	1	1	1		0.277
3	2	1	1	1	1		0.231
3	3	1	1	1	1		0.185
2	2	1	2	2	1		0.443
1	5	1	1	1	1	$[\mathrm{H_3O^+}] = 1\,M$	0.333
1	6	1	2	1	1	$[\mathrm{H_3O^+}] = 1\,M$	0.309

In each equation, $[\text{Red}_1]_{max}$ is the maximum allowable concentration of Red_1 compatible with a quantitative conversion of Red_1 to Ox_1 at the titration equivalence point. Similar equations can be developed relative to redox reactions of the type

$$n_2 c\text{Red}_1 + n_1 a\text{Ox}_2 + n_1 y\text{H}_3\text{O}^+ \rightleftharpoons n_2 d\text{Ox}_1 + n_1 b\text{Red}_2 + n_1 w\text{H}_2\text{O}$$

using rearrangements of Eq. (155) or (156), and assigning some specific value to $[\text{H}_3\text{O}^+]$.

If we assume a titration of reductant and oxidant under conditions where the $[\text{Ox}_1]$ concentration at the equivalence point is approximated as 0.0500 M, and if we set the maximum allowable concentration of Red_1 at this point as $10^{-6}\,M$, we can calculate the minimum allowable difference for the equation $(E^\circ_{f,\text{Ox}_2/\text{Red}_2} - E^\circ_{f,\text{Ox}_1/\text{Red}_1})$. Table 9.7 gives the minimum differences for several redox titration characteristics under these conditions.

It should be noted that all of the titrations discussed in Secs. 9.2 and 9.3 are capable, under the conditions described above, of yielding quantitative conversions at their equivalence points.

9.5 OXIDATION-REDUCTION TITRATION INDICATOR SYSTEMS

Apart from the potentiometric method of locating the equivalence point of redox titrations, a method to be discussed in detail in a later chapter, there are several indicator systems capable of application in the location with reasonable accuracy of the equivalence points of such titrations.

9.5.1 THE TITRANT SERVES AS INDICATOR

In several instances where titrant solutions of intense color are used, the end point of the titration can be detected by the color imparted to the solution by the first excess of titrant. The agreement between the end point and the equivalence point will, to a considerable extent, depend on both the intensity of the titrant color and any masking effect introduced by the solution titrated, if it should itself be colored, turbid, etc. A typical case in point is that in which a permanganate solution is used as the titrant. The intensity of the color of this reagent at the concentrations normally encountered when used as a titrant is such that when colorless or slightly colored solutions are titrated, the eye can detect the pink permanganate color at a level of about $5 \times 10^{-6}\,M$ relative to $[\text{MnO}_4^-]$. In the titration of iron(II) with potassium permanganate, outlined briefly in Sec. 9.2.3, the indicator blank corresponding to this concentration of MnO_4^- would be given by

$$5 \times 10^{-6}\,M = \frac{V_{\text{MnO}_4^-} \times 0.0200\,M - \frac{1}{5} \times 50.00 \times 0.1000\,M}{50.00 + V_{\text{MnO}_4^-}}$$

from which

$$V_{\text{MnO}_4^-} = 50.02 \text{ ml}$$

The indicator blank and titration error here are -0.02 ml and $+0.4$ ppt, respectively. In view of the burette-reading uncertainty of ± 0.02 ml per reading, the indicator blank can be considered negligible for a titration of these characteristics. Where the color of the solution being titrated is dark enough to require indicator-blank values of higher magnitude, these can be determined on an experimental basis.

9.5.2 INTERNAL INDICATORS OTHER THAN OXIDATION-REDUCTION INDICATOR SUBSTANCES

It is possible to add a substance to the redox titration solution system such that it will react with a product or reactant of the system at a level of concentration appropriate to that expected at the titration equivalence point. Thiocyanate, for example, can be used as an indicator reacting with Fe^{3+}, the red color of the $Fe(SCN)^{2+}$ complex appearing at values of $[Fe^{3+}]$ in excess of about $5 \times 10^{-6} M$. Starch can be used as the indicator in titrations involving iodine, and it reacts with iodine to form the blue starch-triiodide complex.

Since this latter indicator situation is commonly encountered in iodimetric and iodometric titrations, the following additional details are of importance.

The amylose components of ordinary starch are capable of the reaction

$$\text{amyloses} + I_3^- \rightleftharpoons (\text{amyloses}, I_3^-)$$
$$\quad\text{Colorless} \qquad\qquad \text{Blue}$$

where the triiodide, I_3^-, forms reversibly an intensely blue adsorption complex with the amyloses. The amylopectin components of starch form red complexes with I_3^- in reactions of limited reversibility and are therefore undesirable. The substance commonly referred to as *soluble starch* contains a significantly high amount of the amyloses, and this form of starch is used to prepare the indicator solutions employed in iodimetric and iodometric titrations. It is important to note that the reaction above involves the triiodide ion, indicating that both iodine and iodide must be present if the indicator is to function with adequate sensitivity. Concentrations of iodide ion in solution of not less than $10^{-4} M$ are required in this connection.

Several important factors are involved in the preparation and use of starch indicator solutions. Bacterial action usually results in decomposition of aqueous solutions in a matter of a few days. Starch solutions should, therefore, be prepared where required on a daily basis. The indicator sensitivity decreases with increasing solution temperature, so that titrations should not be carried out at temperatures higher than room temperature. The presence of agents which retard the adsorption of the triiodide by starch, such as gelatin, alcohols, glycerol, etc., should be avoided. Decreases in the sensitivity result with higher $[H_3O^+]$ values in solution, and the indicator should not be used in strongly acid solutions. Since the presence of high concentrations of iodine yields undesirable products with starch, the indicator should be added in iodimetric and iodometric titrations only near the titration end point.

The sensitivity of starch to iodine is such that a visible blue color is given to a solution containing the normal amount of starch (about 0.01 percent) by a value of $[I_2] = 5 \times 10^{-7} M$, where the $[I^-]$ is not less than $10^{-4} M$ and where the solution being titrated is colorless or only faintly colored. For the titration outlined in Sec. 9.2.2 between iodine and sodium thiosulfate, the indicator blank and titration error would be negligible, since the value of $[I_2]$ at the equivalence point is $3.1 \times 10^{-7} M$, while that at the 49.95-ml point of added $Na_2S_2O_3$ solution is $2.5 \times 10^{-5} M$. Thus the blue color due to the starch-triiodide complex would disappear somewhere just before and very close to the equivalence-point volume (the indicator blank is actually less than +0.01 ml). Even where more dilute solutions are titrated, for example, $0.0050 M\ I_2$ vs. $0.0100 M\ Na_2S_2O_3$, the indicator blank, given by

$$5 \times 10^{-7} M = \frac{50.00 \times 0.0050\ M - V_{S_2O_3^{2-}} \times 0.0100\ M \times 0.5}{50.00 + V_{S_2O_3^{2-}}}$$

$$V_{S_2O_3^{2-}} = 49.99\ \text{ml}$$

is only +0.01 ml, again a negligible value. In the case of colored solutions, the appearance or disappearance of the characteristic blue color may be more difficult to detect, and the indicator blanks as determined experimentally may be large enough to require their application.

9.5.3 OXIDATION-REDUCTION INDICATORS

General Certain groups of organic compounds are capable of oxidation-reduction reactions where the color of the oxidized form of the indicator in solution differs significantly from that of the reduced form. Two general oxidation-reduction reaction types can be considered here; these involve either participation or nonparticipation of H_3O^+ directly in the indicator redox reaction. Thus we have

$$\text{In}_{\text{Ox}} + ne \rightleftharpoons \text{In}_{\text{Red}} \tag{226}$$

$$E_{\text{Ox/Red}} = E^{\circ}_{f,\text{Ox/Red}} + \frac{0.059}{n} \log \frac{[\text{In}_{\text{Ox}}]}{[\text{In}_{\text{Red}}]} \tag{227}$$

and

$$\text{In}_{\text{Ox}} + nH_3O^+ + ne \rightleftharpoons \text{In}_{\text{Red}} \tag{228}$$

$$E_{\text{Ox/Red}} = E^{\circ}_{f,\text{Ox/Red}} + \frac{0.059}{n} \log \frac{[\text{In}_{\text{Ox}}][H_3O^+]^n}{[\text{In}_{\text{Red}}]} \tag{229}$$

With both reaction types, $[H_3O^+]$ in solution may affect $E^\circ_{f,Ox/Red}$ through the diverse-ion effect, but under the solution conditions normal in redox titrations, the variations in $[H_3O^+]$ encountered will not usually introduce significant changes in the value of $E^\circ_{f,Ox/Red}$. In the case of indicator substances conforming in their redox reaction to Eq. (228), the $[H_3O^+]$ value in solution will, in addition, affect the value of $E^\circ_{f,Ox/Red}$ to the extent of 0.059 V per unit change in the solution pH. Again, under the conditions normally surrounding redox titrations, the solution pH will not be expected to differ markedly from titration to titration, so that here again the influence of $[H_3O^+]$ may be minimal.

In either case, therefore, for a specific value of $[H_3O^+]$ in solution, the oxidation-reduction indicator compound will show an $E_{Ox/Red}$ value dependent on the concentration ratio $[In_{Ox}]/[In_{Red}]$. The color of the compound in solution will also be dependent on this ratio, and following along the lines of the treatment used in connection with those organic substances used as indicators in volumetric neutralization titrations, we note that when

$$\frac{[In_{Ox}]}{[In_{Red}]} \geq 10$$

the color of only the oxidized form is seen, while for

$$\frac{[In_{Ox}]}{[In_{Red}]} \leq \frac{1}{10}$$

the color of only the reduced form is seen. For Eq. (227) this yields

$$E_{Ox/Red} = E^\circ_{f,Ox/Red} \pm \frac{0.059}{n} \tag{230}$$

and for Eq. (229),

$$E_{Ox/Red} = E^\circ_{f,Ox/Red} \pm \frac{0.059}{n} + 0.059 \log [H_3O^+] \tag{231}$$

and we note that these equations provide the transition ranges for the indicators involved in terms of $E_{Ox/Red}$. The transition interval may not always be symmetrical around $E^\circ_{f,Ox/Red}$ or $(E^\circ_{f,Ox/Red} + 0.059 \log [H_3O^+])$, and the lack of symmetry is often largely dependent on the nature and intensity of the colors for the oxidized and reduced forms.

The use of such organic compounds as indicators in redox titrations means that the solution potential at the various points in the titration will dictate the indicator color at these points, since the $E_{Ox/Red}$ values for the titration half-cell systems and the indicator system will be identical at all titration points. For a given redox titration there will be a specific ΔE_{cell} around the equivalence point. The selection of an indicator with a transition range of $E_{Ox/Red}$ matching this ΔE_{cell} implies that the indicator will change color at or close to the equivalence point.

The degree to which the end point given by the color change and the stoichiometric equivalence point agree will be a function of the extent to which ΔE_{cell} and the indicator transition interval correspond. Indicator blanks and titration errors of significant magnitude may occur where good correspondence is not secured.

Since the indicator transition range involves the term $\pm 0.059/n$ volts, the range is a maximum when $n = 1$, and has a value of about 0.12 V. Since the transition intervals are therefore small, it is usually sufficient to know the value of $E_{Ox/Red}$, called the transition potential, around which the indicator changes color. The suitability of an indicator for a given redox titration is then considered to be adequate where this transition potential fits into or close to the ΔE_{cell} around the titration equivalence point.

It is understood that since the indicator concentration in redox titrations is very low, those indicators which have redox reactions where H_3O^+ participates directly will not, in such titration systems, affect the value of the $[H_3O^+]$ in the solution involved.

A list of oxidation-reduction indicators in frequent use is given in Table 9.8, together with the transition potentials and the oxidized and reduced form colors.

Table 9.8 Oxidation-reduction indicators

Indicator	Oxidized color	Reduced color	Transition potential approximately, pH = 0
Ruthenium tridipyridine dichloride*	Yellow	Colorless	1.33
5-Nitro-1,10-phenanthroline iron(II) (nitro ferroin)	Pale blue	Red	1.25
1,10-Phenanthroline iron(II) (ferroin)	Pale blue	Red	1.11
Erioglaucine A	Red	Green	1.02
2,2'-Dipyridine iron(II)	Pale blue	Red	0.97
4,7-Dimethyl-1,10-phenanthroline iron(II)	Pale blue	Red	0.88
Sodium diphenylamine sulfonic acid and sodium diphenylbenzidine sulfonic acid	Purple	Colorless	0.85
Diphenylamine and diphenylbenzidine	Violet	Colorless	0.76
Methylene blue	Blue	Colorless	0.53
Nile blue	Blue	Colorless	0.41
Indigo tetrasulfonic acid	Blue	Colorless	0.36
Indigo monosulfonic acid	Blue	Colorless	0.26
1,10-Phenanthroline vanadium(II)†	Pale green	Blue	0.15

*J. Steigmann, N. Birnbaum, and S.M. Edwards, *Ind. Eng. Chem., Anal. Ed.*, **14**:30 (1942).
†W. P. Schaefer, *Anal. Chem.*, **35**:1746 (1963).

Oxidation-reduction indicator classifications *The orthophenanthrolines* Organic compounds known as 1,10-phenanthrolines (or orthophenanthrolines) are capable of forming intensely colored, highly stable complexes with iron(II), vanadium(II), and certain other ions. These compounds have the general structure

1,10-Phenanthroline

with the complex with iron(II) showing the structure

and being called *ferroin* or $(Phen)_3 Fe^{2+}$. The indicator reaction involves

$$(Phen)_3 Fe^{3+} + 1e \rightleftharpoons (Phen)_3 Fe^{2+}$$
Pale blue   Red

the transition range of potential being given approximately by 1.06 ± 0.06 V at pH = 0. Because of the low intensity of the blue color due to $(Phen)_3 Fe^{3+}$ and the high intensity of the red color of $(Phen)_3 Fe^{2+}$, in titrations where the indicator changes from the reduced to the oxidized form color, the color change will not be observed until $[(Phen)_3 Fe^{3+}]$ is about nine times $[(Phen)_3 Fe^{2+}]$, or at an $E_{Ox/Red}$ value of approximately 1.11 V, this latter being the transition potential. This is an instance where a lack of symmetry exists for the color shift around the indicator $E°_{f,Ox/Red}$ value. Because of the possibility of decomposition, indicators of the orthophenanthroline type should not be used where the solution temperature exceeds 50°C.

Diphenylamine and associated compounds Diphenylamine undergoes a series of reactions in strong oxidizing media generally:

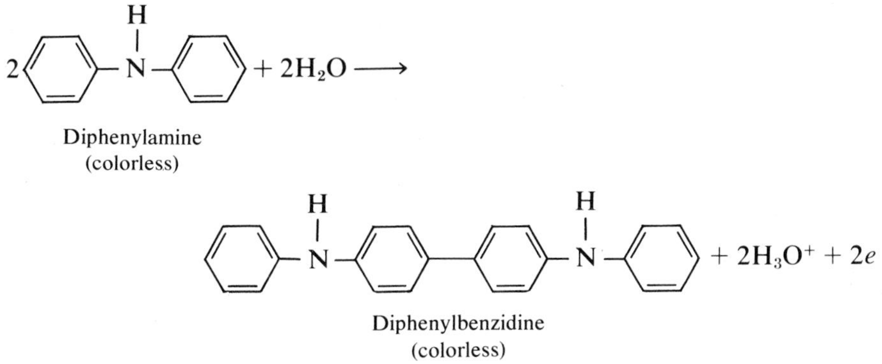

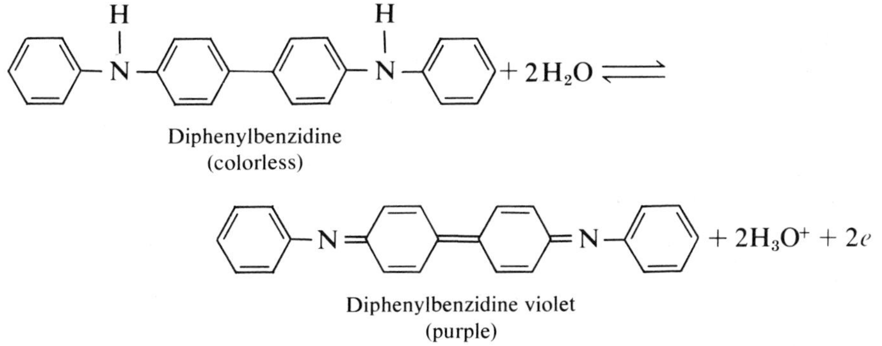

Diphenylamine
(colorless)

Diphenylbenzidine
(colorless)

Diphenylbenzidine
(colorless)

Diphenylbenzidine violet
(purple)

The reaction transforming diphenylamine to diphenylbenzidine is irreversible, so that the use of diphenylbenzidine instead of diphenylamine eliminates this step and avoids the consumption of additional titrant. The indicator reaction is the transformation of diphenylbenzidine to diphenylbenzidine violet. The transition potential is 0.76 V at 25°C and a solution pH = 0.

The low solubility in water of diphenylamine and diphenylbenzidine presents a difficulty which can be avoided by the use of their sulfonic acid derivatives such as sodium diphenylamine sulfonate or sodium diphenylbenzidine sulfonate. The latter is to be preferred, again in order to eliminate the diphenylamine-diphenylbenzidine step. Where either of these sulfonic acid derivatives is used, the transition potential is 0.85 V at 25°C and pH = 0. The diphenylamine and diphenylbenzidine substances form insoluble tungstates and cannot be used in the presence of tungstate.

Applications of oxidation-reduction indicators   The use of 1,10-phenanthroline iron(II) as the indicator in the iron(II)-vs.-cerium(IV) titration described in Sec. 9.2.2 exemplifies an application of oxidation-reduction indicators. The concentration of the indicator commonly used in solution at the equivalence point is approximately $[(Phen)_3 Fe^{2+}] = 1 \times 10^{-5} M$. The transition potential is 1.11 V and the

ΔE_{cell} for ± 0.05 ml around the equivalence point is given by $1.28 - 0.88$ V, the equivalence-point potential being 1.08 V. Thus the indicator changes just at the equivalence point, and the indicator blank, based on the amount of additional titrant required to bring about the indicator transition, is given by

$$1 \times 10^{-5} M = \frac{V_{\text{Ce}^{4+}} \times 0.1000 \ M}{V_{\text{Ce}^{4+}} + 100.00}$$

$$V_{\text{Ce}^{4+}} = 0.01 \text{ ml}$$

the indicator blank of -0.01 ml being negligible relative to the burette-reading uncertainty of ± 0.02 ml per reading. Note that the use of more dilute solutions of titrant will imply larger indicator-blank values, as will the use of indicator concentrations higher than $1 \times 10^{-5} M$ at the equivalence point.

Sodium diphenylbenzidine sulfonate can be used in the titration of iron(II) by potassium dichromate (Sec. 9.2.3). The indicator concentration usually encountered at the equivalence point is approximately $2 \times 10^{-5} M$. The transition potential is 0.85 V, and the ΔE_{cell} for ± 0.05 ml around the equivalence point is given by $1.18 - 0.95$ V, the equivalence-point potential being 1.14 V. The indicator would, under the conditions described for this titration in Sec. 9.2.3, show a color shift appreciably before the equivalence point, in point of fact, at about 48 ml of added potassium dichromate solution.

In order to permit the use of this indicator, titrations such as that between iron(II) and potassium dichromate are carried out in a medium to which sufficient 85% phosphoric acid, H_3PO_4, has been added to provide a concentration of about 0.5 M at the titration equivalence point. The presence of the PO_4^{3-} ion results in the formation of the stable, colorless ferric phosphate complex, which considerably decreases the value of $E^\circ_{f,\text{Fe}^{3+}/\text{Fe}^{2+}}$.

Under these new conditions, the value of $E^\circ_{f,\text{Fe}^{3+}/\text{Fe}^{2+}}$ would be about 0.60 V, and we would have

$$E_{\text{cell}} \text{ at 49.95 ml of added } 0.0167 \ M \text{K}_2\text{Cr}_2\text{O}_7 = 0.78 \text{ V}$$

$$E_{\text{cell}} \text{ at the equivalence point} = 1.12 \text{ V}$$

The indicator now changes color just at the equivalence point, the indicator blank, represented by the amount of additional titrant required to bring about the indicator transition being given by

$$2 \times 10^{-5} = \frac{3 V_{\text{Cr}_2\text{O}_7^{2-}} \times 0.0167 \ M}{V_{\text{Cr}_2\text{O}_7^{2-}} + 100.00}$$

$$V_{\text{Cr}_2\text{O}_7^{2-}} = 0.04 \text{ ml}$$

the indicator blank being -0.04 ml and the titration error $+0.8$ ppt. Again note that the use of more dilute titrant solutions, or the use of indicator concentrations higher than $2 \times 10^{-5} M$ at the equivalence point, will imply larger indicator blanks. It is also important to note that the formation of the colorless iron(III) phosphate complex aids in rendering the indicator color change easier to detect.

9.6 REAGENTS FOR PRETITRATION OXIDATION-REDUCTION PROCESSES

9.6.1 GENERAL

In redox titration work the titration involves either the oxidation of a substance previously quantitatively reduced to the required state for reaction with an oxidant as titrant, or the reduction of a substance previously quantitatively oxidized to the required state for reaction with a reductant as titrant. These pretitration processes of oxidation or reduction require the use of oxidants or reductants possessing certain well-defined characteristics. They should be capable of oxidizing or reducing the substance involved without simultaneously changing the oxidation state of other solution components to forms capable of consuming the titrant to be used. They must be easily removed from the solution, or negated in their effect (by dilution, by boiling, by chemical reaction, etc.) subsequent to their oxidation or reduction function being performed, so that no excess capable of permitting titrant consumption, in one way or another, will exist in the solution. They must perform the oxidation or reduction process rapidly and simply in order to avoid excessive time consumption. The following lists several commonly encountered pretitration oxidants and reductants.

9.6.2 OXIDANTS

Perchloric acid   Perchloric acid, $HClO_4$, is a widely used oxidant. Its oxidizing ability is confined to the hot, concentrated acid. The diluted acid does not possess oxidizing properties, so that dilution alone serves to remove the oxidizing effect. Many substances, such as organic materials, ammonia, bismuth, antimony(III), etc., form explosive mixtures with hot, concentrated perchloric acid. For this reason, the use of the reagent as an oxidant should be confined to proven nonexplosive mixtures, or to mixtures where preliminary or previous testing has demonstrated the absence of possible explosive situations. In the attack of metals and alloys, for example, degreasing with a volatile solvent prior to perchloric acid attack is essential. Because of the possibility of explosive situations developing between perchloric acid fumes and organic materials, fume hoods incorporating organic substances in their construction, such as wood, plastics, etc., cannot be used, and the danger of explosive mixtures with dust and deposited chemical substances, even in stainless steel fume hoods and vents, requires that a water curtain design be mandatory.

Peroxodisulfates   The ammonium, potassium, or sodium peroxodisulfates, $(NH_4)_2S_2O_8$, $K_2S_2O_8$, $Na_2S_2O_8$, are strong oxidizing agents in acid media at elevated temperatures; they can be used for the oxidation of chromium(II) to chromium(VI) (dichromate), manganese(II) to manganese(VII) (permanganate), iron(II) to iron(III), cerium(III) to cerium(IV), etc. The oxidation reactions are relatively slow, even in boiling acid solution, unless catalyzed by the presence of silver(I) ion. The reaction involved is

$$S_2O_8{}^{2-} + 2e \rightleftharpoons 2SO_4{}^{2-}$$

the value of $E^\circ_{f,S_2O_8{}^{2-}/SO_4{}^{2-}}$ in acid media being about 2.0 V. Excesses of the peroxodisulfate ion are conveniently removed by continued boiling over a short period, according to the reaction

$$2S_2O_8{}^{2-} + 6H_2O \rightleftharpoons 4SO_4{}^{2-} + O_2 + 4H_3O^+$$

Periodates Potassium periodate, KIO_4, is frequently used as an oxidant for the oxidation of manganese(II) to manganese(VII) (permanganate) in the spectrophotometric determination of manganese. Because of difficulties relative to the removal of excess KIO_4, the reagent is infrequently used in pretitration treatment. Hot acid media is required to further the oxidation processes based on KIO_4, and the low solubility of the salt in water prevents the use of high concentrations in solution. The reaction in acid media is

$$H_4IO_6{}^- + 2H_3O^+ + 2e \rightleftharpoons IO_3{}^- + 5H_2O$$

with the value of $E^\circ_{f,IO_4{}^-/IO_3{}^-}$ in acid media being about 1.6 V.

Bismuthate Sodium bismuthate, $NaBiO_3$, is often used in the oxidation of manganese(II) to manganese(VII) (permanganate) and chromium(III) to chromium(VI) (dichromate). The procedure involves boiling acid media, and an advantage is the low solubility of $NaBiO_3$, which allows the removal of any excess by filtration. The reaction is

$$BiO_3{}^- + 6H_3O^+ + 2e \rightleftharpoons Bi^{3+} + 9H_2O$$

with $E^\circ_{f,BiO_3{}^-/Bi^{3+}}$ about 1.7 V.

Peroxides Hydrogen peroxide, H_2O_2, is used as an oxidizing agent in acid media for the processes iron(II) to iron(III) and tin(II) to tin(IV). In alkaline media it can be used to oxidize manganese(II) to manganese(IV) (manganese dioxide, MnO_2). Either solutions of H_2O_2 or additions of solid sodium peroxide, Na_2O_2, may be used. The reaction in acid media is

$$H_2O_2 + 2H_3O^+ + 2e \rightleftharpoons 4H_2O$$

the $E^\circ_{f,H_2O_2/H_2O}$ value being about 1.8 V. The excess in acid solution is easily removed by boiling the solution

$$2H_2O_2 \rightleftharpoons 2H_2O + O_2$$

Bromine Bromine is sometimes used as an oxidizing agent in ammoniacal solution for conversions such as manganese(II) to manganese(IV) (hydrous manganese dioxide). The reaction is

$$Br_2 + 2e \rightleftharpoons 2Br^-$$

and the $E^{\circ}_{f,Br_2/Br^-}$ value is about 1.1 V. Excesses of bromine can be removed by boiling the solution subsequent to the oxidation process.

9.6.3 REDUCTANTS

Metals The low reduction potentials of many pure metals indicate that such substances would serve as excellent reductants. Prominent among these are zinc, silver, cadmium, lead, aluminum, and nickel. Where the metal is used in the granulated or powder form as a direct addition to the solution, the excess is subsequently removed by filtration and washing. Where the stick, sheet, or coil form is used, this can be lifted out of the solution and washed off. In many cases of the latter usage, filtration may also be required. The reducing ability of the individual pure metals can be secured from a table of standard reduction potentials.

A more convenient method of using the metals as reductants is to employ them in a *reductor*. Figure 9.9 shows a simple device of this type.

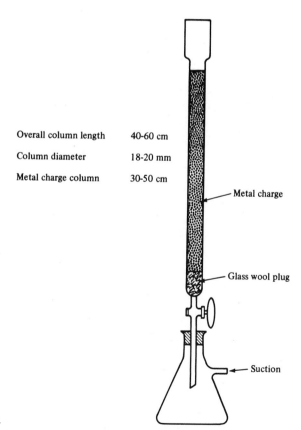

Overall column length	40-60 cm
Column diameter	18-20 mm
Metal charge column	30-50 cm

Metal charge

Glass wool plug

Suction

Fig. 9.9 A typical reductor.

The Jones reductor: zinc amalgam While pure zinc granules can be used as the reductant charge in a Jones reductor, the rate of dissolution of zinc in the acid solutions customarily run through the reductor very significantly limits the life of the charge. In addition to this, the rate of hydrogen evolution under these conditions seriously retards the rate of passage of the solution. The use of zinc granules coated with an amalgam of zinc and mercury, because of the high overvoltage for hydrogen at the mercury surface, provides for an extended charge life. The method of preparing the amalgamated zinc is quite simple. About 300 g of pure zinc, 20-30 mesh, is treated with 300 ml of a solution 2 percent to mercuric nitrate and containing 1 to 2 ml of concentrated nitric acid. The mixture is stirred vigorously for about 10 min, and the solution is decanted off. The coated zinc is then washed four or five times by decantation. The reductor column is filled with water, and the amalgamated zinc is slowly added until the column is filled to the required depth. Subsequently, the charge is washed, using suction, with about 1000 ml of water. Between periods of use, the reductor column is kept filled with water to prevent the formation of basic salts with attendant clogging. A cadmium amalgam, prepared in a similar manner, can also be used.

Reduction processes in zinc or cadmium amalgam reductors are usually carried out in solutions 1 to 2 M to either hydrochloric or sulfuric acid, with the latter acid being preferred. Copper, nickel, tin, arsenic, antimony, and other metals capable of reduction to the elemental state should be absent. Nitrates or nitric acid must be removed by double or triple evaporation of the solution to sulfuric acid fumes prior to passage through the reductor. Before use, a blank determination should be run. For example, 200 ml of 1 M H_2SO_4 should be run through the reductor, and the solution titrated with 0.02 M $KMnO_4$. A blank of not more than 0.05 ml should result, or the test should be repeated until such a blank is secured.

The silver reductor Silver powder for the silver reductor charge is prepared by reducing a silver nitrate solution with copper metal. The reductor is similar in design to that shown in Fig. 9.9, but it is usually only about 10 cm in column length. Between periods of use the column is kept filled with 1 M hydrochloric acid. The reduction processes carried out in the silver reductor involve hydrochloric acid media, usually at either 2 or 4 M strength. The acidity is critical, if certain reduction processes are to yield the required products. Table 9.9 provides a list of reduction reactions typical of the zinc amalgam and silver reductors. Note the difference in the reduction process for certain substances.

Other reductors Reductor column packings involving lead or cadmium have been used for specific applications.

Stannous chloride Stannous chloride, $SnCl_2$, can be used in acid solution as a

Table 9.9 Typical zinc amalgam and silver reductor reactions*

Zinc amalgam, $1 M$ H_2SO_4	Silver, acidity as indicated	
$Fe^{3+} + 1e \rightleftharpoons Fe^{2+}$	$Fe^{3+} + 1e \rightleftharpoons Fe^{2+}$	$1 M$ HCl
$MoO_4^{2-} + 8H_3O^+ + 3e \rightleftharpoons Mo^{3+} + 12H_2O$	$MoO_4^{2-} + 8H_3O^+ + 3e \rightleftharpoons Mo^{3+} + 12H_2O$	$5 M$ HCl
	$MoO_4^{2-} + 4H_3O^+ + 1e \rightleftharpoons MoO_2^+ + 6H_2O$	$2 M$ HCl
$TiO^{2+} + 2H_3O^+ + 1e \rightleftharpoons Ti^{3+} + 3H_2O$	Not reduced	
$Cr^{3+} + 1e \rightleftharpoons Cr^{2+}$	Not reduced	
$VO_4^{3-} + 8H_3O^+ + 3e \rightleftharpoons V^{2+} + 12H_2O$	$VO_4^{3-} + 6H_3O^+ + 1e \rightleftharpoons VO^{2+} + 9H_2O$	$1 M$ HCl
$UO_2^{2+} + 4H_3O^+ + 2e \rightleftharpoons U^{4+} + 6H_2O$	$UO_2^{2+} + 4H_3O^+ + 2e \rightleftharpoons U^{4+} + 6H_2O$	$4 M$ HCl
$UO_2^{2+} + 4H_3O^+ + 3e \rightleftharpoons U^{3+} + 6H_2O$*		
$Cu^{2+} + 2e \rightleftharpoons Cu$	$Cu^{2+} + 1e \rightleftharpoons Cu^+$	$2 M$ HCl

*In the Jones reductor both U^{4+} and U^{3+} are produced by the reduction of UO_2^{2+}. After reduction, passage of air through the solution for a few minutes oxidizes U^{3+} to U^{4+}. U^{4+} is unaffected during this treatment.

pretitration reductant, a particular application being the reduction of iron(III) to iron(II). The reaction for tin(II) is

$$Sn^{2+} - 2e \rightleftharpoons Sn^{4+}$$

the $E^\circ_{f,Sn^{4+}/Sn^{2+}}$ value in $1 M$ HCl being 0.139 V. The excess of Sn^{2+} is easily removed by reaction with added mercuric chloride, $HgCl_2$, and we have

$$Sn^{2+} + 2HgCl_2 \rightleftharpoons Sn^{4+} + Hg_2Cl_2(s) + 2Cl^-$$

The slightly soluble mercurous chloride can be left in the solution, since it does not consume the usual oxidants used in the titration of iron(II). Care should be taken to avoid too large an excess of Sn^{2+} in the reduction process Fe^{3+} to Fe^{2+}, since a large excess permits in part the reaction

$$Sn^{2+} + HgCl_2 \rightleftharpoons Sn^{4+} + Hg + 2Cl^-$$

the finely divided grey or black precipitate of mercury being capable of consuming oxidant titrants such as permanganate, cerium(IV), and dichromate.

Gases Certain gases, such as sulfur dioxide, SO_2, and hydrogen sulfide, H_2S, can be used in acid solution as reductants. Both gases can, for example, be used in the reduction of small amounts of iron(III) to iron(II). The process is time-consuming, however, often requiring the bubbling of the gas through the solution for one-half hour or more. Excesses of either gas are easily removed by subsequent boiling of the solution.

9.7   REAGENTS FOR OXIDATION-REDUCTION TITRATION METHODS

No attempt will be made in this section to outline in detail the preparation and standardization of the oxidant and reductant titrants discussed, nor will detailed methods of analysis involving these titrants be given. The purpose here will be to present the general principles underlying the use of various oxidants and reductants as titrants. Solution preparation and standardization, and methodic details, are in many instances provided in the laboratory experiments chapter of the text.

9.7.1 METHODS INVOLVING PERMANGANATE

General Permanganate solutions are usually prepared using potassium permanganate. Because of a tendency to instability, even when stored in the dark, potassium permanganate solutions, particularly those more dilute than $0.02\,M$, should be restandardized at predetermined intervals. The successful use of dilute hydrochloric acid media for permanganate titrations depends on the particular environment and its effect on the rate of reaction between chloride and permanganate. In some cases the reaction rate is so low that no difficulties arise; in others, for example, where iron(II) vs. permanganate is involved, the induced reaction rate is high enough to create a serious problem relative to permanganate consumption by chloride. Most oxidation reactions where permanganate is involved proceed rapidly even at room temperature. In some cases, elevated temperature and/or catalytic action is required to provide a sufficiently rapid reaction rate. The reaction betweeen oxalate and permanganate must be carried out at elevated temperature, with manganese(II) as the catalyst; that between arsenious oxide, As_2O_3, and permanganate must be catalyzed by traces of iodate or iodide.

Determination of iron Iron(II) can be titrated in acid medium with standard potassium permanganate solution. A general discussion of this oxidation-reduction system was given in Sec. 9.2.3. The pretitration reduction of iron(III) to iron(II) usually involves either the Jones reductor or stannous chloride solution. Where the Jones reductor is used, titanium, vanadium, and chromium are also reduced and subsequently consume titrant, so that this reduction technique should not be applied in the presence of these elements unless one is prepared to correct the total volume of permanganate solution used in the titration. Where stannous chloride is used, titanium and chromium are not reduced, and the method can be applied in the presence of these elements. Vanadium is reduced, however, and will subsequently consume titrant. Where vanadium is present, the solution of iron and vanadium can be reduced by stannous chloride, followed by titration at the proper acidity with dichromate rather than permanganate, iron(II) alone being oxidized under these conditions.

While reduction in the Jones reductor can involve either sulfuric or hydrochloric acid media, reduction by stannous chloride requires a hydrochloric acid medium. The titration of iron(II) with permanganate in hydrochloric acid medium results in an induced reaction whereby chloride is partially oxidized to either chlorine or hypochlorous acid, resulting in excessive consumption of permanganate. The presence of manganese(II) in solution prevents this induced reaction, and in the titration of iron(II) by permanganate in a hydrochloric acid medium, manganese(II) is added prior to titration in the form of the Zimmerman-Reinhardt reagent. This reagent consists of a solution of manganous sulfate, $MnSO_4$, in dilute sulfuric-phosphoric acids.

Even where sulfuric acid media are involved, the addition of phosphoric acid prior to titration is an accepted procedure aimed at sharpening the end-point color change. As discussed in Sec. 9.5.3, phosphate ion forms a stable colorless complex ferric phosphate. The color change is thus from colorless to pink, where permanganate serves as the end-point indicator, rather than from yellow or yellow-green to pink, where phosphoric acid is not present and the solution prior to the end point has the yellow or yellow-green color due to the relatively high concentration of Fe^{3+}.

In the titration of iron(II) vs. permanganate the titrant usually serves as the indicator, the first excess providing a pink color in solution (see Sec. 9.5.1).

Determination of oxalate or calcium Permanganate reacts with oxalate in acid media according to the reaction

$$5C_2O_4^{2-} + 2MnO_4^- + 16H_3O^+ \rightleftharpoons 10CO_2 + 2Mn^{2+} + 24H_2O$$

The medium usually used is a dilute solution of sulfuric acid (about 1 M). The reaction is slow at room temperature, and must be carried out at elevated temperatures of 50 to 60°C.

The first few drops of permanganate solution added do not decolorize rapidly, and no further addition should be made until this first addition is decolorized. The presence of manganese(II) catalyzes the reaction above, and until the first trace of manganese(II) is formed by reduction of the initial permanganate addition, the reaction will be slow. Once this first trace has been formed, further additions of permanganate will be rapidly decolorized.

Because of a tendency of permanganate to autodecomposition, the concentration of MnO_4^- should not be allowed to become too high during the titration. This requires that the titration be conducted at a reasonable speed, with efficient stirring or mixing.

The reaction provides the basis for one of the standardization procedures for permanganate solutions.

Several metals form slightly soluble oxalates and can be quantitatively separated by this precipitation technique. Where such a slightly soluble oxalate is involved, dissolving the salt in dilute sulfuric acid subsequent to separation can be followed by titration of the resulting solution with standard permanganate solution, the titration reaction being outlined in the foregoing. A determination of the metal involved can thus be made. The method is frequently used for the determination of calcium. Separation of the calcium oxalate is accomplished by the addition of ammonium oxalate solution to a dilute hydrochloric acid solution of calcium(II). This is followed immediately by neutralization of the acid with dilute ammonia. Such a precipitate of calcium oxalate will be almost free from coprecipitated ammonium oxalate, in fact to the extent that the subsequent titration will provide an error of only about $+2$ ppt. Deviation from the proper procedure of separation of calcium oxalate, using, for example, ammoniacal instead of hydrochloric acid media at the time of addition of ammonium oxalate solution, can result in the introduction of significant error.

Certain interfering substances, such as titanium and manganese, should be absent when calcium oxalate is precipitated.

Determination of arsenic   Arsenic(III) reacts in acid solution with permanganate to yield

$$5H_3AsO_3 + 2MnO_4^- + 6H_3O^+ \rightleftharpoons 5H_3AsO_4 + 2Mn^{2+} + 9H_2O$$

Arsenious oxide, As_2O_3, can be dissolved in 3 M sodium hydroxide solution and subsequently acidified with hydrochloric acid to provide a solution of H_3AsO_3. To this solution of arsenic(III) a drop or two of 0.002 M potassium iodate is added to catalyze the reaction, and the titration is then carried out using standard permanganate solution. Although the reaction will go quite well in hot hydrochloric acid solution, use of the catalyst and room-temperature titration is to be preferred. When the same titration is carried out in a sulfuric acid medium, the formation of complex manganese arsenates results in serious solution discoloration prior to the end point (brown or green solutions result).

The reaction with arsenic(III) is also a standardization technique for solutions of permanganate.

Other determinations   Permanganate can also be used in the determination of antimony, ferrocyanide, nitrites, hydrogen peroxide and other peroxides, vanadium, iodine, bromine, tin, tungsten, uranium, titanium, sulfites, molybdenum, sodium (indirectly), and potassium (indirectly).

9.7.2   METHODS INVOLVING CERIUM(IV)

General   Cerium(IV) solutions can be used as substitute titrants in almost all the titrations involving permanganate. Standard solutions of cerium(IV) must be prepared in dilute acid media (not less than about pH 1) in order to prevent hydrolysis of cerium(IV). As opposed to permanganate solutions, cerium(IV) solutions in dilute sulfuric acid are stable over extended periods and do not usually require restandardization over the normal period of solution life. The stability of dilute hydrochloric acid solutions is excellent at room temperature but decreases with increasing temperature. Nevertheless, hot dilute hydrochloric acid media can be used in the titration of iron(II) with cerium(IV), since no reaction occurs between chloride and cerium(IV) unless the $[Cl^-]$ exceeds about 3 M. This is a definite advantage over the use of permanganate under similar circumstances. Dilute nitric acid or perchloric acid solutions of Ce^{4+} are of poor stability.

Cerium(IV) cannot be used as a titrant in the presence of phosphates or fluorides, since phosphate results in the precipitation of ceric phosphate and fluoride results in the separation of ceric fluoride, both reactions consuming Ce^{4+} solution. Note that this requires that no addition of phosphoric acid be made in the case of titrations involving iron(II) and cerium(IV). The addition of Zimmerman-Reinhardt reagent to dilute hydrochloric acid media titrations is, of course, not required, since no induced reaction between chloride and cerium normally occurs.

Customarily 1,10-phenanthroline iron(II) is used as the indicator in many cerium(IV) titrations.

Determinations involving cerium(IV) In general, cerium(IV) solutions can be used in determinations where permanganate solutions are used, and the list of determinable substances follows very closely that already given for permanganate. The use of cerium(IV) as a titrant for iron(II) was described in Secs. 9.2.2 and 9.5.3.

9.7.3 METHODS INVOLVING DICHROMATE

General Potassium dichromate can be substituted in many titrations where permanganate is the titrant, although dichromate has a significantly lower reduction potential. Solutions of potassium dichromate are extremely stable, and restandardization is not ordinarily required over the normal period of solution life. Dilute sulfuric, hydrochloric, or perchloric acid media may be used for dichromate titrations. There is no reaction with chloride, even in hot solutions, at the normal hydrochloric acid concentrations of 1 or 2 M, so that the Zimmerman-Reinhardt reagent addition is not required. This is an advantage over the use of permanganate as the titrant.

The indicator used in dichromate titrations is often sodium diphenylbenzidine sulfonate; it should be used in preference to diphenylamine, diphenylbenzidine, or sodium diphenylamine sulfonate where possible. In order to obtain the best correspondence between the indicator end point and the equivalence point in the iron(II)-vs.-dichromate titration, phosphoric acid must be added to the solution before titration, as discussed in Sec. 9.5.3.

Determinations involving dichromate In general, dichromate solutions can be used in determinations where permanganate or cerium(IV) solutions are used. The use of dichromate in the titration of iron(II) was outlined in Secs. 9.2.3 and 9.5.3.

9.7.4 METHODS INVOLVING POTASSIUM BROMATE

General Potassium bromate, $KBrO_3$, can be used in titration analysis in two rather different approaches. One of these involves the use of potassium bromate as an oxidant in the determination by volumetric methods of certain organic compounds. The use of $KBrO_3$ in inorganic analysis normally involves the substance directly as a titrant. Such titrations are carried out in acid media—and most frequently in solutions not less than 1 M to hydrochloric acid.

Where strong reductants are titrated under such conditions with $KBrO_3$ solutions, the reaction of the oxidant is given by

$$BrO_3^- + 6H_3O^+ + 6e \rightleftharpoons Br^- + 9H_2O$$

the value $E^\circ_{f,BrO_3^-/Br^-}$ being about 1.44 V. Where the reductant is a stronger reducing agent than the bromide ion, no bromine appears during the titration until the reductant has been quantitatively reacted with $KBrO_3$, at which point the first excess of $KBrO_3$ results in the reaction

$$5Br^- + BrO_3^- + 6H_3O^+ \rightleftharpoons 3Br_2 + 9H_2O$$

which stems from the lower reduction potential of the bromine-bromide system reaction

$$Br_2 + 2e \rightleftharpoons 2Br^-$$

with a value of $E^\circ_{f,Br_2/Br^-}$ of about 1.09 V.

Where reductants of reduction potentials higher than that of the bromine-bromide system are involved, mixtures of bromide and bromine may arise before the titration equivalence point. The addition of mercury(II) to such solutions before titration often permits the reduction of bromate to bromide alone by the formation of the soluble species $HgBr_2$ and $HgBr_4^{2-}$.

Potassium bromate solutions are extremely stable and usually do not require restandardization over the normal period of solution life.

Since the appearance of bromine in solution generally signals the titration end point, certain organic dye substances of the azo type, which can be easily brominated and by this process decolorized to pale-yellow products, can serve as indicators in $KBrO_3$ titrations. Methyl red and methyl orange can, for example, be so used. Such indicator reactions are irreversible, so that no bromine must be formed in solution until immediately after the titration equivalence point. The titration of reducing substances of higher reduction potential than that of the bromine-bromide system should be avoided because of the possibility of bromine release before the equivalence point. Where mercury(II) additions have been used to minimize this difficulty, titration may be possible. Regardless of this situation, it is evident that all such $KBrO_3$ titrations must be conducted so as to avoid localized high concentrations of $KBrO_3$, with consequent premature bromine release. This implies the slow addition of titrant, with rapid and efficient stirring of the solution. A few redox indicators of reversible color change can be used in $KBrO_3$ titrations; these include p-ethoxychrysoidin (red to colorless), quinoline yellow (yellow-green to colorless) and α-naphthoflavone (yellow to orange-brown).

Determinations involving potassium bromate Potassium bromate can be used in inorganic redox titrations for the determination in part of arsenic, antimony, thallium, copper, tin, and selenium.

9.7.5 METHODS INVOLVING IODIMETRY AND IODOMETRY

General There are two aspects to the application of iodine chemistry to oxidation-reduction titrations. One of these involves the use of iodine directly as an oxidant, where a standard solution of iodine is used as the titrant, ordinarily in slightly acid to slightly basic media. Such techniques are classed as *iodimetric* methods, called *iodimetry*. In the second aspect, iodide is used as a reductant, and the iodine liberated in the associated reaction is titrated, usually in neutral to slightly acid media, with a standard solution of a reductant such as sodium thiosulfate or sodium arsenite. Such techniques are classed as *iodometric* methods, called *iodometry*. Both groups of methods are based on the fact that iodine in the presence of iodide yields the triiodide ion and the iodine-triiodide half reaction. Thus we have

$$I_2 + I^- \rightleftharpoons I_3^- \quad \text{and} \quad I_3^- + 2e \rightleftharpoons 3I^-$$

where $E^\circ_{I_3^-/I^-} = 0.534$ V at 25°C. Customarily, for convenience in writing reactions and making calculations involving equilibria, the half reaction is shown as

$$I_2 + 2e \rightleftharpoons 2I^-$$

with $E^\circ_{I_2/I^-} = 0.535$ V at 25°C.

The reduction potential of the iodine-triiodide system is not particularly high, and relatively few substances are capable of being oxidized by iodine. Thus there are few iodimetric methods. On the other hand, many substances are capable of oxidizing the iodide ion, and there are correspondingly many indirect or iodometric methods. The indicator most commonly used in iodimetric or iodometric titrations is soluble starch. In iodimetry the color change is usually from the solution color (hopefully reasonably colorless) to blue, with the reverse change in iodometry. The student is referred to Sec. 9.5.2 for a discussion of the use of starch as an indicator.

Iodine titrations can not be carried out in very basic media because of the disproportionation reactions

$$I_2 + OH^- \rightleftharpoons HOI + I^-$$

$$3HOI + 3OH^- \rightleftharpoons 2I^- + IO_3^- + 3H_2O$$

and a pH value of about 9 represents the basicity limit for these titrations. Iodide can be oxidized by air in the reaction

$$4I^- + 4H_3O^+ + O_2 \rightleftharpoons 2I_2 + 6H_2O$$

The reaction is very slow in neutral or slightly basic media but increases in reaction rate with increasing acidity and is greatly accelerated by the action of direct sunlight. Various substances catalyze the oxidation reaction; these include copper(I), nitrite, and nitrogen oxide, NO. In certain cases cyclic situations occur. For example, copper(I) is easily oxidized by air to form copper(II) which, in turn, reacts with iodide to form more iodine and copper(I) and so on. Nitrite, when present, results in the reaction

$$2NO_2^- + 2I^- + 4H_3O^+ \rightleftharpoons 2NO + I_2 + 6H_2O$$

with NO being oxidized by air to form higher oxides, such as NO_2, which then react with iodide to produce iodine and NO, and so on. Such cyclic situations can prevent the achievement of a permanent end-point color in iodometric titrations. Titrations conducted in acid media should be completed as soon as possible after the addition of iodide and the liberation of iodine.

Iodine is appreciably volatile where its solutions are concerned. Its volatility is considerably decreased in the presence of an excess of iodide, since this permits the formation of the triiodide ion. Iodine solutions should not, however, be left exposed to the atmosphere for appreciable periods, and standard solutions of iodine should be kept in well-stoppered bottles. Again, on the basis of iodine volatility, iodometric titrations should be conducted as soon as possible after the addition of iodide and the liberation of iodine.

Iodimetry   Direct titration of arsenic(III) by iodine can be used as a method for the determination of arsenic or for the standardization of iodine solutions. The half reaction for arsenic is

$$H_3AsO_3 + 3H_2O - 2e \rightleftharpoons H_3AsO_4 + 2H_3O^+$$

with the value of $E^\circ_{f,As(V)/As(III)}$ being 0.58 V in 1 M HCl at 25°C. Thus, in acid solution, the equilibrium for the above reaction lies well to the left. The value of $E^\circ_{I_2/I^-}$ is 0.535 V, indicating that arsenic(III) cannot be oxidized to arsenic(V) by the iodine-triiodide system in acid solution. Lowering the $[H_3O^+]$ drives the arsenic half reaction to the right and, as developed in an example in Chap. 4, Part 5, at a solution pH of 8 the value of $E_{As(V)/As(III)}$ is -0.10 V. In a medium of pH 8, therefore, the separation of $E_{As(V)/As(III)}$ and $E^{\circ}_{I_2/I^-}$ is large enough to permit the quantitative conversion of arsenic(III) to arsenic(V) by iodine. The reaction in such a medium would be shown as

$$H_3AsO_3 + I_2 + 5H_2O \rightleftharpoons HAsO_4^{2-} + 2I^- + 4H_3O^+$$

Although the oxidation reaction is quantitative from about pH 4 to pH 9, it is appreciably slower at pH values less than 7, and titrations of arsenic(III) with iodine are usually conducted in media buffered to pH 8. Such buffered solutions can be obtained by the addition of sodium bicarbonate, $NaHCO_3$, in excess to a hydrochloric or sulfuric acid solution of arsenic(III).

Antimony can be determined by titration with iodine in a somewhat similar technique. Precautions are taken to avoid the separation of basic antimony salts during the pretitration neutralization procedures, and this usually involves the addition of tartaric acid and the formation of the soluble antimony tartarate complex, $SbOC_4H_4O_6^-$. The titration reaction is

$$SbOC_4H_4O_6^- + I_2 + 3H_2O \rightleftharpoons SbO_2C_4H_4O_6^- + 2I^- + 2H_3O^+$$

A common method for the determination of tin involves the titration in acid solution of tin(II) by iodine in the reaction

$$Sn^{2+} + I_2 \rightleftharpoons Sn^{4+} + 2I^-$$

Tin(IV) is reduced prior to the titration with a metallic reductant such as lead or nickel. Because of the ease of oxidation of tin(II) by air in acid solution, the reduction process and all subsequent steps, including the iodine titration, are carried out under an inert gas such as nitrogen or carbon dioxide.

The analysis of sulfide-containing substances is often carried out iodimetrically. The evolved hydrogen sulfide gas is absorbed in a measured volume of acidified standard iodine solution with the reaction

$$H_2S + I_2 + 2H_2O \rightleftharpoons S + 2I^- + 2H_3O^+$$

The excess of iodine is subsequently titrated with a standard sodium thiosulfate solution.

Additional iodimetric methods include

Thiosulfate: $2S_2O_3^{2-} + I_2 \rightleftharpoons S_4O_6^{2-} + 2I^-$

Sulfites or sulfur dioxide: $SO_3^{2-} + I_2 + 3H_2O \rightleftharpoons SO_4^{2-} + 2I^-$
$$+ 2H_3O^+$$

Hydrazine: $N_2H_4 + 2I_2 \rightleftharpoons N_2 + 4HI$

Iodometry The methods of iodometry depend on the oxidation of iodide to iodine and the titration of the liberated iodine with a standard solution of a reducing agent. Sodium thiosulfate is the commonest reductant used in this connection. The reaction between sodium thiosulfate and iodine was discussed in Sec. 9.2.2. Since the oxidant is always reacted with a reasonable excess of iodide, usually potassium iodide, the subsequent titration of the liberated iodine takes place in the presence of iodide ion.

The hydrolysis of iodine in the reaction

$$I_2 + OH^- \rightleftharpoons HOI + I^-$$

becomes significant at a pH of about 8. This is usually unimportant where iodimetric titrations are conducted at this pH, since the iodine added as the titrant is almost immediately reacted with the reductant in solution. Where iodometric work is concerned, however, the liberation of iodine at this solution pH results in appreciable hydrolysis, with the reaction of thiosulfate and hypoiodite being

$$4HOI + S_2O_3^{2-} + 7H_2O \rightleftharpoons 2SO_4^{2-} + 4I^- + 6H_3O^+$$

thus permitting a departure from the expected stoichiometry of the iodine-thiosulfate reaction. In order to avoid this difficulty, titrations of iodine and sodium thiosulfate are generally conducted in solutions having a pH of not higher than 6.

Sodium thiosulfate solutions tend to show a certain degree of instability, due largely to bacterial action and/or the action of light. These actions result in the decomposition of thiosulfate, one of the products being elemental sulfur. Solutions of thiosulfate as sodium thiosulfate at about 9 to 10 pH appear to exhibit the highest stability, and the addition of small amounts of sodium carbonate to such solutions will provide this pH and thereby increased stability. When such additions are made, care should be taken to ensure that the iodine solutions titrated are acid enough to accept without significant change in pH the effect of minor amounts of sodium carbonate added with the sodium thiosulfate solution.

A list of the more commonly encountered iodometric titrations is given in Table 9.10. Only the determination of copper will be discussed here.

In the determination of copper, a weakly acid solution of copper(II) is reacted with an excess of potassium iodide according to

$$2Cu^{2+} + 4I^- \rightleftharpoons 2CuI(s) + I_2$$

The relatively low solubility of cuprous iodide and the excess of iodide ion allows the reaction to go to completion. An excess of iodide of not less than $0.7\ M$ should be used. The liberated iodine is titrated with standard sodium thiosulfate solution.

Too high a pH value at the time of addition of potassium iodide can result in the hydrolysis of the copper(II) ion and an incomplete reaction of copper(II) with iodide. Too low a pH value at this point can permit interference from such elements as arsenic and antimony, if these are present and in the $+5$ state. Iron, if present in the $+3$ state, will react with iodide to form iodine:

$$2Fe^{3+} + 2I^- \rightleftharpoons 2Fe^{2+} + I_2$$

Table 9.10 Iodometric determinations

Substance determined	Reaction
Copper(II)	$2Cu^{2+} + 4I^- \rightleftharpoons 2CuI(s) + I_2$
Gold(III)	$Au^{3+} + 3I^- \rightleftharpoons AuI(s) + I_2$
Arsenic(V)	$H_3AsO_4 + 2I^- + 2H_3O^+ \rightleftharpoons H_3AsO_3 + I_2 + 3H_2O$
Antimony(V)	$H_3SbO_4 + 2I^- + 2H_3O^+ \rightleftharpoons H_3SbO_3 + I_2 + 3H_2O$
HClO	$HClO + 2I^- + H_3O^+ \rightleftharpoons Cl^- + I_2 + 2H_2O$
Cl_2	$Cl_2 + 2I^- \rightleftharpoons I_2 + 2Cl^-$
Br_2	$Br_2 + 2I^- \rightleftharpoons I_2 + 2Br^-$
ClO_3^-	$ClO_3^- + 6I^- + 6H_3O^+ \rightleftharpoons Cl^- + 3I_2 + 9H_2O$
BrO_3^-	$BrO_3^- + 6I^- + 6H_3O^+ \rightleftharpoons Br^- + 3I_2 + 9H_2O$
IO_3^-	$IO_3^- + 5I^- + 6H_3O^+ \rightleftharpoons 3I_2 + 9H_2O$
IO_4^-	$IO_4^- + 8H_3O^+ + 7I^- \rightleftharpoons 4I_2 + 12H_2O$
I^-	$I^- + 3Cl_2 + 9H_2O \rightleftharpoons IO_3^- + 6Cl^- + 6H_3O^+$
	$IO_3^- + 5I^- + 6H_3O^+ \rightleftharpoons 3I_2 + 9H_2O$
NO_2^-	$2HNO_2 + 2I^- + 2H_3O^+ \rightleftharpoons 2NO + I_2 + 4H_2O$
MnO_4^-	$2MnO_4^- + 10I^- + 16H_3O^+ \rightleftharpoons 2Mn^{2+} + 5I_2 + 24H_2O$
$Cr_2O_7^{2-}$	$Cr_2O_7^{2-} + 6I^- + 14H_3O^+ \rightleftharpoons 2Cr^{3+} + 3I_2 + 21H_2O$
CrO_4^{2-}	$2CrO_4^{2-} + 6I^- + 16H_3O^+ \rightleftharpoons 2Cr^{3+} + 3I_2 + 24H_2O$
Iron(III)	$2Fe^{3+} + 2I^- \rightleftharpoons 2Fe^{2+} + I_2$
H_2O_2	$H_2O_2 + 2I^- + 2H_3O^+ \rightleftharpoons I_2 + 4H_2O$
MnO_2	$MnO_2 + 2I^- + 4H_3O^+ \rightleftharpoons Mn^{2+} + I_2 + 6H_2O$

and the addition of ammonium bifluoride, NH_4HF_2, prior to the addition of potassium iodide is commonly used in order to complex the iron(III) as FeF_6^{3-}. The use of NH_4HF_2 also permits proper buffering of the solution at a pH of about 3 to 4. This pH is high enough to prevent the oxidation of iodide by arsenic(V) and antimony(V) and low enough to prevent loss of copper(II) by hydrolysis. Arsenic(III) and antimony(III) could consume iodine in this pH zone, but the oxidizing method of attack used in copper determinations ensures that these elements are present in the $+5$ state. The addition of the starch indicator is made near the end point, and the color change is from blue to yellow-white. Where much lead is present the color change may be from blue to yellow.

The presence of cuprous iodide results in appreciable adsorption of iodine at the precipitate surface. Slow release of this adsorbed iodine near the end point can lead to a return of the blue color subsequent to the apparent end point being achieved. This situation can be relieved by the addition, near the end point, of a small amount of potassium thiocyanate. A reaction between KSCN and CuI at the precipitate surface,

$$CuI + SCN^- \rightleftharpoons CuSCN(s) + I^-$$

results in the release of adsorbed iodine.

9.8   OXIDATION-REDUCTION TECHNIQUES APPLIED TO ORGANIC QUANTITATIVE ANALYSIS

9.8.1   GENERAL

A characteristic of oxidation-reduction reactions as these apply generally in inorganic quantitative analysis is their rapidity, which permits their use in oxidation-reduction titration techniques. The slow rate which characterizes oxidation-reduction reactions associated with organic compounds prevents in nearly every instance their application in direct titration procedures. Where oxidation-reduction reactions can be applied in the quantitative determination of organic compounds, the normal procedure is to add a measured excess of the oxidant or reductant, allowing the reaction to proceed to completion over a relatively extended period. The unreacted portion of the oxidant or reductant is then determined by some well-established method. Subsequently, the amount of oxidant or reductant determined as having reacted with the organic compound becomes the basis for the quantitative estimation of the compound.

Several problems may arise in connection with such analytical techniques. The oxidants and reductants which can be used are seldom specific, and certain mixtures of organic compounds can not be analyzed by the simple method outlined. Again, organic reactions of the types involved are usually irreversible and do not lend themselves to the application of the customary methods associated with equilibrium calculations. In addition, the oxidation or reduction processes applied to organic substances often result in side reactions and a variety of products, so that an inconsistent stoichiometry is frequently encountered. Despite all these possible drawbacks, many oxidation-reduction procedures involving organic compounds can be applied under controlled and reproducible conditions, so that adequately consistent results can be secured.

9.8.2   PROCESSES INVOLVING THE PERMANGANATE ION

In alkaline media, and most commonly sodium hydroxide media, permanganate ion, MnO_4^-, can be reduced to either manganate ion, MnO_4^{2-}, or manganese dioxide, MnO_2. Reaction conditions that favor the production of MnO_4^{2-} rather than MnO_2 are to be preferred, since the former reaction proceeds easily, cleanly, and rapidly. The use of a fairly concentrated solution of sodium hydroxide (about 2 M) and a reasonable excess of permanganate ion (about two times the theoretical amount) provides a proper basis for the permanganate-to-manganate reaction.

For example, a determination of glycerol can be be made by adding a measured excess of permanganate ion to a 2 M solution of sodium hydroxide containing the glycerol to be estimated. The reaction is allowed to proceed to completion, which requires about 30 min at room temperature and involves the following:

$$\begin{array}{l} H_2C-OH \\ \quad | \\ HC-OH + 14MnO_4^- + 20OH^- \rightarrow 3CO_3^{2-} + 14MnO_4^{2-} + 14H_2O \\ \quad | \\ H_2C-OH \end{array}$$

The excess of MnO_4^- may be determined by one of two methods.

1. The solution may be acidified with sulfuric acid after the reaction is complete, warmed to about 50°C, and the excess MnO_4^-, together with the MnO_4^{2-} produced in the above reaction, reduced to Mn^{2+} by the addition of an excess of a standard solution of sodium oxalate. The excess of sodium oxalate is then back-titrated with a standard solution of potassium permanganate to the appearance of a pink color. The difference between the calculated volume of reductant solution for the amounts of permanganate solutions used and the actual titration volume allows calculation of the amount of permanganate consumed in the reaction with glycerol and, of course, the amount of glycerol. This difference is usually quite small, so that the determination is subject to a fairly large relative error.

It can be shown as a general equation that

$$g = \frac{\{5(vm + tp) - 2wy\}M}{x} \tag{232}$$

where g = weight of the organic compound, mg

v = volume of MnO_4^- added to the 2 M solution of NaOH containing the organic compound, ml

m = molarity of the MnO_4^- solution

x = number of moles of MnO_4^- consumed per mole of organic compound oxidized

w = volume of standard sodium oxalate ($C_2O_4^{2-}$) solution added, ml

y = molarity of the sodium oxalate ($C_2O_4^{2-}$) solution

t = volume of standard potassium permanganate solution used in the back-titration, ml

p = molarity of the potassium permanganate solution

M = molecular weight of the organic compound, g

and the organic compound is oxidized to CO_2 and H_2O only. The method permits a sharp end point but a high relative error of about 10 ppt.

2. The second method requires the addition of excess barium(II) to the solution after the reaction is complete, followed by titration of the excess permanganate with a standard solution of sodium formate in accordance with

$$H \cdot COO^- + 2MnO_4^- + 3Ba^{2+} + 3OH^- \rightleftharpoons$$
$$2BaMnO_4(s) + BaCO_3(s) + 2H_2O$$

Barium ions react with both CO_3^{2-} and MnO_4^{2-} resulting from the above reaction and from the glycerol reaction, allowing the reaction with formate to proceed to completion without the reduction of MnO_4^{2-} to MnO_2. The end point is usually taken as the disappearance of the pink color due to the permanganate ion, but it is somewhat difficult to see on account of the masking effect of the green precipitate of $BaMnO_4$.

Examples of other organic substances which can be determined in a similar manner are methanol, formate ion, formaldehyde, pentoses, hexoses, phenol, glycollic acid, tartaric acid, ethylene glycol, and salicylic acid. Where the amount of MnO_4^- consumed per mole of organic compound is large (about 10 mol or more per mole of organic compound) the acidification and reduction titration method is usually applied; where it is small, the barium(II) addition followed by titration with standard sodium formate is often preferred.

The technique of determination by oxidation with MnO_4^- is somewhat reduced in value by the lack of specificity of MnO_4^-, so that the method is best applied where only one reaction can take place, and that stoichiometrically (e.g., the determination of ethylene glycol or glycerol in water solutions).

Stamm* provides additional details relative to the use of alkaline permanganate in the quantitative determination of organic compounds.

9.8.3 PROCESSES INVOLVING THE CERIUM(IV) ION

Certain organic compounds can be determined as the result of oxidation by the dichromate ion in processes similar to those involving the permanganate ion. For example, an acid solution of dichromate will quantitatively oxidize ethylene glycol in the reaction

$$3 \begin{array}{c} H_2C-OH \\ | \\ H_2C-OH \end{array} + 5Cr_2O_7^{2-} + 40H_3O^+ \rightarrow 6CO_2 + 10Cr^{3+} + 69H_2O$$

A heating period in excess of 90 min, at approximately 95°C, is required however, resulting in an oxidation process too slow to be of value as a quantitative technique.

Where the cerium(IV) ion is used in sulfuric acid medium, the oxidation of appropriate organic compounds generally proceeds slowly and very often with an inconsistent stoichiometry, even at elevated temperatures. Some improvement in reaction rate and a definite improvement in stoichiometry can often be secured through the use of sulfuric acid media at 10 to 12 M instead of the usual 3 to 5 M. Even here, the slow reaction rates (boiling and refluxing for one hour or more) tend to detract from the value of the procedure.

In perchloric acid medium, however, the acid strength being about 4 M, certain organic compounds can be oxidized rapidly, often at room temperature, and with a consistent stoichiometry. The attack results in cleavage of the bond between the carbon atoms for any of the following groups adjacent to each other: $-COOH, -COH, >CO, >CHOH, -CH_2OH$, and the active methylene group

*H. Stamm in W. Böttger (ed.), "Newer Methods of Volumetric Analysis," R. E. Oesper, trans., pp. 55–65, Van Nostrand, Princeton, N.J., 1938.

$>CH_2$. Each of the resulting fragments is oxidized to the appropriate saturated carboxylic acid, except in the case of fragments already containing a carboxyl group, where the carboxyl group is then oxidized to CO_2.

Examples

$$
\begin{array}{l}
COOH \\
| \qquad + 2Ce^{4+} + 2H_2O \rightarrow 2CO_2 + 2Ce^{3+} + 2H_3O^+ \\
COOH
\end{array}
$$
Oxalic
acid

$$
\begin{array}{l}
COOH \\
| \\
HC-OH \\
| \qquad + 6Ce^{4+} + 8H_2O \rightarrow 2CO_2 + 2H\cdot COOH + 6Ce^{3+} + 6H_3O^+ \\
HC-OH \qquad\qquad\qquad Formic \\
| \qquad\qquad\qquad\qquad\quad acid \\
COOH
\end{array}
$$
Tartaric
acid

$$
\begin{array}{l}
H_2C-OH \\
| \\
HC-OH + 6Ce^{4+} + 9H_2O \rightarrow 3H\cdot COOH + 6Ce^{3+} + 6H_3O^+ \\
| \qquad\qquad\qquad\qquad\qquad Formic \\
HC=O \qquad\qquad\qquad\qquad\quad acid
\end{array}
$$
Glyceraldehyde

$$
\begin{array}{l}
H_3C-C-C-CH_3 + 2Ce^{4+} + 4H_2O \rightarrow 2CH_3\cdot COOH + 2Ce^{3+} + 2H_3O^+ \\
\quad\;\; \| \;\; \| \qquad\qquad\qquad\qquad\qquad Acacid \\
\quad\;\; O \;\; O
\end{array}
$$
Biacetyl

$$
\begin{array}{l}
H_2C-OH \\
| \\
HC-OH + 8Ce^{4+} + 11H_2O \rightarrow 3H\cdot COOH + 8Ce^{3+} + 8H_3O^+ \\
| \qquad\qquad\qquad\qquad\qquad Formic \\
H_2C-OH \qquad\qquad\qquad\quad acid
\end{array}
$$
Glycerol

$$
\begin{array}{l}
COOH \\
| \\
CH_2 + 6Ce^{4+} + 8H_2O \rightarrow 2CO_2 + H\cdot COOH + 6Ce^{3+} + 6H_3O^+ \\
| \qquad\qquad\qquad\qquad\qquad Formic \\
COOH \qquad\qquad\qquad\qquad acid
\end{array}
$$
Malonic
acid

$$
\begin{array}{l}
H_3C-C-CH_2-C-CH_3 + 6Ce^{4+} + 10H_2O \rightarrow 2CH_3\cdot COOH + \\
\quad\;\; \| \qquad\qquad \| \qquad\qquad\qquad\qquad\qquad Acetic \quad H\cdot COOH + 6Ce^{3+} + 6H_3O^+ \\
\quad\;\; O \qquad\qquad O \qquad\qquad\qquad\qquad\qquad acid \qquad Formic \\
Acetyl\ acetone \qquad\qquad\qquad\qquad\qquad\qquad\qquad\qquad acid
\end{array}
$$

The common technique is to add a measured excess of cerium(IV) ion in $4\,M$ perchloric acid to the organic compound, allowing the reaction to proceed to completion at room temperature or, where required, at some carefully selected elevated temperature. The Ce^{4+} remaining unreacted is determined by reduction with a standard solution of either sodium oxalate or iron(II), using an appropriate indicator such as 5-nitro-1,10-phenanthroline iron(II) (nitro ferroin).

Smith and Duke* have investigated the reactions of several organic compounds with Ce^{4+} in perchloric acid media; the student is referred to their work for further information.

9.8.4 PROCESSES INVOLVING POTASSIUM BROMATE

Bromine reacts with appropriate organic compounds by substitution or addition. The former is typified by the reaction of phenol with bromine:

In such aromatic ring compound reactions, bromine substitutes for hydrogen in the ring structure. Addition reactions occur with olefins, involve the double bonds, and are typified by the reaction of ethylene and bromine:

Where reactions of these types take place rapidly and on a quantitative basis, the technique can be applied for analytical purposes. One method would be to use as the titrant for the solution or dispersion of the organic compound a standard solution of bromine, detecting as the titration end point the first indication of unreacted bromine. A second method, useful particularly where there is a slow reaction rate between bromine and the organic compound, would be to add a measured excess of standard bromine solution, allowing the reaction to go to completion, and subsequently back-titrating the unreacted bromine with a standard solution of a reductant such as arsenious acid.

Standard solutions of bromine are inconvenient to use, since the volatility of bromine from its aqueous solutions renders it impossible to maintain a fixed concentration level over short periods. On the other hand, a solution of potassium bromate reacts with bromide ion in the presence of about $0.1\,M$ hydrogen ion to

*G. F. Smith and F. R. Duke, *Ind. Eng. Chem., Anal. Ed.*, **15**: 120 (1943).

yield bromine:

$$BrO_3^- + 5Br^- + 6H_3O^+ \rightleftharpoons 3Br_2 + 9H_2O$$

A solution of bromate and bromide thus becomes, in the presence of adequate hydrogen ion, an excellent source of controlled amounts of bromine. Two general techniques may be applied in quantitative organic analysis.

In the direct titration method, a standard solution of potassium bromate is prepared. This solution, if neutral, may be prepared to contain an excess of potassium bromide. Solution neutrality will guarantee that the concentration of bromine will be negligible and the solution highly stable. This solution would then be used to titrate an acidified (about 0.1 M [H_3O^+]) solution or dispersion of the organic substance. Alternatively, the standard potassium bromate solution may be prepared free from potassium bromide, an excess of this latter salt being added to the acidified solution or dispersion of the organic substance. Titration with the standard potassium bromate solution then follows. In either case, bromine is produced during the titration and the bromination of the organic compound proceeds. As soon as the organic compound has reacted completely, the next addition of potassium bromate solution results in unreacted bromine. An appropriate indicator responds to this first bromine excess and signals the titration end point.

In the indirect method, used basically with slow reaction rate systems, a measured excess of standard potassium bromate solution containing an excess of potassium bromide is added to the acidified solution or dispersion of the organic substance. After the reaction has proceeded to completion, the excess bromine is back-titrated with a standard solution of arsenious acid according to

$$H_3AsO_3 + Br_2 + 3H_2O \rightleftharpoons H_3AsO_4 + 2Br^- + 2H_3O^+$$

An alternative to this method of back-titration is to add an excess of potassium iodide to the solution when the bromination process is complete. The liberated iodine is then titrated in the usual manner with a standard sodium thiosulfate solution.

Bromination procedures, particularly those conducted by the addition of an excess of potassium bromate–bromide solution, should be allowed to take place in stoppered flasks in order to prevent loss of bromine.

The indicators capable of being used in direct bromination titrations were discussed generally in Sec. 9.7.4. These include certain azo-type dye substances, such as methyl red or methyl orange, which are brominated easily to pale-yellow products with the first unreacted bromine. Such indicator reactions, being irreversible, require avoidance during titration of localized high concentrations of potassium bromate–bromide solution. Slow addition of the titrant, with rapid and efficient stirring, is essential. Several indicators showing reversible reactions with bromine are available, and these include p-ethoxychrysoidin, quinoline yellow, and α-naphthoflavone.

For the indirect titration process, one of the available indicators having a reversible reaction with bromine should be used in the back-titration procedure involving standard arsenious acid solution. Where potassium iodide in excess has been added, followed by titration of the liberated iodine with standard sodium thiosulfate solution, starch serves as the indicator in the usual manner.

Several organic compounds can be determined by the potassium bromate–bromide method. These include aromatic ring compounds with ring substituents which strongly influence ortho and/or para position bromination. Examples are phenol, salicylic acid, sulfonic acid, *m*-toluidine, aniline, etc. The degree of unsaturation of olefin mixtures can also be determined, the method being applied largely in the determination of the degree of unsaturation of petroleum products, fats, greases, oils, etc.

A particularly interesting example of the technique is the titration of 8-hydroxyquinoline as a means of determining metals which form insoluble 8-hydroxyquinolates. For example, aluminum will react with 8-hydroxyquinoline according to

$$Al^{3+} + 3C_9H_7ON + 3H_2O \rightleftharpoons Al(C_9H_6ON)_3(s) + 3H_3O^+$$

The precipitate can be filtered off, washed, and redissolved in strong acid, returning the 8-hydroxyquinoline to solution. Potassium bromide in excess is then added, and the 8-hydroxyquinoline is titrated with a standard solution of potassium bromate using methyl red as the indicator. The reaction proceeds according to

8-Hydroxyquinoline

The determination of aluminum, as well as other cations precipitated by 8-hydroxyquinoline, can be made in this way.

9.8.5 PROCESSES INVOLVING PERIODIC ACID OR ITS SALTS

Periodic acid and periodates are capable of selectively oxidizing organic compounds having hydroxyl groups on adjacent carbon atoms. This situation was first investigated by Malaprade.* These substances will, in general, oxidize hydroxyl, aldehyde, or ketone groups with adjacent carbon atoms, as well as primary and secondary amine groups on carbon atoms adjacent to one carrying a hydroxyl group or associated with a carbonyl group.

*L. Malaprade, *Compt. Rend.*, **186**:382 (1928).

Aqueous solutions of iodine in the $+7$ state are complex and contain the species H_5IO_6 (paraperiodic acid), HIO_4 (metaperiodic acid), $H_4IO_6^-$, $H_3IO_6^{2-}$, IO_4^-, etc. The distribution of these species depends largely on the solution pH. Most of the reactions of interest in quantitative organic analysis using periodates occur most rapidly at a solution pH of 4. Since we have, for H_5IO_6,

$$H_5IO_6 + H_2O \rightleftharpoons H_3O^+ + H_4IO_6^- \qquad K_1 = 5.1 \times 10^{-4}$$

$$H_4IO_6^- + H_2O \rightleftharpoons H_3O^+ + H_3IO_6^{2-} \qquad K_2 = 2.0 \times 10^{-7}$$

and for HIO_4,

$$HIO_4 + H_2O \rightleftharpoons H_3O^+ + IO_4^- \qquad K_a = 2.30 \times 10^{-2}$$

the implication is that the species $H_4IO_6^-$ and/or IO_4^- are the active oxidants in the organic reactions of interest.

Several forms of periodate may be used, the most common being paraperiodic acid, H_5IO_6, potassium metaperiodate, KIO_4, and sodium metaperiodate, $NaIO_4$. Prepared solutions of paraperiodic acid or the metaperiodates in water may lose strength rapidly if the water contains organic material. The method of preparation and storage should avoid this possibility.

Periodic acid or periodate solutions may be standardized by titration against standard arsenic(III) solutions, such as a standard sodium arsenite solution. In this technique a measured volume of the periodate solution is buffered to a neutral or slightly alkaline condition by the addition of a concentrated solution of either bicarbonate (as $NaHCO_3$) or borate (as $Na_2B_4O_7$). An excess of potassium iodide is added and iodine is liberated according to

$$IO_4^- + 3I^- + H_2O \rightleftharpoons IO_3^- + I_3^- + 2OH^-$$

or

$$H_4IO_6^- + 3I^- \rightleftharpoons IO_3^- + I_3^- + 2OH^- + H_2O$$

The liberated iodine (as the triiodide ion) is titrated in the usual manner with a standard solution of arsenic(III) using starch as the indicator.

Alternatively, a measured volume of the periodate solution may be added to an acid solution containing an excess of potassium iodide. Iodine is then liberated according to

$$HIO_4 + 11I^- + 7H_3O^+ \rightleftharpoons 4I_3^- + 11H_2O$$

or

$$H_5IO_6 + 11I^- + 7H_3O^+ \rightleftharpoons 4I_3^- + 13H_2O$$

The liberated iodine is titrated with a standard solution of sodium thiosulfate using starch as the indicator.

The reactions of periodate with appropriate organic compounds are usually carried out in aqueous media, although limited solubility of the organic substance in water may dictate the use of ethanol, methanol, acetic acid, etc., as the solvent. Most oxidation reactions involving periodate show maximum rates at a solution pH of about 4, so that where possible this solution pH is maintained. The product of periodate is usually the iodate ion, IO_3^-. Since the use of elevated temperatures tends to introduce undesirable side reactions and an inconsistent stoichiometry, most periodate oxidations of organic substances are allowed to take place at or below room temperature. Reaction rates vary from a few minutes to one hour or more, and excesses of periodate are used to hasten the reaction process.

The following summarizes the oxidation processes relative to periodate and organic compounds.

1. Hydroxyl groups on adjacent carbon atoms result in cleavage of the bond between the carbon atoms. The hydroxyl groups are oxidized to carbonyl groups.

$$
\begin{matrix}
H_2C-OH \\
| \\
H_2C-OH
\end{matrix}
+ H_4IO_6^- \rightarrow 2H_2C=O + IO_3^- + 3H_2O
$$

Ethylene glycol   Formaldehyde

2. Carbonyl groups involving adjacent carbon atoms, or a carbonyl group adjacent to a carbon atom carrying a hydroxyl group, result in cleavage of the bond between the carbon atoms. The carbonyl group is oxidized to a carboxyl group and the hydroxyl group to a carbonyl group.

$$
\begin{matrix}
H_3C-C-C-CH_3 \\
\| \| \\
O O
\end{matrix}
+ H_4IO_6^- \rightarrow 2CH_3\cdot COOH + IO_3^- + H_2O
$$

Biacetyl   Acetic acid

$$
\begin{matrix}
H_3C-CH_2-C-CH-CH_3 \\
\| | \\
O OH
\end{matrix}
+ H_4IO_6^-
$$

2-Hydroxy-3-pentanone

$$\rightarrow H_3C-CH_2-COOH + CH_3\cdot CH=O + IO_3^- + 2H_2O$$

Propionic acid   Acetaldehyde

3. A primary amine group attached to a carbon atom adjacent to one carrying a hydroxyl group, or adjacent to a carbonyl group, results in cleavage of the bond between the carbon atoms. The amine group is lost as ammonia, and its carbon group is oxidized to an aldehyde. The hydroxyl group is oxidized to a carbonyl group and the carbonyl group to a carboxyl group.

$$\begin{array}{c} H_2C-CH_2 \\ |\quad| \\ OH\ NH_2 \end{array} + H_4IO_6^- \rightarrow 2H_2C{=}O + NH_3 + IO_3^- + 2H_2O$$

Formaldehyde

Ethanolamine

4. A secondary amine group attached to a carbon atom adjacent to one with a hydroxyl group, or adjacent to a carbonyl group, results in cleavage of the bond between the carbon atoms. The secondary amine group becomes a primary amine and its carbon atom group is oxidized to an aldehyde. The hydroxyl group and carbonyl group are oxidized as indicated in the foregoing.

$$\begin{array}{c} H_2C-CH_2 \\ |\quad| \\ OH\ NH(CH_3) \end{array} + H_4IO_6^- \rightarrow 2H_2C{=}O + CH_3NH_2 + IO_3^- + 2H_2O$$

Formaldehyde   Methylamine

N-Methyl-ethanolamine

Reaction products may be predicted by assuming that attack and oxidation starts at appropriate points at one end of the molecule and proceeds in steps of one appropriate carbon-carbon bond to the other end. For example, in the case of glycerol, we assume the following steps:

1.
$$\begin{array}{c} H_2C-OH \\ | \\ HC-OH \\ | \\ H_2C-OH \end{array} + H_4IO_6^- \rightarrow H_2C{=}O + \begin{array}{c} HC-CH_2 \\ \|\quad| \\ O\ OH \end{array} + IO_3^- + 3H_2O$$

Glycerol   Formaldehyde   Glycolic aldehyde

2.
$$\begin{array}{c} HC-CH_2 \\ \|\quad| \\ O\ OH \end{array} + H_4IO_6^- \rightarrow H{\cdot}COOH + H_2C{=}O + IO_3^- + 2H_2O$$

Glycolic aldehyde   Formic acid   Formaldehyde

Two analytical approaches may be used. The first involves the addition of a measured excess of periodate solution to the solution or dispersion of the organic substance usually at a pH of about 4. The reaction is then allowed to go to completion at room temperature, and the excess periodate is titrated by either the standard arsenic(III) solution or the standard sodium thiosulfate solution method. The second involves the addition of excess periodate solution (the actual excess need not be measured), with the eventual determination of one or more of the reaction products, from which the quantity of the original organic compound may be estimated.

A mixture of ethylene glycol and glycerol may be analyzed by the addition of a measured excess of paraperiodic acid solution subsequently allowing the reaction to go to completion. The oxidation of glycerol produces 2 mol of formaldehyde and 1 mol of formic acid per mole of glycerol, while the ethylene glycol oxidation results in 2 mol of formaldehyde per mole of glycol. Titration of the reaction mixture with standard sodium hydroxide solution yields a titration involving the formic and iodic acids produced, as well as the excess of paraperiodic acid. If the end point is determined using methyl red as the indicator, any excess of paraperiodic acid will be titrated to the first-stage equivalence point only. If the resulting solution is made weakly alkaline by the addition of a concentrated solution of sodium bicarbonate, the addition of an excess of potassium iodide, followed by titration with standard arsenic(III) solution, will permit determination of the excess paraperiodic acid. A measured excess of paraperiodic acid solution, identical to the volume used to initiate the oxidation of glycerol and ethylene glycol, is now titrated alone with the same standard sodium hydroxide solution, using methyl red as the indicator. The volume for this titration, relative to the volume consumed in the titration conducted on the reaction mixture, allows the determination of formic acid and, therefore, of glycerol. The arsenic(III) titration can now be used, together with the original measured excess of paraperiodic acid, to calculate the ethylene glycol content.

A mixture of ethylene glycol and 1,2-propylene glycol, $CH_2OH-CHOH-CH_3$, reacts with periodate:

$$\begin{array}{l} H_2C-OH \\ \quad | \qquad + H_4IO_6^- \rightarrow 2H_2C=O + IO_3^- + 3H_2O \\ H_2C-OH \end{array}$$

$$\begin{array}{l} H_2C-OH \\ \quad | \\ HC-OH + H_4IO_6^- \rightarrow H_2C=O + CH_3 \cdot CH=O + IO_3^- + 3H_2O \\ \quad | \\ CH_3 \end{array}$$

yielding both formaldehyde and acetaldehyde. Such a mixture of reaction products can be analyzed by separating the two aldehydes and determining one of them. A typical technique is to pass carbon dioxide through the solution resulting from the oxidation reaction with periodate. The formaldehyde and acetaldehyde are carried off by the carbon dioxide gas into a closed system, first through a solution of glycine to absorb the formaldehyde and then through a solution of sodium bisulfite to absorb the acetaldehyde. The latter absorption results in the reaction

$$\begin{array}{c} \qquad\qquad\qquad SO_3H \\ \qquad\qquad\qquad | \\ CH_3 \cdot CH=O + NaHSO_3 \rightleftharpoons CH_3-C-O-Na^+ \\ \qquad\qquad\qquad | \\ \qquad\qquad\qquad H \end{array}$$

The excess sodium bisulfite in the solution is destroyed, the acetaldehyde–sodium bisulfite addition compound is attacked to release sodium bisulfite, which is then determined by iodimetric titration to yield the quantity of acetaldehyde and, of course, 1,2-propylene glycol. Note that mixtures of ethylene glycol, glycerol, and 1,2-propylene glycol could be analyzed by a combination of glycerol determined by the formic acid–sodium hydroxide titration technique, 1,2-propylene glycol by the bisulfite titration method, and ethylene glycol by the determination of the excess of paraperiodic acid after completion of the oxidation process and correction of the periodate consumed for the amounts relative to glycerol and 1,2-propylene glycol.

9.9   DETERMINATION OF WATER BY THE KARL FISCHER METHOD

This technique, of very extensive application, is used in the determination of water, particularly in organic substances.* The method is based on the use as a titrant of an anhydrous methanol solution containing iodine, sulfur dioxide, and pyridine. When the reagent is added to a methanol solution containing water, the reactions

$$C_5H_5N \cdot I_2 + C_5H_5N \cdot SO_2 + C_5H_5N + H_2O$$
$$\rightarrow 2C_5H_5N \cdot H^+I^- + C_5H_5N \cdot SO_3$$

$$C_5H_5N \cdot SO_3 + CH_3OH \rightarrow C_5H_5N \cdot H^+CH_3SO_4^-$$

take place. A large excess of methanol is required to prevent the reaction

$$C_5H_5N \cdot SO_3 + H_2O \rightarrow C_5H_5N \cdot H^+HSO_4^-$$

which also consumes water, from taking place. Excess sulfur dioxide and pyridine are also present, so that the reaction with water is determined on the basis of the iodine value of the reagent.

The titrant can be purchased already prepared from most supply houses. It is somewhat unstable and should be standardized frequently against a standard solution of water in methanol. The reagent should be protected at all times against contamination by water. The sample to be analyzed for water content is dissolved in methanol.

The end point can be taken as the first appearance of an excess of the reagent, the solution color change being from the yellow of the reaction products to the brown of the reagent. End-point detection by potentiometric means is to be preferred, a "dead stop" end point being particularly sharp. Such techniques are discussed in a later section on potentiometry and potentiometric titrations (Chap. 12, Sec. 12.2).

*J. Mitchell and D. M. Smith, "Aquametry," Interscience, New York, 1948.

PROBLEMS

1. Using the table of *standard* reduction potentials given in Appendix I, indicate which of the following reactions might be expected to occur spontaneously:

(a) $Ti^{3+} + Fe^{3+} + 3H_2O \rightleftharpoons TiO^{2+} + Fe^{2+} + 2H_3O^+$

(b) $Cu^+ + Ce^{4+} \rightleftharpoons Cu^{2+} + Ce^{3+}$

(c) $2Fe^{2+} + Br_2(aq) \rightleftharpoons 2Fe^{3+} + 2Br^-$

(d) $Pb^\circ + 2Cr^{3+} \rightleftharpoons Pb^{2+} + 2Cr^{2+}$

(e) $2Ce^{3+} + H_3AsO_4 + 2H_3O^+ \rightleftharpoons 2Ce^{4+} + HAsO_2 + 4H_2O$

2. The following titration reactions will occur spontaneously with the E_f° values given. The concentration of the first component on the right, $[Ox_1]$, will be 0.0500 M at the equivalence point in each case if the conversion is quantitative. Again, if the conversion is quantitative, the concentration of the first component on the left, $[Red_1]$, should not exceed 10^{-6} M at the equivalence point in each case. Indicate for each reaction whether or not the conversion of Red_1 to Ox_1 will be quantitative at the titration equivalence point.

(a) $5VO^{2+} + MnO_4^- + 33H_2O \rightleftharpoons 5VO_4^{3-} + Mn^{2+} + 22H_3O^+$
$$E_{f,VO_4^{3-}/VO^{2+}}^\circ = 1.00 \text{ V}$$
$$E_{f,MnO_4^-/Mn^{2+}}^\circ = 1.51 \text{ V}$$

(b) $Ti^{3+} + Fe^{3+} + 3H_2O \rightleftharpoons TiO^{2+} + Fe^{2+} + 2H_3O^+$
$$E_{f,TiO^{2+}/Ti^{3+}}^\circ = 0.10 \text{ V}$$
$$E_{f,Fe^{3+}/Fe^{2+}}^\circ = 0.73 \text{ V}$$

(c) $Sn^{2+} + I_2 \rightleftharpoons Sn^{4+} + 2I^-$
$$E_{f,Sn^{4+}/Sn^{2+}}^\circ = 0.14 \text{ V}$$
$$E_{f,I_2/I^-}^\circ = 0.54 \text{ V}$$

(d) $H_3AsO_3 + I_2 + 4H_2O \rightleftharpoons H_2AsO_4^- + 2I^- + 3H_3O^+$
$$E_{f,As(V)/As(III)}^\circ = 0.38 \text{ V}$$
$$E_{f,I_2/I^-}^\circ = 0.54 \text{ V}$$

3. Iron(III) will react with iodide ion in the reaction

$$2I^- + 2Fe^{3+} \rightleftharpoons I_2 + 2Fe^{2+}$$

(a) What is the equilibrium constant for this reaction?

(b) When equivalent amounts of Fe^{3+} and I^- are added together in solution, will the reaction be quantitative relative to $Fe^{3+} \rightarrow Fe^{2+}$?

(c) Suggest conditions under which this reaction could be made sufficiently quantitative to provide an iodometric method for the determination of iron.

(d) For $[Fe^{2+}] = 5.0 \times 10^{-2} M$ at the close of the reaction and $[Fe^{3+}] = 10^{-6} M$ maximum, indicate the conditions required.

4. For a redox reaction where $n_1 = n_2 = 1$ and $a = b = c = d = 1$, the minimum difference $E_{Ox_2/Red_2}^\circ - E_{Ox_1/Red_1}^\circ$ for a quantitative conversion of Red_1 to Ox_1 is 0.555 V at 25°C. What would the minimum difference be at (a) 50°C and (b) 75°C.   *Ans.* (a) 0.602; (b) 0.649

5. In any redox titration reaction system,

$$n_2c Red_1 + n_1a Ox_2 \rightleftharpoons n_2d Ox_1 + n_1b Red_2$$

At what point of titration completion (10 percent, 25 percent, etc.) will the value of E_{cell} vs. SHE be given by $E_{f,Ox_1/Red_1}^\circ$ for the system being titrated? At what point of titration completion will the value of E_{cell} be given by $E_{f,Ox_2/Red_2}^\circ$ for the titrant system? Assume for simplicity that $n_1 = n_2 = a = b = c = d$ = 1.   *Ans.* 50 percent complete, 200 percent complete

6. (*a*) Determine the value of E_{cell} vs. SHE at the equivalence point for the following titration reactions:

(1) $Ti^{3+} + Fe^{3+} + 3H_2O \rightleftharpoons TiO^{2+} + Fe^{2+} + 2H_3O^+$
$$[H_3O^+] = 1\,M$$

(2) $Fe^{2+} + VO_4^{3-} + 6H_3O^+ \rightleftharpoons Fe^{3+} + VO^{2+} + 9H_2O$
$$[H_3O^+] = 5\,M$$

(3) $U^{4+} + 2VO_4^{3-} + 8H_3O^+ \rightleftharpoons UO_2^{2+} + 2VO^{2+} + 12H_2O$
$$[H_3O^+] = 1\,M$$

Given

$$E^\circ_{f,TiO^{2+}/Ti^{3+}} = 0.100\ V$$

$$E^\circ_{f,Fe^{3+}/Fe^{2+}} = 0.700\ V$$

$$E^\circ_{f,VO_4^{3-}/VO^{2+}} = 1.000\ V$$

$$E^\circ_{f,UO_2^{2+}/U^{4+}} = 0.334\ V$$

(*b*) From Table 9.8 select the indicators best suited for use with the above titrations.

7. The following solutions were prepared and standardized against the substances indicated. Calculate the molarity and normality of each of the solutions in the standardization reactions involved.

(*a*) A solution of $K_2Cr_2O_7$ standardized against $0.2468 \pm 0.0002\,g$ of iron wire 99.4 percent pure requires 42.34 ± 0.04 ml of the solution to reach a sodium diphenylbenzidine sulfonate indicator end point. Phosphoric acid is added to the iron solution before the titration.

(*b*) A solution of $KMnO_4$ standardized against $0.3000 \pm 0.0002\,g$ of pure $Na_2C_2O_4$ requires 37.89 ± 0.04 ml to reach the pink MnO_4^- color end point.

(*c*) A solution of $Na_2S_2O_3$ is standardized as follows. $0.3000 \pm 0.0002\ g$ of $K_2Cr_2O_7$ is dissolved in 50 ml of water. 50 ml of a solution at pH 5, containing HCl and 3 g of KI, is then added and mixed thoroughly. After 5 min, the solution is diluted to 250 ml and the liberated iodine titrated with the $Na_2S_2O_3$ solution. Starch solution is added near the end point as the indicator, and the blue color disappears at 28.95 ± 0.04 ml of $Na_2S_2O_3$ solution.

8. A solution of potassium permanganate is prepared and standardized. The concentration of the solution was determined as $0.01894 \pm 0.00001\ M$. Show the normality and titer of this solution relative to the following titration reactions:

(*a*) $5Fe^{2+} + MnO_4^- + 8H_3O^+ \rightleftharpoons 5Fe^{3+} + Mn^{2+} + 12H_2O$
Titer for iron

(*b*) $5Na_2C_2O_4 + 2MnO_4^- + 16H_3O^+ \rightleftharpoons 10CO_2 + 2Mn^{2+} + 10Na^+ + 24H_2O$
Titer for sodium oxalate

(*c*) $5H_3AsO_3 + 2MnO_4^- + 6H_3O^+ \rightleftharpoons 5H_3AsO_4 + 2Mn^{2+} + 9H_2O$
Titer for arsenic

(*d*) $3Mn^{2+} + 2MnO_4^- + 6H_2O \rightleftharpoons 5MnO_2 + 4H_3O^+$
Titer for manganese

9. A $0.5201 \pm 0.0002\,g$ sample of impure As_2O_3 is dissolved and eventually secured in a solution of pH 8. The arsenic(III) is titrated with $0.0946 \pm 0.0002\ M$ iodine, using starch as the indicator. At the end point 24.89 ± 0.04 ml of iodine solution is added. What is the percent As_2O_3 in the impure material? *Ans.* 44.8%

10. A $5.041 \pm 0.001\ g$ sample of common grey iron is treated with 1:1 HCl and the evolved gases, including H_2S, absorbed in 50.00 ± 0.04 ml of acidified $0.00480 \pm 0.00001\ M$ iodine solution. The excess iodine is then titrated with $0.01000 \pm 0.00005\ M$ $Na_2S_2O_3$ solution, using starch as the indicator. 9.85 ± 0.04 ml of $Na_2S_2O_3$ solution is required to reach the end point. What is the percent sulfur in the iron, as determined by the evolution method?

11. An iron-bearing material contains vanadium as the only substance other than iron capable of being reduced by stannous chloride. The reduction reactions are

$$Sn^{2+} + 2Fe^{3+} \rightleftharpoons Sn^{4+} + 2Fe^{2+}$$

$$Sn^{2+} + 2VO_4^{3-} + 12H_3O^+ \rightleftharpoons Sn^{4+} + 2VO^{2+} + 18H_2O$$

Neither Fe^{2+} nor VO^{2+} are reoxidized by mercuric chloride in the subsequent oxidation of the excess Sn^{2+}. Fe^{2+} and VO^{2+} are oxidized by $KMnO_4$, but only Fe^{2+} is oxidized by $K_2Cr_2O_7$ when the titration is carried out at the proper acidity. A 1.0182 ± 0.0004 g sample of the material is carried through to a titration with 0.01864 ± 0.00002 M $KMnO_4$. The end-point volume corrected for the indicator blank is 24.98 ± 0.04 ml. A 1.1658 ± 0.0004 g sample is now carried through to a titration with 0.01653 ± 0.00002 M $K_2Cr_2O_7$ under conditions where only the Fe^{2+} is oxidized. The end-point volume after indicator blank correction is 20.65 ± 0.04 ml. What are the percentages of iron and vanadium in the material?

12. A powder contains, together with inert substances, both As_2O_3 and As_2O_5. A 1.0168 ± 0.0002 g sample is dissolved in acid solution, buffered to pH 8, and titrated with 0.0506 ± 0.0001 M iodine solution. The starch indicator end point occurs at 30.47 ± 0.04 ml. The solution is now made strongly acid to HCl and an excess of KI is added. The liberated iodine is titrated with 0.3045 ± 0.0002 M $Na_2S_2O_3$ solution, 17.10 ± 0.04 ml being required to reach the starch indicator end point. What are the percentages of As_2O_3 and As_2O_5 in the powder?   *Ans.* As_2O_3, 15.00%; As_2O_5, 12.00%

13. The reaction

$$2Cu^{2+} + 2I^- \rightleftharpoons 2Cu^+ + I_2$$

shows an equilibrium constant of

$$\frac{[Cu^+]^2[I_2]}{[Cu^{2+}]^2[I^-]^2} = 10^{-12.8}$$

indicating that iodide should not be capable of reducing copper(II). However, iodide acts also as a precipitant of copper(I):

$$Cu^+ + I^- \rightleftharpoons CuI(s)$$

the CuI showing a very low solubility as indicated by the K_{sp} (CuI) value of 1.0×10^{-12}. The overall reaction is then

$$2Cu^{2+} + 4I^- \rightleftharpoons 2CuI(s) + I_2$$

(a) Calculate the $[Cu^{2+}]$ remaining in the solution when 0.080 mol of KI is added to 100.00 ml of a 0.0200 M solution of Cu^{2+}.

(b) Calculate the $[Cu^{2+}]$ remaining in the solution when 0.020 mol of KI is added to 100.0 ml of a 0.0200 M solution of Cu^{2+}.

(c) What are your conclusions relative to a and b?

14. A sample weight of 0.2891 ± 0.0002 g of a bronze alloy is dissolved in 1:1 HNO_3. This solution, after preparation, is treated with KI so as to liberate I_2 and leave a residual $[I^-]$ of 0.8 M. The liberated iodine is titrated with 0.1065 ± 0.0002 M $Na_2S_2O_3$ solution. The starch indicator end point occurs at 31.89 ± 0.04 ml. What is the percent copper?

15. (a) Give an explanation of why starch indicator solution is added only near the end of an iodometric titration with $Na_2S_2O_3$ solution but *could* be added at the beginning of an iodimetric titration.

(b) When starch indicator is added at the beginning of an iodimetric titration, what special titration conditions must be observed?

16. A mineral contains calcium oxide and inert substances. A 0.5689 ± 0.0002 g sample is dissolved in an appropriate medium, and the calcium is then precipitated under the best conditions as calcium oxalate. This precipitate is dissolved in dilute sulfuric acid, and the resulting solution is titrated with 0.02081 ± 0.00004 M $KMnO_4$. The titration requires a total of 20.31 ± 0.04 ml of $KMnO_4$ solution, the indicator blank being determined as -0.04 ml. What is the percentage of calcium in the mineral? *Ans.* 7.43%

17. The two half reactions

(a) $MnO_4^- + 4H_3O^+ + 3e \rightleftharpoons MnO_2 + 6H_2O$

(b) $MnO_4^- + 2H_2O + 3e \rightleftharpoons MnO_2 + 4OH^-$

are possible. Note that half reaction *a* is H_3O^+ dependent. From this dependency calculate the $E°$ value for half reaction *b* given $E°$ for half reaction *a* as 1.679 V.

18. A material contains only Na_2S, As_2S_3 and inert substances. A 0.4000 ± 0.0002 g sample is dissolved in acid and the solution is boiled and subsequently buffered to pH 8. A titration of this solution with 0.0750 ± 0.0001 M iodine yields an end-point volume with starch indicator of 18.56 ± 0.04 ml. A 0.2000 ± 0.0002 g sample is dissolved in acid and the evolved H_2S is absorbed in 50.00 ± 0.04 ml of acidified iodine solution of the same strength as the previous titrant. The excess of iodine is then titrated with 0.1515 ± 0.0002 M $Na_2S_2O_3$ solution, 27.33 ± 0.04 ml being required to reach a starch indicator end point. What are the percentages of Na_2S and As_2S_3 in the material?

19. A bleaching powder for industrial use contains calcium hypochlorite, $CaCl(OCl)$. When treated with dilute sulfuric acid the reaction

$$CaCl(OCl) + 2H_3O^+ \rightleftharpoons Ca^{2+} + Cl_2 + 3H_2O$$

takes place, the Cl_2 representing the "available chlorine." When both iodide ion and dilute sulfuric acid solution are used together, we have

$$CaCl(OCl) + 2I^- + 2H_3O^+ \rightleftharpoons Ca^{2+} + I_2 + 2Cl^- + 3H_2O$$

and the liberated iodine is equivalent to the available chlorine. 5.000 ± 0.001 g of bleaching powder is ground in a mortar with a small amount of water until a smooth paste completely free from lumps is secured. This paste is transferred carefully to a 500-ml volumetric flask, made up to the mark with water, and the flask and contents are shaken until a homogeneous suspension is obtained. 50.00 ± 0.02 ml is immediately pipetted into a flask. 2.0 g of KI and 15.0 ml of 10.0% H_2SO_4 solution is added. The liberated iodine is titrated, using starch as the indicator, with 0.2039 ± 0.0001 M $Na_2S_2O_3$. The end point is noted at 29.92 ± 0.04 ml of $Na_2S_2O_3$ solution. Determine

(a) The percent available chlorine

(b) The percent $CaCl(OCl)$

20. A sample of low-grade solder is analyzed for lead. A 0.7589 ± 0.0002 g sample is dissolved in acid, resulting in a solution of Pb^{2+}. An excess of K_2CrO_4 is added and the lead precipitated as $PbCrO_4$. No other chromate separates. The precipitate is filtered off and washed and then dissolved in acid. This solution is treated with an excess of KI and the iodine liberated by the reaction of KI with CrO_4^{2-} is titrated with 0.0509 ± 0.0001 M $Na_2S_2O_3$. Starch is used as the indicator, but because of the yellow color of the precipitated lead iodide, a blank of -0.10 ml must be applied. The total volume at the end point is 11.22 ± 0.04 ml of the $Na_2S_2O_3$ solution. What is the percentage of lead in the solder? *Ans.* 5.15%

21. Develop the general formula

$$g = \frac{\{5(vm + tp) - 2wy\}M}{x}$$

for the determination of the amount of organic compound oxidized by an alkaline excess of MnO_4^- to CO_2 and H_2O, with the formation of MnO_4^{2-}.

22. To 25.00 ± 0.04 ml of a solution containing ethylene glycol is added 50.00 ± 0.04 ml of $0.0450 \pm 0.0001\ M$ $KMnO_4$ which is 3 M to NaOH. The reaction is allowed to proceed for 30 min at room temperature. The solution is then acidified with sulfuric acid and heated to 50°C and the excess MnO_4^- and the MnO_4^{2-} produced by the reaction is reduced to Mn^{2+} by the addition of 10.00 ± 0.04 ml of $0.3000 \pm 0.0003\ M$ sodium oxalate ($Na_2C_2O_4$). The excess of sodium oxalate is then back-titrated with $0.0450 \pm 0.0001\ M$ $KMnO_4$ solution, 2.30 ± 0.04 ml being required to reach the pink end-point color. What is the weight in milligrams of ethylene glycol in the 25-ml sample taken?   *Ans.* 35.8 mg

23. From the details of Sec. 9.8 alone, suggest a method of analysis which would permit the determination individually of formic acid and tartaric acid in a solution containing both substances.

24. A 1.0058 ± 0.0002 g sample of a material containing aluminum and inert substances is dissolved and the aluminum precipitated as aluminum 8-hydroxyquinolate. The precipitate is filtered off, washed, and then redissolved in 0.1 M HCl. An excess of KBr is added and bromination of 8-hydroxyquinoline allowed to go to completion as the solution is titrated with $0.1185 \pm 0.0002\ M$ $KBrO_3$ solution. The end point, using methyl red as the indicator, occurs at 42.28 ± 0.04 ml of bromate solution. What is the percentage of aluminum in the material?

25. An organic substance contains sorbitol, $CH_2OH(CHOH)_4CH_2OH$, as the only compound capable of being oxidized by periodate. 0.465 ± 0.001 g is dissolved and diluted to exactly 100.00 ± 0.08 ml. A 25.00 ± 0.04 ml aliquot is pipetted into 25.00 ± 0.04 ml of $0.0350 \pm 0.0002\ M$ $NaIO_4$ at a pH of 4. The reaction is allowed to take place over a 45-min period, at the end of which the solution is made slightly alkaline by the addition of a concentrated solution of $NaHCO_3$. An excess of KI is added, and the liberated iodine titrated with $0.0360 \pm 0.0002\ M$ arsenious acid. Starch is used as the indicator, and the end point is noted at 6.22 ml.

   (*a*) What is the percentage of sorbitol in the organic substance?

   (*b*) Write the reaction between sorbitol and periodate.

26. The previous problem can be approached differently. Suppose that a second 25.00 ± 0.04 ml aliquot of the prepared solution of the organic substance is added to 25.00 ± 0.04 ml of $0.0350 \pm 0.0002\ M$ $NaIO_4$ solution without any adjustment of pH. After a reaction period of 45 min, the solution is titrated with $0.0765 \pm 0.0002\ M$ NaOH, using phenol red as the indicator. Assuming that only the formic acid produced by the oxidation reaction is titrated, indicate what volume of NaOH solution would be required to reach the end point.   *Ans.* 6.81 ml

27. A liquid contains only glycerol, ethylene glycol, and inert substances. A 2.000 ± 0.001 g sample is diluted to 250.00 ± 0.12 ml in a volumetric flask. A 50.00 ± 0.05 ml aliquot is added to 50.00 ± 0.05 ml of $0.0516 \pm 0.0002\ M$ paraperiodic acid, and the reaction is allowed to continue for 45 min at room temperature. The solution is then titrated with $0.0955 \pm 0.0002\ M$ NaOH using methyl red as the indicator. The end point occurs at 35.46 ± 0.04 ml of NaOH solution. The solution is then made slightly alkaline by the addition of a concentrated solution of $NaHCO_3$. An excess of KI is added, and the liberated iodine is titrated with a $0.0400 \pm 0.0002\ M$ solution of arsenious acid. The starch end point is noted at 13.44 ± 0.04 ml of arsenious acid solution. A 50.00 ± 0.05 ml portion of the 0.0516 M paraperiodic acid solution is now titrated with the 0.0955 M NaOH solution, the methyl red end point occurring at 28.40 ± 0.04 ml. What are the percentages of glycerol and ethylene glycol in the liquid?

28. Suppose that an organic mixture contains only serine, $CH_2OHCH(NH_2)COOH$, and inert substances. 1.0056 ± 0.0004 g of the mixture is dissolved and added to 50.00 ± 0.05 ml of $0.0496 \pm 0.0002\ M$ paraperiodic acid in a closed system. After the reaction is complete, the released NH_3 is distilled off and absorbed in 50.0 ± 0.1 ml of $0.0286 \pm 0.0002\ M$ HCl. The absorbing solution is titrated with $0.0238 \pm 0.0002\ M$ NaOH to a phenol red end point at 13.40 ± 0.04 ml.

   (*a*) What is the percentage of serine in the mixture?

   (*b*) Write the oxidation reaction for serine and paraperiodic acid.   *Ans.* (*a*) 11.6%

29. A mixture contains serine and threonine, $CH_3CHOHCH(NH_2)COOH$, with inert substances. 2.000 ± 0.002 g of the mixture is dissolved and diluted to 50.00 ± 0.05 ml and a 25.00 ± 0.02 ml aliquot is added to 25.00 ± 0.04 ml of 0.0555 ± 0.0002 M paraperiodic acid in a closed system. After the reactions are completed, the released NH_3 is distilled off and absorbed in 25.00 ± 0.04 ml of 0.0508 ± 0.0002 M HCl. The absorbing solution is titrated with 0.0250 ± 0.0002 M NaOH to a chlorophenol end point at 7.20 ± 0.04 ml.

A second 25.00 ± 0.02 ml aliquot is treated with the same volume of the paraperiodic acid solution, again in a closed system. After the reactions are completed, the formaldehyde and acetaldehyde produced are drawn off by a stream of CO_2 gas passed through the solution. The acetaldehyde is then absorbed in a solution of $NaHSO_3$. The excess of $NaHSO_3$ is destroyed, and the acetaldehyde-bisulfite addition compound is attacked so as to release the sodium bisulfite, which is then titrated with 0.0530 ± 0.0002 M iodine solution to an end point at 10.00 ± 0.05 ml.

(*a*) What are the percentages of serine and threonine in the mixture?

(*b*) Write the oxidation reaction for threonine and paraperiodic acid.

REFERENCES

1. Böttger, W. (ed.): "Newer Methods of Volumetric Analysis," trans., R. E. Oesper, Van Nostrand, Princeton, N.J., 1938.
2. Kolthoff, I. M., P. J. Elving, and E. B. Sandell: "Treastise on Analytical Chemistry," Interscience-Wiley, New York, 1959 (continuing series).
3. Kolthoff, I. M., V. A. Stenger, and R. Belcher: "Volumetric Analysis," 2d ed., 3 vols., Interscience, New York, 1957.
4. Laitinen, H. A.: "Chemical Analysis," McGraw-Hill, New York, 1960.
5. Lingane, J. J.: "Electroanalytical Chemistry," 2d ed., Interscience, New York, 1958.
6. Smith, T. B.: "Analytical Processes: A Physico-Chemical Interpretation," 2d ed., E. Arnold, London, 1952.

10

Acid-base Titrations in Nonaqueous Solvents

10.1 GENERAL

It was indicated in Chap. 4, Part 2 that the hydrogen ion is solvated in aqueous solution, existing as the hydrated proton H_3O^+. In a nonaqueous solvent, such as ethanol or glacial acetic acid, the hydrogen ion is also solvated, at least to some degree. In ethanol, we have

$$H^+ + C_2H_5OH \rightleftharpoons C_2H_5OH_2^+$$

and in glacial acetic acid,

$$H^+ + CH_3 \cdot COOH \rightleftharpoons CH_3 \cdot COOH_2^+$$

A solvent can display to a degree acidic or basic properties, or the ability to donate or receive a proton. Water is, for example, a less acidic solvent than glacial acetic acid, which is to say that H_3O^+ donates the proton much less readily than does $CH_3 \cdot COOH_2^+$. This situation is reflected in the effect of such solvents on the ability of a substance such as ammonia to function as a base. NH_3 acts as a stronger base when reacted with hydrogen ion in glacial acetic acid than when so reacted in water; this reflects the more acidic nature of glacial acetic acid as a solvent.

The titration of ammonia by a strong acid in water and in glacial acetic acid well exemplifies this situation. We have

$$NH_3 + H_3O^+ \rightleftharpoons H_2O + NH_4^+$$

and

$$NH_3 + CH_3 \cdot COOH_2^+ \rightleftharpoons CH_3 \cdot COOH + NH_4^+$$

Because of the more acidic nature of glacial acetic acid, NH_3 appears to act as a stronger base in this solvent, so that the equilibrium for the reaction of NH_3 and strong acid in glacial acetic acid lies significantly further to the right than does that for the corresponding reaction in water. Under these conditions it is apparent that a base too weak to be titrated in water by a strong acid, with reasonable accuracy of equivalence-point location, might be so titrated successfully providing that the titration is carried out in a solvent sufficiently acidic in nature. It is possible, for example, to titrate successfully in a properly selected nonaqueous solvent bases with K_b values less than 10^{-7}, a situation not possible in aqueous medium. Conversely, an acid too weak to be titrated successfully in water with a strong base might be titrated satisfactorily in a solvent sufficiently more basic than water. Such titrations are of particular interest in the case of weak acid or base substances which are either insoluble or of limited solubility in water.

10.2 SOLVENTS AND SOLVENT PROPERTIES

10.2.1 SOLVENTS

Solvents may generally be divided into three classes with respect to their acid-base properties. The amphiprotic solvents are those which show both acidic and basic properties and undergo self-dissociation or autoprotolysis, one molecule acting as an acid reacting with a second molecule acting as a base. Water is the most common example of this class, although other solvents are capable of similar dissociation reactions.

1. $H_2O + H_2O \rightleftharpoons H_3O^+ + OH^-$
2. $CH_3OH + CH_3OH \rightleftharpoons CH_3OH_2^+ + CH_3O^-$
3. $C_2H_5OH + C_2H_5OH \rightleftharpoons C_2H_5OH_2^+ + C_2H_5O^-$
4. $CH_3 \cdot COOH + CH_3 \cdot COOH \rightleftharpoons CH_3 \cdot COOH_2^+ + CH_3 \cdot COO^-$
5. $NH_3 + NH_3 \rightleftharpoons NH_4^+ + NH_2^-$
6. $H_2N-CH_2CH_2-NH_2 + H_2N-CH_2CH_2-NH_2 \rightleftharpoons$
                   $H_2N-CH_2CH_2-NH_3^+ + H_2N-CH_2CH_2-NH^-$

Methanol (2) and ethanol (3) show acid-base properties similar to those of water; glacial acetic acid (4) shows stronger acid properties than water; ammonia (5) and ethylenediamine (6) show basic properties stronger than that of water.

A second group of solvents, called aprotic or inert solvents, show no obvious acidic or basic properties, have no dissociable proton, and show no tendency to donate or accept a proton. Typical of this group are solvents such as benzene, chloroform, and carbon tetrachloride.

The third group includes solvents which show basic properties but no obvious acidic properties. Such solvents, typified by ethers, ketones, and pyridine, can react by accepting a proton,

$$\text{Pyridine} + H_2O \rightleftharpoons \overset{\cdot\cdot}{\underset{H^+}{N}} + OH^-$$

but show no reaction where a proton is donated.

10.2.2 THE PROPERTIES OF SOLVENTS

Acidic or basic properties of solvents   It has been already noted as a general statement that a solvent with strong acidic properties tends to permit a solute to show stronger basic properties. Thus, the dissociation constant for a weak base attains higher values in such solvents than in water, a solvent of weak acidic and basic properties.

Conversely, a solvent with strong acidic properties tends to decrease the acidic properties of a solute. It should be noted that in water, both hydrochloric and perchloric acids act as strong acids, being dissociated to an extent so great that it is not possible to determine any difference between their degrees of dissociation. In glacial acetic acid, however, while neither acid attains the degree of dissociation reached in water, that for perchloric acid can be determined as being significantly higher than that for hydrochloric acid. In effect then, the use of a solvent with strong acidic properties allows a differentiation of the strengths of strong acid solutes. Solutes that are weak acids in water display lower dissociation constants in strongly acidic solvents than are secured in water.

With solvents which show strong basic properties, the opposite effect is noted, weak acids showing higher dissociation constants and weak bases showing lower dissociation constants than those typical where water is the solvent.

The autoprotolysis constant of solvents   The degree of completion of an acid-base titration reaction, at the equivalence point, in an amphiprotic solvent, is a function, in part, of the dissociation constant for the solute being titrated and the dissociation constant or, as it is called, the autoprotolysis constant of the solvent. The dissociation constant for a particular solute will, of course, be related to the nature of the solvent involved.

Consider a solute B with weak basic properties but capable in some degree of titration by a strong acid in water as the solvent. We have

$$B + H_3O^+ \rightleftharpoons BH^+ + H_2O$$

and the relationship

$$\frac{[BH^+]}{[B][H_3O^+]} = \frac{[BH^+][OH^-]}{[B]} \frac{1}{[H_3O^+][OH^-]} = \frac{K_b}{K_w} = K_{eq} \tag{233}$$

The equilibrium constant for the titration reaction is thus given by K_b/K_w; and the extent to which the reaction goes to completion will in part be given by the magnitude of K_b/K_w. Here K_b is the dissociation constant for the weak base B in water, with K_w being the dissociation or autoprotolysis constant for water.

The titration of the same weak base with a strong acid in glacial acetic acid as the solvent yields

$$B + CH_3 \cdot COOH_2^+ \rightleftharpoons BH^+ + CH_3 \cdot COOH$$

and we have

$$\frac{[BH^+]}{[B][CH_3 \cdot COOH_2^+]} = \frac{[BH^+][CH_3 \cdot COO^-]}{[B]} \frac{1}{[CH_3 \cdot COOH_2^+][CH_3 \cdot COO^-]}$$
$$= \frac{K_b'}{K_{auto}} = K_{eq} \tag{234}$$

Here the equilibrium constant is given by K_b'/K_{auto}, where

$K_b' = $ the dissociation constant for the weak base B in glacial acetic acid
$$= \frac{[BH^+][CH_3 \cdot COO^-]}{[B]} \tag{235}$$

$K_{auto} = $ the autoprotolysis constant for glacial acetic acid
$$= [CH_3 \cdot COOH_2^+][CH_3 \cdot COO^-] \tag{236}$$

In a similar manner, for the titration of a weak acid HA by a strong base in water, we have

$$HA + OH^- \rightleftharpoons A^- + H_2O$$

and

$$\frac{[A^-]}{[HA][OH^-]} = \frac{[H_3O^+][A^-]}{[HA]} \frac{1}{[H_3O^+][OH^-]} = \frac{K_a}{K_w} = K_{eq} \tag{237}$$

with the equilibrium constant for the titration reaction being given by K_a/K_w.

The same titration carried out in ethanol, using as the titrant the strong base sodium ethoxide, C_2H_5ONa, yields

$$HA + C_2H_5O^- \rightleftharpoons A^- + C_2H_5OH$$

with

$$\frac{[A^-]}{[HA][C_2H_5O^-]} = \frac{[C_2H_5OH_2^+][A^-]}{[HA]} \frac{1}{[C_2H_5OH_2^+][C_2H_5O^-]}$$
$$= \frac{K_a'}{K_{auto}} = K_{eq} \tag{238}$$

the equilibrium constant being given by K_a'/K_{auto}, where

$K_a' =$ the dissociation constant for HA in ethanol
$$= \frac{[C_2H_5OH_2^+][A^-]}{[HA]} \tag{239}$$

$K_{auto} =$ the autoprotolysis constant for ethanol
$$= [C_2H_5OH_2^+][C_2H_5O^-] \tag{240}$$

For the weak base B in titration, the value of K_b' will generally be somewhat greater than K_b, because of the strong acidic properties of glacial acetic acid relative to those of water. The autoprotolysis constant for glacial acetic acid, approximately 3.5×10^{-15}, is somewhat lower than that for water. These two factors will result in a value for $K_b'/K_{auto} > K_b/K_w$. If these were the only two factors to be considered, the titration in glacial acetic acid medium should show a more complete titration reaction at the equivalence point, a larger $\pm\Delta$pH for comparable titration volumes around the equivalence point, and the possibility of a more accurate location of this point. Even where the K_b' value is similar to, or even lower than, K_b, the choice of a solvent with a very low K_{auto} value could improve the degree of completion at the equivalence point and the accuracy of its location. Table 10.1 gives the value of K_{auto} for some common solvents.

The dielectric constant of solvents and its effect on solvent behavior Another factor affecting the strength of the acidic or basic characteristics of a solute is the dielectric constant of the solvent. This constant is representative of the amount of work required to be done to separate, in the dissociation process, the two oppositely charged particles of a solute in the solvent involved. It therefore represents the ease of dissociation of a solute in the particular solvent.

Table 10.1 K_{auto} **for some common solvents**

Solvent	°C	K_{auto}
Formic acid, H·COOH	25	6.3×10^{-7}
Water, H_2O	25	1.0×10^{-14}
Acetic acid, CH_3·COOH	25	3.5×10^{-15}
Ethylenediamine, $H_2N-CH_2CH_2-NH_2$	25	5.0×10^{-16}
Methanol, CH_3OH	25	2.0×10^{-17}
Ethanol, C_2H_5OH	25	8.0×10^{-20}

In any solvent SH, an *uncharged* solute B with basic properties dissociates according to

$$B + SH \rightleftharpoons BH^+ + S^-$$

while an *uncharged* solute HA with acidic properties dissociates according to

$$HA + SH \rightleftharpoons SH_2^+ + A^-$$

Both reactions require a transfer of charge (the ionization step) and the *separation* of the charged particles (the dissociation step). In water, which has a high dielectric constant ($D_{water} = 78.5$), the amount of work required to separate the charged particles is minimal. In a solvent such as glacial acetic acid, with a low dielectric constant ($D_{acetic} = 6.1$), a greater amount of work is required to accomplish the same situation.

It will be apparent that a reaction of dissociation for an uncharged solute proceeds further to the right with a solvent of high dielectric constant than with one of low value for the constant. As a case in point, we note that the dissociation constant for acetic acid, as a weak acid solute, is 10^{-5} in water ($D_{water} = 78.5$) and 10^{-10} in ethanol ($D_{ethanol} = 24.3$).

Where the solute is a *charged* weak acid or base the situation is different, and we have, for solutes of acidic or basic properties,

$$BH^+ + SH \rightleftharpoons SH_2^+ + B$$

$$A^- + SH \rightleftharpoons HA + S^-$$

Here we have the transfer of charge but *no separation of charge*. Under these circumstances the magnitude of the dielectric constant of the solvent has little or no effect on the dissociation constant for the solute involved. For example, the weak acid solute NH_4^+ shows a value of 1.0×10^{-9} for K_a in water and a value of 1.0×10^{-10} for K_a' in ethanol.

Table 10.2 provides dielectric constant values for some common solvents.

10.3   THE CHOICE OF A SOLVENT FOR ACID-BASE TITRATIONS

It is apparent that the choice of a solvent for a given titration, leaving aside as obvious the need for an operational solubility situation for the solute, will be decided by several factors. The most important of these are as follows.

1. The acidic or basic properties of the solvent, that is, its ability to donate or receive a proton, will be of importance. Where a weakly acid solute is to be titrated, a solvent of strong basic properites is to be preferred, and vice versa.
2. The magnitude of the autoprotolysis constant for the solvent will be significant, a solvent of low K_{auto} value being preferred. The lower the value of K_{auto}, the better the possibility of locating the equivalence point with reasonable accuracy.

3. The dielectric constant of the solvent will have an influence on the degree of completion of the titration reaction. High values of the constant will permit greater degrees of completion of the reaction at the equivalence point.

Many combinations of these factors may occur, not all of them resulting in the most favorable titration circumstances. The following provides some idea of how these combinations can be considered.

Suppose that a weak base B is to be titrated by a strong acid in a choice of water or formic acid as the solvent. The acidic properties of formic acid, being much greater than those for water, would be an advantage, as would be the relatively high dielectric constant ($D_{formic} = 58.5$). On the other hand, the value of K_{auto} for formic acid is about 6.3×10^{-7}, being much higher than that for water. This high value would tend to offset the advantages of the strong acidic properties and the high dielectric constant. Formic acid, in effect, would not provide sufficient advantage over water as a solvent for this titration.

If the same titration were considered, projecting the use of glacial acetic acid instead of water, the advantages with respect to glacial acetic acid would depend on the strong acidic properties and the K_{auto} value, which at approximately 3.5×10^{-15} is slightly lower than K_w for water. The dielectric constant of glacial acetic acid ($D_{acetic} = 6.1$) is extremely low, however. This would tend to offset to some extent the advantages; but, all in all, glacial acetic acid does have definite advantages as a solvent for weak base titrations.

Again, using the same titration as an example and considering the use of ethanol as the solvent, we find that an advantage is associated with the low K_{auto} value of about 8.0×10^{-20}. The relatively low dielectric constant ($D_{ethanol} = 24.3$), and the more or less neutral properties of ethanol, tend to offset this advan-

Table 10.2 Dielectric constants of some common solvents

Solvent	°C	D
Water, H_2O	25	78.5
Formic acid, $H \cdot COOH$	16	58.5
Formic acid, $H \cdot COOH$	25	62
Methanol, CH_3OH	25	32.6
Ethanol, C_2H_5OH	25	24.3
Acetone, $CH_3 \cdot CO \cdot CH_3$	25	20.7
Isopropanol, $CH_3 \cdot CHOH \cdot CH_3$	25	18.3
Methyl isobutyl ketone, $(CH_3)_2 \cdot CH \cdot CH_2 \cdot CO \cdot CH_3$	25	13.1
Ethylenediamine, $H_2N \cdot CH_2 \cdot CH_2 \cdot NH_2$	25	12.5
Pyridine, C_5H_5N	25	12.3
Acetic acid, $CH_3 \cdot COOH$	25	6.1
n-Butylamine, $CH_3 \cdot CH_2 \cdot CH_2 \cdot CH_2NH_2$	25	5.3
Benzene, C_6H_6	20	2.3

tage. An example of this offsetting tendency can be noted by observing that the dissociation constants for most weak acids are 10^{-5} to 10^{-6} times the values in water. This reduces the ratio K_a'/K_{auto} very frequently to a level quite similar to that for K_a/K_w, so that only minor advantages are obtained when ethanol is used as the solvent for weak acid titrations.

Where charged weak acids are to be titrated by a strong base, it would be distinctly advantageous to conduct such titrations in a solvent such as ethanol rather than in water. With no charge-separation situation involved the low dielectric constant for ethanol would not seriously reduce the value of the dissociation constant of the weak acid, allowing practically full advantage to be taken of the low K_{auto} value. In order to provide a clear picture of this situation, we shall consider the titration of $0.100\ M$ NH_4^+ (a weak charged acid) with $0.100\ M$ NaOH in water, as compared with the titration of $0.100\ M$ NH_4^+ with $0.100\ M$ sodium ethoxide, C_2H_5ONa, a strong base, in ethanol. The values for the various constants are

$$K_{auto,water} = K_w = 1.0 \times 10^{-14}$$
$$K_{auto,ethanol} = 8.0 \times 10^{-20}$$
$$K_a\,(NH_4^+)\ \text{in water} = 1.0 \times 10^{-9}$$
$$K_a'(NH_4^+)\ \text{in ethanol} = 1.0 \times 10^{-10}$$

In water The reaction is $NH_4^+ + OH^- \rightleftharpoons NH_3 + H_2O$. The value of K_a/K_w is 1.0×10^5. At the equivalence point of the titration we have

$$[NH_3] + [NH_4^+] = 0.0500\ M$$

and

$$[NH_4^+] = [OH^-]$$

so that

$$\frac{[NH_3]}{[NH_4^+][OH^-]} = \frac{[NH_3]}{[NH_4^+]^2} = \frac{K_a}{K_w} = 1.0 \times 10^5$$

and

$$[NH_3] = 1.0 \times 10^5\,[NH_4^+]^2$$

Therefore, at the equivalence point, we have

$$1.0 \times 10^5\,[NH_4^+]^2 + [NH_4^+] = 0.0500\ M$$

from which

$$[NH_4^+]\ \text{at the equivalence point} = 7.0 \times 10^{-4}\ M$$

In ethanol The reaction is $NH_4^+ + C_2H_5O^- \rightleftharpoons NH_3 + C_2H_5OH$. The value of K_a'/K_{auto} is 1.2×10^9. At the equivalence point of this titration we have

$$[NH_3] + [NH_4^+] = 0.0500 \ M$$

and

$$[NH_4^+] = [C_2H_5O^-]$$

so that

$$\frac{[NH_3]}{[NH_4^+][C_2H_5O^-]} = \frac{[NH_3]}{[NH_4^+]^2} = \frac{K_a'}{K_{auto}} = 1.2 \times 10^9$$

and

$$[NH_3] = 1.2 \times 10^9 [NH_4^+]^2$$

At the equivalence point we have, therefore,

$$1.2 \times 10^9 [NH_4^+]^2 + [NH_4^+] = 0.0500 \ M$$

from which

$$[NH_4^+] \text{ at the equivalence point} = 6.5 \times 10^{-6} \ M$$

Note that the degree of completion of the titration reaction at the equivalence point is much greater in ethanol than in water. The water-base titration would indeed be subject to a considerable error in the location of the equivalence point; that in ethanol would yield a relatively small error only. For further consideration of this point the student is referred to Chap. 6, Sec. 6.9.

Aprotic or inert solvents, or solvents with basic but no acidic properties, show extremely low values of the autoprotolysis constant and, therefore, the possibility of advantage with respect to the degree of completion of the titration reaction. This advantage may be offset to a significant extent, where a charged weak acid or base is concerned, if the dissociation constant is appreciably lowered. The use of aprotic solvents very often involves the disadvantage of low solubility of the solute to be titrated and, as a consequence, many methods recommending solvent mixtures of aprotic and amphiprotic solvents are to be found in the literature.

The equations required for the calculation of values for the hydrogen-ion activity, or the activities of other species, at various points in nonaqueous solvent acid-base titrations are no more difficult to derive than are those applied in the calculation of hydrogen-ion concentrations and the concentrations of other species in aqueous solvent titrations. An excellent example, with reference to the quantitative treatment of equilibria and titrations involving the nonaqueous solvent ethylenediamine, is given by Fischer and Peters.*

*Robert B. Fischer and Dennis G. Peters, "Quantitative Chemical Analysis," 3d ed., pp. 341-353, Saunders, Philadelphia, 1968.

10.4 THE DETECTION OF THE EQUIVALENCE POINT IN NONAQUEOUS TITRATIONS

The detection of the equivalence point or end point in nonaqueous acid-base titrations, using indicators of the same types as those applied in aqueous solution titrations, is rendered difficult as the result of certain circumstances which severely limit their applicability. In many cases these indicators, where they can be used, change color in nonaqueous solvents over pH ranges or, more suitably, over ranges of hydrogen-ion activity that are frequently quite different from those normal to aqueous media. In addition to this, the nature of the color change in such solvents may for some indicators be quite unlike that found for water. Another factor important to consider is that the low dielectric constant for most nonaqueous solvents commonly used results in little separation of charge or dissociation where these indicators are concerned. Ionization takes place, but without charge separation, the charged particles existing largely as ion pairs subject to equilibria relationships different from those peculiar to ionization and dissociation in aqueous medium.

Some indicators, such as methyl red, phenolphthalein, thymolphthalein, etc., can be used for end-point detection in certain nonaqueous solvents and associated titrations. In the main, however, equivalence-point or end-point detection is best accomplished by potentiometric means. The general technique is discussed in detail in Sec. 12.2. The procedure can be briefly outlined as follows. Special electrodes, capable of potentiometric measurement in nonaqueous solvents, are sometimes although not always used. Such an electrode immersed in solution will report the changing potential due to changes in the solution hydrogen-ion activity (or concentration) as the titration progresses. These potential changes of the indicator electrode are measured by a potentiometer in reference to a reference electrode which is also immersed in the solution. Such changes are most pronounced in the neighborhood of the equivalence point, permitting equivalence-point location by graphic analysis of the titration curve of potential vs. titrant volume, or from a tabulation of potential change increments relative to titrant volume increments ($\Delta E/\Delta V_T$ or $\Delta^2 E/\Delta V_T^2$) as obtained from titration data. A general discussion of the graphic and tabulation techniques was given in Chap. 5, Sec. 5.2.

10.5 ACID-BASE TITRATIONS IN NONAQUEOUS SOLVENTS

No attempt will be made here to provide a detailed discussion of various acid-base nonaqueous solvent titrations; only a general indication is given of the applicability of several types of such titrations. The sensitivity of these titrations to the presence of water should be mentioned briefly. There is some ability to tolerate water in very small amounts with respect to some nonaqueous solvent titrations. In general, however, even traces of water should be avoided. Methods of preparing and storing nonaqueous solvents and the titrants employed with them should be such as to eliminate the possibility of contamination by water, carbon dioxide, or other impurities of an acidic nature.

10.5.1 TITRATIONS IN ETHYLENEDIAMINE

Ethylenediamine is an amphiprotic solvent with distinctly basic properties, a relatively low K_{auto} of 5.0×10^{-16} and a relatively low dielectric constant of 12.5. It has definite advantages as a solvent for the titration of weak acid substances such as phenols, amino acids, and most weak carboxylic acids, where the dissociation constant in water is so low as to preclude titration in this medium. The strong base titrant used in such titrations is sodium aminoethoxide, $H_2N-CH_2CH_2-O^-Na^+$, and the equivalence point is usually located potentiometrically.

Because of the low dielectric constant, most electrolytes exist in solution in ethylenediamine as ion pairs rather than as dissociated electrolytes, and this is true even of a strong electrolyte such as sodium bromide. While very weak acids such as phenol (K_a in water $= 1.28 \times 10^{-10}$) provide breaks in the titration curve around the equivalence point adequate to the location of this point with reasonable accuracy when titrated in ethylenediamine, acids such as benzoic acid, a weak acid with $K_a = 6.46 \times 10^{-5}$ in water, act as strong acids in ethylenediamine, yielding very large titration curve breaks around the equivalence point.*

10.5.2 TITRATIONS IN GLACIAL ACETIC ACID

Glacial acetic acid is an amphiprotic solvent with distinctly acidic properties, a relatively low K_{auto} of 3.5×10^{-15} and a low dielectric constant of 6.1. It has advantages in its use as a solvent for the titration of weak base substances such as amines, amino acids, the sodium, potassium, or ammonium salts of carboxylic acids that do not yield strongly basic solutions in water, and the sodium or potassium salts of certain inorganic acids, such as Na_2SO_4, $NaBr$, $NaCN$, Na_2CO_3, $NaHCO_3$, etc. The usual titrant employed is perchloric acid dissolved in glacial acetic acid and standardized against potassium acid phthalate in glacial acetic medium.

While the tolerable amount of water present in such titrations is appreciable compared with that permissible in many other nonaqueous solvent titrations, the limit decreasing with increasing weakness of the base to be titrated, water should not be present. The problem is easily avoided in glacial acetic acid titrations by the addition of acetic anhydride to the solvent, the anhydride reacting with any water present to form acetic acid. While it is possible to use certain indicators such as methyl violet, despite the complexity of their color changes, the potentiometric method of equivalence-point location is to be preferred.

---

* M. L. Moss, J. H. Elliott, and R. T. Hall, *Anal. Chem.*, **20**: 784 (1948).

10.5.3 TITRATIONS IN METHANOL OR ETHANOL

Methanol and ethanol are amphiprotic solvents of neutral properties, relatively low K_{auto} values of 2.0×10^{-17} for methanol and 8.0×10^{-20} for ethanol, and relatively low dielectric constants of 32.6 and 24.3, respectively. These solvents have some advantages in the titration of weak acid solutes but show distinct advantages in the titration of charged weak acid substances such as NH_4^+. The usual titrants for weak acid solutes are the strong bases sodium methoxide, CH_3ONa, and sodium ethoxide, C_2H_5ONa. Again, the potentiometric method of location of the equivalence point is to be preferred.

10.5.4 TITRATIONS IN METHYL ISOBUTYL KETONE

Titrations in methyl isobutyl ketone, $(CH_3)_2 \cdot CH \cdot CH_2 \cdot CO \cdot CH_3$, have been investigated by Bruss and Wyld.* An advantage of this solvent is its nearly neutral characteristics. While not truly an aprotic or inert solvent, since it displays some basic properties, it is capable of being used as the solvent in a range of both weak acid and weak base titrations. It has moreover the capability of titration differentiation between a wide range of acid substances. The titrants usually employed are tetrabutyl ammonium hydroxide, $(C_4H_9)_4N^+OH^-$, in isopropanol, $CH_3 \cdot CHOH \cdot CH_3$, for the titration of acid solutes, and perchloric acid in dioxane for the titration of basic solutes. Equivalence-point location is accomplished potentiometrically.

As examples of the ability of this solvent to differentiate between acids the following should be noted.* It is possible to titrate solutions of

1. Sulfuric acid with a break for each stage
2. Hydrochloric and perchloric acids with a break for each acid
3. Perchloric acid and sulfuric acid with a break for perchloric and for each sulfuric acid stage
4. Phenol and the acids perchloric, hydrochloric, salicylic, and acetic with a break for each substance

PROBLEMS

1. Write the equation for the reaction which takes place upon mixing the following:
   (a) Perchloric acid with glacial acetic acid
   (b) Benzoic acid with ethanol
   (c) Ammonia with methanol
   (d) Acetic acid in ethanol

2. Suppose that a solution of benzoic acid in ethanol is to be titrated with sodium ethoxide, C_2H_5ONa. Show the titration reaction and derive the expression for K_{eq}.

*D. B. Bruss and G. E. A. Wyld, *Anal. Chem.*, **29**:232 (1957).

3. Determine the value of pH $\{-\log\ [C_2H_5OH_2{}^+]\}$ and pH $\{-\log\ [CH_3\cdot COOH_2{}^+]\}$ for pure ethanol and pure glacial acetic acid.   *Ans.* Ethanol 9.55

4. Suppose that 50.00 ml of 0.1000 M benzoic acid is to be titrated in water with 0.1000 M NaOH and in ethanol with 0.1000 M C_2H_5ONa. The value of K_a in water is 6.4×10^{-5}, while that of K'_a in ethanol is 1.0×10^{-10}. Show the value of pH $\{-\log\ [H_3O^+]\}$ and pH $\{-\log\ [C_2H_5OH_2{}^+]\}$ for the following titration points (in milliliters):

       0.00   25.00   45.00   49.00   49.90   50.00   50.10   51.00   55.00

What are your conclusions?   *Ans.* 45.00 ml pH: water 5.2, ethanol 11.0

5. Suppose that 50.00 ml of 0.1000 M NH_4Cl is to be titrated in water with 0.1000 M NaOH and in ethanol with 0.1000 M C_2H_5ONa. Show the values of pH $\{-\log\ [H_3O^+]\}$ and pH $\{-\log\ [C_2H_5OH_2{}^+]\}$ for the following titration points (in milliliters). Take K_a as 1.0×10^{-9} and K'_a as 1.0×10^{-10}.

       45.00   49.00   49.90   50.00   50.10   55.00

What are your conclusions?

REFERENCES

1. Fritz, J. S.: "Acid-Base Titrations in Non-aqueous Solvents," G. F. Smith Chemical Company, Columbus, Ohio, 1952.
2. Huber, W.: "Titrations in Nonaqueous Solvents," Academic, New York, 1967.
3. Kolthoff, I. M., P. J. Elving, and E. B. Sandell: "Treatise on Analytical Chemistry," Interscience-Wiley, New York, 1959 (continuing series).
4. Kucharsky, J., and L. Safarik: "Titrations in Non-aqueous Solvents," Elsevier, Amsterdam, 1963.

11
Theory and Procedures of Gravimetric Analysis

11.1 GENERAL

The procedures of gravimetric analysis are based on the determination of a constituent through some form of weighing operation. This may involve the weighing of the required constituent in either its pure form or in the form of some compound of known composition. In the majority of such cases, the constituent is separated by precipitation from solution of a slightly soluble compound involving the constituent. In others, the slightly soluble compound separated may be related indirectly rather than directly to the constituent. In either case, the method is completed by a weight determination relative to the separated compound or the one devolving directly from it. In a few instances, methods of separation by volatilization of the constituent, or of a compound directly or indirectly related to the constituent, are applied, followed by determination on a basis of weight loss.

The method of precipitation is by far the most common of the procedures leading to determination by gravimetric analysis; and it is therefore important that the various aspects of the precipitation process be thoroughly understood. In the most general terms, for a method of precipitation to be adaptable to gravimetric analysis, several fundamental characteristics must be evidenced.

1. The precipitate involved should be so slightly soluble that the required constituent is left in solution at a concentration level such that it would not be detectable within the weighing limits of the analytical balance. This implies a level in the total volume of the solution of 0.0001 g or less; and, in most cases, this condition will be met by requiring that the desired constituent be left in solution, subsequent to the precipitation process completion, to an extent not exceeding $10^{-6}\,M$. For those conditions surrounding the solubility of slightly soluble salts, and associated equilibria, the student is referred to the details outlined in Chap. 4, Part 3.
2. The precipitate secured should be free from contamination by any of the other constituents in the solution from which it originates. This includes contamination by soluble constituents, a phenomenon occurring to a much greater extent than is often appreciated. Again, the presence of other constituents must not hinder the precipitation process in any way.
3. The precipitate should be easily and rapidly filtered and capable of being easily washed free from impurities of a soluble nature.
4. The weighing form of the precipitate (after drying, ignition, etc.) should be of a known and constant composition.

11.2 THE MECHANISM OF PRECIPITATION

11.2.1 GENERAL

In any process of precipitation the ultimate average size of the particles, their nature—whether crystalline or amorphous—and the degree of purity of the precipitate will depend on several factors. The most important of these are

1. The conditions under which precipitation takes place
2. The characteristics of the precipitating substance
3. The treatment of the precipitate subsequent to precipitation

We shall deal, in one way or another, with each of these factors in the sections following.

11.2.2 NUCLEATION

The formation of precipitate particles occurs from a supersaturated solution. Nuclei, or extremely small particles of precipitate, much less than microscopic in size, provide centers at which subsequent growth takes place. The formation of nuclei can be thought of as taking place as the result of one of two processes. Where relatively low degrees of supersaturation exist, nuclei can be thought of as originating by the clustering of ions at centers involving very fine suspended foreign matter in the solution (i.e., dust), or impurities adhering to the walls of the glass vessel. Such nucleation processes are said to be heterogeneous, and within the limits of the relatively low degrees of supersaturation involved, the number of such nuclei will be but slightly affected by the degree of supersaturation. Where relatively high degrees of supersaturation exist, the nuclei are thought to originate mainly as the result of the coming together by chance of ions in clusters or groups involving a minimum number of ions arranged according to a suitable but definite pattern. This nucleation process is said to be homogeneous, and for relatively high degrees of supersaturation, there is the likelihood that there will be an increase in the number of nuclei formed with increasing degree of supersaturation. It is possible, of course, to imagine nucleation by both heterogeneous and homogeneous processes taking place in solutions of high degree of supersaturation.

Ostwald* investigated the question of solubility and temperature relative to a solid in a solvent. Figure 11.1 indicates the general situation. The solubility of the solid with temperature is shown as the line UU_1. Beyond the line MM_1 the solution becomes unstable and subject to spontaneous nucleation and precipitation. Between the two lines the solution is said to be metastable or supersaturated, and while nucleation and precipitation do not take place spontaneously, these processes can be induced by the action of "seeding," that is, the introduction into the solution of a crystal or crystals of the solid. Thus, for a given temperature T, the slow addition of the precipitating agent to the solution will result in a continuous increase in the concentration of the solute until saturation of the solution is reached at the point S on the line UU_1. Further addition of precipitating agent results in supersaturation, until the point Q on the line MM_1 is reached, at which point nucleation and precipitation begin spontaneously. The degree of supersaturation is given by the term $Q - S$, while the relative supersaturation is given by the ratio $(Q - S)/S$.

*W. Ostwald, *Z. physik. Chem., Leipzig*, **22**:289 (1897).

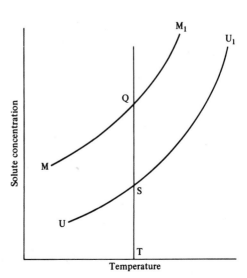

Fig. 11.1   Solubility of a solid with temperature.

The question of the importance of the degree of supersaturation on the particle size of precipitates was investigated by von Weimarn.* The initial rate of precipitation of the solute was taken as being given by

$$\text{Initial rate of precipitation} = K \frac{Q - S}{S} \tag{241}$$

where Q = the concentration of the solute in the supersaturated solution

   S = the equilibrium solubility of the solute (i.e., the solubility of macro particles)

   K = a constant

The higher the value of $Q - S$, the greater the number of nuclei formed initially and, consequently, the finer the particle size of the precipitate. The larger the value of S (and note that this generally increases with increasing temperature), the smaller the value of $(Q - S)/S$ and the less the number of nuclei formed, with consequent relatively large particle size of precipitate. It is apparent that, in the interests of obtaining ease of filtering and washing, larger particles are to be preferred, and the analyst should adjust the precipitation conditions, where possible, to obtain the minimum value for $(Q - S)/S$.

The von Weimarn relationship can at best be only an approximation, since the value of S used refers to the solubility of large crystals. Since the expression deals with the initial formation of precipitate particles, when these are extremely small, the value of S should have reference to microcrystal solubility rather than to that for macrocrystals. Values for S for microcrystals are not generally known, but with most slightly soluble substances, the solubility increases very significantly with particle sizes of two microns (2×10^{-4} cm) and less. Nevertheless, the von Weimarn expression can be used as a guide relative to determining the general conditions under which precipitation should take place.

A more recent study of supersaturation in solution can be found in the work of LaMer and Dinegar.†

11.2.3 PARTICLE SIZE AND GROWTH

The process of nucleation, once started, should proceed to the precipitation of the slightly soluble substance. This would imply particle growth from nucleation size to the average particle size for the initial precipitate. This process of growth will normally be one, for crystalline precipitates, of the spatial extension of the crystal lattice of each nucleus by the orderly addition of ions of the solute from the supersaturated solution. The process of particle growth can, however, be represented by stages of particle-size increase as

Ions (10^{-8} cm) → nucleation clusters (10^{-8} to 10^{-7} cm) → colloidal particles (10^{-7} to 10^{-4} cm) → precipitate particles ($> 10^{-4}$ cm)

---

*P.P. von Weimarn, *Chem. Revs.*, **2**:217 (1926).
†U. K. LaMer and R. H. Dinegar, *J. Amer. Chem. Soc.*, **73**:380 (1951).

Thus all precipitates, during their growth from nucleation centers, pass through the colloidal particle size range, so that a discussion of the colloidal state and its general characteristics will be of fundamental importance.

The colloidal state   Colloids are essentially very small particles ranging from about 10^{-7} to 10^{-4} cm. Because of their small size they exhibit certain properties which, in their turn, serve to characterize the colloidal state.

The extremely small size of these particles results in their continuous and erratic motion in their supporting solution under the influence of collision with solvent molecules. This characteristic motion, visible under the ultramicroscope, is referred to as the brownian movement, and is typical of colloidal suspensions.

Again, the small size implies large surface-area-to-volume ratio; in other words, the number of ions at the surface is large relative to the total number of ions in the particle. In a crystalline particle having a simple cubic lattice structure we note, from Fig. 11.2, that ions of B^+ or A^- below the surface are each surrounded by six ions of opposing charge. Surface ions, on the other hand, are surrounded by only five ions of opposing charge. Such surface ions have, therefore, the ability to attract oppositely charged ions to the surface of the particle and, in turn, to be attracted by oppositely charged ions in the solution or by solvent molecules where these are polar in nature.

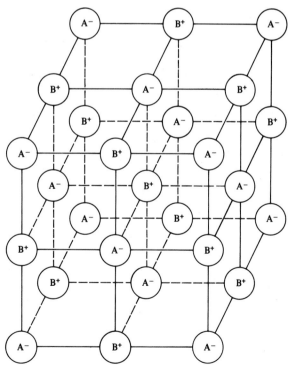

Fig. 11.2   Simple cubic lattice structure.

   The higher solubility of very small particles can then be explained on the basis of the attraction of the surface ions to the polar molecules of solvents such as water. Such ions show a tendency to leave the surface and enter the solution, and with the large surface development of small particles, this results in a high solubility relative to that exhibited by larger particles with less surface development.

   As indicated above, this same situation permits the attraction, to the surface of small particles, of oppositely charged ions from the solution—an action described as adsorption. The tendency will be for such small crystals to preferentially adsorb an ion from the solution common to the crystal lattice, where such an ion is available. Where a common ion is not available, the tendency will be to preferentially adsorb that ion from solution capable of yielding with one of the lattice ions a compound of least solubility. By this means colloidal particles acquire charge, each particle in a given colloidal suspension acquiring a charge of the same sign. A case in point may serve to better illustrate the situation.

   Consider a solution of silver nitrate, $AgNO_3$, to which a very small addition of sodium chloride, NaCl, has been made. The small particles of AgCl are initially formed in a medium containing large amounts of Ag^+ and NO_3^-, with small amounts of Na^+, H_3O^+, and OH^-. These particles will preferentially adsorb Ag^+ ions and will thus all acquire a positive charge. This layer of adsorbed Ag^+ is called the primary layer, and the primary layer will always be found on or very close to the surface of the colloidal particle. This primary layer then attracts a loosely held layer of opposing charge in the form mainly of NO_3^- ions, although some OH^- ions may also be attracted. This loosely held ion layer is called the secondary layer, and while it is not held as closely to the surface as the primary layer, it is held in the immediate neighborhood. Figure 11.3 illustrates the general

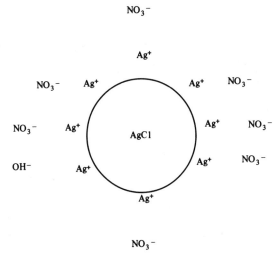

Fig. 11.3 AgCl colloidal particle formed in a solution containing $AgNO_3$.

process, the particle charge here being positive. In the reverse circumstance, where a small volume of $AgNO_3$ solution is added to a solution of NaCl, the primary layer around the AgCl particles will consist of adsorbed Cl^- ions and the secondary layer largely of Na^+ ions with a small contribution of H_3O^+ ions. The AgCl particles will thus all acquire a negative charge.

As an illustration of the process of acquiring charge where an ion common to the precipitate lattice is not available in solution, we may consider very small particles of AgCl in a solution containing $NaNO_3$ and $CH_3 \cdot COONa$. The acetate ion, $CH_3 \cdot COO^-$, will be adsorbed as the primary-layer ion in preference to the nitrate ion, NO_3^-, since this ion provides, with the silver ion of the AgCl lattice, a compound, silver acetate, of lower solubility than silver nitrate. The secondary layer will be mainly Na^+ with a small H_3O^+ contribution. Na^+ will not be adsorbed as the primary-layer ion, since the compound NaCl has a relatively high solubility.

This process of charge acquisition, a necessary preliminary to establishing the colloidal suspension state, results in the particles repelling each other because all have the same charge, preventing the accumulation of particles into larger aggregates and resultant particle size growth to sizes permitting separation by precipitation. The nonsettling of colloidal particles under the influence of gravity is due largely to this effect and to that underlying the brownian movement.

Nucleation processes which, for one reason or another, are in part or in total followed by the establishment of a colloidal suspension state result in either incomplete precipitation or failure to produce a precipitate entirely. Since colloidal particles will pass through the pores of ordinary filters, being retained only by ultrafilters, the method of separation by precipitation fails in such instances.

Although all precipitates pass through the colloidal particle size range during the growth process, duration in this state is seldom prolonged in the majority of cases, since the charge around the colloidal particles is removed as a result of the precipitation process itself. This charge removal results in the discharged particles clustering together into larger aggregates, eventually reaching particle sizes large enough for settling and precipitation. The process of charge removal followed by subsequent growth in particle size is called coagulation or flocculation and, in the normal approach to the process of precipitation, is brought about during the continued addition of precipitant solution to and beyond the level of equivalence of precipitant and substance being precipitated. As will be recalled from Chap. 4, Part 3, it is customary to add an excess of the precipitating ion during the precipitation process in order to reduce to a low level, through the common-ion effect, the concentration in solution of the ion being precipitated. Thus, in the case of the addition of NaCl solution to a solution of $AgNO_3$, continued addition of NaCl results in the separation of colloidal AgCl particles with adsorbed Ag^+ as the primary layer. Very near the point where an amount of Cl^- equivalent to the Ag^+ originally present has been added, the further addition of Cl^- results in the removal of the adsorbed Ag^+ because of the strong attraction between Cl^- and Ag^+. The AgCl particles are thus discharged, permitting particle aggregation to larger particle sizes and subsequently to sizes sufficiently large to permit separation by precipitation.

Ions other than those common to the precipitate lattice can be used in the flocculation process, and a colloidal suspension of AgCl carrying an adsorbed Cl^- primary layer can be flocculated by the addition of soluble salts such as KNO_3, $Ca(NO_3)_2$, or $Al(NO_3)_3$. The minimum quantity of a salt required to flocculate a given quantity of a colloid is called the flocculation value, and the flocculation value will decrease with increasing charge for that salt ion active in initiating the flocculation process. Thus, for a given quantity of the AgCl colloid carrying a primary layer of Cl^-, the flocculation value will decrease relative to the use, in order, of KNO_3, $Ca(NO_3)_2$. or $Al(NO_3)_3$. Table 11.1 shows the mean flocculation values (FV), in millimoles per liter, for various ion charges in relationship to colloids having positive or negative primary-layer charges.

Colloids, upon flocculation, may yield precipitates of two quite different types. Where the precipitate carries down with it a considerable quantity of the solvent, the colloid from which it originated is called lyophilic, and where water is the solvent, the term hydrophilic is applied. Such a precipitate is called a gel; it is typified by precipitates such as $Fe(OH)_3$, $Al(OH)_3$, etc. Gels do not, upon ignition, lose this water readily, and prolonged ignition at high temperature is

Table 11.1   Mean flocculation values (millimoles per liter) for sols of negative and positive charges*

			Negative sols			
Charge of cation	As₂S₃ FV	Ratio†	Au FV	Ratio	AgI FV	Ratio
1+	55	1	24	1	142	1
2+	0.69	0.013	0.38	0.016	2.43	0.017
3+	0.091	0.0017	0.006	0.0003	0.068	0.0005

		Positive sols		
Charge of anion	Fe₂O₃ FV	Ratio†	Al₂O₃ FV	Ratio
1−	11.8	1	52	1
2−	0.21	0.018	0.63	0.012
3−			0.080	0.0015
4−			0.053	0.0010

*H. R. Kruyt (ed.), "Colloid Science," vol. 1, Tables 2 and 4, pp. 308–309, Elsevier, Amsterdam, 1952.
†The ratio refers to the flocculation value (FV) of the monovalent ion.

required. The ignited residues obtained will often exhibit hygroscopicity and must be cooled after ignition in a dessicator carrying an appropriate dessicant. Where the precipitate carries with it but little of the solvent, the originating colloid is called lyophobic, or hydrophobic in the case of water being the solvent. Precipitates from such colloids are called sols and are typified by AgCl, CuS, etc. Such water as is carried down by sols is easily removed, even by drying at temperatures around 110°C, and the dried precipitate or ignited residue is not usually hygroscopic. Although sols and gels represent the two basic forms of precipitate which result from the flocculation of colloidal suspensions, some precipitates exhibit properties somewhere between those for gels and sols.

Even where flocculation processes are successful in separating a colloidal suspension in precipitate form, the adsorptive ability of colloidal particles implies the possibility of serious precipitate contamination by adsorbed foreign ions. This situation will be explored in some depth in a subsequent section.

11.2.4 THE PHYSICAL NATURE OF THE PRECIPITATE

It has been indicated, in the course of the development of some of the basic ideas in the previous sections, that particle size growth eventually results in sizes large enough to settle out as precipitate material. Such an initial precipitate can be termed a primary precipitate, in order to differentiate it from the final form of the precipitate just prior to filtration, which final form may be the end result of such procedures as digestion, aging, etc., carried out on the primary precipitate.

It has been indicated again, in a general way, that precipitate particles will show a crystalline structure. While this is certainly true in nearly every instance of a precipitate just prior to the filtration step, the primary precipitate may or may not show a crystalline structure. We must, of course, be careful to indicate exactly what we mean by a crystal structure. Certain primary precipitates, such as those normally secured in the precipitation of barium sulfate from highly dilute solutions, show a crystal structure clearly visible to the eye, with well-marked crystal faces and a relatively large crystal size. Other primary precipitates, such as silver chloride, show a curdy appearance, with no external evidence of a crystalline form. Examination of such primary precipitates by x-rays presents, however, direct evidence of an internal crystal structure, the crystals being extremely small and resulting in no crystal planes visible to the eye. Still other primary precipitates—gels such as ferric hydroxide for example—show neither visual nor x-ray evidence of a crystal form. Such primary precipitates are said to be amorphous. Upon aging even these precipitates eventually develop an internal crystal structure detectable by x-ray examination.

Two basic factors underlie the development of the physical nature of the primary precipitate. According to Haber*, these factors are the aggregation velocity and the orientation velocity. The aggregation velocity represents the rate at which the primary particles of the precipitate aggregate together to form larger

*F. Haber, *Ber.*, **55**:1717 (1922).

particles. To a very considerable extent the aggregation velocity is a function of the degree of supersaturation at the time of nucleation—the greater the degree of supersaturation, the higher the aggregation velocity. The orientation velocity, on the other hand, represents the rate at which random groupings or aggregates of precipitate ions orient themselves into the orderly array typical of the crystal-lattice structure. The orientation velocity is largely a function of the nature of the precipitate, and strongly polar salts such as barium sulfate show high orientation velocities.

Thus, the physical nature of the primary precipitate will reflect the relative values of the aggregation and orientation velocities. Where the former is strongly dominant, the primary precipitate will tend to show an amorphous or curdy nature; where the latter is dominant, an obviously crystalline form may be the result.

Since the aggregation velocity can be accelerated by increasing the degree of supersaturation, it is apparent that a high value for $(Q - S)/S$ could produce amorphous or curdy precipitates, even in the case of primary precipitates normally found to be obviously crystalline. This is well exemplified when highly concentrated solutions of barium(II) and sulfate ion are mixed together, where the value of $(Q - S)/S$ is extremely high, whereupon barium sulfate in the form of a gel is obtained.

It is important to note, therefore, the significance of the degree of supersaturation on *both* particle size and physical nature of the primary precipitate.

11.2.5 THE CONDITIONS UNDER WHICH PRECIPITATION TAKES PLACE

The preliminary discussions on the degree of supersaturation of solution and its general relationship to primary precipitate particle size and physical nature indicated that certain conditions under which the precipitation process could take place might permit a primary precipitate of relatively large particle size and with a good possibility of displaying a crystalline structure. Where the value of $(Q - S)/S$ could be kept at a comparatively low value, relatively few nuclei would be formed, and these would subsequently grow to provide a primary precipitate of reasonable particle size.

Low values for $(Q - S)/S$ imply low values for Q or high values for S or, preferably, a combination of both. Low values of Q can be obtained by using a dilute solution of the substance to be precipitated and by adding to it a dilute solution of the precipitant—in other words, by maintaining a low concentration of both ions forming the precipitate. At the same time, the slow addition of the precipitant solution, with efficient stirring, assists in this process by minimizing the possibility of high instantaneous or localized ion concentrations.

High values of S, the precipitate solubility, can usually be obtained by precipitation from hot solutions, since most slightly soluble salts show increased solubility with increasing temperature. High values of S can also be obtained in specific instances by adjustment of the solution pH just before or just after the addition of the precipitant, with slow readjustment after thorough mixing. Other techniques, applicable in specific cases, such as complexation procedures, can also be applied.

In general, therefore, in order to facilitate the securing of precipitates of reasonably large primary particle size and with crystalline structures, precipitation should take place as the result of the slow addition, with efficient stirring, of dilute precipitant solutions to hot dilute solutions of the reacting ion. Since the rate of crystal growth increases with increasing temperature, precipitation from hot solutions also aids the process of particle growth by crystal-lattice extension.

Precipitation from homogeneous solution   In the previous section the possibility of increasing S by adjustment of the solution pH either immediately before or after the addition of the precipitant, followed by the slow readjustment of the pH, was mentioned briefly. Such a procedure could, for example, be applied in the case of a precipitate which is soluble (high S value) in acid solution. The acid solution of the substance to be precipitated and the precipitant are subjected to the action of a reagent, or reaction process, which removes hydrogen ions slowly, gradually readjusting the solution pH to a zone where the precipitate is highly insoluble. Such a precipitation process, called precipitation from homogeneous solution, maintains a low degree of supersaturation at all times and permits the formation of relatively few precipitate nuclei that subsequently grow to relatively large size by the slow accumulation of additional precipitate. This slow growth process assists in securing, as well, a crystalline structure. Two cases in point will serve to illustrate the general application of the procedure.

Urea will slowly hydrolyze when acid solutions are boiled, the reaction being

$$CO(NH_2)_2 + H_2O \rightleftharpoons 2NH_3 + CO_2$$

The carbon dioxide is expelled by the boiling solution, and the slow generation of ammonia results in a gradual increase in the pH of the solution. The method was recommended by Willard and Tang* for the separation of calcium as calcium oxalate. Urea is added to an acid solution of calcium (II) and oxalate ion, the solution is boiled gently and calcium oxalate precipitates slowly, as the pH increases, according to

$$Ca^{2+} + HC_2O_4^- + NH_3 \rightleftharpoons CaC_2O_4(s) + NH_4^+$$

Even in the presence of magnesium ion or phosphate ion the calcium oxalate precipitate is generally completely free from contamination and of large enough particle size to permit easy filtration.

Sodium thiosulfate slowly decomposes by hydrolysis to yield sulfide ions when acid solutions of the salt are boiled. The addition of sodium thiosulfate to an acid solution of copper(II), followed by gentle boiling of the solution, results in a copper sulfide precipitate of relatively large particle size.

The technique of increasing the value of S by complexation processes,

*H. H. Willard and N. K. Tang, *J. Am. Chem. Soc.*, **59**:1190 (1937); *Ind. Eng. Chem., Anal. Ed.*, **9**:357 (1937).

where these are applicable, also provides a method of precipitation from homogeneous solution. In such instances a substance capable of forming a complex with the ion to be precipitated is added to solution, followed by the addition of the precipitant. The stability of the complex prevents precipitation—in effect, increases the value of S. The subsequent addition of a reagent capable of slowly forming a more stable complex with the complexing agent results in a slow precipitation process where, again, relatively large particle size is obtained.

11.2.6 INCREASING PARTICLE SIZE AFTER PRECIPITATION

The primary precipitate secured as the result of precipitation from hot dilute solution will, except in the case of slightly soluble salts yielding amorphous precipitates, generally be in the form of small crystals of reasonably perfect lattice structure. Primary precipitates secured from concentrated solutions will generally be either amorphous gels or will consist of very small crystals, very imperfect as to structure. It should be noted here that we refer only to particle size and structure and not to the level of precipitate purity. In both instances an average crystal or particle size is implied, and there will be a considerable variation in particle size for any given primary precipitate.

Where such a primary precipitate is allowed to stand for some time in contact with the solution from which it was obtained, usually at elevated temperature, the tendency will be for the smaller particles, which exhibit the higher solubility, to dissolve and to allow the solution to become supersaturated with respect to the larger particles. The dissolved material then restores the solution equilibrium by deposition on the larger particles. This process, known variously as digestion, aging, or "Ostwald ripening" of the precipitate, permits a general and significant increase in the average particle size. To some extent, and under the proper conditions, the process of digestion also results in improvements in the perfection of the crystal-lattice structure.

Most precipitates tend, for a variety of reasons, to carry down with them impurities from the solution. While the process of digestion may, in certain instances, permit some order of precipitate purification by returning carried-down impurities back to the solution during particle re-solution and redeposition actions, the problem of precipitate purity involves many complex variables. A discussion of the more important of these follows, together with some general observations on the various methods of treating precipitates so as to permit some degree of purification.

11.3 THE PURITY OF PRECIPITATES AND THE TREATMENT OF PRECIPITATES SUBSEQUENT TO PRECIPITATION

11.3.1 GENERAL

Separation by precipitation generally results in a precipitate contaminated to a greater or lesser degree. Frequently the use of certain approaches to the overall process of separation by precipitation can assist significantly in improving precipitate purity. For example, careful selection and control of the method of precipitation, the application of certain techniques in the treatment of the precipitate after precipitation and before filtration, and the washing of the precipitate after filtration with a wash liquid containing a well-chosen electrolyte at some optimum concentration can each contribute in this direction. Obviously what is ultimately required is a sound knowledge of the precipitation and precipitate contamination processes, together with the accumulation of experience in precipitation work, in order to be able to anticipate problems relative to precipitate purity, to pursue methods of separation and treatment capable of minimizing the extent to which precipitate contamination might occur, and to apply and, where necessary, devise techniques for the purification of contaminated precipitates.

A precipitate may be contaminated as the result of simultaneous precipitation, coprecipitation, or postprecipitation. It is apparent that these effects may act separately or, under the proper conditions, in combination.

Simultaneous precipitation is the contamination of a precipitate by an impurity or impurities normally insoluble in the solution under the conditions pertaining during the precipitation process. As a simple case in point, we can consider the separation of chloride as silver chloride by the addition of a silver ion solution to a solution containing chloride ion. When the chloride ion solution contains a significant quantity of bromide ion, the silver chloride precipitate is contaminated by silver bromide, both silver chloride and silver bromide being insoluble under the precipitation conditions described. The contamination of a precipitate by simultaneous precipitation is basically a problem involving an improper approach to the analytical problem on the part of the analyst. No method of separation by precipitation should be applied without a prior knowledge of the general nature of the matrix under treatment. Where necessary, qualitative analysis should be used to achieve this end. In addition to this, even where there is prior knowledge of the nature of the matrix, the analyst must be fully aware of any possibility of simultaneous precipitation phenomena where the matrix substances, the sought-for component, and the precipitation method are concerned.

Where a precipitate is contaminated by a substance or substances normally soluble in the solution medium, the effect is known generally as coprecipitation. For example, the addition of a solution of barium chloride to one of sodium sulfate results in a precipitate of barium sulfate having an analytically significant contamination by sodium sulfate and sodium bisulfate, despite the fact that these sodium salts are easily soluble in water.

Coprecipitation effects may occur as the result of two mechanisms. These are

1. The adsorption of foreign ions at the surfaces of precipitate particles during the process of precipitation
2. The occlusion of foreign ions within the precipitate particles during precipitate formation and particle growth

Coprecipitation may frequently occur without the analyst being aware of it, and since parallel analyses conducted under identical conditions will often provide for an almost identical degree of contamination by coprecipitation, the resulting precision of the analytical method tends to obscure the fact of its inaccuracy.

Postprecipitation is the separation of an impurity from its supersaturated solution some appreciable time after the appearance of a primary precipitate involving the sought-for substance. Where such a primary precipitate does not appear, the impurity may precipitate with extreme slowness. Impurities which postprecipitate in this manner and which undergo a very gradual process of precipitation in the absence of some primary precipitate are capable of yielding very stable supersaturated solutions. In the presence of some primary precipitate, the primary precipitate particles serve as nuclei for the postprecipitation of the impurity from its supersaturated solution. A case in point is the separation of calcium from magnesium by the addition of an acid solution of ammonium oxalate to the acidified solution of calcium and magnesium, followed by the addition of urea and boiling and the further addition of ammonium oxalate (see Sec. 11.2.5). Calcium oxalate separates fairly rapidly under these conditions, and the solution will also be supersaturated to magnesium oxalate. The separation of magnesium oxalate takes place on the primary precipitate of calcium oxalate very slowly, however, and where the amount of magnesium is not too high, the contamination by postprecipitation will be analytically insignificant if the calcium oxalate precipitate is filtered off not more than 3 h after its separation. Where the amount of magnesium is high, the first precipitate of calcium oxalate may have to be filtered off and redissolved and the calcium oxalate reprecipitated under the same conditions.

In view of the importance of coprecipitation as a source of precipitate contamination, this phenomenon will be discussed in somewhat greater detail.

11.3.2 COPRECIPITATION

By adsorption The ability of the surface ions of precipitate particles to adsorb ions of opposing charge from the solution was discussed in Sec. 11.2.3. The tendency is for such particles to adsorb preferentially a lattice ion when one is available in the solution at a sufficient concentration. Where a lattice ion is not available, the precipitate particle surface will adsorb ions foreign to the crystal lattice. The foreign ion most strongly adsorbed, from among those available in the solution, will be that capable of forming with the lattice ion of opposite charge the salt of least solubility, although other foreign ions of a similar charge nature may also be adsorbed to a degree decreasing with increasing solubility of the foreign ion–lattice ion salt combination (Paneth-Fajans-Hahn rule). Such adsorbed ions, whether common or foreign to the lattice, form the primary layer around the precipitate particle involved. Solution ions of opposite charge, in accordance generally with the same solubility rule, are then attracted close to these adsorbed ions in the establishment of the secondary layer.

In the formation of these primary and secondary layers, specific situations leading to contamination of the precipitate by adsorption coprecipitation may arise. For example, during a normal process of precipitation, the slow addition with stirring of a solution of the cation to be precipitated to a solution of the anion to be precipitated results in the initial precipitate particles adsorbing the anion and becoming negatively charged. The secondary layer, or counter ions, will then consist of foreign cations from the solution. Flocculation here results in contamination by coprecipitation of foreign cations. The opposite situation occurs where the precipitate results from the reverse process, that is, the addition of the anion solution to the cation solution, the contamination then being in the form of foreign anions. It is important to note, at this point, that the method of adding the two solutions can often minimize the extent to which contamination by adsorption coprecipitation may occur. Where one of the solutions contains an ion very likely to contaminate by this form of coprecipitation, this solution should then be added to the second solution in order to maintain as low a concentration of the contaminating ion as possible during the precipitation process. Where serious coprecipitation of this nature cannot be avoided, regardless of the method of solution addition, it may be necessary to remove the contaminating ion by some separation process before carrying out the process of precipitation.

During the flocculation of colloidal suspensions, or indeed during the general process of precipitation, the original primary- and secondary-layer ions may be brought down to a considerable extent by the flocculated colloid or precipitate.

The seriousness of such contamination will sometimes depend on the nature of the adsorbed and counter ions. For example, where arsenious sulfide, As_2S_3, has been precipitated by passing H_2S gas into an acid solution of arsenious trichloride, $AsCl_3$, sulfide ions will be adsorbed at the surface of the As_2S_3 particles, with H_3O^+ forming the principal secondary-layer ion. Contamination of the As_2S_3 precipitate by S^{2-} and H_3O^+ represents a reasonably minor source of difficulty, since whatever portion of the S^{2-} and H_3O^+ ions is not removed during the subsequent digestion and filter washing procedures will, because of the volatile nature of both H_2S and H_2O, be removed during the precipitate ignition or drying procedure. On the other hand, where As_2S_3 has been precipitated as the result of passing H_2S gas into an acid solution containing only $AsCl_3$ and $ZnCl_2$, even under conditions of zinc(II) concentration which do not normally permit the precipitation of ZnS, the primary-layer ion will be S^{2-}, with Zn^{2+} ions forming the bulk of the secondary layer. Such contamination in the form of ZnS, being of a nonvolatile nature, represents a much more serious analytical problem.

Where a colloidal suspension has been flocculated by the use of an electrolyte involving strong flocculating ions, replacement of the primary- and/or secondary-layer ions originally adsorbed and counterattracted by those of the flocculating electrolyte may take place, and these replacement ions may then be brought down to a significant extent by the flocculated colloid or precipitate. Where the flocculating ion or ions provide for a more serious contamination than the original primary- and secondary-layer ions, a worsening of the analytical situation may have resulted. A case in point would be the use of barium chloride as a flocculant for a suspension of As_2S_3 which had originally adsorbed S^{2-} and the counter ion H_3O^+ Here the secondary-layer ion H_3O^+ would be largely replaced by Ba^{2+}, the resultant barium sulfide carried down by the flocculated As_2S_3 representing a more serious form of contamination than that which would have resulted from H_2S and H_2O, the products of the original primary- and secondary-layer ions.

Digestion of precipitates contaminated by adsorption coprecipitation removes the contaminating ions to a greater or lesser degree. Since the extent to which this form of contamination takes place is a reflection of the surface development of the precipitate, digestion, by increasing the average particle size and decreasing the overall surface area, decreases the adsorptive effect and the extent of contamination by adsorption. In general, however, the efficiency of the process of precipitate digestion is a function of the nature of the precipitate involved. Crystalline and curdy precipitates can be so purified to a significant extent, where the impurities are surface adsorbed. Gelatinous precipitates or gels do not respond well to the digestion process, the value of S for such precipitates being so low that redissolving and reprecipitation actions are insufficient to permit much in the way of particle size increase and associated precipitate purification.

Again, washing the precipitate thoroughly after filtration often serves, in addition to its other functions, to remove such contaminating ions. This procedure will be effective for crystalline and curdy precipitates where the contaminating impurities are almost exclusively confined to the precipitate surfaces. Where adsorbed impurities have been occluded during crystal growth (and this refers largely to situations involving crystalline rather than curdy precipitates), such washing actions may not be very effective. For gelatinous precipitates, such as ferric and aluminum hydroxides, coprecipitated adsorbed impurities, particularly cations, may not be so well removed by filter washing procedures, although a varying degree of removal is always secured.

Such filter washing procedures rarely involve the solvent (usually water) alone, since during the washing procedure, the concentration of the flocculating electrolyte will fall below the flocculation value, introducing the possibility of the precipitate partly or totally resuming the colloidal suspension state. Such an action, called peptization, is avoided by the use of a wash liquid containing a concentration of an appropriate electrolyte sufficient to prevent peptization. The electrolyte used is chosen so as to avoid introducing a source of contamination. For example, AgCl precipitates can be washed with dilute solutions of nitric acid, a volatile electrolyte at normal precipitate drying temperatures. Many precipitates can be washed with dilute solutions of appropriate ammonium salts volatile at subsequent ignition temperatures. In several instances, a selection of the wash solution electrolyte is made with the aim of not only preventing peptization but of permitting an exchange of nonvolatile adsorbed and/or counter ions for the volatile ions of the electrolyte.

As mentioned previously, many precipitates cannot be sufficiently purified with respect to coprecipitated adsorbed impurities by selection of the precipitation method, flocculation technique, digestion process, or filter wash solution treatment. In such cases, and where the precipitate is easily redissolved, redissolving the precipitate and reprecipitation often provide an efficient method of purification. Some points of importance should be noted with respect to the questions of adsorbed impurities and reprecipitation efficiency in precipitate purification. First of all, the amount of an impurity adsorbed by a precipitate at a specific temperature is a function of the concentration of the impurity in the solution from which the precipitate is secured. This relationship is given by

$$x = kc^{1/n} \tag{242}$$

where x = amount of impurity adsorbed per unit weight of precipitate

k = proportionality constant

c = concentration of impurity in the solution

n = a constant (often about 2)

Thus, a precipitate adsorbs only a small proportion of the impurity available in the solution. Upon redissolving, the concentration of the impurity in the new solution is very much less than that which pertained in the initial solution, and the amount of impurity adsorbed during the reprecipitation process will be considerably less than that adsorbed by the initial precipitate. The effectiveness of the removal of the impurity decreases with successive reprecipitations, but in many cases, the level of impurity is reduced to analytical insignificance as the result of a single redissolving and reprecipitating process.

By occlusion Contamination of a precipitate by occlusion of foreign ions may occur as the precipitate particles increase in size from that of the primary particles. Two general mechanisms may be involved, which may occur simultaneously or separately.

Foreign ions which fit into the crystal lattice of the precipitate Here the solution contains foreign ions which are capable of substituting in the crystal lattice for the regular precipitate anions and/or cations. While substitutions of this nature usually take place with ions having the same general size and the same charge as the ions for which they substitute, they can also involve ions of a similar size but different charge value, providing the chemical formulas are similar. There are many examples of the first situation, such as lead sulfate and barium sulfate, barium sulfate and barium chromate, etc. An example of the second situation is given by the case of barium sulfate and potassium permanganate.

Crystals which are formed under conditions in which such substitution actions can take place are called mixed crystals or solid-solution crystals, and they form over concentration ranges that are specific for each type. Where this form of contamination by occlusion can take place, purification of the resulting precipitate is usually not possible to any worthwhile degree. Processes of digestion result in little in the way of improvement, since such contaminating ions as are released to the solution during redissolving actions are returned and reoccluded during reprecipitation actions. Redissolving the precipitate and going through a process of reprecipitation does not generally result in any significant purification, since the solution obtained by redissolving such a precipitate will itself yield a similarly contaminated precipitate, the formation of solid-solution crystals being on a different basis of impurity incorporation than that involved in the process of adsorption. Contamination which might originate in this way can be most easily avoided by removing the interfering ion or ions before the precipitation process, or by converting them prior to precipitation to some noninterfering form.

Foreigns ions which do not fit into the crystal lattice of the precipitate   Foreign ions of this type are adsorbed in the usual way at the surfaces of the primary precipitate particles and are, during the subsequent stage of rapid crystal growth, covered and enclosed within the crystal structure. Since such foreign ions do not fit into the lattice, that is, they are not capable of substituting for the normal precipitate anions and/or cations, their occlusion results in the formation of imperfections in the crystal structure of the precipitate. An important factor should be recognized here. Such occlusion actions will occur usually with precipitates of a crystalline nature, where the orientation-aggregation velocities relationship is such as to promote crystal growth to relatively large size.

For curdy, gelatinous, or so-called amorphous precipitates, the tendency to remain in the form of agglomerations of extremely small crystals results in insufficient growth to allow for significant occlusion of foreign ions. Thus, contamination by occlusion of foreign ions which do not fit into the crystal lattice occurs to a significant extent in many crystalline precipitates but is of lesser general importance with curdy or gelatinous precipitates.

Where occlusion occurs, many factors such as the type of foreign ion, its charge, the rate of growth of the primary particles, etc., contribute to the degree to which such ions are occluded. To a certain extent, and often very efficiently, occluded ions can be removed during the digestion process normally following precipitation. Dissolving actions during digestion take place, not only where the smaller particles are concerned but also at irregularities on the surfaces of the larger particles. Reprecipitating actions then not only increase the average particle size but also perfect and compact the larger crystals. Both these mechanisms tend to reduce greatly the surface area of the precipitate. Occluded impurities returned to the solution during redissolving actions are not reoccluded during the process of reprecipitation during digestion, since the speed of crystal growth is very low and tends to minimize occlusion actions. Surface adsorption of foreign ions is, of course, considerably reduced as the result of the significant reduction in the precipitate surface exposed to the solution. In many cases, such contamination by occlusion is reduced by almost 100 percent by the application of proper precipitation and digestion procedures.

Where digestion procedures do not sufficiently purify the precipitate, a process of redissolving and reprecipitating can often be applied to advantage.

11.4 SUMMARY OF THE GENERAL CONDITIONS SURROUNDING PRECIPITATION AND THE TREATMENT OF PRECIPITATES

No hard and fast rules can be laid down relative to the precipitation process and the treatment of the precipitate after precipitation. The following does, however, summarize the *general* procedures which provide the best probability of securing precipitates of reasonable purity and ease of handling.

1. Removal from the solution, prior to the precipitation process, and where indicated, of foreign ions which could contaminate the precipitate by coprecipitation involving occlusion by the formation of mixed or solid-solution crystals. The method of removing these ions can involve any efficient technique, including conversion to some noninterfering form.
2. Precipitation by the slow addition, with efficient stirring, of the dilute precipitant solution to the hot dilute solution of the reacting ion. The order of adding the two solutions (i.e., reacting ion to precipitant, or vice versa) may be important from the point of view of minimizing coprecipitation by adsorption. The order used should be the most advantageous.
3. Where a flocculating electrolyte must be added, care should be taken to choose this on the basis of avoiding the introduction of serious adsorption contamination of the precipitate by electrolyte ions.
4. Dilution followed by digestion at elevated temperature, subsequent to precipitation, should normally be carried out to increase the average particle size and to allow a degree of purification of the precipitate. This is particularly important with crystalline precipitates. In the case of curdy precipitates, the digestion process does little to increase average particle size but does permit some crystal perfectioning and significant precipitate purification. Where gelatinous precipitates are involved, digestion is carried out only long enough to permit precipitate coagulation before filtration.
5. Where postprecipitation phenomena are anticipated, the length of time in digestion may have to be carefully restricted in order to minimize contamination originating from this source.
6. Reprecipitation procedures can be used to purify precipitates contaminated by coprecipitation resulting from adsorption or some forms of adsorption with occlusion. The relative degree of purification decreases with successive reprecipitations. Reprecipitation cannot be used to remove contamination

resulting from coprecipitation based on the occlusion of foreign ions in the formation of mixed or solid-solution crystals.

7. The use of electrolytes in proper concentrations where wash solutions are concerned is required in order to prevent peptization, although this is of real importance only where curdy and gelatinous precipitates are to be washed. The electrolyte may also be chosen with the additional aim of the advantageous exchange of adsorbed ions.

11.5 ORGANIC PRECIPITANTS

11.5.1 GENERAL

The method of separation by precipitation can, by drying or ignition after filtering and washing, lead to a determination by gravimetric means. Where the precipitate is separated and redissolved, a volumetric titration technique can often be applied. The ideal situation relative to this method of separation would be achieved where an absolutely specific precipitant was available for each substance to be separated. While this arrangement is a long way from realization, and indeed not likely to be ever fully realized, the area of organic precipitants perhaps approaches it most closely. While specific selectivity cannot be obtained in every case, judicious selection of the organic precipitant, or sequence of precipitants, and control of the conditions surrounding precipitation (solution pH, etc.) can yield a high degree of selectivity.

The precipitates resulting from the use of organic precipitants are often very insoluble in water, so that the degree of separation is then highly quantitative. Quite frequently the precipitate has a high equivalent weight relative to the metal ion being separated, so that a small weight of the sought-for metal yields a large precipitate weight, allowing high sensitivity in the detection of the metal involved. As a further advantage, most organic precipitates have a physical nature (particle size, etc.) which permits ease of filtering and washing. There is often a tendency to bulkiness, but this can frequently be reduced by proper control of the conditions of precipitation.

Certain disadvantages exist where certain organic precipitants are used. For example, the action of drying organic precipitates sometimes results in a dried product of uncertain composition. This may be due to processes such as partial decomposition or volatilization taking place during the drying operation. Ignition to an inorganic form of constant composition is often preferable where volatilization phenomena do not interfere, although the advantage of high equivalent weight may then be negated. In certain instances, where neither drying nor ignition results in the achievement quantitatively of a constant composition compound involving the sought-for substance, the organic precipitant may be used for separation purposes rather than for gravimetric determination. The possibility then for the application of a technique involving redissolving, followed by volumetric titration, should not be overlooked. Again, the low solubility of organic precipitates in water is often matched by a low solubility in water for the organic precipitant. Where such a situation exists, a solution of the precipitant in some organic solvent is usually used, but excesses of the precipitant may result in contamination of the precipitate. Cost is also a factor where many organic precipitants are concerned, and the high cost of some of these reagents may restrict or even prohibit their general use.

Organic precipitants are commonly of two types, chelating precipitants and ion-association precipitants.

11.5.2 CHELATING PRECIPITANTS

Chelates have already been discussed briefly in Chap. 4, Part 4. These are basically heterocyclic ring compounds formed by the reaction of a metal ion, as the central ion, with two or more functional groups of the same ligand. The uncharged or neutral compound so formed is usually of low solubility in aqueous solution. The chelating precipitant will contain both acidic groups (hydrogen replaceable) and coordinating groups (atoms with unshared electron pairs). The metal ion will be of a size, oxidation number, and coordinating number appropriate to the formation of the ring compound. Indicated below are acidic groups and coordinating groups commonly found as chelating precipitant components.

Acidic groups	Coordinating groups
$-COOH$, carboxyl	$-NO_2$, nitro
$-OH$, hydroxyl	$-NH_2$, amino
$-SO_3H$, sulfonic	$-NH-$, imino
$-SH$, mercapto	$=N-$, cyclic nitrogen
	$>CO$, carbonyl

Where chelating precipitants are concerned, the ring compounds formed with metal ions are five- or six-membered. An example is given by the substance dimethylglyoxime:

$$
\begin{array}{ccc}
H_3C & & CH_3 \\
\diagdown & & \diagup \\
& C-C & \\
& \| \quad \| & \\
& N \quad N & \\
\diagup & & \diagdown \\
HO & & OH
\end{array}
$$

The specificity of this reagent is high; it reacts in acid solution with palladium only and in slightly acid to slightly basic solution with nickel only. It forms soluble complexes with ions such as copper(II) and cobalt(II), so that in the presence of these ions large excesses of the precipitant should be added.

The reaction with nickel forms nickel dimethylglyoximate according to the following:

$$
\text{Ni}^{2+} + 2 \quad
\begin{array}{c}
\text{H}_3\text{C} - \text{C} = \text{N} - \text{OH} \\
| \\
\text{H}_3\text{C} - \text{C} = \text{N} - \text{OH}
\end{array}
\quad \rightleftharpoons \quad
\text{Nickel dimethylglyoximate structure} \quad + 2\text{H}^+
$$

Dimethylglyoxime

Nickel dimethylglyoximate

A second chelating precipitant is 8-hydroxyquinoline (oxine), a reagent which is capable of precipitating a large number of metal ions. While the different precipitates involved do show variations in solubility with solution pH, selectivity is secured on this basis to a relatively minor extent only. Oxine can be used, therefore, for the separation of groups of metals rather than for the determination of individual metals. Aluminum and magnesium are exceptions to this situation, however, being frequently determined gravimetrically by precipitation under controlled solution conditions with 8-hydroxyquinoline. Aluminum, for example, yields

$$
\text{Al}^{3+} + 3 \quad \text{(8-Hydroxyquinoline)} + 3\text{H}_2\text{O} \rightleftharpoons \quad \text{(Al-oxine complex)} + 3\text{H}_3\text{O}^+
$$

8-Hydroxyquinoline

Table 11.2 provides a list of some of the commoner chelating precipitants and their general functions.

11.5.3 ION-ASSOCIATION PRECIPITANTS

Certain organic precipitants ionize in aqueous solution to form anions and cations. These anions and/or cations are then capable of reacting with some oppositely

Table 11.2　Common chelating organic precipitants

Precipitant	*Reactants*
$CH_3-C=N-OH$ $CH_3-C=N-OH$ Dimethylglyoxime	Precipitates Ni(II) in solutions of pH 5 to 9 and Pd(II) in solutions of pH 1; used mainly for the determination of nickel
 8-Hydroxyquinoline	Precipitates almost all metals except the alkali metals; group separations can be obtained by pH control; used largely in the determination of aluminum and magnesium
 1-Nitroso-2-naphthol	Precipitates Cu(II), Fe(III), Pd(II), and Co(II); used mainly in the separation of cobalt from nickel; in the reaction, Co(II) is oxidized to Co(III), with the Co(III) being precipitated
 Salicylaldoxime	Precipitates several cations, and group separation possible through pH control; Cu(II) precipitates at about pH 2.5, at which value Co(II), Cd(II), Zn(II), Pb(II), and Fe(II) do not separate
 Ammonium nitrosophenolhydroxylamine (cupferron)	Precipitates Fe(III), Ti(IV), Sn(IV), Ce(IV), Zr(IV) from acid solutions of about 10% HCl or H_2SO_4; mainly used for the separation of iron and titanium from aluminum
 α-Benzoinoxime (cupron)	Used for the separation and determination of copper and molybdenum; Cu(II) can be separated from ammonium tartarate solutions, Mo(IV) from solutions of mineral acids
 Quinaldic acid	Used for the separation and determination of cadmium, copper, and zinc

charged species in solution to yield slightly soluble compounds where the structure is not ringlike and the bond between the precipitant and reacting ions is mainly ionic in character. In several cases the solubility of the resulting salt is low enough to permit a quantitative separation of the sought-for substance; in others, the solubility is too high for a satisfactory separation. A typical example of this type of organic precipitant is given by the use of tetraphenylarsonium chloride as a precipitant for mercury(II). The tetraphenylarsonium chloride dissociates in aqueous solution to yield the tetraphenylarsonium ion, $(C_6H_5)_4As^+$, and the reaction with mercury(II) proceeds according to

$$2 \ (C_6H_5)_4As^+ + HgCl_4{}^{2-} \rightleftharpoons \{(C_6H_5)As\}_2HgCl_4$$

Table 11.3 indicates several common ion-association precipitants.

11.6   THE WASHING OF PRECIPITATES

Subsequent to filtration, the precipitate is washed with a suitable wash liquid to remove impurities adsorbed by the precipitate and contained in the volume of original solution carried down by the precipitate. The wash liquid (water is the commonest solvent) is chosen on the basis of

1. High solubility for impurities and low solubility for the precipitate substance.
2. An electrolyte added to prevent peptization, particularly where the precipitate is a flocculated colloid.
3. An electrolyte added to provide an ion-exchange effect, where adsorbed ions of a nonvolatile nature are exchanged for ions volatile in the subsequent drying or ignition procedure. Frequently the added electrolyte is capable of performing both the antipeptization and ion-exchange functions.

The method of washing the precipitate is important, and the washing cycle should observe the following.

1. The precipitate on the filter should be thoroughly stirred up, using a jet of wash liquid.
2. After disturbing the precipitate, the jet should be used to wash the body and edges of the filter paper thoroughly (where this is the form of filter employed).
3. Total volume for total volume, a large number of small-volume washes is much more efficient in the removal of impurities than is a small number of large-volume washes. The reasoning is quite simple. Consider a single wash of 40-ml volume as opposed to two successive washes of 20-ml volume each. The total volume in each case is the same, so that any precipitate loss by solubility effects will be identical. Suppose the precipitate retains at any time 1 ml of the liquid with which it was last in contact. In the case of a single

Table 11.3   Common ion-association organic precipitants

Cation precipitants	Reactants
$C_6H_5-As\begin{array}{c}\nearrow OH\\ =O\\ \searrow OH\end{array}$ Phenylarsonic acid	Precipitates tetravalent ions such as Ti(IV), Th(IV), Zr(IV), Sn(IV), and Ce(IV) from strongly acid solutions
$\begin{array}{c}O=C-OH\\ \mid\\ O=C-OH\end{array}$ Oxalic acid	Precipitates many metallic ions; used mainly for the separation and determination of calcium and magnesium
$Na^+B(C_6H_5)_4{}^-$ Sodium tetraphenol boron	Precipitates K(I) for determination by gravimetric or volumetric means

Anion precipitants	Reactants
$(C_6H_5)_4As^+Cl^-$ Tetraphenylarsonium chloride	Precipitates Hg(II) as $HgCl_4{}^{2-}$ and the chlorocomplexes of the metals Tl, Cd, Zn, Sn(IV), Ir, Au, Pt, etc; precipitates Mn as $MnO_4{}^-$ and other oxy-anions such as $IO_4{}^-$, $ClO_4{}^-$, $MoO_4{}^{2-}$, etc.
$H_2N-\!\!\bigcirc\!\!-\!\!\bigcirc\!\!-NH_2$ Benzidine	Precipitates $SO_4{}^{2-}$, $PO_4{}^{3-}$, and $WO_4{}^{2-}$; although used frequently for the separation of $SO_4{}^{2-}$, the solubility of benzidinium sulfate is quite high
$Cl-\!\!\bigcirc\!\!-\!\!\bigcirc\!\!-NH_2$ 4-Chloro-4′-aminodiphenyl	Precipitates $SO_4{}^{2-}$ with a solubility approximately one-seventh that of the benzidinium sulfate precipitate

wash with 40 ml, the amount of original liquid, and therefore impurities, remaining would be $\frac{1}{40}$, or 2.5 percent. With two successive washes of 20 ml each, the first wash leaves behind an impurity level of $\frac{1}{20}$, or 5 percent, and the second a level of $1/(20)^2$, or 0.25 percent. Note that two washings of 20 ml each reduced the impurity level in the precipitate to one-tenth the value secured after one 40-ml washing. The general expressions are given by

$$\text{Percent of original impurity} = \frac{100v}{(V)^n} \tag{243}$$

where v = solution retained by the precipitate, ml
V = wash liquid for each washing, ml
n = number of washings

The improvement of n washings over x washings, were $x < n$ and where the same total volume is involved, is given by

$$\frac{\text{Relative amount of impurity after } n \text{ washings}}{\text{Relative amount of impurity after one washing}} = \frac{n}{(V)^{n-1}} \qquad (244)$$

The improvement of n washings over x washings, where $x < n$ and where the same total volume is involved, is given by

$$\frac{\text{Relative amount of impurity after } n \text{ washings}}{\text{Relative amount of impurity after } x \text{ washings}} = \frac{(n/x)^x}{(V)^{n-x}} \qquad (245)$$

where V = wash liquid for each washing in the n wash technique, ml

11.7 DRYING AND IGNITION OF PRECIPITATES

Once a precipitate has been separated, filtered, and washed, it must then undergo two final operations.

1. It must be dried or ignited to some constant, known composition compound.
2. The dried or ignited residue must be cooled under the proper conditions and then accurately weighed.

We shall consider only the first of these final stages in the process of analysis by the gravimetric method. In so doing we shall assume that the precipitate is free from all contamination except that by water (from the precipitation and/or the washing procedure) and any electrolyte added to the wash water to prevent the action of peptization.

Many precipitates can be dried, and their water contents removed, by heating in a drying oven at temperatures in the relatively low range of 100 to 150°C. This requires that the water content be held extremely loosely, and not in the form of strongly adsorbed or occluded water. It requires also that when the wash water contains an antipeptization electrolyte, this should be entirely volatile at the applied drying temperature. Where these conditions are met, the weighing form is nearly always identical to the precipitate form. A precipitate such as silver chloride, for example, washed with a volatile electrolyte such as a highly dilute nitric acid solution, can be dried at about 120°C and, for all normal analytical purposes, the water and electrolyte totally removed. Here the precipitate and weighing forms will be the same. This situation occurs where many precipitates originating with the use of organic precipitants are concerned [i.e., when nickel(II) and dimethylglyoxime are reacted, nickel dimethylglyoximate is both the precipitate and weighing form].

It is perhaps more common to encounter techniques where the washed precipitate is ignited at some elevated temperature and thereby usually converted from the precipitate form to some other form involving a constant known composition. Such ignition procedures are employed for a number of reasons not the least important of which are inability to secure a constant composition form of the precipitate by drying alone, the use of an antipeptization electrolyte volatile at some elevated temperature, but not at the normal drying range of about 100 to 150°C, and the need to guarantee the removal of precipitate water held by occlusion or strong adsorption. Gelatinous precipitates such as those involving the hydrated oxides of aluminum, iron, and silicon contain occluded and absorbed water which cannot be completely removed except by prolonged heating at temperatures in the neighborhood of 1100°C. Several important factors associated with the ignition of precipitates are noted below.

1. Over the passage of many years of experience, the process of trial and error has devised optimum ignition temperatures for the conversion of precipitates to some desirable known and constant composition. In recent years, the use of the thermobalance has resulted in an investigation of such empirical ignition temperatures and, in many cases, a much clearer picture of the temperature required for a given ignition process and the range of temperature over which the desired process can be expected to take place. This instrument is capable of weighing accurately a precipitate as its temperature is increased from room temperature to some elevated temperature (depending on the substance) in excess of 1000°C. A pyrolysis curve of precipitate weight vs. temperature is obtained. Such a curve is shown in Fig. 11.4, and covers the

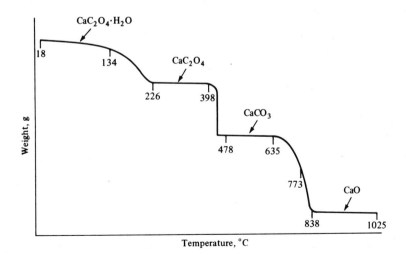

Fig. 11.4 A pyrolysis or thermogravimetric curve. (*Adapted from C. Duval, "Inorganic Thermogravimetric Analysis," 2d ed., Elsevier, Amsterdam, 1963.*)

ignition of calcium oxalate, $CaC_2O_4 \cdot H_2O$, the precipitated form. It will be noted that the pyrolysis curve shows plateaus at the various conversion stages (that is, $CaC_2O_4 \cdot H_2O$ to CaC_2O_4, CaC_2O_4 to $CaCO_3$, $CaCO_3$ to CaO). A typical ignition procedure would be to convert the hydrated calcium oxalate precipitate to calcium oxide by heating to constant weight at 1000°C.

2. Where an ignition procedure is planned it may be necessary, under certain circumstances, to filter the precipitate using a medium other than filter paper. This will be required, for example, where there is a strong tendency for the precipitate to be reduced during ignition by carbon from the filter paper. Where precipitated AgCl is to be ignited, it should be filtered using a previously ignited filter crucible. Even in the case of precipitates with much less tendency to react with carbon during ignition, care should be taken when preparing for ignition filter paper retentions of such precipitates. This will involve placing the properly folded filter paper and precipitate in an appropriate crucible and drying at about 120°C. This will expel easily removed water and help to avoid a rapid conversion of precipitate and paper water to steam during the initial ignition step, with possible loss of precipitate material. The dried paper and precipitate are then heated over a burner at low flame to char the paper completely, this without allowing the paper to flame. Finally, ignition is carried out, preferably in a muffle furnace of adequate capacity as to both size and temperature range, at the required temperature.

3. The ignition temperature should be held at the optimum value. Use of a higher temperature does not necessarily imply faster and more thorough conversion to the desired form. Indeed, the use of some temperature higher than that recommended may result in precipitate loss by volatilization, sublimation, or decomposition.

4. Many ignited residues do not show any marked tendency to absorb atmospheric moisture or carbon dioxide during the cooling cycle that precedes the weighing operation. A significant number do, however, and this is particularly the case with respect to the absorption of moisture. Ignited residues from gelatinous precipitates often show a strong tendency in this direction, Al_2O_3 being a typical example. Such residues must always be cooled in a dessicator containing an efficient dessicant capable of preventing significant water absorption during the cooling period. Where the reaction of an ignited residue with carbon dioxide is a problem, dessicators with atmospheres free from carbon dioxide can be used for cooling purposes.

Example In the formula $(Q - S)/S$, the value of Q can be taken as the instantaneous concentration of the precipitate upon the addition of reactant and precipitant. The value of S represents the molar solubility of microparticles of initial precipitate. An idea of the relative number of precipitate nuclei formed initially will then be given by $(Q - S)/S$, and this, in turn, will provide some idea of the form of the precipitate (crystalline, curdy, etc.) and, very generally, a picture of the relative size of the precipitate particles when precipitation is complete. The following example illustrates the general approach. Consider 100 ml of a solution 0.10 M to SO_4^{2-} and 100 ml of a solution 0.10 M to Cl^- To the former is added 0.050 ml of 0.10 M Ba^{2+}, and to the latter, 0.050 ml of 0.10 M Ag^+. The K_{sp} values at 25°C for the precipitates are

$$K_{sp}(AgCl) = 1.56 \times 10^{-10}$$

$$K_{sp}(BaSO_4) = 1.1 \times 10^{-10}$$

The molar solubility of micro- and macrocrystals of AgCl is almost identical. The molar solubility of microcrystals of $BaSO_4$ is about 1000 times that for macrocrystals.

AgCl *precipitation*:

$$Q = \frac{0.050 \times 0.10}{100.05} = 5.0 \times 10^{-5} M$$

$$[Ag^+] = S(AgCl) = \frac{K_{sp}(AgCl)}{[Cl^-]} = \frac{1.56 \times 10^{-10}}{10^{-1}}$$

$$= 1.56 \times 10^{-9} M$$

$$\frac{Q - S}{S} = \frac{(5.0 \times 10^{-5}) - (1.56 \times 10^{-9})}{1.56 \times 10^{-9}} = 3.2 \times 10^4$$

$BaSO_4$ *precipitation*:

$$Q = \frac{0.050 \times 0.10}{100.05} = 5.0 \times 10^{-5} M$$

$$[Ba^{2+}] = S(BaSO_4) = \frac{K_{sp}(BaSO_4)}{[SO_4^{2-}]} = \frac{1.1 \times 10^{-10}}{10^{-1}}$$

$$= 1.1 \times 10^{-9} M$$

$S(BaSO_4)$ for microcrystals $= 1000 \times 1.1 \times 10^{-9} M = 1.1 \times 10^{-6} M$

$$\frac{Q - S}{S} = \frac{(5.0 \times 10^{-5}) - (1.1 \times 10^{-6})}{1.1 \times 10^{-6}} = 4.4 \times 10$$

Note that the relative number of AgCl nuclei formed is very high. This implies

1. Only minor particle size growth during the subsequent precipitation to completion, since further precipitation occurs at the surfaces of a very large number of primary particles
2. The acquiring of adsorbed charge by these small particles and the establishment of the colloidal state
3. Eventual flocculation in the form of curdy or gelatinous small-particle aggregates

Note the relatively small number of $BaSO_4$ nuclei. This implies

1. Rapid growth to larger particle size during the subsequent precipitation to completion
2. Time at colloidal particle size not appreciable
3. Separation as a crystalline precipitate of relatively large particle size

Example In a solution of Zn^{2+} and HCl at a pH of 1.0, saturation of the solution with H_2S gas results in a concentration of H_2S of about $0.1M$. Using the following values,

$$K_{sp}(ZnS) = 1.6 \times 10^{-24}$$

$$K_1(H_2S) = 9.1 \times 10^{-8}$$

$$K_2(H_2S) = 1.1 \times 10^{-15}$$

we note that

$$\frac{[H_3O^+][HS^-]}{[H_2S]_{un}} = K_1 \qquad \frac{[H_3O^+][S^{2-}]}{[HS^-]} = K_2$$

so that

$$[S^{2-}] = \frac{K_1K_2[H_2S]_{un}}{[H_3O^+]^2} = 1.0 \times 10^{-21}M$$

from which

$$[Zn^{2+}] = \frac{K_{sp}(ZnS)}{[S^{2-}]} = \frac{1.6 \times 10^{-24}}{1.0 \times 10^{-21}} = 1.6 \times 10^{-3}M$$

implying that, where $[Zn^{2+}]$ is less than $1.6 \times 10^{-3}M$, precipitation of ZnS will not occur. However, in a solution at pH = 1.0, $[Zn^{2+}] = 1.0 \times 10^{-3}M$, and $[Hg^{2+}] = 1.0 \times 10^{-2}M$, saturation with H_2S will result in a precipitate of HgS contaminated with ZnS to an analytically significant extent. Can this be explained?

The particles of HgS precipitate acquire charge by adsorbing the common ion S^{2-} from the solution. Because of the resulting relatively high concentration of S^{2-} in the neighborhood of the HgS precipitate, as well as the low solubility of ZnS, the counter ion (secondary-layer ion) adsorbed will be Zn^{2+} rather than H_3O^+. The flocculation of such a primary precipitate yields a final precipitate of HgS carrying with it an appreciable and significant amount of zinc.

Example A solution of nickel and HCl showing $[Ni^{2+}] = 0.050\ M$ and pH = 3.0 is saturated with H_2S, resulting in $[H_2S] = 0.1\ M$. We have

$$K_{sp}(NiS) = 3.0 \times 10^{-21}$$

$$K_1(H_2S) = 9.1 \times 10^{-8}$$

$$K_2(H_2S) = 1.1 \times 10^{-15}$$

so that

$$[S^{2-}] = \frac{K_1K_2[H_2S]_{un}}{[H_3O^+]^2} = 1.0 \times 10^{-17}\ M$$

and

$$[Ni^{2+}] = \frac{3.0 \times 10^{-21}}{1.0 \times 10^{-17}} = 3.0 \times 10^{-4}\ M$$

so that, under the conditions of precipitation, a heavy precipitate of NiS should be the result. Even after saturation, however, with prolonged agitation, no precipitate appears. Can this be explained?

It is apparent that, since the $K_{sp}(NiS)$ is exceeded under the conditions indicated, NiS must be capable of forming an extremely stable supersaturated solution, very probably the result of a very high solubility relative to microparticles of NiS.

PROBLEMS

1. Equal volumes of solutions of Ba^{2+} and SO_4^{2-}, at various concentration levels, were mixed together rapidly and the form of the resultant precipitate of $BaSO_4$ noted.

Solutions of Ba^{2+} and SO_4^{2-}		$BaSO_4$ precipitate
3	M each	Gelatinous
1	M each	Curdy
0.1	M each	Tiny crystallites
0.01	M each	Very small crystals
0.001	M each	Small crystals

Assume that macro- and microcrystals have approximately the same solubility and derive a very general relationship between the form of the precipitate and the relative number of nuclei formed. Explain the difference in the precipitate form on this basis.

2. Determine the relative surface area per mass unit in each case of a crystalline substance where
   (a) The substance is a single cube with a 1-cm edge
   (b) The cube in a is divided into eight equal cubes
   (c) Each cube in b is divided into eight equal cubes
Note the effect of crystal size on the surface area per unit of mass. What does this situation imply in general with respect to the possible purity of the initial precipitate?

3. Explain thoroughly why small particles of a precipitate are more soluble than larger particles.

4. Where a precipitation procedure results in the colloidal state, explain why the choice of ions capable of yielding the primary layer is much more restricted than that for ions forming the secondary layer.

5. Why would it be poor analytical practice to add an unrestrictedly large excess of the common ion during a precipitation process?

6. Where a controlled excess of precipitant does not by itself completely flocculate a colloidal suspension, what factors should be given consideration with respect to the selection of a flocculating electrolyte?

7. Of the two precipitates, AgCl and $Fe(OH)_3$, which can be expected to retain the larger amount of water upon separation? Explain.

8. Outline clearly the individual and combined effects of the two main factors which govern the physical nature of a precipitate (crystalline, curdy, etc.).

9. Explain how the particle size and purity of a precipitate are affected by the technique of homogeneous precipitation.

10. If you wished to secure a reasonably crystalline, relatively large particle size precipitate of zinc sulfide, ZnS, would you elect to conduct the separation by passing H_2S gas into an acid or an alkaline solution of Zn^{2+}? Explain the basis of your selection.

11. Using an example, discuss the action of postprecipitation.

12. Under certain circumstances it can be important as to whether the reactant solution is added to the precipitant solution, or vice versa. Explain this situation and give a case in point.

13. Digestion of a precipitate aids significantly in the removal of adsorbed impurities where crystalline and many curdy precipitates are involved. The procedure yields little advantage with gelatinous precipitates. Explain.

14. Where flocculated colloids are washed subsequent to filtration, it is essential that the wash liquid contain an electrolyte. Discuss the reasons for this, and explain what other advantages might be secured as the result of a judicious selection of the electrolyte used. What dangerous situation, analytically speaking, might be introduced by improper selection of the wash liquid electrolyte?

15. It has been indicated that the amount of impurity adsorbed per unit weight of precipitate is given by

$$x = kc^{1/n}$$

Where $k = 0.005$ and $n = 2$ and we consider a unit concentration of the impurity ($c = 1$) for the original solution from which the precipitate was secured, show what improvement in precipitate purity would be gained as the result of one, two, or three reprecipitations. What are your conclusions relative to the value of multiple reprecipitations?

16. The example at the end of Chap. 11 provided an illustration of a form of contamination by coprecipitation involving the adsorption of ZnS by a precipitate of HgS. Suppose that we consider a solution of pH = 1 which contains only $[Ba^{2+}] = 1.0 \times 10^{-2}M$ and $[Zn^{2+}] = 1.0 \times 10^{-3}M$ and is saturated with H_2S gas. As might be expected, no precipitate would be secured. Suppose now that the Ba^{2+} is precipitated as $BaSO_4$ by the addition of SO_4^{2-}. Might we expect a contamination of the $BaSO_4$ precipitate similar to that which occurred with the precipitate of HgS? Explain.

17. The separation of barium as barium sulfate in the presence of small amounts of lead results in contamination of the precipitated $BaSO_4$ by occlusion of lead ions in the barium sulfate lattice. Discuss this form of contamination and indicate why reprecipitation would not result in significant precipitate purification.

18. Nickel(II) can be precipitated as nickel dimethylglyoximate, separated by filtration, washed, dried, and weighed as this compound. The precipitated nickel dimethylglyoximate could also be filtered, washed, ignited to nickel oxide, NiO, and weighed as such. Which method would you prefer? Give sound reasons for your choice based on example calculations.

19. A precipitate has been separated by filtration. It retains with it at all times 2 ml of the liquid with which it was last in contact. It is now to be washed with a wash liquid. Indicate the improvement secured as the result of a choice of

    (*a*) One washing of 50-ml volume
    (*b*) Two washings of 25-ml volume each
    (*c*) Five washings of 10-ml volume each
    (*d*) Ten washings of 5-ml volume each

REFERENCES

1. Flagg, J. F.: "Organic Reagents Used in Gravimetric and Volumetric Analysis," Interscience, New York, 1948.
2. Gordon, L., M. L. Salutsky, and H. H. Willard: "Precipitation from Homogeneous Solution," Wiley, New York, 1959.
3. Kolthoff, I. M., E. B. Sandell, E. J. Meehan, and Stanley Bruckenstein: "Quantitative Chemical Analysis," 4th ed., Macmillan, London, 1969.
4. Laitinen, H. A.: "Chemical Analysis," McGraw-Hill, New York, 1960.
5. Salutsky, M. L.: Precipitates: Their Formation, Properties and Purity, in I. M. Kolthoff and P. J. Elving (eds.), "Treatise on Analytical Chemistry," pt. 1, vol. 1, pp. 733–766, Interscience-Wiley, New York, 1959.
6. Welcher, F. J.: "Organic Analytical Reagents," 4 vols., Van Nostrand, Princeton, N.J., 1959.

12
Electrochemical Methods of Analysis

12.1 GENERAL

Electrochemistry is the relationship between electrical properties and chemical reactions. As applied in the area of analytical chemistry, this usually involves the measurement of some electrical property under conditions which, directly or indirectly, permit an association between the magnitude of the property measured and the concentration of some particular chemical species. Such measurements are nearly always made in solution environments.

The electrical properties most commonly measured involve either voltage, current, or resistance or combinations of these. In some instances the electrical property may be a function of time, whereupon time may also be a variable to be measured.

Electrochemical techniques are very frequently applied in the field of titrimetry, the end point or equivalence point of the titration being located by a consideration of the change in titrant volume relative to the change in the electrical property being measured. Such techniques include the measurement of

1. The changing potential between an indicator electrode and a reference electrode (potentiometric titration)

2. The changing resistance or conductance of the solution under titration (conductometric titration)
3. The changing diffusion-controlled limiting current between an indicator electrode and a reference electrode where a fixed potential is maintained between the electrodes (amperometric titration)

Such electrochemical techniques are indirect, since they are used to locate the critical volume of titrant, from which a quantitative estimation of the species involved is then made.

Somewhat less frequently encountered, but also of importance, are those electrochemical methods of a nontitrimetric or direct application. A procedure used commonly here is to control one of the electrical properties mentioned, and to measure one or both of the remaining properties. Direct calculation using appropriate formulations, or a calibration technique, then provides the relationship between the property measured and the concentration of a specific solution species. Such more direct methods include

1. The measurement of the potential between an indicator electrode and a reference electrode under conditions of negligible flow of current (potentiometry)
2. The separation of solution species at one or both electrodes by electrode reactions carried out under predetermined conditions of applied voltage and current (electrolytic separations)
3. The measurement of the quantity of electricity required to yield a quantitative electrode reaction involving a particular species in solution (coulometry)
4. The measurement of the diffusion-controlled limiting current between an indicator electrode and a reference electrode where a fixed potential is maintained between the electrodes (voltammetry and polarography)

Some considerable discussion has already been given relative to half-cell potentials and their measurement and calculation, the Nernst equation, reference and indicator electrodes, galvanic cells, etc. This was presented in Chap. 4, Part 5, on oxidation-reduction equilibria, and will be drawn from and expanded in this chapter.

12.2 POTENTIOMETRY AND POTENTIOMETRIC TITRATIONS

12.2.1 MEASUREMENT OF POTENTIAL AND VOLTAGE

Since the techniques of both potentiometry and potentiometric titrations involve the measurement of the emf between an indicator electrode system and a reference electrode system, it is essential that some additional discussion be given on the measurement of cell emf.

It has been indicated that it is not possible to measure half-cell or single electrode potentials in a direct manner. The customary procedure is to measure the potential of the half-cell or single electrode system relative to some reference elec-

trode system of known or accepted potential. This requires the setting up of a galvanic cell consisting of the half-cell systems involved properly connected, the determination of the galvanic cell emf, and the estimation from this value, together
with the accepted value of potential for the reference electrode, of the potential for
the half-cell or single electrode system under investigation. The value thus determined will be relative to the reference system used and the ambient temperature
under which the measurement was made.

While half-cell or single electrode potentials are usually referred to the normal
or standard hydrogen electrode (SHE), such potentials can be expressed relative
to any acceptable reference electrode system, such as the saturated KCl calomel
electrode (SCE) or the saturated KCl silver–silver chloride electrode. The conversion of single electrode potentials, relative to one reference electrode system,
to any other reference system is easily accomplished. The general method was
outlined in Sec. 4.19.3.

The term "potential," when used correctly in connection with any half-cell,
refers to the potential of that half-cell or single electrode system when the electrode reaction takes place in the reduction direction. In the broader interpretation
applicable to potentiometric work, potential can be used to describe the emf of a
galvanic cell involving any two single electrode or half-cell systems connected
properly, and it should be measured between the electrodes under conditions
where no current flows through the cell. Methods of measurement of cell emf
which permit the flow of significant current through the cell during the
measurement process result in voltage values appreciably lower than the potential
value of which the cell is capable. This derives principally from two factors. The
passage of current causes the appropriate electrode reactions to take place, thereby changing the concentrations or activities of critical solution species and lowering the cell potential. Secondly, an ohmic drop in potential over the internal resistance of the cell solution results from the flow of current, providing for measured
emf values lower than the potential expected under conditions of no current flow.

Since no method of measurement of potential can provide for a zero flow of
current through the cell, it is apparent that what is obtained in such measurements
is a voltage or emf value. Where the current flow is insignificant, the voltage value
determined may very closely approximate the cell potential, providing that no
other interfering influences exist. The potentiometric method of measurement
which, in brief, involves balancing the cell output emf against an identical and
measured externally applied emf, under a current flow through the cell of 10^{-12} A
or less, is generally applied in the measurement of cell potential.

Whereas the potential of a cell system can be calculated on a theoretical
basis by the techniques given in Chap. 4, Part 5, the measured values of cell emf
usually show significant differences from these calculated values. It was shown,
for example, that the potential of a galvanic cell could be determined from

$$E_{cell} = E_{Ox_2/Red_2} - E_{Ox_1/Red_1}$$

where the values for E_{Ox_2/Red_2} and E_{Ox_1/Red_1} are calculated from Nernst-type relationships. In such a cell, the system 1 half-cell electrode is the negative pole and that for the system 2 half-cell the positive pole, when the overall cell reaction as written is spontaneous. Where a positive value for E_{cell} is obtained, the overall cell reaction as written will be spontaneous; a negative value for E_{cell} indicates a nonspontaneous overall cell reaction as written.

Where the value of E_{cell} so determined is to be compared with the value obtained by direct measurement, it would be assumed that for a reasonably accurate correlation.

1. The Nernst-type relationships would yield the true values of the half-cell potentials involved.
2. The current flow through the cell would be negligible over the period of measurement.
3. The electrode reactions would be reversible.
4. The electrode reactions would be rapid.
5. No other source of potential, other than those involved in the Nernst-type relationships, would contribute to the measured value.

For the calculation of $E_{Ox/Red}$, the Nernst relationship will yield a representative value for the half-cell potential only where the activity of the ion involved in the half-cell reaction is used. Where the molar concentration for such an ion is used, a customary procedure in most analytical work, this may lead to calculated potential values which differ to a more or less significant extent from the true values.

Cells and cell reactions may be reversible or irreversible. A reversible cell is one where the electrode reactions are reversed by changing the direction of current flow through the cell. Thus, in a reversible cell the electrode reactions when the cell is electrolyzed will be exactly the opposite to those taking place when the cell functions as a galvanic cell. Irreversible cells are those where, at one or both electrodes, entirely different reactions result from a reversal of the direction of current flow. In the main, in order to obtain consistent measurement values, cells important from the point of view of potentiometry and potentiometric titrations are reversible cells.

Where either one or both of the cell electrode reactions are not rapid what is called overvoltage often appears. Generally speaking, this means that for the electrode reaction involved, the emf developed frequently differs from the potential as determined from the Nernst-type relationship. While overvoltage values are not contributory to most analytically important potentiometric work, such as that involved in pH measurement or in potentiometric titrations, they can be of considerable significance in electrolytic separations, where current flow through the cell promotes electrode reactions.

Even where negligible current flows and the cell reactions are both rapid and reversible, measured potentials may differ appreciably from calculated potentials. One of the sources of such differences is the potential which may be developed across the interface between two solutions exhibiting differences in concentration, or composition, and ion mobilities. Such potentials are called junction potentials and may, under the proper circumstances, take on relatively high values. Suppose, for example, that we have two solutions of a substance AB, where

$$AB \rightleftharpoons A^+ + B^-$$

and where the solution concentrations are 1 and 0.01 M relative to AB. Suppose further that the ion A^+ is much more mobile than the ion B^-. The two solutions are now separated by a permeable membrane. Because of the concentration difference, the rates of diffusion of A^+ and B^- from the concentrated to the dilute solution will be higher than the comparable rates in the reverse direction. Again, because of the higher mobility of A^+, the rate of diffusion of A^+ will be much higher than that of B^-. Thus, the dilute solution will gradually acquire a positive charge relative to the concentrated solution. Eventually the charge difference will reach a value capable of offsetting the difference in the ion mobilities, and an equilibrium will be established. The charge difference at this point may represent a very significant potential across the interface.

Any similar situation which involves two solutions of different concentration, or composition, and ion mobilities will be capable of giving rise to a potential across the junction represented by the interface between the two solutions. For example, a galvanic cell consisting of a silver electrode in a solution of silver nitrate connected to a calomel reference electrode by means of a potassium nitrate salt bridge will yield a junction potential at the bridge and silver solution interface and a second junction potential at the bridge and calomel electrode solution interface. Such junction potentials, since they contribute to the emf detected by the measuring system, may provide for measured values of cell emf substantially different from the calculated cell potential. In many instances, junction potentials can be minimized by the use of properly selected conductive bridges between the two solutions involved. Such conductive bridges involve concentrated salt solutions, and the concentration of the salt should be as high as possible in order to provide comparatively high diffusion rates of the salt ions into the two solutions connected by the salt bridge. The mobilities of the ions of the salt comprising the bridge should be as similar as possible. Where, for example, a saturated KCl salt bridge is used, the very slight difference between the mobilities of the K^+ and Cl^- ions provides for junction potentials which, under favorable circumstances, may amount to not more than a millivolt or two. Obviously the most favorable circumstances will not be achieved where the solutions connected by the bridge show total ion species concentrations approaching the salt concentration of the bridge.

Where junction potentials are known or can be determined, and are of a magnitude sufficient to warrant consideration in direct potentiometry, the expression given previously for E_{cell} may be modified to

$$E_{cell} = (E_{Ox_2/Red_2} + E_{j_2}) - (E_{Ox_1/Red_1} + E_{j_1}) \tag{246}$$

12.2.2 METHODS INVOLVING DIRECT POTENTIOMETRY

Advantages and limitations   An advantage of the analytical application of direct potentiometry is the fact that such investigations can be carried out on very small solution volumes, as little as a milliliter or less being required under favorable conditions. In addition to this, the method does not involve the destruction of or compositional change in the sample investigated. The use of a two-electrode probe permits the continuous analysis of liquid systems in motion and is highly adaptable to the monitoring and automated controlling of industrial solution processes of many types. Where microprobe systems are used, the monitoring ability extends to the study of certain biochemical processes taking place in living tissues.

   Although many types of electrode systems may involve limitations peculiar to themselves, there are limitations which, in a general way, restrict the value of direct potentiometry as an analytical tool. One of these involves the interpretation of the measured cell potential, between the indicator and reference electrodes, in terms of the activity of the ion under investigation. For Eq. (246), where indicator and reference electrodes constitute the cell, this may be shown, under the proper circumstances, as

$$E_{cell} = (E_{ind} + E_{j_{ind}}) - (E_{ref} + E_{j_{ref}}) \tag{247}$$

For an indicator electrode showing the general electrode reaction

$$M^+ + 1e \rightleftharpoons M$$

we have

$$E_{ind} = E_{M^+/M} = E^{\circ}_{M^+/M} + 0.059 \log \alpha_{M^+} \qquad \text{at } 25°C \tag{248}$$

so that

$$\alpha_{M^+} = \text{antilog} \frac{1}{0.059} (E_{cell} - E^{\circ}_{M^+/M} - E_{j_{ind}} + E_{ref} + E_{j_{ref}}) \tag{249}$$

An absolute error of only ± 1 mV in the value of the term in parentheses in Eq. (249) results in an error of ± 4 percent in the value of α_{M^+} as determined. For divalent and trivalent ions the errors are, respectively, about ± 8 and ± 12 percent on the same basis. The value of E_{cell} as measured and referred to E_{ind} may differ from the true value of E_{ind} on the basis of variations and uncertainties in the junction potentials or reference electrode potential, variations in the indicator electrode potential due to response systems other than the ideal, temperature variations, etc. Junction potential uncertainties alone may indeed amount to several millivolts, so that direct potentiometric measurement of E_{cell} and the use of this value in the calculation of α_{M^+} is capable of yielding results with a high error probability.

   The Nernst-type relationship, in its more suitable application, involves the activity of the ion. In analytical work the interest centers mainly around ion concentrations. Ion activity coefficient data relating activity and molar concentration are rarely available for most analytical situations, and this, coupled with the variation in activity coefficient with variation in the solution concentrations of other species, renders it difficult to obtain worthwhile concentration data from direct potentiometric measurements. Again, the direct method at its best will report only the concentration of a species relative to the concentration of the species ion reporting at the indicator electrode. Where the species may exist in solution to an appreciable extent in an undissociated or complexed form, the total species concentration can not be directly obtained. For example, in a solution of a weak acid, the measurement of pH will provide information as to the concentration of hydrogen ion but will not directly provide data relative to the concentration of weak acid. Finally, in many cases, the indicator electrode may respond to concentrations of species in solution other than that for which it was presumed to be specific. Where this occurs, the values obtained for the ion activity or concentration in question may be seriously in error.

To some extent, and particularly in the case of pH measurement, the use of calibration methods instead of calculations may offset some of the sources of error, such as those associated with junction and reference electrode potential variations and uncertainties, and activity-concentration relationships for the ion species under investigation. Where this method is applied, the general technique is to prepare a bulk solution which is of the same ionic strength, for concentrations of individual ion species other than that to be determined, as that expected for the unknown solutions to be analyzed. To portions of this bulk solution known concentrations of the ion species to be determined are added, and the solution potentials are measured potentiometrically between an appropriate indicator-reference electrode combination. The potential values are then plotted against the known concentrations. Since the extrapolation of the curve for such a plot may lead to unsuspected error, the range of the plot as to concentration is planned to encompass the range expected for the unknown solutions to be tested.

Metallic indicator electrodes   Where a half-cell reaction involves a metal electrode immersed in a solution of its own ions, we have a situation typified by

$$M^{n+} + ne \rightleftharpoons M$$

and we would have

$$\alpha_{M^+} = \text{antilog} \frac{n}{0.059} (E_{\text{ind}} - E^{\circ}_{M^{n+}/M}) \qquad \text{at } 25°C$$

and, with a lesser degree of accuracy,

$$[M^{n+}] = \text{antilog} \frac{n}{0.059} (E_{\text{ind}} - E^{\circ}_{M^{n+}/M}) \qquad \text{at } 25°C$$

In addition to the general difficulties already mentioned with respect to obtaining an accurate value for E_{ind}, there are limitations more or less specific to the use of metal electrodes in direct potentiometry or in potentiometric titrations.

It is true that certain metals can be used as electrodes in solution situations involving their own ions. The use of a massive silver electrode immersed in a solution containing silver(I) ion is a typical case in point. In the majority of instances, however, two factors tend to rule against the application in analytical chemistry of such metal electrode systems. First, many metallic substances react with solution species other than the metal ion directly involved. Typical of such reactions are those with molecular oxygen dissolved in the solution medium, the formation on the metal surface of basic compounds, such as $Cu(OH)NO_3$ where a copper electrode is involved, and the general attack at the metal surface resulting from unfavorable solution conditions (acidity, etc.). Such reactions establish potential response systems which permit deviations from the ideal relationship, with associated errors arising out of the calculation or calibration systems used to determine ion activity or concentration. Second, many metals, upon being subjected to strain or distortion, show crystal-structure deformations with resultant significant changes in internal energy. Such internal energy changes result in electrode potential changes apart from those associated with the changes in the activity or concentration of the metal ion. For metals such as iron or nickel the sensitivity to strain or distortion effects is quite pronounced.

While not entirely free from the sources of error outlined in the foregoing, the massive silver electrode can be used for the direct potentiometric determination of silver(I) ion activity or concentration, although a more common application is that associated with potentiometric titrations involving silver ion. The concentration of silver ion is given by

$$[Ag^+] = \text{antilog} \frac{1}{0.059} (E_{cell} - E^{\circ}_{Ag^+/Ag} + E_{ref} + E_{j_{ref}}) \qquad \text{at } 25°C$$

Where the calibration method is used, E_{cell} is plotted against the concentration of Ag^+ for solutions of known Ag^+ concentration and of ionic background similar to that expected for the unknown solutions to be investigated. The technique suffers generally from the difficulties outlined under limitations of direct methods.

The measurement of pH using the hydrogen electrode   The hydrogen electrode would appear to offer excellent possibilities as an electrode for the measurement of the hydrogen activity, hydrogen-ion concentration, or pH of solutions. The general form of this electrode was described in Sec. 4.16. Because the solution in which it was immersed showed $\alpha_{H^+} = 1$, the electrode was then described as a normal or standard hydrogen electrode. Obviously, where the electrode proper is immersed in solutions of varying α_{H^+}, the potential will respond to the value of the hydrogen-ion activity in solution. In conjunction with a reference electrode, the hydrogen electrode should be capable of the direct potentiometric determination of pH.

The hydrogen electrode is difficult to use, particularly with respect to the open flow of hydrogen gas. In addition, the electrode shows a potential response not only to the hydrogen ion but also to oxidizing and reducing agents in the same solution. A further problem arises relative to the platinum black coating of the electrode, this coating becoming in time less capable of promoting the equilibrium between hydrogen gas and hydrogen ion and therefore requiring periodic replacement.

The measurement of pH using the glass-membrane electrode   The glass electrode or, more commonly and properly, the glass-membrane electrode, is by far the most extensively used of the electrodes sensitive to hydrogen ion. It is generally unaffected by oxidizing or reducing agents in solution, is resistant to chemical attack by nearly all solution media, and shows a rapid response to changes in the solution concentration of hydrogen ion. These characteristics render the glass-membrane electrode extremely useful in applications involving both direct potentiometry and potentiometric titrations. While the form of the electrode may vary somewhat with the application (extremely small test volumes, continuous flow testing, etc.), a general form is shown in Fig. 12.1. This consists essentially of a silver–silver chloride electrode system immersed in a solution contained within a glass membrane sensitive to hydrogen ion. A glass tube contains a silver wire sealed through one end and protruding into a thin-walled glass bulb (the glass membrane) of a special glass sensitive to hydrogen ion, the glass bulb being fused to the glass tube. The silver wire inside the bulb is coated with silver chloride and is immersed in the solution, usually 0.1 M HCl, contained in the bulb. The opposite end of the wire terminates in a shielded lead wire which connects to the pH meter or potentiometer. The wire in the glass tube is enclosed in either resin or

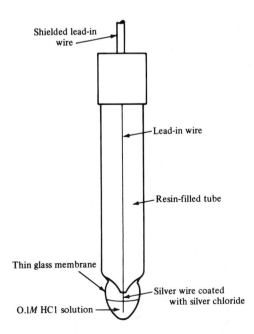

Fig. 12.1   The glass-membrane electrode.

wax. Unless the glass-membrane electrode is of the combination type, incorporating as a single unit both the glass electrode and a reference electrode, it is used in conjunction with a separate reference electrode. The reference electrode in either case is typically a saturated KCl calomel electrode. The glass forming the hydrogen-ion—sensitive bulb or membrane is of a special type, a sodium-calcium silicate glass being commonly used, with a lithium-calcium silicate glass being used to avoid what is known as "alkaline error" where solutions of high sodium ion concentration are to be investigated.

The immersion in water of the glass-membrane electrode results, after a brief period, in the hydration of the glass membrane. The hydration reaction occurs by absorption of water by the membrane interface in contact with the solution, followed by the exchange of univalent cations of the glass for hydrogen ions from the water or solution environment. For a sodium-calcium silicate glass we have then

$$H^+_{solution} + Na^+_{glass} \rightleftharpoons Na^+_{solution} + H^+_{glass}$$

This hydration reaction has also, of course, taken place at the membrane interface adjacent to the internal hydrochloric acid solution. The glass membrane does not function correctly except where hydrated properly. Where the electrode has been removed from aqueous medium for some time, it will show poor response until after immersion for a brief period in water. Similarly, the immersion of the electrode in solutions of strong dehydrating agents, such as absolute ethanol, will result in poor response until proper functioning is restored by immersion for some time in water.

After proper hydration, the glass membrane cross section, usually about 100 μm total thickness, will show an inner and outer layer of approximately 0.1 μm where the substitution of hydrogen ion for sodium ion is significant. Each of these layers consists of a surface immediately adjacent to the solution involved where substitution of H^+ for Na^+ is complete. The replacement level becomes progressively less with extension into the 0.1-μm layer, and at the junction of each hydrated zone with the central or dry zone of the membrane, there is no substitution of H^+ for Na^+. Figure 12.2 shows schematically the general arrangement after hydration of the membrane. Current flow across the membrane is the result of processes which vary with the zone involved. In the interface zones of the membrane immediately adjacent to the solutions, charge transfer is by movement of H^+ ions, in the hydration zones it is by movement of H^+ and Na^+ ions and in the dry zone it is by movement of Na^+ ions. Depending on the nature of the external solution (its hydrogen-ion activity), the net movement may be that of hydrogen ions from the external solution to the internal solution, or vice versa.

Where the quantity of hydrogen ion in the glass-membrane surface immediately adjacent to either the external or internal solution is taken as being identical, the boundary potential over each interface will be a function of the activity of hydrogen ion in the solution at that interface. Thus the potential between the two interfaces will be the resultant of the two boundary potentials. Since the α_{H^+} for the

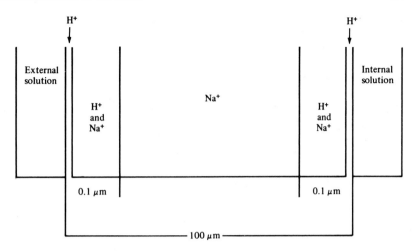

Fig. 12.2 Properly hydrated glass membrane.

internal solution is a constant, the potential between the inner and outer interfaces will be a function of the α_{H^+} of the external solution. The potential is then given by

$$E = k - 0.059 \log \alpha_{H^+} \qquad \text{at } 25°C \tag{250}$$

The value of k accounts for several factors, including the activity of hydrogen ion in the internal solution. Also included in k is the asymmetry potential of the membrane. This potential can be demonstrated to exist by placing on either side of a glass membrane identical solutions and reference electrodes. A zero difference in potential would be expected under these conditions; a small potential of a few millivolts is nevertheless usually noted. This potential arises out of strains in the glass membrane originating during the manufacturing process, chemical attack at the external membrane surface, scratching or general deterioration of the external surface as the result of handling, adsorption of foreign ions at the external surface, dirty surfaces, etc. The asymmetry potential varies slowly in time for a given electrode as well as varying from electrode to electrode.

When a glass-membrane electrode and a reference electrode such as the SCE are immersed in a solution of H^+, the potential of the resultant galvanic cell arises out of several sources. These are

1. The potential of the silver–silver chloride electrode
2. The potential of the reference electrode
3. The junction potential at the interface between the reference electrode and the solution
4. The potential across the interface between the glass membrane and the internal solution

5. The potential across the interface between the glass membrane and the external solution
6. The asymmetry potential of the glass membrane

The values of 1, 2, and 4 are constant at constant temperature. The junction potential may vary according to the ionic strength and composition of the unknown solution; but for most normally dilute solutions it will be reasonably constant for saturated KCl salt bridge junctions. All in all, however, the galvanic cell potential changes result from changes in the values originating with 3, 5, and 6. It can be shown that we have a relationship between the galvanic cell potential E_{cell} as measured potentiometrically and the solution pH, where

$$E_{cell} = K + 0.059\,pH \qquad \text{at } 25°C \tag{251}$$

where K now involves k and includes those factors arising out of 1, 2, 3, 4, and 6. Remember that the contributory sources to E_{cell} are temperature sensitive, so that compensating calculations or instrumental circuits must be available relative to correcting for the effect of varying ambient temperature of testing. Since variation in the values, for example, of the junction and asymmetry potentials can occur, the value of K can not be precisely determined, and the method of measuring the pH of an unknown solution requires a calibration procedure where a comparison of the pH for the unknown solution with that of a standard buffer solution of known pH is made. Thus we have

$$E_{cell,std} = K + 0.059\,pH_{std} \qquad \text{at } 25°C$$

$$E_{cell,unk} = K + 0.059\,pH_{unk} \qquad \text{at } 25°C$$

so that

$$pH_{unk} = pH_{std} + \frac{E_{cell,unk} - E_{cell,std}}{0.059} \qquad \text{at } 25°C \tag{252}$$

Two factors are important here. The value of pH for the buffer standard must be accurately known, and the value of K is assumed to be identical over both measurements. Buffer solutions of pH values known to ± 0.005 unit are readily available commercially, and the K values over both measurements can be assumed to be identical when the standard buffer used has a pH value very closely approximating that of the unknown solution. This latter condition will be valid, however, only where the ionic strength and composition of the unknown and the buffer do not differ sufficiently to introduce a difference in the junction potentials and therefore a difference in the value of K. Under ideal conditions, the error or uncertainty in the value of $E_{cell,unk} - E_{cell,std}$ should not exceed a millivolt, with a resulting error or uncertainty in the pH determined of approximately ± 0.02 unit. Under nonideal conditions, the uncertainty in the pH may be significantly greater. Thus, even where pH meters are capable of reading uncertainties as low as ± 0.005 unit, the derived pH value will not usually be more certain than ± 0.02 unit.

While the calibration method can be applied potentiometrically (using Eq. (252) for the calculation of the pH), in actual practice the procedure is simplified by the use of modified potentiometers in the form of pH meters. Such instruments provide a scale which interprets the millivolt relationship of hydrogen-ion concentration as pH. A temperature-compensation circuit is included which can be set to the temperature at the time of testing, thereby compensating for the variation with temperature of the instrumental interpretation of millivolts as pH. The proper buffer standard is then tested and the pH meter dial is set to read the exact pH of the buffer for the testing temperature involved. The unknown solution is then tested, the pH value being read directly from the dial.

It is apparent that since the substitution of H^+ ions for Na^+ ions requires first the absorption of water by the membrane interface surface, the activity of water in the solution in contact with the outer interface will have some influence on the extent to which these absorption and substitution actions take place. In strongly acidic solutions, where the activity of water is substantially less than unity, the value of pH observed errs by being higher than the actual value. In point of fact, the same error situation would occur in *any* solution where the activity of water is appreciably less than unity, such as solutions of high salt concentration or high concentration of nonaqueous solvents. Although the name of "acid error" has been given to this phenomenon because it is commonly encountered in connection with the testing of strongly acid solutions, it is a misnomer. Where acid solutions are involved, the phenomenon is not of great significance, since its effects are not noted except at solution pH values of zero or lower.

When the hydrogen-ion concentration of a solution is less than $10^{-9} M$ (pH less than 9) other cations, when present in the same solution at concentrations 10^9 times as great or more, tend to compete with hydrogen ion for exchange sites in the hydrated outer layer of the glass membrane. Sodium ions are particularly effective in this connection, although potassium and lithium ions also exert a comparatively strong influence. The effect of this situation is to yield an observed pH value lower than the actual value, the extent of this "alkaline error," as it is called, varying with the concentration and nature of the cation involved and with the concentration of hydrogen ion. Since sodium ion is by far the most serious offender, electrode membrane glasses have been developed which tend to reduce the alkaline error due to high sodium-ion concentrations in solution. Such glasses usually involve the substitution of lithium for the sodium content of ordinary membrane glass, although many lithium glasses also contain barium and cesium and/or lanthanum in place of calcium. Figures 12.3 and 12.4 show the effect of acid error and alkaline error relative to glass-membrane electrodes involving Corning 015 sodium-calcium silicate glass.

Glass-membrane electrodes sensitive to cations other than hydrogen ion The situation relative to alkaline error led to the investigation by Eisenman and others* of glass-membrane electrodes sensitive to cations other than hydrogen

*G. Eisenman, "Advances in Analytical Chemistry and Instrumentation," vol. 4., Wiley, New York, 1965.

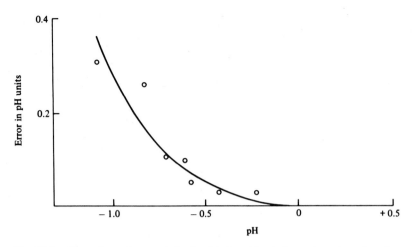

Fig. 12.3 Glass-electrode error in hydrochloric acid solutions. *(Based on data from M. Dole, "The Glass Electrode," Wiley, New York, 1944, p. 140.)*

ion. Membrane glass of the general composition 22% Na_2O, 6% CaO, and 72% SiO_2 shows excellent response to hydrogen ion for solutions ranging in pH from 0 to 9. Where the pH value exceeds 9, the electrode becomes responsive to other cations, particularly to univalent cations such as Na^+, Li^+, and K^+, the response depending on the pH and the concentration of such cations. For such a membrane, the response is especially strong to Na^+ ions. By varying the composition of the glass, it is possible to enhance such response systems. Membrane glass of approximately 27% Na_2O, 5% Al_2O_3, and 68% SiO_2 is responsive in particular to K^+, whereas a 15% Li_2O, 25% Al_2O_3, and 60% SiO_2 glass responds strongly to Li^+. Electrodes with these and other special glass membranes are now available commercially for the determination of Na^+, Li^+, K^+, Ag^+, NH_4^+, and several other univalent cations. Where these electrodes are used, depending upon the nature of the membrane, there are limitations to the concentration of hydrogen ion that can be present in the solution under test. In addition to this, an electrode responsive to a particular cation may be affected by other cations where these are present in solution in similar concentrations. Such interfering actions are, of course, based on competition for exchange sites in the external hydrated layer of the glass membrane.

Ion-selective electrodes or specific-ion electrodes It is possible to develop electrodes for the measurement of the activity or concentration of certain anion types. Pungor et al.* developed such an electrode for the measurement of the activity of iodide ion in aqueous solution. A monomeric silicone rubber was polymerized in the presence of a fine dispersion of silver iodide particles. A por-

---

*E. Pungor, J. Havas, and K. Toth, *Z. Chem.*, **5**:9 (1965).

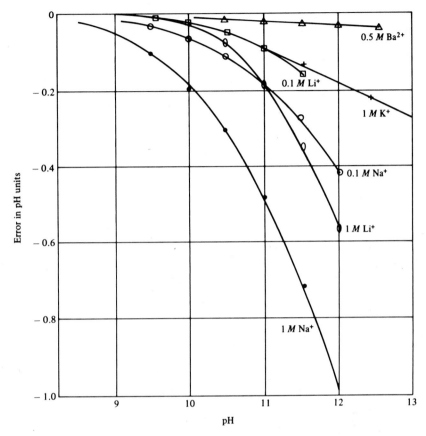

Fig. 12.4 Corning 015 glass-electrode error in alkaline solution with several cations. *(Based on data from M. Dole, "The Glass Electrode," Wiley, New York, 1941, p. 128, Table 7.2.)*

tion of the resultant polymer was used as a membrane to seal off the bottom of a glass tube. Potassium iodide solution was added to the tube and a silver electrode in contact with this solution was sealed into the opposite end of the tube. Immersion of this electrode into a solution containing iodide ion results in the preferential adsorption of iodide ions by the silver iodide particles at the solution-membrane interface (see Sec. 11.2.3). The extent of such adsorption will be a function, at either interface, of the activity of iodide ions in the respective solutions, but it will be constant for the inner interface. Boundary potentials over the membrane inner and outer interfaces will result. The electrode potential will thus reflect the overall boundary potential across the membrane and can, through a Nernst-type relationship, be associated with the activity of iodide ion in the solution in contact with the outer membrane interface. Where a reference electrode is used in conjunction with this electrode, a calibration procedure involving E_{cell} and known activity or concentration solutions of iodide ion can be applied. Similar membrane electrodes can be used for the determination of chloride and bromide. Fluoride may be determined by a somewhat related technique, using an electrode with a rare earth-treated single-crystal membrane.

A fairly recent development in the field of ion-selective electrodes is the use of liquid ion-exchange membranes, such liquids being commonly nonvolatile, water-insoluble organic substances involving chelating or other functional groups. The construction of an electrode embodying such a membrane is shown in Fig. 12.5, and it is compared with that of a typical glass-membrane electrode for pH measurement. Such ion-selective electrodes can be applied in the direct potentiometric determination of several anions and polyvalent cations. The electrode consists, in general, of a liquid ion-exchange substance held between two porous glass or polymeric disks. These disks allow contact, but prevent physical mixing, of the ion-exchange liquid with the inner and outer aqueous solutions involved. The inner solution has a definite concentration of the ion to which the electrode is selective. Immersed in this solution, and connected to an external lead, is a silver–silver chloride electrode. Before use, the ion-selective electrode is immersed for some time in a solution of the ion to which it is responsive.

The potential developed by the electrode has its basis in situations similar to those which apply in the case of the hydrogen-ion–sensitive glass-membrane electrode. Boundary potentials are established across the two solution–membrane interfaces; that across the outer interface being a function of the ion activity in the external solution; that across the inner interface being a function of the ion activity in the internal solution and, therefore, constant at constant temperature. The potential developed by the electrode will be related to the two boundary potentials and can be shown to be related to the activity of the ion in the external solution

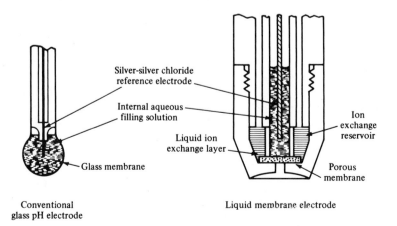

Conventional
glass pH electrode

Liquid membrane electrode

Silver-silver chloride
reference electrode

Internal aqueous
filling solution

Liquid ion
exchange layer

Glass membrane

Ion
exchange
reservoir

Porous
membrane

Fig. 12.5 Glass-membrane electrode and typical liquid membrane electrode. *(Courtesy of Orion Research, Inc., Cambridge, Mass.)*

under test by

$$E = \text{constant} + \frac{0.059}{n} \log \alpha_{M^{n+}} \qquad \text{at } 25°C \qquad (253)$$

Where, specifically, the calcium-ion–sensitive electrode is involved, the liquid ion-exchange substance is an organic derivative of phosphoric acid, and the inner solution is a $CaCl_2$ solution of fixed concentration. Prior to use the electrode is immersed for some time in a solution of Ca^{2+}. This results in the formation in the liquid membrane, by ion exchange of Ca^{2+} for H^+, of the calcium salt or calcium chelate of the organophosphate. Repeated treatment results in the ion-exchange liquid being mainly in the desired salt or chelate form. Subsequent immersion of the electrode in a solution of Ca^{2+} then results, depending on the activity of Ca^{2+} in this solution, in migration of Ca^{2+} from the membrane interface to the solution, or vice versa. Because of their very large size, the organophosphate anions do not migrate into the solution to any significant extent. Migration of Ca^{2+} ions into or out of the membrane interface in contact with the inner solution will, of course, occur in the opposite sense.

Boundary potentials are thus established, and electrode response occurs. In actual use, the prepared electrode and a reference electrode are immersed in solutions of known Ca^{2+} activity and concentration, and a calibration plot of E_{cell} vs. activity and/or concentration is obtained. This calibration plot is then used to determine the activity and/or concentration of Ca^{2+} in unknown solutions. Figure 12.6 shows a typical calibration plot for Ca^{2+} activity and concentration. The

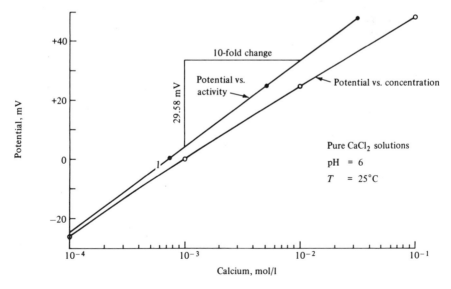

Fig. 12.6 Activity and concentration calibration plots for the calcium-ion-sensitive electrode. *(Courtesy of Orion Research, Inc., Cambridge, Mass.)*

selectivity of the calcium-ion—sensitive electrode is based on the affinity of the organophosphoric acid ion-exchange substance for Ca^{2+} over other ions such as Mg^{2+} or Na^+. Calibration solutions should, of course, be of an ionic background similar to that expected for the unknown solutions to be tested. Ion-sensitive electrodes may be independent of solution pH within certain ranges of pH only. The calcium-ion—sensitive electrode is, for example, independent of pH between about 5 and 11 pH. At higher values of hydrogen-ion concentration in the solution to be investigated, the possibility of H^+ exchange with the calcium salt or chelate of the organophosphate exists, with corresponding H^+ as well as Ca^{2+} sensitivity.

Ion-selective electrodes of this type have been developed for use in the direct potentiometric determination of several anions and polyvalent cations.

12.2.3 METHODS INVOLVING POTENTIOMETRIC TITRATIONS

General By far the most fruitful use of potentiometry in analytical chemistry lies in its application to the area of titrimetry. Such potentiometric titrations, as they are called, can be applied in connection with neutralization, precipitation, complexation, oxidation-reduction, and nonaqueous solvent systems. In the potentiometric titration, the end point or equivalence point is determined, in general, by plotting the emf value of an appropriate galvanic cell against the volume of titrant. The galvanic cell consists of two half-cell systems, a reference half-cell or electrode and an indicator half-cell or electrode. The indicator electrode is responsive to the changes in the activity and/or concentration of a particular ion species that occurs as the titration progresses. In its simplest application, the method requires only the usual titration equipment, plus a potentiometer (usually a dual potentiometer and pH meter), a magnetic stirring device, and suitable indicator-reference electrode pairs. Figure 12.7 indicates a typical potentiometric titration setup.

The advantages of potentiometric titration stem from several sources. It should be remembered that the methods of direct potentiometry yield at best the activity and/or concentration of a specific ion species in the solution tested. Where the substance involved exists in solution to a significant extent in a form other than the ionic state, that portion will not be determinable from data obtained by direct potentiometric measurement. For example, the determination of the pH of an aqueous solution of a weak acid reports only the hydrogen-ion concentration. Since the major portion of the acid is present in the undissociated form, direct potentiometry can not provide data yielding the *total* concentration of the acid. Potentiometric titration involves titrating the acid solution with a standard solution of a base, plotting the solution pH vs. the volume of standard base, determining the equivalence point of the titration, and calculating the weak acid concentration from this stoichiometric data. Again, in direct potentiometry, an accurate and precise measurement of E_{cell} must be made, and accurate values of E_j and E_{ref} known, in order to evaluate E_{ind} and obtain through the Nernst relationship an

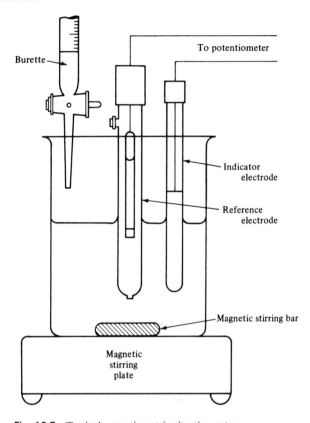

Fig. 12.7 Typical potentiometric titration setup.

accurate value for the ion activity. The relationship between α_{ion} and [ion] for the particular medium investigated should be known in order to determine with worthwhile accuracy the value of [ion]. Even where the calibration method is applied, the results become highly doubtful where the ionic strength or background of the calibrating solutions does not duplicate that of the unknown. In a potentiometric titration, the position of the E_{cell}-vs.-titrant-volume curve relative to the vertical axis (the E_{cell} axis) may vary with variations in E_j, E_{ref}, $E°$, etc.; but for a particular titration, its position relative to the horizontal axis (the titrant-volume axis) does not vary. The equivalence point for a particular titration is thus fixed relative to the titrant-volume axis, regardless of its position with respect to the E_{cell} axis. If the same titration is repeated under varying conditions of ionic background for species not participating in the titration reaction, the resultant titration curves will yield the same equivalence-point volume, although they may differ from each other relative to their E_{cell}-axis position. In a potentiometric titration, it is not even necessary to set or calibrate the potentiometer at any particular reading of E for the start of the titration. It is only necessary to ensure that the *range* of E_{cell} over the course of the entire titration lies within the scope of the potentiometer dial readings. Varying the starting value of the E reading on the potentiometer dial can only vary the vertical positioning of the titration curve; it can not affect the position of the equivalence point with respect to the volume axis. Thus a knowledge of E_j, E_{ref}, or even $E°$, is not really important in potentiometric titration, nor is the determination of E_{cell} required to be accurate, or even as precise, as is required for direct potentiometry.

The method of potentiometric titration possesses several important possible advantages over standard titrimetric procedures. Where solutions under titration are colored, internal indicators which change color at the end point or equivalence point may be unsuitable or, if used, may introduce significant titration errors. Even where such internal indicators can be applied under ideal conditions of solution color, the interpretation of the color change coinciding with the titration end point is a matter of subjective decision and may therefore be open to error. Again, it is not unusual to find that the end point given by the color change of the indicator does not coincide with the stoichiometric equivalence point, thereby necessitating the application of blank calculations made on a subjective basis. In certain cases, suitable internal indicators may not be available for the titration system involved. In all such situations, the potentiometric titration method offers distinct advantages. Finally, because the equivalence point is interpreted electrochemically, on the basis of large changes in E_{cell} in the neighborhood of the equivalence-point volume, the potentiometric titration method is highly adaptable to automated techniques, such as automatically plotted titration curves, titrations automatically stopped at the equivalence point, etc.

Certain disadvantages are associated with the technique, not the least of which are that (1) it is not always possible to find an indicator electrode adequately responsive for the titration system under consideration, and (2) the time required to complete a potentiometric titration may be somewhat longer than that for the comparable internal indicator titration. This latter situation is often less significant in the case of automated potentiometric titrations.

Equivalence-point location In Sec. 5.2, it was pointed out that the plots of pR vs. V_T, $\Delta pR/\Delta V_T$ vs. V_T, or $\Delta^2 pR/\Delta V_T^2$ vs. V_T could be used to locate the point of inflection, or its attributes, of the titration curve, and that proper applications of curve analysis techniques could serve to locate the titration equivalence point, even where this point and the point of inflection, or its attributes, did not coincide. Since the value of pR and the E_{cell} value are related in all properly conducted potentiometric titrations through some form of the Nernst relationship, the plots of E_{cell} vs. V_T, $\Delta E_{cell}/\Delta V_T$ vs. V_T, or $\Delta^2 E_{cell}/\Delta V_T^2$ vs. V_T can serve in the same way to locate the titration equivalence point. Thus we have, in the determination of chloride by titration with a standard solution of Ag^+, the relationship

$$E_{ind} = E_{AgCl/Ag} = E°_{AgCl/Ag} - 0.059 \log [Cl^-] \qquad \text{at } 25°C$$
$$= E°_{AgCl/Ag} + 0.059 \, pCl$$

and, since we have from Eq. (247)

$$E_{cell} = (E_{ind} + E_{j_{ind}}) - (E_{ref} + E_{j_{ref}})$$

we have

$$E_{cell} = \text{constant} + 0.059 \log \text{pCl} \quad \text{at } 25°C$$

since changes in E_{ref}, $E°_{AgCl/Ag}$, and E_j are not of importance in potentiometric titrations. Changes in pCl during the titration are directly reflected by changes in E_{cell}.

It was stated in Sec. 5.2 that the method of hand-drawing such curves is rarely applied in the case of manual (nonautomatic) potentiometric titrations because of the latitude for inaccuracy in locating the equivalence point and the general tedium and time consumption involved in obtaining sufficient data. Where automatic potentiometric titration techniques and equipment are used, and the plot of pH vs. V_T, $\Delta pH/\Delta V_T$ vs. V_T, E_{cell} vs. V_T, or $\Delta E_{cell}/\Delta V_T$ vs. V_T is obtained by automatic continuous curve plotting, such techniques of graphical location of the equivalence point may be applied. The curves obtained under such circumstances are secured rapidly and automatically, and they are smooth enough to permit reasonably accurate equivalence-point-volume location by the various curve-analysis techniques.

By far the simplest and most rapid method of equivalence-point location applicable in the case of manual potentiometric titrations is the tabulation technique described in Sec. 5.2.5 involving the calculation from E_{cell} and V_T values of $\Delta^2 E_{cell}/\Delta V_T^2$ data for titration points around the equivalence point. The method is based on the addition of identical small-volume increments of titrant before and after the equivalence-point volume. The implication here is that the equivalence-point volume will at least be approximately known. This is certainly true in the majority of instances involving commercial control analysis. Where the approximate volume required is not known, this can be determined by a rapid preliminary titration using relatively large increments of V_T.

It is most important to recall that in the case of asymmetrical titration curves, where there is a lack of coincidence between the equivalence point and the point of inflection of the titration curve, some discrepancy may exist between the equivalence point located through graphical or tabulation methods and the true equivalence point. This point was discussed in Sec. 5.2.6.

It is possible to conduct a manual potentiometric titration under conditions where the titration is stopped at a value of E_{cell} associated with the equivalence point. Of greater practical interest, however, is the fact that several manufacturers of potentiometric titration equipment supply titrators capable of automatically stopping the titration at the equivalence point. In either case, for such an application to be of real value, the E_{cell} (or more properly the potentiometer dial E reading) corresponding to the equivalence point must be determined by titration of a similar system carried out previously, and the conditions under which subsequent titrations of such systems are carried out must be such as to permit this E_{cell} or E value to be reproduced at each equivalence point. Calculation of the theoretical value of E_{cell} at the equivalence point, although it can be made, is usually unnecessary. What is more important is that the instrumental parameters once established by the preliminary titration must not be altered for subsequent titrations.

In the case of potentiometric titrations, it is of particular importance to ensure that the speed of the titration is compatible with the reaction rate and the rate at which the electrode system establishes an equilibrium. The system should be allowed to come to equilibrium conditions after each addition of titrant and before the potentiometer E value is read. For manual titrations this ordinarily implies that after each addition of titrant, the potentiometer E value is read when the potential drift is negligible (a millivolt or two). For automatic titrators, where the speed of the titration can be set by a control on the instrument, it will be important to select a titration speed suited to the reaction and electrode conditions. In all potentiometric titrations, efficient stirring is an essential factor.

Neutralization titrations Such titrations are carried out through the use of a suitable glass-membrane electrode, a reference electrode, and a pH meter. Where manual methods are used, the titration technique may involve a plot of pH vs. V_T, or one of $\Delta pH/\Delta V_T$ vs. V_T, for points around the equivalence point, with the usual curve-analysis approach being applied for equivalence-point-volume location. Alternatively $\Delta^2 pH/\Delta V_T^2$ vs. V_T data may be accumulated for points around the equivalence point, and the equivalence-point volume located by the tabulation method. Where automatic curve plotting titrators are used, a complete titration curve of pH vs. V_T or $\Delta pH/\Delta V_T$ vs. V_T will be available for curve analysis.

The glass-membrane electrode is *usually* calibrated using a known buffer solution of pH 7.00 prior to such titrations. Remember that it will not be possible to extract from such a titration *accurate* pH data for titration points with pH values which are significantly different from 7.00. This does not, however, affect the ability to locate the titration equivalence-point volume.

The general advantages of the potentiometric method over the usual internal indicator procedures have already been noted. Where neutralization titrations are concerned, particularly those carried out with automatic curve plotting titrators, additional advantages include the following.

1. A complete titration curve can be obtained for mixtures of acids or bases, or for multicomponent systems in general. The analysis of such a curve will generally permit a somewhat better location of the various equivalence points than would be obtained with internal indicators.
2. The complete titration curve of, for example, the titration of the salt of a dibasic acid and a strong base by a strong acid solution can be used to determine the values of K_1 and K_2 for the dibasic acid. The value of K_2 is obtained from $[H_3O^+] \approx K_2 C_{s,1}/C_{s,2}$, which holds at that point in the titration exactly halfway between the start and the first-stage equivalence point. The value of K_1 is then obtained from $[H_3O^+] \approx \sqrt{K_1 K_2}$, which holds at the

first-stage equivalence point. The accuracy achieved will be best, although always somewhat limited, where the glass electrode has been previously calibrated for a pH value between the midway point pH and that at the first-stage equivalence point.

Many examples of neutralization titrations involving pH data were given in Chap. 6. These should be reexamined relative to equivalence-point-volume location.

Example Table 6.13 showed pH-vs.-V_B data for the titration of the dibasic acid, maleic acid, by a standard NaOH solution. The titration curve was given in Fig. 6.9. From the former a first-stage equivalence point pH of 4.12 is obtained. From the latter, a value of about 6.2 pH is found for the titration point midway between the first- and second-stage equivalence points. We have

$$pK_2 \approx 6.2$$
$$\tfrac{1}{2}(pK_1 + pK_2) \approx 4.12$$

so that

$$K_1 \approx 9.1 \times 10^{-3} \quad \text{and} \quad K_2 \approx 6.3 \times 10^{-7}$$

These values are reasonably close to the tabled values of K_1 at 1.2×10^{-2} and K_2 at 6.0×10^{-7}.

Precipitation titrations Subject to the restrictions already noted for metal electrode systems, the indicator electrode used in potentiometric precipitation titrations may be a metal electrode representing the same species as the reacting cation. In several instances it is possible, and indeed convenient, to use as the indicating electrode an appropriate ion-sensitive electrode of the liquid-membrane type discussed previously. Either electrode type is used in conjunction with a suitable reference electrode. Where the reference electrode solution is capable of reacting with the solution environment during the titration (e.g., a saturated KCl calomel electrode immersed in a solution containing silver(I) ion), an inert salt bridge is used to connect the reference electrode to the solution environment. Thus a saturated KNO_3 salt bridge can be used to connect the SCE reference electrode to the solution involving the titration of chloride ion by standard silver nitrate solution. Figure 12.8 illustrates such an arrangement.

Consider further, as an example, the potentiometric titration of chloride ion by a standard silver(I) ion solution. If a silver electrode, with a properly bridged SCE, is used, we have, after the first addition of silver(I) ion solution and the formation of some AgCl precipitate,

$$E_{ind} = E^{\circ}_{AgCl/Ag} + 0.059 \text{ pCl} \qquad \text{at } 25\,°C$$
$$\begin{aligned} E_{cell} &= E_{ind} - E_{ref} \\ &= E^{\circ}_{AgCl/Ag} - E_{ref} + 0.059 \text{ pCl} \\ &= -0.023 + 0.059 \text{ pCl} \qquad \text{at } 25°C \end{aligned} \qquad (254)$$

where $E^{\circ}_{AgCl/Ag}$ and E_{ref} (SCE in reduction) at 25°C have values, respectively, of 0.222 and 0.245 V. As the titration progresses and pCl increases, the value of E_{cell} increases. Alternatively, the silver electrode may be considered responsive

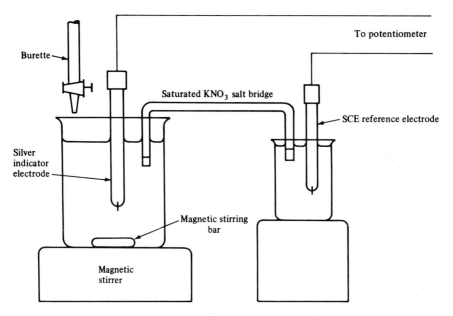

Fig. 12.8   Typical arrangement for the potentiometric titration of chloride ion by silver(I) ion.

to $[Ag^+]$ in solution (as it actually is) and we have

$$E_{ind} = E^\circ_{Ag^+/Ag} - 0.059 \text{ pAg} \qquad \text{at } 25°C$$
$$E_{cell} = E^\circ_{Ag^+/Ag} - E_{ref} - 0.059 \text{ pAg}$$
$$= 0.555 - 0.059 \text{ pAg} \qquad \text{at } 25°C \qquad (255)$$

with a 25°C value of $E^\circ_{Ag^+/Ag}$ at 0.800 V. Note that since

$$\text{pAg} = pK_{sp}(AgCl) - \text{pCl}$$
$$= 9.80 - \text{pCl} \qquad \text{at } 25°C$$

Eq. (255) yields

$$E_{cell} = 0.555 - 0.578 + 0.059 \text{ pCl}$$
$$= -0.023 + 0.059 \text{ pCl} \qquad \text{at } 25°C$$

as before. Both expressions, for given values of pCl and pAg, yield the same E_{cell} value. Table 12.1 shows the data for a potentiometric titration involving a silver indicating electrode and an SCE reference electrode, where 50.00 ml of 0.1000 M NaCl is titrated with 0.1000 M AgNO$_3$. The application of the tabulation technique for locating the equivalence-point volume as 49.99 ml of the AgNO$_3$ solution is given. In consideration of the burette-reading uncertainty, this would be reported as 50.00 ml.

Table 12.1 Potentiometric titration of chloride ion*

50.00 ml of 0.1000 M NaCl vs. 0.1000 M AgNO$_3$
Silver indicating electrode and SCE reference electrode

Volume AgNO$_3$, *ml*	pCl	E_{cell} vs. SCE, V	$(\Delta E_{cell}/\Delta V_{Ag}) \times 10^3$ 0.1 *ml* = ΔV_{Ag}	$(\Delta^2 E_{cell}/\Delta V_{Ag}^2) \times 10^3$ 0.1 *ml*/0.1 *ml*
0.00	1.00	0.036		
			0.11 (5.00)	
10.00	1.18	0.047		
			0.11 (15.00)	
20.00	1.37	0.058		
			0.13 (25.00)	
30.00	1.60	0.071		
			0.22 (35.00)	
40.00	1.96	0.093		
			0.38 (42.50)	
45.00	2.28	0.112		
			0.80 (46.50)	
48.00	2.69	0.136		+0.05 (47.50)
			1.8 (48.50)	
49.00	3.00	0.154		+0.24 (48.88)
			3.6 (49.25)	
49.50	3.30	0.172		+1.0 (49.45)
			7.7 (49.65)	
49.80	3.70	0.195		+5.1 (49.75)
			18 (49.85)	
49.90	4.00	0.213		+35 (49.90)
			53 (49.95)	
50.00	4.90	0.266		+1 (50.00)
			54 (50.05)	
50.10	5.81	0.320		−37 (50.10)
			17 (50.15)	
50.20	6.11	0.337		−4.5 (50.25)
			8.0 (50.35)	
50.50	6.51	0.351		−1.1 (50.55)
			3.6 (50.75)	
51.00	6.81	0.379		−0.25 (51.12)
			1.7 (51.50)	
52.00	7.10	0.396		
60.00	7.85	0.440		

*pCl values are not corrected for the dilution effect. E_{cell} values are calculated from $E_{cell} = -0.023 + 0.059$ pCl. The values in parentheses after each $\Delta E_{cell}/\Delta V_{Ag}$ and $\Delta^2 E_{cell}/\Delta V_{Ag}^2$ value associate the average slope, or rate of change of the slope, with the average volume value between the titration points involved.

$$\text{Equivalence-point volume} = 49.90 + \left(\frac{35}{35 + 37} \times 0.20 \text{ ml}\right)$$
$$= 49.99 \text{ ml}$$

Complexation titrations Since the more important complexation titrations analytically speaking are those where a metal ion species is titrated with an
aminopolycarboxylic acid such as EDTA, it would appear that a metal electrode
of the same species as the reacting cation could serve as an indicator electrode in a
situation corresponding to the potentiometric form of such a titration. However,
because of the previously mentioned restrictions surrounding the use of such electrodes, and the possibility of electrode response to situations other than that involving only the activity of the cation in question, very few of such metal electrode systems are really properly operative in potentiometric complexation titrations.

In recent years there has been a growing tendency to apply ion-sensitive
liquid-membrane electrodes as indicator electrodes in the potentiometric form of
complexation titrations. In this connection the application should be quite obvious.

An interesting form of metal indicator electrode system for potentiometric titrations involving soluble complexes is the general use of a mercury electrode to
follow EDTA titrations where the reacting cation forms a soluble complex with
EDTA appreciably less stable than the mercury(II)-EDTA complex.* In such
applications the mercury electrode, as a mercury pool in a cup immersed in the titration vessel solution, acts as the indicator electrode, with an SCE as the reference electrode. Figure 12.9 illustrates a typical arrangement.

We shall consider the operation of this electrode relative to its use in a titration of magnesium(II) ion by EDTA solution. Before the titration starts, the solution contains magnesium(II) ion. A small volume of a solution of mercury(II)-
EDTA complex is now added, the amount being sufficient to yield a solution concentration of about $10^{-4}\ M$. Once the titration has been started, the solution will
contain

Magnesium(II) ion	Mg^{2+}
Magnesium(II)-EDTA complex	$Mg(Y)^{2-}$
Mercury(II) ion	Hg^{2+}
Mercury(II)-EDTA complex	$Hg(Y)^{2-}$
EDTA anion	Y^{4-}

The mercury(II)-EDTA complex dissociates according to

$$Hg(Y)^{2-} \rightleftharpoons Hg^{2+} + Y^{4-}$$

and we have

$$[Hg^{2+}] = \frac{[Hg(Y)^{2-}]}{[Y^{4-}]\ K_{stab}\ (Hg(Y)^{2-})}$$

*C. N. Reilley and R. W. Schmid, *Anal. Chem.*, **30**:947 (1958).

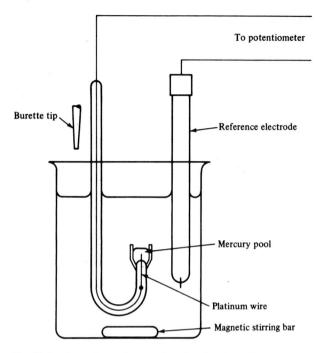

Fig. 12.9 Potentiometric complexation titration using a mercury-indicator electrode.

so that

$$E_{ind} = E_{Hg^{2+}/Hg} = E^{\circ}_{Hg^{2+}/Hg} + \frac{0.059}{2} \log [Hg^{2+}]$$

$$= E^{\circ}_{Hg^{2+}/Hg} + \frac{0.059}{2} \log \frac{[Hg(Y)^{2-}]}{[Y^{4-}] K_{stab} (Hg(Y)^{2-})}$$

In conjunction with the reference electrode we have

$$E_{cell} = E_{ind} - E_{ref}$$

$$= E^{\circ}_{Hg^{2+}/Hg} - E_{ref} + \frac{0.059}{2} \log \frac{[Hg(Y)^{2-}]}{[Y^{4-}] K_{stab} (Hg(Y)^{2-})} \qquad (256)$$

The magnesium-EDTA system, because the concentration of magnesium(II) ion is very much higher than that for mercury(II), dictates the concentration of Y^{4-} in the solution up to and including the titration equivalence point, with the excess of EDTA solution providing the $[Y^{4-}]$ after this point. We have, therefore,

$$Mg^{2+} + Y^{4-} \rightleftharpoons Mg(Y)^{2-}$$

so that

$$[Y^{4-}] = \frac{[Mg(Y)^{2-}]}{[Mg^{2+}] \, K_{stab} \, (Mg(Y)^{2-})} \tag{257}$$

Substitution of Eq. (257) for $[Y^{4-}]$ into Eq. (256) now yields

$$E_{cell} = E^\circ_{Hg^{2+}/Hg} - E_{ref} + \frac{0.059}{2} \log \frac{K_{stab} \, (Mg(Y)^{2-}) \, [Hg(Y)^{2-}]}{K_{stab} \, (Hg(Y)^{2-}) \, [Mg(Y)^{2-}]}$$
$$+ \frac{0.059}{2} \log [Mg^{2+}]$$

$$= E^\circ_{Hg^{2+}/Hg} - E_{ref} + \frac{0.059}{2} \log \frac{K_{stab} \, (Mg(Y)^{2-}) \, [Hg(Y)^{2-}]}{K'_{stab} \, (Hg(Y)^{2-}) \, [Mg(Y)^{2-}]}$$
$$- \frac{0.059}{2} \, pMg \tag{258}$$

The stability of the mercury(II)-EDTA complex is significantly greater than that for the magnesium(II)-EDTA complex (6.3×10^{21} vs. 4.9×10^{8}), so that the value of $[Hg(Y)^{2-}]$ will remain reasonably constant as the titration progresses. Again, around the equivalence point of the titration, the value of $[Mg(Y)^{2-}]$ will change to a negligible extent only. In the neighborhood of the equivalence point, therefore, $[Hg(Y)^{2-}] / [Mg(Y)^{2-}]$ will be reasonably constant, so that Eq. (258) becomes

$$E_{cell} = constant - \frac{0.059}{2} \, pMg \tag{259}$$

The value of E_{cell} will then vary with pMg, and the large changes in pMg around the equivalence point will be reflected by associated large changes in E_{cell}. Note that as the titration progresses and pMg increases, E_{cell} will decrease. Table 12.2 provides a general idea of the course of such a titration.

Substances determinable by potentiometric complexation titration with EDTA, using the mercury electrode as the indicator electrode, include those cations capable of yielding soluble EDTA complexes appreciably less stable than the mercury(II)-EDTA complex.

Oxidation-reduction titrations Potentiometric oxidation-reduction titrations are usually conducted using a platinum-wire indicator electrode. This so-called inert electrode is inert in the sense that it does not react with the components of the reactant or titrant solutions, or the products of the titration reaction, but acts as a collector for electron transfer in the general reaction

$$Ox + ne \rightleftharpoons Red$$

where Ox and Red represent differently charged states for a particular species. Such a reaction is typified by

$$Fe^{3+} + 1e \rightleftharpoons Fe^{2+}$$

Table 12.2   Potentiometric complexation titration of magnesium(II)*
50.00 ml of 0.1000 M Mg^{2+} vs. 0.1000 M EDTA at pH = 10.0
Mercury indicating electrode and SCE reference electrode

Volume EDTA, ml	pMg	E_{cell} $vs.$ SCE, V	$(\Delta E_{cell}/\Delta V_{EDTA}) \times 10^3$ 0.1 $ml = \Delta V_{EDTA}$	$(\Delta^2 E_{cell}/\Delta V^2_{EDTA}) \times 10^3$ 0.1 $ml/0.1$ ml
0.00	1.00	0.100		
5.00	1.09	0.098		
10.00	1.18	0.095		
20.00	1.37	0.090		
30.00	1.60	0.083		
40.00	1.96	0.072		
			−0.33 (44.50)	
49.00	3.00	0.042		−0.031 (46.88)
			−1.8 (49.25)	
49.50	3.30	0.033		−0.55 (49.45)
			−4.0 (49.65)	
49.80	3.70	0.021		−2.5 (49.75)
			−9.0 (49.85)	
49.90	4.00	0.012		−12 (49.89)
			−18 (49.92)	
49.95	4.30	0.003		−20 (49.95)
			−28 (49.98)	
50.00	4.77	−0.011		0.0 (50.00)
			−28 (50.02)	
50.05	5.25	−0.025		+20 (50.05)
			−18 (50.08)	
50.10	5.55	−0.034		+13 (50.11)
			−8.0 (50.15)	
50.20	5.85	−0.042		
50.50	6.26	−0.055		
51.00	6.55	−0.063		
60.00	7.55	−0.093		

*[Hg(Y)$^{2-}$] at the start of the titration = 1.0×10^{-4} M. pMg values are not corrected for the dilution effect. Around the equivalence point, [Hg(Y)$^{2-}$] is taken as 5.0×10^{-5} M and [Mg(Y)$^{2-}$] is taken as 5.0×10^{-2} M.

$$K_{stab}(Hg(Y)^{2-}) = 6.3 \times 10^{21} \qquad K_{stab}(Mg(Y)^{2-}) = 4.9 \times 10^8$$

E_{cell} *calculated* from (not secured from an actual titration)

$$E_{cell} = E°_{Hg^{2+}/Hg} - E_{ref} + \frac{0.059}{2} \log \frac{4.9 \times 10^8 \times 5.0 \times 10^{-5}}{6.3 \times 10^{21} \times 5.0 \times 10^{-2}} - \frac{0.059}{2} pMg$$

$$= 0.850 - 0.245 - 0.475 - \frac{0.059}{2} pMg$$

$$= 0.130 - \frac{0.059}{2} pMg$$

The values in parentheses after each $\Delta E_{cell}/\Delta V_{EDTA}$ and $\Delta^2 E_{cell}/\Delta V^2_{EDTA}$ value associate the average slope, or rate of change of the slope, with the average volume value between the titration points involved. The equivalence-point volume = 50.00 ml of 0.1000 M EDTA solution.

Thus the inert electrode reports the potential of the iron(III)-iron(II) system:

$$E_{Fe^{3+}/Fe^{2+}} = E^\circ_{Fe^{3+}/Fe^{2+}} + 0.059 \log \frac{[Fe^{3+}]}{[Fe^{2+}]}$$

without itself taking part chemically in the establishment of this potential. Other inert electrodes, such as gold, carbon, and palladium, are sometimes used as indicator electrodes in specific circumstances.

A suitable reference electrode, such as the saturated KCl calomel electrode, is used in conjunction with the platinum electrode. Ordinarily the reference electrode is immersed in the solution to be titrated, although a salt bridge may be required where the electrode solution and the solution under titration, or the titrant solution, have mutually reactive components. The potential reported by the indicator-reference electrode couple is given by

$$E_{cell} = E_{ind} - E_{ref}$$
$$= E_{Ox/Red} - E_{ref}$$

where $E_{Ox/Red}$ is greater than E_{ref} in reduction. The value of E_{cell} is, of course, relative to the particular reference electrode employed. As with all potentiometric titrations, the absolute value of E_{cell} need not be recorded, only the potentiometer dial E reading being required, since the relative changes in E as the titration progresses are alone of significance.

While direct potentiometry with respect to more or less irreversible systems is not possible, such systems can be titrated potentiometrically without difficulty, providing that the reactant and titrant substances react with adequate speed and in stoichiometric proportions. The indicator electrode must be capable of responding to the changing potential of at least one of the reacting systems involved in the titration. In the majority of instances, the indicator electrode will be capable of responding to either system, but in an oxidation-reduction titration, where equilibrium is established between both systems after each addition of titrant, each system achieves the same equilibrium potential.

Tabulated data for various oxidation-reduction titrations were given in Chap. 9, these tabulations involving E_{cell} values relative to the SHE. The E_{cell} value relative to the SCE can readily be approximated through the use of the equation

$$E_{cell} \text{ vs. SCE} = E_{cell} \text{ vs. SHE} - 0.245 \text{ V} \qquad \text{at } 25°C$$

The method of calculating the equivalence-point volume from $\Delta^2 E_{cell}/\Delta V_T^2$ data is obvious. Remember that the tabulation technique will not, for asymmetrical titration reactions, yield the exact equivalence-point volume. This will be noted for the asymmetrical titration reactions described in detail in Chap. 9. In the majority of cases, however, the discrepancy will often lie within the burette-reading uncertainty, commonly ±0.02 ml per reading. The circle-fit method of curve analysis for the plot of E_{cell} vs. volume of titrant will usually locate the equivalence-point volume accurately for both symmetrical and asymmetrical titration reactions.

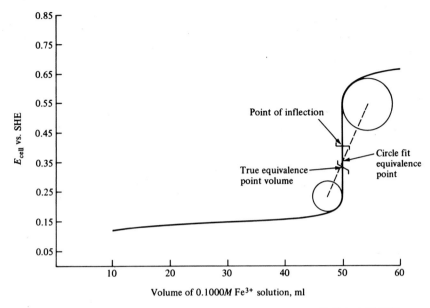

Fig. 12.10   Titration of 50.00 ml of 0.0500 M Sn^{2+} vs. 0.1000 M Fe^{3+} in 2 M HCl.

Figures 12.10 and 12.11 show this technique as applied to the asymmetrical titration curves for the titration of 0.0500 M Sn^{2+} vs. 0.1000 M Fe^{3+} in 2 M HCl,
and 0.1000 M Fe^{2+} vs. 0.0200 M KMnO$_4$ in a solution of pH $= 1.00$.

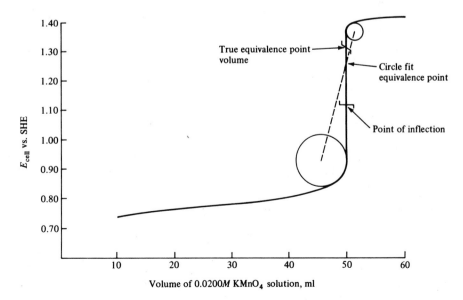

Fig. 12.11   Titration of 50.00 ml of 0.1000 M Fe^{2+} vs. 0.0200 M KMnO$_4$ at pH $= 1.00$.

Nonaqueous titrations   Acid-base titrations in nonaqueous solvents can general-ly be conducted potentiometrically by the use of an appropriate glass-membrane indicator electrode and a saturated KCl calomel reference electrode.   The millivolt scale of the pH meter, or the potentiometer scale, is used rather than the pH scale, since the pH scale calibrated by aqueous solution buffers has no mean-ing relative to nonaqueous solvent titrations.   In order to avoid dehydration of the glass membrane by the nonaqueous solvent, as well as malfunctioning of the refer-ence electrode, both electrodes should be stored in water between titration applications.

12.2.4   CALCULATIONS IN POTENTIOMETRY AND POTENTIOMETRIC TITRATIONS

Although the value of the factor $2.303RT/nF$ in the Nernst equation has been taken as $0.059/n$, it should be remembered that this is an approximation suitable for our present calculation purposes and specific for reactions carried out at 25°C. In direct potentiometric measurements, where potential may on occasion be measured to four significant figures at 25°C, the more significant figure of $0.05915/n$ should be used for calculation purposes, particularly where activities rather than concentrations are being determined.   In potentiometric titrations, where only the changes in E or pH as the titration progresses around the equiva-lence point are important, the situation takes on less significance.   A picture of the variation of the factor $2.303RT/nF$ with temperature is given by the fact that this has a value of $0.05816/n$ at 20°C, $0.05915/n$ at 25°C, and $0.06014/n$ at 30°C.

12.3   ELECTROGRAVIMETRY AND ELECTROLYTIC SEPARATIONS

12.3.1   GENERAL

The method of direct potentiometry involves the flow negligible current during the period over which the potential between the two electrodes comprising the couple is measured.   Thus the typical half-cell and overall cell reactions do not take place to any significant extent, and the solution composition is unaffected by the poten-tial measuring process.   Similarly, in potentiometric titrations, although the solu-tion composition changes as the titration progresses, no compositional changes are associated with the process of measuring the potential changes between the electrode systems involved as the titration proceeds.

The methods of electrogravimetry and electrolytic separation, which in es-sence provide for significant current flow through the solution as the result of a po-tential applied between the electrodes and originating with an external source, result in solution compositional changes.   Such changes are the result of driving the electrochemical reaction involved to a point of quantitative completion.   For example, the application of an adequate potential difference between two large platinum gauze electrodes immersed in an acid solution of copper(II) ion can result in the deposition of copper on the cathode or negative electrode on the basis of the half reaction

$$Cu^{2+} + 2e \rightleftharpoons Cu$$

Under the proper conditions of applied potential, this reversible reaction can be driven to the right to the extent that the removal of copper(II) ion from the solution is in time quantitatively complete. The weight of copper deposited can then lead to an accurate determination of the amount of copper originally present in the solution and, of course, in any sample relationship established on the solution basis. Such a separation technique may be applied in two different but related ways in quantitative analysis. In one approach, that of electrogravimetry, the weight of material separated as an element or compound may be evaluated and a quantitative determination of the substance required obtained. The separation and determination of copper just described is a common electrogravimetric procedure. In the second approach, that of electrolytic separation, an element, a compound, a group of elements, or a group of compounds may be separated from solution on a quantitative basis. Such a separation may be conducted solely for the purpose of removing from the solution a substance or group of substances capable of interfering in a projected method of analysis to be carried out on the solution. The removal of iron, nickel, manganese, and zinc by deposition electrolytically on a mercury cathode is, for example, frequently carried out in order to prevent interference by these elements in the separation of aluminum by precipitation as aluminum hydroxide using aqueous ammonia as the precipitant. Again, as a somewhat different approach, such a quantitative electrolytic separation may be carried out for the purpose of separating a group of substances from the solution, the separation itself being subsequently subjected to an analytical investigation to establish the quantitative aspect of each component. An example of such a procedure would be the separation simultaneously of copper and silver by deposition on a platinum cathode, with a quantitative determination of each of these elements arising out of a subsequent investigation of the deposit.

There are distinct advantages associated with analytical techniques involving electrogravimetry and electrolytic separation. The method of electrogravimetry is based on a weight measurement, and providing that no influences capable of adversely affecting the purity and quantitativeness of the separation exist, the accuracy and precision will be largely dictated by the reading uncertainty of the balance. This allows excellent accuracy and precision over a large range of quantity deposited. The relative accuracy and precision, for example, in the determination of 0.01 g of copper should not be worse than about ±2 percent; that for 1.0 g of copper not worse than about ±0.02 percent.

Although the process of separation by electrolysis usually takes place over an appreciable period (1 to 3 h being common), the method has merit in the fact that it is automatic and once started does not require the attention of the analyst until completed. The time over which electrolysis takes place can thus be used in other directions. Again the general (although not always) lack of such procedures as solution preparation prior to precipitation, filtration, washing, etc., renders the techniques of electrolytic separation relatively free from manipulative complications, very much lowering the possibility of the poor precision and accuracy sometimes associated therewith.

12.3.2 DECOMPOSITION POTENTIAL

The process of solution electrolysis obviously involves an overall cell reaction that is not spontaneous. This cell reaction, in reverse, would constitute theoretically or practically a galvanic cell reaction. As a galvanic cell it would be capable of developing a potential. It is apparent that this potential, sometimes called the "back emf," must be matched by the applied potential and, for any electrolytic action of a significant nature to take place, must be exceeded to some degree.

Where a slow reaction rate is associated with one or both electrode reactions, overvoltage values may also be encountered. Where such overvoltages exist, the applied potential must exceed the sum of the back emf and overvoltage values if significant electrolytic action is to be achieved.

The value of the applied potential that matches the back emf plus overvoltage values is called the decomposition potential. Values of applied potential exceeding the decomposition potential for a given system will be capable of initiating significant current flow through the solution. Since the solution resistance will be a factor requiring consideration, any demand for a definite current flow over this resistance will require that the applied potential exceed the decomposition potential by some definite margin.

As will be shown shortly, the decomposition potential will not be clearly defined for any cell system, since decomposition or electrolytic action will increase gradually over a zone of applied potential. Two situations should be noted. Since the applied potential is capable of promoting a nonspontaneous cell reaction, it will carry a negative sign. Where the applied potential is said to increase, or to exceed a certain value, it is becoming increasingly negative, or is more negative than the value involved.

We shall consider as an example the application of a gradually increasing applied potential between two large platinum gauze electrodes immersed in a solution 0.05 M to cupric sulfate and 1.0 M to sulfuric acid. The potential will be applied from an external source and the solution will be stirred as efficiently as possible by means of a magnetic stirrer. The current flow through the solution will be measured and plotted against the applied potential. Figure 12.12 indicates the general arrangement.

Where the solution is not stirred, or is inefficiently stirred, the electrode reactions associated with the flow of current through the solution result generally in the solution layer just adjacent to either electrode becoming almost immediately

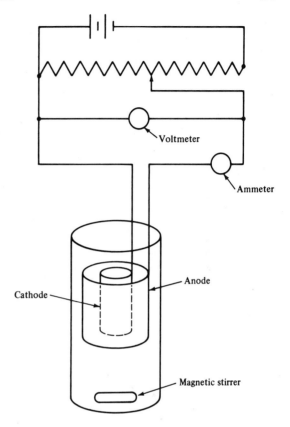

Fig. 12.12   Simple electrolysis circuit.

depleted with respect to the ion or substance oxidized or reduced at the electrode involved. Thus the composition of such solution layers becomes different from that of the bulk of the solution. This phenomenon is called concentration polarization, and where it occurs, the current becomes a function of the mass transfer rate rather than of the electrode reaction rate based on the electrode potential. The Nernst equation and associated calculations involving electrode potentials is based on the electrode reaction rate situation and on the assumption of solution homogeneity. The existence of concentration polarization zones in the immediate neighborhood of either electrode would constitute a failure of this assumption. As the electrolytic process progresses and depletion of ion concentration in the general solution occurs, concentration polarization situations do develop and the current tends to depend on the mass transfer rate rather than on the electrode potential, regardless of the stirring efficiency.

For the copper-oxygen system under discussion, it is possible to attempt a prior calculation of the applied potential that should be required to just match the back emf. The electrode reactions for electrolytic action will be

At the cathode: $Cu^{2+} + 2e \rightleftharpoons Cu$

At the anode: $H_2O - 2e \rightleftharpoons \frac{1}{2}O_2 + 2H^+$

with the overall cell reaction under electrolysis being

$$H_2O + Cu^{2+} \rightleftharpoons \frac{1}{2}O_2 + 2H^+ + Cu$$

Note that these reactions in reverse constitute the half-cell and cell reactions to be expected for the corresponding galvanic cell and as such provide half-cell and cell potential values given by

$$E_{Cu^{2+}/Cu} = E^\circ_{Cu^{2+}/Cu} + \frac{0.059}{2} \log [Cu^{2+}]$$
$$= 0.340 + 0.0295 \log (5.0 \times 10^{-2}\ M)$$
$$= 0.302\ V \qquad\qquad\qquad \text{at } 25°C$$

$$E_{O_2/H_2O} = E^\circ_{O_2/H_2O} + \frac{0.059}{2} \log \frac{[O_2]^{1/2}[H^+]^2}{[H_2O]}$$

Assuming a value of oxygen at 1 atm (unit activity) and the activity of water as unity we have

$$E_{O_2/H_2O} = E^\circ_{O_2/H_2O} + 0.059 \log [H^+]$$
$$= 1.229 + 0.059 \log 1\ M$$
$$= 1.229\ V \qquad\qquad \text{at } 25°C$$

The value of E_{cell} for the galvanic couple would then be

$$E_{cell} = E_{O_2/H_2O} - E_{Cu^{2+}/Cu}$$
$$= 1.299 - 0.302$$
$$= 0.927\ V \qquad\qquad \text{at } 25°C$$

On the basis of the calculations so far, we would expect the cell in question to show no electrolytic action until this back emf value of 0.927 V was exceeded. We would not expect to see copper deposited at the cathode and oxygen liberated at the anode until the applied potential assumed a value more negative than -0.927 V. Again, on the basis of the *calculation*, we would assume a definite decomposition potential of -0.927 V.

The solid line of Fig. 12.13 indicates what the plot of current vs. applied potential might be *expected* to be on the basis of these calculations. Note that the *expected* plot shows no current flow through the solution up to an applied potential equal to the decomposition potential. Once the decomposition potential has been exceeded, the current is expected to increase linearly with increasing applied potential.

Where the experiment is conducted on a practical basis, the dotted line of Fig. 12.13 shows in general the actual plot of current vs. applied potential. The

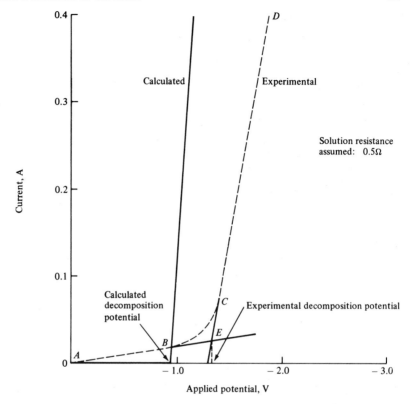

Fig. 12.13 Plot of current vs. applied potential 0.05 M $CuSO_4$ and 1.0 M H_2SO_4.

explanations covering the obvious differences between this plot and the expected plot follow.

The first very low values of applied potential, in the dotted line zone AB, result in very low current values; and, in general, the current flow increases slowly but linearly with the applied potential in this zone. Two factors contribute to this situation. The current, called the residual current, stems largely from current flow associated with solution impurities capable of being reduced at the cathode and reoxidized at the anode. Contributing reactions to residual current are those associated with iron(III) ion in solution,

At the cathode: $\quad Fe^{3+} + 1e \rightleftharpoons Fe^{2+}$

At the anode: $\quad Fe^{2+} - 1e \rightleftharpoons Fe^{3+}$

and those associated with dissolved oxygen,

At the cathode: $\quad O_2 + 2H^+ + 2e \rightleftharpoons H_2O_2$

At the anode: $\quad H_2O_2 - 2e \rightleftharpoons O_2 + 2H^+$

This latter situation represents a very significant contribution to the total residual current. The second source of residual current stems from a rather interesting development. When the Nernst equation was used to calculate the value of $E_{Cu^{2+}/Cu}$, the assumption was made that the activity of solid copper was unity. However, before any potential is applied, the platinum cathode will have no deposit of copper, and it can be shown here that the activity of solid copper is extremely low. As the applied potential is increased, even at values very much less than the *calculated* decomposition potential, small amounts of copper are deposited and the activity of solid copper gradually increases as more of the platinum surface becomes covered with copper. When the entire surface is covered to a depth of a few copper atoms, the activity of solid copper achieves a unity value. Several points of importance here are the following.

1. The deposition of copper starts well before any *calculated* decomposition potential.
2. The residual current, in part, results from the deposition of copper before the calculated decomposition potential.
3. The decomposition potential will not be, for a given system, a single clearly defined value.

Over the section represented by BC, copper continues to be deposited on the cathode, but the rate of deposition increases very rapidly with increasing applied potential. This zone of applied potential encompasses the *experimental* decomposition potential. Over the section CD, this decomposition potential has been exceeded and, in general, the current increases linearly with applied potential.

For this experimental plot of current vs. applied potential, the decomposition potential for the system $Cu^{2+} \rightarrow Cu$ and $H_2O \rightarrow \frac{1}{2}O_2 + 2H^{2+}$, in a medium 0.05 M to $CuSO_4$ and 1.0 M to H_2SO_4, which we shall denote as $E_d(Cu-O_2)$, can be approximated by extrapolating the straight-line sections AB and DC to intersect at the point E. A line from E perpendicular to the applied potential axis yields the approximate value of $E_d(Cu-O_2)$.

It will be noted that $E_d(Cu-O_2)$ so obtained is about -1.33 V, that is, it is appreciably more negative than the calculated value of -0.927 V. This difference arises out of the slow reaction rate for the electrode reaction

$$H_2O - 2e \rightleftharpoons \tfrac{1}{2}O_2 + 2H^+$$

which is quite irreversible and results in an overvoltage at this electrode (the anode). This overvoltage represents the additional energy required in the form of applied potential in order to permit the reaction to occur at an appreciable rate. Although a brief discussion of overvoltage will be given later, it will be sufficient for the moment to assign here a value of about -0.40 V as representative of the overvoltage for the discharge of oxygen on smooth platinum under the very low current density conditions existing around the decomposition potential. Thus we have, for E_{cell},

$$E_{cell} = (1.229 + 0.40) - 0.302$$
$$= 1.33 \text{ V} \qquad\qquad \text{at } 25°C$$

requiring an applied potential of -1.33 V to match the combination of back emf and overvoltage. This, in effect, is the approximate decomposition potential $E_d(Cu-O_2)$ for this cell. Admittedly, for electrolysis to take place at an appreciable rate, a significant current must flow through the solution. For any applied potential exceeding the decomposition potential, the current through the solution effective in the deposition of copper will be given by

$$i - i_r = \frac{E_d(Cu-O_2) - E_{app}}{R} \qquad\qquad (260)$$

where

$$i = \text{total current at the applied potential, A}$$
$$i_r = \text{residual current at the applied potential, A}$$
$$E_d(Cu-O_2) = \text{decomposition potential for the conditions involved, V}$$
$$E_{app} = \text{applied potential, V}$$
$$R = \text{solution resistance, } \Omega$$

For a total current of 1 A, yielding a current density of say 0.01 A/cm², the overvoltage for the discharge of oxygen on smooth platinum will be about -0.85 V. If the solution resistance is 0.5 Ω, and a residual current of 0.01 A exists, this implies an applied potential of

$$E_{app} = E_d(Cu-O_2) - R(i - i_r)$$
$$= -1.78 - 0.5(1 - 0.01)$$
$$= -2.28 \text{ V} \qquad\qquad \text{at } 25°C$$

Note that with the change in the oxygen-discharge overvoltage resulting from the higher current density, the value of $E_d(Cu-O_2)$ under these conditions is -1.78 V. The value of i_r is usually negligible in normal electrogravimetric or electrolytic separation processes, where reasonably high total current flow values are maintained.

Suppose, for the same system, that we now consider a continued increase in the applied potential beyond the decomposition potential for the copper system. We can now calculate a value for the cell potential of the hydrogen-oxygen system. This will involve the electrode and cell reactions in electrolysis.

At the cathode: $2H^+ + 2e \rightleftharpoons H_2$

At the anode: $H_2O - 2e \rightleftharpoons \frac{1}{2}O_2 + 2H^+$

Overall cell: $H_2O \rightleftharpoons \frac{1}{2}O_2 + H_2$

and we have

$$E_{2H^+/H_2} = E^\circ_{2H^+/H_2} + 0.059 \log [H^+] \quad \text{at } 25°C$$
$$= 0.000 \text{ V}$$
$$E_{O_2/H_2O} = 1.229 \text{ V (as before)} \quad \text{at } 25°C$$

The hydrogen reaction also occurs at a slow rate, and the overvoltage value for hydrogen *discharge* on *copper* (since at this point the platinum cathode will be completely covered by the copper deposit) will be about -0.50 V at the low current density for the hydrogen system around its decomposition potential. This value, together with an oxygen-discharge overvoltage of about -0.85 V, compatible with the high current density for the copper-oxygen system associated with an applied potential more negative than $E_d(\text{Cu}-\text{O}_2)$, leads to an E_{cell} of*

$$E_{\text{cell}} = (1.229 + 0.85) - (0.000 - 0.50)$$
$$= 2.58 \text{ V} \quad \text{at } 25°C$$

providing an approximate value of -2.58 V for $E_d(\text{H}_2-\text{O}_2)$.

The following points should now be noted concerning initial electrolysis conditions for the system under discussion.

1. As the applied potential is increased gradually, a significant copper deposition and oxygen liberation will be obtained once the applied potential exceeds the value of $E_d(\text{Cu}-\text{O}_2)$.
2. Increases in the applied potential between $E_d(\text{Cu}-\text{O}_2)$ and $E_d(\text{H}_2-\text{O}_2)$ will result in an increase in the rates of copper deposition and oxygen liberation.
3. Applied potential values beyond $E_d(\text{H}_2-\text{O}_2)$ will result in the simultaneous deposition of copper and liberation of hydrogen at the cathode, with liberation of oxygen at the anode. The total current through the solution here will be the sum of the appropriate residual current, the current effective in copper deposition, and that effective in hydrogen liberation.
4. As the value of the applied potential is increased beyond $E_d(\text{H}_2-\text{O}_2)$, the rate of copper deposition will increase slightly, with the rate of liberation of hydrogen increasing considerably. This will be based mainly on the availability of water and, of course, hydrogen ion. The hydrogen ion discharged at the cathode is replaced by the anodic reaction.
5. If the system should be extended to include metal ions capable of yielding $E_d(\text{metal}-\text{O}_2)$ values less than $E_d(\text{Cu}-\text{O}_2)$, where these metals are capable of deposition on the cathode from aqueous solution, such metals will be so deposited at appropriate applied potentials less than that for $E_d(\text{Cu}-\text{O}_2)$.

* If the solution consisted of 1.0 M H_2SO_4 alone, the oxygen-discharge overvoltage would be -0.40 V, that associated with a very low current density for the hydrogen-oxygen system. The hydrogen-discharge overvoltage on *platinum* would be about -0.02 V at this low current density.

6. If the system should be extended to include metal ions capable of yielding E_d (metal—O_2) values between E_d(Cu—O_2) and E_d(H_2—O_2), where such metals are capable of deposition on the cathode from aqueous solution, these metals will successively come into the deposition process as the applied potential is increased within this range.

7. If the system should be extended to include metal ions capable of yielding E_d(metal—O_2) values higher than E_d(H_2—O_2), such metals will not be deposited at the cathode even where the applied potential is more negative than E_d(metal—O_2). Here increases in the applied potential beyond E_d(H_2—O_2) result in $\{E_d(H_2$—$O_2) - E_{app}\}$ values directed mainly towards increases in the current flow and reaction rate associated with the cell reaction

$$H_2O \rightleftharpoons \tfrac{1}{2}O_2 + H_2$$

Thus the cathode does not achieve potential values sufficiently negative to permit the discharge of such metal ions.

12.3.3 OVERVOLTAGE

As mentioned elsewhere, overvoltage values are associated with slow electrode reaction rates and represent the additional electric energy required to permit significant rates for such reactions. The magnitude of the overvoltage value depends on a number of variables. These are listed on a very generalized basis below.

1. The type of metal used as the electrode. In general, overvoltage values increase with increasing "softness" of the metal used, being low for platinum and high for mercury.
2. The physical state of the discharged substance affects the overvoltage value. Again in general, the overvoltage will be high for gaseous discharges and low for metallic discharges. Values of overvoltage as associated with the discharge of metals are often negligibly small. With certain metals, however, commonly called "hard" metals, such as chromium, molybdenum, tungsten, etc., overvoltage values may be very significant.
3. Overvoltage in general increases rapidly with increasing current density—that is, the current through the solution per square centimeter of electrode surface.
4. The solution temperature, which affects the overvoltage value, causing it to decrease with increasing temperature. Here part of the additional energy required to promote a more rapid reaction rate is provided thermally.

Table 12.3 provides a general picture of overvoltage values for hydrogen and oxygen discharge against various electrodes and under varying conditions of current density.

Table 12.3 Overvoltage values for hydrogen and oxygen discharge on various metal electrodes
25°C; dilute H_2SO_4*

| | *Overvoltage* | | | | | | | |
| | Current density, 0.001 A/cm² | | Current density, 0.01 A/cm² | | Current density, 0.1 A/cm² | | Current density, 1 A/cm² | |
Electrode	H_2	O_2	H_2	O_2	H_2	O_2	H_2	O_2
Smooth platinum	−0.024	−0.72	−0.068	−0.85	−0.29	−1.3	−0.68	−1.5
Platinized platinum	−0.015	−0.35	−0.030	−0.52	−0.041	−0.64	−0.048	−0.76
Bismuth	−0.78	· · · ·	−1.05	· · · ·	−1.1	· · · ·	−1.23	· · · ·
Copper	−0.48	−0.42	−0.58	−0.58	−0.70	−0.66	−1.27	−0.79
Gold	−0.24	−0.67	−0.39	−0.96	−0.59	· · · ·	−0.80	−1.63
Iron	−0.40	· · · ·	−0.56	· · · ·	−0.82	· · · ·	−1.3	· · · ·
Lead	−0.52	· · · ·	−1.09	· · · ·	−1.2	· · · ·	−1.3	· · · ·
Mercury	−0.9	· · · ·	−1.1	· · · ·	−1.1	· · · ·	−1.2	· · · ·
Nickel	−0.56	−0.35	−0.75	−0.52	−1.1	−0.64	−1.24	−0.85
Tin	−0.86	· · · ·	−1.08	· · · ·	−1.2	· · · ·	−1.2	· · · ·
Zinc	−0.72	· · · ·	−0.75	· · · ·	−1.1	· · · ·	−1.2	· · · ·

*The overvoltage values for hydrogen discharge are, very generally, about −0.1 to −0.3 V additional in alkaline media. Hydrogen discharged on copper metal in 1 M NaOH shows an overvoltage of about −0.80 V at approximately 0.01 A/cm² current density. The overvoltage value for oxygen discharge on smooth platinum in 1 M NaOH is about −1.40 V at very low current density and about −1.80 V at a current density of approximately 0.01 A/cm².

12.3.4 RELATIONSHIP BETWEEN DECOMPOSITION POTENTIAL, APPLIED POTENTIAL, AND ION CONCENTRATION

Before discussing the various techniques of electrogravimetry and electrolytic separation, we shall explore the general relationship between decomposition potential, applied potential, and ion concentration, which is used to investigate the quantitative nature of such separations. The copper system described previously will serve here. We shall not, at the moment, consider the situation as involving applied potential values exceeding $E_d(H_2{-}O_2)$. We had shown for this system a value of −1.33 V for $E_d(Cu{-}O_2)$ and −2.58 V for $E_d(H_2{-}O_2)$. Suppose that an applied potential of −2.00 V is maintained. Under these conditions an appreciable current through the solution will be established initially. The overvoltage for oxygen discharge will change from −0.40 V to a more negative value on the basis of this higher current density. Ignoring the residual current value, the current effective in copper deposition will be given by Eq. (260) as

$$i = \frac{-\{(E_{O_2/H_2O} + \text{oxygen overvoltage}) - E_{Cu^{2+}/Cu}\} - E_{app}}{R}$$

assuming negligible overvoltage in the discharge of copper. Here we have the value of $-\{(E_{O_2/H_2O} + \text{oxygen overvoltage}) - E_{Cu^{2+}/Cu}\}$ representing $E_d(Cu{-}O_2)$

for this situation. As the electrolysis progresses, copper is deposited and the concentration of Cu^{2+} in the bulk of the solution decreases. The result of this decrease in $[Cu^{2+}]$ is that it permits a situation to develop in which the availability of Cu^{2+} is not sufficient to maintain the earlier current flow. The current thus decreases as the electrolysis progresses at the applied potential of -2.00 V. The current decrease results in changes in the oxygen-discharge overvoltage to less negative values. Since the applied potential remains fixed, the decrease in i and decrease in oxygen overvoltage in Eq. (260) can only be met by an increase in E_{O_2/H_2O} and/or a decrease in $E_{Cu^{2+}/Cu}$. The value of E_{O_2/H_2O} will increase slightly on the basis of the increase in hydrogen-ion concentration resulting from the electrolytic cell reaction

$$H_2O + Cu^{2+} \rightleftharpoons \tfrac{1}{2} O_2 + 2H^+ + Cu$$

This increase in $[H^+]$ will, however, be only from 1 to 1.1 M, even considering all of the copper to be deposited; and the corresponding increase in E_{O_2/H_2O} will be minimal. Because of the continuing availability of water, no other factor will bring about an increase in E_{O_2/H_2O}. The real change is represented by a decrease in the value of $E_{Cu^{2+}/Cu}$, and this occurs on the basis of the decrease in $[Cu^{2+}]$ in the solution. When the electrolysis is carried out to equilibrium at the applied potential involved, the current through the solution effective in copper deposition will be practically zero, and the value of the expression

$$-\{(E_{O_2/H_2O} + \text{oxygen overvoltage}) - E_{Cu^{2+}/Cu}\}$$

or $E_d(Cu-O_2)$, will be equal to the applied potential. At this very low current density the overvoltage for oxygen discharge will again be about -0.40 V, and we have

$$-\{(E_{O_2/H_2O} + \text{oxygen overvoltage}) - E_{Cu^{2+}/Cu}\} = -2.00 \text{ V}$$

With E_{O_2/H_2O} at this point given approximately by

$$E_{O_2/H_2O} = 1.229 + 0.059 \log 1.1 \ M = 1.231 \text{ V} \qquad \text{at } 25°\text{C}$$

we have

$$(1.231 + 0.40) - E_{Cu^{2+}/Cu} = 2.00 \text{ V}$$

$$E_{Cu^{2+}/Cu} = -0.369 \text{ V} \qquad \text{at } 25°\text{C}$$

This leads to a value of $[Cu^{2+}]$ at equilibrium under an applied potential of -2.00 V of

$$[Cu^{2+}] = \text{antilog} \frac{2}{0.059}(-0.369 - E°_{Cu^{2+}/Cu})$$

$$\approx 10^{-24} \ M \qquad\qquad \text{at } 25°\text{C}$$

Note that the system started with an $E_d(Cu-O_2)$ of -1.33 V, with the value of $E_d(Cu-O_2)$ gradually increased as the deposition of copper continued and the system finished at equilibrium with an $E_d(Cu—O_2)$ value of -2.00 V. The oxygen-discharge overvoltage for $E_d(Cu—O_2) = -1.33$ V was -0.40 V, the same value applying at equilibration with $E_d(Cu—O_2) = -2.00$ V, since both of the decomposition potential situations are associated with minimal current density. Again, since the system started and finished (at equilibrium) with the same oxygen overvoltage value, all changes in $E_d(Cu—O_2)$ were based on changes in $E_{Cu^{2+}/Cu}$, the changes in E_{O_2/H_2O} being negligible by comparison.*

We can now generalize this situation for similar electrolytic actions. Suppose that the starting concentration of a metal ion M^{n+}, to be deposited on the cathode as M, is given as $[M^{n+}]_s$, and the value at equilibrium under a given applied potential is $[M^{n+}]_f$. The decomposition potential at the start and finish of electrolysis will be given by

$$E_d(M—O_2)_s = -\{(E_{O_2/H_2O} + O_2 \text{ overvoltage})_s - (E_{M^{n+}/M} + M \text{ overvoltage})_s\}$$

$$E_d(M—O_2)_f = -\{(E_{O_2/H_2O} + O_2 \text{ overvoltage})_f - (E_{M^{n+}/M} + M \text{ overvoltage})_f\}$$

The value of $(E_{O_2/H_2O} + O_2 \text{ overvoltage})$ will not change significantly from start to equilibrium during normal electrogravimetric or electrolytic separation processes conducted in acid medium, so that

$$E_d(M—O_2)_f - E_d(M—O_2)_s = (E_{M^{n+}/M} + M \text{ overvoltage})_f - (E_{M^{n+}/M} + M \text{ overvoltage})_s$$

The value of overvoltage for the process $M^{n+} \to M$ will usually be negligible or, at least, will not change appreciably over the electrolytic process, and we have

$$E_d(M—O_2)_f - E_d(M—O_2)_s = E_{M^{n+}/M}^\circ + \frac{0.059}{n} \log [M^{n+}]_f - E_{M^{n+}/M}^\circ$$
$$- \frac{0.059}{n} \log [M^{n+}]_s$$
$$= \frac{0.059}{n} \log \frac{[M^{n+}]_f}{[M^{n+}]_s} \qquad \text{at 25°C} \qquad (261)$$

Since the electrolysis is carried out to equilibrium at the applied potential involved, we have at equilibrium

$$E_d(M—O_2)_f = E_{app}$$

and

$$E_{app} - E_d(M—O_2)_s = \frac{0.059}{n} \log \frac{[M^{n+}]_f}{[M^{n+}]_s}$$

---

*The situation occurs only because of the high starting concentration of hydrogen ion at $1\ M$. If the system had started with $[Cu^{2+}] = 0.05\ M$ and $[H^+] = 10^{-6}\ M$, the increase in $[H^+]$ from 10^{-6} to $0.1\ M$ from start to complete copper deposition would have appreciably influenced the value of E_{O_2/H_2O}.

so that

$$[M^{n+}]_f = [M^{n+}]_s \, 10^{\left(\dfrac{E_{app} - E_d(M-O_2)_s}{0.059/n}\right)} \quad \text{at } 25°C \qquad (262)$$

We can not guarantee the distribution of the applied potential between cathode and anode, except in controlled cathode potential electrolysis. Since we are usually interested in the cathodic portion of the applied potential, we can show alternatively that

$$E_{app} - E_d(M-O_2)_s = E_{M^{n+}/M,f} - E_{M^{n+}/M,s}$$
$$= E_{app, \text{cathode},f} - E_{app, \text{cathode},s}$$

and

$$[M^{n+}]_f = [M^{n+}]_s \, 10^{\left(\dfrac{E_{app,\text{cathode},f} - E_{app,\text{cathode},s}}{0.059/n}\right)} \quad \text{at } 25°C \qquad (263)$$

where the $E_{app,\text{cathode}}$ values are the portions of the applied potential specific at the cathode. Note that Eq. (262) is applicable only in situations where the value of $(E_{O_2/H_2O} + O_2 \text{ overvoltage})$ is not appreciably different at the start and finish of the electrolytic process.

   No such process of electrolysis would be conducted at a minimum applied potential or cathodic applied potential calculated to yield a quantitative separation of the metal involved, since uncertainties in setting the applied potential and in the calculations (stemming from overvoltage approximations, the use of concentrations rather than activities, etc.) would lead to serious errors. One would apply the maximum potentials possible, due attention being paid to such factors as interferences arising out of the discharge of other metal ions and the simultaneous liberation of hydrogen at the cathode. This would permit the maximum effective current in the deposition of the metal involved. In order to avoid the extended time interval often required to reach equilibrium at such an applied potential, the solution would be tested in some way at appropriate periods to determine that the metal-ion concentration had achieved a value of $10^{-6} M$ or less.

Example   In the system just described the value of $[Cu^{2+}]$ at the start was $5.0 \times 10^{-2} M$ and that of $[H^+]$ was $1.0 M$. The applied potential was -2.00 V. The oxygen-discharge overvoltage was -0.40 V at the start and at equilibrium under -2.00 V applied potential. Determine the value of $[Cu^{2+}]$ at equilibrium for $25°C$.

$$E_d(Cu-O_2)_s = -\{(1.229 + 0.40) - (0.302)\}$$
$$= -1.33 \text{ V}$$

$$[Cu^{2+}]_f = 5.0 \times 10^{-2} M \, 10^{\left(\dfrac{-2.00 + 1.33}{0.059/2}\right)}$$
$$= 1.0 \times 10^{-24} M$$

Example   A solution contains $0.0250 M$ $Cu(NO_3)_2$ and $[H^+] = 10^{-6} M$. An electrolysis is carried out between smooth platinum gauze electrodes to equilibrium at an applied potential of -2.40 V. The overvoltage for oxygen discharge on smooth platinum at the start and at equilibrium can be taken as -0.40 V. The overvoltage for copper discharge can be taken as negligible. Determine the value of $[Cu^{2+}]$ at equilibrium for 25°C.

E_{app} at equilibrium $= -\{(E_{O2/H2O} + O_2 \text{ overvoltage})_f - (E_{Cu^{2+}/Cu})_f\}$
$E_{O2/H2O}$ at equilibrium $= 1.229 + 0.059 \log (5.0 \times 10^{-2} M)$
$= 1.152$ V
$-2.40 = -\{(1.152 + 0.40) - (E_{Cu^{2+}/Cu})_{(f)}\}$
$(E_{Cu^{2+}/Cu})_f = -0.848$ V
$[Cu^{2+}]_f = 5.4 \times 10^{-41} M$

12.3.5 ELECTROLYSIS AT CONSTANT APPLIED POTENTIAL

This technique of electrogravimetry and electrolytic separation involves the use of a constant or fixed applied potential over the period of the electrolysis. Although not necessarily restrictive in this sense, we shall only consider the technique in association with applied potential values less negative than $E_d(H_2-O_2)$ for the systems electrolyzed. Electrolysis involving applied potentials more negative than $E_d(H_2-O_2)$, resulting in the simultaneous cathodic processes of metal deposition and hydrogen evolution, will be dealt with under constant current electrolysis.

The discussion just completed relative to decomposition potential, applied potential, and ion-concentration relationships covered many of the factors of importance with respect to electrolysis at constant applied potential. It is apparent that in an aqueous solution containing only *one* depositable metal capable of yielding $E_d(\text{metal}-O_2)$ *appreciably* less negative than $E_d(H_2-O_2)$, the method might permit the quantitative separation and determination of the metal involved. In the case of the $CuSO_4-H_2SO_4$ system given previously as an example, the use of a constant applied potential of -2.00 V resulted in time in a quantitatively complete deposition of copper without simultaneous evolution of hydrogen, since the cathodic applied potential did not apparently attain the value of -0.50 V required for hydrogen-ion discharge. The time interval for the quantitative deposition of copper might have been conveniently shortened from that required to reach equilibrium at -2.00 V applied potential by termination of the electrolysis on the basis of a solution test capable of indicating a solution value of $[Cu^{2+}]$ less than $10^{-6} M$. Although the current efficiency relative to copper deposition will be 100 percent, neglecting the very small residual current, the current will decrease exponentially with time. Any such system will perform best with respect to time interval for a quantitative separation where we have $E_d(\text{metal}-O_2) - E_d(H_2-O_2)$ as a high value—in other words, where the decomposition potential for the metal system is significantly less negative than that for the hydrogen system. When this condition is obtained, a constant applied potential appreciably more negative than $E_d(\text{metal}-O_2)$ can be used, resulting in a current of reasonable magnitude and a time interval of practical duration to attain quantitative separation.

Cases may arise where the $E_d(\text{metal}-O_2)$ value is quite close to

$E_d(H_2-O_2)$ but still shows what appears to be a difference adequate for quantitative separation of the metal. The need to limit the applied potential to a value less negative than $E_d(H_2-O_2)$, the changes in the oxygen-discharge overvoltage with changes in the current density and the decreases in the cathodic applied potential as the electrolysis progresses may then seriously limit the value of $\{E_d(\text{metal}-O_2) - E_{app}\}$ and therefore the current effective in the deposition of the metal involved. Under such circumstances the effective current may be so low, even at the start of the electrolysis, that the time to achieve a quantitative separation may be prohibitively long.

A really serious situation arises where this form of electrolysis is considered in connection with solutions that contain more than one depositable metal capable of yielding $E_d(\text{metal}-O_2)$ less negative than $E_d(H_2-O_2)$. Consider the case of 100 ml of a solution containing 0.050 M $AgNO_3$, 0.050 M $Cu(NO_3)_2$, and 1.0 M HNO_3. The solution shows a resistance of 2Ω and is to be considered for the separation of silver quantitatively by electrolysis between large smooth platinum gauze electrodes at constant applied potential. The galvanic cell potentials for the silver and copper systems are given by

$$
\begin{aligned}
E_{cell} &= (E_{O_2/H_2O} + O_2 \text{ overvoltage}) - E_{Ag^+/Ag} \\
&= (1.229 + 0.40) - 0.723 = 0.91 \text{ V} \qquad \text{at } 25°C \\
E_{cell} &= (E_{O_2/H_2O} + O_2 \text{ overvoltage}) - E_{Cu^{2+}/Cu} \\
&= (1.229 + 0.40) - 0.302 = 1.33 \text{ V} \qquad \text{at } 25°C
\end{aligned}
$$

with the overvoltages for silver and copper discharge being taken as negligible. The values of $E_d(Ag-O_2)$ and $E_d(Cu-O_2)$ are therefore -0.91 and -1.33 V, respectively. Suppose that a constant applied potential of -1.20 V is used. At equilibrium the value of $[Ag^+]$ is given by

$$
\begin{aligned}
[Ag^+]_f &= 5.0 \times 10^{-2} \, M \quad {}_{10}\left(\frac{-1.20 + 0.91}{0.059}\right) \\
&= 6.1 \times 10^{-7} \, M
\end{aligned}
$$

On this basis it would appear that silver might be separated from copper and determined quantitatively. Several factors should be noted, however. First of all, the current effective in the deposition of silver initially would be, neglecting the residual current, given by

$$
\begin{aligned}
i &= \frac{E_d(Ag-O_2) - E_{app}}{R} \\
&= \frac{-0.91 + 1.20}{2} = 0.15 \text{ A}
\end{aligned}
$$

This effective current will not be achieved, however, since the establishment of a significant current density will alter the oxygen-discharge overvoltage to a more negative value, resulting in a more negative value of $E_d(Ag-O_2)$ and decreasing the effective current for silver deposition to appreciably less than 0.15 A. The time required for a quantitative separation might, under such circumstances, be prohibitive for practical purposes. Again, the uncertainties in calculating and setting and maintaining the required applied potential value, together with the changes in oxygen overvoltage with current density changes, could lead to cathodic applied potentials capable of permitting significant deposition of copper. Where the two metals involved in a similar system show cathodic potentials for metal discharge in close proximity to each other, separation on a quantitative basis becomes impossible.

Because of the foregoing difficulties, the method of electrolysis at constant applied potential is rarely encountered in quantitative analytical chemistry.

12.3.6 ELECTROLYSIS AT CONSTANT CURRENT

Electrolysis carried out at constant current involves the application of an applied potential capable of yielding some initial value of current effective in the deposition of the metal affected. As the concentration of the metal ion decreases, this current will tend to decrease if the applied potential is maintained at its initial value. Increasing the applied potential to a more negative value results in a decreased cathodic potential (more negative) and an increase in the electrode reactions rates, returning the effective current to its initial level. As the metal-ion concentration continues to fall off, further changes in the applied potential are made in order to maintain the current level. Eventually a concentration of metal ion will be reached where the metal-oxygen system will no longer be able to sustain the required current, at which point a further change in the applied potential results in a decrease in the cathodic potential applied to the level at which hydrogen ion is discharged. The introduction of a current effective in hydrogen discharge reestablishes the initial current as a total current value. At this point, because of the availability of water, the cathodic potential will be stabilized. Further deposition of the metal will result in a continued decrease in the current effective in metal deposition, the stabilized cathodic potential will be directed more and more to the discharge of hydrogen ion, and the current effective in hydrogen-ion discharge will increase correspondingly. Thus, once the applied potential is more negative than $E_d(H_2-O_2)$, with the cathodic potential stabilized at the value for hydrogen discharge, the total current will remain constant, with less contribution arising in time, out of metal deposition and more out of hydrogen discharge.

It is apparent immediately that the following constitute important factors to be considered relative to constant-current electrolysis.

1. The technique will be of little value in electrogravimetry except where only one metal with $E_d(metal-O_2)$ less negative than $E_d(H_2-O_2)$ exists in the solution.
2. The necessity of maintaining a reasonable constant current, in order to obtain a

quantitative deposition in a practical time interval, requires almost without exception that the value of the applied potential will eventually become more negative than $E_d(H_2-O_2)$. This implies the simultaneous deposition of the metal and evolution of hydrogen at the cathode. While the simultaneous discharge of hydrogen does not affect the quantitative nature of the metal deposition, the continuous evolution of hydrogen gas bubbles can affect the physical nature of the deposit, rendering it porous and less adherent to the electrode. Increased porosity and associated increased surface development can result in an increase in the extent to which oxidation of the deposit may take place during electrolysis (entrained oxygen bubbles from the anodic reaction) and during subsequent electrode washing and drying operations. Poor adhesion can permit deposit loss during electrolysis and during manipulative operations. These factors are important but should not be unnecessarily emphasized. A standard method used in industry for the determination of copper in copper-base alloys involves constant-current electrolysis with an applied potential significantly more negative than $E_d(H_2-O_2)$, and its continuing success attests to the need to avoid unduly stressing the effect of hydrogen evolution on metal deposit quality.

3. Where electrogravimetry by constant-current electrolysis is carried out, the solution customarily contains only one depositable metal with $E_d(\text{metal}-O_2)$ less negative than $E_d(H_2-O_2)$ and a separation between the cathodic potentials for metal deposition and hydrogen discharge large enough to guarantee in time a quantitative separation of the metal. The applied potential is usually set initially at a value significantly more negative than $E_d(H_2-O_2)$ and capable of providing a comparatively high total current value. The subsequent electrolysis is conducted at what amounts to constant total current, with the deposition of the metal contributing in time less to this total current and hydrogen discharge contributing more and more.*

4. It has been observed already that where $E_d(\text{metal}-O_2)$ for a system is more negative than $E_d(H_2-O_2)$, such a metal can not be deposited at the cathode, since the cathodic applied potential eventually becomes stabilized at the discharge value for hydrogen ion and does not attain a value negative enough for metal-ion discharge. In some instances, however, a decrease in the $[H^+]$ of the solution can result in a change in the value of $E_d(H_2-O_2)$ so that it becomes more negative than $E_d(\text{metal}-O_2)$—and consequently in the possibility of a quantitative metal deposition from this lower $[H^+]$ solution. Consider a solution $0.050\,M$ to Zn^{2+} and $1.0\,M$ to H^+. Electrolysis is projected between large smooth platinum gauze electrodes, with the cathode

---

*The current will be constant only with respect to the ammeter reading, not with respect to either system being electrolyzed. Even the current reported by the ammeter may not be truly constant because of solution compositional changes, overvoltage changes, etc., as the electrolysis progresses.

protected by a prior copper deposit.* The E_d values are given by

$$E_d(H_2-O_2) = -\{(1.229 + 0.40) - (0.000 - 0.50)\}$$
$$= -2.13 \text{ V} \qquad\qquad\qquad \text{at } 25°C$$

$$E_d(Zn-O_2) = -\{(1.229 + 0.40) - (-0.801 - 0.0)\}$$
$$= -2.43 \text{ V} \qquad\qquad\qquad \text{at } 25°C$$

and it is evident that zinc can not be deposited at the cathode from this acid solution. If the solution were now neutralized with solid NaOH, and an excess added to yield a 1.0 M value of NaOH, we would have a solution $0.050 M$ to Zn^{2+} and $10^{-14} M$ to H^+. Using the proper overvoltage values from Table 12.3 for hydrogen and oxygen discharge in 1 M NaOH we have

$$E_d(H_2-O_2) = -\{(0.403 + 1.40) - (-0.826 - 0.80)\}$$
$$= -3.43 \text{ V} \qquad\qquad\qquad \text{at } 25°C$$

$$E_d(Zn-O_2) = -\{(0.403 + 1.40) - (-0.801 - 0.0)\}$$
$$= -2.60 \text{ V} \qquad\qquad\qquad \text{at } 25°C$$

and zinc may be deposited quantitatively from a 1-M NaOH solution. Obviously, where copper and zinc are present in an acid solution, constant current electrolysis of the acid solution will yield a separation and determination of copper without interference from zinc. After the copper separation, the solution may be made alkaline to NaOH, the electrodes returned to the solution (the cathode with its previous copper deposit as a protective measure), and the zinc separated and determined by a second constant-current electrolysis.

5. For electrolytic separations under constant-current electrolysis, with applied potentials more negative than $E_d(H_2-O_2)$, all depositable metals in solution with $E_d(\text{metal}-O_2)$ less negative than $E_d(H_2-O_2)$ may be deposited at the cathode. The removal of each metal will be quantitative, provided that the cathodic potential for a quantitative separation of each metal is sufficiently less negative than that at which hydrogen is discharged. As already noted, care must be taken when conducting such separations with a platinum cathode to protect the electrode against the possibility of reaction with metals such as zinc or bismuth.

Electrolytic separations may also be conducted through the use of a mercury cathode and a platinum anode, the more negative overvoltage for hydrogen discharge on mercury providing for the deposition in acid medium of certain metals not depositable on a platinum cathode in such media. As an example, the acid solution of Zn^{2+} described in 4 showed, with platinum

---

*This protective copper deposit on the cathode is required to prevent the zinc deposit from alloying with platinum, thus permitting platinum loss during the subsequent electrode deposit stripping process using dilute nitric acid. Such protective measures are required where metals such as zinc and bismuth are to be deposited.

electrodes, $E_d(\text{Zn}-\text{O}_2)$ more negative than $E_d(\text{H}_2-\text{O}_2)$. With a mercury cathode–platinum anode we have a more negative overvoltage for hydrogen discharge, with still what amounts to a negligible overvoltage for zinc discharge. Thus we have

$$E_d(\text{H}_2-\text{O}_2) = -\{(1.229 + 0.40) - (0.000 - 1.1)\}$$
$$= -2.73 \text{ V} \qquad\qquad \text{at } 25°C$$

$$E_d(\text{Zn}-\text{O}_2) = -\{(1.229 + 0.40) - (-0.801 - 0.0)\}$$
$$= -2.43 \text{ V} \qquad\qquad \text{at } 25°C$$

and this permits the quantitative removal of zinc(II) ion from solution by constant-current electrolysis.*

This process of mercury cathode electrolysis at constant current is commonly applied in the quantitative separation from acid solution of metals such as zinc, iron, manganese, nickel, cobalt, etc., from metals such as aluminum, vanadium, titanium, etc. A typical mercury cathode electrolysis system is shown in Fig. 12.14. It is customary to use very negative applied potentials (in the range of -8 to -10 V), resulting in high total current values and the simultaneous separation of the various metals affected. It should be pointed out again that the mercury cathode separation method is not intended to be used for electrogravimetric purposes; it is applied almost exclusively in electrolytic separation techniques.

12.3.7 TIME INTERVAL FOR QUANTITATIVE DEPOSITION UNDER CONSTANT–CURRENT ELECTROLYSIS†

It is important to gain some idea of the time interval required to achieve quantitative separation of a given metal under constant-current electrolysis, with the initial applied potential more negative than $E_d(\text{H}_2-\text{O}_2)$ and the metal involved being the only metal in solution with $E_d(\text{metal}-\text{O}_2)$ less negative than $E_d(\text{H}_2-\text{O}_2)$. The following can be considered as both an approximate and a general approach.

Consider a solution containing a metal ion at a concentration of $[\text{M}^{n+}]$, *in equivalents per liter*, and a hydrogen-ion concentration of $[\text{H}^+]$ where $E_d(\text{metal}-\text{O}_2)$ is significantly less negative than $E_d(\text{H}_2-\text{O}_2)$. The efficiently stirred solution will be electrolyzed between large smooth platinum gauze electrodes at an applied potential more negative than $E_d(\text{H}_2-\text{O}_2)$ and capable of yielding a total current (ignoring the residual current) of I amperes. We shall assume that the number of metal ions reduced at the cathode over any brief time interval will be proportional to the $[\text{M}^{n+}]$ in solution, and that the number of hydrogen ions reduced over the same time interval will be proportional to the $[\text{H}^+]$ in

---

*In actual practice, the solution pH would be adjusted, before the start of electrolysis, to a value very significantly higher than zero. Why?

†Adapted from L. Meites and H. C. Thomas, "Advanced Analytical Chemistry," pp. 196–197, McGraw-Hill, New York, 1958.

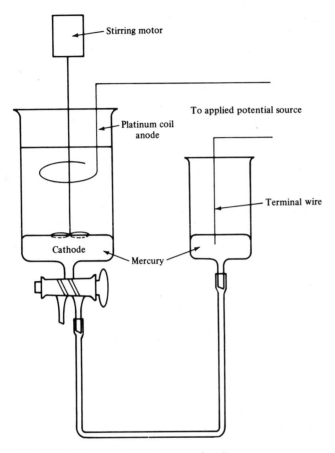

Fig. 12.14 Mercury cathode electrolysis cell.

solution.* Thus we will have the current due to metal reduction, $i_{M^{n+}}$, proportional to $[M^{n+}]$, and that for hydrogen reduction, i_{H^+}, proportional to $[H^+]$. We then have

$$i_{M^{n+}} + i_{H^+} = I$$
$$K_{M^{n+}}[M^{n+}] + K_{H^+}[H^+] = I \qquad (264)$$

Assuming for the moment that the electrolysis goes to equilibrium at the applied potential, for the metal-oxygen system, the value of $[M^{n+}]$ at this point will be practically zero. The concentration of hydrogen ion here will be

$$[H^+]_s + [M^{n+}]_s$$

---

*This assumption is not quite valid; the current efficiency will be somewhat greater relative to metal deposition than the assumption allows. However, the situation will be adequate for our purpose here.

where $[H^+]_s$ and $[M^{n+}]_s$ are, respectively, the concentrations in equivalents per liter of each ion at the start of electrolysis. We then have, from Eq. (264), at equilibrium

$$K_{H^+}([H^+]_s + [M^{n+}]_s) = I$$

$$K_{H^+} = \frac{I}{[H^+]_s + [M^{n+}]_s} \tag{265}$$

At the start of electrolysis Eq. (264) yields

$$K_{M^{n+}}[M^{n+}]_s + K_{H^+}[H^+]_s = I$$

Substitution of the value just determined for K_{H^+} gives

$$K_{M^{n+}} = \frac{I}{[H^+]_s + [M^{n+}]_s} \tag{266}$$

Suppose that a number of equivalents of M^{n+} are to be removed in time dt. This will require a quantity of electricity given by

$$\frac{i_{M^{n+}} dt}{F}$$

where $i_{M^{n+}}$ is the current effective in metal deposition over time dt and F is the faraday. Suppose that over this same time interval dt, the removal of metal ion from solution is given by $-Vd[M^{n+}]$, where V is the volume of the solution in liters and $d[M^{n+}]$ is the change in concentration in equivalents per liter. The negative sign indicates decreasing concentration. We then have

$$Vd[M^{n+}] = -\left(\frac{i_{M^{n+}} dt}{F}\right) \quad \text{and} \quad \frac{d[M^{n+}]}{[M^{n+}]} = -\left(\frac{K_{M^{n+}} dt}{VF}\right)$$

Integration of this latter expression leads to

$$\ln[M^{n+}] = -\left(\frac{K_{M^{n+}} t}{VF}\right) + C \quad \text{and} \quad \ln\frac{i_{M^{n+}}}{K_{M^{n+}}} = -\left(\frac{K_{M^{n+}} t}{VF}\right) + C$$

When $t = 0$ (the start of the electrolysis), the constant of integration is

$$C = \ln\frac{i_{M^{n+}}^\circ}{K_{M^{n+}}}$$

where $i_{M^{n+}}^\circ$ is the current effective in metal deposition at the start of the electrolysis. Substitution of this value for C into the previous expression gives

$$\ln\frac{i_{M^{n+}}}{K_{M^{n+}}} = -\left(\frac{K_{M^{n+}} t}{VF}\right) + \ln\frac{i_{M^{n+}}^\circ}{K_{M^{n+}}} \quad \text{and} \quad \ln\frac{i_{M^{n+}}}{i_{M^{n+}}^\circ} = -\left(\frac{K_{M^{n+}} t}{VF}\right)$$

with

$$\log\frac{i_{M^{n+}}}{i_{M^{n+}}^\circ} = -\left(\frac{0.4343 K_{M^{n+}} t}{VF}\right)$$

and, finally

$$\frac{i_{M^{n+}}}{i^{\circ}_{M^{n+}}} = {}_{10} - \left(\frac{0.4343\, K_{M^{n+}}\, t}{VF}\right) \tag{267}$$

with $K_{m^{n+}}$ given by Eq. (266). Note that Eq. (267) provides a means of calculating the value of $i_{M^{n+}}$ at any time t. Note also that Eq. (267) can be modified to yield

$$\frac{[M^{n+}]_f}{[M^{n+}]_s} = {}_{10} - \left(\frac{0.4343\, K_{M^{n+}}\, t}{VF}\right) \tag{268}$$

where $[M^{n+}]_f$ is the concentration of M^{n+} at any time t, and $[M^{n+}]_s$ is the starting concentration. Note that when the ratio of $[M^{n+}]_f$ to $[M^{n+}]_s$ is used, the concentrations can be of any homogeneous form.

Equation (267) or (268) can be rewritten to yield a value of t. For example, Eq. (268) gives

$$t = -\left(\frac{VF}{0.4343\, K_{M^{n+}}}\right) \log \frac{[M^{n+}]_f}{[M^{n+}]_s} \tag{269}$$

or, in more absolute terms, using the equality given by Eq. (266),

$$t = -\left(\frac{VF([H^+]_s + [M^{n+}]_s)}{0.4343\, I}\right) \log \frac{[M^{n+}]_f}{[M^{n+}]_s} \tag{270}$$

Example   1.0000 g of a copper-base alloy containing 58.97 percent of copper, and no interfering substances, was dissolved in 20 ml of 1:1 nitric acid. The solution was boiled to expel NO_2 and diluted to 150 ml. The approximate $[H^+]$ is now 0.5 M. Constant-current electrolysis is now conducted between large smooth platinum gauze electrodes at a total current of 5 A. Calculate the approximate time required to reach the point where the solution contains not more than 0.0001 g of undeposited copper. Calculate the approximate current effective in copper deposition at this point.

$$[Cu^{2+}]_s = \frac{0.5897}{63.54 \times 0.150} = 0.061^9\, M$$

$$t \approx -\left(\frac{0.150 \times 96490(0.5 + 0.124)}{0.4343 \times 5}\right) \log \frac{0.0001}{0.5897}$$

$$\approx 4.4\,\text{h}$$

$$K_{Cu^{2+}} = \frac{5}{0.5 + 0.124} = 8$$

$$i^{\circ}_{Cu^{2+}} = 8.0 \times 0.124$$
$$= 0.99^2\,\text{A}$$

$$\frac{i_{Cu^{2+}}}{i^{\circ}_{Cu^{2+}}} = \frac{i_{Cu^{2+}}}{0.99^2} = \frac{[Cu^{2+}]_f}{[Cu^{2+}]_s} = \frac{0.0001}{0.5897}$$

so that $i_{Cu^{2+}}$ when 0.0001 g copper maximum is left in solution is

$$i_{Cu^{2+}} \approx 0.000168\,\text{A} \approx 0.168\,\text{mA}$$

While the time interval may seem excessive, it should be remembered that the value of t can be decreased, without changing the applied potential, by such expedients as increasing the solution temperature, increasing the area of the electrodes, decreasing the separation distance between the electrodes, etc.

12.3.8 ELECTROLYSIS AT CONTROLLED CATHODE POTENTIAL

It has been noted that constant potential electrolysis, with the applied potential less negative than $E_d(H_2-O_2)$, is rarely used for analytical purposes. The problems encountered in maintaining a cathodic applied potential less than E_d (hydrogen), and capable of avoiding simultaneous deposition where more than one metal with E_d (metal$-O_2$) less negative than $E_d(H_2-O_2)$ exists in solution, in the face of overvoltage changes, uncertainties in the applied potential, and concentration polarization effects, are in the practical sense insurmountable. Where constant-current electrolysis is concerned, the technique is satisfactory for electrogravimetry where only one depositable metal with E_d(metal$-O_2$) sufficiently less negative than $E_d(H_2-O_2)$ exists in the solution, and where the applied potential is more negative than $E_d(H_2-O_2)$. It is also satisfactory for the electrolytic separation simultaneously of all depositable metals in solution showing E_d values sufficiently less negative than $E_d(H_2-O_2)$. The situation relative to the separation of two or more depositable metals from each other, where the cathodic potentials for quantitative deposition are adequately separated and sufficiently less negative than that at which hydrogen is discharged, is perfectly feasible theoretically speaking. The minimum required separation of the cathodic potentials (and therefore of the E_d(metal$-O_2$) values is given by Eq. (263):

$$\frac{10^{-6}\ M}{10^{-1}\ M} = \left(\frac{E_{app,cathode,f} - E_{app,cathode,s}}{0.059/n} \right)_{10}$$

with the value of $[M^{n+}]_s$ assumed to be 0.1 M for the general case. Thus we have

$$E_{app,cathode,f} - E_{app,cathode,s,min} = \frac{-0.295}{n}$$

so that for $n = 1$, the minimum difference is about -0.30 V, and for $n = 2$ and $n = 3$, it is -0.15 and -0.10 V, respectively.

Maintaining such potential differences is virtually impossible in any process of electrolysis where the cathodic applied potential is not rigidly controlled. Even where the cathodic potentials for two metal systems show a difference considerably greater than the minimum required for quantitative separation, the changes in the cathodic potential occurring as the result of concentration polarization effects, overvoltage changes, and solution resistance changes during electrolysis render it almost impossible to obtain a separation without some interference where the technique does not involve cathode potential control. Again, even where some possibility of a satisfactory separation exists, on the basis of a very pronounced difference in the cathodic potentials, the need to avoid interference requires the use of restrictive applied potentials, with a corresponding low effective deposition current and prolonged separation times.

In order to ensure that the cathodic applied potential maintains a value capable of permitting quantitative separation of the metal involved, but does not attain negative values large enough to result in some simultaneous deposition of other metals present in solution, it is necessary to control the applied potential at the cathode. To exert such control the cathodic potential must be monitored. This is accomplished by measuring the potential relative to some reference electrode. An electrolysis carried out under a system of measuring and controlling the cathodic potential is called controlled cathode potential electrolysis. Figure 12.15 shows a general arrangement for a manually controlled electrolysis of this type. The reference electrode, commonly a saturated KCl calomel electrode, is connected to the solution by means of a nonreactive salt bridge, where this is required by mutually reactive electrode solution and electrolysis solution components. The cathode and reference electrode are linked through a potentiometer, with the cathode and anode being connected to a source of applied potential in the customary manner.

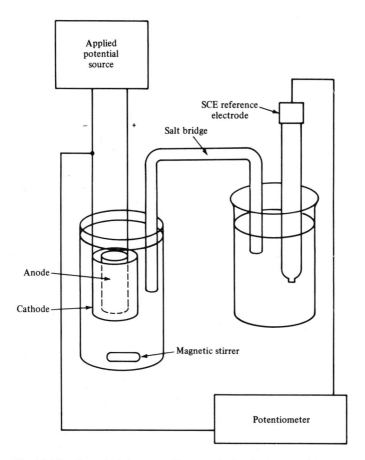

Fig. 12.15 Electrolysis by manually controlled cathode potential.

Suppose, for example, that silver is to be quantitatively separated from copper by electrolysis of 100 ml of a solution with 0.050 M AgNO$_3$, 0.050 M Cu(NO$_3$)$_2$, and 1.0 M HNO$_3$. The cathodic applied potential at which the concentration of Ag$^+$ will be reduced to 10^{-7} M at equilibrium is given by

$$E_{\text{app,cathode}} = 0.800 + 0.059 \log 10^{-7}\ M$$
$$= 0.387\ \text{V} \qquad\qquad \text{at } 25°\text{C}$$

while the value of cathodic applied potential at which copper will be just ready to deposit will be

$$E_{\text{app,cathode}} = 0.340 + \frac{0.059}{2} \log (5.0 \times 10^{-2}\ M)$$
$$= 0.302\ \text{V} \qquad\qquad \text{at } 25°\text{C}$$

If the cathodic applied potential is maintained at a value of 0.38 V relative to the SHE, this requires a value vs. the SCE of

$$E \text{ vs. SCE} = E \text{ vs. SHE} - E_{\text{SCE,red}}$$
$$= 0.38 - 0.245$$
$$= 0.14\ \text{V} \qquad\qquad \text{at } 25°\text{C}$$

and a quantitative deposition of silver without interference from copper would result.

To start the electrolytic process, the applied potential is gradually increased until the potentiometer linking the cathode and SCE shows a value of 0.14 V. It is apparent that at the start of the electrolysis, a relatively high current is obtained, and the silver-ion concentration will decrease rapidly in the early stages, resulting in concentration polarization effects, overvoltage changes, etc., with corresponding changes in the cathodic potential requiring adjustment of the applied potential to less negative values. As the electrolysis progresses, the magnitude of such cathodic potential changes diminishes, and the associated applied potential adjustments become minor. Since the current is effective in the deposition of silver alone (ignoring the residual current), it can be expected to decrease exponentially with time. A current of almost zero will indicate that the separation of silver is quantitatively complete. Alternatively, the application of a chemical test to small portions of the solution at appropriate intervals can be used to indicate quantitative removal (10^{-6} M residual [Ag$^+$] in solution) and therefore practical completion of the electrolysis.

Manual control of the adjustments required to be made to the applied potential and the overseeing by an analyst can be avoided by automating the equipment. A simple form of automation is shown in Fig. 12.16. As long as the cathodic potential, relative to the SCE, maintains the value originally set on the potentiometer, no current flows in the potentiometer circuit and the current sensing attachment. When the cathodic potential starts to shift, a very small current flows, the direction depending on the directional change in the cathodic potential. The

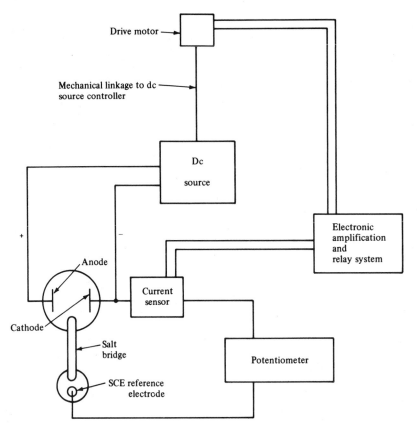

Fig. 12.16   Schematic diagram for automated controlled cathode potential electrolysis.

current sensing attachment alerts the electronic amplification and relay device, which is usually set to respond when the potentiometer circuit current exceeds some limiting value. Operation of the relay actuates a drive motor at the source of the applied potential, increasing or decreasing the potential until the cathodic potential is returned to the required value. Automated controlled cathode or anode potential electrolysis equipment is available commercially.

The time required for a quantitative separation under controlled cathode potential electrolysis will be discussed in general in Sec. 12.4.2.

12.3.9   FACTORS AFFECTING ELECTROLYTIC DEPOSITION

In general, metal depositions for electrogravimetric purposes should be fine-grained, dense, smooth, and free from oxides and should adhere firmly to the electrode. Deviations from these characteristics may result in a nonquantitative determination. Several factors can affect the physical nature of the deposit.

The simultaneous evolution of large volumes of gas, particularly hydrogen gas at the cathode, can lead to porous and poorly adherent deposits. The evidence is quite strongly against unnecessary emphasis in this connection, as has already been pointed out. Moderate hydrogen gas evolution will infrequently exert a harmful effect on deposit quality.

Efficient stirring is required to minimize concentration polarization effects, particularly since such effects are very significant in the early stages of any electrolysis. Stirring by means of a magnetic stirrer is to be preferred, although stirring by the use of a mechanically rotated central electrode is also acceptable, providing that care is taken to prevent contact between the electrodes during electrolysis and during initial washing procedures. Stirring by passage of nitrogen or air into the solution should be avoided, if for no other reason than that of the possibility of mechanical loss of solution.

Electrolysis at elevated temperature usually provides for lower overvoltage values, decreased concentration polarization effects, lower solution resistance, and less time to reach quantitative completion. Elevated temperature can also lead to the possibility of significant oxidation of the deposit during the electrolytic process. Optimum temperatures for electrolysis are best determined experientially for each system.

It was indicated that a fine-grained structure was desirable. Such deposits are frequently obtained where the current density is maintained at a reasonable level. Low current densities tend to lead to coarse-grained deposits, while high current densities yield uneven, rough deposits with relatively poor adhesive properties. Current densities may vary from 0.005 to 0.1 A/cm^2, or 0.5 to 10 A current for an electrode of 100 cm^2 area.

The presence of complexing agents can reduce the $[M^{n+}]$ to a very low level. The electrolysis of such a complex-ion solution of the metal requires an applied potential at the cathode appreciably more negative than that required where the metal is not complexed. The deposit obtained by the electrolysis of a complex-ion solution has a tendency to be of superior quality with respect to its grain size, smoothness, and property of adhesion to the electrode. The use of a complexing agent capable of complexing one metal in solution and not another, or of forming a much more stable complex with one metal than another, often permits selective deposition procedures. It is possible, for example, to separate cadmium quantitatively from copper by controlled cathode potential electrolysis of a solution of their cyanide complexes as cadmium(II) and copper(I). In acid solution and in the absence of a complexing agent, copper may be separated quantitatively from cadmium by controlled cathode potential electrolysis.

Several substances may be separated for electrogravimetric purposes at the anode. Lead as $Pb(NO_3)_2$ in nitric acid solution can be separated at the anode according to the reaction

$$Pb^{2+} + 2H_2O - 2e \rightleftharpoons PbO_2 + 4H^+$$

the lead dioxide being attracted to and held firmly at the platinum anode. A subsequent weight estimation of the lead dioxide leads to an accurate and precise determination of lead.

12.4 COULOMETRY AND COULOMETRIC TITRATIONS

12.4.1 GENERAL

The methods of coulometry and coulometric titration are based on the measurement of the quantity of electricity involved in an electrochemical electrolysis reaction. This quantity is expressed in coulombs; it represents the product of the current in amperes by the duration of the current flow in seconds. The quantity of electricity thus determined represents, through the laws of Faraday, the equivalents of reactant as associated with the electrochemical reaction taking place at the electrode of significance in the electrochemical cell action resulting from electrolysis. The equivalents of reactant may then be directly or indirectly related to the quantity of the unknown species under investigation. Essentially we have then

$$W = \frac{W_m Q}{nF} \tag{271}$$

where W = weight of reactant, g

   W_m = gram-molecular weight of reactant

   n = electron transfer for the reactant in the electrode reaction involved

   F = faraday (96,490 C/equiv)

   Q = quantity of electricity consumed, C

The relationship of the reactant to the unknown species establishes the weight in grams of the unknown species involved.

In the direct or coulometric method, the electrode reaction generally involves the unknown species directly in one form or another or, more infrequently, some substance directly related chemically to the unknown; and the electrochemical reaction occurring at the electrode is brought about by the application of a controlled potential at that electrode. The reaction desired may be one of oxidation (anodic) or reduction (cathodic), and the associated current will, as has been noted in Sec. 12.3.8, decrease exponentially with time. The time required to attain a reaction position that is quantitatively complete can be established at the point where the current has decreased to a value indicative of the reactant concentration in solution having reached a level of 10^{-6} M residual or less, the concentration level here depending on the requirement for quantitative completion. Alternatively, the critical time may be determined when some indicator system shows the required reactant concentration level to have been attained. The quantity of electricity consumed may be evaluated by a coulometer, or by a current-time integrator, and is then applied directly in the estimation of the quantity of reactant and unknown species originally present. Where current-time integrator techniques are used, it is apparent that the value of Q will be obtained from

$$Q = \int i \, dt \tag{272}$$

with

$$i = i^\circ \, e^{-kt} \tag{273}$$

where i° = current at the start of electrode reaction, A

k = a constant specific for the electrode reaction

t = time (usually time to reach reaction completion level), s

Q then represents the area under the current-vs.-time curve between the start and the time to reach reaction completion.

With respect to the indirect or coulometric titration method, the electrode reaction generally involves the production anodically or cathodically of a substance capable of reacting chemically and stoichiometrically with a substance related to the unknown species. The substance so generated, called the titrant, is the product of a constant-current situation rather than a controlled potential application. The quantitative completion of the reaction between the generated titrant and the reactant is signaled by an indicator system that may involve a standard visual change indicator technique or a system of end-point detection based on potentiometric, absorptiometric, etc., methods. In any case, the end-point signal permits the determination of the value of Q as the product of the constant current and the time in seconds to reach the end point. Thus we have

$$Q = it \tag{274}$$

and the stoichiometry of the titration reaction allows the determination of the quantity of the unknown species originally present by a proper treatment of W in Eq. (271).

It is at once apparent that in all coulometric methods, it is of prime importance that the quantity of electricity consumed be related exclusively to the particular electrode reaction involved. The current, in other words, must be 100 percent efficient relative to the specific electrochemical reaction involved. Participation of any fraction of the current in secondary reactions leads to error in the final result. In instances where coulometric titration techniques are applied, and where the unknown species participates in a secondary reaction with the product of the electrochemical reaction, the stoichiometry of the secondary reaction must be rigidly established, and the reaction must be of an exclusive nature.

12.4.2 COULOMETRY AT CONTROLLED POTENTIAL

The methods of coulometry at controlled potential resemble those of controlled cathode potential electrolysis in electrogravimetry, although both cathode and anode controlled potential techniques are found in the coulometric approach. The methods of controlled cathode potential coulometry are, however, often applicable to systems which do not lend themselves to the electrogravimetric application. This is particularly the case when the separation of metallic substances proceeds readily relative to a mercury cathode, when certain metals may be separated at the mercury cathode in acid solution but not so separated at a platinum cathode, when deposit quality presents a problem in electrogravimetry, and when the quantity to be determined is too small to be accurately estimated, if at all, by the electrogravimetric method. The mercury cathode does not, of course, lend itself to the electrogravimetric process of analysis, but in a similar coulometric approach, the measurement of the quantity of electricity and its relationship to the weight of metal discharged at the cathode avoids these difficulties. Coulometric techniques involving controlled mercury cathode potential include the determination of nickel and cobalt when present together, the determination of copper and bismuth in the same solution, and the determination of lead in the presence of cadmium.[*]

The coulometric method can be applied in the case of the determination of the halides chloride, bromide, and iodide through the use of controlled anode potential at a silver anode, the cathode being of the platinum gauze type. These separations involve the generalized reaction

$$Ag + X^- - 1e \rightleftharpoons AgX$$

where Ag is converted to Ag^+ as the anodic reaction, the resulting Ag^+ reacting with the halide to form the slightly soluble silver halide salt. Where the solution contains only one of the halides, no problems of a significant nature are encountered. With mixtures of halides, certain limitations relative to allowable concentrations exist, and close anodic potential control is required. Iodide may be determined in the presence of chloride, even where the chloride concentration is $1\ M$. Iodide may be determined in the presence of bromide up to a bromide concentration of approximately $0.05\ M$. The determination of bromide in the presence of chloride presents some problem, in that the allowable concentration of chloride is somewhat less than $0.005\ M$.[†]

Electrochemical reactions which do not result in depositable products may also be made use of in coulometric techniques, and the controlled potential coulometric determination of iron from the reaction Fe^{2+} to Fe^{3+} and arsenic from the reaction As^{3+} to As^{5+} are commonly applied. In several instances, the method may be applied in the determination of organic substances, particularly in cases where such substances are capable of quantitatively complete reduction reactions at the mercury cathode under controlled potential conditions. The determination of picric acid and trichloracetic acid by such methods have been demonstrated to be feasible by Meites and Meites.[‡]

[*]J. J. Lingane, "Electroanalytical Chemistry," 2d ed., pp. 462–464, Interscience, New York, 1958.
[†]Data from J. J. Lingane, "Electroanalytical Chemistry," 2d ed. pp. 465–471, Interscience, New York, 1958.
[‡]T. Meites and L. Meites, *Anal. Chem.*, 27: 1531(1955); 28: 103(1956).

In the direct or coulometric method, there are two factors of prime importance. There is, first of all, the need to guarantee to the highest degree possible 100 percent current efficiency for the electrode reaction involved. This necessitates the minimization of the residual or background current. The importance of this factor can not be overemphasized. Where small quantities of reactants at the electrode are to be determined, the magnitude of unminimized residual currents may be extremely significant relative to the total current through the system. The judicious selection of the supporting electrolyte can frequently assist in this minimization process; but of greater general importance is the removal of oxygen dissolved in the solution by flushing with an inert gas such as nitrogen prior to the coulometric procedure. This minimizes that significant residual or background current stemming from the electrode reactions

$$O_2 + 2H^+ + 2e \rightleftharpoons H_2O_2 \quad \text{cathode}$$

$$H_2O_2 - 2e \rightleftharpoons O_2 + 2H^+ \quad \text{anode}$$

Other sources of residual current exist, of course, such as that originating with hydrogen-ion reduction when the technique involves controlled cathode potential coulometry. Where some residual current persists, even subsequent to minimization efforts, electrolysis of the basic electrolyte, in the absence of the reactive species to be determined, and under identical methodic conditions, may serve to establish the level of this residual current. An assumption that this current will remain identical and constant during the coulometric process involving the reactive species will permit total current values to be corrected accordingly.

The second important factor has to do with the determination of the current-time relationship during the coulometric process and, of course, the subsequent determination of the quantity of electricity consumed. There are several methods which can be applied here. A rather simple technique involves Eq. (273). Here we have

$$\log i = \log i° - 0.4343\, kt \qquad (275)$$

If i is determined for various values of t during the coulometric procedure, a plot of $\log i$ vs. t will yield a straight line. The value of $\log i°$ and $i°$ may be obtained from the intercept of the line with the $\log i$ axis at zero time, with k being obtained from the line slope $-0.4343\, k$. This estimation of $i°$ and k now permits the determination of Q_{max}, the quantity of electricity for the coulometric process carried to theoretical completion. The value of Q_{max} is obtained from

$$Q_{max} = \int_0^\infty i° e^{-kt}\, dt$$

$$= \frac{i°}{k} \qquad (276)$$

Thus, the amount of the reacting species may be estimated from Eq. (271) without the necessity of carrying the electrochemical reaction involved to quantitative completion. The values of i determined for various values of t may be corrected for the residual current, if this latter has been shown by previous tests to attain a significant level.

While the method appears to be attractive, it requires, for a reasonably accurate estimate of Q_{max}, multiple determinations of i and t, with a corresponding expenditure of operator time. The accuracy of the estimation of the reactive species may be limited to about ± 1 or ± 2 percent (± 10 or ± 20 ppt), particularly where electrolyte temperature fluctuations result from the passage of fairly high currents (macrodeterminations) and where current variations may occur during the coulometric process because of poor applied potential control. This latter situation is not at all uncommon where manual rather than automatic applied potential control systems are used.

The lack of precision and accuracy, together with the extended operator involvement, characteristic of the foregoing method, can be avoided by the use of a coulometer for the measurement of the quantity of electricity consumed. A coulometer is a device, placed in series with the controlled potential coulometric apparatus, through which flows the same current. The electrochemical reaction taking place in the coulometer is one which utilizes this current flow with 100 percent efficiency. One of the most venerable of such devices involves a platinum cathode and a silver anode immersed in a well-stirred solution of silver perchlorate. The passage of current through this system under the proper conditions results in the deposition of silver on the cathode, which deposition is carried on until a quantitatively complete electrode reaction has occurred in the controlled potential coulometric apparatus. The weight of silver deposited, in relationship to Faraday's law, permits the determination of the species reactive in the coulometric process. A second and more efficient coulometer is the hydrogen-oxygen coulometer. This consists basically of an inverted tube filled with 0.5 M K_2SO_4 solution and containing two sealed-in platinum electrodes. When connected in series with the controlled potential coulometric apparatus, oxygen and hydrogen are liberated at the anode and cathode. Both gases are collected together under known temperature and pressure conditions. The total volume of the combined gases is measured and, after correction for water vapor pressure and calculation to standard temperature and pressure conditions, is used through the faradian relationship to determine the quantity of the reactive species in the controlled potential process.

Many techniques of direct coulometry now make use of current-time integrators to determine the area under the current-time curve and therefore the quantity of electricity consumed in the coulometric process. In such cases, the value of Q at any time t is given by

$$Q_t = \int_0^t i° e^{-kt} \, dt = -\frac{i°}{k} (e^{-kt} - 1) \tag{277}$$

The value of Q_{max} was given by Eq. (276) as

$$Q_{max} = \frac{i^\circ}{k}$$

so that the relative error in stopping the coulometric process at any time t is given by

$$\text{Relative error in } Q_t = \frac{Q_{max} - Q_t}{Q_{max}} = e^{-kt} = 10^{-0.4343kt} \qquad (278)$$

It is possible to determine in this way the relative error in Q_t or in the determination of the reacting species resulting from stopping the process at any time t. Conversely, it is possible to calculate the time at which the process should be stopped in order to obtain a desired level of relative error in the final result.

Very frequently in controlled applied potential coulometry, it is essential to prevent the required electrode reaction from undergoing a reversal at the opposing electrode, or to prevent reaction products from the opposed electrode interfering with the required electrode reaction. This is usually accomplished by compartmentizing the opposing electrode through the use of a porous or sintered-glass disk. A typical controlled applied potential coulometry apparatus is shown in Fig. 12.17.

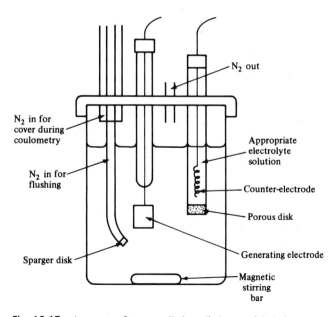

Fig. 12.17 Apparatus for controlled applied potential coulometry.

Example A solution of picric acid was electrolyzed at a controlled applied potential at a mercury cathode. The picric acid was reduced to triaminophenol under conditions where the current efficiency was 100 percent.

150.0 s after the start of the electrolysis, the recorded current was 0.2136 A; 600.0 s later the current had decreased to 0.0122 A. Determine the number of milligrams of picric acid in the original solution.

$$\log i \text{ at } 150.0 \text{ s} = \log 0.2136 = -0.6704$$

$$\log i \text{ at } 750.0 \text{ s} = \log 0.0122 = -1.9136$$

$$-0.6704 = \log i^\circ - 0.4343\, k\ 150.0$$

$$-1.9136 = \log i^\circ - 0.4343\, k\ 750.0$$

$$i^\circ = 0.436^9 \text{ A}$$

$$k = 0.00477^1/\text{s}$$

$$Q_{max} = i^\circ/k$$

and we have

$$W = \frac{W_m (i^\circ/k)\ 1000}{nF} \text{ mg}$$

$$= 12.08 \text{ mg}$$

Example Suppose in the previous problem that the system had been placed in series with a hydrogen-oxygen coulometer. When a quantitative level of conversion of picric acid to triaminophenol was secured, the coulometer showed 17.23 ± 0.04 ml of mixed hydrogen and oxygen gases at 24.8°C and 789.65 mmHg. If the vapor pressure of water is 23.48 mmHg at 24.8°C, calculate the quantity of picric acid in milligrams in the unknown solution.

$$\text{At STP, } 11,207 \text{ ml of } H_2 = 1 \text{ equiv} = 96,490 \text{ C}$$

$$\text{At STP, } 5603.5 \text{ ml of } O_2 = 1 \text{ equiv} = 96,490 \text{ C}$$

$$\text{At STP, } 16,810.5 \text{ ml of mixed gases} = 1 \text{ equiv} = 96,490 \text{ C}$$

$$\text{At STP, } 1 \text{ ml of mixed gases} = \frac{96,490}{16,810.5} \text{ C}$$

$$\text{Gas collected} = 17.23 \pm 0.04 \text{ ml at } 789.65 \text{ mmHg and } 24.8°C$$

Water vapor pressure correction yields

$$\text{Gas collected} = 17.23 \pm 0.04 \text{ ml at } 766.17 \text{ mmHg and } 24.8°C$$

$$\text{Gas collected at STP} = \frac{17.23 \times 766.17 \times 273.2}{760.0 \times 298.0} \text{ ml}$$

$$Q = \frac{17.23 \times 766.17 \times 273.2 \times 96,490}{760.0 \times 298.0 \times 16,810.5} \text{ C}$$

$$W(\text{picric acid}) = \frac{229.11 \times 17.23 \times 766.17 \times 273.2 \times 96,490 \times 1000}{18 \times 96,490 \times 760.0 \times 298.0 \times 16,810.5}$$

$$= 12.06 \text{ mg}$$

12.4.3 COULOMETRY AT CONSTANT CURRENT: COULOMETRIC TITRATIONS

General   In the ideal application of controlled applied potential coulometry, a single electrode reaction is guaranteed by a judicious selection of the electrolyte composition and by the use of the proper applied and controlled potential at the electrode involved. The current decays exponentially as the electrode reaction proceeds, and the quantity of electricity consumed in carrying out the desired electrochemical reaction to quantitative completion is determined by integration using a chemical coulometer, a current-time integrator, or some other means capable of yielding the area under the current-time curve.   In constant-current coulometry, a selected constant current is forced through the electrolyte, and the quantity of electricity required to carry the electrochemical reaction to some point of quantitative completion is obtained by multiplying the constant current by the time to reach completion.   Where the product of the electrochemical reaction reacts stoichiometrically and rapidly with a second species, the quantity of the second species may be easily determined.

It will be recalled from Sec. 12.3.6 that electrolytic processes conducted under constant-current conditions eventually result in concentration polarization effects being observed at the electrode involved, with a corresponding shift in the electrode potential and the subsequent appearance of an electrode reaction other than that intended. The appearance of this second reaction will result in less than 100 percent current efficiency relative to the prime reactive species, except where a product of the second reaction can react rapidly and stoichiometrically with the reactive species involved in the first reaction.

Consider, for example, an attempt to determine cerium quantitatively by reacting Ce^{4+} at a platinum cathode according to

$$Ce^{4+} + 1e \rightleftharpoons Ce^{3+}$$

under constant-current coulometry and using a compartmentized platinum anode as the counterelectrode. If the solution starts off as $0.001\ M\ Ce^{4+}$ and $1\ M\ H^+$, the cathodic decomposition potentials for the reactions $Ce^{4+} + 1e \rightleftharpoons Ce^{3+}$ and $2H^+ + 2e \rightleftharpoons H_2$ will be about 1.46 and -0.50 V, respectively. Where the solution is well stirred and where dissolved oxygen has been removed by flushing with pure nitrogen gas, an idealized plot of cathodic applied potential vs. current through the solution would yield curve 1 of Fig. 12.18. It should be noted that the concentration polarization effect results in a leveling off of the current at a value of i_1. This represents what is called the *limiting current* for $0.001\ M\ Ce^{4+}$ under the relevant solution conditions. The current values i_2 and i_3 for curves 2 and 3 represent the limiting currents due to concentration polarization effects for the same base solution but with one-half and one-quarter the Ce^{4+} concentration, respec-

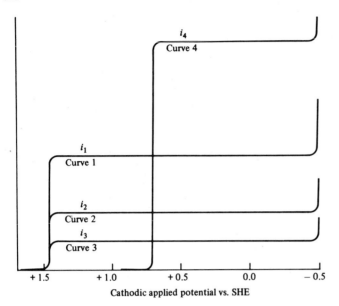

Fig. 12.18   Limiting current curves for Ce^{4+} solutions with $1\ M\ [H^+]$.

tively (that is, $0.0005\ M$ and $0.00025\ M$).   Under ideal conditions the values of i_2 and i_3 are, respectively, one-half and one-quarter the value of i_1. For the solution concentration of $0.001\ M\ Ce^{4+}$, the application of a constant current of i_2 will result in the reduction of Ce^{4+} to Ce^{3+} at the cathode with 100 percent efficiency until the concentration of Ce^{4+} is reduced to about $0.0005\ M$.   At this level, because of concentration polarization effects, Ce^{4+} ions can no longer arrive at the cathode in sufficient quantity to sustain the current requirement of i_2. The cathode potential will therefore drift to more negative values until it reaches a value of about $-0.50\ V$ vs. SHE, at which potential hydrogen ion is discharged at the cathode.   At this point the current efficiency for the reaction $Ce^{4+} + 1e \rightleftharpoons Ce^{3+}$ becomes less than 100 percent and decreases continuously as the reactions go on, with the reaction $2H^+ + 2e \rightleftharpoons H_2$ assuming a larger share of the current i_2. It is at once apparent that the coulometric determination of cerium by such an application of the constant-current method would be impossible on a quantitative basis.

Suppose, however, that a species such as Fe^{3+} is introduced into the solution of Ce^{4+}, before the coulometric procedure, at a concentration level about twice that of Ce^{4+}. The ion mobilities for Ce^{4+} and Fe^{3+} are approximately the same, so that their limiting currents will be similar for similar concentrations.   Curve 4 of Fig. 12.18 represents ideally the plot of cathodic potential vs. current for Fe^{3+} alone at about $0.002\ M$ concentration.   The cathodic decomposition potential for the reaction $Fe^{3+} + 1e \rightleftharpoons Fe^{2+}$ is approximately $0.70\ V$ vs. SHE in $1\ M\ H^+$. The value of i_4 represents the limiting current for Fe^{3+} under these conditions. Consider now the application to the mixed solution, 0.001 M Ce^{4+} and 0.002 M Fe^{3+}, of a constant current of i_2. As before, the current efficiency for the reaction $Ce^{4+} + 1e \rightleftharpoons Ce^{3+}$ will be 100 percent down to a value of $[Ce^{4+}]$ of about 0.0005 M. Fe^{3+} will not be reduced during this initial period, since the cathodic potential, at about 1.46 V vs. SHE, is insufficiently negative. Concentration polarization effects relative to the cerium reduction reaction will appear when $[Ce^{4+}]$ is about 0.0005 M, resulting in the cathode potential drifting to more negative potential values. Eventually the cathode potential will reach a value of approximately 0.70 V vs. SHE, at which point the reaction $Fe^{3+} + 1e \rightleftharpoons Fe^{2+}$ will start to take place at the cathode. This means that from this point, the current efficiency for the reaction $Ce^{4+} + 1e \rightleftharpoons Ce^{3+}$ will become increasingly less than 100 percent. However, the product of the Fe^{3+} reaction, Fe^{2+}, reacts immediately and stoichiometrically in the stirred solution with Ce^{4+} according to

$$Fe^{2+} + Ce^{4+} \rightleftharpoons Fe^{3+} + Ce^{3+}$$

so that the $[Fe^{3+}]$ in solution remains constant, and the limiting current for the $Fe^{3+} + 1e \rightleftharpoons Fe^{2+}$ reaction remains constant, as long as the solution contains Ce^{4+}. The cathode potential, under these conditions, does not achieve a sufficiently negative value for the reaction $2H^+ + 2e \rightleftharpoons H_2$. The net reaction is the reduction of Ce^{4+} quantitatively under conditions where one equivalent of Ce^{4+} is reduced for every 96,490 C (1 faraday) of electricity passing through the solution.

It is apparent that (1) the constant-current value applied should not exceed i_4 and, indeed, should be appreciably less than this value, and (2) the concentration of Fe^{3+} should be much higher than that of Ce^{4+}. It is also apparent that, where the constant current applied is greater than i_1 but less that i_4, the method is one of coulometric titration, the Ce^{4+} in the main being titrated by Fe^{2+} generated electrolytically and coulometrically. Note that the added species, Fe^{3+}, was required to be of a nature such that the reduction potential was appreciably more negative than that for the reaction $Ce^{4+} + 1e \rightleftharpoons Ce^{3+}$ and appreciably more positive than that for the interfering electrode reaction (in this case $2H^+ + 2e \rightleftharpoons H_2$), as well as such that the cathode reaction product, Fe^{2+}, was capable of reacting rapidly and stoichiometrically with Ce^{4+} to produce the same product, Ce^{3+}, as resulted from the cathodic reaction $Ce^{4+} + 1e \rightleftharpoons Ce^{3+}$.

A constant-current coulometric process somewhat more representative of the coulometric titration form is that where the constant current is used to generate electrolytically, with 100 percent efficiency, a substance which acts immediately as the titrant in a titration reaction with the desired reactive species in the solution. A typical example is the generation of iodine by constant-current coulometry between two platinum electrodes immersed in an oxygen-free solution of potassium iodide. The generating reaction is anodic:

$$3I^- - 2e \rightleftharpoons I_3^-$$

with the cathode being compartmentized. The buffering of the iodide solution to a
pH of about 8 allows the titration of As^{3+} in the solution according to

$$H_3AsO_3 + I_3^- + 5H_2O \rightleftharpoons HAsO_4^{2-} + 3I^- + 4H_3O^+$$

Note that the iodide concentration remains constant and the limiting current for
the iodide-triiodide system remains unchanged throughout the course of the titration. The use of an appreciable iodide concentration in the solution, and the
application of a constant current less than the limiting current associated with the
iodine-triiodide system, permits 100 percent current efficiency. The quantity of
arsenic can be calculated from the constant current, the time to reach a quantitative completion of the triiodide-arsenic(III) reaction, and the stoichiometry of this
latter reaction.

It has been indicated that in many instances, it is necessary to compartmen-
tize the opposing electrode in order to prevent interference in the coulometric
process. It has also been indicated that where the generating electrode is capable,
under certain electrolyte solution conditions, of undergoing side reactions, there
will result a current efficiency of less than 100 percent for the desired electrode
reaction. Both situations may often be avoided by a procedure providing for the
generation of the titrant external to the solution containing the reactant. A typical
arrangement for such external generation of titrant is shown in Fig. 12.19. The

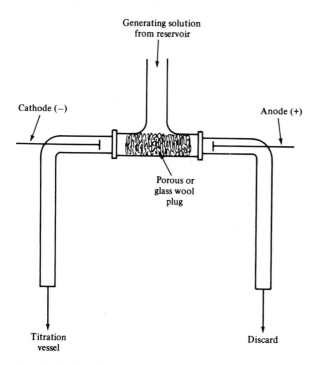

Fig. 12.19   Cathodic generation externally of titrant.

slow flow of titrant generating electrolyte from the reservoir is divided into two streams by passage along the arms of the T tube. One arm contains the cathode, the other the anode. If the cathodic reaction results in the required titrant, the flow from this arm is directed to the titration vessel, and that from the anodic arm is discarded. Flow from the reservoir continues briefly whenever the generating current ceases, but only long enough to flush out residual titrant from the system, whereupon it is then cut off. The anodic reaction product, and the anode itself, cannot interfere in the required titration reaction, nor can the generating electrode become involved in side reactions with respect to any of the components of the reactant solution in the titration vessel.

Titration methods The coulometric titration technique can be applied in neutralization titrations involving strong acids or weak acids. The reduction of water at a platinum cathode results in the production of hydroxyl ion according to

$$2H_2O + 2e \rightleftharpoons H_2 + 2OH^-$$

the hydroxyl ion produced titrating the strong or weak acid substance. Since the cathodic reduction of water is ordinarily accompanied by its anodic oxidation at a platinum anode according to

$$2H_2O - 4e \rightleftharpoons O_2 + 4H^+$$

the anode, in the case of internal generation of the titrant, must be compartmentized. It is possible, however, to provide for an anodic reaction other than that of water oxidation and the production of hydrogen ion. The addition of bromide or chloride ion to the reactant solution prior to coulometric titration, and the use of a silver rather than a platinum anode, results in the anodic reaction

$$Ag + Br^- - 1e \rightleftharpoons AgBr(s)$$

and an end product that does not interfere in the coulometric titration of acids by the generated hydroxyl ion. Where a system of external generation of hydroxyl ion is used, interference by possible side reactions at the cathode or by anodic reaction products, is of no moment. An important factor associated with the coulometric titration of acids is that of the elimination of the need to protect standard base solution (i.e., sodium hydroxide solutions) from the absorption of atmospheric carbon dioxide. Any carbon dioxide in the reactant solution and generating solution may be removed prior to titration by nitrogen or other inert gas flushing. The titration of strong or weak bases may be similarly accomplished, using anodically generated hydrogen ion, and the same general reasoning applies as surrounded the acid titration approach.

Precipitation and complexation titration methods are often easily adapted to the coulometric titration technique. The titration of the halides chloride, bromide, and iodide may be accomplished through the use of a heavy silver anode and a platinum cathode. The generation anodically of silver(I) ion serves to titrate the halide involved. Similar methods of coulometric precipitation titrimetry have been developed using the generation of mercury(I) ion at a mercury anode. Where an ammonia–ammonium nitrate buffered solution of mercury(II)-EDTA-ammine complex is subjected to constant current coulometry between a platinum anode and a mercury cathode of the pool type, the complex reacts cathodically according to

$$HgNH_3(Y)^{2-} + NH_4^+ + 2e \rightleftharpoons Hg + 2NH_3 + HY^{3-}$$

and HY^{3-} is released to the solution.* This liberated HY^{3-} ion will react rapidly with metal-ion species in solution, such as Ca^{2+}, Zn^{2+}, Cu^{2+}, and Pb^{2+}. The chelate formed by the metal-ion species involved must, of course, be appreciably less stable than the mercury chelate, so that reaction prior to HY^{3-} generation will not occur. Since the opposing electrode reaction at the platinum anode results in the liberation of oxygen according to

$$2H_2O - 4e \rightleftharpoons O_2 + 4H^+$$

the anode should be compartmentized in order to prevent the liberated oxygen from reacting cathodically.

Oxidation-reduction reactions are generally adaptable in one way or another to the methods of coulometric titrimetry, and very many applications involving such reactions have been made. In addition to those discussed earlier in this section, Table 12.4 gives an outline of common coulometric oxidation-reduction and other titration methods.

*C. N. Reilley and W. W. Porterfield, *Anal. Chem.*, **28**:443(1956).

Table 12.4   Common coulometric titration methods

Generated titrant	Generating electrode reaction	Reactive species
Cl_2	$2Cl^- + 2e \rightleftharpoons Cl_2$	I^-, As(III)
Br_2	$2Br^- + 2e \rightleftharpoons Br_2$	I^-, As(III), Sb(III), U(IV), Tl(I), SCN^-, NH_3, 8-hydroxyquinoline, aniline, phenol, salicylic acid
I_2	$3I^- - 2e \rightleftharpoons I_3^-$	As(III), Sb(III), $S_2O_3^{2-}$, Se(IV), H_2S, SO_2
Ce^{4+}	$Ce^{3+} - 1e \rightleftharpoons Ce^{4+}$	Ti(III), U(IV), Fe(II), As(II), I^-, $Fe(CN)_6^{4-}$
Ag^{2+}	$Ag^+ - 1e \rightleftharpoons Ag^{2+}$	Ce(III), V(IV), As(III), $C_2O_4^{2-}$
Mn^{3+}	$Mn^{2+} - 1e \rightleftharpoons Mn^{3+}$	$C_2O_4^{2-}$, Fe(II), As(III)
Fe^{2+}	$Fe^{3+} + 1e \rightleftharpoons Fe^{2+}$	Cr(VI), Mn(VII), Ce(IV), V(V)
U^{4+}	$UO_2^{2+} + 4H^+ + 2e \rightleftharpoons U^{4+} + 2H_2O$	Cr(VI), Ce(IV)

Indicator techniques in coulometric titrimetry　In the main, the titration reactions of the coulometric titration method are identical to those of standard volumetric titrimetry. The basic difference in the methods lies in the use of an electrically generated titrant rather than a prepared standard titrant solution. The indicator techniques used in volumetric titrimetry apply therefore generally where the detection of the end point or equivalence point in coulometric titrimetry is concerned. Visual color change indicators are used extensively in coulometric titrations, particularly where the determination of reasonably large quantities of reactant is involved. The adaptability of the method to the determination of low and very low concentrations of reactant requires, however, the use of other than visual color change indicator methods since these do not lend themselves to the precise and accurate location of the equivalence point where significantly dilute reactant solutions are involved (see Chap. 6). In such cases, instrumental methods for the location of the equivalence point are usually employed. Such methods include potentiometric, conductometric, polarographic, and absorptiometric techniques, the technique applied being customarily dictated by its adaptability relative to the reaction and solution conditions surrounding the particular coulometric titration method involved. The potentiometric technique for equivalence-point location was discussed in Sec. 12.2.3. The techniques associated with the polarographic and conductometric methods will be discussed in subsequent sections of this chapter, with the absorptiometric method being discussed in Chap. 13.

Apparatus for coulometric titrations　The constant-current generators used in coulometric titrimetry must be capable of providing precise and accurate constant currents ranging in steps from about 1 to 20 mA. Generators with more extended current ranges, or with a greater number of current selections in more restricted ranges, are also used. Electronic instrumentation is available commercially which is capable of providing constant currents with a precision and accuracy of about ± 0.1 to ± 0.2 percent for the lower current values of 0.1 to 20 mA, and about ± 0.01 percent for the higher current values of 100 to 200 mA. The timer for such a constant-current generator is normally an electrically operated stop clock, driven by a synchronous motor and incorporating an electric braking device to minimize errors arising out of lag in starting and coasting after stopping. The timer should be capable of providing time intervals precise and accurate to not worse than about ± 0.1 percent, although ac line frequency fluctuations may provide for a lower order of precision and accuracy. The timer and generator are, of course, actuated and stopped through a linked switch or relay system. Under ideal circumstances, the time interval required to complete a coulometric titration, based on the reactant concentration and an appropriately selected constant-current level, should not be less than approximately 100 s. With a precision and accuracy of about ± 0.1 percent each for the constant current and the time interval, this would permit a maximum possible relative error level for the derived result of approximately and theoretically 2 ppt.

The generating electrode for a coulometric titration will have an area to some extent dependent on the reactant concentration, being about 0.5 to 1 cm² for coulometric titrimetry involving microquantities and up to 25 cm² for macroquantities. Figure 12.20 shows a typical constant-current coulometric titration arrangement as equipped for the potentiometric location of the equivalence point.

The adaptability of the coulometric titration technique to automation is of importance where the equivalence point is located by instrumental means. We shall use as an example the potentiometric method in the coulometric titration of arsenic(III) ion by triiodide ion generated anodically in a solution of potassium iodide. The potentiometer E value for the potentiometric electrode pair, and for the quantitative completion of the titration reaction

$$H_3AsO_3 + I_3^- + 5H_2O \rightleftharpoons HAsO_4^{2-} + 3I^- + 4H_3O^+$$

is determined experimentally. An electronic control device between the potentiometer and the constant-current generator and timer is incorporated where, with

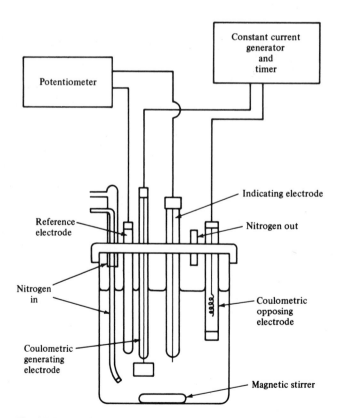

Fig. 12.20 Coulometric titration with potentiometric location of equivalence point.

this E value set on the potentiometer, the generator-timer is deactivated for this value of E for the potentiometer electrode pair, and is activated for E values of the electrode pair prior to the equivalence point. The addition to the equilibrated system of a small measured volume of a standard arsenic(III) ion solution, of the same general concentration as the unknown solution, results in the generator-timer being set into operation. When the equivalence point E value is obtained for the potentiometer electrode pair, the generator-timer is automatically shut off and a printout mechanism prints out t_s, the time interval for the standard solution reaction. The addition now of the same measured volume of an unknown solution allows the process to be repeated, and the printout then yields t_u, the time interval for the unknown solution reaction. Where the electrolyte contains an adequately high concentration of potassium iodide, and is properly buffered, and where the volume of unknown or standard solution added is small relative to the electrolyte solution volume, multiple unknowns may be successively treated through the use of an indexing table which adds the identical volume for each unknown solution. The indexing table of such an automated titrimeter can be loaded with one standard solution and 50 to 100 unknown solutions, these being automatically titrated one after the other, with the printout yielding the values t_s, t_{u_1}, t_{u_2}, etc. It is apparent that since the experimental conditions are generally the same, we have for each of the unknown solutions

$$\text{Concentration unknown} = \frac{(\text{concentration standard})(t_u)}{t_s}$$

General advantages of coulometric titrimetry   The technique is particularly capable of determining small quantities of a reactant species precisely and, all other factors being equal, accurately. For example, the passage of 1 mA of current for 100 s by no means represents even an approach to the lower limits of detection. Yet this involves only 0.1 C of electricity, or the corresponding quantity of reacting species of approximately 10^{-6} equiv. If the electrolyte solution contained 10 ml of the original unknown solution, this would represent an unknown solution of 0.0001 N. The titration of such a highly dilute solution by standard volumetric titrimeter would lead to a high relative error in the location of the equivalence-point volume and, therefore, in the determination of the amount of the unknown. The implied relative error in the titration by coulometric means would be about a part or two per thousand.

The techniques of coulometric titrimetry eliminate the need for the preparation and standardization, and subsequent periodic restandardization, of a large variety of titrant solutions. In addition to this, the method permits the use of titration reactions involving titrant substances normally unstable enough to render impossible, or at least most difficult, their preparation, storage, and use in standard volumetric titrimetry.

Example   1.0000 g of a plant pesticide was dissolved in a water solution medium and eventually diluted to 100.0 ml. It was assumed that no interfering substances would be present. 10.00 ml of this solution was added to 50 ml of an electrolyte solution involving 5% KI and buffered to a pH of 8.0. This solution was coulometrically titrated using a compartmentized platinum cathode and a platinum generating anode for the generation of the triiodide ion. A constant current of 10.00 ± 0.01 mA was forced through the solution, a starch solution was added to provide the indicator, and the end-point color change to blue was noted after 724.7 ± 0.1 s. Neglecting any indicator-blank corrections in time, calculate the percentage of sodium arsenite, $NaAsO_2$, in the pesticide.

The molecular weight of $NaAsO_2$ is 129.91, and the electron transfer for As(III) to As(V) is $n = 2$. The percentage of $NaAsO_2$ is given by

$$\% \, NaAsO_2 = \frac{724.7 \times 0.01000 \times 129.91 \times 10 \times 100}{96{,}490 \times 2 \times 1.0000} = 4.88$$

12.5 VOLTAMMETRY, POLAROGRAPHY, AND AMPEROMETRIC TITRATIONS

12.5.1 GENERAL

Voltammetry is the study of the entire course of the applied-potential-vs.-current curve as this exists for an electrode of very small area, or microelectrode, and a reference electrode of relatively large area. Where such a study involves a dropping mercury electrode as the microelectrode, this particular aspect of voltammetry is called *polarography*. Where the principles of voltammetry and polarography are applied to plot the current-vs.-volume-of-titrant relationship during a titration, in order to locate the titration equivalence-point volume, the technique is described as an *amperometric titration*.

Since the area of the microelectrode used in these applications is extremely small, any current flowing through the solution will also be extremely small—in the microampere range. Because of the resulting very low value for the iR drop through the solution, the applied potential will be virtually equal to the potential between the microelectrode and the reference electrode. The potential of the reference electrode will not be affected by such low values of current flow. During the course of increasing the applied potential between such a microelectrode and, say, an SCE reference electrode pair, it is apparent that where the applied potential is less negative than the value of E for the galvanic cell, and where the cell reaction is reversible (although this latter condition is not essential), the galvanic cell half reactions

$$R - ne \rightleftharpoons O \qquad \text{microelectrode, negative}$$

$$Hg^+ + 1e \rightleftharpoons Hg \qquad \text{SCE reference, positive}$$

will take place. Here we have O and R, the oxidized and reduced states of the ion species, providing the negative electrode reaction, and Hg^+ and Hg the oxidized and reduced states for the positive reference electrode reaction. The current flow in the external circuit in this case will be from the microelectrode to the reference electrode. Where the applied potential is exactly equal to E, zero current will flow; and this represents the point in the current-voltage relationship at which potentiometric measurements are made. Thus the value of the applied potential at this point, in conjunction with the known reference electrode potential, will yield the indicator or microelectrode potential which, in turn, through the Nernst relationship

$$E_{\text{ind}} = E^{\circ}_{f,\text{O/R}} + \frac{0.059}{n} \log \frac{[\text{O}]}{[\text{R}]} \qquad \text{at } 25°C$$

will yield the ratio value for [O]/[R]. Where the applied potential is slightly more negative than the value of E, the electrolysis cell reactions

$$O + ne \rightleftharpoons R \qquad \text{microelectrode, cathode}$$

$$Hg - 1e \rightleftharpoons Hg^+ \qquad \text{SCE reference, anode}$$

will take place. Under these three very general sets of conditions, and considering only the microelectrode, we have successively the cathodic applied potential less negative than E_{ind} $\{E^{\circ}_{f,\text{O/R}} + 0.059/n \log ([\text{O}]/[\text{R}])\}$, equal to E_{ind} and more negative than E_{ind}.

Where galvanic cell or electrolysis cell reactions are concerned, and in particular where electrolysis reactions go on, it is apparent that the current flow through the solution will be proportional to the applied potential at the cathode, providing that the ion species O and R can be immediately brought to and removed from the electrode surface. Where O and R cannot be so transferred to and from the surface, the current will depend largely on the transport situations for these ions. The ability to move the ion species O to the electrode surface—and we shall concentrate on this aspect for the moment—will depend on the migration effect as related to the electrostatic attraction of the cathode for the species O, the convection effect as related to the efficiency of solution mixing or stirring, and the diffusion effect as related to the rate at which O diffuses to the electrode surface on the basis of the difference in the concentration of O in the bulk of the solution and at the electrode surface. These effects yield, respectively, the migration current, convection current, and diffusion current contributions to the total current through the solution. In voltammetry and polarography, the objective is to have the rate of arrival of O at the electrode surface basically dependent on the diffusion effect. To this end the migration and convection effects are eliminated, or at least minimized, so as to represent negligible contributions. The convection effect is largely removed by keeping the electrodes static, by not stirring the solution in any way, and by maintaining an electrolysis time of not more than a minute or two. The migration effect is minimized by providing in the solution, for cathodic microelectrode reduction processes, a concentration of a positively charged ion that will be attracted to the cathode but, in the solution medium involved, not discharged or reduced. The concentration of the salt providing such an ion, called the *supporting electrolyte*, will be high relative to the concentration of O in the bulk of the solution. An example would be the use of the supporting electrolyte KNO_3, where K^+ ions accumulate at the cathodic microelectrode surface but are not discharged in aqueous solution. This accumulation of K^+ ions at the interface in amounts such that their positive charge is approximately equal to the negative charge at the cathode reduces to practical insignificance the electrostatic attraction of the microelectrode for the species O. This situation is well exemplified by the fact that the transference number $t_{Cd^{2+}}$ for a 10^{-3} M solution of $Cd(NO_3)_2$ is about 0.50, the value decreasing to 0.009 in the presence of 0.1 M KNO_3 and to 0.0009 in 1 M KNO_3.

In voltammetric and polarographic studies, where these two conditions are maintained, the species O will reach the electrode surface almost exclusively by a process of mass transfer involving diffusion over the concentration difference between [O] at the electrode surface and [O] in the bulk of the solution. In the explanation to follow, we shall consider only a cathodic microelectrode reaction in electrolysis of $O + ne \rightleftharpoons R$, defining [O] in the bulk of the solution as C_0 and [O] at the microelectrode surface as $C_0(x = 0)$, where x is the distance into the solution from the electrode surface. For this discussion we shall consider that the solution is aqueous, that the convection and migration effects have been by appropriate means reduced to insignificance, that the microelectrode has a plane surface and that any residual or background current can be, for the moment, ignored.

Where the applied potential at the cathode attains a value slightly more negative than the equilibrium potential for the original solution, the reaction $O + ne \rightleftharpoons R$ will take place at the electrode surface, with the value of $C_0(x = 0)$ becoming less than C_0. Under these conditions, the current i through the solution will be controlled by the rate of diffusion of O to the electrode surface. The rate of diffusion of O will be proportional to the concentration gradient $\Delta C_0/\Delta x$, and we have

$$\text{Rate of diffusion of O} = D_0 \frac{\Delta C_0}{\Delta x} \tag{279}$$

where rate of diffusion of O = moles of O diffusing to the electrode surface per cm² per s

D_0 = the proportionality constant or *diffusion coefficient*, cm²/s

$\Delta C_0 = C_0 - C_0(x = 0)$, mol/cm³

Δx = distance from electrode surface to point x in solution where [O] equals C_0; cm

The value of D_0 would be, of course, dependent on the nature of the reaction $O + ne \rightleftharpoons R$. The value of Δx will increase as the time of electrolysis increases, since the front of the depleted [O] zone will move further into the solution with time. As the cathodic applied potential becomes more negative, the value of $C_0(x = 0)$ decreases further, and with a value of applied potential sufficiently negative to ensure immediate reduction of O upon arrival at the electrode surface, the value of $C_0(x = 0)$ becomes much less than C_0 and approaches a zero value. The electrode is now completely concentration polarized. Under these conditions, the value of $\Delta C_0 = C_0 - C_0(x = 0)$ becomes dependent only on C_0. The value of Δx will, as mentioned previously, increase with increasing time of electrolysis. Obviously the rate of diffusion will be directly proportional to C_0 and inversely proportional to Δx. In consequence, the current through the solution will be directly governed by the rate of diffusion, and will be the diffusion current i_d. It can be shown that for the conditions stipulated, the number of moles of O arriving at the electrode surface per square centimeter of surface area and per second will be given by N in the expression

$$N = C_0 \sqrt{\frac{D_0}{\pi t}} \tag{280}$$

where C_0 is in moles per cubic centimeter, D_0 is in centimeters squared per second, π has the conventional value, and t is in seconds and represents the time from the start of the electrolysis reaction. The current through the solution will be the diffusion current for the system. It will be given by i_d in the expression

$$i_d = nFAC_0 \sqrt{\frac{D_0}{\pi t}} \tag{281}$$

where i_d is the diffusion current in amperes, n is the electron transfer for the reaction $O + ne \rightleftharpoons R$, F is the faraday at 96,490 C/equiv, A is the microelectrode surface area in square centimeters, and C_0, D_0, and t have the terms and units as previously outlined. Where the value of C_0 is expressed in millimoles per liter, the value of i_d will be given in microamperes.

We have noted that i_d decreases with increasing t. If i_d could always be measured at a constant value of t, it would be directly proportional to C_0. A plot of i_d vs. C_0, for solutions of variable but known C_0 secured under these conditions, would yield in general a straight-line relationship that could be used, for solutions of identical background but unknown C_0, to determine the value of C_0 from the measured value of i_d. We can in no simple way, however, guarantee that the measurement of i_d will be made at precisely the same value of t for successive systems, so that the situation developed so far, relative to a plane-surface microelectrode, would not be adaptable to quantitative analytical purposes.

It should be noted that Eq. (281) implies that for $t = 0$, the start of the electrolysis, the value of i_d would be infinite. This results from the fact that we have not taken the solution resistance into consideration. At the start of electrolysis, the current will be largely governed by the solution resistance, although this factor becomes negligible immediately after the start of electrolysis.

12.5.2 THE DROPPING MERCURY ELECTRODE, OR DME

General The dropping mercury electrode, as a microelectrode, permits to a significant extent relief of the time dependency of i_d noted for the plane-surface microelectrode. Where this electrode is used, the voltammetric technique is called polarography, and the same *general* conditions relative to migration and other effects apply as applied in the discussion just completed. The fundamental principles of polarography in its application to analytical chemistry were developed some considerable time ago by Heyrovsky.* Essentially, the dme consists of a reservoir of pure mercury connected to a glass capillary tube about 10 cm in length, 7 mm in external diameter, and 0.03 to 0.04 mm in internal radius, The capillary tube is fixed vertically in the solution, and the head of mercury (the distance from the level in the reservoir to the capillary tip) is such as to yield a drop life of from 2.5 to 6 s. Drop life is the time in seconds from the start of drop growth to the parting of the drop from the capillary tip. The drop-life range indicated will normally require a head of mercury, H, of 40 to 70 cm. The drop size at the moment of dislodgement will generally be about 0.5 mm radius.

The dme differs from the plane-surface electrode in that (1) the surface is spherical, (2) the surface area increases during drop life, and (3) the surface of the drop moves into the solution during drop life. It can be shown that where the drop life is small (between 2.5 and 6 s) and where the drop radius at dislodgement is about 0.5 mm, the rate of diffusion of O to the spherical drop surface will be approximately but adequately given by Eq. (280). The increasing surface area during drop life can be shown to be given by

$$A = 4\pi\left(\frac{3}{4\pi d}\right)^{2/3} m^{2/3} t^{2/3} \tag{282}$$

where A = drop surface area, cm² in time t

d = density of mercury, g/cm³

m = rate of flow of mercury into the drop, g/s

t = time during drop life, s

Finally, it can be shown that the effect of the drop surface moving into the solution can be compensated for by multiplication of the rate of diffusion by the factor $\sqrt{\frac{7}{3}}$. A combination of these factors leads to the dme expression for i_d as

$$i_d = nF4\pi\left(\frac{3}{4\pi d}\right)^{2/3} m^{2/3} t^{2/3} \sqrt{\frac{7}{3}} \sqrt{\frac{D_0}{\pi t}} C_0 \qquad \text{at 25°C}$$

$$= 708 nm^{2/3} t^{1/6} D_0^{1/2} C_0 \tag{283}$$

where i_d is given in microamperes when C_0 is expressed in millimoles per liter and the rate of flow m is given in milligrams per second. The value of 708 holds for 25°C, and includes the density of mercury as 13.53 g/cm³ at 25°C. The faraday has the usual value, and D_0 and t are expressed respectively in square centimeters per second and seconds. This expression is due to Ilkovič.† The important features involved in this interpretation of the diffusion current at the dme are those relative to the direct relationship between i_d and C_0. At an applied cathodic potential permitting immediate reduction of O upon arrival at the electrode sur-

*J. Heyrovsky, *Chem. listy*, **16**:256(1922).
†D. Ilkovič, *Collection Czech. Chem. Commun.*, **6**:498(1934).

face, and in the very early stages of drop life, the relative rate of increase of the electrode surface area is very high and the movement of the surface into the solution rapid enough to prevent any worthwhile concentration gradient zone from being established at the electrode surface. The current increases rapidly here because of these conditions. As the drop grows larger, the relative rate of surface area increase and movement into the solution decreases and, at a short but appreciable period before drop dislodgement, a well-defined concentration gradient zone is established where $C_0(x = 0) <<< C_0$ and approximately equal to zero. This establishes a value of current which is the diffusion current and dependent only on the value of C_0. This diffusion current value, i_d, now persists virtually unchanged until drop dislodgement, since the effect of the small further increase in t involves only its one-sixth power. Subsequent to dislodgement, the concentration gradient zone formed around the drop is destroyed by the inrushing solution and the succeeding drop forms and dislodges, attaining the same value of i_d as previously. Since the amount of O reduced per drop will be negligible, and the stirring action of the drop fall insignificant, repetitive drops will provide identical values of i_d at drop dislodgement. The theoretical repetitive pattern is shown in Fig. 12.21. The current grows rapidly in the early stages of drop life. The pla-

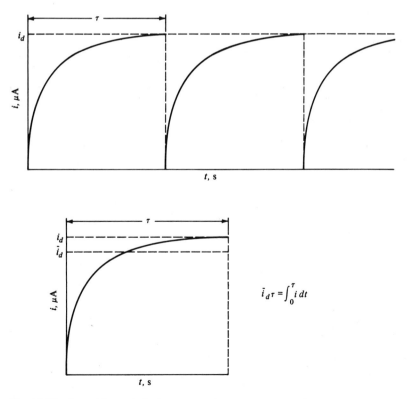

Fig. 12.21   Repetitive and single mercury drops: current vs. time.

teau sections of current represent the i_d value as based on C_0. Lower or higher values of C_0 will yield correspondingly lower or higher values of i_d. Figure 12.21 also shows the development of the current i with time for a single drop. The value of τ is the drop life in seconds. Note that the plateau current i_d is established appreciably before drop dislodgement. The area under the i-vs.-t curve to the value of $t = \tau$ yields the quantity of electricity consumed in the reduction of O to R over this time interval. The very rapid increase in i taking place during the early stages of drop life can not be adequately followed, even by very rapid undamped electronic current sensing devices and recorders, although the current changes in the latter stages of growth can be so followed. In point of fact, *undamped* electronic current sensing devices and recorders will provide the maximum diffusion current value i_d. Where instruments involving only galvanometric current sensing devices, or *damped* electronic devices and recorders, are used, the average of the minimum and maximum oscillations of the galvanometer or recorder with the growth and decay of current for each drop will provide the time-average diffusion current $\bar{i}_d$ over the drop life. This time-average current can be related to Eq. (283) on the basis of

$$\bar{i}_d\tau = \int_0^\tau i\, dt \quad = \int_0^\tau 708 nm^{2/3}t^{1/6}D_0^{1/2}C_0\, dt$$

$$\bar{i}_d = 607 nm^{2/3}\tau^{1/6}D_0^{1/2}C_0 \qquad \text{at } 25°C \qquad (284)$$

so that the time-average current provided by galvanometric and damped instrumentation is also proportional to C_0. Note that $\bar{i}_d = \frac{6}{7}i_d$ $(t = \tau)$.

Factors influencing the diffusion current other than C_0 It is apparent that changes in the values of m and τ can result where different capillaries are used. Such changes must be properly compensated for, when necessary, where a capillary is replaced. This will require the experimental determination of the new values of m and τ, and the subsequent comparison of i_d values for the new capillary with those for the replaced unit.

The head of mercury has an influence on the diffusion current value. All other factors being constant, changes in the head of mercury, H, will affect both m and τ. The result of these influences is to provide a dependency of $\bar{i}_d$ on H where $\bar{i}_d$ is proportional to $\sqrt{H}$. This effect can be reduced to a negligible level by the use of a mercury reservoir of large diameter. If, for example, H has a value of 50 cm, a variation of ± 1 mm would result in a relative error in $\bar{i}_d$ of ± 0.1 percent. Since each drop at dislodgement has a radius of approximately 0.5 mm, the drop volume at this point is about 5.2×10^{-4} cm³. With a reservoir of about 7-cm diameter, a change of 1 mm in the value of H would require the formation of about 7400 drops. At a drop life of, say, 4 s, a continuous operation of about 8 h is implied.

Any variation in the value of the applied potential at the dme affects the surface tension of mercury and, in turn, the interfacial tension force between mercury and the solution. This latter force is matched by the effect of gravity on drop mass at the time of dislodgement. Any change in the interfacial tension force therefore results in a change in drop life. With increasing negative charge to the dme, the drop life increases to a maximum at the electrocapillary maximum potential for the solution involved, decreasing thereafter as the applied potential becomes more negative.* The value of m is much less affected, although it decreases to a minimum at the electrocapillary maximum potential and subsequently increases as the applied potential becomes increasingly negative. It is important in highly precise and accurate work to determine the value of τ and $\bar{\imath}_d$ at exactly the same applied potential. However, since the contribution of τ in Eq. (284) involves only the one-sixth power, the effect on the value of $\bar{\imath}_d$ of applied potential induced variations in τ can be ignored in routine work.

The temperature of the system at the time of measurement of i_d is of primary importance, since changes in temperature result in changes relative to all the components of Eq. (283) or (284), with the exception of n. These changes are based on 607 (density of mercury), m(viscosity of mercury), τ(interfacial tension force through the surface tension of mercury), D_O(energy of diffusing particles), and C_O (solution volume). The variation of D_O with temperature is by far the most significant of these temperature dependencies; it can be considered as the prime source of the temperature dependency of $\bar{\imath}_d$. It can be shown that around 15 to 50°C, the relative error introduced into the value of $\bar{\imath}_d$ by temperature variation is about 1 to 2 percent per degree Celsius.† Thus, temperature control of the system is essential in polarographic work. This can be achieved by immersion of the polarographic cell in a constant-temperature water bath or by the use of water-jacketed cells which allow the circulation of water under controlled temperature. By such means, solution temperature control is easily achieved at a level of ± 0.1°C, with a corresponding relative error in $\bar{\imath}_d$ of about ± 0.1 percent.

12.5.3 CURRENT–POTENTIAL CURVES: DROPPING MERCURY ELECTRODE

General Apart from its applicability to analytical chemistry, which will be outlined in a subsequent section, the polarographic technique has provided a means for the investigation of reaction characteristics. A further discussion of current-potential curves for reversible reaction systems will serve to illustrate the point.

Reversible reactions: product R is a soluble species Consider the reversible electrode reaction $O + ne \rightleftharpoons R$, where the product R is a species soluble in the solution (that is, $Cr^{3+} + 1e \rightleftharpoons Cr^{2+}$). The Nernst relationship is

$$E_{O/R} = E_{O/R}^{\circ} + \frac{0.059}{n}\log\frac{C_O}{C_r}\,(x = 0) \qquad \text{at } 25°C \qquad (285)$$

*L. Meites, "Polarographic Techniques," 2d ed., p. 135, Interscience, New York, 1965.
†I. M. Kolthoff and J. J. Lingane, "Polarography," 2d ed., pp. 52–53 and 90–93, Interscience, New York, 1952.

where $E_{O/R}$ is the cathodic applied potential and C_O and C_R are the concentrations [O] and [R] at the electrode surface. The Ilkovič relationship yields*

$$\bar{\imath} = 607 nm^{2/3} \tau^{1/6} D_O^{1/2} \{C_O - C_O(x = 0)\} \qquad \text{at } 25°C \qquad (286)$$

where $C_O(x = 0)$ is slightly less than C_O at an applied potential slightly more negative than the equilibrium potential for the original solution. This leads to

$$\bar{\imath} = \bar{\imath}_d - 607 nm^{2/3} \tau^{1/6} D_O^{1/2} C_O(x = 0) \qquad \text{at } 25°C \qquad (287)$$

Since $C_O - C_O(x = 0) = C_R(x = 0)$, we have

$$\bar{\imath} = 607 nm^{2/3} \tau^{1/6} D_R^{1/2} C_R(x = 0) \qquad \text{at } 25°C \qquad (288)$$

The ratio of C_O/C_R at $(x = 0)$ is now given by Eqs. (287) and (288) as

$$\frac{C_O}{C_R}(x = 0) = \frac{D_R^{1/2}}{D_O^{1/2}} \frac{\bar{\imath}_d - \bar{\imath}}{\bar{\imath}}$$

from which, through Eq. (285), we have

$$E_{O/R} = E_{O/R}^\circ + \frac{0.059}{n} \log \frac{D_R^{1/2}}{D_O^{1/2}} \frac{\bar{\imath}_d - \bar{\imath}}{\bar{\imath}}$$

$$= E_{1/2\ O/R} + \frac{0.059}{n} \log \frac{\bar{\imath}_d - \bar{\imath}}{\bar{\imath}} \qquad \text{at } 25°C \qquad (289)$$

where $E_{1/2\ O/R} = E_{1/2} = E_{O/R}^\circ + 0.059/n \log (D_R^{1/2}/D_O^{1/2})$. A polarographic curve may be obtained by plotting $\bar{\imath}$ vs. $E_{O/R}$, but it may be just as fully represented by a plot of $\bar{\imath}/\bar{\imath}_d$ vs. $E_{O/R} - E_{1/2}$. This latter form is somewhat more convenient for this development. Figure 12.22 shows such a plot. The form is typically that of a *polarographic wave*, and it should be noted that, for $n = 1$, we have the following.

1. When $E_{O/R}$ is about 0.1 V less negative than $E_{1/2}$, $\bar{\imath}$ is about $0.02\bar{\imath}_d$ or much less than $\bar{\imath}_d$.
2. When $E_{O/R}$ is about 0.1 V more negative than $E_{1/2}$, $\bar{\imath}$ is about $0.98\bar{\imath}_d$ or approximately equal to $\bar{\imath}_d$.
3. When $E_{O/R}$ is identical to $E_{1/2}$, $\bar{\imath}$ is exactly one-half $\bar{\imath}_d$.

The plot of $\bar{\imath}$ vs. $E_{O/R}$ yields the same wave form, the plateau of the wave then corresponding to the condition $\bar{\imath} = \bar{\imath}_d$. Condition 3 shows that $E_{1/2}$ is identified as the value of $E_{O/R}$ at which $\bar{\imath} = \frac{1}{2}\bar{\imath}_d$, or at the half-wave portion of the curve. The value of $E_{1/2}$ is therefore called the *half-wave potential* for the system and is independent of the concentration for most reversible reactions. For such reactions, the value of $E_{1/2}$ will be constant for a given reaction in a specific solution medium, and can serve to identify the species O. Note from Fig. 12.22 that as the value of n increases, the wave form becomes steeper and more clearly defined between the limits of about ± 0.1 V around $E_{1/2}$. For reversible reactions in general, a plot of

*J. Heyrovsky and D. Ilkovič, *Collection Czech. Chem. Commun.*, 7:198(1935).

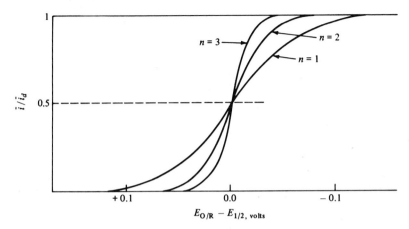

Fig. 12.22   The polarographic curve or wave.

$E_{O/R}$, the applied cathodic potential, vs. log $\{(\bar{i}_d - \bar{i})/\bar{i}\}$ will yield a straight line from which $E_{1/2}$ may be determined as the intercept with the $E_{O/R}$ axis for log $\{(\bar{i}_d - \bar{i})/\bar{i}\} = 0$, with n being determined from the slope of the line.

Reversible reactions: hydrogen-ion participates in the electrode reaction   The general reaction here at the dme is $O + mH^+ + ne \rightleftharpoons R + wH_2O$, where the degree of reaction completion depends on $[H^+]$ and where R is a solution-soluble species.   Here we have

$$E_{O/R} = E_{O/R}^\circ + \frac{0.059}{n} \log \frac{[H^+]^m \, C_O}{C_R}(x = 0)$$

$$= E_{O/R}^\circ - \frac{0.059m}{n} pH + \frac{0.059}{n} \log \frac{C_O}{C_R}(x = 0)$$

$$= E_{1/2} + \frac{0.059}{n} \log \frac{(\bar{i}_d - \bar{i})}{\bar{i}} \qquad \qquad \text{at } 25°C \qquad (290)$$

where

$$E_{1/2} = E_{O/R}^\circ + \frac{0.059}{n} \log \frac{D_R^{1/2}}{D_O^{1/2}} - \frac{0.059m}{n} pH \qquad \text{at } 25°C \qquad (291)$$

Where the reaction is reversible, Eq. (290) indicates that $E_{1/2}$ for a specific pH may be determined in the usual way by plotting $E_{O/R}$ vs. log $\{(\bar{i}_d - \bar{i})/\bar{i}\}$, with n being determined from the straight-line slope. $E_{1/2}$ may be determined for different solution values of pH.   Equation (291) then indicates that a plot of $E_{1/2}$ vs. pH will yield a straight line from which, using the determined value of n, the value of m may be evaluated from the slope.

Reversible reactions: metal-ion reduction with amalgam formation   Here O is represented by the metal ion M^{n+} and R by the metal M.   The metal is capable of amalgam formation with mercury, and when the reaction $M^{n+} + ne \rightleftharpoons M$ takes place at the dme surface, the metal diffuses into the drop, D_R being the diffusion coefficient for this process. For reversible reactions of this type, Eq. (289) again in general holds, and we have

$$E_{O/R} = E_{1/2,a} + \frac{0.059}{n} \log \frac{(\bar{\imath}_d - \bar{\imath})}{\bar{\imath}} \qquad \text{at 25°C} \tag{292}$$

where

$$E_{1/2,a} = E^{\circ}_{a,M^{n+}/M} + \frac{0.059}{n} \log \frac{D_M^{1/2}}{(D_{M^{n+}})^{1/2}} \qquad \text{at 25°C} \tag{293}$$

The value of E°_a may be significantly different from E°, with $E_{1/2,a}$ similarly different from $E_{1/2}$. As before, $E_{1/2,a}$ and n may be determined generally for reversible reactions by plotting $E_{O/R}$ vs. $\log \{(\bar{\imath}_d - \bar{\imath})/\bar{\imath}\}$.

Reversible reactions: metal complex ion reduction with amalgam formation   Here we have the metal complex ion $M(L)_x^{(n-xp)}$ of the ligand L^{p-} dissociating:

$$M(L)_x^{(n-xp)} \rightleftharpoons M^{n+} + xL^{p-}$$

with M^{n+} being reduced to M at the dme with amalgam formation. $M(L)_x^{(n-xp)}$ must then diffuse to the electrode surface. We have

$$\frac{[M^{n+}][L^{p-}]^x}{[M(L)_x^{(n-xp)}]} = K_{inst} \tag{294}$$

for the dissociation of the metal complex ion

$$E_{O/R} = E^{\circ}_{a,O/R} + \frac{0.059}{n} \log [M^{n+}] \qquad \text{at 25°C} \tag{295}$$

for the reduction of the metal ion, where O and R represent M^{n+} and M and

$$E_{O/R} = E^{\circ}_{c,O/R} + \frac{0.059}{n} \log [M(L)_x^{(n-xp)}] \qquad \text{at 25°C} \tag{296}$$

for the reduction of the metal complex ion, where O and R represent $M(L)_x^{(n-xp)}$ and M. The value of E°_c is the standard reduction potential for the reduction of the metal complex ion. Lingane and von Stackelberg* have shown independently the relationship between E°_a and E°_c. From Eqs. (294), (295), and (296) we have

$$E^{\circ}_a + \frac{0.059}{n} \log [M^{n+}] = E^{\circ}_c + \frac{0.059}{n} \log \frac{[M^{n+}][L^{p-}]^x}{K_{inst}}$$

*J. J. Lingane, *Chem. Rev.*, **29**:1(1941); M. von Stackelberg, *Z. Elektrochem.*, **45**:466(1939).

so that

$$E_c^\circ = E_a^\circ - \frac{0.059}{n} \log \frac{[L^{p-}]^x}{K_{inst}} \qquad \text{at } 25°C \qquad (297)$$

The values of $E_{1/2,a}$ and $E_{1/2,c}$ will be given by

$$E_{1/2,a} = E_a^\circ + \frac{0.059}{n} \log \frac{(D_M)^{1/2}}{(D_{M^{n+}})^{1/2}} \qquad \text{at } 25°C \qquad (298)$$

$$E_{1/2,c} = E_c^\circ + \frac{0.059}{n} \log \frac{(D_M)^{1/2}}{(D_{M(L)_x^{(n-xp)}})^{1/2}} \qquad \text{at } 25°C \qquad (299)$$

and, of course, Eq. (289) will hold in general for reversible reactions. From Eq. (297) we now have

$$E_{1/2,c} = E_a^\circ - \frac{0.059}{n} \log \frac{[L^{p-}]^x}{K_{inst}} + \frac{0.059}{n} \log \frac{(D_M)^{1/2}}{(D_{M(L)_x^{(n-xp)}})^{1/2}}$$

$$= E_{1/2,a} - \frac{0.059}{n} \log \frac{[L^{p-}]^x}{K_{inst}} + \frac{0.059}{n} \log \frac{(D_{M^{n+}})^{1/2}}{(D_{M(L)_x^{(n-xp)}})^{1/2}}$$

$$= E_{1/2,a} - \frac{0.059x}{n} \log [L^{p-}] + \frac{0.059}{n} \log K_{inst} + \frac{0.059}{n} \log \frac{(D_{M^{n+}})^{1/2}}{(D_{M(L)_x^{(n-xp)}})^{1/2}}$$

$$\text{at } 25°C \qquad (300)$$

Note that the magnitude of $E_{1/2,c}$ depends on the magnitude of K_{inst} and the concentration of the ligand. Where the metal complex ion is very stable, the value of $E_{1/2,c}$ will be very much more negative than $E_{1/2,a}$, a difference of 1 to 2 V more negative being common for stable complexes. For reversible reactions, $E_{1/2,a}$ and n may be determined in the usual manner from a plot of $E_{O/R}$ vs. log $\{(\bar{i}_d - \bar{i})/\bar{i}\}$ for the system $M^{n+} + ne \rightleftharpoons M$ in the absence of the ligand. Again for reversible reactions involving the metal complex ion, $M(L)_x^{(n-xp)} + ne \rightleftharpoons M + xL^{p-}$, we may determine the value of $E_{1/2,c}$ for each of several solutions of the metal complex ion with different excess ligand concentrations by plotting the applied potential $E_{O/R}$ vs. log $\{(\bar{i}_d - \bar{i})/\bar{i}\}$ for each of these solutions. If we now plot the values of $E_{1/2,c}$ against log $[L^{p-}]$, we shall for reversible reactions obtain a straight line. Using the determined value of n, the slope of this line will permit evaluation of x, the coordination number for the metal complex ion. For a fixed value of $[L^{p-}]$, using the determined values of n, x, and $E_{1/2,a}$, the determined value of $E_{1/2,c}$ for the value of $[L^{p-}]$ involved, and assuming that $(D_{M^{n+}})^{1/2}/(D_{M(L)_x^{(n-xp)}})^{1/2}$ will approximate unity, Eq. (300) can be applied to determine the value of K_{inst} for the metal complex ion.

It is important to note that where the values of $E_{1/2,1}$ and $E_{1/2,2}$ for two uncomplexed metal ion species M_1^{n+} and M_2^{n+} may be too close for polarographic separation and their subsequent identification, induced complexation reactions in which one species is not complexed or yields a much less stable complex than the other can result in $E_{1/2,c,1}$ and $E_{1/2,c,2}$ values sufficiently separated to permit identification.

Irreversible reactions   In the majority of cases involving reversible reactions, the value of $E_{1/2}$ is independent of the concentration of O. With irreversible reactions, $E_{1/2}$ is dependent on [O] and, as determined from the polarographic curve, is more negative than would have been anticipated if the reaction had been reversible. The polarographic wave itself is less clearly defined, as can be noted from Fig. 12.23. The value of $\bar{\imath}_d$ in such cases remains dependent on the value of [O], however, so that irreversible reactions are also adaptable to the analytical applications of polarography.

12.5.4   THE RESIDUAL CURRENT

General   In our discussions so far we have ignored the fact that in the absence of any species O, the current through the solution with increasingly negative applied potentials should remain at zero, until hydrogen ion is discharged, but is frequently of very significant magnitude. Figure 12.24 shows an idealized pair of polarograms, one of a solution of the metal ion M^{n+} and the other of the same solution in the absence of the metal ion. It should be noted that a *residual current* exists in the absence of the metal ion, and that this residual current would yield an average value $\bar{\imath}_r$ contributing to the magnitude of the limiting current as measured at any specific but appropriate applied potential. The residual current i_r originates

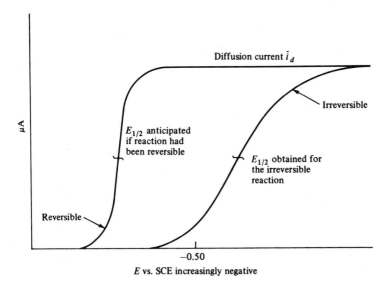

Fig. 12.23   Reversible opposed to irreversible polarographic waves.

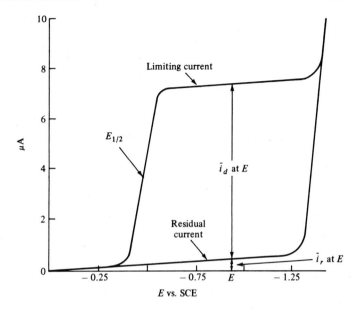

Fig. 12.24 Idealized limiting, diffusion, and residual current relationships.

from two sources. One of these, the *faradian current* i_f, involves the reduction of traces of impurities, including dissolved oxygen, in the solution. The other, the *charging* or *condenser current* i_c, involves the so-called electrical double layer formed at the dme surface-solution interface.

Residual current: faradian current component   Impurities in solution contributing to the faradian current may originate with heavy-metal and other impurities in the distilled water and the supporting electrolyte salt. The use of distilled water and supporting electrolyte salts of the highest available purity is essential. In the majority of cases, however, the main contributor to the faradian current component of the residual current is dissolved oxygen in the solution. Here we have the reduction of oxygen at the dme to hydrogen peroxide, with a further reduction possibility of hydrogen peroxide to water. The reactions involved are

$$O_2 + 2H^+ + 2e \rightleftharpoons H_2O_2$$
$$H_2O_2 + 2H^+ + 2e \rightleftharpoons 2H_2O$$

the former with an $E_{1/2}$ of about -0.1 V vs. SCE, and the latter with an $E_{1/2}$ of about -0.9 V vs. SCE. Apart from errors introduced by the addition of the oxygen and hydrogen peroxide reactions currents to the diffusion currents determined for analytical purposes in the quantitative sense, the influence of the polarographic waves for oxygen and hydrogen peroxide in rendering less clear the definition of the polarographic waves for other reducible species with $E_{1/2}$ values between zero and -1 V vs. SCE is obvious. The removal of dissolved oxygen by bubbling an inert gas, commonly pure nitrogen, through the solution for several minutes prior to polarographic examination, and over the solution during such examination, is required in order to prevent the interference of this substance in qualitative and quantitative analytical polarography.

Residual current: condenser or charging current component   For a given supporting electrolyte, the electrocapillary maximum potential is that applied potential at the dme which just counterbalances the tendency of the drop to adsorb at its surface the anions of the electrolyte. The electrocapillary maximum varies with the nature of the the supporting electrolyte and is, for example, -0.46 V vs. SCE for a 0.1 M KCl solution. At applied potentials less negative than this, electrons are repelled from the drop surface by adsorbed Cl^- ions, and as each drop forms and falls, electron flow away from the dme is established (negative condenser current flow). At applied potentials more negative than -0.46 V vs. SCE, electrons move to the dme surface, Cl^- ions at the interface are repelled, and the value of $[K^+]$ at the interface will be slightly higher than $[Cl^-]$. This is the electrical double layer, and as each drop forms and falls, removing electrons, an electron flow to the dme is established (positive condenser current). When the applied potential is equal to -0.46 V vs. SCE, the condenser current is zero. The magnitude of the value of the average condenser current during drop life is the limiting factor relative to the concentration that can be detected by normal polarography. Where $\bar{i}_c$ and $\bar{i}_d$ are similar, the concentration level for this $\bar{i}_d$ value then represents the lower level of detection for the method. It can be shown that this lower level of detection, for *normal* polarography (without $\bar{i}_c$ compensation, etc.) is about 10^{-5} to $10^{-6} M$.

12.5.5   CURRENT MAXIMA AND MAXIMUM SUPPRESSORS

A continuous curve of $\bar{i}$ vs. applied potential may often show a continual increase in $\bar{i}$ to a high level, with rapid return to a plateau or $\bar{i}_d$ value. Such a curve is shown in Fig. 12.25. Although the theoretical background of such maxima is not yet thoroughly understood, the effect is known to be related to the streaming of the solution past the drop surface. Since such maxima can obviously cause considerable difficulty relative to the identification of $E_{1/2}$ values and the evaluation of $[O]$ from measured $\bar{i}_d$ values, the maxima effect is eliminated or minimized by the addition to the solution of small quantities of a maximum suppressor of reasonably high molecular weight, such as gelatin, gum arabic, methyl red and other dyes, nonionic detergents such as Triton X-100, camphor, etc. These materials, by adsorption at the dme surface, retard the streaming motion of the solution past the drop surface.

12.5.6   ANODIC AND ANODIC–CATHODIC CURRENT–POTENTIAL WAVES

It is possible to obtain anodic polarographic waves where the dme functions as an anode and the electrode reaction involves $R - ne \rightleftharpoons O$. The limitation in this area

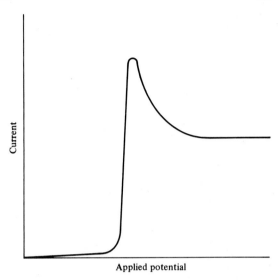

Fig. 12.25   Typical maxima effect.

is the fact that at applied potential values more positive than about 0.35 V vs. SCE, the oxidation of mercury takes place. It is also possible to obtain anodic-cathodic polarographic waves where the electrode reaction (and of course the current flow) is reversed during the applied potential increase to more negative values. Anodic-cathodic waves may also be obtained where an anodic wave for species R(1) may be followed by a cathodic wave for species O(2). Figure 12.26 shows the polarogram for a solution containing Fe^{2+} and Cd^{2+}, where the anodic wave for $Fe^{2+} - 1e \rightleftharpoons Fe^{3+}$ is followed by the cathodic wave for $Cd^{2+} + 2e \rightleftharpoons Cd(Hg)$.

12.5.7 ANALYTICAL APPLICATIONS OF POLAROGRAPHY

General   The polarographic wave for a solution containing a single reducible species may be analyzed to determine the half-wave potential, and with a consideration of the supporting electrolyte involved, the species may be identified. Mixtures of reducible species may also be so analyzed, and the various species identified in the same way, providing that the half-wave potentials are separated by about 0.2 to 0.3 V, even where only one of the adjacent half-waves represents a reaction with $n = 1$, and by 0.1 to 0.2 V where the adjacent species half-waves represent reactions each with $n = 2$. Although the qualitative aspect of polarography should not be overlooked, by far the most important analytical application of the technique is that having to do with the quantitative estimation of a species or the components of a mixture of suitable species. The Ilkovič relationship is the basis of the quantitative aspect since, all other components of the expression being accounted for, the value of $\bar{i}_d$ will be directly proportional to the

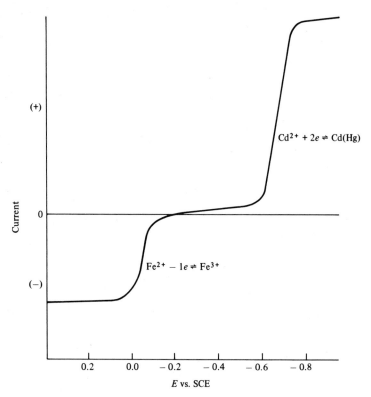

Fig. 12.26 Polarographic waves (anodic-cathodic) for a solution of $[Fe^{2+}]$ $= 10^{-3}\,M$ and $[Cd^{2+}] = 10^{-3}\,M$ in 0.25 M citrate at pH = 4.

concentration of the reducible (cathodic) or oxidizable (anodic) species. In order to evaluate concentration by the polarographic technique, it is necessary to determine the plateau or diffusion current as the time-average diffusion current $\bar{\imath}_d$.

Measurement of diffusion current   A typical polarogram is shown in Fig. 12.27. The oscillations represent the growth and decay of current during drop life and have been instrumentally damped to reduce their magnitude, this situation generally applying whether a recorded polarogram is obtained or readings are taken from a galvanometer for a specific value of applied potential. The average of the maximum and minimum oscillations for a given applied potential represent a very close approximation to the time-average diffusion current $\bar{\imath}_d$ for this potential. Such an average is easily obtained, whether for the oscillations at a specific applied potential or over the entire plateau of a recorded wave. Such a measured value of $\bar{\imath}_d$ will actually represent the time-average limiting current, since it will include the value of $\bar{\imath}_r$; and where $\bar{\imath}_r$ is significant relative to the total current, this latter should be corrected for the value of $\bar{\imath}_r$. The most accurate method of correction is to produce two polarograms, one in the presence of the single reducible

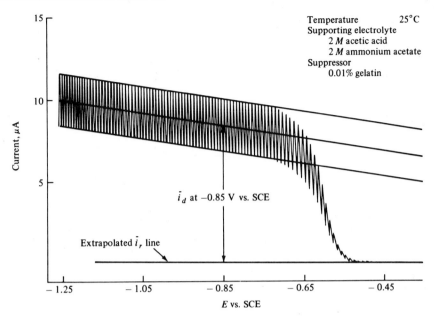

Fig. 12.27 Polarographic current-potential curve: Cadmium(II), $10^{-3} M$.

species and the other, with the identical solution background, in its absence. The limiting current at any appropriate potential may then be determined and corrected for the $\bar{i}_r$ value at this potential and the value of $\bar{i}_d$ thus determined. Figure 12.24 shows the general technique.

The foregoing method is time-consuming in that two complete polarograms must be taken. Where the value of $\bar{i}_r$ is sufficiently small relative to the total current, a situation which occurs where the concentration of the species is relatively high (around $10^{-3} M$), it suffices to determine a value of current at a potential close to but before the start of the wave for the species. The value of the limiting current is then determined for an applied potential corresponding to a well-defined plateau region, and this value, corrected for the first value, will closely approximate the required $\bar{i}_d$ value. For situations where $\bar{i}_d$ for different solution concentrations of the species involved must be compared, the current values must be determined for the same applied potential values. This technique will apply generally with single reducible species solutions. For a mixture of species, where the half-wave potentials are adequately separated, a somewhat similar method can be applied. Here a single polarogram is taken. That portion of the current-potential curve before the first reducible species wave is extrapolated as a continuous residual current line and, at each appropriate applied potential, the limiting current determined, corrected for the associated $\bar{i}_r$ and, where necessary, for the $\bar{i}_d$ values for the previous waves, will yield the required $\bar{i}_d$. Figure 12.28 indicates the general technique. It should be pointed out that several sources of error may exist

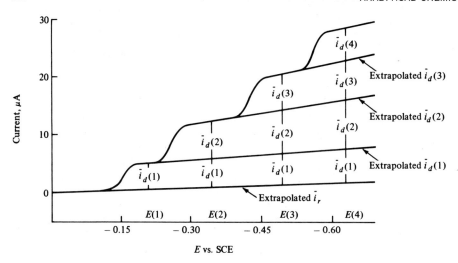

Fig. 12.28 Typical multiple-species polarogram: interpretation of $\bar{i}_d$ values.

here (changes in $\bar{i}_c$ with applied potential, etc.), but the method is reasonably accurate where $\bar{i}_r$ is always small compared with the limiting current.

Methods of inorganic quantitative analysis   Most metal ions can be reduced at the dme, either to the metallic state with amalgam formation or to a lower soluble oxidation state, by the proper choice of applied potential and supporting electrolyte.   Selections of supporting electrolytes, with tabulations of the associated half-wave potentials for most metallic cations, are available.   An excellent source of data in this connection is Meites.* Many anions may also be determined quantitatively by the polarographic method, and again, the same source provides data relative to supporting electrolytes and half-wave potentials.

Although the *absolute method* may be applied in the quantitative determination of a species, and this involves the use of the *diffusion current constant* $607nD_0^{1/2}$ and the evaluation of m and τ for the applied potential involved, by far the more commonly applied methods are those involving *comparison techniques*. The simplest of these requires the preparation of a series of standard solutions of known concentration of the species to be determined.   The range of concentration should encompass the range expected for the unknown concentration solutions, and the compositional background as to supporting electrolyte, suppressor, etc., should be identical to that for the unknowns.   The value of $\bar{i}_d$ should be determined for all solutions at the same appropriate value of applied potential.   A calibration curve of $\bar{i}_d$ vs. concentration will then show the linearity, or lack of linearity, for this relationship.   Where the relationship is linear, the analysis of unknowns may be based on a ratio technique using $\bar{i}_d$ and concentration for a standard solution

*L. Meites, "Polarographic Techniques," 2d. ed., pp. 615–670, Interscience, New York, 1965.

and $\bar{\imath}_d$ for the unknown. Where the relationship is nonlinear, the calibration curve may be used. In all applications of the method, the temperature at the time of current measurement must be a repetitive constant value, and this is particularly important where a calibration curve is used.

A second comparison method of approach is to make a measured volume addition of an appropriate standard concentration solution of the species involved. Here we have

$$C_0 = \frac{\bar{\imath}_{d,1} V_s C_{0,s}}{(V_s + V)\bar{\imath}_{d,2} - \bar{\imath}_{d,1} V} \tag{301}$$

where $\bar{\imath}_{d,1}$ = corrected diffusion current for the unknown solution, μA

$\bar{\imath}_{d,2}$ = corrected diffusion current for the solution after the addition of the standard solution, μA

V = volume of unknown solution, ml

V_s = volume of standard solution added, ml

$C_{0,s}$ = concentration of standard solution, mM

C_0 = concentration of the unknown solution, mM

Analysis of organic substances Most organic electrode reactions are irreversible, and the majority of these take place in a stepwise fashion. In very many instances these reactions are pH dependent. Where H^+ is consumed or produced by the electrode reaction, the solution must be adequately buffered to prevent undesirable $[H^+]$ values at the interface as the reaction proceeds. In most cases solvents other than water must be used, and this often requires that corrections be made because of high solution resistance and associated high iR drop through the solution. Because of these characteristics the polarographic data obtained is sometimes difficult to interpret. Nevertheless, many organic compounds may be treated polarographically with respect to qualitative identification, quantitative estimation, and structural elucidation. Several functional groups react at the dme, and compounds containing these groups may often be treated polarographically. Such groups include organic halogen compounds; carbonyl compounds including ketones, sugars, aldehydes, quinones, and hydroquinones; some carboxylic and dicarboxylic acids, excepting the simple aromatic and aliphatic monocarboxylic acids; some keto and aldehydo acids; nitro, nitroso, azo, and amine oxide group compounds; oxygen-oxygen, sulfur-sulfur and carbon-sulfur bond substances.

12.5.8 AMPEROMETRIC TITRATIONS

General In any titration process the concentration of the reactant decreases, that of the titrant increases, and those of the products increase as the titration progresses. Where one or more of these substances is capable of reduction or oxidation at a microelectrode, the voltammetric or polarographic technique may be used to follow the course of the titration and to provide a means of locating the equivalence-point volume. Where such a method is employed, the titration is called an *amperometric titration*, and a plot of diffusion current or limiting current vs. volume of titrant will yield a curve from which the equivalence-point volume may be evaluated. The amperometric titration technique is subject to less error than the corresponding voltammetric or polarographic method. With the polarographic titration method, for example, the supporting electrolyte and capillary characteristics exert less influence, and the temperature, while it should remain constant during the titration, is not of such fundamental importance. Again, as noted in the foregoing, the reactant itself need not be reactive at the microelectrode, and the method may be based on a reactive titrant or reaction product. The method of plotting the curve involves data accumulated on either side of but well removed from the equivalence-point zone, so that titration reactions which are incomplete to a considerable degree at the equivalence point may be adapted to the amperometric technique. The degree of dilution of the reactant and/or titrant is not as important as it is in standard titration work, since the method is basically suited to low concentration media. As we shall see, however, the *relative* concentrations of titrant and reactant solutions may, under certain conditions, be of significance.

Polarographic amperometric titrations The general forms for amperometric titration curves where (*a*) the reactant is dme reactive, (*b*) the titrant is reactive (*c*) a product of the titration reaction is reactive, and (*d*) both reactant and titrant are reactive are shown in Fig. 12.29. The applied potential at the dme is in each case, of course, appropriate to the plateau current zone for the electrode reaction involved. The equivalence-point-volume location technique is illustrated for each titration form.

In the amperometric titration method, it is necessary to correct each current reading, usually the limiting current value, for the volume dilution effect of the titrant. Here the starting volume of the reactant solution must be known, and the correction involves multiplying each current value by the factor $(V + v)/V$, where V is the starting volume and v is the volume of titrant solution added to the particular point in the titration. Alternatively, the concentration of the titrant solution may be made at least 20 times that of the reactant solution, so that the volume v at any point is always very small compared with V. In this way, the use of the correction factor may be avoided with little in the way of error introduced. Since, where this method is applied, the titrant volume at the equivalence point will only be a milliliter or two, a microburette should be used to avoid high relative burette-reading uncertainties.

Amperometric titrations with the rotating platinum microelectrode The dme has the disadvantage of the mercury being susceptible to attack by oxidizing agents of various types and by certain other substances. Where this situation may

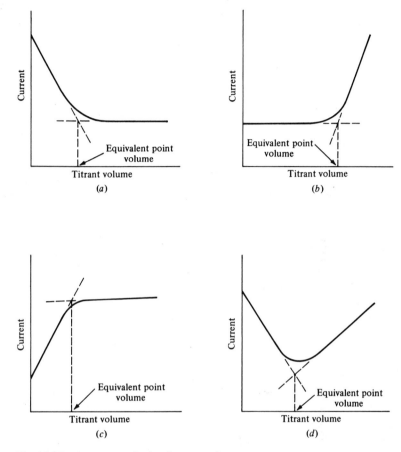

Fig. 12.29   Amperometric titration curve forms.

exist, and where the required applied potential does not exceed that at which hydrogen is discharged at platinum (a value appreciably less negative than that for
mercury), the rotating platinum electrode may be used. This microelectrode consists of a very short length of platinum wire fused into the side of a glass tube and
connected, through a mercury column inside the tube, to the applied potential
source. The tube is rotated, without precession, at a constant rpm of about 600 by
a synchronous motor. Although certain fundamental differences exist between
the solution currents for this system and the polarographic system, the method of
equivalence-point-volume location is generally similar.

12.5.9 POLAROGRAPHIC EQUIPMENT

Although a polarograph may be put together from fairly simple and basic components, excellent equipment for analytical polarography and amperometric titrations is available from a large number of commercial suppliers. We shall avoid

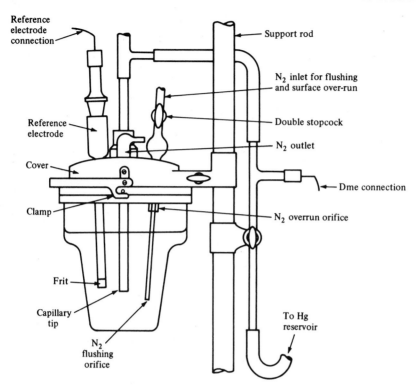

Fig. 12.30 Polarographic cell. *(Courtesy Metrohm Ltd., Herisau, Switzerland.)*

any lengthy description and discussion of the basic instrumentation. Figure 12.30 shows a typical commercially available arrangement. The potential is applied and measured potentiometrically, and the current through the solution is measured galvanometrically or electronically through a recorder. The current oscillations over the drop life are damped to various selective degrees. The capillary characteristics have already been discussed. It should be emphasized that the capillary tip should be cut flat and should be parallel to the solution. The capillary itself should be rigidly perpendicular to the solution and should not be tilted. The entire arrangement must be located, for obvious reasons, in a vibration-free area. The flow of mercury through the capillary, when the investigation is completed, should be maintained at some low level until the tip has been washed with water and immersed below the surface of pure mercury in a storage vessel. The tip should not be removed from the storage vessel until a mercury flow has been initiated.

12.6 CONDUCTOMETRY AND CONDUCTOMETRIC TITRATIONS

12.6.1 GENERAL

Three of the electroanalytical techniques so far discussed have dealt with conditions involving current flow through the solution, and in two of these —electrolytic separations and coulometry—appreciable current flow was involved. Where current flow through an electrolyte solution takes place, the conduction of current is accomplished by the migration of anions and cations to the appropriate electrodes. The end result of such migrations, where a continuing flow of current is established unidirectionally, will be the proper electrode reactions as associated with the migrating species. The ability of an electrolyte solution to carry current will be related to the number of charged particles in the solution, their respective mobilities in the medium, and the magnitude of the applied potential between the electrodes. Where the solution contains a single electrolyte species, the *conductance* will be related to the concentration of this species when a constant applied potential is involved. Under the same conditions of constant applied potential, a solution containing several electrolyte species will display a conductance associated with the relative concentrations of the various species involved, and each species will contribute a well-defined fraction of the total current flow.

Obviously, where a solution contains a single electrolyte species, the conductance and electrolyte concentration relationship may be established through the use of known concentration solutions, fixed temperature and applied potential conditions, a specific conductance cell, etc. Once established, the plotted relationship may be used, under the same conditions, to evaluate the concentration for an unknown solution involving only the same single species. Such methods are called *conductometric methods* or, in general, *conductometry*. In the presence of more than one electrolyte species, the method provides no simple solution to the concentration question, since the dependency of conductance on the contributions of *all* conducting species renders the technique nonspecific. The applicability of the method of direct conductometry is extremely limited relative to analytical chemistry, and apart from its use in the determination of the purity of water, it is rarely encountered.

Where the conductance of a solution under titration is measured, and where the titration reaction results in conductance changes that are clearly detectable, even against a background of conductance arising out of the presence of other electrolytes in the solution, the conductance values may be plotted against the corresponding titrant volumes. The resulting curve will permit the location of the equivalence-point volume of the titration. Titrations carried out under such conditions are called *conductometric titrations*, and the method is much more adaptable to analytical situations than is the direct conductometric technique. The general process of conductometric titration, first investigated by Dutoit,* has several advantages, which include its applicability to titration reactions that are incomplete at the equivalence point to a significant degree (an advantage shared with amperometric and absorptiometric titrations) and its ability to deal with highly diluted solutions. The titration of weak acids or bases with K_a or K_b values less than 10^{-7} is feasible, as is the titration of many weak acids by the weak base ammonia. Again, the titration of a strong acid—weak acid mixture by conductometric means yields much more clearly defined equivalence-point locations than

*P. Dutoit, *J. Chim. Phys.*, **8**:12(1910).

the corresponding indicator or potentiometric titration. Certain inherent disadvantages are associated with the method, which include a general lack of application to oxidation-reduction titrations and an increasing loss of sensitivity with increasing background electrolyte concentration.

12.6.2 ELECTROLYTIC CONDUCTANCE AND ELECTROLYTE CONCENTRATION

The conductance of a solution is the reciprocal of its resistance. Where the resistance is expressed in ohms, the conductance is expressed in mhos or ohm^{-1}. The resistance of a column of solution A square centimeters in cross-sectional area between two electrodes separated by a distance of l centimeters is given by

$$R = \rho \frac{l}{A} \tag{302}$$

where ρ is the *specific resistance* in ohm centimeters. The conductance is given by

$$L = \frac{1}{R} = \kappa \frac{A}{l} \tag{303}$$

with κ the *specific conductance* in ohm^{-1} centimeters^{-1}. The quantity l/A is called the *cell constant* and is specific for a given conductance cell. It is measured by determining the conductance for the cell using solutions (usually potassium chloride solutions) of accurately known specific conductance.

The *equivalent conductance* Λ is defined as the conductance of a solution containing 1 gram equivalent of solute between electrodes separated by a distance of 1 cm. Where the solution has a concentration of C gram equivalents per liter (very close to 1000 cm^3), the value of Λ is given by

$$\Lambda = \frac{1000 \, \kappa}{C} \tag{304}$$

and, with κ given by Eq. (303), we have

$$\Lambda = \frac{1000Ll}{CA} \tag{305}$$

whence we have

$$L = \frac{A}{1000l} \Lambda C \tag{306}$$

The value of Λ may be determined from Eq. (304) where the value of κ is known for a solution of specific concentration C. Λ varies with the concentration, increasing with decreasing concentration. Decreases in the interionic attraction and repulsion forces with increasing dilution is the principal underlying cause of this phenomenon. For strong electrolytes, the increase in Λ with dilution is linear at low concentrations; it can be represented by

$$\Lambda = \Lambda° - A\sqrt{C} \tag{307}$$

where Λ° is the equivalent conductance at infinite dilution and A is a constant which includes factors involving the temperature and the characteristics of the solute and solvent. The value of Λ° can be obtained by the extrapolation to zero concentration of the straight line obtained by plotting Λ vs. $\sqrt{C}$ for low C values. For weak electrolytes, the degree of dissociation increases with increasing dilution, this providing for a nonlinear relationship, and Λ° must be determined by indirect methods.

The value of Λ° is given by

$$\Lambda^\circ = \lambda_+^\circ + \lambda_-^\circ \tag{308}$$

for a single electrolyte solute, λ_+° and λ_-° being the equivalent ionic conductances at infinite dilution for the cation and anion. The values of λ_i° for various ions are given in Table 12.5. The prefixes $\frac{1}{2}$, $\frac{1}{3}$, etc., refer to the fact that concentration in equivalents per liter is involved. The relationship

$$\Lambda = \lambda_+ + \lambda_- \tag{309}$$

also holds for the equivalent conductance and equivalent ionic conductances at finite dilution. Where more than one electrolyte solute is involved, it can be shown, using basically Eqs. (306) and (309), that

$$L = \frac{A}{1000\,l}\,\Sigma\,C_i\,\lambda_i \tag{310}$$

where C_i and λ_i represent the concentration and equivalent ionic conductance for each ion species in the solution. While the value of λ_i at finite concentrations is

Table 12.5 Values of equivalent ionic conductance at infinite dilution*
25°C

Cations	λ_+°	Anions	λ_-°
H^+	349.82	OH^-	198.0
K^+	73.52	$\frac{1}{2}SO_4^{2-}$	79.8
NH_4^+	73.4	Br^-	78.4
$\frac{1}{2}Pb^{2+}$	73	I^-	76.8
$\frac{1}{2}Ba^{2+}$	63.64	Cl^-	76.34
Ag^+	61.92	NO_3^-	71.44
$\frac{1}{2}Ca^{2+}$	59.50	$\frac{1}{2}C_2O_4^{2-}$	70
$\frac{1}{2}Sr^{2+}$	59.46	ClO_4^-	68.0
$\frac{1}{2}Cu^{2+}$	54	HCO_3^-	44.48
$\frac{1}{2}Mg^{2+}$	53.06	CH_3COO^-	40.9
Na^+	50.11		
Li^+	38.69		

*Generally from J. J. Lingane, "Electroanalytical Chemistry," 2d. ed., p. 181, Interscience, 1958.

not entirely independent of its environment, the value of λ_i° is. For most conductometric titration purposes λ_i° may be used as a substitute for λ_i.

The values of solution conductance are measured using an ac source of potential in order to avoid the compositional changes resulting in solution when dc sources are used. A Wheatstone bridge is usually used to obtain the conductance measurements. Figure 12.31 shows both a bridge and a common type of conductometric titration conductance cell. The conductance cell used for titration work is not required to have a known value of the cell constant (l/A), since it is assumed that the cell constant will remain unchanged during any given titration.

12.6.3 CONDUCTOMETRIC TITRATIONS

General   The conductance of the solution is measured during the course of the titration, enough values being obtained on either side of the equivalence-point zone

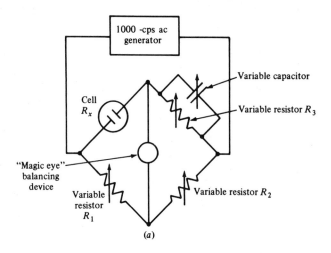

(a)

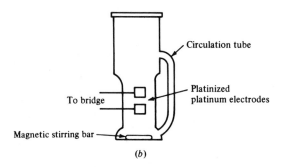

(b)

Fig. 12.31   (a) Wheatstone bridge and (b) conductometric titration cell.

to permit accurate location of the equivalence-point volume within the limitations of the titration reaction. These values are not taken too close to the equivalence-point zone, and since the departure from linearity in this zone becomes more extended with increasing degree of reaction incompletion, care must be taken relative to the titration points at which conductance is measured. The corrected conductance values are plotted against the volume of titrant.

As was the case with amperometric titrations, the titration curve may take one of several forms, the particular form reflecting the general conductance characteristics of the reactant, titrant, and product species. Figure 12.32 shows an assembly of titration forms. Again as was the case with amperometric titrations, each measured conductance value will require multiplication by the titrant volume dilution correction factor $(V + v)/V$, except in instances where the concentration of the titrant solution is not less than 20 times that of the reactant solution. This latter alternative will necessitate the use of a microburette.

In all conductometric titration work the temperature must remain constant during the course of the titration, although a *specific* temperature is not required. It should be noted that whereas a plot of corrected conductance vs. titrant volume is generally linear, for appropriate titrations, in those zones adequately removed from that of the equivalence point, a plot of solution resistance vs. volume of titrant is nonlinear both before and after the equivalence point. Such plots cannot be applied in the location of the equivalence point.

Neutralization titrations   In the titration of a strong acid by a strong base, say HCl by NaOH, the concentration of H^+ decreases during the titration, being replaced by an equivalent amount of Na^+. The difference between $\lambda^{\circ}_{H^+}$ at 349.82 and $\lambda^{\circ}_{Na^+}$ at 50.11 results, before the equivalence point, in a decrease in the solution conductance. The value of $[Cl^-]$ remains fairly constant during this portion of the titration, except for the dilution effect of the titrant volume, and the value of $[OH^-]$ and its contribution to the conductance is negligible. At the equivalence point the conductance is basically due to $[Na^+]$ and $[Cl^-]$. After the equivalence point the $[OH^-]$ increases with the excess of base, and since the value of $\lambda^{\circ}_{OH^-}$ is 198.0, the conductance increases quite rapidly. The plot of conductance vs. titrant volume yields a V-shaped curve for this titration, and the method of equivalence-point location is shown in Fig. 12.32a. Since $\lambda^{\circ}_{OH^-}$ is significantly lower than $\lambda^{\circ}_{H^+}$, the slope of the curve after the equivalence point is less than that before.

Where a weak acid is titrated by a strong base such as NaOH (and the same reasoning as that to follow will apply to weak base–strong acid titrations), the feasibility of the titration will depend on the weak acid concentration and the value of K_a for the acid involved. Where weak acids of K_a value less than about 10^{-5} are titrated, the $[H^+]$ at the start will be fairly low, decreasing rapidly with the establishment of buffered solution conditions. The conductance decreases over this relatively short titration zone. The establishment of the buffer solution condition stabilizes the $[H^+]$ to a degree, but with further additions of base up to the equiv-

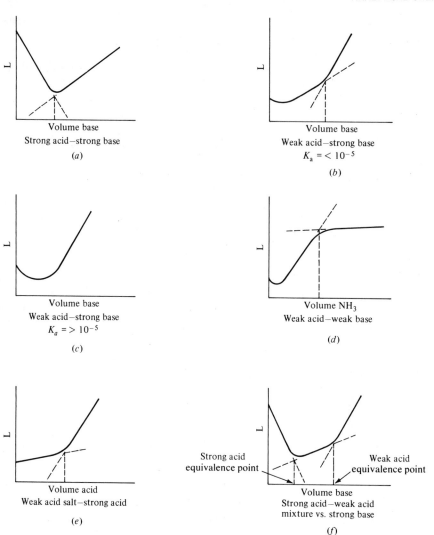

Fig. 12.32 Conductometric titration forms.

alence point, the conductance will gradually increase because of the increase in $[Na^+]$ and [acid anion]. After the equivalence point a rapid increase in conductance occurs as the $[OH^-]$ increases. Figure 12.32b shows a typical curve of this type. Here, because of the high value of $\lambda^{\circ}_{OH^-}$ compared with $\lambda^{\circ}_{Na^+}$, the slope of the curve after the equivalence point is greater than that before. For a weak acid of K_a value between about 10^{-5} and 2×10^{-2}, titration with NaOH does not yield accurate location of the equivalence point. Here the weak acid dissociates to a fairly high degree. The $[H^+]$ and the conductance decrease slowly as a buffered solution condition is established, but the subsequent increase in conductance due to increasing $[Na^+]$ and [acid anion] and decreasing $[H^+]$ is offset to some extent by the relatively high degree of dissociation for the remaining acid. The curve obtained is therefore nonlinear before the equivalence point, preventing accurate extrapolation and location of the equivalence point. A typical curve for a titration of this type is shown in Fig. 12.32c.

The titration of weak acids of K_a value higher than 10^{-5} and as low as about 10^{-8} is feasible, however, where a solution of NH_3 is used as the titrant. Here the low conductance of the weak base titrant provides for a much reduced angle between the curve portions before and after the equivalence point, allowing greater definition in the location of the equivalence point. Figure 12.32d illustrates such a titration.

Solutions of the salts of weak acids and strong bases may be titrated with a strong acid (or those of weak bases and strong acids by strong bases) using conductometric means. For example, the titration of sodium acetate solution by a strong acid results in the formation, before the equivalence point, of increasing amounts of the undissociated weak acid and the replacement of the acetate ion by the anion of the strong acid. Depending on the degree of weakness of the acid and the relative ionic conductances of the weak and strong acid anions, the solution conductance may show some condition between a gradual increase to a gradual decrease. After the equivalence point, the excess of strong acid provides for increasing $[H^+]$ and significant increases in the conductance. Salts of reasonably strong weak acids may be titrated in this way, with less limitation as to the accuracy of locating the equivalence point than is associated with the corresponding indicator or potentiometric titration. Figure 12.32e shows a typical form of weak acid salt vs. strong acid titration.

The titration of strong acid—weak acid mixtures conductometrically generally proceeds with better equivalence-point location than is permitted by the parallel titration as conducted using the indicator or potentiometric technique. Figure 12.32f shows a typical mixed acid titration curve.

Precipitation and complexation titrations   Because of the high equivalent ionic conductances for H^+ and OH^-, neutralization titrations provide for the maximum sensitivity and adaptability relative to the use of the conductometric method. The technique can be applied to precipitation and complexation titrations, as exemplified by Fig. 12.33 showing the titration curve for a titration of lead as lead acetate by a standard solution of sulfuric acid. Before the equivalence point the acetate ion is converted to the relatively undissociated acetic acid, with Pb^{2+} being precipitated as lead sulfate. The conductance thus decreases in this zone. After the equivalence point the excess of H_2SO_4 provides for increasing $[H^+]$ and rapid increases in the conductance. The reaction for the formation of $PbSO_4$ is quite incomplete at the equivalence point, due to the relatively high solubility of $PbSO_4$, but the technique permits quite accurate location of the equivalence point.

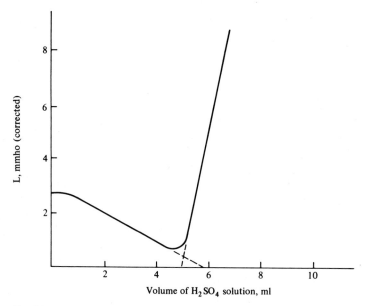

Fig. 12.33   Titration of 0.05 M lead acetate by 0.5 M H_2SO_4 at 25°C.

While many precipitation and complexation titration reactions can be adapted to the conductometric technique, the generally much lower equivalent ionic conductances of the ions involved, compared with $\lambda^{\circ}_{H^+}$ and $\lambda^{\circ}_{OH^-}$, leads to less in the way of significant changes in conductance during the titration process than are found for neutralization titrations. In addition, the generally slower reaction rate for such titrations, and the possibility of interfering actions such as coprecipitation, contribute to a lower order of accuracy in the location of the equivalence point.

Oxidation-reduction titrations   Such titrations are not normally adaptable to the conductometric method since (1) the equivalent ionic conductances of the reactant, titrant, and product ions are frequently insufficiently different to permit easily detectable conductance changes during the titration and (2) most of such titrations are carried out in the presence of significant $[H^+]$, which provides for a high background conductance against which the small conductance changes occurring as the result of the titration reaction are not detectable.

PROBLEMS

Secs. 12.1 to 12.2

1. E_{cell} may be found from

$$E_{cell} = E_{ind} - E_{ref}$$

Considering the Nernst equation from the point of view of ion concentration rather than activity, and taking all values of $E°$ with an uncertainty of ± 0.001 V, calculate the ion concentration for each of the following aqueous solutions at 25°C.

(a) $[Ag^+]$ for a solution of $AgNO_3$, where a silver electrode and an SCE reference electrode (saturated KNO_3 salt bridge) are used to obtain an E_{cell} of 0.424 ± 0.002 V at 25°C. E_{SCE} in reduction $= 0.241 \pm 0.001$ V at 25°C.

(b) $[Cu^{2+}]$ for a solution of $Cu(NO_3)_2$, where a copper electrode and an SCE reference electrode yield an E_{cell} of -0.165 ± 0.002 V at 25°C. E_{SCE} in reduction $= 0.245 \pm 0.001$ V at 25°C.

(c) $[Hg^{2+}]$ for a solution of $Hg(NO_3)_2$, where a mercury electrode and an SCE reference electrode (saturated KNO_3 salt bridge) are used to obtain an E_{cell} of 0.437 ± 0.002 V at 25°C. The value of E_{SCE} in reduction $= 0.241 \pm 0.001$ V at 25°C.

(d) What is the meaning of the negative value of E_{cell} in b? How would you write E_{cell} relative to E_{ind} and E_{ref} to conform to a positive value for E_{cell} in b?

2. A silver electrode is immersed in an aqueous solution of silver(I) ion in conjunction with an SCE reference electrode (saturated KNO_3 salt bridge). E_{SCE} in reduction at 25°C $= 0.241 \pm 0.001$ V. The value of $E°_{Ag^+/Ag}$ at 25°C is 0.800 ± 0.001 V. If the couple gives an emf read from the potentiometer dial with a reading uncertainty of ± 0.001 V, determine the maximum relative uncertainty in the calculated value of $[Ag^+]$.   *Ans.* $\pm 12\%$.

3. A hydrogen electrode (1 atm hydrogen gas pressure)–SCE reference electrode couple immersed in a weak monobasic acid solution exactly 0.100 M yielded an emf of 0.390 V at 25°C. What is the value of K_a for the weak acid? E_{SCE} in reduction $= 0.245$ V at 25°C.

4. Given the value of $E°_{Ag^+/Ag}$ as 0.800 V and $K_{sp}(AgBr)$ as 5.0×10^{-13}, both at 25°C, calculate the value of $E°_{AgBr/Ag}$ at 25°C.   *Ans.* 0.074 V

5. 50.00 ± 0.02 ml of a solution of weak monobasic acid was titrated by 0.0981 ± 0.0002 M NaOH, the titration being followed by a glass-membrane electrode–SCE reference electrode couple. The following values were secured.

Volume NaOH, ml	pH
30.00	5.30
35.00	5.77
36.00	5.94
37.00	6.21
37.50	6.45
37.70	6.59
37.90	6.85
38.10	7.21
38.30	9.91
38.50	10.48
38.70	10.72
38.90	10.87
39.50	11.15
40.00	11.29
41.00	11.48
42.00	11.60

Determine
(a) The curve of pH vs. V_{NaOH}

(b) The curve of $\Delta pH/\Delta V_{NaOH}$ vs. V_{NaOH}

(c) The equivalence-point volume from a and b by curve analysis

(d) The tabulation of $\Delta^2 pH/\Delta V^2_{NaOH}$ vs. V_{NaOH} and the equivalence-point volume by the tabulation method

(e) The estimated value of pH at the equivalence point

(f) The concentration of the weak acid at the start of the titration (using the tabulation-method equivalence-point volume)

(g) The estimated value of K_a for the weak acid (using the simplest equation)

6. 0.5624 ± 0.0002 g of an alloy containing silver and no interfering substances is weighed out, dissolved in dilute nitric acid, and finally diluted to about 100 ml. A silver electrode–SCE reference electrode (saturated KNO_3 salt bridge) couple is used to follow the course of the titration of this solution by $0.0845 \pm 0.0002 M$ NaCl. The E values are as shown.

Volume NaCl, ml	E vs. SCE
0.00	0.445
2.00	0.429
4.00	0.421
6.00	0.409
8.00	0.388
8.50	0.380
9.00	0.365
9.20	0.355
9.40	0.341
9.50	0.328
9.60	0.302
9.70	0.245
9.80	0.214
9.90	0.201
10.10	0.185
10.50	0.169
11.00	0.157
12.00	0.143

(a) Plot a curve of E vs. V_{NaCl} and determine the equivalence-point volume by curve analysis.

(b) Tabulate $\Delta^2 E/\Delta V^2_{NaCl}$ vs. V_{NaCl} and determine the equivalence-point volume by the tabulation method.

(c) Use the value from b to determine the percent silver in the alloy.

(d) Estimate E at the equivalence-point volume and, from this value, estimate $K_{sp}(AgCl)$.

7. The value of $K_{sp}(AgCl)$ estimated from the data in Prob. 6 will likely be only approximate. Indicate how you would go about trying to determine, by potentiometric titration of Ag^+ by Cl^- solution, a more accurate value of $K_{sp}(AgCl)$.

8. A material contains KI and no interfering substances. A 1.0652 ± 0.0002 g sample is dissolved in water and diluted to 100.00 ± 0.04 ml. This solution is then titrated with $0.0546 \pm 0.0002 M$ $AgNO_3$, using a couple involving a silver electrode and an SCE reference electrode (saturated KNO_3 salt bridge). The following data were recorded.

Volume $AgNO_3$, ml	E vs. SCE
2.00	−0.243
4.00	−0.236
6.00	−0.229
8.00	−0.220
10.00	−0.205
12.00	−0.170
12.20	−0.161
12.30	−0.155
12.40	−0.148
12.50	−0.137
12.60	−0.118
12.70	+0.239
12.80	+0.307
12.90	+0.323
13.00	+0.333
13.50	+0.358
14.00	+0.370

(a) Determine the equivalence-point volume by the tabulation method.

(b) Determine the percent KI in the material.

(c) Evaluate the relative error that would have resulted in the determination of KI if the titration had been stopped at an E-vs.-SCE value of −0.100 V.

Ans. (a) 12.65 ml; (b) 10.76%

9. An iron-titanium mineral is to be analyzed for iron and titanium. 1.0675 ± 0.0002 g of the powdered mineral is dissolved and treated so as to obtain the iron and titanium as Fe^{2+} and Ti^{3+}. This solution is diluted to about 100 ml, the final solution being $1 M$ to H_2SO_4. The solution is now titrated with $0.2132 \pm 0.0002 M$ cerium(IV) ion solution, the titration being followed by a platinum electrode —SCE reference electrode couple. The titration yielded the following data.

Volume Ce^{4+}, ml	E vs. SCE	Volume Ce^{4+}, ml	E vs. SCE
5.00	−0.130	54.00	+0.527
10.00	−0.097	55.00	+0.543
15.00	−0.039	55.90	+0.576
15.50	−0.019	56.00	+0.585
15.70	−0.004	56.10	+0.598
15.80	+0.010	56.20	+0.625
15.90	+0.029	56.30	+1.035
16.00	+0.284	56.40	+1.062
16.10	+0.312	56.50	+1.075
16.20	+0.325	56.60	+1.084
16.30	+0.334	57.00	+1.103
.	.	58.00	+1.126
.	.		
.	.		

Determine the percent values for iron and titanium in the mineral.   *Ans.* 15.26% Ti, 45.0% Fe

Sec. 12.3

10. A solution with 0.060 M Au^{3+} and 0.050 M HNO_3 is to be considered for electrolysis between smooth platinum gauze electrodes of 100 cm² area each. Assume that the electrolysis is feasible.

(*a*) Calculate the back emf of the gold-oxygen system.

(*b*) The oxygen-discharge overvoltage on smooth platinum at very low current density is −0.40 V; that for the discharge of gold on platinum is negligible. Calculate the theoretical decomposition potential for the gold-oxygen system.

(*c*) Assume that the electrolysis takes place with an initial current of 1.0 A over a solution resistance of 1.5Ω. At this initial current density, the oxygen-discharge overvoltage will be −0.85 V. Ignore the residual current and calculate the initial applied potential required. What is the initial current density?

(*d*) Assume the oxygen-discharge overvoltage to be −0.85 V, and that for hydrogen discharge on the gold-plated cathode to be −0.10 V. Calculate the theoretical decomposition potential for the hydrogen-oxygen system.

(*e*) Describe the cathodic and anodic potential conditions, and the solution composition conditions, as the electrolysis progresses with efficient stirring to equilibrium at a constant applied potential of −2.10 V.

(*f*) Determine the value of [Au^{3+}] when the electrolysis reaches equilibrium at a constant applied potential of −2.10 V. You may consider the value of E_{O_2/H_2O} to remain virtually unchanged from the start to equilibrium.

11. A solution contains 0.030 M Au^{3+}, 0.050 M Ag^+, 0.060 M Cu^{2+}, and 0.100 M HNO_3. Although gold and silver are best deposited from solutions of their cyanide complexes, electrolysis of this solution between large platinum gauze electrodes is proposed. Disregard any difficulties relative to cathodic potential variations arising out of concentration polarization phenomena occurring during electrolysis. Accept the oxygen-discharge overvoltage on platinum as −0.40 V, and that of hydrogen discharge on copper as −0.50 V. Consider the discharge overvoltage values for gold, silver, and copper on platinum or on each other as being negligible.

(*a*) Calculate the theoretical values for $E_d(Au-O_2)$, $E_d(Ag-O_2)$, $E_d(Cu-O_2)$, and $E_d(H_2-O_2)$.

(*b*) Show the order of discharge for the various ions Au^{3+}, Ag^+, Cu^{2+}, and H^+.

(*c*) Determine the theoretical quantitativeness and feasibility of the separation of gold, silver, and copper from each other and from the solution without the simultaneous discharge of hydrogen ion. Accept applied potential values for the separation of each ion 0.10 V less negative than the E_d value for the next ion to successively discharge.

12. (*a*) A solution contains 0.100 M Cu^{2+}, 0.100 M Zn^{2+}, and 0.100 M H^+. The discharge overvoltage values for oxygen, hydrogen, copper, and zinc can be taken as

O_2 on the smooth platinum	−0.40 V
H_2 on copper	−0.50 V
Copper on smooth platinum	Negligible
Zinc on copper	Negligible

Calculate the theoretical decomposition potentials $E_d(Cu-O_2)$, $E_d(Zn-O_2)$, and $E_d(H_2-O_2)$.

(*b*) The efficiently stirred solution in Prob. 12*a* is electrolyzed between large platinum gauze electrodes at an applied potential appreciably more negative than $E_d(Zn-O_2)$. Will zinc be deposited at the cathode? Give the reasoning underlying your answer.

13. The solution in Prob. 12a is, without volume change, made alkaline to NH_3 and an excess is added. The reactions

$$Cu^{2+} + 4NH_3 \rightleftharpoons Cu(NH_3)_4{}^{2+}$$

$$Zn^{2+} + 4NH_3 \rightleftharpoons Zn(NH_3)_4{}^{2+}$$

take place. Again without volume change KCN is added in large excess and the complexation reactions

$$2Cu(NH_3)_4{}^{2+} + 7CN^- + H_2O \rightleftharpoons 2Cu(CN)_3{}^{2-} + CNO^- + 6NH_3 + 2NH_4{}^+$$

$$Zn(NH_3)_4{}^{2+} + 4CN^- \rightleftharpoons Zn(CN)_4{}^{2-} + 4NH_3$$

take place. Note that in the complexation reaction with CN^-, the copper(II) is reduced to copper(I). The following data are relevant.

$[NH_3]$ in the final solution	About 1 M
$[OH^-]$ in the final solution	$1.0 \times 10^{-3} M$
$[CN^-]$ in the final solution	$1.0 M$
$K_1 K_2 K_3$ for $Cu(CN)_3{}^{2-}$	3.9×10^{28}
$K_1 K_2 K_3 K_4$ for $Zn(CN)_4{}^{2-}$	1.6×10^{20}
O_2-discharge overvoltage on platinum	-1.40 V
H_2-discharge overvoltage on zinc or copper	-0.90 V
Copper or zinc discharge overvoltage on steel	Negligible

(*a*) Calculate the theoretical values for $E_d(Cu\text{—}O_2)$, $E_d(Zn\text{—}O_2)$, and $E_d(H_2\text{—}O_2)$.

(*b*) Suppose that a steel casting is used as the cathode, with a large platinum gauze electrode as the anode. What will now be the effect of an applied potential more negative than $E_d(Cu\text{—}O_2)$ and $E_d(Zn\text{—}O_2)$?

(*c*) What prevents any appreciable change in either E_{2H^+/H_2} or E_{O_2/H_2O} as the copper and zinc are deposited?

The cathodic deposit here is an alloy of copper and zinc called brass. The procedure is essentially similar to that used in commercial brass plating, although in the industrial application cheap inert metals or alloys would be used anodically in place of platinum.

Ans. (*a*) -3.20 V, -3.37 V, -3.53 V

14. Electrolysis with efficient stirring is to take place between large platinum gauze electrodes immersed in a solution with $0.100 M$ Cu^{2+}, $0.100 M$ Cd^{2+}, and $1.0 M$ H^+. Assume

O_2-discharge overvoltage on platinum	-0.40 V
H_2-discharge overvoltage on cadmium	-0.50 V
Copper- or cadmium-discharge overvoltage on platinum or each other	Negligible

(*a*) What are the theoretical values of $E_d(Cu\text{—}O_2)$, $E_d(Cd\text{—}O_2)$, and $E_d(H_2\text{—}O_2)$?

(*b*) What is the order of discharge at the cathode for cadmium, copper, and hydrogen?

(*c*) Ignore any difficulties arising out of cathodic potential variations during electrolysis, use E_d values from *a* and applied potential values for each substance equal to the E_d value for the next successive substance discharged and show

(1) If it is theoretically possible to separate copper from cadmium quantitatively

(2) If it is theoretically possible to separate cadmium from the solution quantitatively under any conditions of applied potential

15. Suppose that the solution in Prob. 14 was made alkaline to NH_3, with a large excess added. Suppose further that KCN was added in large excess, and that both additions resulted in no volume change for the original solution. Copper(II) will be reduced to copper(I) during the complexation reaction with cyanide ion. The copper will exist almost exclusively in the form $Cu(CN)_3{}^{2-}$, while the cadmium will be almost completely in the form $Cd(CN)_4{}^{2-}$. The following data are pertinent.

$[NH_3]$ in the final solution	About 1 M
$[OH^-]$ in the final solution	$1.0 \times 10^{-3} M$
$[CN^-]$ in the final solution	$1.0 M$
$K_1 K_2 K_3$ for $Cu(CN)_3{}^{2-}$	3.9×10^{28}
$K_1 K_2 K_3 K_4$ for $Cd(CN)_4{}^{2-}$	5.8×10^{18}
O_2-discharge overvoltage on platinum	-1.40 V
H_2-discharge overvoltage on copper or cadmium	-0.90 V
Copper- or cadmium-discharge overvoltage on platinum or each other	Negligible

(a) What are the theoretical values of $E_d(\text{Cu}-\text{O}_2)$, $E_d(\text{Cd}-\text{O}_2)$, and $E_d(\text{H}_2-\text{O}_2)$?

(b) What is the order of discharge at the cathode for copper, cadmium, and hydrogen?

(c) Is it theoretically possible to separate cadmium quantitatively from copper? {Use an applied potential value equal to $E_d(\text{Cu}-\text{O}_2)$.}

(d) Is it theoretically possible to separate copper quantitatively from the solution without simultaneous liberation of hydrogen? {Use an applied potential value 0.10 V less negative than $E_d(\text{H}_2-\text{O}_2)$.}

(e) Discuss generally the effect of complexation on the electrolytic separation of metals, using as examples material from Probs. 12, 13, 14, and 15.

16. 1.0654 ± 0.0002 g of a brass alloy is dissolved in dilute nitric acid. The solution is boiled to expel NO_2 and diluted to 100 ml. The diluted solution is noted to be clear and is then further diluted to 180 ml. The $[H^+]$ at this point is about $0.10 M$. The solution is electrolyzed with efficient stirring between two large platinum gauze electrodes at a constant current of 3.0 A. The cathode and anode weigh initially 24.6732 ± 0.0002 g and 12.6984 ± 0.0002 g, respectively. After a test of the solution shows a copper(II) ion concentration of less than $10^{-6} M$, the electrodes are washed and dried and found to weigh 25.2894 ± 0.0002 g for the cathode and 12.7299 ± 0.0002 g for the anode. The deposit of lead dioxide is stripped from the anode, the cathode retaining its copper deposit. The solution is made alkaline to NaOH, with about a $1 M$ excess. The electrodes are replaced in this solution and a further electrolysis with efficient stirring carried out at an applied potential more negative than $E_d(\text{H}_2-\text{O}_2)$. When a test of the solution shows $[\text{Zn}^{2+}] < 10^{-6} M$, the electrodes are washed and dried. The cathode is now found to weigh 25.7006 ± 0.0002 g. What is the composition of the alloy relative to copper, lead, and zinc?   *Ans.* 57.84% Cu, 2.56% Pb, 38.59% Zn

17. A 200-ml volume of solution contains $[H^+] = 1.00 \times 10^{-4} M$ and $[\text{Cd}^{2+}] = 2.50 \times 10^{-2} M$. Electrolysis is conducted with efficient stirring between two large platinum gauze electrodes, a constant current of 3.0 A being maintained by an applied potential more negative than $E_d(\text{H}_2-\text{O}_2)$. Assume that the concentration of Cd^{2+} can be reduced under these conditions to less than $10^{-6} M$ (a quantitative separation). Calculate the approximate elapsed time in hours to allow the $[\text{Cd}^{2+}]$ to be reduced to $10^{-6} M$.   *Ans.* 0.90 h approximately

18. (*a*) Calculate the theoretical applied cathodic potential vs. SHE required in a controlled cathode potential electrolysis to separate quantitatively (10^{-6} *M* residual) the metal in each of the following solutions:

    (1) 0.0200 *M* Cu^{2+} in aqueous solution
    (2) 0.0100 *M* Ag^+ in aqueous solution
    (3) 0.0200 *M* copper in a solution 1.00 *M* to NH_3 at equilibrium
    (4) 0.0100 *M* silver in a solution 1.00 *M* to NH_3 at equilibrium
    (5) 0.0200 *M* copper in a solution 1.00 *M* to CN^- at equilibrium
    (6) 0.0100 *M* silver in a solution 1.00 *M* to CN^- at equilibrium
    (7) 0.0200 *M* copper in a solution 0.0100 *M* to EDTA $\{[Y^{4-}] = 0.0100\ M\}$ at equilibrium
    (8) 0.0100 *M* silver in a solution 0.0100 *M* to EDTA $\{[Y^{4-}] = 0.0100\ M\}$ at equilibrium

    (*b*) Determine the values of the applied cathodic potentials in *a* relative to (1) the SCE and (2) the saturated KCl Ag/AgCl electrode.

19. It is possible to separate *individually* Cl^-, Br^-, and I^- from solution by controlled anode potential electrolysis of the solution between a large platinum cathode and a massive silver anode. The typical anode reaction is given by

$$Cl^- + Ag - 1e \rightleftharpoons AgCl$$

Suppose that we have solutions of the following compositions:

    (1) 0.100 *M* each in Cl^- and Br^-
    (2) 0.100 *M* each in Br^- and I^-
    (3) 0.100 *M* each in Cl^- and I^-

    (*a*) Indicate in each case if the quantitative separation (10^{-6} *M* residual concentration of the ion separating first) is feasible.

    (*b*) Where a separation is feasible, indicate the controlled anode potential vs. SCE which permits the most quantitative separation.

20. 200 ml of solution contains $[Cu^{2+}] = 0.100\ M$, $[Bi^{3+}] = 0.050\ M$, and $[H^+] = 1.0 \times 10^{-2}\ M$. Indicate whether or not copper may be separated quantitatively from bismuth by controlled cathode potential electrolysis, using an assumed cathode potential control of ± 0.010 V. Consider the separation to be quantitative when the residual copper in solution is less than 0.0001 g.

21. Prepare a plot showing the approximate decay in time of the current effective in the deposition of copper, when constant-current electrolysis takes place between large smooth platinum gauze electrodes relative to an acid solution of copper(II) ion, where the following conditions exist.

$[Cu^{2+}]$ at the start	0.050 *M*
$[H^+]$ at the start	1.0 *M*
E_{app} more negative than $E_d(H_2-O_2)$ and $E_d(Cu-O_2)$	
Total current at the start	1 A
Solution volume	100 ml

22. Determine for the solution characteristics given in Prob. 21 the time required to reach a $[Cu^{2+}]$ in solution of 10^{-6} *M* (a quantitative separation of copper). What will be the approximate current effective in copper deposition at this point in the electrolysis?   *Ans.* 32 h, 1.8 μA

23. Prepare data and plot curves showing the effect of each of the following on the time interval required to reach 10^{-6} *M* residual $[Cu^{2+}]$ in the solution involved when constant-current electrolysis is involved. The starting solution characteristics are given in each case.

The effect of	Starting solution characteristics			
	V, ml	[Cu²⁺], M	[H⁺], M	I, A
Total current	100	0.050	0.50	...
Solution volume	...	0.050	0.50	4
Starting [H⁺]	100	0.050	...	4
Starting [Cu²⁺]	100	...	0.50	4

Sec. 12.4

24. A sample procured from an ore concentrating process contained both lead and cadmium but no other substances capable of interfering in the subsequent analytical procedure. A 0.8765-g sample of the material was taken, dissolved, and eventually secured in 100.0 ml of a 0.5-M solution of KCl. The dissolved oxygen was removed by nitrogen gas flushing, and the solution was eventually subjected to direct coulometry between a mercury cathode maintained at a controlled potential of -0.50 V vs. SCE and a compartmentized anode. At this applied potential, Pb(II) ion is reduced to Pb(0) but the Cd(II) ion is not affected. A water coulometer was placed in series with the coulometric apparatus, and when the current through the solution had decreased to the residual or background current previously determined for a 0.5-M KCl solution alone, it was noted that 24.65 ± 0.04 ml of mixed gases had been collected at 25.0°C and 745.00 mmHg. If the water vapor pressure at 25.0°C is 23.76 mmHg, calculate the percentage of lead in the concentrate. *Ans.* 15.07%

25. A silver-copper alloy used for brazing purposes is to be analyzed for silver only. A sample weight of 0.9743 g is dissolved, the solution composition and volume adjusted, and the resulting solution electrolyzed at a controlled cathodic potential which allowed only silver deposition at the cathode. The following current values were noted at the time intervals outlined.

Current, mA	Elapsed time, s
281	300
158	600
89	900

(a) Determine the percentage of silver in the alloy.

(b) On a theoretical basis, assuming 100 percent current efficiency, calculate the relative error in the determination of silver based on a current-time integrator value of Q if the electrolysis was stopped in 20 min (1200 s).

26. Lactic acid can be titrated coulometrically with generated hydroxyl ion. Under a constant current of 15.00 mA, 20.00 ml of a sample solution containing the acid requires 185.6 s to complete the titration. Calculate the number of milligrams of lactic acid in the 20.00 ml of sample solution tested.

27. Three samples of a contaminant containing arsenic, and no interfering compounds, are dissolved and treated in a manner resulting in the arsenic being available in the As(III) form. Each solution involved 0.1000 g of the material, and the solutions were diluted to exactly 100.0 ml. 5.00 ml of a standard solution containing 0.1000 g of As(III) per liter was added to a base solution containing 5 percent KI and sufficient $NaHCO_3$ to buffer the solution to about 8 pH. This solution was coulometrically titrated with anodically generated iodine, using potentiometric means to locate the equivalence point. The titration reaction is

$$As^{3+} + I_2 \rightleftharpoons As^{5+} + 2I^-$$

The constant current applied was 10.00 mA and the equivalence point occurred at 128.8 s. Successively, 5.00 ml of each unknown solution was added to the base solution, the times to reach the equivalence points being

Sample 1   98.9 s
Sample 2   101.6 s
Sample 3   105.3 s

Show the arsenic content for each of the three samples. Assume 100 percent current efficiency.

Ans. Sample 1: 7.68%; Sample 2: 7.89%; Sample 3: 8.18%

28. Consider again the Example in Sec. 12.4.2. Suppose that the coulometric process had been halted at 1000.0 s. Calculate the relative error in the determination of picric acid for this electrolysis period. What is the value in milligrams of picric acid that would be obtained?

29. Magnesium may be precipitated with oxine (8-hydroxyquinoline) according to the reaction

After separation from the precipitation medium, and adequate washing, the precipitate may be used in a gravimetric determination of magnesium. Alternatively, the precipitate may be dissolved in acid and the magnesium determined volumetrically by titration of the liberated oxine with bromine, using a standard solution of $KBrO_3$ in the presence of excess KBr. The titration reaction involves

Where very small amounts of magnesium are concerned, neither the gravimetric nor the volumetric procedures are satisfactory. The titration may be performed coulometrically, however, using anodically generated bromine.

   A 1.0048-g sample of an industrial waste is dissolved, the magnesium precipitated with oxine, and the precipitate separated, washed, and dissolved in an acid solution. The liberated oxine is titrated coulometrically, after the addition of excess KBr, under a constant current of 10.00 mA. The end point was noted in 189.6 s. Assuming 100 percent current efficiency, determine the percentage of magnesium in the industrial waste.

30. Using the general technique described in the Example in Sec. 12.4.2 it was noted that 6.99 mg of pure picric acid required 53.0 C, as measured by a coulometer, to be quantitatively converted to triaminophenol at a controlled potential mercury cathode. What is the electron transfer in the reduction process?

31. Phenol may be determined by a titration involving a standard solution of $KBrO_3$ in an acidified solution containing an excess of KBr. The reaction involved is

A pharmaceutical product is known to contain phenol but no compounds capable of interference in the above reaction. The density of the compound (a liquid) at 25°C is 0.896 g/cm³. 1.000 ± 0.005 ml of the product is taken and diluted to exactly 250 ml. 5.00 ± 0.01 ml of this solution is added to an acidified solution containing excess KBr, and the phenol is titrated by anodically generated bromine, using a constant current of 15.00 mA. The end point is detected at 543.6 s. Assume a current efficiency of 100 percent, and calculate the percent phenol by weight in the product. *Ans. 7.40%*

32. A metallic casting used in a corrosive environment started to develop a coating of a corrosion product. The presence of iron in the product was suspected. A 0.1348-g sample of the product was dissolved and the iron content isolated in a solution in the Fe(II) state. A compartmentized cathode and a platinum anode were immersed in the solution, which also contained a supporting electrolyte. The Fe(II) was oxidized to Fe(III) at the anode under a controlled potential of 1.05 V vs. a saturated KCl silver–silver chloride electrode. A series-connected chemical coulometer was used, where iodine was generated at the anode. The generated iodine representing the quantitative conversion of Fe(II) to Fe(III) was titrated with 0.0156 M $Na_2S_2O_3$ solution, and the starch end point occurred at 15.65 ± 0.04 ml. A test conducted on the same background solution and the same volume, using the same applied anodic potential for an identical time period, but in the absence of any *added* Fe(II), yielded a titration volume of 0.54 ± 0.01 ml of the $Na_2S_2O_3$ solution.

(*a*) What is the percentage of iron in the corrosion product?

(*b*) Discuss the possible sources of the current which yielded the value of 0.54 ml of $Na_2S_2O_3$ solution in the absence of added Fe(II).

33. A hydrocarbon fuel source is periodically checked for sulfur content in order to control the SO_2 content of the stack gas. A sample involving 0.1055 g of the fuel is pyrolyzed, and the combustion product of interest, SO_2, is absorbed in a solution containing excess iodine and iodide. The iodine-iodide solution is under precise potentiometric control. The reaction

$$SO_2 + I_3^- + 2H_2O \rightleftharpoons SO_4^{2-} + 4H^+ + 3I^-$$

takes place with a corresponding change in the potential for the I_3^-/I^- system. This change in potential, through the potentiometric control device, actuates a coulometric generating electrode pair in the solution, generating iodine anodically at a constant current of 5.000 mA until the potential returns to its original value. For the sample in question, a return to the original potential occurs after a period of 124.3 s. What is the percentage of sulfur in the fuel? *Ans. 0.0979%*

34. A copper recovery plant is engaged in burning off the insulating coating from discarded electrical copper wiring. Much of the coating material involves chlorine-containing plastic materials, such as polyvinyl chloride. The gases issuing from the furnace are monitored for chlorine content as HCl in order to control a scrubber operation designed to remove the HCl from the furnace gases before escape to the atmosphere.

A 100-ml sample of the furnace gas product is automatically taken every 20 min, and the HCl content is absorbed and reacted by 100 ml of a dilute solution of NaOH. The NaOH solution is monitored by a very precise potentiometric device. The absorption of HCl results in a decrease in the $[OH^-]$ of the solution, and this causes the potentiometric device to activate a platinum electrode pair connected to a constant-current supply. OH^- is generated cathodically until the potential returns to its original value. A particular sample of 100 ml of gas requires 387.3 s at a constant current of 20.00 mA to restore equilibrium. What is the percent of Hcl in the furnace gas product? Use the relationship 1 mol HCl provides 22.414 liters of HCl vapor.

35. The calcium content of a common pharmaceutical product is checked by dissolving a 250-mg tablet in 50.00 ml of water. 5.00 ml of this solution is added to a properly buffered solution containing an excess of $HgNH_3(Y)^{2-}$. HY^{3-} is then generated at a mercury cathode (see Sec. 12.4.3) under a constant current of 12.50 mA. 92.4 s is required to reach the end point of the titration. What is the milligram content of calcium per tablet?

Sec. 12.5

36. A polarographic diffusion current $\bar{\imath}_d$ was measured as 5.86 μA under a head of mercury of 50.0 cm. What will be the relative error in the current when measured at a mercury head of 55.0 cm? Do not take into consideration the back-pressure effect of the solution.

37. The change in concentration for a species O involved in the dropping mercury electrode reaction

$$O + 1e \rightleftharpoons R$$

is negligible. Suppose that we have a drop life of 4.35 s, a flow rate of 1.00 mg/s of mercury, and a D_O value of 3.6×10^{-5} cm^2/s. The solution is initially 1.00 mM with respect to O. If the current is assumed to be constant over a 2-h period of continuous examination, calculate the decrease in [O] for a solution whose volume is 25.00 ml.   *Ans.* Decrease 0.014 mM

38. An electrode reaction at the dropping mercury electrode yields the following polarographic data.

$E_{applied}$ *vs.* SCE, V	*Average diffusion current,* μA
-0.653	0.45
-0.660	0.75
-0.667	1.20
-0.675	1.75
-0.683	2.50
-0.691	3.18
-0.697	3.60
-0.704	4.00
-0.718	4.40

The average diffusion current $\bar{\imath}_d$ at plateau value is found to be 4.64 μA. Use a graphic technique to show, within the graphic limits, whether or not the electrode reaction is reversible. If the reaction is proven to be reversible, determine the value of $E_{1/2}$ and n from the graph, and compare these values with those obtained by solving simultaneous equations involving selected data from the above.

39. A hydrogen-ion concentration dependency exists for the polarographic dropping mercury electrode reaction

$$O + mH^+ + ne \rightleftharpoons R + wH_2O$$

The following values of $E_{1/2}$ for the reaction were obtained by making separate plots of $E_{applied}$ vs. log $\{(\bar{\imath}_d - \bar{\imath})/\bar{\imath}\}$ for the system as buffered to various pH values. These curves indicated reaction reversibility, and n was determined from the slope as 6.

$E_{1/2}$ *vs.* SCE, V	pH
1.03	0.5
0.96	1.0
0.91	1.4
0.85	1.8
0.82	2.0

Determine the value of $E_{1/2}$ for a solution of pH $= 0$.   Determine the value of m in the reaction indicated.   *Ans.* $E_{1/2} = 1.10$ V vs. SCE: $m = 14$

40. A particular metal complex ion may be reduced with amalgam formation at the dropping mercury electrode. The metal ion in a solution without the complexing ligand yields a plot of $E_{applied}$ vs. log $\{(\bar{i}_d - \bar{i})/\bar{i}\}$ which indicates that the reaction is reversible and gives a value of $E_{1/2,a}$ of 0.050 V vs. SCE, with a value of $n = 2$. The following values of $E_{1/2,c}$ were obtained for metal complex ion solutions of different excess ligand concentrations by plotting separately $E_{applied}$ vs. log $\{(\bar{i}_d - \bar{i})/\bar{i}\}$.

$E_{1/2,c}$ *vs.* SCE, V	*Ligand concentration, M*
-0.658	1.0
-0.632	0.6
-0.611	0.4
-0.540	0.1

From a plot of $E_{1/2,c}$ vs. log ligand concentration determine the value of the coordination number for the metal complex ion. Determine the value of $K_{instability}$ for the metal complex ion.

41. The atmosphere of a foundry specializing in the production of copper-base alloy castings containing lead must be tested daily to determine the lead content for health reasons. A small electrostatic precipitator treats 0.10 m^3 of air per minute, precipitating the contained solid matter against the walls of a tube. The deposit is eventually washed off and treated and the lead content is subsequently determined polarographically. A 5-h sampling period in a given location in the foundry is involved, and the deposit is eventually dissolved and secured in exactly 100.0 ml of solution. 10.00 ml of this solution is added to 30 ml of 1 M KCl, and the resulting solution is diluted to exactly 50.0 ml. After removal of the dissolved oxygen by nitrogen degassing, the average diffusion current $\bar{i}_d$ is measured as 0.98 μA at an applied potential of -0.50 V vs. SCE. 10.00 ml of 0.000300 M Pb^{2+} solution is now added, the solution degassed again, and the average diffusion current determined as 1.12 μA. 30.00 ml of 1 M KCl, with 10.00 ml of "blank" solution (no lead from atmospheric sampling) is diluted to 50.0 ml and degassed and the average diffusion current is read as 0.05 μA under the same conditions. What is the atmospheric lead content in this area in milligrams per 10 m^3 of air?

42. A solution containing Cu^{2+} and Bi^{3+} is to be tested for [Cu^{2+}] and [Bi^{3+}] polarographically. In a supporting electrolyte of 2 M acetic acid–2 M ammonium acetate the half-wave potentials are -0.07 and -0.25 V vs. SCE for Cu(II) and Bi(III), respectively. In order to provide good resolution of the two waves at variable concentrations of the two elements, the $E_{1/2,a}$ values should have a somewhat larger separation. The addition of a ligand complexes the Bi^{3+}, resulting in a complex with a $K_{stability}$ of 10^{20}. Cu^{2+} is not affected. Calculate the separation now of $E_{1/2,a}$ for Cu(II) and $E_{1/2,c}$ for the bismuth(III) complex ion.   *Ans.* -0.57 V

43. 50.00 ml of a solution containing Pb^{2+} is to be titrated amperometrically with a standard solution of K$_2$CrO$_4$. The titration reaction is

$$Pb^{2+} + CrO_4^{2-} \rightleftharpoons PbCrO_4(s)$$

For comparison purposes 50.00-ml portions of the lead solution are titrated with
      (1) 0.0500 M K$_2$CrO$_4$
      (2) 0.5000 M K$_2$CrO$_4$
after degassing with nitrogen. After each addition of titrant nitrogen is again bubbled through the solution to degas and stir prior to making the average diffusion current measurement. The current is

measured at an applied potential of -1.00 V vs. SCE at the dropping mercury electrode, at which potential both Pb^{2+} and CrO_4^{2-} are reduced. The following titration data are obtained.

	Titration 1		Titration 2	
Volume K_2CrO_4, ml (± 0.04)	Current $\bar{i}_d, \mu A\ (\pm 0.3)$		Volume K_2CrO_4, ml (± 0.004)	Current $\bar{i}_d, \mu A\ (\pm 0.3)$
0.00	157.6		0.000	157.6
4.00	124.4		0.400	132.6
8.00	95.9		0.800	108.0
12.00	71.0		1.200	83.8
16.00	49.1		1.600	60.0
20.00	29.7		2.000	36.5
24.00	12.4		2.400	13.4
28.00	26.3		2.800	34.1
32.00	43.8		3.200	62.2
36.00	59.7		3.600	89.8
40.00	74.2		4.000	116.9

Determine:

(a) The curve for titration 1 from the recorded data

(b) The curve for titration 1 from the microampere recorded data corrected for the titrant volume dilution effect

(c) The curve for titration 2 from the recorded data

(d) The curve for titration 2 from the microampere recorded data corrected for the titrant volume dilution effect

(e) The concentration of lead in the original solution in moles per liter from the curves for b, c, and d

(f) Your comments relative to the methods of titration involved

Sec. 12.6

44. Consider the following conductometric titrations. Use the equivalent ionic conductance values given in Table 12.5. The cell constant will be 0.200 cm^{-1} in each case. Calculate the conductance values for the titration points shown. Where the titrant solution is less than 20 times the concentration of the reactant solution, make the necessary titrant volume dilution effect corrections. Draw each titration curve, determine the equivalence-point volume from the curve and compare the value with the theoretical value.

(a) 50.00 ml of 0.0108 M HCl vs. 0.0145 M NaOH. Consider the titration points

0.00	5.00	10.00	15.00	20.00	25.00	30.00	35.00	40.00	45.00
50.00	55.00	60.00							

Ans. 5.00 ml $= 0.0208$, 45.00 ml $= 0.0096$

(b) 50.00 ml of 0.0100 M CH$_3$·COOH vs. 0.2000 M NaOH. Take K_a as 2.00×10^{-5} and consider the titration points

0.000	0.500	1.000	1.500	2.000	2.500	3.000	3.500	4.000
4.500	5.000							

Ans. 1.000 ml $= 0.00187$, 4.500 ml $= 0.0145$

(c) 50.00 ml of 0.0180 M $CH_3 \cdot COOH$ vs. 0.500 M NH_3. Take K_a as 2.00×10^{-5} and K_b as 1.00×10^{-5}. Consider the titration points

0.000	0.400	0.800	1.200	1.600	2.000	2.400	2.800
3.200	3.600	4.000					

Ans. 0.800 ml = 0.00462, 3.200 ml = 0.01081;

(d) 50.00 ml of 0.0500 M $CH_3 \cdot COONa$ vs. 0.500 M $HClO_4$. Take K_a as 2.00×10^{-5} and consider the titration points

0.000	0.500	1.000	2.000	3.000	4.000	5.000	6.000
7.000	8.000	9.000					

Ans. 1.000 ml = 0.0241, 8.000 ml = 0.0922;

(e) 50.00 ml of 0.0125 M $(CH_3 \cdot COO)_2$ Pb vs. 0.498 M H_2SO_4. Titration points

0.000	0.200	0.400	0.600	0.800	1.000	1.200	1.400
1.600	1.800	2.000	2.200	2.400			

Ans. 0.400 ml = 0.0105, 1.600 ml = 0.0129

REFERENCES

1. Bates, R. G.: "Determination of pH," Wiley, New York, 1964.
2. Delahay, P.: "Instrumental Analysis," Macmillan, New York, 1956.
3. Heyrovsky, J.: "Polarographie," Springer, Vienna, 1941.
4. Kolthoff, I. M., and J. J. Lingane: "Polarography," 2d. ed., 2 vols., Interscience, New York, 1952.
5. Lingane, J. J.: "Electroanalytical Chemistry," 2d. ed., Interscience, New York, 1958.
6. Meites, L.: "Advanced Analytical Chemistry," McGraw-Hill, New York, 1958.
7. Meites, L.: "Polarographic Techniques," 2d. ed., Interscience, New York, 1965.
8. Milner, G. W. C., and G. Phillips: "Coulometry in Analytical Chemistry," Pergamon, Oxford, 1967.
9. Purdy, W. C.: "Electroanalytical Methods in Biochemistry," McGraw-Hill, New York, 1963.
10. Rechnitz, G. A.: "Controlled-potential Analysis," Macmillan, New York, 1963.
11. Ross, J. W.: Conductometric Titrations, in F. J. Welcher (ed.), "Standard Methods of Chemical Analysis," pp. 297–322, Van Nostrand, New York, 1966.
12. Zuman, P.: "Organic Polarographic Analysis," Pergamon, Oxford, 1964.

13
Absorptiometry and Absorptiometric Methods of Analysis

13.1 GENERAL

A large body of analytical techniques is based on the ability of substances to emit or absorb electromagnetic radiation. Many of these methods are of fairly recent origin, although the general principles of the processes have been understood for many years. A relationship can be established relative to the intensity of a given radiation wavelength as emitted by a substance and the concentration of the substance in the medium under examination. Similarly, a relationship may be established between the ability of a substance to absorb radiation of a given wavelength and the concentration of the substance in the matrix involved. Our interest will center around methods of quantitative analysis which are based on relationships involving absorptivity. Such methods are, in general, identified as *absorptiometric methods*.

In this particular context, what we shall measure is the radiation intensity for a given wavelength as transmitted by the medium containing the absorbing substance. Since for a primary or incident beam of constant intensity, the transmitted intensity will depend on the extent to which the primary beam is absorbed and therefore on the concentration of the absorbing substance in the medium, a rela-

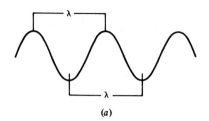

(a)

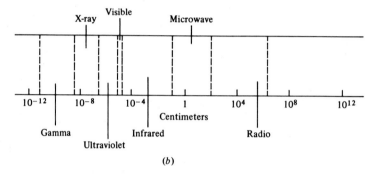

(b)

Fig. 13.1 (*a*) A wavelength; (*b*) the electromagnetic spectrum.

tionship can be established which associates the transmitted intensity and the concentration. Although absorptiometric methods are not restrictive with respect to the wavelength of the examining radiation, and this may range over the entire electromagnetic spectrum, our discussions will be directed in the main to *absorptiometry* as associated with that very small portion of the spectrum which encompasses the visible light zone.

13.2 ELECTROMAGNETIC RADIATION AND ITS PROPERTIES

13.2.1 THE ELECTROMAGNETIC SPECTRUM

Electromagnetic radiation represents the transfer and propagation of energy by alternating associated electric and magnetic force fields. Its wavelike properties may be described by mutually perpendicular electric and magnetic vectors, both vectors being perpendicular to the direction of propagation of the radiation wave. Such waves are characterized by the properties of wavelength, velocity of propagation, frequency, etc. The *wavelength* refers to the distance separating adjacent wave peaks or troughs, as indicated by Fig. 13.1. The *wave number*, which represents the number of waves per unit of length, is the reciprocal of the wavelength. The speed at which the wave front advances is called the *velocity*. The number of waves passing a given point in unit time is described as the *frequency*. The relationship between these properties is given by

$$\frac{1}{\lambda} = \bar{\nu} = \frac{\nu}{v} \tag{311}$$

where λ = wavelength
ν = frequency
$\bar{\nu}$ = wave number
v = velocity in the medium involved

The velocity of electromagnetic radiation in a vacuum is given by c at about 3×10^{10} cm/s, and the velocity in any other medium is less than the value of c. It should be noted that the frequency of a given radiation is fixed by its source. Since the velocity will vary according to the medium, it is apparent that radiation of a given frequency will show varying wavelength according to the medium through which it travels.

The terms of expression of the wavelength often depend on the spectral region being described. The angstrom unit, Å (10^{-7} mm), is commonly used to describe x-ray radiation wavelength, the nanometer, nm (10^{-6} mm), for ultraviolet and visible wavelength, and the micron, μm (10^{-3} mm), for infrared wavelength. The wave number in centimeter^{-1} units is also frequently encountered as a descriptive term. The spectrum of electromagnetic radiation encompasses a very wide range. A very generalized picture of this spectrum, together with the descriptive terms for the various wavelength zones, is given by Fig. 13.1. Where the visible zone is concerned, the eye receives the impression of light and color. The color seen depends on the wavelength or wavelength combinations of the light reaching the eye. Where well-distributed radiation covering the entire visible spectrum (about 400 to 750 nm) reaches the eye, the color interpretation of "white" is made. Where such radiation passes first through a medium which transmits some wavelengths and absorbs others, the eye will record the color of the medium in terms of the transmitted wavelengths. A solution will appear blue, for example, where the green, yellow, and red wavelength components (about 500 to 750 nm) are largely absorbed by the solution, with the blue wavelength components (about 400 to 500 nm) being transmitted.

13.2.2 THE INTERACTION OF RADIATION ENERGY AND MATTER

While the interpretation of electromagnetic radiation as a wave motion permits the explanation of such optical phenomena as reflection, refraction, interference, diffraction, etc., certain effects are more readily explained by assuming that radiation is represented by a stream of particles or bundles of energy, each bundle being called a *photon* or a *quantum*. The amount of energy per quantum varies from one frequency of radiation to another but is constant for radiation of a specific frequency. The quantum energy relationship to frequency is given by

$$E = h\nu = \frac{hc}{\lambda} \tag{312}$$

where E is the energy per quantum (ergs) and h is Planck's constant (6.626×10^{27} erg·s). Thus, the quantum energy varies directly with the frequency and, for light propagation in a vacuum, inversely as the wavelength. The *total energy* of a molecule is made up of a number of components, and three of these are of importance relative to the absorptiometric aspect of analytical chemistry. These are

1. The *rotational energy*, which involves the rotation of the molecule
2. The *vibrational energy*, which involves interatomic vibrational motion within the molecule
3. The *electronic energy*, which involves the energy levels occupied orbitally by the electrons of the atoms of the molecule

For a given molecule, changes in the energy levels for each of these components may take place, but only within the limitations of certain allowable energy levels. For example, a molecule has certain permitted levels of rotational, vibrational, and electronic energy. To bring about a transition from a lower to a higher level, for any of these energies, a *specific* quantity of energy is required, and the transition can not be brought about by smaller or greater energy quantities. This "quantization" of energy for permissible transitions in rotational, vibrational, or electronic energy levels, together with the frequency and energy per quantum relationship for electromagnetic radiation, leads to the basis of the interaction of matter and radiant energy.

Where radiation falls on molecular aggregates, effects relative to changes in rotational, vibrational, or electronic energy levels will take place only where the radiation includes frequencies providing quantum energies exactly matching the transition energy requirements. The irradiation of a molecular species by polychromatic radiation will result in the absorption of those frequencies alone which have quantum energies capable of bringing about permissible transitions for the molecular species involved. All other frequencies will be transmitted. The quantum energies required to initiate rotational, vibrational, and electronic transitions vary very considerably, being relatively high for electronic shifts (x-ray, ultraviolet, and visible radiation), lower for vibrational shifts (infrared radiation), and much lower for rotational shifts (far infrared and microwave radiation).

Where white light is used to irradiate an appropriate molecular species, electronic energy transitions will take place, and these will, in general, dictate the region of the visible spectrum where absorption takes place. It might be expected that for a single type of electronic transition, this would involve the absorption of radiation of an isolated wavelength. However, the electronic energy level transitions will be accompanied by vibrational and rotational energy transitions, and these latter will be superimposed on the absorption region for the electronic transitions. Thus, the absorption region will not be a very narrowly restricted spectral zone but will involve wavelength bands on either side of the optimum wavelength absorbed for the electronic transitions. These multiple bands are not resolved

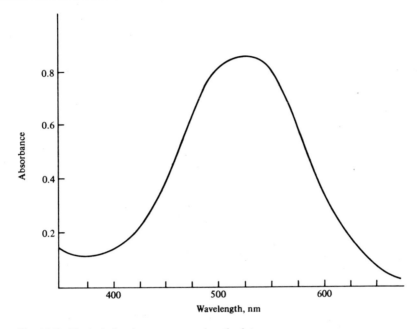

Fig. 13.2   Typical absorbance-vs.-wavelength plot.

under normal instrumental conditions, so that the absorbance-vs.-wavelength pattern takes on generally a relatively broad-based form. Figure 13.2 gives some idea of such an absorbance-vs.-wavelength relationship.

13.3   ABSORPTION OF RADIATION: QUANTITATIVE BASIS

13.3.1   BOUGUER-LAMBERT LAW

Figure 13.3 shows a schematic representation of the arrangement for the irradiation of a solution containing an absorbing species, using monochromatic radiation in the visible zone. The Bouguer-Lambert law states that the ratio of the transmitted light intensity or radiant power P to the incident radiant power P_0 will be a

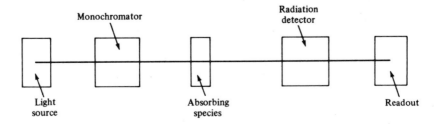

Fig. 13.3   The irradiation of an absorbing species.

function of the path length b through the absorbing medium. It can be shown that

$$\Delta P = -k_1 P \, \Delta b \tag{313}$$

where ΔP is a small increment of change in radiant power brought about by the small path-length increment Δb. The minus sign indicates that the change involves a decrease in radiant power. Thus we have

$$-\frac{\Delta P}{P} = k_1 \, \Delta b \qquad \text{and} \qquad -\frac{dP}{P} = k_1 \, db \tag{314}$$

and, by integration between P and P_0 for the path lengths b and zero, we have

$$-\int_{P_0}^{P} \frac{dP}{P} = \int_0^b k_1 \, db$$

$$\ln \frac{P_0}{P} = k_1 b \qquad \text{and} \qquad \log \frac{P_0}{P} = k_2 b \tag{315}$$

with

$$\frac{P_0}{P} = 10^{k_2 b} \tag{316}$$

Thus the transmitted radiant power P varies exponentially as b and directly as P_0. Note that in its strictest sense, the law holds for monochromatic radiation only.

13.3.2 BEER'S LAW

Beer approached the situation from the point of view of the effect of the concentration of the absorbing species under a constant path length b. He indicated that the transmitted radiant power P varied exponentially as the concentration c and directly as the incident radiant power P_0. The mathematical expression of this law, derived in the same manner as outlined for the Bouguer-Lambert law, is

$$\log \frac{P_0}{P} = k_3 c \tag{317}$$

with

$$\frac{P_0}{P} = 10^{k_3 c} \tag{318}$$

It can be easily shown mathematically that the two laws may be combined to yield

$$\log \frac{P_0}{P} = K b c \tag{319}$$

with

$$\frac{P_0}{P} = 10^{K b c} \tag{320}$$

These expressions are often generally referred to as Beer's law. This law, as given by Eq. (319) or (320), is again in the strictest sense valid for monochromatic radiation only. Where c is expressed in grams per liter and b in centimeters, the constant K is replaced by a, and a is called the *absorptivity*. Where c is expressed in moles per liter, with b in centimeters, K is replaced by ϵ, the *molar absorptivity*. We shall, in general, make use of Beer's law in the gram per liter concentration form as

$$\log \frac{P_0}{P} = abc \quad \text{and} \quad \frac{P_0}{P} = 10^{abc} \tag{321}$$

Several terms relative to absorptiometry should now be noted. These are

1. $\log P_0/P = abc = A$ A is called the *absorbance*
2. $P/P_0 = 10^{-abc} = T$ T is called the *transmittance*
3. $P/P_0 \times 100 = \% \, T$ $\% \, T$ is called the *percent transmittance*

A plot of absorbance A vs. concentration c for a set path length b will yield, within the solution concentration range where Beer's law is obeyed, a straight line of slope ab. A plot of transmittance T vs. c under similar conditions will yield an exponential curve, but a plot of $-\log T$ vs. c will yield a straight line of slope ab. A plot of T vs. c made on semilog graph paper, with T being plotted on the logarithmic axis, will yield a straight line of slope $-ab$. Most absorptiometric instruments incorporate percent transmittance and absorbance scales, the former being linear since $\% \, T$ varies directly with P, the latter being logarithmic since A varies logarithmically with P.

13.4 DEVIATIONS FROM BEER'S LAW

13.4.1 CHEMICAL SOURCES OF DEVIATION

With systems where Beer's law applies, the linearity of A vs. c is usually restricted to certain fairly well defined concentration ranges. The law is not capable of being applied to some systems. In certain cases the system shows apparent deviations from Beer's law on the basis of chemical reactions occurring in the solution. For example, we can imagine a fairly general analytical absorptiometric system where the substance to be determined reacts to produce a compound BA which dissociates

$$BA \rightleftharpoons B^+ + A^-$$

where BA is 100 percent dissociated regardless of concentration, and A^- is the absorbing species in the absorptiometric system. Here the value of absorbance will be directly related to the concentration of A^-, which, in turn, is directly related to the quantity of the substance to be determined. On the other hand, if the degree of dissociation of BA is a function of its concentration, as it would be for weak electrolyte BA substances, and if the degree of dissociation over the concentration ranges anticipated for analytical purposes changes significantly, then the absorbance will be a function of the concentration of A^-, itself a function of both the quantity of the substance to be determined and the degree of dissociation of BA. The plot of A vs. c, where c is the concentration of the unknown, will yield a nonlinear curve of the type shown in Fig. 13.4. Similar situations will occur where the concentration of the absorbing species is affected by complexation or hydration (pH dependency) reactions.

13.4.2 INSTRUMENTAL SOURCES OF DEVIATION

Beer's law, as noted previously, can be strictly applied only where monochromatic incident radiation is used, the particular wavelength involved being normally that capable of providing the maximum change in absorbance per unit change in concentration. Where polychromatic radiation is used, certain wavelengths will be much more strongly absorbed than others. At each successive solution layer Δb, less of the strongly absorbed wavelengths will be available, so that each successive layer will absorb a slightly lower fraction of the total radiant power incident at that layer. This is in direct opposition to the Bouguer-Lambert and Beer laws, which demand that at each successive layer, the same fraction of the radiant

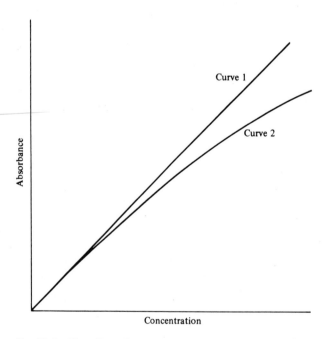

Fig. 13.4 The effect of varying degree of dissociation on the plot of absorbance vs. concentration. Curve 1: Beer's law obeyed rigidly. Curve 2: Beer's law deviation based on degree of dissociation.

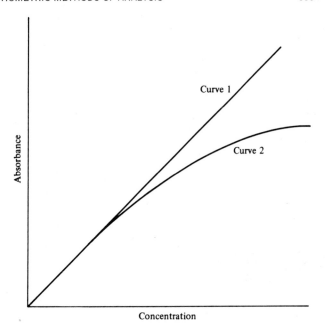

Fig. 13.5 The effect of bandwidth for monochromated light on the plot of absorbance vs. concentration. Curve 1: effective bandwidth about 2 nm. Curve 2: effective bandwidth about 50 nm.

power incident at each layer be absorbed. It can be shown mathematically that the greater the difference between the minimum and maximum wavelength values for the polychromatic incident radiation, the greater will be the introduced deviation from Beer's law.* Figure 13.5 shows the general effect of polychromatic incident light on a plot of A vs. c. Note the departure from linearity.

13.5 INSTRUMENTATION IN VISIBLE LIGHT ABSORPTIOMETRY

13.5.1 VISIBLE LIGHT SOURCES

The most commonly used visible light absorptiometers, or, as they are called, *photometers* or *spectrophotometers*, use as a light source a glass-enclosed tungsten filament light bulb. The filament temperature, at normal operating voltages of about 6 to 12 V, is about 2400°C; and the emitted radiation is well distributed through the visible spectrum, although there is some concentration of radiant power in the near infrared. This radiant power concentration manifests itself as heat, and the heat output of the tungsten lamp is often sufficient to require that a water jacket should surround the lamp housing, or that at least a water cell should

* L. Meites and H. C. Thomas, "Advanced Analytical Chemistry," pp. 255–256, McGraw-Hill, New York, 1958.

be placed in the optical path immediately in front of the lamp housing. Since the lamp output of total radiant power varies significantly with the input voltage, the better instruments incorporate a voltage control device which limits input ac line voltage variations to the transformer to output variations of a few millivolts.

13.5.2 MONOCHROMATORS

The function of the *monochromator* is to remove from the radiation output from the source wavelengths other than that required for the examination of the particular species involved. The critical characteristics of a monochromator are

1. The effective bandwidth, which is expressed as the width of the transmittance band at one-half the maximum transmittance value
2. The value of the wavelength at the center of the transmittance band
3. The value of transmittance at this central wavelength value

In order, these characteristics express the degree of monochromation, the effective wavelength transmitted by the monochromator, and the radiant power transmitted at this wavelength. Figure 13.6 shows a typical percent-transmittance-vs.-wavelength plot for a monochromator, with the characteristics clearly shown.

    Monochromators for visible light absorptiometers may be the transmission filter type, the interference filter type, a prism, or a diffraction grating. The trans-

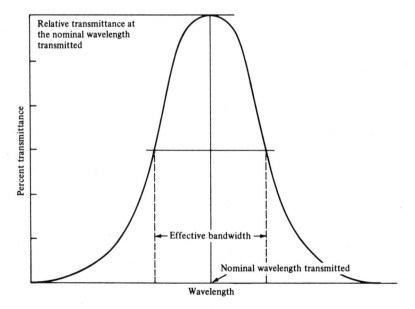

Fig. 13.6 Percent transmittance vs. wavelength for a monochromator.

mission filter is a relatively simple device which absorbs strongly at certain wavelengths and weakly at others. The monochromation ability of such filters depends to a considerable extent on their nature. Colored-glass filters will ordinarily yield transmitted radiation with an effective bandwidth of 20 to 50 nm. It will be noted that this is by no means monochromatic light, and transmission filter photometers often show Beer's law deviations relative to absorbance-vs.-concentration plots. The interference filter consists normally of a thin sheet of clear dielectric substance, such as calcium fluoride, enclosed by two thin glass sheets the inner surfaces of which have been coated with a reflecting but semitransparent metallic material such as silver. Interference effects, originating with the reflection of light from the two metallic coatings, result in cancellation of out-of-phase wavelengths and reinforcement of in-phase wavelengths, the latter being the radiation wavelengths exiting from the filter. The thickness of the dielectric sheet is carefully controlled and dictates the wavelength value provided by the filter. Such filters are capable of effective bandwidths of about 5 to 10 nm.

Prism monochromators function on the basis of variation in the index of refraction of the prism material with variation in the incident light wavelength. Red light (longer wavelengths) is thus refracted at a lesser angle than blue light (shorter wavelengths), so that a prism is able to disperse incident white light into its visible spectrum. The use of a slit at the focal plane of the spectrum allows the isolation of the wavelength required for absorptiometric purposes. Since the spectral dispersion is greater (in say nanometers per millimeter of spectral length) at the blue than at the red end, a slit, in order to provide approximately the same effective bandwidth along the spectrum, must be adjustable to a degree, being narrower at the red end than at the blue end. The narrowness of the slit is, of course, limited by the necessity of avoiding slit edge diffraction effects and the need to pass sufficient radiant power to activate the absorptiometric system. The effective bandwidth of prism monochromators, usually of optical glass for visible light spectrophotometers, is about 1 to 2 nm.

Diffraction grating monochromators involve glass (transmission) or polished metal (reflecting) surfaces ruled with a very large number of parallel lines very closely spaced (1000 to 2000 lines per millimeter). For light from the source incident to such a grating, the areas between the lines function as prime sources of light, and interference effects relative to the light emitted by these sources result, for specific emergent angles relative to the optical axis of the incident light, in cancellation of certain wavelengths and reinforcement of others. The reinforced wavelength for a given grating will be a function of the emergent angle, so that a visible spectrum is thus obtained from incident white light. A slit traversing this spectrum will permit isolation of the required wavelength value. Since the dispersion of a diffraction grating is almost identical throughout the spectrum, the slit need not be adjustable with respect to its width, although avoidance of slit edge diffraction effects imposes limitations, as before, relative to the minimum width. Grating instruments are capable of yielding an effective bandwidth of about 1 to 2 nm.

13.5.3 DETECTORS

For visible light photometers and spectrophotometers the detector, which evaluates the radiant power of the light transmitted to it, must show a high response to the visible spectral region, fast response time, capability of response amplification, and a linear response to changes in transmitted radiant power. Two types of photoelectric detectors are commonly used. One of these devices, the phototube, is shown in Fig. 13.7. It consists of an evacuated glass bulb containing a metallic cathode large enough in area to intercept the complete transmitted beam. This cathode is coated with a layer of a metal capable of emitting electrons when light ranging over the wavelengths of the visible spectrum impinges upon it. A potassium coating on a silver cathode will respond adequately to a wavelength range of about 250 to 700 nm. The electrons emitted are attracted to a wire anode, establishing a current proportional to the radiant power of the beam falling on the cathode. This current may be amplified and the amplified current read directly by a microammeter or measured by a potentiometer.

Alternatively, a photomultiplier tube may be used as the detector. This device allows greater sensitivity, and since the current multiplication effect is carried out within the tube, external amplification circuits may in many instances be eliminated. Such a tube contains a primary cathode of the same general nature as that used in the phototube, as well as several electrodes or dynodes each more positively charged than the preceding one. The electrons emitted by the cathode

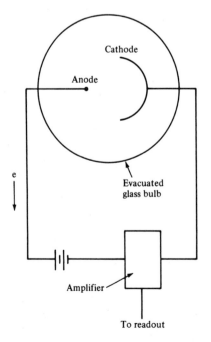

Fig. 13.7   The phototube.

are accelerated to and focused on the first dynode, which then releases many more secondary electrons. The accumulated electrons are then accelerated and focused on the next dynode, and so on. The final collecting anode then receives an electron flow which may be as much as 10^8 times that originating from the primary cathode. Such larger currents are often used for measuring purposes without amplification, although under certain circumstances external amplification circuits are also employed.

A number of phototubes or photomultiplier tubes may be used interchangeably, each detector being highly sensitive to a specific spectral zone. This type of operation allows maximum sensitivity for the monochromated radiation being used for analytical purposes.

13.5.4   THE SAMPLE CELL

The solution to be examined absorptiometrically is contained in a sample cell or cuvette. These cells, for work involving the visible spectrum, are made of glass. Ideally, sample cells should have incident and emergent surfaces that are perfectly flat (to avoid reflection losses), parallel to each other, and free from scratches or etching effects; and each cell of a given type should provide an identical path length. Cylindrical cells, with curved incident and emergent surfaces, may also be used, and indeed are often used in the interests of economy; but care should be taken that each cell is marked so that its insertion in the cell holder of the instrument always provides for the same incident and emergent surfaces.

13.5.5   ABSORPTIOMETRIC INSTRUMENTATION

General   Photometers normally involve fairly simple instrumentation, consisting essentially of a light source, a filter monochromator, a sample cell holder, a photoelectric detector, and a current measuring device. Spectrophotometers, commonly more sensitive instruments, involve a light source, a prism or grating monochromator, a slit, a sample cell holder, a photoelectric detector, and a current measuring device. In our descriptions we shall restrict ourselves to spectrophotometric instrumentation.

Single-beam instruments   Figure 13.8 shows a typical single-beam spectrophotometer. The intensity of the light from the source can be controlled by a supply voltage regulator, and the source light is then passed to a collimator. The light is monochromated by a diffraction grating and slit device and passed to the sample cell and the transmitted radiant power is evaluated by a photoelectric detector. The current from the detector may or may not be amplified prior to being measured directly by a microammeter. The use of single-beam instruments involves, typically, the use of an opaque shutter to cut off the detector from the optical path. The small current (dark current) then flowing in the detector due to low-level thermal emission of electrons is canceled out by a balancing device, so that the meter

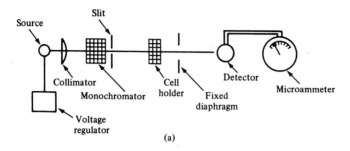

(a)

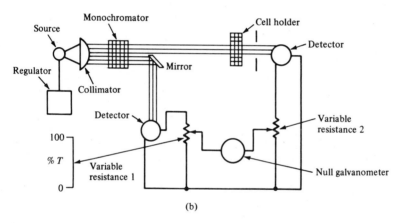

(b)

Fig. 13.8 (*a*) Single-beam spectrophotometer; (*b*) simple double-beam spectrophotometer.

scale reads zero transmittance (infinite absorbance). The monochromator is then set to transmit the wavelength of maximum absorbance for the analytical system involved. The method of identifying this wavelength will be discussed in a later section. A reference solution (pure solvent, an analytically blank solution, etc.) is then placed in the sample cell, the opaque shutter is raised, and a meter reading of 100 percent transmittance (zero absorbance) is obtained by adjusting the light source voltage regulator and/or the amplification gain. Substitution in the sample cell successively of solutions of known concentration of absorbing substance then provides the associated values of percent transmittance and absorbance. A plot of absorbance vs. concentration will yield, for systems obeying Beer's law, a linear relationship. The concentration of an unknown solution, where the concentration lies within the limits of the calibrating solution concentrations, may be obtained by relating the associated absorbance to concentration on this plot. As an alternative, and where the system has been shown to obey Beer's law for the concentration range of interest, the absorbance and concentration for a standard solution may be used, with the absorbance for the unknown solution, in a ratio method for the determination of the concentration of the unknown. The concentration for the unknown solution must, of course, lie within that range of concentrations for which conformity to Beer's law has been established.

Simple double-beam instruments   The basic difficulty with the single-beam spectrophotometer stems from the possibility that over a series of measurements, the intensity of the light source may vary as the result of ac input voltage variations. The incorporation of a voltage control device to ensure constant voltage to the light source for a given setting is expensive and not feasible relative to comparatively inexpensive instrumentation. A simple double-beam spectrophotometer provides an economical means of minimizing the effect of input line voltage fluctuations. Figure 13.8 also shows a typical instrument of this type. The monochromated light is divided into two beams of approximately the same radiant power. One beam passes through a sample cell and is transmitted to a detector. The second beam is passed directly to a similar detector. Each detector circuit involves an adjustable resistance, the two resistances being interconnected, directly on one hand, and through a suitable galvanometer functioning as a null current device on the other. The variable resistance (1) is calibrated as the percent transmittance (or absorbance) scale. In actual use, the transmittance scale is set to read zero, and with the detectors closed off by opaque shutters, the difference in the two "dark currents" is adjusted by means of an auxiliary resistor (not shown) until the null device shows no current flow. The monochromator is then set to the required wavelength for the analytical system involved, a reference solution is placed in the sample cell, the transmittance scale is set to read 100 percent, and the opaque shutters are lifted. The variable resistance (2) is now adjusted until the null indicator again shows no current flow. The reference solution is replaced successively by solutions of known concentration of the absorbing substance, and for each solution, the transmittance scale resistance is adjusted until the null device shows no current flow, the percent transmittance (or absorbance) being recorded in each case. A plot of absorbance vs. concentration will then yield, for systems obeying Beer's law, a straight-line relationship. Unknown solutions may then be analyzed, either with reference to the graph of absorbance vs. concentration or by the ratio method using a known concentration solution. The continuous monitoring of the output radiation from the source and the monochromator by the second detector minimizes the effect of variations in input voltage at the light source.

13.5.6 MINIMUM RELATIVE ERROR
Errors in absorptiometric techniques may arise out of a variety of sources. Some of these can be avoided simply by adequate care and the application of some very simple common-sense rules. Sample cells should be scrupulously clean,

unscratched, unetched, and positioned properly and repetitively in the cell holder. The instrument should be regularly inspected for cleanliness relative to collimator, mirror, detector face, etc., surfaces. The monochromator should be calibrated at regular intervals and circuit drifts or instabilities should be corrected as soon as observed. The solutions examined should be clear and free from gas bubbles and of a range of concentration where Beer's law is obeyed for the analytical system involved.

Apart from these obvious sources of error, it can be shown that a distinct source of relative error is associated with the measurement of absorbance or transmittance, and that this relative error will be high where the transmittance is high (low absorbance), will decrease with decreasing transmittance to a minimum value at a *fixed value* of transmittance (or absorbance), and will increase again with decreasing transmittance beyond this value. The relative error here is related to the instrumental uncertainty in the measurement of transmittance (or absorbance) and, for systems obeying Beer's law, is independent of the nature of the system. Using the expression

$$A = \log \frac{P_0}{P} = abc$$

and differentiating with respect to A, we obtain

$$dA = -\frac{0.4343}{P} dP$$

Dividing by A, we have

$$\frac{dA}{A} = -\frac{0.4343}{AP}\frac{dP}{} = -\frac{0.4343}{AP_0 10^{-A}}\frac{dP}{} \tag{322}$$

Using finite increments, we now have

$$\frac{\Delta A}{A} = -\frac{0.4343}{AP_0 10^{-A}}\Delta P \tag{323}$$

where ΔP represents the uncertainty in the instrumental measurement of the value of P. Differentiating Eq. (323) again with respect to A with ΔP constant

$$d\left(\frac{\Delta A}{A}\right) = -\frac{0.4343\,\Delta P}{P_0}\left(\frac{10^A \ln 10}{A} - \frac{10^A}{A^2}\right) dA \tag{324}$$

Where the value of $\Delta A/A$ is a minimum, $d(\Delta A/A)$ will be zero, and Eq. (324) will yield

$$-\frac{0.4343\,\Delta P}{P_0}\left(\frac{10^A \ln 10}{A} - \frac{10^A}{A^2}\right) dA = 0 \tag{325}$$

with

$$A = \frac{1}{\ln 10} = 0.4343 \tag{326}$$

This value of absorbance corresponds to a percent transmittance of 36.8. Since the relative error in concentration $\Delta c/c$ corresponds to that in absorbance $\Delta A/A$, it is apparent that for a system obeying Beer's law, the relative error in concentration as determined for this source will be dependent on the transmittance or absorbance value, will be independent of the analytical system, and will be a minimum for $A = 0.4343$ or $\%T = 36.8$. The magnitude of the relative error will depend on the value of ΔP, the *photometric error*, a characteristic of the instrument. For the minimum relative error, for example, we have

$$\frac{\Delta A}{A} = \frac{\Delta c}{c} = -\frac{0.4343\,\Delta P}{AP_0 10^{-A}} = -\frac{0.4343\,\Delta P}{0.4343 P_0 10^{-0.4343}}$$

If we set $P_0 = 1$ to normalize the expression, we have

$$\frac{\Delta c}{c} = -2.72\,\Delta P$$

and the magnitude of the minimum relative error, and indeed the relative error for any value of transmittance or absorbance, increases with increasing ΔP. Table 13.1 shows data relating the relative error to both absorbance and transmittance values for a photometric error of ± 0.01 (± 1 percent). Figure 13.9 shows the plots of relative error vs. percent transmittance for ΔP at ± 0.01 and ± 0.005. Note the necessity, where possible, of avoiding percent transmittance readings higher than 80 or lower than 10.

13.6   ANALYTICAL TECHNIQUES

13.6.1   SELECTION OF THE PROPER WAVELENGTH OF EXAMINATION

The plot of absorbance vs. wavelength for a single absorbing species will normally yield a curve showing a peak value of absorbance at a specific wavelength value or an extremely confined wavelength range. Figure 13.2 illustrates such a relationship. At this wavelength, the change in absorbance per unit change in the concentration is a maximum, and the maximum sensitivity is thus realized. For a given species under investigation, the optimum wavelength for maximum sensitivity is obtained from a plot of absorbance vs. wavelength for a solution containing the species at a concentration about halfway in the range where Beer's law is expected to be obeyed. This sets, for the analytical system involved, the wavelength for the monochromator device (filter, prism or grating setting, etc.).

Where other than the optimum wavelength must be used, because of interference arising out of other solution components and affecting the absorbance curve in the neighborhood of the peak wavelength, the value chosen should yield as high an absorbance change per concentration unit change as possible, and it should not be located in that portion of the absorbance-vs.-wavelength curve where the change in absorbance with small changes in wavelength is excessively high.

Table 13.1 Tabulation of absorbance and percent transmittance vs. relative error: $\Delta P = \pm 0.01$

Absorbance	Relative error, ±	Percent transmittance	Relative error, ±
0.01	44.4	99	102
0.05	9.7	98	51.5
0.10	5.5	97	33.9
0.15	4.1	96	25.7
0.20	3.4	95	20.6
0.25	3.1	90	10.6
0.30	2.9	85	7.2
0.35	2.8	80	5.6
0.40	2.7	75	4.6
0.45	2.7	70	4.0
0.50	2.7	65	3.6
0.55	2.8	60	3.3
0.60	2.9	55	3.0
0.65	3.0	50	2.9
0.70	3.1	45	2.8
0.75	3.3	40	2.7
0.80	3.4	35	2.7
0.85	3.6	30	2.8
0.90	3.8	25	2.9
0.95	4.1	20	3.1
1.00	4.3	15	3.5
1.10	5.0	10	4.3
1.20	5.7	9	4.6
1.30	6.7	8	4.9
1.40	7.8	7	5.4
1.50	9.2	6	5.9
1.60	10.8	5	6.7
1.70	12.8	4	7.8
1.80	15.2	3	9.5
1.90	18.2	2	12.8
2.00	21.7	1	21.7
3.00	144.8		
4.00	1086		

13.6.2 THE GRAPHIC METHOD AND RATIO METHOD

We have already discussed the general technique of quantitative analysis as based on the graphic and ratio methods (see Sec. 13.5.5). Basically, where Beer's law is obeyed, a plot of absorbance vs. concentration will yield a straight line. Where the instrument, with the monochromator providing the optimum wavelength, has been set at zero percent transmittance (infinite absorbance) using the opaque shutter, and at 100 percent transmittance (zero absorbance) using a reference or background solution, the insertion successively of solutions of known concentration will yield the associated values of percent transmittance or absorbance.

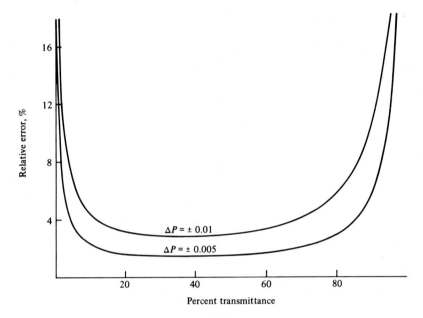

Fig. 13.9   Relative error vs. percent transmittance.

Where a plot of absorbance vs. concentration for these solutions shows conformity with Beer's law, the concentration range is satisfactory. Where a zone of nonconformity to Beer's law is observed, the unknown solutions must be adjusted to yield concentration ranges within the zone of conformity. Where Beer's law is obeyed, subsequent examinations may involve an appropriate known concentration solution and the unknown, whereupon the unknown concentration may be obtained by the ratio method using the absorbance and concentration for the known solution and the absorbance for the unknown.

13.6.3   THE DIFFERENTIAL METHOD

The differential method allows reduction of the relative error in concentration as based on the photometric error of the instrument, thereby increasing the precision and accuracy of the determination. Excellent presentations of this technique are given by Hiskey, and Reilley and Crawford.* We shall show in a very simplified manner the application of the method.

Suppose that we intend to use a spectrophotometer with a photometric error of ±0.01 or ±1 percent. The instrument is calibrated in the usual manner, using the opaque shutter to set the zero transmittance scale point and a reference or background solution to set the 100 percent transmittance scale point. The following situations may be observed.

*C. F. Hiskey, *Anal. Chem.*, **21**:1440(1949); C. N. Reilley and C. M. Crawford, *Anal. Chem.*, **27**:716(1955).

1. The unknown solutions yield %T values higher than 80, resulting in relative errors in concentration higher than ±5.6 percent. In this case, a standard solution providing a % T on the original scale of 80 is now used to set the zero % T scale point, with the 100 percent point being set as usual with reference solution. The new scale relative to the original is shown in Fig. 13.10. The original scale range of 80 to 100% T is now a scale range of 0 to 100% T, and the scale expansion factor is 5. An unknown solution which read 90%T on the original scale, with a relative error in concentration of ±10.6

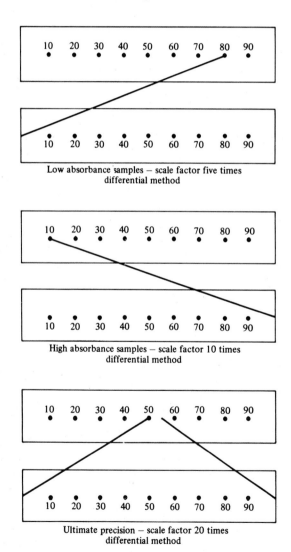

Low absorbance samples – scale factor five times
differential method

High absorbance samples – scale factor 10 times
differential method

Ultimate precision – scale factor 20 times
differential method

Fig. 13.10 The differential method.

percent, now reads 50% T on the new scale, with a relative error which can be shown to be about ±2.1 percent.

2. The unknown solutions yield % T values lower than 10, with relative errors in concentration higher than ±4.3 percent. A standard solution providing a % T value of 10 on the original scale is now used to set the 100% T scale point, with the zero point being set as usual with the opaque shutter. The new scale relative to the original is shown in Fig. 13.10, and the original range of 0 to 10% T is now a scale range of 0 to 100% T, the scale expansion factor being 10. An unknown solution which read 1% T on the original scale, with a relative error of ±21.7 percent, now reads 10% T on the new scale, with a relative error which can be shown to be about ±2.2 percent.

3. The unknown solutions yield % T values between 50 and 55. Standard solutions providing % T values of 50 and 55 on the original scale are now used to set respectively the zero and 100% T points on the new scale. The new and original scales are shown in Fig. 13.10. The original % T range of 50 to 55 is now a scale range of 0 to 100% T, and the scale expansion factor is 20. An unknown solution which read 51% T on the original scale, with a relative error of ±2.9 percent, now reads 20% T on the new scale, with a relative error which can be determined as about ±0.15 percent.

Note that in all of these differential techniques, the relative error associated with the expanded scale readings is that for the original scale divided by the scale expansion factor. The technique can be applied only to systems where Beer's law is obeyed.

13.6.4 THE ANALYSIS OF BINARY MIXTURES

Figure 13.11 shows three possible situations relative to the plot of absorbance vs. wavelength for a solution containing two absorbing components. In the case of Fig. 13.11a, the two components show separate absorbance-vs.-wavelength curves, with no overlapping. In such instances, which are reasonably rare, it is possible to measure the $A_1(\lambda_1)$ and $A_2(\lambda_2)$ and, from two separate calibration curves, or by two applications of the ratio method, determine the concentrations c_1 and c_2 for substances 1 and 2.

In Fig. 13.11b, there is an overlap of one absorption spectrum on the other that is one-sided. System 2 overlaps system 1 at λ_1, but system 1 does not overlap system 2 at λ_2. Here the absorbance as measured at λ_1 will be the sum of the absorbances $A_1(\lambda_1)$ and $A_2(\lambda_1)$. The absorbance as measured at λ_2 will be due to $A_2(\lambda_2)$ only.

Figure 13.11c shows a double spectral overlap, and the absorbance at λ_1 will be the sum of $A_1(\lambda_1)$ and $A_2(\lambda_1)$ with that at λ_2 being the sum of $A_1(\lambda_2)$ and $A_2(\lambda_2)$. In this latter case, we have

$$A \text{ at } \lambda_1 = \{a_1(\lambda_1)c_1 + a_2(\lambda_1)c_2\} b \tag{327}$$

$$A \text{ at } \lambda_2 = \{a_1(\lambda_2)c_1 + a_2(\lambda_2)c_2\} b \tag{328}$$

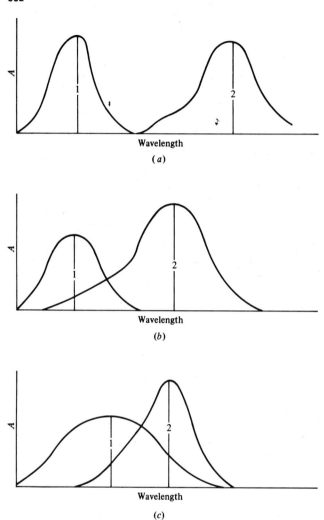

Fig. 13.11   Absorbance-vs.-wavelength curves for binary systems.

The values of $a_1(\lambda_1)$, $a_2(\lambda_1)$, $a_1(\lambda_2)$, and $a_2(\lambda_2)$ can be determined by measuring the absorbance values for pure known concentration solutions of substance 1 and substance 2 at the wavelengths λ_1 and λ_2.   Once determined, these values of absorptivity can be used to solve Eqs. (327) and (328) simultaneously for the values of c_1 and c_2.

The single overlap situation of Fig. 13.11b is merely a special case where we have $a_1(\lambda_2)$ a zero value.

It is possible to analyze in this way solutions containing more than two components.   Highly accurate absorbance values must be determined, however, with highly accurate values of the absorptivities being secured. Where multicomponent systems much beyond the binary are involved, solution of the simultaneous equations is best obtained by computer.

13.7 ABSORPTIOMETRIC TITRATION TECHNIQUES

As pointed out frequently elsewhere, during a titration reaction the concentration of the reactant decreases, with increasing titrant and reaction products concentrations, as the titration progresses. Where, for example,

1. The reactant substance absorbs strongly at a specific wavelength and the background solution, titrant, and reaction products do not.
2. The titrant substance absorbs strongly at a specific wavelength, and the background solution, reactant, and reaction products do not.
3. One of the reaction product substances absorbs strongly at a specific wavelength, but the background solution, reactant, titrant, and other reaction products do not.
4. Both reactant and titrant substances absorb at a specific wavelength, but one much more strongly than the other, and the background solution and reaction products do not.

such titration reactions are amenable to location of the equivalence-point volume by absorptiometric means. Titrations conducted where the absorbance of the solution is monitored for an appropriate wavelength of the light incident on the titration vessel during the course of the titration, with the equivalence-point volume being located from a plot of absorbance vs. titrant volume, are called *absorptiometric titrations*. More specifically, relative to the absorptiometric system used, they may be called *photometric* or *spectrophotometric titrations*.

The titration curves are generally similar in form to those for amperometric titrations, where the absorptiometric system does not involve the plotting of solution absorbance changes associated with secondary titration indicator color change situations. In this latter instance, the plot of absorbance vs. titrant volume will resemble the S-shaped form typical of standard volumetric titration plots of pR vs. V_T.

Several typical absorptiometric titration curve forms are shown in Fig. 13.12, representing titration systems without added indicator substances. The absorbing system must obey Beer's law if satisfactory curves are to be secured. As was the case with amperometric and conductometric titrations, reactions quite incomplete at the equivalence point can be treated by the absorptiometric titration method, providing that precautions are taken to obtain absorbance values at titration points adequately removed from the equivalence-point zone.

As was the case again with amperometric and conductometric titrations, the dilution effect of the titrant volume may be corrected for by multiplication of each

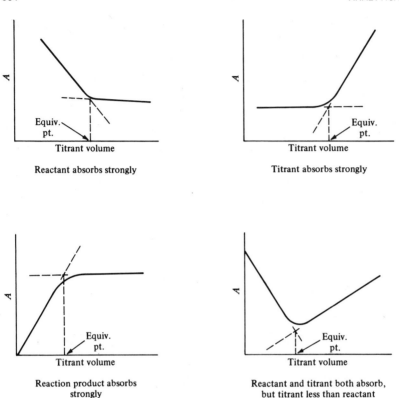

Fig. 13.12 Some typical spectrophotometric titration curves.

absorbance value by the correction factor $(V + v)/V$. Alternatively, the need to apply the correction factor may be eliminated by use of a titrant solution not less than 20 times as concentrated as the reactant solution.

The absorptiometric titration technique is particularly suited to titration reactions where a relatively significant degree of reaction incompletion exists at the titration equivalence point, and it also has the ability to be applied relative to solutions appreciably more dilute than can be handled by potentiometric titration means. As mentioned previously, it is possible to conduct an absorptiometric titration where, at the wavelength of examination, the reactant, titrant, reaction products, and background solution do not absorb appreciably, the wavelength selected being that at which the response is strong relative to absorbance changes associated with the color changes for an indicator substance added to the titration system. For example, in the titration of sodium acetate solution by a standard solution of perchloric acid, using methyl orange as the indicator, the standard volumetric technique gives quite poor results. Similarly, the potentiometric titration approach yields poor equivalence-point-volume location. The absorptiometric titration technique, using methyl orange as the indicator and an appropriate optimum examining wavelength of about 530 nm, gives a significant improvement in accuracy.

PROBLEMS

1. A solution containing 20 ppm of manganese as $KMnO_4$ shows absorbance values of 0.010, 0.700 0.880, 0.300, and 0.180 at wavelengths of 440, 510, 525, 545, and 565 nm, respectively, and a cell path length of 1.00 cm. If the instrument used has a smallest detectable absorbance value of 0.005, show the lowest detectable level of manganese in parts per million at each wavelength. Assume Beer's law to be obeyed. What are your conclusions?

2. Using the data provided in Prob. 1, determine
   (*a*) The absorptivities in liters/cm^{-1}g^{-1}
   (*b*) The molar absorptivities
for manganese as $KMnO_4$ at the wavelengths indicated.

3. Using the data provided in Prob. 1, determine the percent transmittance values corresponding to the absorbance values indicated for each wavelength.

4. An ilmenite ore was analyzed for manganese absorptiometrically. A 1.0612-g ore sample was dissolved and treated to remove or neutralize any interfering elements or compounds and the manganese was oxidized to MnO_4^-. The solution was diluted to exactly 100.0 ml. This solution, in a properly adjusted spectrophotometer, gave an absorbance of 0.450 ± 0.002 for a cell path length of 1.00 cm. A similar solution containing exactly 1.65 mg of manganese per 100.0 ml gave an absorbance of 0.73 ± 0.01 under the same conditions. Assume Beer's law to be obeyed and calculate the percentage of manganese in the ore.

5. An industrial water is analyzed for iron. Two 30.0-ml portions of the water are taken. To one portion (sample 2) is added 5.00 ml of a solution which is 50.0 ppm to Fe^{3+}. The other portion (sample 1) is treated as is. Both samples are now treated with nitric acid and an excess of KSCN, the Fe^{3+} being converted to $FeSCN^{2+}$. Both solutions are now diluted to exactly 50.0 ml. Sample 1 gives an absorbance of 0.40 ± 0.01 and sample 2 an absorbance of 0.61 ± 0.01 when examined at 455 nm using a cell path length of 1.00 cm. Assume conformity to Beer's law and calculate the iron content of the water in ppm.   *Ans.* 15.9 ppm

6. A substance in solution gave an absorbance peak at 610 nm. Solutions containing variable concentrations of this substance were examined at 610 nm in 1.00-cm cells. The following results were obtained.

Absorbance	*Concentration, M,* $\pm 0.1 \times 10^{-6}$ *M*
0.150 ± 0.005	3.1×10^{-6}
0.240 ± 0.005	5.0×10^{-6}
0.380 ± 0.005	7.9×10^{-6}
0.45 ± 0.01	9.4×10^{-6}
0.55 ± 0.01	12.0×10^{-6}
0.61 ± 0.01	14.0×10^{-6}
0.78 ± 0.01	19.0×10^{-6}
0.82 ± 0.02	21.0×10^{-6}

Construct a plot of absorbance vs. concentration.
   (*a*) Is Beer's law obeyed over the concentration range involved?
   (*b*) If not, what is the upper concentration limit for conformity to Beer's law?

(*c*) Within the limits of conformity to Beer's law, what is the molar absorptivity for the substance at 610 nm?

7. Nickel and cobalt both form complexes with 2, 3-quinoxalinedithiol. Nickel forms a blue complex with an absorbance peak at 650 nm. Cobalt forms a red complex with an absorbance peak at 505 nm. Separate solutions were prepared, one with 2.00 ppm of nickel as the complex and the other with 2.00 ppm of cobalt as its complex. Using a 1.00-cm cell, the following data were obtained.

	% T at 650 nm	% T at 505 nm
Nickel solution	30.2 ± 0.1	67.6 ± 0.1
Cobalt solution	89.1 ± 0.1	6.3 ± 0.1

A solution containing unknown amounts of both nickel and cobalt as their complexes gave, under the same conditions, $35.9 \pm 0.1\%$ T at 650 nm and $3.5 \pm 0.1\%$ T at 505 nm. What are the concentrations of nickel and cobalt in ppm in this solution?

8. A vitamin tablet contains iron as $FeSO_4$ and is regularly assayed for this substance. A tablet is dissolved and, after neutralization of any interfering substances, is obtained as 100.0 ml of solution containing all of the iron in the Fe(III) state. 5.00 ml of this solution is taken and the iron is reduced to the Fe(II) state and, after the proper treatment, complexed with 1, 10-phenanthroline. The solution is now diluted to exactly 100.0 ml. The red ferrous phenanthroline complex has an absorbance peak at 510 nm, and the solution absorbance for a 1.00-cm cell is found to be 0.448 ± 0.005. A solution, treated in the same way, contains 0.35 mg of iron per 100.0 ml and gives an absorbance of 0.70 ± 0.01.

(*a*) What is the content of $FeSO_4$ in milligrams per tablet?

(*b*) What is the absorptivity in liters/cm^{-1}g^{-1} for ferrous phenanthroline complex at 510 nm?

Ans. (*a*) 12.2 mg; (*b*) 200

9. The acid form (red) of the indicator methyl red has an absorbance peak at 530 nm, the acid form being represented by HIn. The basic form (yellow), represented by In$^-$ does not absorb at all at this wavelength. The same concentration of the indicator is added to nine buffered solutions, and the absorbance is measured at 530 nm using a 1.00-cm cell. The following values were obtained.

pH	Absorbance
1.00	1.43
2.00	1.43
3.00	1.42
4.00	1.34
5.00	0.875
6.00	0.194
7.00	0.022
8.00	0.002
9.00	0.000

(*a*) What is the dissociation constant K_a for methyl red?

(*b*) What is the approximate pH value which represents the central value for the indicator color change range in pH?

10. An indicator HIn has an acid color with an absorbance peak at 610 nm and slight absorbance at 450 nm. The basic color has an absorbance peak at 450 nm and absorbs slightly at 610 nm. A

1.2×10^{-3} M solution of the indicator in solutions buffered to pH 1.00 and 9.00 yielded, in a 1.00-cm cell, the following.

Solution pH	Absorbance at 450 nm	Absorbance at 610 nm
1.00	0.070	1.46
9.00	0.760	0.051

A low concentration of the indicator is added to a buffered solution at pH 5.00. Examination under the same conditions gave 0.311 and 0.700 absorbance values at 450 and 610 nm, respectively. Calculate the K_a value for the indicator. $\qquad$ *Ans.* $K_a = 7.85 \times 10^{-6}$

11. In the *molar ratio method* a series of solutions is prepared which contain the same molarity of a metal ion, but with varying molar concentrations of the ligand with which it forms a complex ion. The absorbances of these solutions are then measured at a wavelength at which the metal complex ion absorbs, but the ligand and uncomplexed metal ion do not. A plot of absorbance vs. the ratio of ligand molarity to metal ion molarity is then made. The extrapolated straight-line sections of this plot meet at a point which locates on the ratio axis the ratio of ligand to metal ion in the complex (the coordination number). Where the line perpendicular to the ratio axis from the intersection point cuts the plot curve, the corresponding absorbance value A_c, relative to the absorbance value at the straight-lines intersection point A_i, yields $K_{\text{instability}}$ for the metal complex ion from

$$K_{\text{instability}} = \frac{x^x C^x (1 - A_c/A_i)^{(x+1)}}{A_c/A_i}$$

where x = coordination number for the complex
    C = concentration of the metal ion plus metal complex ion (starting concentration of metal ion)

For our purposes we shall assume that only one metal complex ion is formed. To solutions containing 1.00×10^{-4} M metal ion, various ligand concentrations are added. The absorbance is measured at the wavelength where only the metal complex ion absorbs, with the following results.

Ligand, M	Absorbance
3.0×10^{-5}	0.065
6.0×10^{-5}	0.130
9.0×10^{-5}	0.195
1.2×10^{-4}	0.260
1.5×10^{-4}	0.325
1.8×10^{-4}	0.390
2.1×10^{-4}	0.455
2.4×10^{-4}	0.500
2.7×10^{-4}	0.550
3.0×10^{-4}	0.580
3.3×10^{-4}	0.600
3.6×10^{-4}	0.620
3.9×10^{-4}	0.640
4.2×10^{-4}	0.650
4.5×10^{-4}	0.650
4.8×10^{-4}	0.650

Plot carefully absorbance vs. the ratio of ligand molarity to metal ion molarity. The latter is a constant value at 1.00×10^{-4}.

(a) What is the coordination number for the metal complex ion?

(b) What is the value of $K_{instability}$ for this complex?

12. A nickel chelate substance in solution has an absorbance peak at 530 nm. The following data were obtained for solutions of varying nickel content, examined in 1.00-cm cells.

Nickel, ppm	Percent transmittance
0.200	75.2
0.400	56.5
0.600	42.5
0.800	31.9
1.000	24.0
1.200	18.0
1.400	13.6
1.600	10.2
1.800	7.6

The photometric error for the instrument is $\Delta P = \pm 0.01$.

(a) Is Beer's law obeyed within these concentration values?

(b) What is the absorptivity in liters/cm^{-1}mg^{-1} for the nickel chelate at 530 nm?

(c) If it is required to determine nickel in the range of 1.00 to 1.60 ppm with a relative error of not more than ± 2 percent, based on the photometric error, what standard solutions may be used in the differential method to set zero and 100 percent transmittance on the instrument?

(d) An unknown solution containing 1.40 ppm yielded 13.6 percent transmittance on the old scale. Calculate the relative error due to photometric error. What will be the percent transmittance on the new scale devised in c? What will now be the approximate relative error based on photometric error?

13. Silica (SiO_2) may be determined by conversion to the silicomolybdate compound which has a blue color. A series of samples is prepared which contain between 0.030 and 0.100 mg of SiO_2. A standard solution known to contain 0.020 mg of SiO_2 is used to set the instrument at 100 percent transmittance. The zero point is set in the usual way. A second standard solution known to contain 0.100 mg of SiO_2 now yields a percent transmittance of 14.4. One of the unknown solutions yields under the same conditions a transmittance of 31.8 percent. What is the SiO_2 content of the unknown solution in milligrams?   *Ans.* 0.067 mg

REFERENCES

1. Bauman, R. P.: "Absorption Spectroscopy," Wiley, New York, 1962.

2. Headridge, J. B.: "Photometric Titrations," Pergamon, New York, 1961.

3. Mellon, M. G. (ed.): "Analytical Absorption Spectroscopy," Wiley, New York, 1950.

4. Sandell, E. B.: "Colorimetric Determination of Traces of Metals," 2d. ed., Interscience, New York, 1950.

5. Underwood, A. L.: Photometric Titrations, in C. N. Reilley (ed.), "Advances in Analytical Chemistry and Instrumentation," vol. 3, Interscience, New York, 1964.

14
Methods of Separation

14.1 GENERAL

The technique of chemical analysis involves four basic steps. As the first step it is necessary to obtain a representative sample of the material to be analyzed. If the material is homogeneous, obtaining a representative sample presents no real problem. Where the material is heterogeneous, procuring a representative sample may require involved procedures such as bulk sampling, particle size reduction, reduction of the initial sample size by repeated quartering, etc.

As a second step it is necessary to secure the desired substance in a form capable of measurement under conditions where the result of the measurement can be directly and unmistakably associated with the amount of the desired substance. It requires, in general, the application of a series of physical or chemical processes.

The third step involves the measurement of the desired substance. The form in which the desired substance is eventually obtained will be dictated by the method of measurement to be applied (colorimetric, potentiometric, etc.). Few problems can be anticipated relative to this step, providing that the second step has been carried out successfully and any interfering substances either separated or their effects nullified.

The final step is associated with the interpretation of the measurement in terms of the desired substance. In this step calculations are made to allow expression of the final result of the analysis in an acceptable form. Depending on the number of measurements, and their type, this may or may not involve some form of statistical analysis of the data obtained.

Of the sequence of steps outlined above, the most critical, from the point of view of the chemistry of the analytical process, is the second step. Very few single reactions can be used in chemical analysis under conditions where only the desired substance reacts and reports in the measurement system. In the majority of cases, the material to be analyzed will contain other substances which, by undergoing similar reactions or by inhibiting necessary reactions, will increase or decrease the property finally measured, thereby allowing faulty estimates of the amount of desired substance.

Interfering substances may be treated in the second step in two general ways to eliminate or minimize their interfering effects. One approach maintains the interfering substance in the measurement environment but introduces a method of nullifying the interfering action. An example is the complexation of iron by fluoride in the iodometric determination of copper. The second approach involves the application of a physical or chemical process which isolates the interfering substance or substances in one phase while maintaining the desired substance in a second phase. The two phases are then separated, and that containing the desired substance is carried through a further process of analysis to the final measurement. Such methods of separation include the following.

1. Precipitation techniques where the solution is treated with a properly selected precipitant to separate the desired and interfering substances.
2. Extraction techniques where the solution is treated with a properly selected immiscible solvent, or where a solid mixture is treated with a properly selected solvent, in order to permit the separation of substances.
3. Volatilization techniques where interfering and desired substances, or mixtures of desired substances, are separated by volatility differences.
4. Chromatographic techniques where interfering and desired substances, or mixtures of desired substances, are separated on the basis of a difference in their rates of migration through a stationary phase when carried by a moving phase. The stationary phase may be a solid or a liquid, and the moving phase a liquid or a gas.
5. Electrodeposition techniques where properly selected potentials are applied between appropriate electrodes immersed in a solution of the desired and interfering substances, resulting in a separation of the substances by deposition or liberation.
6. Ion-exchange techniques where interfering ions may be separated and exchanged for noninterfering ions.

This chapter will deal very generally with these methods of separation. However, separation by precipitation has already been discussed in Chap. 11, and separation by electrolytic means has been dealt with adequately in Sec. 12.3, so that in the main this chapter will deal with extraction, volatilization, chromatographic, and ion-exchange techniques.

14.2 SEPARATION BY EXTRACTION

14.2.1 GENERAL

The transference of a material from one phase to another is called *partition*. We shall discuss in this section only that process of partition which involves the separation of a solute between two immiscible liquid phases, a process commonly referred to as *extraction*.

The *distribution law*, or law of heterogeneous equilibrium, provides the relationship governing the ratio of the equilibrium activities (or more commonly, concentrations) of a solute which is distributed between two immiscible solvents. Expressed in mathematical form this law is

$$\frac{C_2}{C_1} = K \tag{329}$$

where C_2 and C_1 are the equilibrium concentrations of a solute in solvent 1 and solvent 2. For a given temperature, the value of K is a constant called the *distribution constant* or *distribution coefficient*. There are certain limitations to this law, based mainly on the dissociation or nondissociation of the solute in the solvents. We shall, however, ignore such limitations where the present discussion is concerned.

The amount of a solute found at equilibrium in each of two immiscible solvents will be a function of the volume of each solvent and the solubility of the solute in each solvent. Consider the following example. Suppose that we have 100 ml of solvent 1 in which there is 1.0 g of solute. We now add 100 ml of solvent 2, where solvents 1 and 2 are completely immiscible and where the solute is four times as soluble in solvent 2 as it is in solvent 1. At equilibrium, after thorough mixing by agitation, solvent 2 is separated from solvent 1. The 100 ml of separated solvent 1 is treated with a fresh 100 ml of solvent 2 and, following the previous procedure, solvents 1 and 2 are again separated. The process is continued. After the first extraction, solvent 1 will contain 0.2 g of solute, while the first 100 ml of solvent 2 will contain 0.8 g of solute. After the second extraction, solvent 1 will contain 0.04 g of solute, after the third it will contain 0.008 g, and so on. The decreasing amounts of solute in the 100 ml of solvent 1, 1.0 g, 0.2 g, 0.04 g, 0.008 g, . . ., represent an exponential series involving, in this case, 5^0, 5^{-1}, 5^{-2}, 5^{-3}, etc. A relationship may be developed on this basis which relates the weight of solute left with the distribution coefficient, the weight of solute present originally, the volumes of the two solvents, and the number of extractions. This is shown as

$$W_1 = W_0 \left(\frac{V_1}{KV_2 + V_1} \right)^n \tag{330}$$

where W_1 = weight of solute in solvent 1 after n extractions, g
W_0 = weight of solute in solvent 1 at the start, g
V_1 = volume of solvent 1, ml
V_2 = volume of solvent 2 for each extraction, ml
K = distribution coefficient for the temperature involved
n = number of extractions

Note that for the example given, and *as a general case*, multiple small volume extractions are more efficient than one extraction involving the same total volume. In the example used, four extractions with 25 ml of solvent 2 yields a W_1 value of 0.0625 g, while one extraction of 100 ml yields a W_1 value of 0.2 g.

14.2.2 APPLICATIONS

The extraction method may be used to remove a single solute from a solvent medium. It may also be applied, where the K values differ appreciably, to separate a group of solutes from each other. In many cases only a separatory funnel is required. The method has distinct advantages over comparable precipitation techniques. The time required to obtain a quantitative separation is, for example, much less than that associated with a corresponding precipitation method, since the latter involves the time-consuming processes of precipitation, digestion, filtration, etc. Again, the difficulties often associated with precipitation analysis, such as coprecipitation, postprecipitation, etc., are avoided where the extraction process is applied. Finally, the extraction method permits the handling of trace quantities of substances at levels far below those normally dealt with in precipitation techniques.

The separation of certain groups of metal chlorides from others by ether extraction in hydrochloric acid media has long been applied in analytical chemistry. Iron(III) can, for example, be extracted by ether from 6 M HCl solutions containing such metal ions as cobalt(II), nickel(II), manganese(II), chromium(III), aluminum(III), and titanium(IV). The efficiency of the extraction of iron may be better than 90 percent, the named ions being unaffected in the extraction process. The advantages of the procedure for the removal of the bulk of the iron before conducting analytical processes for the determination of traces of the unaffected metal ions is obvious. The method is particularly useful in the treatment of high-iron content materials such as ferrous alloys and iron ores.

The extraction technique can be used most profitably in the extraction of a metal as the metal chelate substance, where the metal chelate has a high solubility in an immiscible solvent such as chloroform, benzene, etc., relative to its solubility in aqueous media. Partition to the organic solvent phase permits the quantitative separation of the affected metal as its chelate. An example of this situation is the quantitative separation of lead in aqueous solution by extraction with a chloroform solution of diphenylthiocarbazone. In the absence of any interfering action, the relatively high solubility of lead diphenylthiocarbazone chelate in chloroform permits an excellent separation with a minimum of extractions.

Such an application of diphenylthiocarbazone (HDz) can be used to exemplify the general process of extraction as it applies to the separation of metals as their chelates. The reaction of a metal ion M^{n+} with HDz is given generally by

$$M^{n+} + nHDz + nH_2O \rightleftharpoons M(Dz)_n + nH_3O^+$$

The distribution coefficient for the metal chelate will be given by

$$K_{M(Dz)_n} = \frac{[M(Dz)_n]_{org}}{[M(Dz)_n]_{aq}} \tag{331}$$

while the *distribution ratio* will be given by

$$D = \frac{C_{M,org}}{C_{M,aq}} \tag{332}$$

where $C_{M,org}$ and $C_{M,aq}$ represent respectively the *total* metal concentrations in the organic and aqueous phases. Where the organic phase is concerned, we can ignore contributions from M^{n+}, $M(Dz)_{n-1}$, etc. In the aqueous phase we can ignore sources such as $M(Dz)_n$, $M(Dz)_{n-1}$, etc. We have, therefore,

$$D = \frac{[M(Dz)_n]_{org}}{[M^{n+}]_{aq}} \tag{333}$$

Combining Eqs.(331) and (333) yields

$$D = \frac{K_{M(Dz)_n}[M(Dz)_n]_{aq}}{[M^{n+}]_{aq}} \tag{334}$$

In the aqueous phase the chelating reaction is

$$M^{n+} + nDz^- \rightleftharpoons M(Dz)_n$$

and the formation (stability) constant is given by

$$K_{stab} = \frac{[M(Dz)_n]_{aq}}{[M^{n+}]_{aq}[Dz^-]_{aq}^n} \quad \text{with} \quad [M^{n+}]_{aq} = \frac{[M(Dz)_n]_{aq}}{K_{stab}[Dz^-]_{aq}^n} \tag{335}$$

The dissociation of the chelating substance in aqueous medium is given by

$$HDz + H_2O \rightleftharpoons H_3O^+ + Dz^-$$

and we have

$$K_a = \frac{[H_3O^+]_{aq}[Dz^-]_{aq}}{[HDz]_{aq}} \quad \text{with} \quad [Dz^-]_{aq}^n = \frac{K_a^n[HDz]_{aq}^n}{[H_3O^+]_{aq}^n} \tag{336}$$

The distribution coefficient for the chelating substance is given by

$$K_{HDz} = \frac{[HDz]_{org}}{[HDz]_{aq}} \tag{337}$$

A combination of Eqs. (334), (335), (336), and (337) as follows:

$$\frac{[M^{n+}]_{aq}}{[M(Dz)_n]_{aq}} = \frac{1}{K_{stab}[Dz^-]^n_{aq}}$$

$$\frac{K_{M(Dz)_n}}{D} = \frac{[H_3O^+]^n_{aq}}{K_{stab}K_a^n[HDz]^n_{aq}}$$

$$= \frac{[H_3O^+]^n_{aq}K_{HDz}^n}{K_{stab}K_a^n[HDz]^n_{org}}$$

yields an expression for D as

$$D = \frac{K_{M(Dz)_n}K_{stab}K_a^n[HDz]^n_{org}}{[H_3O^+]^n_{aq}K_{HDz}^n} \tag{338}$$

Thus the distribution ratio D is given in terms of values specific for the particular metal ion and chelating substance reacting together (for example, K_{stab}, $K_{M(Dz)_n}$, K_{HDz} and K_a) and in terms of factors controllable during the chelating reaction (for example, $[H_3O^+]$ and $[HDz]_{org}$). Note that the concentration of the metal ion *does not appear* in any form in this expression, which exemplifies the significance of the extraction technique in that it can be applied to very minute quantities (trace amounts) or to very high quantities (macroamounts) of substance.

In the industrial and commercial field, as well as certain types of research work, extraction by the countercurrent distribution principle is frequently applied on a large-scale automatic basis, and is especially suited to the separation of components in a mixture under conditions where the differences in the distribution coefficients are small.*

Extraction by the countercurrent principle requires some additional discussion, since the general theory has application in the area of separation by chromatographic methods. Figure 14.1 will serve to follow the technique. Consider an aqueous solution which contains 1000 mg of solute in 100 ml of solution. This solution is treated with 100 ml of an immiscible organic solvent. We shall, for the moment, assume that the distribution coefficient is 1.00 and that all successive treatments with either the organic solvent or water will involve 100-ml volumes. At each operation the solvents will be thoroughly mixed, equilibrated, and allowed to separate. The first operation, transfer zero ($n = 0$), is carried out in vessel 0. The organic (upper) phase and the aqueous (lower) phase now each contain 500 mg of solute. The organic phase is now transferred to vessel 1 and is treated with water. The aqueous phase is retained in vessel 0, and is treated with

*L. C. Craig and O. Post, *Anal. Chem.*, **21**:500(1949); L. C. Craig and D. Craig, Laboratory Extraction and Countercurrent Distribution, in A. Weissberger (ed.), "Technique of Organic Chemistry," 2d ed., vol 3, pt. 1, pp. 149–332, Interscience, New York, 1956.

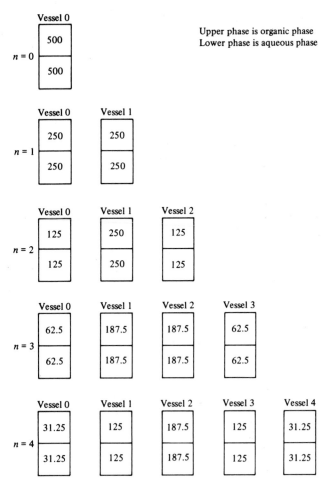

Fig. 14.1 Craig countercurrent separation technique.

organic solvent. This operation, transfer one ($n = 1$), results in 250 mg of solute in each phase in each vessel. The organic phase from vessel 1 is now transferred to vessel 2 and is treated with water. The organic phase from vessel 0 is added to the aqueous phase retained in vessel 1. The aqueous phase retained in vessel 0 is treated with organic solvent. This operation, transfer two ($n = 2$), yields 125, 250, and 125 mg of solute in each phase of vessels 2, 1, and 0, respectively. For transfer three ($n = 3$), we obtain 62.5, 187.5, 187.5, and 62.5 mg of solute in each phase of vessels 3, 2, 1, and 0, respectively. The situation continues as indicated in Fig. 14.1. The following characteristics should be noted and *verified* by the student:

1. With successive transfers the solute becomes distributed throughout the vessels.

2. If the fraction of the original solute is plotted against vessel number for each transfer, the resulting curve will show a peak value for a given vessel, the peak value moving to the right with increasing transfer number.
3. With increasing transfer number the distribution curve decreases in peak height and increases in base length.
4. As the value of the distribution coefficient $K = C_{org}/C_{aq}$ increases beyond 1.00, the rate of movement of the peak to the right increases, and the peak will be found increasingly further to the right of the central vessel. Conversely, with decreasing values of K below 1.00 the rate of movement of the peak to the right decreases and the peak will be found increasingly further to the left of the central vessel.

It is apparent that in this technique, it will be important to determine the number of transfers required to effect a given separation efficiency, and the number of vessels from which the phases should be collected to achieve such a separation.

Suppose we consider that for any vessel, the fraction of the solute in the organic phase will be given by x, with that in the aqueous phase given by y. It is apparent that for any vessel, the sum of x and y will be 1.00, while the ratio of x to y will be K, the distribution coefficient, where equal volumes of both phases are involved. Suppose we have W grams of solute at the start. For $(n = 0)$ we have xW grams of solute in the organic phase and yW grams of solute in the aqueous phase. Thus we have vessel 0 containing a fraction of the original solute W grams given by:

$$f_{0,0} = 1.00$$

For $(n = 1)$ we have

For vessel 1	xW grams to be partitioned
with	x^2W grams in the organic phase
and	xyW grams in the aqueous phase
For vessel 0	yW grams to be partitioned
with	xyW grams in the organic phase
and	y^2W grams in the aqueous phase

and we have the fractions of the original solute given by

Vessel 1: $f_{1,1} = x$
Vessel 0: $f_{1,0} = y$

For $n = 2$ we have

For vessel 2	x^2W grams to be partitioned
with	x^3W grams in the organic phase
and	x^2yW grams in the aqueous phase

For vessel 1 $2xyW$ grams to be partitioned
  with $2x^2yW$ grams in the organic phase
  and $2xy^2W$ grams in the aqueous phase
For vessel 0 y^2W grams to be partitioned
  with xy^2W grams in the organic phase
  and y^3W grams in the aqueous phase

and we have the fractions of the original solute given by

Vessel 2: $f_{2,2} = x^2$
Vessel 1: $f_{2,1} = 2xy$
Vessel 0: $f_{2,0} = y^2$

Carrying out this same procedure we can show that

For $(n = 3)$: $f_{3,3} = x^3$
$f_{3,2} = 3x^2y$
$f_{3,1} = 3xy^2$
$f_{3,0} = y^3$
For $(n = 4)$: $f_{4,4} = x^4$
$f_{4,3} = 4x^3y$
$f_{4,2} = 6x^2y^2$
$f_{4,1} = 4xy^3$
$f_{4,0} = y^4$

and that, for a given transfer number n, each vessel will contain a fraction of the solute given by a term in the corresponding binomial expansion $(x + y)^n$. For any term r in such an expansion we have.

$$f_{n,r} = \left\{ \frac{n!}{r!(n - r)!} \right\} x^r y^{(n-r)} \tag{339}$$

as the fraction of the original solute in the rth vessel for n transfers. We have already noted that, for any vessel,

$x + y = 1$ and $x/y = K$ where the phase volumes are identical

so that

$x = yK$ and $y = \dfrac{1}{1 + K}$

Substitution of these relationships in Eq. (339) yields

$$f_{n,r} = \left\{ \frac{n!}{r!(n - r)!} \right\} \left\{ \frac{1}{1 + K} \right\}^n K^r \tag{340}$$

Where the phase volumes are not equal, it is apparent that

$x + y = 1$ $\dfrac{x}{y} = K\dfrac{V_o}{V_a}$ and $x = yK\dfrac{V_o}{V_a}$ $y = \dfrac{1}{1 + K(V_o/V_a)}$

and from Eq. (339) we have

$$f_{n,r} = \left\{\frac{n!}{r!(n-r)!}\right\} \left\{\frac{1}{1 + K(V_o/V_a)}\right\}^n K^r \left(\frac{V_o}{V_a}\right)^r \tag{341}$$

where V_o and V_a are the volumes respectively of organic and aqueous phases, V_o and V_a being repetitive for each transfer.

While mathematical tables can be used to provide the values of the factorials in the expressions above, with increasing value of n the arithmetic operations become tedious. Although the distribution is a binomial one, where n is large (in the neighborhood of 50 or greater), it tends to approach the gaussian form, and the value of $f_{n,r}$ can then be found from

$$f_{n,r} = \frac{1}{\sqrt{2\pi nK/(1+K)^2}} \exp\left\{-\frac{(r_{max}-r)^2}{2nK/(1+K)^2}\right\} \tag{342}$$

where r_{max} is the vessel number which contains the maximum solute content. Where n is large,

$$r_{max} \approx n\left(\frac{1}{1+K}\right) K \tag{343}$$

and we have

$$f_{n,r max} = \frac{1}{\sqrt{2\pi nK/(1+K)^2}} \tag{344}$$

Equations (342), (343), and (344) apply where V_o and V_a are identical.

14.3   SEPARATION BY VOLATILIZATION

14.3.1   GENERAL

Separation by volatilization may take one of several forms, but the basis of the technique is the separation of substances on the basis of the conversion of one or more of them to a volatile state. In certain instances, the volatile substance or a volatile compound of a substance may be expelled from the mixture without recovery, the purpose here being to remove an undesirable substance. In other cases, the loss of weight in the process may be measured for analytical purposes. In still others, the volatile substance may be expelled and recovered, and the technique may be used for the determination of the volatile substance or a mixture component related to this volatile substance.

14.3.2   APPLICATIONS

Water may be expelled from a substance or a mixture by a heating process for the purpose of drying the material only. On the other hand, the determination of the loss of weight as water may form a critical part of an analytical scheme. Where the loss of weight may be only partially due to water loss, the condensation or absorption of the expelled water can lead to a determination of the water component of the total loss of weight factor.

Carbon is frequently determined in organic compounds and in steels by oxygen-stream combustion methods, the evolution and absorption of CO_2 providing the basis for the determination. Carbonates in certain mixtures or materials may be determined by treatment of the material with acid, or by heating, the expelled CO_2 again providing the determination basis.

Nitrogen is often determined in organic compounds, and in inorganic ammonia compound mixtures, by the evolution and absorption of ammonia. The Kjeldahl method is a typical example.

Certain metal ions in solution may be separated by a variety of volatilization methods. Arsenic(III) may be separated from antimony(III) and tin(IV) by distillation, under the proper conditions, of a solution of their chlorides. The more volatile $AsCl_3$ distills over first, and may be recovered by absorption in a proper liquid medium with subsequent quantitative determination. Chromium may be separated readily from substances such as iron, nickel, cobalt, etc., by the addition of sodium chloride to a boiling perchloric acid solution of iron(III), nickel(II), cobalt(II), and chromium(VI). The volatile CrO_2Cl_2 is absorbed in an appropriate alkaline solution.

The conversion of the sulfur in hydrocarbon fuels to SO_2 at elevated temperature, and the absorption of the evolved SO_2, leads to the quantitative determination of the hydrocarbon sulfur content. Finally, the evolution of volatile SiF_4 by the treatment of SiO_2 with HF and H_2SO_4 leads to a weight-loss situation important in the determination of SiO_2 and Si in many industrial and commercial products.

14.4 SEPARATION BY CHROMATOGRAPHY

14.4.1 GENERAL

The basic principles of chromatography were first coherently described by Tswett* in 1906 when he reported the separation of several colored natural substances (chlorophyll, xanthophyll, etc.) by means of a calcium carbonate column technique. The various pigmented substances, retained at different levels of the column, gave rise to a color-banded appearance. Tswett described the column as a *chromatogram* and the method as a *chromatographic method*. The more general definition of the process was given in Sec. 14.1. There are thus many types of chromatography, some of which are listed in Table 14.1 according to the nature of the stationary and moving phases involved. We shall make no attempt to describe all of these chromatographic processes in detail, but shall confine our discussion to the more general aspects of several different but typical forms.

---

* M. Tswett, *Ber. Deut. Botan. Ges.*, **24**:384(1906).

Table 14.1 Chromatographic methods

Stationary phase	Moving phase	Type of chromatography
Solid	Liquid	Adsorption chromatography, thin layer chromatography, ion-exchange methods
	Gas	Gas-solid chromatography (GSC)
Liquid	Liquid	Paper chromatography, partition chromatography
	Gas	Gas-liquid chromatography (GLC)

In general, the progress of a chromatographic method of analysis involves

1. Adsorption on the stationary phase of the substance or substances to be separated
2. Separation of the adsorbed substances by a continuous flow of the moving phase
3. Recovery of the separated substances progressively by elution
4. Qualitative and/or quantitative analysis of the separated and recovered substances

In many of the methods of chromatography the governing principle will be the distribution law in a form identical to or similar to that of Eq. (329).

14.4.2 ADSORPTION CHROMATOGRAPHY: ELUTION ANALYSIS

By adsorption chromatography we shall imply here the use of a packed column in a technique similar to that employed by Tswett. The stationary phase is then a column packed with an adsorbent substance such as finely divided silica, alumina, calcium carbonate, etc., the nature of the column packing being largely dependent on the materials to be separated. The moving phase is a liquid solvent. The distribution coefficient, often called the *partition coefficient*, is given by

$$D = \frac{\text{concentration of solute in the moving phase}}{\text{concentration of solute in the stationary phase}}$$

and the various values of D for the components to be separated governs the degree of separation. In use, the column is first thoroughly wetted with the solvent. The sample involving the solvent and the solute components to be separated is then introduced at the top of the column. Fresh portions of pure solvent are now added at the top of the column, and the components become distributed between the surface of the stationary phase (the packing) and the solvent, moving down the column at different rates according to their D values. Where D is large, the solute component remains basically in the moving or solvent phase, passing rapidly through the column. Where D is small, the solute component remains

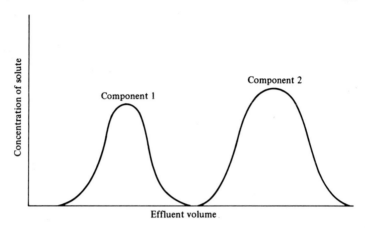

Fig. 14.2 Adsorption chromatography: elution peaks.

largely with the stationary phase and moves slowly down the column. The process of passing the moving phase through the stationary phase is called *elution*.

Where the solute components are differently colored, their separation and removal may be conducted on a visual basis. Where they are not colored, testing of the eluted solvent will serve to detect the separation and removal of each component. Figure 14.2 gives a rather idealized but typical picture of the separation of two components. Where overlapping of the two peaks occurs, this may often be minimized by a gradual change in the composition of the solvent during elution (the addition of increasing quantities of a more polar solvent). Where such a technique is applied, the process is called *gradient elution*.

14.4.3 THIN-LAYER CHROMATOGRAPHY

A slurry of the required adsorbent (silica gel, cellulose, alumina, etc.) containing a binder such as starch is spread in a uniform thin (0.15 to 0.25 mm) layer over a glass plate approximately 20×20 cm. The slurry is allowed to dry and the adsorbent layer is activated by heating for an appropriate period at approximately 110°C. In use, a few microliters of solvent containing 10 to 100 μg of the solutes to be separated is placed on the adsorbent layer at one end of the plate and allowed to dry. The plate is placed vertically with the solute spotted end dipping into a suitable solvent (the moving phase). Solvent then moves up the adsorbent layer (the stationary phase). As it moves, the sample solutes are carried up at rates dependent on their distribution coefficients and expressed as

$$R_F = \frac{\text{distance traveled by a solute from the origin}}{\text{distance traveled by the solvent front from the origin}}$$

After the solvent has migrated some specific distance, say 10 cm, the plate is removed and examined. If the solute components are colored, direct visual methods of analysis may be applied. Where they are not colored, the application of selective reagents may result in color development. Again, certain solutes may be identified by their ability to fluoresce under ultraviolet light. Solutes are often characterized by their R_F values, thus permitting identification on this basis. Quite accurate quantitative analysis results can be obtained by scraping a specific spot or area off the plate, eluting the solute from the solid material, and performing a quantitative estimation, very often by spectrophotometric means.

14.4.4 GAS-LIQUID CHROMATOGRAPHY (GLC)

In gas-liquid chromatography the stationary phase is a nonvolatile liquid coating on particles of a porous solid. The moving phase is a gas. The components of a mixture carried through the stationary phase by the gas distribute themselves between the two phases in accordance with their respective distribution or partition coefficients. The techniques applied include elution, displacement, and frontal analysis. The elution technique is most commonly applied and will be, in general, the method described.

The adsorbing column is mounted in an oven and the liquid sample containing the solutes to be separated is injected and vaporized at the head of the column. A small quantity of air is injected shortly thereafter. The vaporized solutes are carried into the column by the moving phase *carrier gas*. The solutes distribute themselves between the gas and the static liquid phases according to their D values. Separation of the solutes occurs where the respective D values differ sufficiently. With continued flow of the carrier gas, each component leaves the column separately in a small volume of the gas and is detected by a detector system which then records the elution peaks and the time since injection. The resulting chromatogram is illustrated in Fig. 14.3. The injection peak represents the time of injection of the sample and the air peak the time of injection of the small volume of air. On the chromatogram the distance from the injection peak to each component peak is called the *retention time*, while that from the air peak to each component peak is called the *adjusted retention time*. These times are characteristic for given compounds under specific conditions of examination and can serve to identify the compounds. The separation of the components in a mixture is a function of the ratio of their retention times, and the degree of separation is called the *resolution*.

Figure 14.4 gives a very general idea of the gas-liquid chromatographic system. The carrier gas used will depend on the type of column liquid, the detection system, and the nature of the compounds to be separated. Hydrogen, helium, nitrogen, argon, and carbon dioxide are among some of the common gases employed. The column is commonly of copper, stainless steel, or aluminum tubing containing a supporting material such as a diatomaceous earth derivative, and the stationary liquid phase on this support may be a variety of substances, depending on the specific conditions involved, such as squalene, polyethylene glycol, silicone oil, etc. The detector system may involve differential thermal conductivity, flame ionization, or β-ray ionization. While these are the commoner types of detector, infrared and mass spectrometric detection methods are often employed.

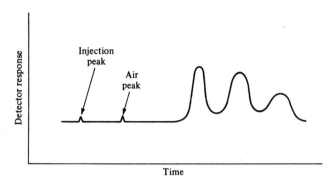

Fig. 14.3 A typical gas-liquid chromatogram.

The detector system reports to a recorder, time being one of the axes of the strip-chart.

The *retention volume* V_R of any component is the volume of gas required to carry the component peak value through the column to elution. It is related to the retention time t_R by the expression

$$V_R = t_R F_c \tag{345}$$

and we have

$$V_G = t_G F_c$$

where F_c is the outlet flow rate of the carrier gas, and V_G and t_G are the retention volume and retention time for the carrier gas alone. The *corrected retention volume* V_R° is the retention volume corrected relative to the average gas pressure in the column, and can be shown to be given by*

$$V_R^0 = j t_R F_c \qquad \text{with} \qquad V_G^0 = j t_G F_c \tag{346}$$

where the value of j is given by

$$j = \frac{3\{(p_i/p_o)^2 - 1\}}{2\{(p_i/p_o)^3 - 1\}} \tag{347}$$

*A. T. James and A. J. P. Martin, *Analyst*, **77**:915 (1952).

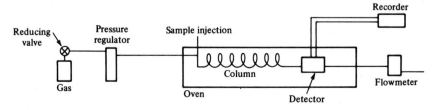

Fig. 14.4 Diagram of a gas-liquid chromatograph.

with p_i and p_o the inlet and outlet column gas pressures. The *retardation factor* R_F represents the rate of movement of the component peak value through the column relative to the rate of movement of the carrier gas, so that

$$R_F = \frac{t_G}{t_R} \tag{348}$$

We have therefore, from Eqs. (346) and (348),

$$R_F = \frac{V_G^0}{V_R^0} \tag{349}$$

The distribution of the component between the liquid and gas phases is given by

$$D_R = \frac{\text{concentration of component in liquid phase}}{\text{concentration of component in gas phase}}$$

and we can show that

$$R_F = \frac{A_G}{A_G + D_R A_L} \tag{350}$$

where A_G and A_L are, respectively, the gas phase and liquid phase cross-sectional areas. For a given column length, we have

$$R_F = \frac{V_G^0}{V_G^0 + D_R V_L} \tag{351}$$

where V_L is the total volume of liquid in the column (column length $\times A_L$). From Eqs. (349) and (351) we have

$$V_R^0 = V_G^0 + D_R V_L \tag{352}$$

and, combining Eqs. (352) and (346) we obtain

$$t_R = \frac{V_G^0 + D_R V_L}{j F_c} \tag{353}$$

Thus, the retention time for a given component is a function of D_R where the values of j, F_c, V_L, and V_G^0 are constant.

Qualitative analysis in its simplest sense involves the determination of the retention time or volume for each peak and the identification of the compound from this data. Refinements of the qualitative procedure involve compound identification from infrared spectra or mass spectrometer data. Quantitative estimation is based generally on the determination of the area under each peak and the ratio of this area for each compound to the total area for all of the peaks obtained. While the technique appears quite simple, various problems arise relative to the method of area measurement, the need for a reproducible technique, and the requirement for internal standards.

14.5 ION-EXCHANGE METHODS

Certain types of materials display the ability to exchange an ion in a contacting solution for an ion of the material. Both natural and synthetic materials can be found which show this ability. Where analytical purposes are involved, the synthetic exchange materials are preferred. Such synthetic materials are called ion-exchange resins. They involve an organic polymer matrix with introduced ionizable groups of specific types. The polymer, such as, for example, polystyrene with added divinylbenzene, is cross-linked between polymer units. The degree of three-dimensional cross-linking can be varied, as can the nature of the impregnating ionizable groups, so as to modify the resin properties with respect to exchange capacity, selectivity, and rate of exchange.

The exchange process is actually a form of solid stationary phase–moving liquid phase chromatography; and, in general, the distribution law applies relative to such ion-exchange mechanisms.

Where such resins contain ionizable groups such as the anions of sulfonic or carboxylic acids, they will function as cation-exchange resins. If these groups are in the acid form, their hydrogen ion is exchanged for cations in the contacting solution. For example, the passage through the resin of a solution involving Na^+ would result in the exchange of Na^+ for H^+, the eluted solution being then free from Na^+. Subsequently the passage through the resin of a solution containing Ca^{2+} or Mg^{2+} would result in the exchange of Na^+ and H^+ for these ions.

Where the resin contains groups such as substituted amines or quaternary ammonia compounds, exposure to aqueous media results in the formation of NH_3^+ groups according to

$$\text{Resin–NH}_2 + H_2O \rightleftharpoons \text{Resin–NH}_3^+ \cdot OH^-$$

The hydroxyl ion may then be exchanged for ions such as Cl^-, NO_3^-, SO_4^{2-}, etc. Such resins are called anion-exchange resins.

Ion-exchange resins are applied frequently in the purification of water, and their abilities in this connection are quite phenomenal. They may also be used in specific analytical procedures. For example, the total salt content of a solution may be determined by passing the solution through a cation-exchange resin in the hydrogen form. Here the salt cations are exchanged for hydrogen ions. The resulting eluted solution may then be titrated with a standard base solution and the total salt content determined. In a similar manner, the chloride content of a solution containing only the chloride ion may be determined by passing the solution through an anion-exchange resin, the chloride ions being exchanged for hydroxyl ions and the eluted solution subsequently titrated with a standard acid solution.

Again, in the analytical process, interfering ions may be removed from a solution by passage through a properly selected ion-exchange resin. The removal of phosphate ion by such procedures is particularly valuable.

Finally, trace amounts of ions in solutions may be removed by the ion-exchange method. Treatment of the resin substance by a small volume of water or an appropriate liquid subsequently yields an eluted solution of relatively high concentration of the removed ions, resulting in a concentration process for trace substances.

PROBLEMS

1. In a simple extraction procedure iron(III) is separated from chromium(III) in aqueous solution by treatment with ether. The distribution coefficient K (org/aq) has a value of 50. 100 ml of the aqueous solution containing 1.00 g of iron as Fe^{3+} is treated once with 200 ml of ether. Indicate the weight of iron left in the aqueous layer.

2. In a simple extraction process, 100 ml of an aqueous solution of substance A is treated three times with 10.0-ml portions of an immiscible organic solvent. The efficiency of extraction of A from the aqueous layer finally resulting is found to be 95 percent. What is the value of the distribution coefficient K (org/aq)?

3. 100 ml of an aqueous solution containing 100 mg of substance A is to be treated with an immiscible organic solvent with a K (org/aq) of 20.0. Not more than 60.0 ml of the organic solvent is to be used. Show

(a) The extraction efficiency for a single extraction with 60.0 ml of the solvent
(b) The extraction efficiency for five extractions of 12.0 ml each of the solvent
(c) The minimum number of identical volume extractions to yield not more than 1 mg of substance A in the aqueous layer

4. In a Craig countercurrent extraction procedure, the value of K (org/aq) is 10.0. 50.00 ml of an aqueous solution contains 100 mg of solute A. The treatment always involves equal volumes (50.0 ml) of either water or the organic solvent. Determine

(a) The weight of A in vessel 0 when $n = 10$.
(b) Which vessel (or vessels) is the peak vessel.
(c) What weight of solute A is contained in the peak vessel(s)?
(d) From what vessels must the organic layer be collected in order to recover not less than 90 percent of solute A?

$Ans.$ (a) 3.86×10^{-9} mg; (b) vessels 9 and 10; (c) 77.2 mg; (d) vessels 7, 8, 9 and 10

5. A metal ion in aqueous solution is extracted by a chelating substance HL, dissolved in an immiscible organic solvent, according to

$$M^{2+}_{aq} + 2HL_{org} \rightleftharpoons M(L)_{2,org} + 2H^+_{aq}$$

The formation constant for the reaction from left to right has a value given by 1.5×10^{-1}. Suppose that 20.0 ml of an aqueous solution of M^{2+} is extracted with 10.0 ml of organic solvent which contains 2.0×10^{-2} M HL. Show the efficiency of extraction of the metal to the organic layer at solution pH values of 1.00, 1.50, 2.00, 2.50, 3.00, and 3.50.

6. Two metal ions Ma^{2+} and Mb^{2+} are capable of extraction from aqueous solution by the chelating substance HL, dissolved in an immiscible organic solvent, in accordance with

$$Ma^{2+}_{aq} + 2HL_{org} \rightleftharpoons Ma(L)_{2, org} + 2H^+_{aq}$$
$$Mb^{2+}_{aq} + 2HL_{org} \rightleftharpoons Mb(L)_{2 org} + 2H^+_{aq}$$

The formation constants are respectively $K_{Ma} = 1.0 \times 10^{-2}$ and $K_{Mb} = 1.0 \times 10^{-7}$. 20.0 ml of an aqueous solution containing both metal ions is treated with 10.0 ml of the solvent which contains 2.0×10^{-2} M HL. Show the percent of each metal extracted to the organic layer at pH 2.00, 2.50, 3.00, 3.50, 4.00, 4.50, 5.00, 5.50, 6.00, and 6.50. Indicate the pH at which you would separate the two metals most efficiently.

REFERENCES

1. Morrison, G. H., and H. Frieser: "Solvent Extraction in Analytical Chemistry," Wiley, New York, 1957.
2. Pecsok, R. L., and L. D. Shields: "Modern Methods of Chemical Analysis," Wiley, New York, 1968.
3. Morris, C. J. O. R., and P. Morris: "Separation Methods in Biochemistry," Interscience, New York, 1963.
4. Nogare, S. Dal, and R. S. Juvet, Jr.: "Gas-Liquid Chromatography," Interscience, New York, 1962.
5. Browning, D. R. (ed.): "Chromatography," McGraw-Hill, London, 1969.

15
Laboratory Experiments

15.1 NEUTRALIZATION TITRATIONS*

15.1.1 PREPARATION OF APPROXIMATELY 0.1 *M* SODIUM HYDROXIDE

Boil 1.2 l of distilled water in a 1500-ml Erlenmeyer flask for 5 min to expel CO_2. Cover with a watch glass and cool. Carefully wash out a 1-l polyethylene bottle. Transfer to the bottle, using a graduated pipette and rubber bulb, 6.0 ml of *clear* 1:1 NaOH solution. Wash out the pipette into the bottle using the boiled water, and make up to approximately 1 l with the boiled water. Mix well. The 1:1 NaOH solution should be available from the instructor. It is prepared by adding *carefully* and *slowly* 50 g of NaOH pellets to 50 ml of distilled water. The solution should have been standing for several days to allow Na_2CO_3 to precipitate and settle out. Only the *clear* solution is used to prepare the standard NaOH solution.

15.1.2 STANDARDIZATION OF APPROXIMATELY 0.1 *M* SODIUM HYDROXIDE

Place about 5 g of pure potassium acid phthalate (standardization grade) in a weighing bottle or dish and dry in the oven for 1 h at 110°C. Cool in the dessica-

*See Chap. 6.

tor. Weigh 0.70- to 0.90-g samples into three clean, dry, numbered Erlenmeyer flasks. Record your weights to the nearest 0.1 mg. Add 50 ml of boiled and cooled distilled water to each flask, and then add 2 drops of phenolphthalein indicator solution (0.1% in 80% alcohol). Rinse a *clean* 50-ml burette with the NaOH solution to be standardized, and fill the burette with the solution. The volume required for a titration will be 30 to 45 ml. Titrate the first solution to the first permanent pink color (color must persist for 30 s). Record the volume required to the nearest 0.01 ml. Titrate the remaining two solutions in the same way and record the volumes. Calculate the molarity (normality) of the NaOH solution for each titration. Inspect for result rejection using the Q test. Average the acceptable results and determine the average deviation. This latter value should not exceed about 2 ppt. Label the bottle properly with the molarity (normality) of the NaOH solution.

15.1.3 DETERMINATION OF THE PURITY OF POTASSIUM ACID PHTHALATE

Unless otherwise advised, place the unknown sample given to you in a weighing bottle or dish and dry in the oven for 1 h at 110°C. Cool in the dessicator. Weigh out four samples of an appropriate weight (see the instructor) into four clean, dry, numbered Erlenmeyer flasks. Record your weights to the nearest 0.1 mg. Add 75 ml of boiled and cooled distilled water to each flask and dissolve the samples. Gentle heat may be applied where required, and the solution subsequently cooled. Add 2 drops of phenolphthalein indicator solution. Rinse and fill a *clean* 50-ml burette with standard NaOH solution. Titrate the first solution to a permanent pink color using 2-ml increments of NaOH solution. This locates the approximate end-point volume for this sample weight. Calculate the approximate end-point volumes for the remaining three solutions and titrate these carefully to the *first* permanent pink color. Record the volumes for these three solutions to the nearest 0.01 ml. Calculate the percentage of potassium acid phthalate in the unknown for each sample. Inspect for result rejection using the Q test. Average the acceptable results and determine the average deviation. Report *all* data, including any rejected value.

15.1.4 DETERMINATION OF ACETIC ACID IN VINEGAR

Other acid substances may be present in commercial vinegar, but in the analytical procedure following, these are reported as acetic acid.

Obtain a sample of commercial vinegar from the instructor. Pipette the required volume (see the instructor) into a 250-ml volumetric flask. Dilute to the mark with distilled water and mix thoroughly. Transfer a 50-ml aliquot by pipette to a clean, numbered Erlenmeyer flask, and add 50 ml of distilled water. Repeat the entire process (sampling, diluting, aliquoting, etc.) with three other pipetted samples from your unknown. *Do not use* as replicates aliquots from one sample.

To each of the prepared solutions add 2 drops of phenolphthalein indicator solution. Rinse and fill a *clean* 50-ml burette with standard NaOH solution (Sec. 15.1.2). Titrate the first solution to a permanent pink color using 2-ml increments of standard NaOH. With this approximate end-point volume as a guide, titrate the remaining three solutions to the *first* permanent pink color. Record your volumes for these three solutions to the nearest 0.01 ml. Assume all the contained acid in the vinegar to be acetic acid, and calculate for each of the last three solutions the acetic acid content of the vinegar in grams per 100 ml. Report the percentage by weight of acetic acid on the basis of a density value of 1.000 for the vinegar. Inspect for result rejection using the Q test. Average the acceptable values and determine the average deviation. Report *all* data, including any rejected value.

15.1.5   PREPARATION OF APPROXIMATELY 0.1 M HYDROCHLORIC ACID

Carefully wash out a 1-l glass bottle with a glass stopper. Add 50 ml of distilled water and transfer 8.5 ml of concentrated HCl (about 12 M) to the bottle, using a graduated pipette and a rubber bulb. Make up to approximately 1 l with distilled water. Mix thoroughly.

15.1.6   STANDARDIZATION OF APPROXIMATELY 0.1 M HYDROCHLORIC ACID WITH SODIUM CARBONATE

Place about 3 g of pure sodium carbonate (standardization grade) in a weighing bottle or dish and dry in the oven for 1 h at 200°C. Cool in the dessicator. Weigh out 0.20- to 0.25-g samples into four clean, dry, numbered Erlenmeyer flasks. Record your weights to the nearest 0.1 mg. Add 50 ml of distilled water and swirl gently to dissolve. Add 2 drops of methyl red indicator solution (0.1% of the sodium salt in water). The indicator will assume the *yellow* color of the basic form. Rinse and fill a *clean* 50-ml burette with the HCl solution to be standardized. Titrate the first solution, using 2-ml increments of the HCl solution, until the indicator turns a distinct *red* (the acid-form color). This will locate the approximate second-stage end-point volume. Calculate the approximate end-point volumes for the remaining three solutions. Titrate the first of these until the indicator turns distinctly red. Add an excess approximate but measured volume of HCl solution of about 2 ml. Remove the CO_2 by boiling the solution gently for not less than 5 min. The color should remain red. If it does not, add an additional measured volume of HCl solution. Cool to room temperature and back-titrate the excess HCl with standard NaOH solution (Sec. 15.1.2). Record the total HCl and NaOH solution volumes to the nearest 0.01 ml. Calculate, from the NaOH–HCl solution relationship, the volume of HCl solution required to attain the end point. Repeat the procedure for the remaining two solutions. Calculate the molarity (normality) of the HCl solution for each of the last three titrations. Inspect for result rejection using the Q test. Average the acceptable values and determine the average deviation. This latter should not exceed 4 ppt. Label the bottle properly with the molarity (normality) of the HCl solution.

15.1.7 DETERMINATION OF SODIUM CARBONATE IN SODA ASH

Soda ash, a commercial and impure form of sodium carbonate, may contain small quantities of sodium bicarbonate and sodium hydroxide. The method of analysis to follow determines the total akalinity of soda ash and reports this as sodium carbonate.

Unless otherwise advised, place the unknown sample of soda ash given to you in a weighing bottle or dish and dry in the oven for 1 h at 200°C. Cool in a dessicator. Weigh four samples of the appropriate weight (see the instructor) into four clean, dry, numbered Erlenmeyer flasks. Record your sample weights to the nearest 0.1 mg. Add 50 ml of distilled water, swirl to dissolve, and add 2 drops of methyl red indicator solution. Rinse and fill a *clean* 50-ml burette with standard HCl solution. Titrate the first solution, using 2-ml increments of HCl solution, to a distinct red color. This locates the approximate second-stage end-point volume for this sample weight. Calculate the approximate end-point volumes for the remaining three samples. Titrate these samples exactly in accordance with the procedure in Sec. 15.1.6. Record the total volumes of HCl and NaOH solutions used to the nearest 0.01 ml. Use the NaOH–HCl solution relationship and calculate the volume of HCl solution required to attain the end point in each case. Calculate the percentage of sodium carbonate in the soda ash for each sample. Inspect for result rejection using the Q test. Average the acceptable values and determine the average deviation. Report *all* data, including any rejected value.

15.1.8 DETERMINATION OF AMMONIA IN DOMESTIC AMMONIA

Obtain a sample of commercial ammonia used for domestic purposes from the instructor. Rinse a clean graduated pipette with the sample and pipette the required volume (see the instructor) into a clean, previously weighed weighing bottle, using a rubber bulb for the rinsing and pipetting operations. Stopper the bottle and weigh. Record the weight of the ammonia sample to the nearest 0.1 mg. Transfer the sample to a 250-ml volumetric flask, rinsing the weighing bottle three times into the flask with distilled water. Dilute to the mark with distilled water and mix thoroughly. Keep stoppered until ready for use. Repeat the entire process (sampling, weighing, diluting, etc.) with three other samples pipetted from your unknown. *Do not use* as replicates aliquots from one sample. Transfer from each diluted sample a 25-ml aliquot by pipette to clean, numbered Erlenmeyer flasks. Add 2 drops of methyl red indicator solution and 25 ml of distilled water. Rinse and fill a *clean* 50-ml burette with standard HCl solution (Sec. 15.1.6). Titrate the first solution, using 2-ml increments of HCl solution, to a distinct red color. With this approximate end-point value as a guide, titrate the remaining three solutions to the *first* appearance of the red color. Record the volumes for these three solutions to the nearest 0.01 ml. Calculate the percent by weight of ammonia in the commercial product for each sample. Inspect for result rejection using the Q test. Average the acceptable results and determine the average deviation. Report *all* data, including any rejected value.

15.2 PRECIPITATION TITRATIONS*

15.2.1 PREPARATION OF APPROXIMATELY 0.1 *M* SILVER NITRATE

Carefully wash out with distilled water a 500-ml brown-glass bottle with a glass stopper. Weigh out 8.5 g of pure silver nitrate and dissolve in 100 ml of distilled water. Transfer to the bottle and make up to approximately 500 ml with distilled water. Protect from direct light as much as possible.

15.2.2 STANDARDIZATION OF APPROXIMATELY 0.1 *M* SILVER NITRATE

Place about 2 g of pure sodium chloride (standardization grade) in a weighing bottle or dish and dry in the oven for 1 h at 200°C. Cool in the dessicator. Weigh 0.20- to 0.25-g samples into three clean, dry, numbered Erlenmeyer flasks. Record the weights to the nearest 0.1 mg. Add 40 ml of distilled water, swirl gently to dissolve, and add 1 ml of 0.1 *M* acetic acid–0.1 *M* sodium acetate solution. Now add 10 drops of dichlorfluorescein indicator solution (0.1% sodium dichlorfluoresceinate in water or 0.1% dichlorfluorescein in 80% alcohol). Finally add 10 ml of 1% dextrin solution. Rinse and fill a *clean* 50-ml burette with the AgNO₃ solution to be standardized. The volume required for the titration will be about 35 to 45 ml. Titrate the first solution, with an efficient swirling action, until the color of the AgCl precipitate changes from yellow-white to the first persistent pink. Titrate the remaining two solutions in the same way. Record the end-point volumes to the nearest 0.01 ml. Calculate the molarity (normality) of the AgNO₃ solution for each titration. Inspect for result rejection using the Q test. Average the acceptable results and determine the average deviation. This latter should not exceed about 3 ppt. Label the bottle properly with the molarity (normality) of the AgNO₃ solution.

15.2.3 DETERMINATION OF CHLORIDE BY FAJAN'S METHOD

Obtain an unknown chloride sample from the instructor and, unless advised otherwise, place in a weighing bottle or dish and dry in the oven for 1 h at 170°C. Cool in the dessicator. Weigh four samples of appropriate weight (see the instructor) into four clean, dry, numbered Erlenmeyer flasks. Record your weights to the nearest 0.1 mg. Dissolve each sample in 40 ml of distilled water, add 1 ml of 0.1 *M* acetic acid–0.1 *M* sodium acetate solution, 10 drops of dichlorfluorescein indicator, and 10 ml of 1% dextrin solution. Rinse and fill a *clean* 50-ml burette with standard AgNO₃ solution (Sec. 15.2.2). Titrate the first solution, using 2-ml increments of AgNO₃ solution and efficient swirling, until the AgCl precipitate shows a distinct pink color. This locates the approximate end-point volume for this sample weight. Calculate the approximate end-point volumes for the remaining three samples, and titrate these carefully to the *first* persistent pink precipitate color. Record the end-point volumes for these titrations to the nearest 0.01 ml. Calculate the percentage of chloride in the unknown for each sample. Inspect for

---

*See Chap. 7.

result rejection using the Q test. Average the acceptable results and determine the average deviation. Report *all* data, including any rejected result.

15.2.4 PREPARATION OF APPROXIMATELY 0.05 *M* POTASSIUM FERROCYANIDE

Weigh 21.1 g of pure potassium ferrocyanide trihydrate ($K_4Fe(CN)_6 \cdot 3H_2O$) into a clean 400-ml beaker. Add 250 ml of distilled water and a stirring bar, and stir until dissolved. Wash out a glass bottle with distilled water. Transfer the solution to the bottle and dilute to approximately 1 l with distilled water.

15.2.5 STANDARDIZATION OF APPROXIMATELY 0.05 *M* POTASSIUM FERROCYANIDE

The reaction of potassium ferrocyanide with zinc involves

$$3Zn^{2+} + 2K_4Fe(CN)_6 \rightleftharpoons K_2Zn_3(Fe(CN)_6)_2(s) + 6K^+$$

Weigh 1.5 g of pure zinc metal into a clean 250-ml beaker. Record the weight to the nearest 0.1 mg. Add 40 ml of 1:3 HCl. Cover with a watch glass and warm to dissolve. Cool and transfer to a 250-ml volumetric flask. Dilute to the mark with distilled water and mix thoroughly. Pipette 25-ml portions into three clean, numbered, 500-ml Erlenmeyer flasks. Add 30 ml of 1:1 H_2SO_4 and dilute to 150 ml with distilled water. Add 25 ml of 20% ammonium sulfate $\{(NH_4)_2SO_4\}$ solution, 10 ml of 10% potassium fluoride (KF) solution, 4 drops of freshly-prepared 1% potassium ferricyanide ($K_3Fe(CN)_6$) solution, and 10 drops of 1% sodium diphenylbenzidine sulfonic acid indicator solution (1% sodium diphenylbenzidine sulfonic acid in water). Let the solutions stand until a deep blue color develops. Rinse and fill a *clean* 50-ml burette with the potassium ferrocyanide solution. The titration volume will be approximately 30 ml. Titrate the first solution carefully, swirling strongly after each addition of titrant and waiting to see if the blue color reappears. The blue color may change to violet near the end point. Continue the titration. Each addition of titrant will cause the blue or violet color to disappear, but until the true end point is reached, the color will reappear. Continue to titrate carefully until the blue or violet color disappears and the solution is an apple-green color which persists for several minutes. Titrate the remaining two solutions in the same manner. Record the end-point volumes to the nearest 0.01 ml. Calculate the molarity of the potassium ferrocyanide solution for each titration. Inspect for result rejection using the Q test. Average the acceptable results and determine the average deviation. Label the bottle properly with the molarity of the potassium ferrocyanide solution.

15.2.6 DETERMINATION OF ZINC IN MAGNESIUM-BASE ALLOYS

Certain commercial magnesium-base alloys contain about 5% zinc and 1% zirconium. The following method permits the determination of zinc in these alloys or in solutions containing zinc.

Obtain a sample of magnesium-base alloy from the instructor or, alternatively, a solution of unknown zinc concentration. Weigh or pipette appropriate quantities (see the instructor) into four clean, numbered 500-ml Erlenmeyer flasks. If the alloy is involved, record the sample weights to the nearest 0.1 mg. For magnesium-base alloy samples add 75 ml of distilled water, then add *slowly* and *carefully* 6 ml of 1:1 H_2SO_4 per gram of sample weight (Note 1). When the sample is dissolved, add a 30-ml excess of 1:1 H_2SO_4 and dilute to 150 ml with distilled water. For liquid samples add 30 ml of 1:1 H_2SO_4 and dilute to 150 ml. Now add 25 ml of 20% ammonium sulfate, 10 ml of 10% potassium fluoride, 4 drops of freshly prepared 1% potassium ferricyanide, and 10 drops of 1% sodium diphenylbenzidine sulfonic acid solution. Allow the solutions to stand until they darken. They may not at this point turn blue or violet, but they must darken before the titration is started. Rinse and fill a *clean* 50-ml burette with standard potassium ferrocyanide solution (Sec. 15.2.5). Titrate the first solution, using 2-ml increments of the titrant and following the general procedure outlined in Sec. 15.2.5, until a persistent apple-green color is obtained. Using this approximate end-point volume as a guide, titrate the remaining three solutions carefully, again following the procedure given in Sec. 15.2.5, until a persistent apple-green color is obtained. Record the end-point volumes for these three titrations to the nearest 0.01 ml. If the unknown is a magnesium-base alloy, calculate for each sample the percentage of zinc in the alloy. If the unknown is a zinc solution, calculate for each sample the zinc content in grams per 100 ml of the original solution. Inspect for result rejection, in either case, by the Q test. Average the acceptable results and determine the average deviation. Report *all* data, including any rejected result.

Note 1   Magnesium-base alloys react violently with sulfuric acid. In order to avoid loss of sample, add the acid in small quantities until the sample is dissolved, waiting for the action to cease after each addition.

15.2.7 DETERMINATION OF PHENOBARBITAL IN TABLETS

Phenobarbital is barbituric acid, 5-ethyl-5-phenyl-, and in the method of analysis to follow the reactions are

2-Phenobarbital + Na_2Co_3 $\rightleftharpoons$ 2-phenobarbital Na + H_2CO_3
Phenobarbital Na + Ag^+ $\rightleftharpoons$ phenobarbital Ag(s) + Na^+

The technique is, in point of fact, a potentiometric titration, but it is included here because the titration reaction is one of precipitation.

A potentiometer with a silver electrode and a KNO_3 salt bridge SCE reference electrode is required. A combination silver–reference electrode system may also be used. Follow the instructions of the laboratory instructor relative to the operation of the instrument.

Prepare and standardize an approximately 0.05-M solution of silver nitrate, following in general the details given in Secs. 15.2.1 and 15.2.2. Obtain from the instructor a sample of tablets containing phenobarbital but no stearates (Note 1). Crush enough tablets to yield a sample containing 0.10 to 0.20 g of phenobarbital (see the instructor). Three such samples will be required, and these should be transferred to three clean, numbered, 250-ml beakers. Add 50 ml of 3% sodium carbonate solution and 25 ml of acetone and dilute to 100 ml with distilled water. Add a stirring bar. Rinse and fill a *clean* 50-ml burette with standard silver nitrate solution. Insert the electrodes and adjust the potentiometer according to the advice of the instructor. Stirring efficiently, titrate carefully, recording the volume values to the nearest 0.01 ml and the potential values to the reading uncertainty permitted by the potentiometer scale. Use 0.1-ml increments of titrant when near the equivalence point, and continue these increments until well past this point. Repeat the procedure for the remaining two solutions. Using the $\Delta^2 E/\Delta T^2$ tabulation method (Sec. 5.2) determine the equivalence-point volume for each titration. Calculate for each titration the weight of phenobarbital in milligrams per tablet. Average the values, and determine the average deviation. Report *all* data.

Note 1   The instructor may provide you with a solution containing an unknown amount of phenobarbital. In this case, follow the instructor's advice concerning the volume of solution to be pipetted for each titration, and report the phenobarbital content of the solution in the manner requested.

15.3   COMPLEXATION TITRATION*

15.3.1   PREPARATION OF APPROXIMATELY 0.01 *M* DISODIUM DIHYDROGEN EDTA

Place about 5 g of pure disodium dihydrogen EDTA dihydrate in a weighing bottle or dish and dry in the oven for 1 h at 70°C. Cool in the dessicator. Weigh 4.0 g of the salt and 0.1 g of magnesium chloride hexahydrate ($MgCl_2 \cdot 6H_2O$) into a 600-ml beaker (Note 1). Add 400 ml of distilled water and a stirring bar, and stir until the salts are dissolved. Wash out with distilled water a 1-l glass bottle and glass stopper. Transfer the EDTA solution to the bottle, and make up to approximately 1 l with distilled water. Mix thoroughly.

Note 1   The magnesium is added to permit the use of Eriochrome black T as the indicator in the calcium carbonate titrations to follow.

15.3.2   STANDARDIZATION OF APPROXIMATELY 0.01 *M* DISODIUM DIHYDROGEN EDTA WITH CALCIUM CARBONATE

Place 0.5 g of pure calcium carbonate (standardization grade) in a weighing bottle or dish and dry in the oven for 1 h at 100°C. Cool in the dessicator. Weigh exactly 0.4000 g of calcium carbonate into a clean, dry 500-ml volumetric flask. Add 100 ml of distilled water, being sure to wash down any of the salt clinging to the sides of the flask. Add dropwise 1:1 HCl until effervescence ceases and the solution is clear. Be *most* careful to avoid any unnecessary excess of 1:1 HCl. Dilute to the mark with distilled water and mix thoroughly. Pipette 50-ml aliquots into each of three clean, numbered Erlenmeyer flasks. Add 5 ml of pH 10 buffer

*See Chap. 8.

solution (Note 1) and 5 drops of Eriochrome black T indicator solution (Note 2). Rinse and fill a *clean* 50-ml burette with the EDTA solution to be standardized. The titration volume should be about 35 to 45 ml. Titrate the first solution carefully until the color changes from red to a blue without a tinge of red. Titrate the remaining two solutions in the same manner. Record the end-point volumes to the nearest 0.01 ml. Calculate the molarity of the EDTA solution, and its calcium carbonate titer in milligrams per milliliter, for each of the titrations. Inspect for result rejection using the Q test. Average the acceptable results and determine the average deviation. This latter should not exceed about 2 ppt. Label the bottle properly with the molarity of EDTA and the calcium carbonate titer in milligrams per milliliter.

Note 1 The buffer consists of 6.8 g of ammonium chloride and 57 ml of concentrated ammonia diluted to 100 ml

Note 2 Dissolve 0.5 g of reagent grade Eriochrome black T in 100 ml of alcohol. Date the solution label on the bottle and do not use if older than 6 weeks.

15.3.3 DETERMINATION OF THE TOTAL HARDNESS OF NATURAL WATER

Most natural waters from ground sources will contain calcium and magnesium. The method of analysis to follow reports both magnesium and calcium together as total hardness in terms of calcium carbonate.

Obtain from your instructor a sample of a natural water from a ground source. Pipette appropriate volumes (see the instructor) into four clean, numbered Erlenmeyer flasks. Add 5 ml of pH 10 buffer solution and 5 drops of Eriochrome black T indicator solution. The volume at this point should be about 50 ml. If it is less, adjust to 50 ml with distilled water. Rinse and fill a *clean* 50-ml burette with standard EDTA solution (Sec. 15.3.2). Titrate the first solution, using 2-ml increments of EDTA solution, until a clear blue color with no trace of red is obtained. Using this approximate end-point volume as a guide, carefully titrate the remaining three solutions until the red color changes to the first clear blue without any trace of red. Record the end-point volumes for these titrations to the nearest 0.01 ml. Calculate for each of the last three titrations the total hardness of the water sample in parts per million of calcium carbonate. Inspect for result rejection using the Q test. Average the acceptable results and determine the average deviation. Report *all* data, including any rejected result.

15.4 OXIDATION-REDUCTION TITRATIONS*

15.4.1 PREPARATION OF APPROXIMATELY 0.02 *M* (0.1 *N*) POTASSIUM PERMANGANATE

Weigh 3.2 g of reagent grade potassium permanganate ($KMnO_4$) into a clean 2-l beaker. Add about 1 l of distilled water and dissolve. Heat to just below boiling

*See Chap. 9.

and maintain at this temperature for 1 h. Cover with a watch glass and let stand and cool for 24 h. Filter by suction through a sintered glass crucible and transfer to a clean brown-glass 1-l bottle. Store in a dark place when not being used.

15.4.2 STANDARDIZATION OF APPROXIMATELY 0.02 M (0.1 N) POTASSIUM PERMANGANATE

Place about 1.5 g of sodium oxalate ($Na_2C_2O_4$) (standardization grade) in a weighing bottle or dish and dry in the oven for 1 h at 120°C. Cool in the dessicator. Weigh 0.25- to 0.30-g portions into three clean, numbered Erlenmeyer flasks. Record the weights to the nearest 0.1 mg. Add 100 ml of 1.5 N H_2SO_4 (20 ml of concentrated H_2SO_4 in 400 ml of water). Heat the solutions to 80 to 90°C. Place a stirring bar and a thermometer in the solution. Rinse and fill a *clean* 50-ml burette with the potassium permanganate solution. The titration volume will be about 35 to 45 ml. Titrate the first solution by adding about 0.2 to 0.5 ml of titrant and stirring until the pink color disappears. Continue to titrate slowly until a first pink color that persists for not less than 30 s is obtained. If the solution temperature falls below 60°C, reheat the solution. Titrate the remaining two solutions in the same manner. Record the end-point volumes to the nearest 0.01 ml. Calculate for each titration the molarity and normality of the potassium permanganate solution. Inspect for result rejection using the Q test. Average the acceptable values and determine the average deviation. This latter value, for the expression of normality, should not exceed about 2 ppt. Label the bottle properly with the molarity and normality of the potassium permanganate solution.

15.4.3 DETERMINATION OF OXALATE AS SODIUM OXALATE

Obtain from the instructor a sample of unknown sodium oxalate content. Unless otherwise advised, place the sample in a weighing bottle or dish and dry in the oven for 1 h at 120°C. Weigh appropriate portions (see the instructor) into four clean, numbered Erlenmeyer flasks. Record the sample weights to the nearest 0.1 mg. Add 100 ml of 1.5 N H_2SO_4 and heat to 80 to 90°C. Rinse and fill a *clean* 50-ml burette with standard potassium permanganate solution (Sec. 15.4.2). Add to the first solution a stirring bar and a thermometer. Titrate this solution by adding first 0.2 to 0.5 ml of titrant and stirring until the pink color disappears. Continue the titration, adding slowly 2-ml increments of titrant, until a permanent pink or red color is obtained. With this end-point volume as a guide, titrate the remaining three solutions in the manner outlined in Sec. 15.4.2, until a first pink color which persists for not less than 30 s is obtained. Record the three titration end-point volumes to the nearest 0.01 ml. Calculate for each titration the percentage of sodium oxalate in the unknown. Inspect for result rejection using the Q test. Average the acceptable results and determine the average deviation. Report *all* values, including any rejected value.

15.4.4 PREPARATION OF APPROXIMATELY 0.0167 M (0.1 N) POTASSIUM DICHROMATE

Place about 5.5 g of pure potassium dichromate ($K_2Cr_2O_7$) in a weighing bottle or dish and dry in the oven for 1 h at 100°C. Cool in the dessicator. Weigh 4.9 g into a 250-ml beaker and dissolve in 100 ml of distilled water. Transfer to a clean 1-l glass bottle and dilute to approximately 1 l.

15.4.5 STANDARDIZATION OF APPROXIMATELY 0.0167 *M* (0.1 *N*) POTASSIUM DICHROMATE

Weigh 0.20-g portions of rust-free pure iron wire into three clean, numbered 250-ml beakers. Record the weights to the nearest 0.1 mg. Add 20 ml of 1:1 HCl, cover with a watch glass, and warm to dissolve. Heat to just under boiling and reduce the iron by the dropwise addition of stannous chloride solution (Note 1) until the yellow color has disappeared and the solution is colorless or faintly green. Add 2 drops of stannous chloride solution in excess. Cool the solution in running water and add at one stroke 20 ml of saturated mercuric chloride solution (Note 2). Allow to stand for 3 min and transfer to a 500-ml Erlenmeyer flask. Add 5 ml of concentrated H_2SO_4, 5 ml of concentrated H_3PO_4, 250 ml of water, and 5 drops of 1% sodium diphenylbenzidine sulfonic acid indicator solution (1% of the salt in water). Rinse and fill a *clean* 50-ml burette with the potassium dichromate solution. Titrate the first solution by adding the titrant slowly, with constant swirling, until the solution changes color from green to a first purple color that persists for several minutes. Titrate the remaining two solutions in the same manner. Record the end-point volumes to the nearest 0.01 ml. Calculate for each titration the molarity and normality of the titrant. Inspect for result rejection by the Q test. Average the acceptable results and determine the average deviation. This latter, for the normality expression, should not be more than 2 to 3 ppt. Label the bottle properly with the molarity and normality of the potassium dichromate solution.

Note 1 The stannous chloride solution is prepared by dissolving 11 g of $SnCl_2 \cdot 2H_2O$ in 25 ml of concentrated HCl and diluting to 100 ml with water.

Note 2 The saturated mercuric chloride solution must be added at one stroke with efficient stirring to avoid the formation of metallic mercury.

15.4.6 DETERMINATION OF IRON IN AN IRON ORE

Obtain a sample of iron ore from the instructor and, unless otherwise advised, dry it in a weighing bottle or dish for 1 h in an oven set at 110°C. Weigh appropriate portions (see the instructor) into four clean, numbered 250-ml beakers. Record the sample weights to the nearest 0.1 mg. Add 20 ml of 1:1 HCl, cover with a watch glass, and heat to just below boiling to dissolve. This may take up to 1 h, and only a white residue of undissolved silica may be left (Note 1). Evaporation losses of 1:1 HCl should be periodically replaced. Reduce the iron with stannous chloride, cool in running water and add saturated mercuric chloride, all as outlined in Sec. 15.4.5. Let stand for 3 min and transfer to a 500-ml Erlenmeyer flask. Add 5 ml of concentrated H_2SO_4, 5 ml of concentrated H_3PO_4, 250 ml of water, and 5 drops of 1 percent sodium diphenylbenzidine sulfonic acid indicator solution. Rinse and fill a *clean* 50-ml burette with standard potassium dichromate solution (Sec. 15.4.5). Titrate the first solution, using 2-ml increments of titrant, until a persistent purple color is obtained. Using this end-point volume as a guide, titrate the remaining three solutions as indicated in Sec. 15.4.5. Record these titration end-point volumes to the nearest 0.01 ml. Calculate for each titration the percentage of iron in the ore. Inspect for result rejection by the Q test. Average the acceptable values and determine the average deviation. Report *all* data, including any rejected value.

Note 1   If a residue exists that is not white, see the instructor. This may require the following procedure. Dilute the solution to 40 ml and filter on a coarse paper. Wash the filter paper twice with 1% HCl and three times with water. Retain the filtrate. Place the filter paper in a small platinum crucible and ignite. Cool and mix into the residue 0.6 g of very fine anhydrous Na_2CO_3. Reheat the crucible and contents to give a clear melt. Cool and add 5 ml of water and, carefully and slowly, 5 ml of 1:1 HCl. Warm to dissolve the melt and transfer the solution to the original filtrate. Evaporate the filtrate to about 20 ml and continue the analytical process.

15.4.7 PREPARATION OF APPROXIMATELY 0.1 M SODIUM THIOSULFATE

Weigh 25 g of pure sodium thiosulfate pentahydrate ($Na_2S_2O_3 \cdot 5H_2O$) and 0.1 g of sodium carbonate into a clean 400-ml beaker. Add 250 ml of previously boiled and cooled distilled water and stir to dissolve. Transfer to a clean 1-l brown-glass bottle and dilute to approximately 1 l with the previously boiled and cooled water. Keep in a dark place when not in use.

15.4.8 STANDARDIZATION OF APPROXIMATELY 0.1 M SODIUM THIOSULFATE

Dry about 1 g of pure potassium dichromate in a weighing bottle or dish for 1 h in an oven set at 170°C. Weigh 0.20-g portions of the salt into three clean 500-ml Erlenmeyer flasks. Record the sample weights to the nearest 0.1 mg. Dissolve in 50 ml of distilled water. Treat each sample separately from now on. Add a freshly prepared solution involving 4 g of potassium iodide (KI), 5 ml of 1:1 HCl, and 50 ml of water. Swirl gently to mix, cover with a watch glass, and let stand for 5 min in a dark place. Add 150 ml of water. Rinse and fill a *clean* 50-ml burette with the sodium thiosulfate solution. The titration will require about 35 to 45 ml. Titrate the first solution until the yellow color of the triiodide ion has almost completely disappeared. Add 5 ml of starch solution (Note 1) and continue the titration carefully until the blue starch-triiodide-complex color changes to the clear green color of the chromous ion. Treat and titrate the remaining two solutions in the same manner. Record the end-point volumes to the nearest 0.01 ml. Calculate the molarity (normality) of the sodium thiosulfate solution for each titration. Inspect for result rejection using the Q test. Average the acceptable results and determine the average deviation. This latter value should not exceed about 2 ppt. Label the bottle properly with the molarity (normality) of the solution.

Note 1   Make a paste of 2 g of soluble starch and 25 ml of water. Add this paste slowly and with constant stirring to 500 ml of boiling water. Continue boiling and add 1 g of boric acid. Heat until a clear solution is obtained (2 to 3 min). Cool and store in a glass-stoppered bottle.

15.4.9 DETERMINATION OF COPPER IN A BRASS ALLOY

This alloy should contain not more than about 2% iron, but may contain normal amounts of tin, lead, and zinc.

Obtain a sample of brass from the instructor. Weigh 0.30-g portions into three clean, numbered, 250-ml Erlenmeyer flasks. Record the sample weights to the nearest 0.1 mg. Add 5 ml of 1:1 HNO_3 and warm under the hood until the sample is dissolved. Cool and add 10 ml of 1:1 H_2SO_4. Evaporate under the hood to copious white fumes of SO_3. Cool and cautiously add 20 ml of water. Boil for 2 min and cool. Swirling continuously, add carefully 1:3 ammonia until the deep blue color of the tetraminecopper(II) complex just appears. Do not add any unnecessary excess of ammonia. Add 5 ml of glacial acetic acid and 2 g of ammonium hydrogen fluoride $\{(NH_4)HF_2\}$. Swirl for 2 min and cool to room temperature. Treat each solution individually from now on. Rinse and fill a *clean* 50-ml burette with standard sodium thiosulfate solution (Sec. 15.4.8). Dissolve 4 g of potassium iodide in 10 ml of water and add to the copper solution. Titrate immediately and rapidly until the triiodide ion color has almost disappeared. Add 5 ml of starch solution and 2 g of potassium thiocyanate (KSCN) and titrate until the blue color disappears and the precipitate is white or grey-white when settled. Treat and titrate the remaining two solutions in the same manner. Record the endpoint volumes to the nearest 0.01 ml. Calculate for each titration the percentage of copper in the brass. Inspect for result rejection using the Q test. Average the acceptable results and determine the average deviation. Report *all* values, including any rejected value.

15.4.10 DETERMINATION OF SODIUM HYPOCHLORITE AS BLEACHING AGENT IN LAUNDRY BLEACH

The reaction of sodium hypochlorite with potassium iodide is

$$NaOCl + 3I^- + 2H_3O^+ \rightleftharpoons NaCl + I_3^- + 3H_2O$$

Obtain a sample of liquid laundry bleach from the instructor. Pipette 15-ml portions into three clean, dry, previously weighed weighing bottles. Stopper and weigh, recording the weights to the nearest 0.1 mg. Transfer the samples to clean 250-ml volumetric flasks, rinsing the weighing bottles into the flasks with a wash bottle stream. Dilute to the mark with distilled water and mix thoroughly. Pipette a 50-ml aliquot from each flask into clean, numbered Erlenmeyer flasks. Treat each sample separately from now on. Rinse and fill a *clean* 50-ml burette with standard sodium thiosulfate solution (Sec. 15.4.8). Add to the sample solution 3 g of potassium iodide dissolved in 10 ml of water, swirl, and add 10 ml of glacial acetic acid. Titrate immediately with sodium thiosulfate solution until the triiodide ion yellow color has almost disappeared. Add 5 ml of starch solution and titrate to the disappearance of the blue color. Treat and titrate the remaining two solutions in the same manner. Record the end-point volumes to the nearest 0.01 ml. Calculate the percentage of sodium hypochlorite in the laundry bleach for each sample. Inspect for result rejection by the Q test. Average the acceptable values and calculate the average deviation. Report *all* results, including any rejected result.

15.4.11 PREPARATION OF APPROXIMATELY 0.05 *M* (0.1 *N*) IODINE SOLUTION

Weigh 12.7 g of reagent grade iodine into a 250-ml beaker. Add 40 g of iodate-free potassium iodide and 50 ml of distilled water. Stir to dissolve and transfer to a clean glass bottle with a glass stopper. Dilute to approximately 1 l with water.

15.4.12 STANDARDIZATION OF APPROXIMATELY 0.05 *M* (0.1 *N*) IODINE SOLUTION

Place 1.5 g of pure arsenious oxide (As_2O_3) (standardization grade) in a weighing bottle or dish and dry in the oven for 1 h at 110°C. Weigh 0.20-g portions into three clean, dry, numbered Erlenmeyer flasks, and record the weights to the nearest 0.1 mg. Add 10 ml of 1 *M* NaOH and warm to dissolve. Cool and add 75 ml of distilled water and 2 drops of phenolphthalein indicator solution (0.1% in 80% alcohol). Carefully add 1:1 HCl until the pink color disappears, then add a 1-ml excess of the acid. In small portions add solid sodium bicarbonate ($NaHCO_3$), swirling after each addition, until effervescence ceases. Add an additional 3 g of $NaHCO_3$ and swirl to dissolve. Add 5 ml of starch solution (Note 1, Sec. 15.4.8). Rinse and fill a *clean* 50-ml burette with the iodine solution. The titration volume will be about 35 to 45 ml. Titrate carefully to the appearance of a first blue or violet color which persists for not less than 1 min. Record the end-point volumes for the three titrations to the nearest 0.01 ml, and calculate for each sample the molarity and normality of the iodine solution. Inspect for result rejection by the Q test. Average the acceptable values and calculate the average deviation. Label the bottle properly with the molarity and normality of the iodine solution.

15.4.13 DETERMINATION OF THE PURITY OF ARSENIOUS OXIDE MATERIAL

Obtain a sample of the material from the instructor and, unless otherwise advised, place it in a weighing bottle or dish and dry in the oven for 1 h at 110°C. Weigh appropriate portions (see the instructor) into four clean, dry, numbered Erlenmeyer flasks, recording the weights to the nearest 0.1 mg. Add 10 ml of 1 *M* NaOH and warm to dissolve. Follow exactly the procedure outlined in Sec. 15.4.12. Titrate the first solution, using 2-ml increments of standard iodine solution (Sec. 15.4.12), until a persistent blue color is obtained. Use this end-point volume as a guide in titrating the remaining three solutions. Record the end-point volumes for these three titrations to the nearest 0.01 ml, and calculate for each titration the percentage of arsenious oxide in the sample. Inspect for result rejection using the Q test. Average the acceptable values and calculate the average deviation. Report *all* values, including any rejected result.

15.4.14 PREPARATION OF 0.0200 *M* (0.1200 *N*) POTASSIUM BROMATE

Dry 5 g of reagent grade potassium bromate ($KBrO_3$) in a weighing bottle or dish for 1 h in an oven set at 100°C. Cool in the dessicator. Weigh 3.3000 ± 0.0002 g of the salt into a clean 400-ml beaker. Add 200 ml of distilled water, stir to dissolve, and transfer to a 1000-ml volumetric flask. Make up to the mark with water and mix thoroughly. Transfer to a clean glass-stoppered 1-l bottle and label as 0.0200 *M* and 0.1200 *N* potassium bromate. This nonstandardization procedure will be adequate for the experiment.

15.4.15 PREPARATION AND STANDARDIZATION OF APPROXIMATELY 0.1 *M* (0.1 *N*) SODIUM THIOSULFATE

Prepare 1 l of approximately 0.1 *M* (0.1 *N*) sodium thiosulfate solution in the manner outlined in Sec. 15.4.7. Pipette 25-ml portions of potassium bromate solution (Sec. 15.4.14) into three clean, numbered Erlenmeyer flasks. Treat the solutions separately from here on. Add 2 g of potassium iodide dissolved in 10 ml of water and 5 ml of 1:5 H_2SO_4. Rinse and fill a *clean* 50-ml burette with the sodium thiosulfate solution and titrate until the yellow triiodide ion color has almost disappeared. Add 5 ml of starch solution (Note 1, Sec. 15.4.8) and titrate to the disappearance of the blue color. Repeat the process for the remaining two solutions. Record the titration volumes to the nearest 0.01 ml. Determine the value of the thiosulfate solution in terms of the milliliters of potassium bromate solution per milliliter of sodium thiosulfate solution for each titration. Inspect for result rejection by the *Q* test. Average the acceptable values and determine the average deviation. Label the bottle with the milliliters of potassium bromate solution per milliliter of sodium thiosulfate solution.

15.4.16 DETERMINATION OF PHENOL

The reactions involved in this analytical process are

$$KBrO_3 + 5KBr + 6H_3O^+ \rightleftharpoons 3Br_2 + 9H_2O + 6K^+$$

$$C_6H_5OH + 3Br_2 \rightleftharpoons C_6H_2Br_3OH + 3HBr$$

$$2KI + Br_2 \rightleftharpoons I_2 + 2KBr$$

$$2S_2O_3{}^{2-} + I_2 \rightleftharpoons S_4O_6{}^{2-} + 2I^-$$

Obtain from the instructor a liquid sample containing phenol. Pipette appropriate volumes (see the instructor—but each volume should contain 3 to 4 mmol of phenol) into three clean, dry, numbered, and previously weighed weighing bottles. Record the weights to the nearest 0.1 mg. Transfer, with careful rinsing, to 250-ml volumetric flasks and dilute to the mark with distilled water. Mix thoroughly and pipette a 25-ml aliquot from each flask into clean, numbered, *glass-stoppered* 250-ml Erlenmeyer flasks. Pipette exactly 25 ml of potassium bromate solution (Sec. 15.4.14) into each flask. Add 0.5 g of potassium bromide (KBr) and 5 ml of 1:5 H_2SO_4. Stopper the flask *immediately* after the addition of the acid to avoid loss of Br_2. Mix and let stand for 15 min. Quickly add 2 g of potassium iodide dissolved in 10 ml of water and re-stopper the flask. Swirl to mix. Rinse and fill a *clean* 50-ml burette with standard sodium thiosulfate solution (Sec. 15.4.15) and titrate until the yellow color of the triiodide ion has almost disappeared. Add 5 ml of starch solution and continue the titration to the disappearance of the blue color. Repeat the procedure for the remaining two solutions. Record the titration volumes to the nearest 0.01 ml. For each titration calculate the percentage of phenol in the sample. Inspect for result rejection using the Q test. Average the acceptable results and calculate the average deviation. Report *all* values, including any rejected results.

15.5 NONAQUEOUS SOLVENT TITRATIONS*

15.5.1 PREPARATION OF APPROXIMATELY 0.1 *M* (0.1 *N*) PERCHLORIC ACID

Using a graduated pipette and a rubber bulb, pipette 8.5 ml of 72% perchloric acid ($HClO_4$) into a clean, dry 2-l beaker containing 250 ml of glacial acetic acid. Stir in 20 ml of acetic anhydride and allow to stand for 1 h. Dilute to approximately 1 l with glacial acetic acid. Cool and transfer to a clean, dry, 1-l glass-stoppered bottle. Let stand for 24 h to allow complete reaction between the acetic anhydride and the water content of the perchloric acid.

15.5.2 PREPARATION OF APPROXIMATELY 0.1 *M* (0.1 *N*) SODIUM ACETATE

Weigh 8.2 g of pure anhydrous sodium acetate ($CH_3 \cdot COONa$) into a clean, dry, 400-ml beaker and dissolve in 300 ml of glacial acetic acid. Transfer to a clean, dry 1-l glass-stoppered bottle and dilute to approximately 1 l with glacial acetic acid.

15.5.3 DETERMINATION OF THE CONCENTRATION RELATIONSHIP FOR THE PERCHLORIC ACID AND SODIUM ACETATE SOLUTIONS

Rinse and fill *clean, dry*, 50-ml burettes with the perchloric acid and sodium acetate solutions. Add 35.00 ml of perchloric acid solution to a clean, dry 200-ml Erlenmeyer flask. Add 2 drops of methyl violet indicator solution (0.2% in chlorobenzene) and titrate with the sodium acetate solution to the first violet color. Repeat the procedure two more times. Record the titration volumes to the nearest 0.01 ml. Inspect for rejection by the Q test. Average the acceptable sodium acetate solution volumes. Calculate the milliliters of perchloric acid solution per milliliter of sodium acetate solution. Label the bottles with this data.

15.5.4 STANDARDIZATION OF THE APPROXIMATELY 0.1 *M* (0.1 *N*) PERCHLORIC ACID AND SODIUM ACETATE SOLUTIONS

Place 3 g of pure potassium acid phthalate in a weighing bottle or dish and dry in the oven for 1 h at 110°C. Cool in the dessicator. Weigh 0.50- to 0.60-g portions into three clean, dry, numbered Erlenmeyer flasks. Record the sample weights to the nearest 0.1 mg. Treat each sample separately from here on. Add 60 ml of

*See Chap. 10.

glacial acetic acid and warm cautiously to dissolve the salt. Cool and add 2 drops of methyl violet indicator. Rinse and fill *clean, dry* 50-ml burettes with perchloric acid (Sec. 15.5.1) and sodium acetate (Sec. 15.5.2) solutions. Titrate the sample solution carefully with the perchloric acid solution until the violet color just disappears. Alternatively, titrate with perchloric acid solution until the violet color has disappeared and an excess of perchloric acid has been added. Back-titrate with sodium acetate solution to the first reappearance of the violet color. Titrate the remaining two solutions in the manner selected. Record all volumes to the nearest 0.01 ml. Calculate for each titration the volume of perchloric acid solution required to reach the end point, using if necessary the data from Sec. 15.5.3. Calculate for each titration the molarity (normality) of the perchloric acid solution and, using the data from Sec. 15.5.3, the molarity (normality) of the sodium acetate solution. Inspect for result rejection in each case by the Q test. Average the acceptable values and determine the average deviations. Label the respective bottles properly with the associated molarities (normalties).

15.5.5 DETERMINATION OF AN AMINE

Obtain a sample of an appropriate-named amine from the instructor. Weigh the required portions (see the instructor—but these should contain about 2 to 4 mequiv of the amine) into three clean, dry, numbered Erlenmeyer flasks, and record the weights to the nearest 0.1 mg. Add 50 ml of glacial acetic acid and dissolve. Add 2 drops of methyl violet indicator. Rinse and fill *clean, dry,* 50-ml burettes with standard perchloric acid and sodium acetate solutions (Sec. 15.5.4). Titrate with the perchloric acid solution until the violet color has disappeared and a slight excess of the titrant has been added. Back-titrate with the sodium acetate solution to the first reappearance of the violet color. Repeat the procedure with the remaining two solutions. Record all volumes to the nearest 0.01 ml. Using the concentration relationship (Sec. 15.5.3), calculate for each titration the percentage of the named amine in the sample. Inspect for result rejection by the Q test and average the acceptable results. Determine the average deviation and report *all* data, including any rejected result.

15.5.6 PREPARATION OF APPROXIMATELY 0.025 *M* (0.025 *N*) PERCHLORIC ACID

Use a standard perchloric acid solution as prepared and standardized in Secs. 15.5.1 and 15.5.4. Pipette 25 ml of the standard solution into a clean, dry 100-ml volumetric flask and dilute to the mark with glacial acetic acid. Mix thoroughly. This solution may be taken as one-quarter the molarity (normality) of the original perchloric acid solution and will be approximately 0.025 *M* (0.025 *N*). Label the flask accordingly.

15.5.7 DETERMINATION OF NICOTINE IN TOBACCO*

Nicotine reacts with *two* protons in this analytical procedure. Dry 15 g of tobacco in the oven for 2 h at 70°C. Cool and weigh 3-g portions into three clean, dry,

*M. L. Benston and M. K. Jaatteenmaki, "Quantitative Chemistry," Van Nostrand, New York, 1972. © 1972 by Litton Educational Publishing Co., pp. 172–173.

numbered 250-ml Erlenmeyer flasks. Record the weights to the nearest 0.1 mg. Add 1 g of barium hydroxide and 15 ml of saturated barium hydroxide solution. Swirl the flask until the tobacco is thoroughly wetted, adding more barium hydroxide solution if necessary. Pipette 100 ml of a 9:1 mixture of benzene and chloroform into the flask, stopper tightly, and shake vigorously for 20 min. Add 2 g of Celite, swirling thoroughly until the Celite is dispersed. Allow the two liquid phases to separate. Pour off the top organic layer through a fluted No. 2 filter paper into a clean, dry 125-ml Erlenmeyer flask. Pour carefully so as to leave the aqueous layer behind. Should any aqueous layer be accidentally poured over, decant the organic layer into a second 125-ml Erlenmeyer flask. Add 2 g of anhydrous magnesium sulfate to the organic layer to dry it. Swirl the flask gently, stopper it, and let stand for at least 15 min. Pour the organic liquid through a fluted No. 2 filter paper into a dry, 250-ml beaker. Pipette a 25-ml aliquot into a dry 125-ml Erlenmeyer flask. Add a stirring bar and 2 drops of methyl violet indicator solution (0.2% in chlorobenzene). Rinse and fill a *clean, dry,* 10-ml burette with standard perchloric acid solution (Sec. 15.5.6) and titrate carefully (the volume will be small) until the color changes from violet to green. Repeat the procedure for the remaining two samples. Record the volume values to the nearest 0.01 ml. Calculate for each sample the percentage of nicotine in the tobacco. Inspect for result rejection by the Q test. Average the acceptable values and determine the average deviation. Report *all* data, including any rejected value.

 Determine the average weight of tobacco in a cigarette made from this tobacco and determine the approximate milligram content of nicotine per cigarette.

15.6 GRAVIMETRIC DETERMINATIONS*

15.6.1 DETERMINATION OF CHLORINE

Clean thoroughly three porous-bottom crucibles, either sintered glass or porcelain type, of the "medium" porosity grade. After cleaning, draw through by suction, from both sides, 10 ml of hot 1:1 HNO_3. Finally draw through, from both sides, 50 ml of hot distilled water. Number properly and dry in the oven at 150°C to constant weight (agreement of weighings between heating periods to within 0.3 to 0.4 mg). Finally, hold in the dessicator until required. Obtain from the instructor a sample of unknown chloride content, place in a weighing bottle or dish and, unless otherwise advised, dry in the oven for 2 h at 150°C. Cool in the dessicator. Weigh appropriate portions (see the instructor) into three clean, dry, numbered 400-ml beakers, and record the weights to the nearest 0.1 mg. Dissolve in 150 ml of distilled water and add 1 ml of concentrated HNO_3. Heat the solution to just under the boiling point. Add 0.1 M silver nitrate solution (17 g of $AgNO_3$ in 1 l of distilled water) from a pipette in small portions while stirring efficiently. After each addition allow the precipitate of AgCl to settle somewhat, observing for the

_____
* See Chap. 11.

subsequent addition whether or not a further precipitation occurs. Continue this process until an addition of silver nitrate results in no further precipitation. At this point add a 10% excess volume of silver nitrate solution. Keep hot for about 20 min, or until the precipitate has coagulated properly. Remove from the heat, cover with a watch glass, and cool in a dark place to room temperature. When cooled, remove and add a few additional drops of silver nitrate solution to the clear supernatant liquid to test for completeness of precipitation of AgCl. Filter by suction through the appropriately numbered porous-bottom crucible, retaining the bulk of the precipitate in the beaker. Wash the precipitate by decantation twice with 25-ml portions of 0.01 M HNO_3 (2 drops of concentrated HNO_3 in 100 ml of distilled water), decanting the washings into the crucible. Wash the precipitate into the crucible with a stream of 0.01 M HNO_3 from a wash bottle, using a rubber policeman to detach and scrub off any precipitate adhering to the walls and bottom of the beaker. Wash the precipitate four times in the crucible with small portions of 0.01 M HNO_3, allowing the suction to drain the wash liquid each time. Collect the last washing and test it for the absence of silver ion by adding one drop of concentrated HCl. Place the three crucibles in a large beaker provided with glass rim hooks, set a watch glass on the hooks, and dry in the oven for 2 h at 150°C. Cool in the dessicator and weigh to the nearest 0.1 ml. Reheat for 30 min, cool, and reweigh. Repeat the process until successive weighings differ by 0.3 to 0.4 mg or less. Calculate for each analysis the percentage of chloride in the sample. Inspect for result rejection by the Q test. Average the acceptable values and determine the average deviation. Report *all* data, including any rejected value.

15.6.2 DETERMINATION OF NICKEL IN STEEL

Clean thoroughly three porous-bottom crucibles of the medium porosity grade in the manner outlined in Sec. 15.6.1. After weighing as indicated, hold in the dessicator until required. Obtain from the instructor a sample of nickel steel which does not contain significant cobalt or copper. Make sure it is free from grease or oil (see the instructor for cleaning instructions if required). Weigh appropriate portions (see the instructor) into three clean, numbered 400-ml beakers. Add 60 ml of 1:1 HCl and warm until decomposition is complete. Add cautiously 10 ml of 1:1 HNO_3 and boil under the hood until the iron and carbides are oxidized and the brown fumes of NO_2 have been expelled. Dilute to 200 ml with hot water and add 20 ml of tartaric acid solution (25 g per 100 ml of water). Neutralize with ammonia and add a 1-ml excess. Filter on an open-texture filter paper into a clean, numbered 400-ml beaker. Wash the paper several times with a hot solution containing 1% each of ammonia and ammonium chloride, and discard the paper. Add 1:1 HCl to the filtrate until slightly acid, warm to 80°C, and add 20 ml of dimethylglyoxime solution (1% in methanol). See the instructor to ensure that this volume is sufficient. Add concentrated ammonia until slightly alkaline and digest for 30 min at 60°C. Cool to room temperature. Filter with light suction through the appropriately numbered porous-bottom crucible. Do not increase the suction until a mat of the precipitate covers the bottom of the crucible. Use a rubber policeman and a stream of cold water from a wash bottle to transfer all of the precipitated nickel dimethylglyoximate into the crucible. Wash the precipitate in the crucible several times with cold water, letting the liquid drain each time. Add 5 ml of the dimethylglyoxime solution and 0.5 ml of concentrated ammonia to the filtrate and washings. Stir and allow to stand to determine whether or not precipitation was complete. Place the crucibles in a large beaker provided with glass rim hooks, set a watch glass on the hooks, and dry in the oven for 2 h at 150°C. Dry to constant weight by the method outlined in the latter part of Sec. 15.6.1. Calculate for each sample the percentage of nickel in the steel. Inspect for result rejection by the Q test. Average the acceptable values and calculate the average deviation. Report *all* data, including any rejected result.

15.7 ELECTROANALYTICAL METHODS*

15.7.1 CONSTANT CURRENT ELECTROLYSIS: DETERMINATION OF COPPER AND LEAD IN BRASS OR BRONZE ALLOYS

The proper application of the method of analysis to follow requires the recovery of any copper and lead carried down by the initial precipitate of tin as metastannic acid. This recovery procedure will be ignored here without the introduction of overly significant errors in the copper and lead determinations. The technique requires the use of a constant-current electrolysis apparatus and associated equipment. See the instructor for the method of operation.

The analytical process should be carried out in triplicate unless advised otherwise. Obtain from the instructor the necessary sets of platinum gauze electrodes and a sample of the brass or bronze alloy. Clean the electrodes thoroughly (see the instructor) and handle them from here on without contact with bare fingers. Place them in an oven and dry for 1 h at 110°C. Cool and weigh to the nearest 0.1 mg. Weigh appropriate portions of the sample (see the instructor) into three clean, numbered 250-ml beakers. Add 25 ml of 1:1 HNO_3, cover with a watch glass, and warm under the hood to dissolve. When dissolved, boil gently to expel NO_2 brown fumes. Add 50 ml of hot water and digest at just under the boiling point for 30 min. Lower the temperature very slightly and let stand and settle for 30 min. Keeping hot, filter through a close-texture filter paper into a 250-ml beaker numbered appropriately. Wash three times with hot 1:99 HNO_3 and at least four times with hot water. Be sure each time to wash the filter-paper edges. Dilute to 150 ml with water and add a stirring bar. Insert the electrodes, cover with a pair of split watch glasses, and start the stirring action. The choice of anode and cathode as to relative electrode size will be made by the instructor on the basis of the lead content expected. If this exceeds 5 percent, the larger area electrode should be the anode. Electrolyze at a current of 4 A (this assumes an electrode area of about 100 cm²) for 2.5 h. At the end of this period, wash down the split watch glasses and the sides of the beaker, raising the solution level until

*See Chap. 12.

fresh electrode supporting wire sections are clearly immersed. Continue the electrolysis until the deposition of copper is complete, as indicated by failure to plate on a new surface when the solution level is raised. When deposition of copper is complete, stop the stirring action but maintain the flow of current. Lower the beaker with care, while washing the cathode with a gentle stream of water from a wash bottle. Cut off the current, remove the cathode, rinse it in distilled water, and dip it into two successive baths of ethanol or methanol. Dry in the oven for 5 min at 110°C, cool, and weigh to the nearest 0.1 mg. Remove the anode, rinse in distilled water thoroughly, and dry in the oven for 30 min at 110°C. Handle this electrode with care as the deposit of lead dioxide (PbO_2) is fragile. Cool and weigh to the nearest 0.1 mg. For each sample determine the weight of the copper and lead dioxide deposits. Calculate the lead from the lead dioxide, and determine for each sample the percentages of copper and lead in the alloy. Inspect for result rejection by the Q test. Average the acceptable results and determine the average deviation. Report *all* data, including any rejected value.

15.7.2 CONTROLLED CATHODE POTENTIAL ELECTROLYSIS: DETERMINATION OF COPPER AND BISMUTH*

A potentiostatic electrolysis apparatus with the necessary attachments is required. The SCE electrode, connected to the solution in an appropriate manner, is used as the reference electrode. See the instructor for the method of operation.

The analysis should be performed in triplicate unless advised otherwise. Obtain from the instructor the necessary sets of platinum gauze electrodes and a solution containing appropriate concentrations of copper and bismuth. Clean the electrodes thoroughly (see the instructor) and, after cleaning, avoid direct contact with the fingers. Place the cathodes in the oven and dry for 1 h at 110°C. Cool and weigh to the nearest 0.1 mg. Pipette 50-ml aliquots of the solution (see the instructor) into three clean, numbered 250-ml beakers. Dilute to 150 ml with water, and add 1.5 g of urea, 12 g of sodium tartarate dihydrate, 2 g of hydrazine hydrochloride, and 1 g of succinic acid. The 1 g of succinic acid covers 100 mg of copper, and the instructor should be consulted in the event of additional succinic acid being required. Dilute to 200 ml with water, and using a pH meter, adjust the pH to 5.9 using careful additions of 2 M NaOH. Cool if required to less than 30°C. Add a stirring bar, insert the platinum electrodes, and connect the platinum and SCE electrodes to the potentiostat. Start the action and electrolyze to deposit the copper at a cathodic potential of −0.30 V vs. SCE. When the blue color has disappeared, add sufficient water to cover a new portion of the cathode and continue the electrolysis until failure of copper to deposit on a newly immersed platinum section indicates completion of the deposition. Stop the stirrer but maintain the current, and slowly lower the beaker while washing the cathode with a gentle stream of water from a wash bottle. Cut off the current, remove the cathode, and rinse in distilled water and then in two successive baths of ethanol or methanol.

*J. J. Lingane and S. L. Jones, *Anal. Chem.*, **23**:1798 (1951).

Dry on a watch glass for 5 min in the oven at 110°C. Cool and weigh to the nearest 0.1 mg. Reconnect the cathode, with its copper deposit, to the potentiostat, replace the solution, commence stirring, and electrolyze for bismuth deposition at a cathodic potential of -0.40 V vs. SCE. Electrolyze for 1 h and perform the same test as outlined in the foregoing to determine the completion of bismuth deposition. Remove the cathode, rinse, dry, and weigh exactly as described for the copper deposition. Calculate for each sample the weight of copper and bismuth in milligrams per 100 ml of solution provided by the instructor. Inspect for result rejection by the Q test. Average the acceptable values and determine the average deviation in each case. Report *all* data, including any rejected result.

15.7.3 POTENTIOMETRIC TITRATIONS: DETERMINATION OF ACETYLSALICYLIC ACID IN ASPIRIN

A pH meter with a glass electrode and an SCE electrode, or a glass-SCE combination electrode, is required. See the instructor for operating instructions.

Obtain from the instructor a sample of *unbuffered* commercial aspirin tablets. Place two tablets in each of four clean, numbered 400-ml beakers. Add 50 ml of ethanol, break up the tablets with a stirring rod, and stir to dissolve. Add 150 ml of distilled water and a stirring bar. Calibrate the pH meter with a buffer of about 8 pH. Rinse the electrodes thoroughly and immerse in the first solution to be titrated. Rinse and fill a *clean*, 50-ml burette with standard NaOH solution (Sec. 15.1.2). For this first solution titrate, using 1- or 2-ml increments of titrant, and locate very approximately the equivalence-point volume by the $\Delta^2 pH/\Delta T^2$ tabulation method (see Sec. 5.2). Using this approximate equivalence-point volume as a guide, titrate the remaining three solutions carefully, with efficient stirring, adding 0.1-ml increments of titrant when near the equivalence point and continuing with these increments until well after this point. The equivalence-point area shows larger changes in pH per volume addition. Record all volume and associated pH readings to the nearest 0.01 ml and reading uncertainty limit of the pH meter scale. For each of these three titrations, tabulate the $\Delta^2 pH/\Delta T^2$ values (see Sec. 5.2) and determine the equivalence-point volume. Calculate for each titration the milligrams of acetylsalicylic acid per aspirin tablet. Average the values and calculate the average deviation. Report *all* values.

15.7.4 STANDARDIZATION OF APPROXIMATELY 0.1 *M* (0.1 *N*) SILVER NITRATE

A potentiometer with either a silver wire electrode and a KNO_3 salt bridge SCE electrode, or a silver-reference combination electrode, is required. The instructor will explain the operation of the instrument.

Prepare approximately 0.1 *M* (0.1 *N*) silver nitrate solution in the manner outlined in Sec. 15.2.1. Standardize the solution as follows. Place about 2 g of pure sodium chloride in a weighing bottle or dish and dry in the oven for 1 h at 200°C. Cool in the dessicator and weigh 0.20- to 0.25-g portions into three clean, dry, numbered 250-ml beakers. Record the weights to the nearest 0.1 mg. Add 75 ml of distilled water and a stirring bar, and stir to dissolve. Rinse and fill a *clean*, dry 50-ml burette with silver nitrate solution. Insert the electrodes and adjust the potentiometer according to the instructor's advice. The titration will require about 35 to 45 ml. Stir efficiently and titrate carefully, recording the volume values to the nearest 0.01 ml and the potential values to the reading uncertainty permitted by the potentiometer scale. Use 0.1-ml increments of titrant volume when near the equivalence point, and continue these increments until well past this point. Repeat the procedure for the remaining two solutions. Using the $\Delta^2 E/\Delta T^2$ tabulation method (see Sec. 5.2), determine the equivalence-point volume for each titration. Calculate the molarity (normality) of the silver nitrate solution for each titration. Inspect for result rejection by the Q test. Average the acceptable values and determine the average deviation. Label the bottle properly with the molarity (normality) of the silver nitrate solution.

15.7.5 POTENTIOMETRIC TITRATIONS: DETERMINATION OF CHLORIDE

Obtain a sample of unknown chloride content from the instructor. Unless advised otherwise, place the sample in a weighing bottle or dish and dry in the oven for 1 h at 170°C. Cool in the dessicator and weigh appropriate portions (see the instructor) into four clean, dry, numbered 250-ml beakers. Record the weights to the nearest 0.1 mg. Add 75 ml of distilled water and a stirring bar, and stir to dissolve. Rinse out and fill a *clean* 50-ml burette with standard silver nitrate solution (Sec. 15.7.4). Insert the electrodes and adjust the potentiometer as before. Stir and titrate the first solution, using 1- or 2-ml increments of titrant, until the approximate equivalence point can be located from a $\Delta^2 E/\Delta T^2$ tabulation. Using this approximate equivalence-point volume as a guide, titrate the remaining three solutions carefully, using 0.1-ml increments of titrant when near the equivalence point and until well after this point. For each of these titrations use a $\Delta^2 E/\Delta T^2$ tabulation to locate the equivalence-point volume. Calculate for each titration the percentage of chloride in the sample. Inspect for result rejection by the Q test. Average the acceptable values and determine the average deviation. Report *all* values, including any rejected value.

15.7.6 PREPARATION AND STANDARDIZATION OF APPROXIMATELY 0.5 *M* (0.5 *N*) FERROUS AMMONIUM SULFATE SOLUTION

A potentiometer with a platinum-SCE electrode combination is required. The instructor will explain the operation procedure for the instrument.

Weigh 100 g of ferrous ammonium sulfate ($FeSO_4 \cdot (NH_4)_2SO_4 \cdot 6H_2O$) into a clean 1-l beaker. Add 300 ml of distilled water and dissolve. Transfer to a 500-ml volumetric flask and make up to the mark with water. Mix thoroughly and transfer to a clean 500-ml bottle.

Dry 4 g of pure potassium dichromate in the oven for 1 h at 100°C. Cool and weigh 1.0-g portions into three clean, numbered 400-ml beakers. Record the weights to the nearest 0.1 mg. Add 100 ml of distilled water and stir to dissolve. Add, with stirring, 50 ml of 1:1 H_2SO_4. Cool and add 50 ml of water and a stirring bar. Insert the electrodes and adjust the potentiometer according to the instructor's advice. Rinse and fill a *clean* 50-ml burette with the ferrous ammonium sulfate solution. Titrate the first solution, recording the volume values to the nearest 0.01 ml and the potential values to the limit of the reading uncertainty of the potentiometer scale. Use 0.1-ml volume increments of titrant when near the equivalence point and until well past this point. Repeat the procedure with the remaining two solutions. Use a $\Delta^2 E / \Delta T^2$ tabulation method (see Sec. 5.2) to locate the equivalence-point volumes. Calculate the molarity (normality) of the ferrous ammonium sulfate solution for each titration. Inspect for result rejection using the Q test. Average the acceptable values and determine the average deviation. Label the bottle properly with the molarity (normality).

15.7.7 DETERMINATION OF GLUCOSE FRUCTOSE IN COMMERCIAL SYRUP

Only syrups or solutions containing glucose and/or fructose should be used in this procedure. The reaction of glucose and fructose with potassium dichromate is

$$4Cr_2O_7^{2-} + C_6H_{12}O_6 + 32H_3O^+ \rightleftharpoons 8Cr^{3+} + 6CO_2 + 54H_2O$$

Obtain from the instructor a sample of commercial syrup, or a prepared solution, containing glucose and/or fructose. Weigh into four clean, dry, numbered, and previously weighed weighing bottles or dishes appropriate portions (see the instructor—but these should not exceed 4 g each), using a laboratory scale for the initial weighing. Subsequently wash and wipe dry any spills on the outside of the bottle, and make the final weighings to the nearest 0.1 mg on the analytical balance. Transfer the samples from the bottles, with adequate rinsing, into clean numbered 100-ml volumetric flasks, dilute to the mark with distilled water, and mix thoroughly.

Weigh into four clean, dry, appropriately numbered 500-ml Erlenmeyer flasks 3-g portions of pure potassium dichromate, recording the weights to the nearest 0.1 mg. Add 25 ml of water and dissolve. For each sample, rinse a 10-ml pipette with diluted syrup solution and transfer 10 ml to the appropriate Erlenmeyer flask. Swirling continuously add very carefully 10 ml of concentrated H_2SO_4. Avoid losses by frothing and let stand for 2 min. Again with continuous swirling, add very carefully 30 ml of concentrated H_2SO_4 and let stand for 15 min. Add about 100 ml of water to a clean, appropriately numbered 400-ml beaker and, stirring vigorously, transfer cautiously the contents of the Erlenmeyer flask to the beaker. Rinse out the flask into the beaker four times with water from a wash bottle. Cool the solution, insert the electrode combination, add a stirring bar, and adjust the potentiometer according to instructions. Rinse and fill a *clean* 50-ml burette with standard ferrous ammonium sulfate solution (Sec. 15.7.6). Titrate the first solution using 1-ml increments of titrant and locating the approximate equivalence-point volume by the $\Delta^2 E / \Delta T^2$ tabulation method. Using this approximate volume as a guide, titrate the remaining three solutions, after treatment in the same manner, using 0.1-ml titrant increments when near the equivalence point and until well past this point. Locate the equivalence-point volumes for these three titrations by the tabulation method and calculate for each titration the percentage of glucose-fructose in the sample. Inspect for result rejection by the Q test. Average the acceptable results and determine the average deviation. Report *all* data, including any rejected result.

15.7.8 COULOMETRIC TITRATIONS: DETERMINATION OF ARSENIC

Prepare a 0.00200 M (0.00400 N) arsenious oxide solution by weighing 0.1978 g of pure previously dried arsenious oxide into a clean 250-ml beaker. Add 1 g of NaOH dissolved in 10 ml of water and stir until dissolved. Add 50 ml of water and 2 drops of phenolphthalein indicator (0.1% in 80% alcohol). Now add 1:1 HCl until the pink color has disappeared and the solution is just acid. Transfer carefully to a 500-ml volumetric flask, make up to the mark with distilled water, and mix thoroughly. Label this solution 0.00200 M (0.00400 N). This nonstandardization procedure will be adequate for the experiment.

A constant-current coulometric apparatus, with an accurate timer, proper platinum generating electrodes, and a generating vessel will be required. The instructor will explain its operation.

Obtain from the instructor a sample of an unknown arsenic content solution with the arsenic in the arsenic(III) state. Pipette appropriate volumes (see the instructor) into three clean, numbered 100-ml volumetric flasks. Make up to the mark with distilled water and mix thoroughly.

Direct method   Place in the generating cell of the constant-current coulometric apparatus a solution consisting of 50 ml of water, 2 g of potassium iodide, and 2 g of sodium bicarbonate. Add 5 ml of starch solution (Note 1, Sec. 15.4.8), insert the generating electrodes, and add a stirring bar. Start stirring, set the current control for a current of 10 mA, and start the timer-generator together. Stop the timer-generator as soon as a blue color develops. If this color persists for 30 s, the predetermination color development procedure is complete; if not, stop and start the timer-generator until a blue color of this persistency is obtained. Now pipette 5 ml of one of the diluted unknown solutions into the generating cell, start the stirrer and the timer-generator units, and titrate coulometrically until a similar persistent blue color develops. Record the time interval to the limiting reading uncertainty of the timer. Repeat the procedure for the remaining two solutions. The instructor will give you the uncertainty in the 10-mA current applied. Use these values to calculate directly the percentage of arsenic in the original unknown solution for each sample, assuming the solution density to be 1.000. Inspect for result rejection by the Q test, average the acceptable values, and determine the average deviation. Report *all* data, including any rejected value.

Ratio method   With certain types of coulometric generating equipment the current applied may only be approximately known. In such instances, the ratio method may be applied. Use the timer interval values from the previous experiment. Add by pipetting 5 ml of the standard 0.00200 M (0.00400 N) arsenious oxide solution previously prepared, and titrate exactly as before. Use the time interval for the standard solution, the concentration for the standard solution, and the time intervals for the three diluted unknown solutions to calculate by the ratio method, for each sample, the percentage of arsenic in the original unknown solution. Inspect, average, and determine the average deviation as before. Compare the two values obtained.

15.7.9 POLAROGRAPHIC DETERMINATION OF ACETAZOLAMIDE IN TABLETS

Acetazolamide is a carbonic anhydrase inhibitor used in the manufacture of certain pharmaceutical tablet products. The polarographic reaction involves the reduction of:

$$CH_3CONH \diagdown \overset{S}{\underset{N\!-\!N}{\diagup \diagdown}} SO_2NH_2$$

at the dropping mercury electrode.

Polarographic equipment with a temperature controlled cell and an SCE reference electrode is required. The instructor will explain the operation of the instrument.

Weigh 0.0600 g of acetazolamide into a 100-ml beaker.* Add 40 ml of boiling distilled water and heat on a steam bath for about 15 min. Cool and transfer to a clean 100-ml volumetric flask, dilute to the mark with water, and mix thoroughly. Let settle if required, and pipette 3.00-, 2.50-, 2.00-, and 1.50-ml portions of the clear supernatant liquid into clean 25-ml volumetric flasks. Add 2.0 ml of 1 N HCl, dilute to the mark with water, and mix thoroughly. Transfer all or a portion of the strongest of these solutions to a polarographic cell with temperature control (water bath, circulating double-wall cell, etc.) set at 25.0 ± 0.5°C. Deaerate the solution by bubbling purified nitrogen through it for 10 min. Run the gas over the solution during the subsequent procedure. Record the polarogram over the voltage interval from −0.20 to −0.75 V vs. SCE. Determine the diffusion current (i_d or $\bar{i}_d$ depending upon the instrumental capacity) at −0.70 V vs. SCE. Repeat the procedure for each of the remaining solutions. Plot the diffusion current vs. the milligrams of acetazolamide per 25 ml of the test solution.

Obtain from the instructor a sample of tablets containing acetazolamide. Crush 2 tablets to a fine powder and weigh to the nearest 0.1 mg a portion which will contain between 40 to 50 mg of acetazolamide (see the instructor). Transfer to a clean 100-ml beaker, add 40 ml of boiling distilled water, and heat on a steam bath for 15 min. Cool, transfer to a clean 100-ml volumetric flask, make up to the mark with distilled water, and mix thoroughly. Allow any undissolved material to settle, pipette 3.00-ml portions of the clear supernatant liquid into two clean 25-ml volumetric flasks, add 2 ml of 1 N HCl, dilute to the mark with water, and mix thoroughly. Continue exactly as outlined before, and determine the diffusion current at −0.70 V vs. SCE. Repeat for the second sample, refer to the graph, and

* A U.S.P. Acetazolamide Reference Standard can be obtained if required.

calculate for each sample the milligrams of acetazolamide per tablet. Average and report the milligrams of acetazolamide per tablet.

15.7.10 AMPEROMETRIC TITRATION: DETERMINATION OF LEAD

Polarographic equipment and an SCE reference electrode are required. If a rapid drop attachment is available for the dropping mercury electrode, this should be used to facilitate current reading. The instructor will explain the operation of the instrument.

The following solutions should be prepared.

Supporting electrolyte-buffer solution   Dissolve 10 g of KNO_3 and 4.1 g of sodium acetate ($CH_3 \cdot COONa \cdot 3H_2O$) in 200 ml of water containing 10 ml of glacial acetic acid. Dilute to 500 ml and mix well.

0.1 *percent gelatin solution*   Dissolve 0.1 g of gelatin in 100 ml of hot water.

Standard potassium chromate solution   Dissolve 1.942 g of K_2CrO_4 in 50 ml of water. Transfer to a 100-ml volumetric flask, dilute to the mark with water, and mix thoroughly.

Standard lead nitrate solution   Dissolve 1.656 g of pure $Pb(NO_3)_2$ in 50 ml of water, transfer to a 100-ml volumetric flask, dilute to the mark with water, and mix well. Assume this solution to be 0.050 M to lead(II).

Standardization of potassium chromate solution   Pipette 7.00-ml portions of the standard lead nitrate solution into three clean, numbered 50-ml volumetric flasks, using a graduated pipette and a rubber bulb. Add 20 ml of water and 2 ml of gelatin solution and dilute to the mark with supporting electrolyte–buffer solution. Mix well. Transfer one of the solutions to the cell of the polarographic equipment and deaerate by bubbling nitrogen gas for 10 min. Stop the flow of gas and determine the diffusion current at an applied potential of -1.00 V vs. SCE at the mercury drop. Rinse and fill a *clean* 10-ml burette with potassium chromate solution. Titrate by adding 0.4-ml increments of titrant, passing nitrogen gas for 1 min after each addition to mix and deaerate and reading the diffusion current with the gas flow shut off. Continue the additions of potassium chromate solution until at least four points well past the equivalence point have been obtained. Record all volume values to the nearest 0.01 ml and all diffusion current values to the limiting reading uncertainty of the scale. Repeat the process with the remaining two solutions. Correct all diffusion current values for the volume dilution effect of the titrant. Plot for each solution the corrected diffusion current vs. the titrant volume, determine the equivalence-point volume by extrapolating the straight-line sections of the curve, and calculate the molarity of the potassium chromate solution. Inspect for result rejection by the Q test, average the acceptable values, and determine the average deviation. Label the potassium chromate solution bottle properly with the molarity of the solution.

Determination of lead   Obtain from the instructor a solution containing an unknown amount of lead in the lead(II) state. Pipette appropriate volumes (see the instructor) into three clean, numbered 50-ml volumetric flasks. Continue the analytical process exactly as outlined in the previous section on standardization of potassium chromate solution. From the equivalence-point volumes determined, calculate for each sample the concentration of lead in the original solution in milligrams per 100 ml. Inspect for result rejection by the Q test, average the acceptable values, and determine the average deviation. Report *all* data, including any rejected value.

For one of the titrations, plot the uncorrected diffusion current vs. the volume of titrant. Comment on anything that appears significant where this curve is compared with the appropriate curve resulting from the plot of corrected diffusion current vs. titrant volume.

15.7.11   CONDUCTOMETRIC TITRATION: DETERMINATION OF LEAD

A Wheatstone bridge suitable for conductometric titration work and a conductometric titration cell are required. The instructor will explain the operation of the equipment.

Approximately 0.5 *M sulfuric acid solution*   Add 14.0 ml of concentrated H_2SO_4 to 200 ml of distilled water. Cool, transfer to a clean 500-ml glass bottle, and dilute to approximately 500 ml with distilled water.

Standardization of approximately 0.5 *M sulfuric acid solution*   Standardize in triplicate against 50.00-ml portions of 0.1 *M* sodium hydroxide solution (Sec. 15.1.2). Label the bottle properly with the molarity of the sulfuric acid solution.

Determination of lead   Obtain from the instructor a solution containing an unknown concentration of lead as lead acetate, $Pb(CH_3 \cdot COO)_2$. This should lie between about 0.02 to 0.05 *M* relative to lead. Using a rubber bulb, pipette a 50-ml portion into the conductometric titration cell. Add a small stirring bar. Read the conductance (resistance) of the solution. Rinse and fill a *clean* 10-ml burette with the standard sulfuric acid solution. Start the stirring action and add 0.3 ml of titrant. Stir for 15 s, stop stirring, and obtain the conductance (resistance) value. Continue this procedure, using 0.3-ml increments of titrant, until at least four points well past the equivalence point have been obtained. Record all volume values to the nearest 0.01 ml and all conductance (resistance) values to the limiting reading uncertainty dictated by the measuring system. Repeat the procedure with two more solution samples. For each titration, plot the uncorrected conductance vs. the volume of titrant. Examine the curves closely. Is it really necessary to correct the values of conductance for the volume dilution effect of the titrant? If you feel that it is, correct the values and plot the corrected conductance vs. the volume of titrant. In either case, determine the equivalence-point volume of titrant for each titration by extrapolation of the straight-line sections to intersection.

Calculate for each titration the molarity of the original solution relative to lead. Inspect for result rejection by the Q test, average the acceptable values, and determine the average deviation. Report *all* data, including any rejected result.

15.8 SPECTROPHOTOMETRIC DETERMINATIONS*

15.8.1 DETERMINATION OF IRON IN INDUSTRIAL WATER EFFLUENT

Iron reacts with thioglycolic acid in ammoniacal medium to form a red-purple complex. The color intensity is a function of the total iron present. *Caution: Thioglycolic acid is an intense skin irritant. Do not get it on your skin. Do not use mouth pipettes to measure or transfer this substance. Do not open or pour from bottles of thioglycolic acid except in the fume hood. When you have finished with the acid, return it immediately to the instructor.*

A good spectrophotometer with a wavelength range from about 400 to 700 nm, with a cell path length of not less than 1 cm, is required. The instructor will explain the operation of the instrument.

Standard iron solution Weigh exactly 0.0500 g of pure iron wire into a clean 250-ml beaker. Add 25 ml of 1:1 HCl and warm to dissolve. Cool and transfer to a clean 250-ml volumetric flask. Make up to the mark with distilled water and mix thoroughly. The iron value for this solution will be 1 ml = 0.2000 mg of iron.

Preparation of the calibration curve Rinse and fill a *clean* 10-ml burette with the standard iron solution. Transfer 0.50-, 1.00-, 1.50-, 2.00-, and 3.00-ml portions to clean, marked 150-ml beakers. Use a sixth beaker to prepare the reference solution. Add to each beaker 10 ml of distilled water, 2.00 ml of concentrated HCl, and 2.0 ml of 30% hydrogen peroxide solution. Cover with watch glasses, heat to boiling under the hood, and boil for 3 min. Wash the covers into the beakers with 5 ml of distilled water and transfer to clean appropriately marked 100-ml volumetric flasks. Treat each solution separately from here on. The reference solution is treated first. Add 5.0 ml of 25% citric acid solution, 10.0 ml of concentrated ammonia, and 2.0 ml of reagent grade thioglycolic acid. Use pipettes and rubber bulbs for these operations. Dilute immediately to the mark with distilled water and transfer back to the proper 150-ml beaker. Stir well for 30 s. Set the spectrophotometer wavelength at 535 nm and transfer a portion of the solution to a cell. Set the zero and 100 percent transmittance scale points according to the advice of the instructor. Successively prepare the five standard solutions in the same manner, reading the absorbance or percent transmittance value for each solution within the limiting reading uncertainty for the scale point involved. Plot an absorbance-vs.-concentration curve, expressing the concentration in milligrams of iron per 100 ml of solution examined. If a straight line is not obtained, consult the instructor.

* See Chap. 13.

Determination of iron   Obtain from the instructor a water sample which will contain between 1.5 and 5.5 mg of iron per 100 ml of water sample. Pipette 10-ml portions into three clean, numbered 150-ml beakers. Use a fourth beaker to prepare a reference solution. Add to each beaker 10 ml of distilled water, 2.00 ml of concentrated HCl, and 2.0 ml of 30% hydrogen peroxide solution. Cover with watch glasses, heat to boiling under the hood, and boil for 3 min. Wash the covers into the beakers with 5 ml of distilled water and transfer to clean appropriately numbered 100-ml volumetric flasks. Treat each solution separately from here on in the manner outlined under Preparation of the Calibration Curve. Using the calibration plot, determine for each sample the concentration of iron in the original unknown solution in milligrams per 100 ml. Inspect for result rejection by the Q test, average the acceptable values, and determine the average deviation. Report *all* values, including any rejected value.

15.8.2   DETERMINATION OF MANGANESE AND CHROMIUM SIMULTANEOUSLY IN STEEL

A good spectrophotometer with a wavelength range of from about 350 to 650 nm, with a cell path length of not less than 2 cm is required. The instructor will explain the operation of the instrument.

Determination of the absorptivities of manganese and chromium   Obtain from the instructor the following solutions:

1. A standard manganese solution with 1 ml = 0.200 mg of manganese
2. A standard chromium solution with 1 ml = 0.200 mg of chromium
3. A standard iron solution with 1 ml = 10 mg of iron

Using a 10-ml burette, measure 5.00 and 8.00 ml of standard manganese solution into clean, marked 150-ml beakers. Measure into two other beakers the same volumes of standard chromium solution. Set aside a fifth beaker for the preparation of the reference solution. Measure into each beaker 10.0 ml of the standard iron solution, using a pipette for this purpose. Add to each beaker 50 ml of distilled water, 5 ml of concentrated H_2SO_4, 5 ml of 85% phosphoric acid, 4 drops of 0.5 M silver nitrate solution, and 5 g of potassium persulfate. Stir to dissolve, cover with watch glasses, and boil for 5 min under the hood. Cool slightly and add 0.5 g of potassium periodate. Heat to boiling under the hood and boil for 5 min. Cool to room temperature and transfer to clean, marked 100-ml volumetric flasks. Dilute to the mark with distilled water and mix thoroughly. Transfer the reference solution to a spectrophotometer cell and set the wavelength at 440 nm. Following the advice of the instructor, set the zero and 100 percent transmittance scale points. Read the absorbance values for each solution at this wavelength. Replace the reference solution in the instrument, set the wavelength at 545 nm, and set the zero and 100 percent transmittance scale points. Read the absorbance values for each solution at this wavelength. Using the concentrations of manganese and chromium in milligrams per 100 ml of solution tested, determine the absorptivities in accordance with:

$$a_{Mn}(545) = \frac{A(545)Mn}{bC_{Mn}} \qquad a_{Mn}(440) = \frac{A(440)Mn}{bC_{Mn}}$$

$$a_{Cr}(545) = \frac{A(545)Cr}{bC_{Cr}} \qquad a_{Cr}(440) = \frac{A(440)Cr}{bC_{Cr}}$$

where b is the cell path length. It is not necessary to have a numerical value for b since in solving the subsequent simultaneous equations, the term b will disappear. Average the two values obtained for each of the four absorptivities. The value of $a_{Cr}(545)$ will likely be zero, since chromium should not absorb at this wavelength. If it is, this will facilitate the solution of the subsequent equations. The wavelength values used are not the wavelengths for peak absorbance for manganese and chromium. In your report this situation should be explained.

Determination of manganese and chromium simultaneously in steel   Obtain from the instructor a sample of steel which contains appropriate percentages of manganese and chromium. Carry out the analysis to follow in triplicate. Weigh out 1.0-g portions (see the instructor) into three clean, numbered 250-ml Erlenmeyer flasks. Record the weights to the nearest 0.1 mg. Add to each flask 30 ml of mixed acids solution (Note 1). Warm to dissolve. Boil under the hood to expel NO_2 fumes, dilute to 100 ml, and warm, if necessary, to dissolve any salts. Cool to room temperature, transfer to clean 250-ml volumetric flasks, dilute to the mark with distilled water, and mix thoroughly. Pipette a 25-ml aliquot into a 150-ml beaker, add 50 ml of distilled water, 5 ml of concentrated H_2SO_4, 5 ml of 85% H_3PO_4, 4 drops of 0.5 M silver nitrate solution, and 5 g of potassium persulfate. Stir to dissolve, cover with a watch glass, and boil under the hood for 5 min. Cool slightly and add 0.5 g of potassium periodate. Heat to boiling under the hood and boil for 5 min. Cool to room temperature and transfer to a clean 100-ml volumetric flask. Dilute to the mark with distilled water and mix thoroughly. Using the reference solution from the determination of the absorptivities, set the wavelength at 440 nm and set the zero and 100 percent transmittance scale points. Read the absorbance value for each solution at this wavelength. Repeat the procedure with the instrument set at 545 nm. Using the absorbance values obtained and the absorptivities previously determined, solve the equations:

$$A_{(total)}(545) = \{a_{Mn}(545)C_{Mn} + a_{Cr}(545)C_{Cr}\}b$$

$$A_{(total)}(440) = \{a_{Mn}(440)C_{Mn} + a_{Cr}(440)C_{Cr}\}b$$

to determine for each sample the values of C_{Mn} and C_{Cr} in milligrams per 100 ml of the solution examined. Using the proper aliquot factor and the sample weight, calculate for each sample the percentages of manganese and chromium in the steel. Inspect for result rejection using the Q test, average the acceptable values and determine the average deviations. Report *all* data, including any rejected results.

Note 1. The mixed acids solution is prepared by adding 10 ml of concentrated H_2SO_4 slowly to 50 ml of distilled water. Cool and add 12.5 ml of 85% H_3PO_4 and 25 ml of concentrated HNO_3.

15.9 GAS-LIQUID CHROMATOGRAPHY*

Accurate analysis involving gas-liquid chromatography involves considerable time and experience. The following qualitative and semiquantitative experiment will serve to yield some idea of the operation of the instrument and the general aspects of gas-liquid chromatography.

A gas chromatograph is required. Follow the instructor's advice carefully relative to the operation of the instrument.

15.9.1 QUALITATIVE AND SEMIQUANTITATIVE DETERMINATION OF HYDROCARBONS

The instructor will set up and explain the operation of the gas chromatograph, and will supply samples of pure n-pentane, n-hexane, and n-octane, as well as a sample containing a mixture of these three hydrocarbons.

With a microliter syringe inject independently appropriate volumes (see the instructor) of each pure hydrocarbon. Obtain chromatograms for each pure hydrocarbon and examine them carefully. With the instructor's assistance identify the main peaks, the peaks for any minor impurities, the injection peak, and the air peak.

Inject the appropriate volume (see the instructor) of the mixture of three hydrocarbons and obtain the chromatogram. Determine the retention time for each main peak and identify the component. Measure the total area under all of the appropriate peaks according to the instructor's advice. Divide the peak area for each main component by the total area and multiply by 100. This will yield a semiquantitative estimate of the weight percent of each component in the mixture. Report these values.

*See Chap. 14.

appendix A
Relative Atomic Weights

Element	Symbol	Atomic no.	Atomic weight	Element	Symbol	Atomic no.	Atomic weight
Actinium	Ac	89	227*	Mercury	Hg	80	200.59
Aluminum	Al	13	26.9815	Molybdenum	Mo	42	95.94
Americium	Am	95	243*	Neodymium	Nd	60	144.24
Antimony	Sb	51	121.75	Neon	Ne	10	20.183
Argon	Ar	18	39.948	Neptunium	Np	93	237*
Arsenic	As	33	74.9216	Nickel	Ni	28	58.71
Astatine	At	85	210*	Niobium	Nb	41	92.906
Barium	Ba	56	137.34	Nitrogen	N	7	14.0067
Berkelium	Bk	97	247*	Nobelium	No	102	254*
Beryllium	Be	4	9.0122	Osmium	Os	76	190.2
Bismuth	Bi	83	208.980	Oxygen	O	8	15.9994
Boron	B	5	10.811	Palladium	Pd	46	106.4
Bromine	Br	35	79.909	Phosphorous	P	15	30.9738
Cadmium	Cd	48	112.40	Platinum	Pt	78	195.09
Calcium	Ca	20	40.08	Plutonium	Pu	94	244*
Californium	Cf	98	249*	Polonium	Po	84	210*
Carbon	C	6	12.01115	Potassium	K	19	39.102
Cerium	Ce	58	140.12	Praseodymium	Pr	59	140.907
Cesium	Cs	55	132.905	Promethium	Pm	61	145*
Chlorine	Cl	17	35.453	Protactinium	Pa	91	231*
Chromium	Cr	24	51.996	Radium	Ra	88	226*
Cobalt	Co	27	58.9332	Radon	Rn	86	222*
Copper	Cu	29	63.54	Rhenium	Re	75	186.2
Curium	Cm	96	245*	Rhodium	Rh	45	102.905
Dysprosium	Dy	66	162.50	Rubidium	Rb	37	85.47
Einsteinium	Es	99	254*	Ruthenium	Ru	44	101.07
Erbium	Er	68	167.26	Samarium	Sm	62	150.35
Europium	Eu	63	151.96	Scandium	Sc	21	44.956
Fermium	Fm	100	252*	Selenium	Se	34	78.96
Fluorine	F	9	18.9984	Silicon	Si	14	28.086
Francium	Fr	87	223*	Silver	Ag	47	107.870
Gadolinium	Gd	64	157.25	Sodium	Na	11	22.9898
Gallium	Ga	31	69.72	Strontium	Sr	38	87.62
Germanium	Ge	32	72.59	Sulfur	S	16	32.064
Gold	Au	79	196.967	Tantalum	Ta	73	180.948
Hafnium	Hf	72	178.49	Technetium	Tc	43	99*
Helium	He	2	4.0026	Tellurium	Te	52	127.60
Holmium	Ho	67	164.930	Terbium	Tb	65	158.924
Hydrogen	H	1	1.00797	Thallium	Tl	81	204.37
Indium	In	49	114.82	Thorium	Th	90	232.038
Iodine	I	53	126.9044	Thulium	Tm	69	168.934
Iridium	Ir	77	192.2	Tin	Sn	50	118.69
Iron	Fe	26	55.847	Titanium	Ti	22	47.90
Krypton	Kr	36	83.80	Tungsten	W	74	183.85
Lanthanum	La	57	138.91	Uranium	U	92	238.03
Lead	Pb	82	207.19	Vanadium	V	23	50.942
Lithium	Li	3	6.939	Xenon	Xe	54	131.30
Lutetium	Lu	71	174.97	Ytterbium	Yb	70	173.04
Magnesium	Mg	12	24.312	Yttrium	Y	39	88.905
Manganese	Mn	25	54.9380	Zinc	Zn	30	65.37
Mendelevium	Md	101	256*	Zirconium	Zr	40	91.22

*Value of the mass number of the most stable known isotope.

appendix B
Formula Weights

AgBr	187.78	HCl	36.46
AgCl	143.32	$HClO_4$	100.46
Ag_2CrO_4	331.73	HIO_4	191.91
AgI	234.77	HNO_3	63.02
$AgNO_3$	169.88	$H_2C_2O_4 \cdot 2H_2O$	126.07
AgSCN	165.95	H_2O	18.015
Al_2O_3	101.96	H_2O_2	34.01
$Al_2(SO_4)_3$	342.14	H_2S	34.08
As_2O_3	197.84	H_2SO_3	82.08
As_2O_5	229.84	H_2SO_4	98.08
$BaCl_2$	208.25	H_3PO_4	98.00
$BaCl_2 \cdot 2H_2O$	244.27	H_5IO_6	227.94
$BaCO_3$	197.35	$HgCl_2$	271.50
$BaCrO_4$	253.33	HgO	216.59
BaO	153.34	Hg_2Cl_2	472.09
$Ba(OH)_2$	171.36	KBr	119.01
$BaSO_4$	233.40	$KBrO_3$	167.01
Bi_2O_3	465.95	KCl	74.56
$CHCl_3$	119.38	$KClO_3$	122.56
CO_2	44.011	$KClO_4$	138.55
$C_2H_4(OH)_2$ (ethylene glycol)	62.07	$KHC_8H_4O_4$ (phthalic)	204.23
$CaCO_3$	100.09	KCN	65.12
CaC_2O_4	128.10	$KHSO_4$	136.17
CaF_2	78.08	KH_2PO_4	136.09
CaO	56.08	KI	166.01
$Ca_3(PO_4)_2$	310.18	KIO_3	214.00
$Ce(HSO_4)_4$	528.40	$KIO_3 \cdot HIO_3$	389.93
CeO_2	171.12	KIO_4	230.00
$Ce(SO_4)_2$	332.24	$KMnO_4$	158.04
$Ce(NH_4)_2(NO_3)_6$	548.23	KOH	56.11
$Ce(NH_4)_4(SO_4)_4 \cdot 2H_2O$	632.55	KSCN	97.18
Cr_2O_3	151.99	K_2CrO_4	194.20
CuO	79.54	$K_2Cr_2O_7$	294.19
Cu_2O	143.08	K_2O	94.20
CuS	95.60	K_2HPO_4	174.18
$CuSO_4$	159.60	K_2SO_4	174.27
$CuSO_4 \cdot 5H_2O$	249.68	$K_3Fe(CN)_6$	329.26
FeO	71.85	$K_4Fe(CN)_6$	368.36
Fe_2O_3	159.69	$MgCl_2$	95.22
Fe_3O_4	231.54	$MgCO_3$	84.32
FeS	87.91	$Mg(NH_4)(PO_4)$	137.32
$FeSO_4 \cdot (NH_4)_2SO_4 \cdot 6H_2O$	392.15	MgO	40.31
FeS_2	119.97	$Mg_2P_2O_7$	222.57
HBr	80.92	$MgSO_4$	120.37
$HCO_2 \cdot H$ (formic)	46.02	MnO_2	86.94
$HCO_2 \cdot CH_3$ (acetic)	60.05	Mn_2O_3	157.88
$HCO_2 \cdot C_6H_5$ (benzoic)	122.12	Mn_3O_4	228.81

Formula weights (continued)

MoO_3	143.94	$Na_2S_2O_3 \cdot 5H_2O$	248.19
$NH_2CO_2NH_2$	60.05	$Ni(C_4H_7O_2N_2)_2$ (NiDMG)	288.94
NH_3	17.031	P_2O_5	141.95
NH_4Cl	53.49	$PbCl_2$	278.10
NH_4NO_3	80.04	$PbCrO_4$	323.19
NH_4OH	35.05	PbO	223.19
$(NH_4)_2C_2O_4 \cdot H_2O$	142.11	PbO_2	239.19
$(NH_4)_2SO_4$	132.14	$PbSO_4$	303.25
$NaBr$	102.90	Pb_3O_4	685.57
$NaCN$	49.01	SO_2	64.06
$NaCO_2 \cdot CH_3$	82.04	SO_3	80.06
$NaCO_2 \cdot C_6H_5$	144.10	SiO_2	60.09
$NaCl$	58.44	$SnCl_2$	189.61
NaF	41.99	SnO_2	150.69
$NaHCO_3$	84.01	$SrCO_3$	147.63
NaH_2PO_4	119.98	$SrSO_4$	183.68
$NaOH$	40.00	ThO_2	264.04
$NaSCN$	81.07	TiO_2	79.90
Na_2CO_3	105.99	U_3O_8	842.09
$Na_2C_2O_4$	134.00	V_2O_5	181.88
$Na_2H_2EDTA \cdot 2H_2O$	372.24	ZnO	81.37
Na_2O	61.98	ZrO_2	123.22
Na_2SO_4	142.04		

appendix C
Dissociation Constants for Weak Acids

Acid	Formula	°C	Step	K_a	pK_a
Acetic	$CH_3 \cdot COOH$	25	...	1.76×10^{-5}	4.75
Acetoacetic	$CH_3 \cdot COCH_2 \cdot COOH$	18	...	2.62×10^{-4}	3.58
Adipic	$(COOH)(CH_2)_4(COOH)$	25	(1)	3.71×10^{-5}	4.43
	$(COOH)(CH_2)_4(COO^-)$	25	(2)	3.87×10^{-5}	4.41
Arsenic	H_3AsO_4	18	(1)	5.62×10^{-3}	2.25
	$H_2AsO_4^-$	18	(2)	1.70×10^{-7}	6.77
	$HAsO_4^{2-}$	18	(3)	3.95×10^{-12}	11.60
Arsenious	H_3AsO_3	25	...	6.0×10^{-10}	9.23
Benzoic	$C_6H_5 \cdot COOH$	25	...	6.46×10^{-5}	4.19
Boric	H_3BO_3	20	(1)	7.3×10^{-10}	9.14
	$H_2BO_3^-$	20	(2)	1.8×10^{-13}	12.74
	HBO_3^{2-}	20	(3)	1.6×10^{-14}	13.80
Carbonic	H_2CO_3	25	(1)	4.30×10^{-7}	6.37
	HCO_3^-	25	(2)	5.61×10^{-11}	10.25
Chloracetic	$CH_2Cl \cdot COOH$	25	...	1.40×10^{-3}	2.82
Chromic	H_2CrO_4	25	(1)	1.8×10^{-1}	0.74
	$HCrO_4^-$	25	(2)	3.2×10^{-7}	6.49
Citric	$H_3C_6H_5O_7$	18	(1)	8.4×10^{-4}	3.08
	$H_2C_6H_5O_7^-$	18	(2)	1.8×10^{-5}	4.74
	$HC_6H_5O_7^{2-}$	18	(3)	4.0×10^{-6}	5.40
Dichloracetic	$CHCl_2 \cdot COOH$	25	...	3.32×10^{-2}	1.48
EDTA (shown	H_4Y	25	(1)	1.0×10^{-2}	2.00
as)	H_3Y^-	25	(2)	2.1×10^{-3}	2.68
	H_2Y^{2-}	25	(3)	6.9×10^{-7}	6.16
	HY^{3-}	25	(4)	5.5×10^{-11}	10.26
Formic	$H \cdot COOH$	20	...	1.77×10^{-4}	3.75
Glycolic	$HOCH_2 \cdot COOH$	25	...	1.32×10^{-4}	3.88
Hydrocyanic	HCN	25	...	5.00×10^{-10}	9.30
Hydrofluoric	HF	25	...	3.53×10^{-4}	3.45
Hydrosulfuric	H_2S	25	(1)	9.1×10^{-8}	7.04
	HS^-	25	(2)	1.1×10^{-15}	14.96
Hypobromous	$HOBr$	25	...	2.06×10^{-9}	8.69
Hypochlorous	$HOCl$	25	...	2.95×10^{-8}	7.53
Hypoiodous	HOI	25	...	2.3×10^{-11}	10.64
Iodic	HIO_3	25	...	1.69×10^{-1}	0.77
Lactic	$CH_3 \cdot CHOH \cdot COOH$	25	...	1.38×10^{-4}	3.86
Maleic	$C_2H_2(COOH)_2$	25	(1)	1.2×10^{-2}	1.92
	$C_2H_2 \cdot COOH \cdot COO^-$	25	(2)	6.0×10^{-7}	6.22
Nitrous	HNO_2	25	...	4.6×10^{-4}	3.35
Oxalic	$COOH \cdot COOH$	25	(1)	5.90×10^{-2}	1.23
	$COOH \cdot COO^-$	25	(2)	6.40×10^{-5}	4.19
Periodic(meta)	HIO_4	25	...	2.30×10^{-2}	1.64
Periodic(para)	H_5IO_6	25	(1)	5.1×10^{-4}	3.29
	$H_4IO_6^-$	25	(2)	2.0×10^{-7}	6.70

Dissociation constants for weak acids (continued)

Acid	Formula	°C	Step	K_a	pK_a
Phenol	C_6H_5OH	25	...	1.28×10^{-10}	9.89
Phosphoric	H_3PO_4	25	(1)	7.52×10^{-3}	2.12
	$H_2PO_4^-$	25	(2)	6.23×10^{-8}	7.21
	HPO_4^{2-}	25	(3)	4.80×10^{-13}	12.32
Phthalic	$C_6H_4(COOH)_2$	25	(1)	1.3×10^{-3}	2.89
	$C_6H_4 \cdot COOH \cdot COO^-$	25	(2)	3.9×10^{-6}	5.41
Picric	$(NO_2)_3 \cdot C_6H_2OH$	25	...	4.2×10^{-1}	0.38
Sulfuric	H_2SO_4	25	(1)	high	
	HSO_4^-	25	(2)	1.20×10^{-2}	1.92
Sulfurous	H_2SO_3	25	(1)	1.7×10^{-2}	1.77
	HSO_3^-	25	(2)	6.2×10^{-8}	7.21
Tartaric	$H_2C_4H_4O_6$	25	(1)	1.04×10^{-3}	2.98
	$HC_4H_4O_6^-$	25	(2)	4.55×10^{-5}	4.34
Trichloracetic	$CCl_3 \cdot COOH$	25	...	1.29×10^{-1}	0.89

appendix D
Dissociation Constants for Weak Bases

Base	Formula	°C	K_b	pK_b
Ammonium hydroxide	NH_4OH	25	1.79×10^{-5}	4.75
Aniline	$C_6H_5NH_2$	25	4.27×10^{-10}	9.37
Diethylamine	$(C_2H_5)_2NH$	25	1.3×10^{-3}	2.89
Dimethylamine	$(CH_3)_2NH$	25	5.2×10^{-4}	3.28
Ethylamine	$C_2H_5NH_2$	25	5.6×10^{-4}	3.25
Methylamine	CH_3NH_2	25	5.0×10^{-4}	3.30
Hydrazine	$H_2N \cdot NH_2$	25	3.0×10^{-6}	5.52
Hydroxylamine	$H_2N \cdot OH$	25	1.07×10^{-8}	7.97
Pyridine	C_5H_5N	25	1.4×10^{-9}	8.85

appendix E
Solubility Product Constants

Salt	°C	K_{sp}	pK_{sp}	Salt	°C	K_{sp}	pK_{sp}
AgBr	25	5.0×10^{-13}	12.30	CuSCN	25	1.9×10^{-13}	12.72
AgC$_2$H$_3$O$_2$				Fe(OH)$_2$	25	1.4×10^{-15}	14.85
(CH$_3$·COOAg)	25	2.30×10^{-3}	2.64	Fe(OH)$_3$	25	4.5×10^{-37}	36.35
Ag$_2$CO$_3$	25	8.1×10^{-12}	11.09	FePO$_4$	25	1.4×10^{-22}	21.85
AgCN	25	2.0×10^{-12}	11.70	FeS	25	5.0×10^{-18}	17.30
AgCl	20	1.00×10^{-10}	10.00	Fe$_2$S$_3$	25	1.0×10^{-88}	88.00
AgCl	25	1.56×10^{-10}	9.81	Hg$_2$Cl$_2$	25	1.3×10^{-18}	17.89
AgCl	100	2.17×10^{-8}	7.66	Hg$_2$I$_2$	25	4.5×10^{-29}	28.35
Ag$_2$CrO$_4$	25	9.0×10^{-12}	11.05	HgS	25	1.6×10^{-54}	53.80
AgI	25	1.50×10^{-16}	15.82	Hg$_2$SO$_4$	25	7.1×10^{-7}	6.15
AgIO$_3$	25	2.0×10^{-8}	7.70	MgCO$_3$	25	1.0×10^{-5}	5.00
AgOH	25	2.0×10^{-8}	7.70	MgC$_2$O$_4$	25	8.6×10^{-5}	4.07
AgSCN	25	1.07×10^{-12}	11.97	Mg(NH$_4$)PO$_4$	25	2.5×10^{-13}	12.60
Al(OH)$_3$	25	2.0×10^{-32}	31.70	Mg(OH)$_2$	25	2.4×10^{-11}	10.62
BaCO$_3$	25	5.1×10^{-9}	8.29	MnCO$_3$	25	1.8×10^{-11}	10.74
BaC$_2$O$_4$	25	1.5×10^{-8}	7.82	MnC$_2$O$_4$	25	1.0×10^{-15}	15.00
BaSO$_4$	25	1.1×10^{-10}	9.96	Mn(OH)$_2$	25	1.9×10^{-13}	12.72
Bi$_2$S$_3$	25	1.0×10^{-96}	96.00	MnS	25	1.4×10^{-15}	14.85
CaCO$_3$	25	4.8×10^{-9}	8.32	NiCO$_3$	25	6.6×10^{-9}	8.18
CaC$_2$O$_4$	25	2.57×10^{-9}	8.59	NiS	25	3.0×10^{-21}	20.52
CaF$_2$	25	3.95×10^{-11}	10.40	PbCO$_3$	25	1.0×10^{-13}	13.00
Ca(OH)$_2$	25	5.5×10^{-6}	5.26	PbCrO$_4$	25	1.8×10^{-14}	13.74
Ca$_3$(PO$_4$)$_2$	25	1.0×10^{-26}	26.00	PbS	25	2.5×10^{-27}	26.60
CaSO$_4$	25	1.2×10^{-6}	5.92	PbSO$_4$	25	1.70×10^{-8}	7.77
CdCO$_3$	25	5.2×10^{-12}	11.28	Sn(OH)$_2$	25	3.2×10^{-26}	25.50
CdC$_2$O$_4$	25	1.8×10^{-8}	7.74	SnS	25	1.0×10^{-26}	26.00
CdS	25	5.0×10^{-27}	26.30	SrC$_2$O$_4$	25	5.6×10^{-10}	9.25
CoCO$_3$	25	8.0×10^{-13}	12.10	SrSO$_4$	25	2.5×10^{-7}	6.60
CoS	25	5.0×10^{-22}	21.30	ZnCO$_3$	25	2.1×10^{-11}	10.68
CuCO$_3$	25	2.5×10^{-10}	9.60	Zn(OH)$_2$	25	3.3×10^{-17}	16.48
CuI	25	1.0×10^{-12}	12.00	Zn$_3$(PO$_4$)$_2$	25	1.0×10^{-32}	32.00
CuS	25	8.0×10^{-36}	35.10	ZnS	25	1.6×10^{-24}	23.80

appendix F
Stepwise Formation (Stability) Constants of Complexes*

Ligand	°C	Cation	K_1	K_2	K_3	K_4
NH_3	25	Ag^+	2.0×10^3	8.0×10^3		
			3.30	3.90		
	25	Cd^{2+}	3.2×10^2	9.1×10	2.0×10	6.2
			2.51	1.96	1.30	0.79
	25	Cu^{2+}	1.3×10^4	3.2×10^3	8.0×10^2	1.3×10^2
			4.11	3.50	2.90	2.11
	25	Hg^{2+}	6.3×10^8	5.0×10^8	10	6.0
			8.80	8.70	1.00	0.78
	25	Ni^{2+}	4.7×10^2	1.3×10^2	4.1×10	1.2×10
			2.67	2.11	1.61	1.08
	25	Zn^{2+}	1.5×10^2	1.8×10^2	2.0×10^2	9.1×10
			2.18	2.26	2.30	1.96
CN^-	25	Ag^+	$Ag^+ + 2CN^- \rightleftharpoons Ag(CN)_2^-$		$K_1 K_2 = 7.1 \times 10^{19}$	
					19.85	
	25	Cd^{2+}	3.0×10^5	1.4×10^5	3.6×10^4	3.8×10^3
			5.48	5.14	4.56	3.58
	25	Cu^+	$Cu^+ + 2CN^- \rightleftharpoons Cu(CN)_2^-$		3.9×10^4	5.0×10
			$K_1 K_2 = 1.0 \times 10^{24}$		4.59	1.70
			24.0			
	25	Hg^{2+}	1.0×10^{18}	5.0×10^{16}	6.8×10^3	9.6×10^2
			18.00	16.70	3.83	2.98
	25	Zn^{2+}	$Zn^{2+} + 3CN^- \rightleftharpoons Zn(CN)_3^-$			5.0×10^2
			$K_1 K_2 K_3 = 3.2 \times 10^{17}$			2.70
			17.50			
EDTA	20	Ag^+	2.1×10^7			
			7.32			
	20	Ba^{2+}	6.0×10^7			
			7.78			
	20	Ca^{2+}	1.0×10^{11}			
			11.00			
	20	Cd^{2+}	3.9×10^{16}			
			16.59			
	20	Co^{2+}	1.6×10^{16}			
			16.21			
	20	Cu^{2+}	6.2×10^{18}			
			18.79			
	20	Fe^{2+}	2.1×10^{14}			
			14.33			
	20	Fe^{3+}	1.7×10^{24}			
			24.23			
	20	Hg^{2+}	6.3×10^{21}			
			21.80			
	20	Mg^{2+}	4.9×10^8			
			8.69			

*Log $K_{stability}$ is given under each $K_{stability}$ value.

Stepwise formation constants* (continued)

Ligand	°C	Cation	K_1
	20	Ni^{2+}	3.6×10^{18}
			18.56
	20	Pb^{2+}	2.0×10^{18}
			18.30
	20	Zn^{2+}	1.8×10^{16}
			16.26
SCN^-	25	Fe^{3+}	1.38×10^2
			2.14
Trien	25	Cu^{2+}	2.5×10^{20}
			20.40

appendix G
Dissociation Constants for Metallochromic Indicators

Indicator		Step	pK_a*	K_a†
Calcon	H_2In	(1)	7.4	4.0×10^{-8}
	HIn^-	(2)	13.5	3.2×10^{-14}
Calmagite	H_2In	(1)	8.1	7.9×10^{-9}
	HIn^-	(2)	12.4	4.0×10^{-13}
Eriochrome black T	H_3In	(1)	1.6	2.5×10^{-2}
	H_2In^-	(2)	6.3	5.0×10^{-7}
	HIn^{2-}	(3)	11.6	2.5×10^{-12}
Metalphthalein	H_6In	(1)	2.2	6.3×10^{-3}
	H_5In^-	(2)	2.9	1.3×10^{-3}
	H_4In^{2-}	(3)	7.0	1.0×10^{-7}
	H_3In^{3-}	(4)	7.8	1.6×10^{-8}
	H_2In^{4-}	(5)	11.4	4.0×10^{-12}
	HIn^{5-}	(6)	12.0	1.0×10^{-12}
Murexide	H_3In	(1)	0	1.0
	H_2In^-	(2)	9.2	6.3×10^{-10}
	HIn^{2-}	(3)	10.5	3.2×10^{-11}
Pyrocatechol violet	H_4In	(1)	0.3	5.0×10^{-1}
	H_3In^-	(2)	7.8	1.6×10^{-8}
	H_2In^{2-}	(3)	9.8	1.6×10^{-10}
	HIn^{3-}	(4)	12.5	3.2×10^{-13}
1-2-(Pyridylazo)-2-naphthol	H_2In	(1)	<2	
	HIn^-	(2)	12.3	5.0×10^{-13}
Xylenol orange	H_5In	(1)	2.6	2.5×10^{-3}
	H_4In^-	(2)	3.2	6.3×10^{-4}
	H_3In^{2-}	(3)	6.4	4.0×10^{-7}
	H_2In^{3-}	(4)	10.5	3.2×10^{-11}
	HIn^{4-}	(5)	12.3	5.0×10^{-13}

*pK_a values taken from I. M. Kolthoff, E. B. Sandell, E. J. Meehan, and Stanley Bruckenstein, "Quantitative Chemical Analysis," 4th ed., p. 1151, Macmillan, London, 1969.
†The value of K_a shown in association with each pK_a value is that suggested for calculation purposes.

appendix H
Formation (Stability) Constants for Metallochromic Indicators*

Cation	CAL†	CMG†	EBT†	MPT†	MUR†	PYV†	PAN†	XYO†
Al^{3+}	...	...	...	...	...	19.1 1.3×10^{19}		
Ba^{2+}	...	...	...	6.2 1.6×10^6				
Bi^{3+}	...	...	...	...	...	27.5 3.2×10^{27}	...	5.5 3.2×10^5
Ca^{2+}	5.6 4.0×10^5	6.1 1.3×10^6	5.4 2.5×10^5	7.8 6.3×10^7	5.0 1.0×10^5			
Co^{2+}	... ...	...	...	...	...	9.0 1.0×10^9	>12 $>1.0 \times 10^{12}$	
Cu^{2+}	21 1×10^{21}	...	...	...	17.9 7.9×10^{17}	16.5 3.2×10^{16}	16 1×10^{16}	
Fe^{3+}	... ...	...	...	...	...		...	5.7 5.0×10^5
Mg^{2+}	7.6 4.0×10^7	8.1 1.3×10^8	7.0 1.0×10^7	8.9 7.9×10^8	...	4.4 2.5×10^4	8.5 3.2×10^8	
Ni^{2+}	... ...	...	...	...	11.3 2.0×10^{11}	9.4 2.5×10^9	12.7 5.0×10^{12}	
Pb^{2+}	... ...	...	...	...	...	13.3 2.0×10^{13}		
Zn^{2+}	12.5 3.2×10^{12}	...	13.5 3.2×10^{13}	15.1 1.3×10^{15}	...	10.4 2.5×10^{10}	11.2 1.6×10^{11}	6.2 1.6×10^6

The value of $K_{stability}$ shown below each log $K_{stability}$ value is that suggested for calculation purposes. Log $K_{stability}$ for reaction $Me + In \rightleftharpoons MeIn$. Ionic strength about 0.1.

*Log $K_{stability}$ values taken from I. M. Kolthoff, E. B. Sandell, E. J. Meehan, and Stanley Bruckenstein, "Quantitative Chemical Analysis," 4th ed., p. 1151, Macmillan, London, 1969.

†CAL, calcon; CMG, calmagite; EBT, Eriochrome black T; MPT, metalphthalein; MUR, murexide; PYV, Pyrocatechol violet; PAN, 1-2-(pyridylazo)-2-naphthol; XYO, Xylenol orange.

appendix I
Standard Reduction Potentials
and Some Formal Reduction
Potentials at 25°C*

Reaction		Potential, V
$Ag^{2+} + 1e \rightleftharpoons Ag^+$		+1.98
$Ag^{2+} + 1e \rightleftharpoons Ag^+$	(4 M HNO$_3$)	+1.93
$Ag^+ + 1e \rightleftharpoons Ag$		+0.800
Ag/AgCl electrode (saturated KCl)		+0.1956
Ag/AgCl electrode (saturated KCl) + $E_{junction}$		+0.1989
$AgBr + 1e \rightleftharpoons Ag + Br^-$		+0.0713
$AgCN + 1e \rightleftharpoons Ag + CN^-$		−0.02
$AgCl + 1e \rightleftharpoons Ag + Cl^-$		+0.2223
$AgI + 1e \rightleftharpoons Ag + I^-$		−0.1519
$Al^{3+} + 3e \rightleftharpoons Al$	(0.1 M NaOH)	−1.706
$H_3AsO_4 + 2H_3O^+ + 2e \rightleftharpoons HAsO_2 + 4H_2O$	(1 M HCl)	+0.58
$AsO_4^{3-} + 2H_2O + 2e \rightleftharpoons AsO_2^- + 4OH^-$	(1 M NaOH)	−0.08
$Au^{3+} + 3e \rightleftharpoons Au$		+1.42
$H_2BO_3^- + 5H_2O + 8e \rightleftharpoons BH_4^- + 8OH^-$		−1.24
$Ba^{2+} + 2e \rightleftharpoons Ba$		−2.90
$Ba^{2+} + 2e \rightleftharpoons Ba(Hg)$		−1.570
$Bi^{3+} + 3e \rightleftharpoons Bi$		+0.200
$BiCl_4^- + 3e \rightleftharpoons Bi + 4Cl^-$		+0.168
$BiOCl + 2H_3O^+ + 3e \rightleftharpoons Bi + Cl^- + 3H_2O$		+0.1583
$BiO + 2H_3O^+ + 3e \rightleftharpoons Bi + 3H_2O$		+0.32
$Br_2(aq) + 2e \rightleftharpoons 2Br^-$		+1.087
$Br_2(l) + 2e \rightleftharpoons 2Br^-$		+1.065
$Br_3^- + 2e \rightleftharpoons 3Br^-$		+1.05
$BrO_3^- + 6H_3O^+ + 6e \rightleftharpoons Br^- + 9H_2O$		+1.44
$BrO_3^- + 3H_2O + 6e \rightleftharpoons Br^- + 6OH^-$		+0.61
$Ca^{2+} + 2e \rightleftharpoons Ca$		−2.76
Calomel electrode (saturated KCl) (SCE)		+0.2412
Calomel electrode (saturated KCl) (SCE) + $E_{junction}$		+0.2445
Calomel electrode (1 M KCl)		+0.2801
Calomel electrode (1 M KCl) + $E_{junction}$		+0.2830
Calomel electrode (0.1 M KCl)		+0.3337
Calomel electrode (0.1 M KCl) + $E_{junction}$		+0.3356
$Cd^{2+} + 2e \rightleftharpoons Cd$		−0.4026
$Cd^{2+} + 2e \rightleftharpoons Cd(Hg)$		−0.3521

*Note 1: Some formal potential values are also given. Note 2: Reactions such as

$$MnO_4^- + 8H_3O^+ + 5e \rightleftharpoons Mn^{2+} + 12H_2O$$

are also written in the form

$$MnO_4^- + 8H^+ + 5e \rightleftharpoons Mn^{2+} + 4H_2O$$

Standard reduction potentials (continued)

Reaction		Potential, V
$Ce^{3+} + 3e \rightleftharpoons Ce$		-2.335
$Ce^{3+} + 3e \rightleftharpoons Ce(Hg)$		-1.4373
$Ce^{4+} + 1e \rightleftharpoons Ce^{3+}$		$+1.443$
$Ce^{4+} + 1e \rightleftharpoons Ce^{3+}$	$(0.5\ M\ H_2SO_4)$	$+1.459$
$Ce^{4+} + 1e \rightleftharpoons Ce^{3+}$	$(1\ M\ HCl)$	$+1.28$
$Cl_2(g) + 2e \rightleftharpoons 2Cl^-$		$+1.3583$
$ClO_3^- + 6H_3O^+ + 6e \rightleftharpoons Cl^- + 9H_2O$		$+1.45$
$ClO_4^- + 2H_3O^+ + 2e \rightleftharpoons ClO_3^- + 3H_2O$		$+1.19$
$ClO_4^- + 8H_3O^+ + 8e \rightleftharpoons Cl^- + 12H_2O$		$+1.37$
$Co^{3+} + 1e \rightleftharpoons Co^{2+}$	$(2\ M\ HNO_3)$	$+1.842$
$Co^{2+} + 2e \rightleftharpoons Co$		-0.28
$Cr^{3+} + 1e \rightleftharpoons Cr^{2+}$		-0.41
$Cr^{2+} + 2e \rightleftharpoons Cr$		-0.557
$Cr^{3+} + 3e \rightleftharpoons Cr$		-0.74
$Cr^{6+} + 3e \rightleftharpoons Cr^{3+}$	$(2\ M\ H_2SO_4)$	$+1.10$
$Cr^{6+} + 3e \rightleftharpoons Cr^{3+}$	$(1\ M\ NaOH)$	-0.12
$CrO_4^{2-} + 8H_3O^+ + 3e \rightleftharpoons Cr^{3+} + 12H_2O$		$+1.195$
$Cr_2O_7^{2-} + 14H_3O^+ + 6e \rightleftharpoons 2Cr^{3+} + 21H_2O$		$+1.330$
$Cr_2O_7^{2-} + 14H_3O^+ + 6e \rightleftharpoons 2Cr^{3+} + 21H_2O$	$(2\ M\ H_2SO_4)$	$+1.10$
$Cr_2O_7^{2-} + 14H_3O^+ + 6e \rightleftharpoons 2Cr^{3+} + 21H_2O$	$(1\ M\ HCl)$	$+1.00$
$Cu^{2+} + 1e \rightleftharpoons Cu^+$		$+0.158$
$Cu^+ + 1e \rightleftharpoons Cu$		$+0.522$
$Cu^{2+} + 2e \rightleftharpoons Cu$		$+0.340$
$Cu^{2+} + 2e \rightleftharpoons Cu(Hg)$		$+0.345$
$F_2(g) + 2e \rightleftharpoons 2F^-$		$+2.87$
$F_2 + 2H_3O^+ + 2e \rightleftharpoons 2HF(aq) + 2H_2O$		$+3.06$
$Fe^{3+} + 1e \rightleftharpoons Fe^{2+}$		$+0.770$
$Fe^{3+} + 1e \rightleftharpoons Fe^{2+}$	$(1\ M\ HCl)$	$+0.700$
$Fe^{3+} + 1e \rightleftharpoons Fe^{2+}$	$(1\ M\ HClO_4)$	$+0.730$
$Fe^{3+} + 1e \rightleftharpoons Fe^{2+}$	$(0.5\ M\ H_2SO_4)$	$+0.700$
$Fe^{3+} + 1e \rightleftharpoons Fe^{2+}$	$(1\ M\ H_2SO_4)$	$+0.680$
$Fe^{2+} + 2e \rightleftharpoons Fe$		-0.409
$Fe(CN)_6^{3-} + 1e \rightleftharpoons Fe(CN)_6^{4-}$	$(1\ M\ H_2SO_4)$	$+0.69$
$Fe(CN)_6^{3-} + 1e \rightleftharpoons Fe(CN)_6^{4-}$	$(0.1\ M\ HCl)$	$+0.56$
$2H_3O^+ + 2e \rightleftharpoons H_2 + 2H_2O$		0.0000
$2H_2O + 2e \rightleftharpoons H_2 + 2OH^-$		-0.8277
$H_2O_2 + 2H_3O^+ + 2e \rightleftharpoons 4H_2O$		$+1.776$
$2Hg^{2+} + 2e \rightleftharpoons Hg_2^{2+}$		$+0.905$
$Hg^{2+} + 2e \rightleftharpoons Hg$		$+0.850$
$Hg_2^{2+} + 2e \rightleftharpoons 2Hg$		$+0.796$
$I_2(s) + 2e \rightleftharpoons 2I^-$		$+0.535$
$I_3^- + 2e \rightleftharpoons 3I^-$		$+0.5338$
$IO_3^- + 6H_3O^+ + 5e \rightleftharpoons \frac{1}{2}I_2 + 9H_2O$		$+1.195$
$IO_3^- + 3H_2O + 6e \rightleftharpoons I^- + 6OH^-$		$+0.26$
$K^+ + 1e \rightleftharpoons K$		-2.924
$Li^+ + 1e \rightleftharpoons Li$		-3.045
$Mg^{2+} + 2e \rightleftharpoons Mg$		-2.375
$Mn^{2+} + 2e \rightleftharpoons Mn$		-1.029

Standard reduction potentials (continued)

Reaction		Potential, V
$MnO_2 + 4H_3O^+ + 2e \rightleftharpoons Mn^{2+} + 6H_2O$		+1.208
$MnO_4^- + 1e \rightleftharpoons MnO_4^{2-}$		+0.564
$MnO_4^- + 4H_3O^+ + 3e \rightleftharpoons MnO_2 + 6H_2O$		+1.679
$MnO_4^- + 2H_2O + 3e \rightleftharpoons MnO_2 + 4OH^-$		+0.588
$MnO_4^- + 8H_3O^+ + 5e \rightleftharpoons Mn^{2+} + 12H_2O$		+1.510
$Na^+ + 1e \rightleftharpoons Na$		−2.711
$NO_3^- + H_2O + 2e \rightleftharpoons NO_2^- + 2OH^-$		0.00
$Ni^{2+} + 2e \rightleftharpoons Ni$		−0.23
$O_2 + 2H_3O^+ + 2e \rightleftharpoons H_2O_2 + 2H_2O$		+0.682
$O_2 + 2H_2O + 2e \rightleftharpoons H_2O_2 + 2OH^-$		−0.146
$O_2 + 4H_3O^+ + 4e \rightleftharpoons 6H_2O$		+1.229
$O_2 + 2H_2O + 4e \rightleftharpoons 4OH^-$		+0.401
$O_3 + 2H_3O^+ + 2e \rightleftharpoons O_2 + 3H_2O$		+2.07
$Pb^{2+} + 2e \rightleftharpoons Pb$		−0.126
$Pb^{2+} + 2e \rightleftharpoons Pb(Hg)$		−0.120
$PbO_2 + 4H_3O^+ + 2e \rightleftharpoons Pb^{2+} + 6H_2O$		+1.46
$PbO_2 + SO_4^{2-} + 4H_3O^+ + 2e \rightleftharpoons PbSO_4 + 6H_2O$		+1.685
Quinhydrone electrode ($\alpha_H^+ = 1$)		+0.6995
$S_2O_8^{2-} + 2e \rightleftharpoons 2SO_4^{2-}$		+2.01
$S + 2H_3O^+ + 2e \rightleftharpoons H_2S + 2H_2O$		+0.140
$S_4O_6^{2-} + 2e \rightleftharpoons 2S_2O_3^{2-}$		+0.10
$Sb_2O_3 + 6H_3O^+ + 6e \rightleftharpoons 2Sb + 9H_2O$		+0.1445
$Sb_2O_5 + 6H_3O^+ + 4e \rightleftharpoons 2SbO^+ + 9H_2O$		+0.581
$SbO^+ + 2H_3O^+ + 3e \rightleftharpoons Sb + 3H_2O$		+0.212
$Sb^{5+} + 2e \rightleftharpoons Sb^{3+}$	(3.5 M HCl)	+0.75
$Sn^{2+} + 2e \rightleftharpoons Sn$		−0.136
$Sn^{4+} + 2e \rightleftharpoons Sn^{2+}$	(0.1 M HCl)	+0.070
$Sn^{4+} + 2e \rightleftharpoons Sn^{2+}$	(1 M HCl)	+0.139
$Sr^{2+} + 2e \rightleftharpoons Sr$		−2.89
$Ti^{3+} + 1e \rightleftharpoons Ti^{2+}$		−2.0
$TiO_2^+ + 2H_3O^+ + 1e \rightleftharpoons Ti^{3+} + 3H_2O$		+0.10
$UO_2^{2+} + 4H_3O^+ + 2e \rightleftharpoons U^{4+} + 6H_2O$		+0.334
$U^{4+} + 1e \rightleftharpoons U^{3+}$		−0.631
$V(OH)_4^+ + 2H_3O^+ + 1e \rightleftharpoons VO^{2+} + 5H_2O$		+1.000
sometimes given as		
$VO_4^{3-} + 6H_3O^+ + 1e \rightleftharpoons VO^{2+} + 9H_2O$		
$VO^{2+} + 2H_3O^+ + 1e \rightleftharpoons V^{3+} + 3H_2O$		+0.337
$V^{3+} + 1e \rightleftharpoons V^{2+}$		−0.255
$Zn^{2+} + 2e \rightleftharpoons Zn$		−0.763
$Zn^{2+} + 2e \rightleftharpoons Zn(Hg)$		−0.763

appendix J
Logarithms of Numbers

10	0000	0043	0086	0128	0170	0212	0253	0294	0334	0374
11	0414	0453	0492	0531	0569	0607	0645	0682	0719	0755
12	0792	0828	0864	0899	0934	0969	1004	1038	1072	1106
13	1139	1173	1206	1239	1271	1303	1335	1367	1399	1430
14	1461	1492	1523	1553	1584	1614	1644	1673	1703	1732
15	1761	1790	1818	1847	1875	1903	1931	1959	1987	2014
16	2041	2068	2095	2122	2148	2175	2201	2227	2253	2279
17	2304	2330	2355	2380	2405	2430	2455	2480	2504	2529
18	2553	2577	2601	2625	2648	2672	2695	2718	2742	2765
19	2788	2810	2833	2856	2878	2900	2923	2945	2967	2989
20	3010	3032	3054	3075	3096	3118	3139	3160	3181	3201
21	3222	3243	3263	3284	3304	3324	3345	3365	3385	3404
22	3424	3444	3464	3483	3502	3522	3541	3560	3579	3598
23	3617	3636	3655	3674	3692	3711	3729	3747	3766	3784
24	3802	3820	3838	3856	3874	3892	3909	3927	3945	3962
25	3979	3997	4014	4031	4048	4065	4082	4099	4116	4133
26	4150	4166	4183	4200	4216	4232	4249	4265	4281	4298
27	4314	4330	4346	4362	4378	4393	4409	4425	4440	4456
28	4472	4487	4502	4518	4533	4548	4564	4579	4594	4609
29	4624	4639	4654	4669	4683	4698	4713	4728	4742	4757
30	4771	4786	4800	4814	4829	4843	4857	4871	4886	4900
31	4914	4928	4942	4955	4969	4983	4997	5011	5024	5038
32	5051	5065	5079	5092	5105	5119	5132	5145	5159	5172
33	5185	5198	5211	5224	5237	5250	5263	5276	5289	5302
34	5315	5328	5340	5353	5366	5378	5391	5403	5416	5428
35	5441	5453	5465	5478	5490	5502	5514	5527	5539	5551
36	5563	5575	5587	5599	5611	5623	5635	5647	5658	5670
37	5682	5694	5705	5717	5729	5740	5752	5763	5775	5786
38	5798	5809	5821	5832	5843	5855	5866	5877	5888	5899
39	5911	5922	5933	5944	5955	5966	5977	5988	5999	6010
40	6021	6031	6042	6053	6064	6075	6085	6096	6107	6117
41	6128	6138	6149	6160	6170	6180	6191	6201	6212	6222
42	6232	6243	6253	6263	6274	6284	6294	6304	6314	6325
43	6335	6345	6355	6365	6375	6385	6395	6405	6415	6425
44	6435	6444	6454	6464	6474	6484	6493	6503	6513	6522
45	6532	6542	6551	6561	6571	6580	6590	6599	6609	6618
46	6628	6637	6646	6656	6665	6675	6684	6693	6702	6712
47	6721	6730	6739	6749	6758	6767	6776	6785	6794	6803
48	6812	6821	6830	6839	6848	6857	6866	6875	6884	6893
49	6902	6911	6920	6928	6937	6946	6955	6964	6972	6981
50	6990	6998	7007	7016	7024	7033	7042	7050	7059	7067
51	7076	7084	7093	7101	7110	7118	7126	7135	7143	7152
52	7160	7168	7177	7185	7193	7202	7210	7218	7226	7235
52	7243	7251	7259	7267	7275	7284	7292	7300	7308	7316
54	7324	7332	7340	7348	7356	7364	7372	7380	7388	7396

Logarithms of numbers (continued)

55	7404	7412	7419	7427	7435	7443	7451	7459	7466	7474
56	7482	7490	7497	7505	7513	7520	7528	7536	7543	7551
57	7559	7566	7574	7582	7589	7597	7604	7612	7619	7627
58	7634	7642	7649	7657	7664	7672	7679	7686	7694	7701
59	7709	7716	7723	7731	7738	7745	7752	7760	7767	7774
60	7782	7789	7796	7803	7810	7818	7825	7832	7839	7846
61	7853	7860	7868	7875	7882	7889	7896	7903	7910	7917
62	7924	7931	7938	7945	7952	7959	7966	7973	7980	7987
63	7993	8000	8007	8014	8021	8028	8035	8041	8048	8055
64	8062	8069	8075	8082	8089	8096	8102	8109	8116	8122
65	8129	8136	8142	8149	8156	8162	8169	8176	8182	8189
66	8195	8202	8209	8215	8222	8228	8235	8241	8248	8254
67	8261	8267	8274	8280	8287	8293	8299	8306	8312	8319
68	8325	8331	8338	8344	8351	8357	8363	8370	8376	8382
69	8388	8395	8401	8407	8414	8420	8426	8432	8439	8445
70	8451	8457	8463	8470	8476	8482	8488	8494	8500	8506
71	8513	8519	8525	8531	8537	8543	8549	8555	8561	8567
72	8573	8579	8585	8591	8597	8603	8609	8615	8621	8627
73	8633	8639	8645	8651	8657	8663	8669	8675	8681	8686
74	8692	8698	8704	8710	8716	8722	8727	8733	8739	8745
75	8751	8756	8762	8768	8774	8779	8785	8791	8797	8802
76	8808	8814	8820	8825	8831	8837	8842	8848	8854	8859
77	8865	8871	8876	8882	8887	8893	8899	8904	8910	8915
78	8921	8927	8932	8938	8943	8949	8954	8960	8965	8971
79	8976	8982	8987	8993	8998	9004	9009	9015	9020	9025
80	9031	9036	9042	9047	9053	9058	9063	9069	9074	9079
81	9085	9090	9096	9101	9106	9112	9117	9122	9128	9133
82	9138	9143	9149	9154	9159	9165	9170	9175	9180	9186
83	9191	9196	9201	9206	9212	9217	9222	9227	9232	9238
84	9243	9248	9253	9258	9263	9269	9274	9279	9284	9289
85	9294	9299	9304	9309	9315	9320	9325	9330	9335	9340
86	9345	9350	9355	9360	9365	9370	9375	9380	9385	9390
87	9395	9400	9405	9410	9415	9420	9425	9430	9435	9440
88	9445	9450	9455	9460	9465	9469	9474	9479	9484	9489
89	9494	9499	9504	9509	9513	9518	9523	9528	9533	9538
90	9542	9547	9552	9557	9562	9566	9571	9576	9581	9586
91	9590	9595	9600	9605	9609	9614	9619	9624	9628	9633
92	9638	9643	9647	9652	9657	9661	9666	9671	9675	9680
93	9685	9689	9694	9699	9703	9708	9713	9717	9722	9727
94	9731	9736	9741	9745	9750	9754	9759	9763	9768	9773
95	9777	9782	9786	9791	9795	9800	9805	9809	9814	9818
96	9823	9827	9832	9836	9841	9845	9850	9854	9859	9863
97	9868	9872	9877	9881	9886	9890	9894	9899	9903	9908
98	9912	9917	9921	9926	9930	9934	9939	9943	9948	9952
99	9956	9961	9965	9969	9974	9978	9983	9987	9991	9996

Index

Page references in **boldface** indicate Laboratory Experiments